List of Names

dates before the seventeenth century are approximate.)

Thomas Bayes (1702–1761)
Gabriel Cramer (1704–1752)
Leonhard Euler (1707–1783)
Thomas Simpson (1710–1761)
Roger Boscovich (1711–1787)
Alexis-Claude Clairaut (1713–1765)
Jean Le Rond d'Alembert (1717–1783)
Maria Gaetana Agnesi (1718–1799)
Tobias Mayer (1723–1762)
Johann Lambert (1729–1777)
Joseph Louis Lagrange (1736–1813)
Caspar Wessel (1745–1818)
Gaspard Monge (1746–1818)
Pierre-Simon de Laplace (1749–1827)
Adrien-Marie Legendre (1752–1833)
William Playfair (1759–1823)
Silvestre-François Lacroix (1765–1843)
Joseph Fourier (1768–1830)
Sophie Germain (1776–1831)
Carl Friedrich Gauss (1777–1855)
Bernard Bolzano (1781–1848)
Augustin-Louis Cauchy (1789–1857)
Nikolai Ivanovich Lobachevsky (1792–1856)
Charles Babbage (1792–1871)
George Green (1793–1841)
Franz Taurinus (1794–1874)
Gabriel Lamé (1795–1870)

Adolphe Quetelet (1796–1874)
Mikhail Ostrogradsky (1801–1861)
János Bolyai (1802–1860)
Niels Henrik Abel (1802–1829)
William Rowan Hamilton (1805–1865)
Peter Lejeune-Dirichlet (1805–1859)
Augustus De Morgan (1806–1871)
Hermann Grassmann (1809–1877)
Joseph Liouville (1809–1882)
Benjamin Peirce (1809–1880)
Ernst Kummer (1810–1893)
Evariste Galois (1811–1832)
James Joseph Sylvester (1814–1897)
Karl Weierstrass (1815–1897)
Ada Byron King (1815–1852)
George Stokes (1819–1903)
Florence Nightingale (1820–1910)
Arthur Cayley (1821–1895)
Eduard Heine (1821–1881)
Hermann von Helmholtz (1821–1894)
Francis Galton (1822–1911)
Leopold Kronecker (1823–1891)
Ferdinand Gotthold Eisenstein (1823–1852)
Georg Bernhard Riemann (1826–1866)
Richard Dedekind (1831–1916)
Clerk Maxwell (1831–1879)
Ludvig Sylow (1832–1918)

Charles Dodgson (1832–1898)
Eugenio Beltrami (1835–1900)
Emile Mathieu (1835–1890)
Camille Jordan (1838–1922)
Hermann Hankel (1839–1873)
Josiah Willard Gibbs (1839–1903)
Heinrich Weber (1842–1913)
Georg Cantor (1845–1918)
Francis Edgeworth (1845–1926)
Georg Frobenius (1849–1917)
Alfred Kempe (1849–1922)
Felix Klein (1849–1925)
Henri Poincaré (1854–1912)
Walter Dyck (1856–1934)
Karl Pearson (1857–1936)
Giuseppe Peano (1858–1932)
Otto Hölder (1859–1937)
Percy Heawood (1861–1955)
Frank N. Cole (1861–1927)
David Hilbert (1862–1943)
George A. Miller (1863–1951)
George Udny Yule (1871–1951)
Ernst Zermelo (1871–1953)
Bertrand Russell (1872–1970)
Leonard Dickson (1874–1954)
William Gosset (1876–1937)
Joseph H. M. Wedderburn (1882–1948)

Emmy Noether (1882–1935)
George David Birkhoff (1884–1944)
Hermann Weyl (1885–1955)
Louis Mordell (1888–1972)
Ronald Fisher (1890–1962)
Abraham Fraenkel (1891–1965)
Stefan Banach (1892–1945)
Jerzy Neyman (1894–1981)
Egon Pearson (1895–1980)
Carl Siegel (1896–1981)
Alfred Tarski (1901–1983)
Abraham Wald (1902–1950)
John von Neumann (1903–1957)
B. L. van der Waerden (1903–1996)
André Weil (1906–1998)
Heinrich Heesch (1906–1995)
Kurt Gödel (1906–1978)
Claude Chevalley (1909–1984)
Leonid V. Kantorovich (1912–1986)
Alan Turing (1912–1954)
George Dantzig (b. 1914)
Claude Shannon (1916–2001)
David Blackwell (b. 1919)
Daniel Gorenstein (1923–1992)
Jean-Pierre Serre (b. 1926)
Yutaka Taniyama (1927–1958)
Goro Shimura (b. 1928)

Wolfgang Haken (b. 1928)
Walter Feit (b. 1930)
John Thompson (b. 1932)
Kenneth Appel (b. 1932)
Paul Cohen (b. 1934)
Robert Langlands (b. 1936)
Gerhard Frey (b. 1944)
Andrew Wiles (b. 1953)

THE HISTORY OF MATHEMATICS

BRIEF VERSION

VICTOR J. KATZ

University of the District of Columbia

Boston San Francisco New York
London Toronto Sydney Tokyo Singapore Madrid
Mexico City Munich Paris Cape Town Hong Kong Montreal

Publisher: Greg Tobin
Acquisitions Editor: William Hoffman
Associate Editor: RoseAnne Johnson
Editorial Assistant: Mary Reynolds
Managing Editor: Karen Guardino
Production Supervisor: Cindy Cody
Marketing Manager: Pamela Laskey
Marketing Coordinator: Heather Peck
Manufacturing Buyer: Hugh Crawford
Designer: Barbara Atkinson
Text Designer: Leslie Haimes
Cover photo: Charles O'Rear/CORBIS

Library of Congress Cataloging-in-Publication Data

Katz, Victor J.
History of mathematics : brief version / Victor J. Katz.
p. cm.
Includes bibliographical references and index.
ISBN 0-321-16193-9
1. Mathematics–History. I. Title.

QA21.K35 2004
510′.9–dc21

2003056029

The cover image is of an astrolabe, an instrument used for finding altitudes of the sun, moon, and stars, usually in connection with determining one's latitude on earth. The reverse of the device often had tables and charts to assist with the calculations. Astrolabes were perfected in the Islamic world as early as the eighth century and were later taken over by Europeans.

Art Acknowledgements:

Page 5, figure 1.1 © The Rhind Mathematical Papyrus, 1979, National Council of Teachers of Mathematics. All Rights Reserved.
Page 129, figure 5.9 © "The Chinese Connection between the Pascal Triangle and the Solution of Numerical Equations of Any Degree" by Lam Lay-Yong, *Historia Mathematica*, Vol. 7, No. 4, November 1980. © 1980, Academic Press, Inc.
Page 245, figure 9.11, Courtesy of Dept. of Special Collections, Stanford University Libraries.
Page 513, figure 20.3 © Neuhart Donges Neuhart Designers

1 2 3 4 5 6 7 8 9 10 — DOC — 06050403

To Phyllis,
for her help and encouragement
and for everything else

Contents

Preface

Approach and Guiding Philosophy

In *A Call for Change: Recommendations for the Mathematical Preparation of Teachers of Mathematics*, the Mathematical Association of America's (MAA) Committee on the Mathematical Education of Teachers recommends that all prospective teachers of mathematics in schools

> develop an appreciation of the contributions made by various cultures to the growth and development of mathematical ideas; investigate the contributions made by individuals, both female and male, and from a variety of cultures, in the development of ancient, modern, and current mathematical topics; [and] gain an understanding of the historical development of major school mathematics concepts.

According to the MAA, knowledge of the history of mathematics shows students that mathematics is an important human endeavor. Mathematics was not discovered in the polished form seen in our textbooks, but often developed in intuitive and experimental fashion out of a need to solve problems. The actual development of mathematical ideas can be effectively used in exciting and motivating today's students.

My textbook *A History of Mathematics: An Introduction* grew out of the conviction that not only prospective school teachers of mathematics but also prospective college teachers of mathematics need a background in the history of the subject to teach it more effectively to their students. However, many readers felt that this text was too long. It clearly had far more material than could be reasonably covered in the typical one-semester course in the history of mathematics. Therefore, I have prepared a briefer version, one that allows the instructor to reach twentieth-century topics in mathematics in the course. This text, like the longer version, is designed for junior or senior mathematics majors who intend to teach in college or high school and thus concentrates on the history of those topics typically covered in an undergraduate curriculum or in elementary or high school. Because the history of any given mathematical topic often provides excellent ideas for teaching the topic, there is sufficient detail in each explanation of a concept for the future (or present) teacher of mathematics to develop a classroom lesson or series of lessons based on the concept's history. My hope is that the student and prospective teacher will gain from this book a knowledge of how we got here from there, a knowledge that will provide a deeper understanding of many of the important concepts of mathematics.

Distinguishing Features

Flexible Organization

Although the chief organization of the book is by chronological period, the material is organized topically within each period. The chapter headings reflect this organization. Thus, since many instructors believe that a history course should be taught topically, that is, first covering algebra, then geometry, then analysis, and then probability and statistics, it is very easy to use this text in that manner.

Astronomy and Mathematics

Because the development of astronomy is so intimately connected with the development of mathematics, the book contains substantial material on that subject. Ptolemy's geocentric astronomy is discussed in an early chapter, while the work of Copernicus and that of Kepler in developing the heliocentric theory are covered later on. We also discuss in some detail Newton's synthesis of this material, including his derivations of Kepler's laws, and then some of the "translation" of Newton's geometric ideas into Leibnizian analysis.

Non-Western Mathematics

A special effort has been made to consider mathematics developed in parts of the world other than Europe. Thus, there is substantial material on mathematics in China, India, and the Islamic world. The reader will see how certain mathematical ideas have arisen in many places, as people tried to answer similar questions.

Topical Exercises

Each chapter contains many exercises, some of which are simple computations while others help to fill the gaps in the mathematical arguments presented in the text. There are also some open-ended discussion questions, many of which ask students to think about how they would use historical material in the classroom. (Answers to some of the computational exercises are provided in an appendix.) Even if readers do not attempt many of the exercises, they should at least read them to gain a fuller understanding of the material of the chapter.

Additional Pedagogy

Given that a major audience for this text consists of prospective teachers of secondary mathematics, I have provided an appendix giving details on how to use the material of the text in teaching mathematical topics. There is a detailed listing of where the history of the various topics of the secondary curriculum may be found in the text; there are suggestions as to how to organize some of this material for classroom use; and there is a detailed time line which helps to relate the mathematical discoveries to other events happening in the world. Finally, given that students may have difficulty pronouncing the names of some mathematicians, the index has a special feature—a phonetic pronunciation guide.

Prerequisites

A working knowledge of one year of calculus is sufficient to understand most of the text. The mathematical prerequisites for some of the later chapters are more demanding, but the titles of the various sections indicate clearly what kind of mathematical knowledge is required.

What Has Been Left Out

Difficult choices had to be made to shorten the text. A comparison of this version with the longer version will show cuts in every section. But because new discoveries are constantly being made in the history of mathematics, this version also has additional material that did not appear earlier, especially relating to the twentieth century. Among the major cuts were the biographical material and the section on ethnomathematics. However, students can easily find biographical material online, especially at the St. Andrews web site: http://www-history.mcs.st-and.ac.uk/history/Mathematicians. And Marcia Ascher's two books *Ethnomathematics: A Multicultural View of Mathematical Ideas* (Pacific Grove, Calif.: Brooks/Cole, 1991) and *Mathematics Elsewhere* (Princeton: Princeton University Press, 2002) provide a wonderful summary of the ethnomathematics of many cultures around the world.

Acknowledgments

Many people have reviewed sections of the manuscript and have offered suggestions. They include

Richard Davitt, University of Louisville;
Michael J. Kallaher, Washington State University;
Mary Ann McLoughlin, College of Saint Rose;
William Blubaugh, University of Northern Colorado;
Richard W. Carey, University of Kentucky;
Joan Lukas, University of Massachusetts–Boston; and
Christopher A. Terry, Augusta State University.

As always, my wife Phyllis has been very supportive during the preparation of this book. I thank her for her help, both explicit and implicit. In particular, she kept reminding me to keep the needs of the students in mind as I was writing. I owe her much more than I can ever repay.

I also want to thank Bill Hoffman, Greg Tobin, RoseAnne Johnson, and Cindy Cody at Addison Wesley, who worked hard to make this book a reality, as well as Quica Ostrander and Jane Hoover at Lifland et al., Bookmakers for handling the production aspects. I hope that users of this book will continue to send me suggestions for improvement.

CHAPTER ONE

Egypt and Mesopotamia

Accurate reckoning. The entrance into the knowledge of all existing things and all obscure secrets.

—*Introduction to Rhind Mathematical Papyrus*

The opening quotation, taken from one of the few documentary sources on Egyptian mathematics, illustrates some of the difficulties in giving an accurate picture of ancient mathematics. Mathematics certainly existed in virtually every ancient civilization for which there are records. But in every one of these civilizations, mathematics was the domain of specially trained priests and scribes, officials whose job it was to develop and use mathematics for the benefit of the government in such areas as tax collection, measurement, building, trade, calendar making, and ritual practices. Yet, even though the origins of many mathematical concepts stemmed from their usefulness in these contexts, mathematicians always exercised their curiosity by extending these ideas far beyond the limits of practical necessity. And because mathematics was a tool of power, its methods were passed on to only the privileged few, often through an oral tradition. Hence, the written records are generally sparse and seldom provide much detail.

In recent years, however, a great deal of scholarly effort has gone into reconstructing the mathematics of ancient civilizations from whatever clues can be found. Naturally, all scholars do not agree on every point, but there is enough agreement that a reasonable picture can be presented of the mathematical knowledge of the ancient civilizations in Egypt and Mesopotamia. Our discussion of the mathematics of each of these civilizations begins with a brief survey of the society and a description of the sources from which current knowledge of the mathematics is derived.

1.1 Egypt

1.1.1 Introduction

Agriculture emerged in the Nile valley in Egypt nearly 7000 years ago, but the first dynasty to rule both Upper Egypt (the river valley) and Lower Egypt (the delta) dates from about 3100 BCE. The culture at the time of the first pharaohs included an elite of officials and priests, a luxurious court, and, for the kings themselves, a role as intermediaries between mortals and gods. This role fostered the development of Egypt's monumental architecture, including the pyramids, built as royal tombs, and the great temples at Luxor and Karnak. Writing began in Egypt at about the time of this first dynasty, and much of the earliest writing concerned accounting, primarily of various types of goods. There were several different systems of measurement, depending on the particular goods being measured. Since there were only a limited number of signs, however, the same signs meant different things in connection with different measuring systems. From the beginning of Egyptian writing, there were two styles: hieroglyphic writing for monumental inscriptions and hieratic, or cursive, writing, done with a brush and ink on papyrus. Greek domination of Egypt from about 300 BCE was responsible for the disappearance of both of these native Egyptian writing forms. Fortunately, Jean Champollion (1790–1832) was able to begin the process of understanding Egyptian writing early in the nineteenth century with the help of the multilingual inscriptions on the Rosetta stone—in hieroglyphics and in Greek as well as in the later demotic writing, a form of the hieratic writing used on papyri.

Modern knowledge of the mathematics of ancient Egypt comes not from the hieroglyphs in the temples but from two papyri containing collections of mathematical problems with their solutions: the *Rhind Mathematical Papyrus*, named for the Scotsman A. H. Rhind (1833–1863), who purchased it at Luxor in 1858, and the *Moscow Mathematical Papyrus*, purchased in 1893 by V. S. Golenishchev (d. 1947), who later sold it to the Moscow Museum of Fine Arts. The former papyrus was copied by the scribe A'h-mose in about 1650 BCE from an original about 200 years older, and it is approximately 18 feet long and 13 inches high. The latter papyrus dates from roughly the same period and is over 15 feet long but only about 3 inches high. Unfortunately, although a good many papyri have survived the ages because of the generally dry Egyptian climate, papyrus is very fragile. Thus, besides the two papyri mentioned, only a few short fragments of other original Egyptian mathematical papyri are still extant.

1.1.2 Number Systems and Computations

The Egyptians developed two different number systems, one for each of their two writing styles. In the hieroglyphic system, each of the first several powers of 10 was represented by a different symbol, beginning with the familiar vertical stroke for 1. Thus, 10 was represented by ∩, 100 by 𓍢, 1000 by 𓆼, and 10,000 by 𓂭. Whole numbers were then represented by appropriate repetitions of the symbols. For example, to represent 12,643, the Egyptians would write |||∩∩∩∩𓍢𓍢𓍢𓍢𓍢𓍢𓆼𓆼𓂭. (Note that the usual practice was to put the smaller digits on the left.)

The hieratic system, in contrast to the hieroglyphic, is a ciphered system. Here each number from 1 to 9 had a specific symbol, as did each multiple of 10 from 10 to 90 and each multiple of 100 from 100 to 900, etc. A given number was represented with the appropriate combination of symbols; for example, 37 was written by putting the symbol for 7 next to that for 30. Since the symbol for 7 was [hieratic 7] and that for 30 was [hieratic 30], 37 was written [hieratic 37]. Similarly, since 3 was written as [hieratic 3], 40 as [hieratic 40], and 200 as [hieratic 200], the symbol for 243 was [hieratic 243]. Although a zero symbol is not necessary in a ciphered system, the Egyptians did have such a symbol. It did not occur in mathematical papyri, however, but in papyri dealing with architecture, where it is used to denote the bottom leveling line in the construction of a pyramid, and in papyri on accounting, where it is used in balance sheets to indicate that the disbursements and income are equal.

Once there is a system of writing numbers, it is only natural for a civilization to devise rules for computation with these numbers. The particular rules used are closely related to the systems of writing numbers. These rules may be considered some of the earliest algorithms developed.

An **algorithm** is an ordered list of instructions designed to produce an answer to a given type of problem. Ancient peoples created algorithms of all sorts to handle many different problems. In fact, ancient mathematics can be characterized as algorithmic, in contrast to the later Greek emphasis on theory. In most ancient mathematical documents, the author describes a problem to be solved and then proceeds to use an algorithm, either explicit or implicit, that provides the solution. There is little concern in the documents for how the algorithm was discovered, why it works, or what its limitations are. In most cases, there are simply many examples of the use of the algorithm, often in increasingly complex situations. Nevertheless, as we discuss these various algorithms, we will indicate possible origins and justifications for them, the possible answers the scribes gave to their students who asked the eternal question "Why?"

Addition and subtraction are quite simple in a hieroglyphic grouping system: Combine the units, then the tens, then the hundreds, and so on. Whenever you have ten of one type of symbol, replace them with one symbol for the next higher power. Hence, to add 783 and 275, put 𓏺𓏺𓏺𓎆𓎆𓎆𓎆𓎆𓎆𓎆𓎆𓍢𓍢𓍢𓍢𓍢𓍢𓍢 and 𓏺𓏺𓏺𓏺𓏺𓎆𓎆𓎆𓎆𓎆𓎆𓎆𓍢𓍢 together to get 𓏺𓏺𓏺𓏺𓏺𓏺𓏺𓏺𓎆𓎆𓎆𓎆𓎆𓎆𓎆𓎆𓎆𓎆𓎆𓎆𓎆𓎆𓎆𓍢𓍢𓍢𓍢𓍢𓍢𓍢𓍢𓍢. Since there are fifteen 𓎆's, replace ten of them with one 𓍢. This then gives ten of the latter. Replace these with one 𓆼. The final answer is 𓏺𓏺𓏺𓏺𓏺𓏺𓏺𓏺𓎆𓎆𓎆𓎆𓎆𓆼, or 1058. Subtraction is done similarly. Whenever "borrowing" was necessary, one of the symbols would be converted to ten symbols of the next lower power. Such a simple algorithm for addition and subtraction is not possible in the hieratic system. Probably, the scribes who used the hieratic system simply memorized basic addition tables.

The Egyptian algorithm for multiplication was based on a continual doubling process. To multiply the number b by the number a, the scribe would first write down the pair 1, b. He would then double each number in the pair repeatedly, until the next doubling would cause the first element of the pair to exceed a. Then, having determined the powers of 2 that add to a, the scribe would add the corresponding multiples of b to get his answer. For example, to multiply 12 by 13 the scribe would set down the following lines:

ˋ1	12
2	24
ˋ4	48
ˋ8	96

At this point he would stop, because the next doubling would give him 16 in the first column, which is larger than 13. He would then check off those multipliers that added to 13, namely, 1, 4 and 8, and add the corresponding numbers in the other column. The result would be written as:

Totals	13	156

There is no record of how the scribe did the doubling. The answers are simply written down. Perhaps the scribe had memorized an extensive two times table. In fact, there is some evidence that doubling was a standard method of computation in areas of Africa to the south of Egypt and that therefore the Egyptian scribes learned from their southern colleagues. In addition, the scribes were somehow aware that every positive integer could be uniquely expressed as the sum of powers of two. That fact provides the justification for the procedure. How was it discovered? The best guess is that it was discovered by experimentation and then passed down as tradition.

Because division is the inverse of multiplication, a problem such as $156 \div 12$ would be stated as "multiply 12 so as to get 156." The scribe would then write down the same lines for the multiplication problem above. This time, however, he would check off the lines containing numbers in the right-hand column that sum to 156; in this example, those would be 12, 48, and 96. Then the sum of the corresponding numbers on the left, namely, 1, 4, and 8, would give the answer 13. Of course, division does not always "come out even." When it did not, the Egyptians resorted to fractions.

The Egyptians only dealt with unit fractions, or "parts" (fractions with numerator 1), with the single exception of 2/3, perhaps because these fractions are the most "natural." The fraction $1/n$ (the nth part) is represented in hieroglyphics by the symbol for the integer n with the symbol 𓂋 above. In the hieratic system, a dot is used instead. So 1/7 is denoted in the former system by 𓂋||||||| and in the latter by 𝓏̇. The single exception, 2/3, had a special symbol: 𓂎 in hieroglyphic and ϒ in hieratic. (The former symbol is indicative of the reciprocal of $1\frac{1}{2}$.) In what follows, however, the notation $\overline{n}$ will represent $1/n$, and $\overline{\overline{3}}$ will represent 2/3.

Because fractions show up as the result of divisions that do not come out evenly, surely there is a need to be able to deal with fractions other than unit fractions. It was in this connection that the most intricate of the Egyptian mathematical techniques developed, the representation of any fraction in terms of unit fractions. Wherever modern mathematicians would use a nonunit fraction, the Egyptians simply wrote a sum of unit fractions. For example, problem 3 of the *Rhind Mathematical Papyrus* asks how to divide 6 loaves among 10 men. The answer given is that each man gets $\overline{2}\ \overline{10}$ loaves (that is, $\frac{1}{2} + \frac{1}{10}$). The scribe checks this by multiplying this value by 10. We may regard the scribe's answer as more cumbersome than our answer of $\frac{3}{5}$, but in some sense the actual division is easier to accomplish this way. Five of the loaves are divided in half, the sixth one in tenths, and then each man is given one half plus one tenth. It is then clear to all that every man has the same portion of bread. Cumbersome or not, this Egyptian method of unit fractions was used throughout the Mediterranean basin for over 2000 years.

In multiplying whole numbers, the important step is the doubling step. So too in multiplying fractions, the scribe had to be able to express the double of any unit fraction. For example, in the bread problem, the check of the solution was written as follows:

1	$\overline{2}\ \overline{10}$
'2	$1\ \overline{5}$
4	$2\ \overline{3}\ \overline{15}$
'8	$4\ \overline{\overline{3}}\ \overline{10}\ \overline{30}$
10	6

How are these doubles formed? To double $\overline{2}\ \overline{10}$ is easy; because each denominator is even, it is merely halved. In the next line, however, $\overline{5}$ must be doubled. It was here that the scribe had to use a table to get the answer $\overline{3}\ \overline{15}$ (that is, $2 \cdot \frac{1}{5} = \frac{1}{3} + \frac{1}{15}$). In fact, the first section of the *Rhind Papyrus* is a table of the division of 2 by every odd integer from 3 to 101 (Fig. 1.1), and the Egyptian scribes realized that the result of multiplying $\overline{n}$ by 2 is the same as that of dividing 2 by n. It is not known how the division table was constructed, but there are several scholarly accounts giving hypotheses for the scribes' methods. In any case, the solution of *Rhind* problem 3 depends on using that table twice, first as already indicated and second, in the next step, where the double of $\overline{15}$ is given as $\overline{10}\ \overline{30}$ (or $2 \cdot \frac{1}{15} = \frac{1}{10} + \frac{1}{30}$). The final step in this problem involves the addition of $1\ \overline{5}$ to $4\ \overline{\overline{3}}\ \overline{10}\ \overline{30}$, and here the scribe just gave the answer. Again, the conjecture is that an extensive table existed for such addition problems. The *Egyptian Mathematical Leather Roll*, which dates from about 1600 BCE, contains a short version of such an addition table. Several other tables for dealing with

FIGURE 1.1 Transcription and hieroglyphic translation of $2 \div 3$, $2 \div 5$, and $2 \div 7$ from the *Rhind Mathematical Papyrus*. (Source: *The Rhind Mathematical Papyrus*, N.C.T.M.)

unit fractions and a multiplication table for the special fraction 2/3 have survived. It thus appears that the arithmetic algorithms used by the Egyptian scribes involved both extensive knowledge of basic tables for addition, subtraction, and doubling and a definite procedure for breaking multiplication and division problems into steps, each of which could be done using the tables.

Besides the basic procedure of doubling, the Egyptian scribes used other techniques in performing arithmetic calculations. For example, they could find halves of numbers as well as multiply by 10; they could figure out what fractions have to be added to a given mixed number to get the next whole number; and they could determine by what fraction a given whole number needs to be multiplied to get a given fraction. These procedures are illustrated in problem 69 of the *Rhind Papyrus*, which includes the division of 80 by $3\ \overline{2}$ and its subsequent check:

1	$3\ \overline{2}$	‘1	$22\ \overline{\overline{3}}\ \overline{7}\ \overline{21}$
10	35	‘2	$45\ \overline{3}\ \overline{4}\ \overline{14}\ \overline{28}\ \overline{42}$
‘20	70	‘$\overline{2}$	$11\ \overline{3}\ \overline{14}\ \overline{42}$
‘$\overline{2}$	7	$3\ \overline{2}$	80
‘$\overline{\overline{3}}$	$2\ \overline{3}$		
‘$\overline{21}$	$\overline{6}$		
‘$\overline{7}$	$\overline{2}$		
$22\ \overline{\overline{3}}\ \overline{7}\ \overline{21}$	80		

In the second line, the scribe has taken advantage of the decimal nature of his notation to give immediately the product of $3\ \overline{2}$ by 10. In the fifth line, he has used the 2/3 multiplication table mentioned earlier. The scribe then realized that since the numbers in lines three through five of the right-hand column add to $79\ \overline{3}$, he needed to add $\overline{2}$ and $\overline{6}$ in that column to get 80. Thus, because $6 \times 3\ \overline{2} = 21$ and $2 \times 3\ \overline{2} = 7$, it follows that $\overline{21} \times 3\ \overline{2} = \overline{6}$ and $\overline{7} \times 3\ \overline{2} = \overline{2}$, as indicated in the sixth and seventh lines. The check shows several uses of the table of division by 2 as well as great facility in addition.

1.1.3 Linear Equations and Proportional Reasoning

The mathematical problems the scribes could solve, as illustrated in the *Rhind* and *Moscow* papyri, deal with linear equations, proportions, and geometry. For example, the Egyptian papyri present two different procedures for dealing with what modern mathematicians call linear equations.

First, problem 19 of the *Moscow Papyrus* uses the normal current technique to find the number such that if it is taken $1\frac{1}{2}$ times and then 4 is added, the sum is 10. In modern notation, the equation is simply $1\frac{1}{2}x + 4 = 10$. The scribe proceeds as follows: "Calculate the excess of this 10 over 4. The result is 6. You operate on $1\frac{1}{2}$ to find 1. The result is $\frac{2}{3}$. You take $\frac{2}{3}$ of this 6. The result is 4. Behold, 4 says it. You will find that this is correct." Namely, after subtracting 4, the scribe notes that the reciprocal of $1\frac{1}{2}$ is $\frac{2}{3}$ and then multiplies 6 by this quantity. Similarly, problem 35 of the *Rhind Papyrus* asks to find the size of a scoop that fills a 1-hekat measure in $3\frac{1}{3}$ trips. The scribe solves the equation, which would today be written as $3\frac{1}{3}x = 1$ by dividing 1 by $3\frac{1}{3}$. He writes the answer as $\overline{5}\ \overline{10}$ and proceeds to prove that the result is correct.

The Egyptians' more common technique of solving a linear equation, however, is the method of **false position**, the method of assuming a convenient but probably incorrect answer and then adjusting it by using proportionality. For example, problem 26 of the *Rhind Papyrus* asks to find a quantity such that when it is added to $\frac{1}{4}$ of itself, the result is 15. The scribe's solution is as follows: "Assume [the answer is] 4. Then $1\ \overline{4}$ of 4 is 5. ... Multiply 5 so as to get 15. The answer is 3. Multiply 3 by 4. The answer is 12." In modern notation, the problem is to solve $x + \frac{1}{4}x = 15$. The first guess is 4, because $\frac{1}{4}$ of 4 is an integer. But then the scribe notes that $4 + \frac{1}{4} \cdot 4 = 5$. To find the correct answer, he must multiply 4 by the ratio of 15 to 5, namely, 3. The *Rhind Papyrus* has several similar problems, all solved using false position. The step-by-step procedure of the scribe can therefore be considered as an algorithm for the solution of a linear equation of this type. There is, however, no discussion of how the algorithm was discovered or why it works. But it is evident that the Egyptian scribes understood the basic idea of proportionality of two quantities.

This understanding is further exemplified in the solution of more explicit proportion problems. For example, problem 75 of the *Rhind Papyrus* asks for the number of loaves of *pesu* 30 that can be made from the same amount of flour as 155 loaves of *pesu* 20. (*Pesu* is the Egyptian measure for the inverse "strength" of bread and can be expressed as *pesu* = [number of loaves]/[number of hekats of grain], where a hekat was a dry measure approximately equal to $\frac{1}{8}$ bushel.) The problem is thus to solve the proportion $x/30 = 155/20$. The scribe accomplished this by dividing 155 by 20 and multiplying the result by 30 to get $232\frac{1}{2}$. Similar problems occur elsewhere in the *Rhind Papyrus* and in the *Moscow Papyrus*.

On the other hand, the method of false position is also used in the only quadratic equation extant in the Egyptian papyri. On the *Berlin Papyrus*, a small fragment dating from approximately the same time as the other papyri, is a problem in which a square area of 100 square cubits is to be divided into two other squares, where the ratio of the sides of the two squares is 1 to $\frac{3}{4}$. The scribe begins by assuming that in fact the sides of the two needed squares are 1 and $\frac{3}{4}$, then calculates the sum of the areas of these two squares to be $1^2 + (\frac{3}{4})^2 = 1\frac{9}{16}$. But the desired sum of the areas is 100. The scribe realizes that he cannot compare areas directly but must compare their sides. So he takes the square root of $1\frac{9}{16}$, namely, $1\frac{1}{4}$, and compares this to the square root of 100, namely, 10. Since 10 is 8 times as large as $1\frac{1}{4}$, the scribe concludes that the sides of the two other squares must be 8 times the original guesses, namely, 8 and 6 cubits, respectively.

There are numerous more complicated problems in the surviving mathematical papyri. For example, problem 64 of the *Rhind Papyrus* reads: "If it is said to thee, divide 10 hekats of barley among 10 men so that the difference of each man and his neighbor in hekats of barley is $\frac{1}{8}$, what is each man's share?" It is understood in this problem, as in similar problems elsewhere in the papyrus, that the shares are to be in arithmetic progression. The average share is 1 hekat. The largest share could be found by adding $\frac{1}{8}$ to this average share half the number of times as there are differences. However, since there is an odd number (9) of differences, the scribe instead adds half of the common difference ($\frac{1}{16}$) a total of 9 times to get $1\frac{9}{16}$ ($1\ \overline{2}\ \overline{16}$) as the largest share. He finishes the problem by subtracting $\frac{1}{8}$ from this value 9 times to get each share.

A final problem, problem 23 of the *Moscow Papyrus*, is what is often thought of today as a "work" problem: "Regarding the work of a shoemaker, if he is cutting out only, he can do 10 pairs of sandals per day; but if he is decorating, he can do 5 per day. As for the number he can both cut and decorate in a day, what will that be?" Here the scribe notes that the shoemaker cuts 10 pairs of sandals in one day and decorates 10 pairs of sandals in two days, so that it takes three days for him to both cut and decorate 10 pairs. The scribe then divides 10 by 3 to find that the shoemaker can cut and decorate $3\frac{1}{3}$ pairs in one day.

1.1.4 Geometry

The Egyptian scribes certainly knew how to calculate the areas of rectangles, triangles, and trapezoids by normal modern methods. It is their calculation of the area of a circle, however, that is particularly interesting. Problem 50 of the *Rhind Papyrus* reads: "Example of a round field of diameter 9. What is the area? Take away $\frac{1}{9}$ of the diameter; the remainder is 8. Multiply 8 times 8; it makes 64. Therefore, the area is 64." In other words, the Egyptian scribe is using the formula $A = (d - d/9)^2 = (\frac{8}{9}d)^2$. A comparison with the formula $A = (\pi/4)d^2$ shows that the Egyptian value for the constant π in the case of area was $\frac{256}{81} = 3.16049\ldots$. Where did the Egyptians get this value, and why was the answer expressed as the square of $\frac{8}{9}d$ rather than in modern terms as a multiple (here $\frac{64}{81}$) of the square of the diameter?

A hint is given in problem 48 of the same papyrus, in which is shown the figure of an octagon inscribed in a square of side 9 (Fig. 1.2). There is no statement of the problem, however, only a bare computation of $8 \times 8 = 64$ and $9 \times 9 = 81$. If the scribe had inscribed a circle in the same square, he would have seen that its area was approximately that of the octagon. What is the size of the octagon? It depends on how one interprets the diagram in the papyrus. If one believes the octagon to be formed by cutting off four corner triangles, each with area $4\frac{1}{2}$, then the area of the octagon is $\frac{7}{9}$ that of the square, namely, 63. The scribe therefore might have simply taken the area of the circle as $A = \frac{7}{9}d^2$ $(= \frac{63}{81}d^2)$. But since he wanted to find a square whose area was equal to the given circle, he may have approximated $\frac{63}{81}$ by $(\frac{8}{9})^2$, thus giving the area of the circle in the form $[(\frac{8}{9})d]^2$, indicated in problem 50 of the papyrus. On the other hand, in the diagram, the octagon does not look symmetric. So perhaps the octagon was formed by cutting off from the square of side 9 two diagonally opposite corner triangles, each equal to $4\frac{1}{2}$, and two other corner triangles, each equal to 4. This octagon then has area 64, as explicitly written on the papyrus, and thus this may be the square that the scribe wanted, which was equal in area to a circle.

FIGURE 1.2 Octagon inscribed in a square of side 9, from problem 48 of the *Rhind Mathematical Papyrus*

It should be noted that problem 50 is not an isolated problem of finding the area of a circle. In fact, there are several problems in the *Rhind Papyrus* in which the scribe uses the rule $V = Bh$ to calculate the volume of a cylinder where B, the area of the base, is calculated by this circle rule. The scribes also knew how to calculate the volume of a rectangular box, given its length, width, and height.

Because one of the prominent forms of building in Egypt was the pyramid, one might expect to find a formula for the volume of a pyramid. Unfortunately, such a formula does not appear in any extant document. The *Rhind Papyrus* does have several problems dealing with the *seked* (slope) of a pyramid; this is measured as so many horizontal units to one vertical unit rise. The workers building the pyramids, or at least their foremen, had to be aware of this value as they built. Since the *seked* is in effect the cotangent of the angle of slope of the pyramid's faces, one can easily calculate the angles given the values appearing in the problems. It is not surprising that these calculated angles closely approximate the actual angles used in the construction of the three major pyramids at Giza.

The *Moscow Papyrus*, however, does have a fascinating formula related to pyramids, namely, the formula for the volume of a truncated pyramid (problem 14): "If someone says to you: a truncated pyramid of 6 for the height by 4 on the base by 2 on the top, you are to square this 4, the result is 16. You are to double 4; the result is 8. You are to square this 2; the result is 4. You are to add the 16 and the 8 and the 4; the result is 28. You are to take $\frac{1}{3}$ of 6; the result is 2. You are to take 28 two times; the result is 56. Behold, the volume is 56. You will find that this is correct." If this algorithm is translated into a modern formula, with the length of the lower base denoted by a, that of the upper base by b, and the height by h, it gives the correct result $V = (h/3)(a^2 + ab + b^2)$. Although no papyrus gives the formula $V = \frac{1}{3}a^2h$ for a completed pyramid of square base a and height h, it is a simple matter to derive that formula from the given formula by simply setting $b = 0$. We therefore assume that the Egyptians were aware of this result. On the other hand, it takes a higher level of algebraic skill to derive the volume formula for the truncated pyramid from that for the complete pyramid. Still, although many ingenious suggestions involving dissection have been given, no one knows for sure how the Egyptians found their algorithm.

No one knows either how the Egyptians found their procedure for determining the surface area of a hemisphere. But they succeeded in problem 10 of the *Moscow Papyrus*: "A basket with a mouth opening of $4\frac{1}{2}$ in good condition, oh let me know its surface area. First, calculate $\frac{1}{9}$ of 9, since the basket is $\frac{1}{2}$ of an egg-shell. The result is 1. Calculate the remainder as 8. Calculate $\frac{1}{9}$ of 8. The result is $\frac{2}{3}\ \frac{1}{6}\ \frac{1}{18}$ [that is, $\frac{8}{9}$]. Calculate the remainder from these 8 after taking away those [$\frac{8}{9}$]. The result is $7\frac{1}{9}$. Reckon with $7\frac{1}{9}$ four and one-half times. The result is 32. Behold, this is its area. You will find that it is correct." Evidently, the scribe has calculated the surface area S of this basket of diameter $d = 4\frac{1}{2}$ by first taking $\frac{8}{9}$ of $2d$, then taking $\frac{8}{9}$ of the result, and finally multiplying by d. As a modern formula, this result would be $S = 2(\frac{8}{9}d)^2$, or, since the area A of the circular opening of this hemispherical basket is given by $A = (\frac{8}{9}d)^2$, the result could be written as $S = 2A$, the correct answer.

1.2 Mesopotamia

1.2.1 Introduction

The Mesopotamian civilization is older than the Egyptian, having developed in the Tigris and Euphrates river valleys beginning sometime in the fifth millennium BCE. Many different governments ruled this region over the centuries. Initially, there were many small city-states, but then the area was unified under a dynasty from Akkad, which lasted from approximately 2350 to 2150 BCE. Shortly thereafter, the Third Dynasty of Ur expanded rapidly, until it controlled most of southern Mesopotamia. This dynasty produced a very centralized bureaucratic state. In particular, it created a large system of scribal schools to train members of the bureaucracy. Although the Ur Dynasty collapsed around 2000 BCE, the small city-states that succeeded it still demanded numerate scribes. By 1700 BCE, Hammurapi, the ruler of Babylon, one of these city-states, had expanded his rule to much of Mesopotamia and instituted a legal system to help regulate his empire. As is standard in the history of mathematics, we will use the adjective "Babylonian" to refer to the civilization and culture of Mesopotamia, even though Babylon itself was the major city of the area for only a limited time.

Writing began in Mesopotamia, quite possibly in the southern city of Uruk, at about the same time as in Egypt, namely, at the end of the fourth millennium BCE. As in Egypt, writing began with the needs of accountancy—the necessity of recording and managing labor and the flow of goods. Initially, small clay tokens were used for this purpose, often enclosed in clay envelopes, sealed with an official's personal seal. The tokens were marked with pictographs representing the type of goods being considered. Thus, for example, five ovoids represented five jars of oil. But the system eventually evolved to the point where there was only one symbol representing the goods and several symbols representing numbers of various sizes. And if the goods in question were such that it was convenient to have several different units of measure, as in the case of grain, there were different symbols for each type of unit.

Certain of these units of measure, but by no means all, bore the relationship that one "large" unit was equal to 60 "small" units. At some point, then, the system of recording numbers developed to the point where the same digit 1 represented 60 as well. We do not know why the Babylonians decided to have one large unit represent 60 small units and then adapted this method for their numeration system. One plausible conjecture is that 60 is evenly divisible by many small integers. Therefore, fractional values of the large unit could easily be expressed as integral values of the small. But eventually, the Babylonians did develop a sexagesimal (base-60) place value system, which in the third millennium BCE became the standard system used throughout Mesopotamia. By that time, too, writing by using a stylus on a moist clay tablet began to be used in a wide variety of contexts. Thousands of these tablets have been excavated during the past 150 years. It was Henry Rawlinson (1810–1895) who, by the mid-1850s, was first able to translate this cuneiform writing by comparing the Persian and Babylonian inscriptions on a rock face at Behistun (in modern Iran) describing a military victory of King Darius I of Persia (sixth century BCE).

A large number of these tablets are mathematical in nature, containing mathematical problems and solutions or mathematical tables. Several hundreds of these have been copied, translated, and explained. These tablets, generally rectangular but occasionally round, usually fit comfortably into the hand and are an inch or so thick. Some, however, are as small as a postage stamp, and others are as large as an encyclopedia volume. It is fortunate that these tablets are virtually indestructible, because they are the only source for Mesopotamian mathematics. The written tradition that they represent died out under Greek domination in the last centuries BCE and was totally lost until the nineteenth century. The great majority of the excavated tablets date from the time of Hammurapi; some small collections date from the earliest beginnings of Mesopotamian civilization, from the centuries surrounding

1000 BCE, and from the Seleucid period, around 300 BCE. The discussion in this section, however, will generally deal with the mathematics of the "Old Babylonian" period (the time of Hammurapi).

1.2.2 Methods of Computation

The Babylonians at various times used different systems of numbers, but the standardized system that the scribes generally used for calculations in the Old Babylonian period was a base-60 place value system together with a grouping system based on 10 to represent numbers up to 59. Thus, a vertical stylus stroke on a clay tablet (𒁹) represented 1, and a tilted stroke (𒌋) represented 10. By grouping, scribes would, for example, represent 37 by 𒌋𒌋𒌋𒁹𒁹𒁹𒁹𒁹𒁹𒁹. For numbers greater than 59, the powers of 60, the base of this place value system, are represented by "places" rather than symbols: The digit in each place represents the number of each power to be counted. Hence, $3 \times 60^2 + 42 \times 60 + 9$ (or 13,329) was represented by the Babylonians as 𒁹𒁹𒁹 𒌋𒌋𒌋𒌋𒁹𒁹 𒁹𒁹𒁹𒁹𒁹𒁹𒁹𒁹𒁹. (This will be written from now on as 3,42,09 rather than with the Babylonian strokes.) The Old Babylonians did not use a symbol for zero, but often left an internal space if a given number was missing a particular power. There would not be a space at the end of a number, so it is difficult to distinguish $3 \times 60 + 42$ (3,42) from $3 \times 60^2 + 42 \times 60$ (3,42,00). Sometimes, however, they would give an indication of the absolute size of a number by writing an appropriate word, typically a metrological one, after the numeral. Thus, "3 42 sixty" represented 3,42, while "3 42 thirty-six hundred" meant 3,42,00. On the other hand, the Babylonians never used a symbol to represent zero in the sense of "nothingness," as in the modern notation $42 - 42 = 0$.

That the Babylonians used tables in the process of performing arithmetic computations is proved by extensive direct evidence. Many of the preserved tablets are in fact multiplication tables. No addition tables have turned up, however. Because over 200 Babylonian table texts have been analyzed, it may be assumed that addition tables did not exist and that the scribes knew their addition procedures well enough that they could write down the answers when needed. On the other hand, there are many examples of "scratch tablets," on which a scribe has performed various calculations in the process of solving a problem. In any case, since the Babylonian number system was a place value system, the actual algorithms for addition and subtraction, including carrying and borrowing, may well have been similar to modern ones. For example, to add 23,37 (= 1417) to 41,32 (= 2492), a scribe would first add 37 and 32 to get 1,09 (= 69). He would write down 09 and carry 1 to the next column. Then $23 + 41 + 1 = 1{,}05$ (= 65), and the final result is 1,05,09 (= 3909).

Because the place value system was based on 60, the multiplication tables were extensive. Any given table listed the multiples of a particular number, say 9, from 1×9 to 20×9 and then gave 30×9, 40×9, and 50×9 (Fig. 1.3). To find the product 34×9, a scribe simply added the two results $30 \times 9 = 4{,}30$ (= 270) and $4 \times 9 = 36$ to get 5,06 (= 306). For multiplication of two- or three-digit sexagesimal numbers, several such tables were needed. The exact algorithm the Babylonians used for such multiplications—where the

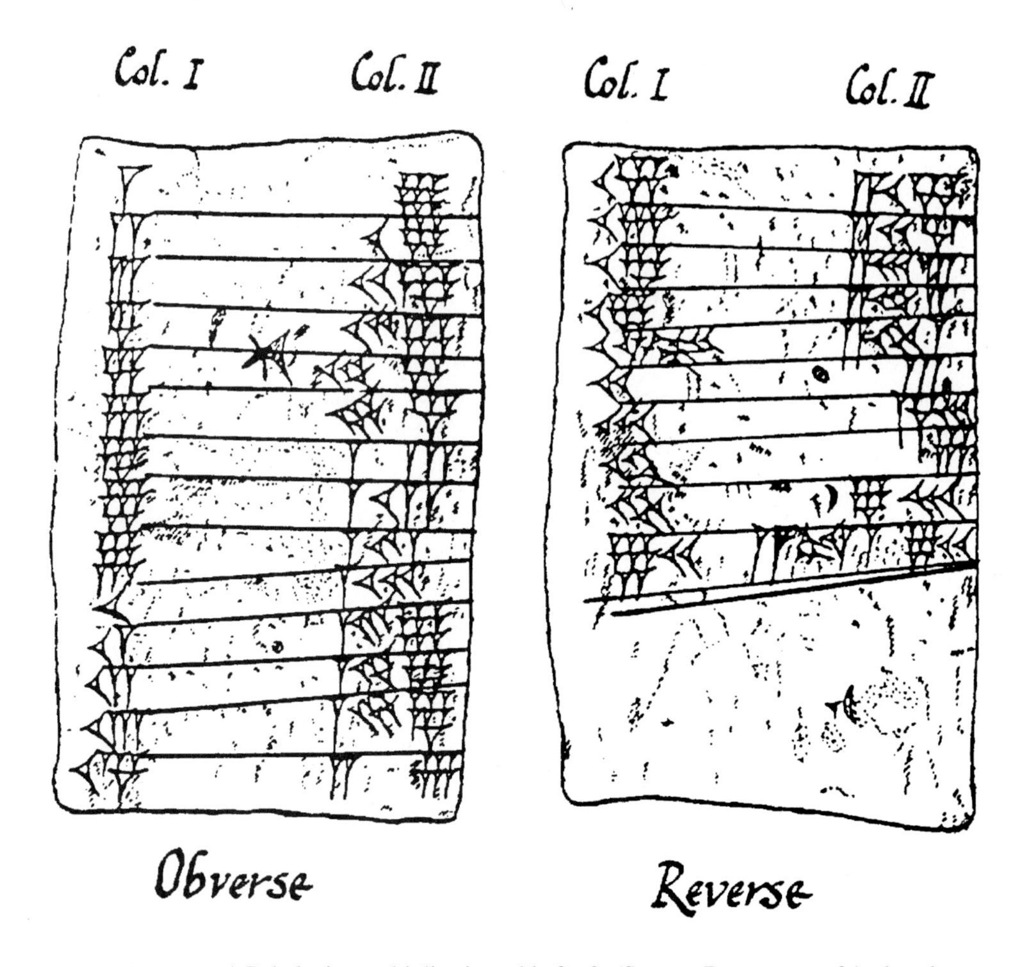

FIGURE 1.3 A Babylonian multiplication table for 9. (Source: Department of Archaeology, University of Pennsylvania)

partial products are written and how the final result is obtained—is not known, but it may well have been similar to our own.

One might think that a complete system of Babylonian tables would include a table for each integer from 2 to 59. Such was not the case, however. In fact, although there are no tables for 11, 13, 17, for example, there are tables for 1,15, 3,45, and 44,26,40. We do not know precisely why the Babylonians made these choices; however, with the single exception of 7, all multiplication tables found so far are for **regular** sexagesimal numbers, that is, numbers whose reciprocal is a terminating sexagesimal fraction. The Babylonians treated all fractions as sexagesimal fractions, analogous to modern decimal fractions. Namely, the first place after the "sexagesimal point" (which in this text will be denoted with a semicolon) represents 60ths, the next place 3600ths, etc. Thus, the reciprocal of 48 is the sexagesimal fraction 0;1,15, which represents $1/60 + 15/60^2$, while the reciprocal of 1,21 (= 81) is 0;0,44,26,40, or $44/60^2 + 26/60^3 + 40/60^4$. Because the Babylonians did not indicate an initial 0 or sexagesimal point, the latter number would be written as 44,26,40. As noted, there exist multiplication tables for this regular number. In such a table, there is no indication of the absolute size of the number, nor is one necessary. When the Babylonians used the table, of course, they realized that, as in today's decimal calculations, the eventual placement of the sexagesimal point depended on the absolute size of the numbers involved, and this placement was made based on context.

Besides multiplication tables, there are also extensive tables of reciprocals, of which a portion is reproduced here. A table of reciprocals is a list of pairs of numbers whose product is 1 (where the 1 can represent any power of 60). Like the multiplication tables, these tables contained only regular sexagesimal numbers.

2	30	16	3,45	48	1,15
3	20	25	2,24	1,04	56,15
10	6	40	1,30	1,21	44,26,40

The reciprocal tables were used in conjunction with the multiplication tables to do division. Thus, the multiplication table for 1,30 (= 90) served not only to give multiples of that number, but also, since 40 is the reciprocal of 1,30, to do divisions by 40. In other words, the Babylonians considered the problem $50 \div 40$ to be equivalent to $50 \times \frac{1}{40}$, or, in sexagesimal notation, to $50 \times 0;1,30$. The multiplication table for 1,30, part of which appears below, then gives 1,15 (or 1,15,00) as the product. The appropriate placement of the sexagesimal point gives 1;15 ($= 1\frac{1}{4}$) as the correct answer to the division problem.

1	1,30	10	15	30	45
2	3	11	16,30	40	1
3	4,30	12	18	50	1,15

1.2.3 Geometry

The Babylonians applied their sexagesimal place value system to a wide range of problems. For example, they developed procedures for determining areas and volumes of various kinds of figures. They worked out algorithms to determine square roots. They solved problems that modern mathematicians would interpret in terms of linear and quadratic equations, problems often related to agriculture or building. In fact, the mathematical tablets themselves are generally concerned with the solution of problems, to which various mathematical techniques are applied. We will look at some of the problems the Babylonians solved and try to figure out what lies behind their methods. In particular we will see that the reasons behind many of the Babylonian procedures come from a tradition different from the accountancy traditions with which Babylonian mathematics began. This second tradition was the "cut-and-paste" geometry of surveyors, who had to measure fields and lay out public works projects. Not only did these manipulations of squares and rectangles develop into procedures for determining square roots and finding Pythagorean triples, but they also developed into what modern mathematicians think of as "algebra."

As we work through the Babylonian problems, keep in mind that, like the Egyptians, the Babylonian scribes did not have any symbolism for operations or unknowns. Thus, solutions are presented with purely verbal techniques. Remember also that the Babylonians often thought about problems in ways that were different from modern approaches. Thus, even though their methods are usually correct, they may seem strange.

As one example of the scribes' different methods, consider their procedures for determining lengths and areas. In general, they presented what we would call "formulas" in terms of what are today called coefficient lists, lists of constants that embody mathematical relationships between certain aspects of various geometrical figures. Thus, the number 0;52,30 ($= \frac{7}{8}$) as the coefficient for the height of a triangle means that the altitude of an equilateral triangle is $\frac{7}{8}$ of the base, while the number 0;26,15 ($= \frac{7}{16}$) as the coefficient for area means that the area of an equilateral triangle is $\frac{7}{16}$ times the square of a side. (Note, of course, that these results are only approximately correct, in that they both approximate $\sqrt{3}$ by $\frac{7}{4}$.) In each case, the idea is that the "defining component" for the triangle is the side.

We too use the length of a side as the "defining component" for an equilateral triangle. But for a circle, we generally use the radius r as the defining component. Thus, we give formulas for the circumference and area in terms of r. The Babylonians, on the other hand, took the circumference as the defining component of a circle. Thus, they gave two coefficients for the circle: 0;20 ($= \frac{1}{3}$) for the diameter and 0;05 ($= \frac{1}{12}$) for the area. The first coefficient means that the diameter is one-third of the circumference; the second means that the area is one-twelfth of the square of the circumference. For example, on the tablet YBC 7302 (Yale Babylonian Collection), there is a circle with the numbers 3 and 9 written on the outside and the number 45 written on the inside (Fig. 1.4). The interpretation of this is that the circle has circumference 3 and that the area is found by dividing $9 = 3^2$ by 12 to get 0;45 ($= \frac{3}{4}$). Another tablet, Haddad 104, illustrates that circle calculations virtually always use the circumference. One problem on this tablet is to find the area of the cross section of a log of diameter 1;40 ($= 1\frac{2}{3}$). Rather than determine the radius, the scribe first multiplies by 3 to find that the circumference is equal to 5, then squares 5 and multiplies by $\frac{1}{12}$ to get the area 2;05 ($= 2\frac{1}{12}$). Note further, of course, that the Babylonian value for π, the ratio of circumference to diameter, is 3; this value produces the value $4\pi = 12$ as the constant by which to divide the square of the circumference to give the area.

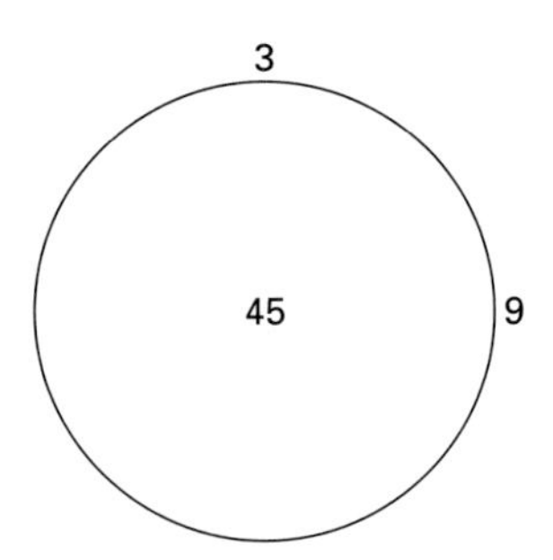

FIGURE 1.4 Circle from Babylonian tablet YBC 7302

There are also Babylonian coefficients for other figures bounded by circular arcs. In particular, the Babylonians calculated areas of two different double bows, the "barge," made up of two quarter-circle arcs, and the "bull's eye," composed of two third-circle arcs (Fig. 1.5). These figures were analogous to the circle, in that their defining component was the arc making up one side. Thus, areas of these two figures are given as $\frac{2}{9}a^2$ and $\frac{9}{32}a^2$, respectively, where in each case a is the length of that arc. These results are accurate under the assumptions that the area of the circle is $C^2/12$ and that $\sqrt{3} = \frac{7}{4}$.

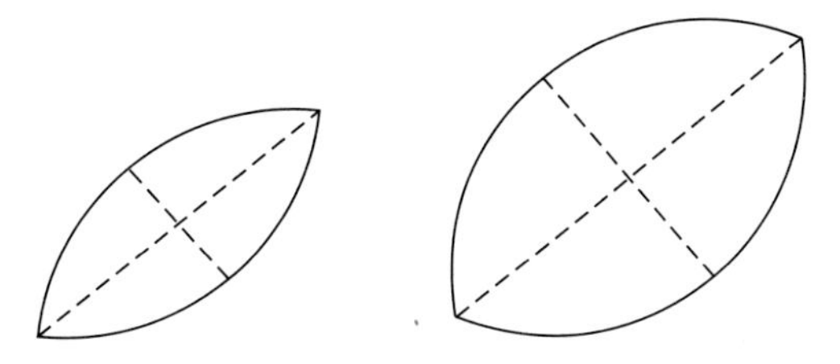

FIGURE 1.5 Babylonian barge and bull's eye

The Babylonians also dealt with volumes of solids. They realized that the volume V of a rectangular block is $V = \ell w h$, and they also knew how to calculate the volume of a prism given the area of the base. But, as in Egypt, there is no document that gives explicitly the volume of a pyramid, even though the Babylonians certainly built pyramidal structures. Nevertheless, on tablet BM 96954 (British Museum), there are several problems involving a grain pile in the shape of a rectangular pyramid with an elongated apex, like a pitched roof (Fig. 1.6a). The method of solution corresponds to the modern formula

$$V = \frac{hw}{3}\left(\ell + \frac{t}{2}\right),$$

where ℓ is the length of the solid, w the width, h the height, and t the length of the apex. Although no derivation of this correct formula is given on the tablet, we can derive it by breaking up the solid into a triangular prism with half a rectangular pyramid on each side (Fig. 1.6b). Then the volume will be the sum of the volumes of these solids. Thus, $V =$ volume of triangular prism + volume of rectangular pyramid, or

$$V = \frac{hwt}{2} + \frac{hw(\ell - t)}{3} = \frac{hw\ell}{3} + \frac{hwt}{6} = \frac{hw}{3}\left(\ell + \frac{t}{2}\right),$$

as desired. Thus, it seems reasonable to assume from the result discussed here that the Babylonians were aware of the correct formula for the volume of a pyramid.

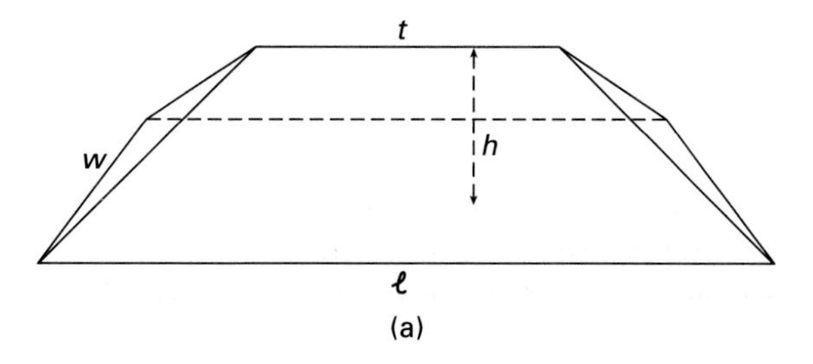

FIGURE 1.6 (a) Babylonian grain pile and dissection

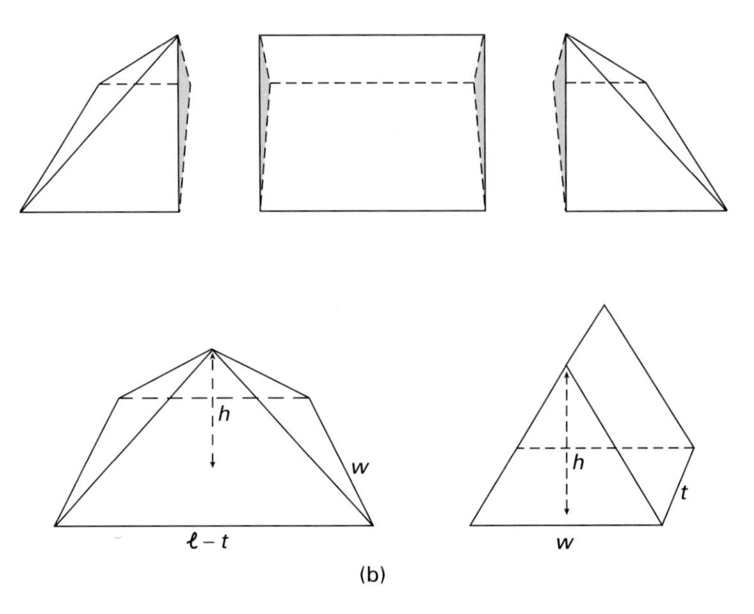

FIGURE 1.6(b) Babylonian grain pile and dissection

That assumption is even more convincing because there is a tablet giving a correct formula for the volume of a truncated pyramid with square base a^2, square top b^2, and height h in the form $V = ([(a+b)/2]^2 + \frac{1}{3}[(a-b)/2]^2)h$. The complete pyramid formula, of course, follows from this by setting $b = 0$. On the other hand, there are tablets where this volume is calculated by the rule $V = \frac{1}{2}(a^2 + b^2)h$, a simple but incorrect generalization of the rule for the area of the trapezoid. It is well to remember, however, that although this formula is incorrect, the calculated answers would not differ much from the correct ones. It is difficult to see how anyone would realize that the answers were wrong in any case, because there was no accurate method for measuring the volume empirically. And because the problems in which these formulas occurred were practical ones, often related to the number of workmen needed to build a particular structure, the slight inaccuracy produced by using this rule would have little effect on the final answer.

1.2.4 Square Roots and the Pythagorean Theorem

Consider another type of Babylonian algorithm, the square root algorithm. Usually, when square roots are needed to solve problems, the problems are arranged so that the square root is one listed in a table of square roots (of which there are many) and is a rational number. But there are cases where an irrational square root is needed, in particular, $\sqrt{2}$. When this particular value occurs, the result is generally written as 1;25 ($= 1\frac{5}{12}$). However, one interesting tablet, YBC 7289, includes a drawing of a square with side indicated as 30 and two numbers, 1;24,51,10 and 42;25,35, written on the diagonal (Fig. 1.7). The product of 30 and 1;24,51,10 is precisely 42;25,35. It is a reasonable assumption that the last number represents the length of the diagonal and that the other number represents $\sqrt{2}$.

Whether $\sqrt{2}$ is given as 1;25 or as 1;24,51,10, there is no record of how the value was calculated. But because the scribes were surely aware that the square of neither of these

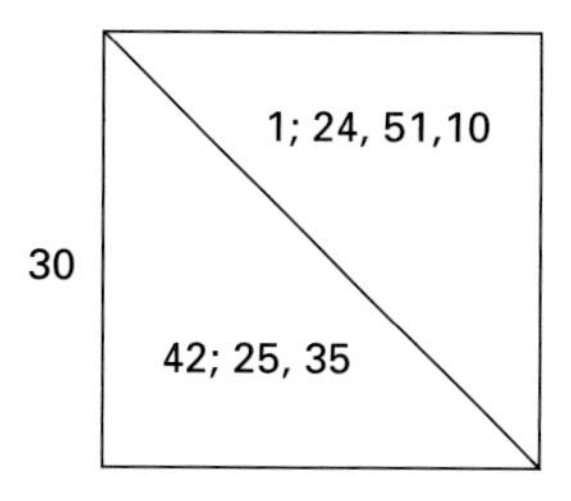

FIGURE 1.7 $\sqrt{2}$ on a Babylonian tablet

was exactly 2, or that these values were not exactly the length of the side of a square of area 2, they must have known that these values were approximations. How were the values determined? One possible method, a method for which there is some textual evidence, begins with the algebraic identity $(x + y)^2 = x^2 + 2xy + y^2$, whose validity was probably discovered by the Babylonians from its geometric equivalent. Given a square of area N for which one wants to know the side $\sqrt{N}$, the first step would be to choose a regular value a close to, but less than, the desired result. Setting $b = N - a^2$, the next step is to find c so that $2ac + c^2$ is as close as possible to b (Fig. 1.8). If a^2 is "close enough" to N, then c^2 will be small in relation to $2ac$, so c can be chosen to equal $(1/2)b(1/a)$, that is, $\sqrt{N} = \sqrt{a^2 + b} \approx a + (1/2)b(1/a)$. (In keeping with Babylonian methods, the value for c has been written as a product rather than a quotient, and, since one of the factors is the reciprocal of a, it is evident why a must be regular.) A similar argument shows that $\sqrt{a^2 - b} \approx a - (1/2)b(1/a)$. In the particular case of $\sqrt{2}$, begin with $a = 1;20$ $(= \frac{4}{3})$. Then $a^2 = 1;46,40$, $b = 0;13,20$, and $1/a = 0;45$, so $\sqrt{2} = \sqrt{1;46,40 + 0;13,20} \approx 1;20 + (0;30)(0;13,20)(0;45) = 1;20 + 0;05 = 1;25$ (or $\frac{17}{12}$). Of course, it isn't really necessary to have a procedure to get $\sqrt{2} \approx 1;25$. One simply needs a good guess or a table in which 1;25 appears as one of the numbers to be squared. After all, $(1;25)^2$ differs from 2 by only $0;0,25 = \frac{1}{144}$.

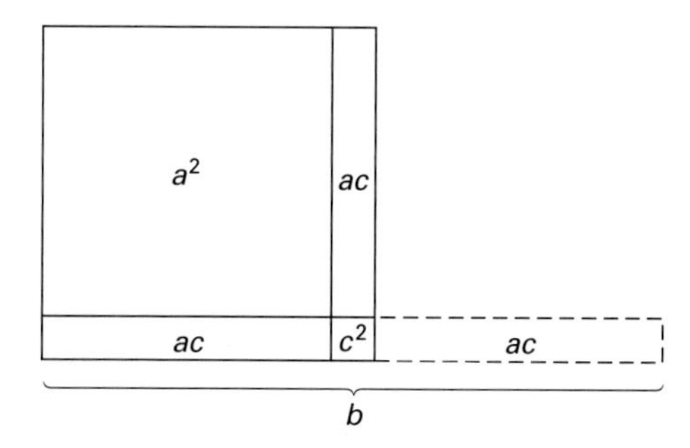

FIGURE 1.8 Geometric version of $\sqrt{N} = \sqrt{a^2 + b} \approx a + 1/2 \cdot b \cdot 1/a$

One of the Babylonian square root problems was connected to the relation between the side of a square and its diagonal. That relation is a special case of the result known as the Pythagorean theorem: In any right triangle, the sum of the areas of the squares on the legs equals the area of the square on the hypotenuse. This theorem, named after the sixth-century BCE Greek philosopher-mathematician, is arguably the most important elementary theorem in mathematics, since its consequences and generalizations have wide-ranging application. It is one of the earliest theorems known to ancient civilizations. In fact, there is evidence that it was known at least 1000 years before Pythagoras.

In particular, there is substantial evidence of interest in Pythagorean triples in the Babylonian tablet Plimpton 322 (in the Plimpton Collection at Columbia University). The extant piece of the tablet consists of four columns of numbers. Other columns were probably broken off on the left. The numbers on the tablet are shown in Table 1.1, reproduced in modern decimal notation with a few recent corrections and with one added column on the right.

TABLE 1.1

Pythagorean Triples from Plimpton 322

$(d/y)^2$	x	d	#	y
1.9834028	119	169	1	120
1.9491586	3367	4825	2	3456
1.9188021	4601	6649	3	4800
1.8862479	12,709	18,541	4	13,500
1.8150077	65	97	5	72
1.7851929	319	481	6	360
1.7199837	2291	3541	7	2700
1.6845877	799	1249	8	960
1.6426694	481	769	9	600
1.5861226	4961	8161	10	6480
1.5625	45	75	11	60
1.4894168	1679	2929	12	2400
1.4500174	161	289	13	240
1.4302388	1771	3229	14	2700
1.3871605	28	53	15	45

It was a major piece of mathematical detective work for modern scholars, first, to decide that Plimpton 322 was a mathematical work rather than a list of orders from a pottery business, and second, to find a reasonable mathematical explanation for the columns of numbers. But find one they did. The columns headed x and d (the headings in the original can be translated as "square-side of the short side" and "square-side of the diagonal") contain in each row two of the three numbers of a Pythagorean triple. It is easy enough to subtract the square of column x from the square of column d. In each case, a perfect square results, whose square root is indicated in the added column y. Finally, the first column on the left represents the quotient $(d/y)^2$.

How and why were these triples derived? One cannot find Pythagorean triples of this size by trial and error. There have been many suggestions over the years as to how the scribe found these and the purpose of the tablet. From a purely mathematical perspective, there are many methods that could generate the table. But since this tablet was written at a particular time and place, probably in Larsa around 1800 BCE, an understanding of its construction and meaning must come from an understanding of the context of the time and how mathematical tablets were generally written. In particular, it is important to note that the first column in a Babylonian table is virtually always written in numerical order (either ascending or descending), while subsequent columns depend on those to their left. It is also important to figure out the words at the top of the column here labeled $(d/y)^2$. This was difficult, because some of the cuneiform wedges had been damaged, but it appears that the heading means "the holding-square of the diagonal from which 1 is torn out so that the short side comes up." The "1" in that heading indicates that the scribe is dealing with reciprocal pairs, very common in Babylonian tables. To relate reciprocals to Pythagorean triples, we note that to find integer solutions to the equation $x^2 + y^2 = d^2$, we can divide by y and first find solutions to $(x/y)^2 + 1 = (d/y)^2$, or, setting $u = x/y$ and $v = d/y$, to $u^2 + 1 = v^2$. This latter equation is equivalent to $(v + u)(v - u) = 1$. That is, we can think of $v + u$ and $v - u$ as the sides of a rectangle whose area is 1 (Fig. 1.9). Now split off from this rectangle one with sides u and $v - u$ and move it to the bottom left after a rotation of 90°. The resulting figure is an L-shaped figure, usually called a **gnomon**, with long sides both equal to v, a figure that is the difference $v^2 - u^2 = 1$ of two squares. Note that this square, whose area, $v^2 = (d/y)^2$, is the entry in the left-most column in Table 1.1 from tablet Plimpton 322, has a gnomon of area 1 torn out so that the remaining square is the square on the short side, as the original column heading says.

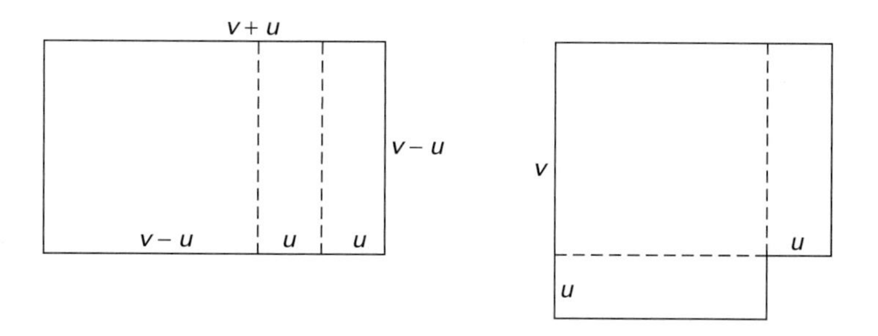

FIGURE 1.9 A rectangle of area 1 turned into the difference of two squares

To calculate the entries on the tablet, it is possible that the scribe began with a value for what we have called $v + u$. Next, he found its reciprocal $v - u$ in a table and solved for $u = \frac{1}{2}[(v + u) - (v - u)]$. The first column in the table is then the value $1 + u^2$. He could then find v by taking the square root of $1 + u^2$. Since u, 1, and v satisfy the Pythagorean identity, the scribe could find a corresponding integral Pythagorean triple by multiplying each of these values by a suitable number y, one chosen to eliminate "fractional" values. For example, if $v + u = 2;15\ (= 2\frac{1}{4})$, the reciprocal $v - u$ is $0;26,40\ (= \frac{4}{9})$. The scribe then finds $u = 0;54,10 = \frac{65}{72}$. We could find v by taking half the sum of $v + u$ and $v - u$, but the scribe found v as $\sqrt{1 + u^2} = \sqrt{1;48,54,01,40} = 1;20,50$, or $\sqrt{1 + u^2} = \sqrt{1.8150077} = 1\frac{25}{72}$. Multiplying the values for u, v, and 1 by $1,12 = 72$ gives the values 65 and 97 for x and d, respectively, shown in line 5 of Table 1.1, as well as the value 72 for y. Conversely, the value of $v + u$ for line 1 of Table 1.1 can be found by adding $\frac{169}{120}$ (= 1;24,30) and $\frac{119}{120}$ (= 0;59,30) to get $\frac{288}{120}$ (= 2;24).

Why were the particular Pythagorean triples on Plimpton 322 chosen? Again, it is impossible to know the answer definitively. But if we calculate the values of $v + u$ for every line of the tablet, we notice that they form a decreasing sequence of regular sexagesimal numbers of no more than four places from 2;24 to 1;48. Not all such numbers are included—there are five missing—but it is possible that the scribe may have decided that the table was long enough without them. He may also have begun with numbers larger than 2;24 or smaller than 1;48 on tablets that have not yet been unearthed. In any case, it is likely that the column of values for $v + u$, in descending numerical order, was one of the missing columns on the original tablet. And the scribe, quite probably a teacher, had thus worked out a list of integral Pythagorean triples, which could be used in constructing student problems for which he would know that the solutions would be possible in integers or finite sexagesimal fractions.

Whether or not the method presented above was the one the Babylonian scribe used to write Plimpton 322, the fact remains that such scribes were well aware of the Pythagorean relationship. And although Table 1.1 offers no indication of a geometrical relationship except for the headings of the columns, there are problems in Old Babylonian tablets that make explicit geometrical use of the Pythagorean theorem. For example, in a problem from tablet BM 85196, a beam of length 30 stands against a wall. The upper end has slipped down a distance 6. How far did the lower end move? In other words, $d = 30$ and $y = 24$ are given, and x is to be found. The scribe calculates x using the theorem: $x = \sqrt{30^2 - 24^2} = \sqrt{324} = 18$. Another slightly more complicated example comes from a tablet found at Susa in modern Iran. The problem is to calculate the radius of a circle circumscribed about an isosceles triangle with altitude 40 and base 60. By considering the right triangle ABC (Fig. 1.10), whose hypotenuse is the desired radius, the scribe derived the equation $r^2 = 30^2 + (40 - r)^2$ from the Pythagorean theorem. He then calculated that $1,20r = 30^2 + 40^2 = 41,40$ and, using reciprocals, that $r = (0;0,45)(41,40) = 31;15$.

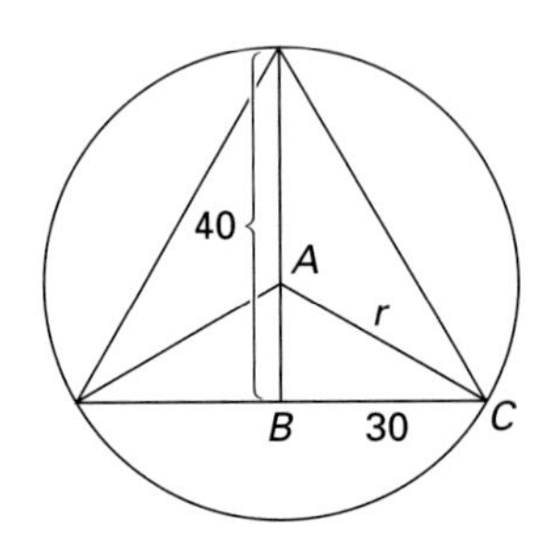

FIGURE 1.10 Circumscribing a circle about an isosceles triangle

1.2.5 Solving Equations

Finally, we will consider equations in Babylonian mathematics. There are few linear equations in extant Babylonian texts, and most of these occur in the context of systems of two linear equations. The Babylonians, like the Egyptians, used the method of false position.

Here is an example from the Old Babylonian text VAT 8389 (Vorderasiatische Abteilung, Tontafeln collection of the Berlin Museum): One of two fields yields $\frac{2}{3}$ *sila* per *sar*, the second yields $\frac{1}{2}$ *sila* per *sar*. [*Sila* and *sar* are measures for capacity and area, respectively.] The yield of the first field was 500 *sila* more than that of the second; the areas of the two fields were together 1800 *sar*. How large is each field? It is easy enough to translate the problem into a system of two equations with x and y representing the unknown areas:

$$\frac{2}{3}x - \frac{1}{2}y = 500$$
$$x + y = 1800.$$

A modern solution might be to solve the second equation for x and substitute the result into the first. But this Babylonian scribe made the initial assumption that x and y were both equal to 900. He then calculated that $\frac{2}{3} \cdot 900 - \frac{1}{2} \cdot 900 = 150$. The difference between the desired 500 and the calculated 150 is 350. To adjust the answers, the scribe presumably realized that every unit increase in the value of x and consequent unit decrease in the value of y gave an increase in the "function" $\frac{2}{3}x - \frac{1}{2}y$ of $\frac{2}{3} + \frac{1}{2} = \frac{7}{6}$. He therefore needed only to solve the equation $\frac{7}{6}s = 350$ to get the necessary increase, $s = 300$. Adding 300 to 900 gave him 1200 for x, and subtracting 300 from 900 gave him 600 for y—the correct answers.

Presumably, the Babylonians also solved single linear equations by the same method, although the few such problems available do not reveal their method. For example, here is a problem from tablet YBC 4652: "I found a stone, but did not weigh it; after I added one-seventh and then one-eleventh [of the total], it weighed 1 *mina* [= 60 *gin*]. What was the original weight of the stone?" We can translate this into the modern equation $(x + x/7) + \frac{1}{11}(x + x/7) = 60$. On the tablet, the scribe just presents the answer, $x = 48\frac{1}{8}$. If he had solved the problem by false position, the scribe would first have guessed that $y = x + x/7 = 11$. Since then $y + \frac{1}{11}y = 12$, instead of 60, the guess must be increased by the factor $\frac{60}{12} = 5$ to the value 55. Then, to solve $x + x/7 = 55$, the scribe could have guessed $x = 7$. This value would produce $7 + \frac{7}{7} = 8$ instead of 55. So the last step would be to multiply the guess of 7 by the factor $\frac{55}{8}$ to get $\frac{385}{8} = 48\frac{1}{8}$, the correct answer.

While tablets containing linear problems are limited, there are very many Babylonian tablets whose problems can be translated into quadratic equations. In fact, many Old Babylonian tablets contain extensive lists of quadratic problems. And in solving these problems, the scribes made full use of the "cut-and-paste" geometry developed by the surveyors. In particular, they applied this approach to various standard problems such as finding the length and width of a rectangle, given the semiperimeter and the area. For example, consider the problem $x + y = 6\frac{1}{2}$, $xy = 7\frac{1}{2}$, from tablet YBC 4663. The scribe first halves $6\frac{1}{2}$ to get $3\frac{1}{4}$. Next, he squares $3\frac{1}{4}$, getting $10\frac{9}{16}$. From this is subtracted $7\frac{1}{2}$, leaving $3\frac{1}{16}$, and then the square root is extracted to get $1\frac{3}{4}$. The length is thus $3\frac{1}{4} + 1\frac{3}{4} = 5$, while the width is given as $3\frac{1}{4} - 1\frac{3}{4} = 1\frac{1}{2}$. A close reading of the wording of the tablets seems to indicate that the scribe had in mind a geometric procedure (Fig. 1.11). (For the sake of generality, the sides in the figure have been labeled in accordance with the generic system $x + y = b$, $xy = c$.) The scribe began by halving the sum b and then constructing the square on it. Since $b/2 = x - (x - y)/2 = y + (x - y)/2$, the square on $b/2$ exceeds the original rectangle of area c by the square on $(x - y)/2$, that is,

$$\left(\frac{x + y}{2}\right)^2 = xy + \left(\frac{x - y}{2}\right)^2.$$

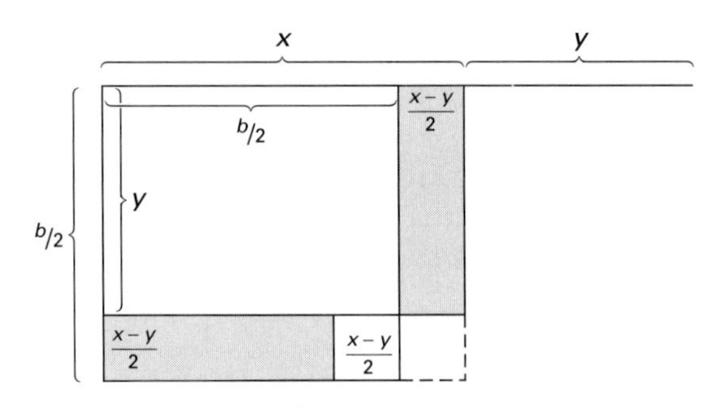

FIGURE 1.11 Geometric procedure for solving the system $x + y = b$, $xy = c$

The figure then shows that adding the side of this square, namely, $\sqrt{(b/2)^2 - c}$, to $b/2$ gives the length x, and subtracting it from $b/2$ gives the width y. The algorithm is therefore expressible in the form

$$x = \frac{b}{2} + \sqrt{\left(\frac{b}{2}\right)^2 - c} \qquad y = \frac{b}{2} - \sqrt{\left(\frac{b}{2}\right)^2 - c}.$$

A geometric interpretation can also be given to the Babylonian solution of what modern mathematicians would consider to be a single quadratic equation. Several such problems are given on tablet BM 13901, including the following: "The sum of the area of a square and $\frac{4}{3}$ of the side is $\frac{11}{12}$. Find the side." In modern terms, the equation to be solved is $x^2 + \frac{4}{3}x = \frac{11}{12}$. For the solution, the scribe tells us to take half of $\frac{4}{3}$, giving $\frac{2}{3}$, square the $\frac{2}{3}$, giving $\frac{4}{9}$, then add this result to $\frac{11}{12}$, giving $1\frac{13}{36}$. This value is the square of $\frac{7}{6}$. Subtracting $\frac{2}{3}$ from $\frac{7}{6}$ gives $\frac{1}{2}$ as the desired side. The Babylonian rule is easily translated into a modern formula for solving $x^2 + bx = c$, namely,

$$x = \sqrt{\left(\frac{b}{2}\right)^2 + c} - \frac{b}{2},$$

which is recognizable as a version of the quadratic formula. The question, however, is how the Babylonians interpreted their procedure. At first glance, it would appear that the statement of the problem is not a geometric one, since we are asked to add a multiple of a side to an area. But the geometric language of the solution seems to indicate that this multiple is to be considered as a rectangle with length x and width b, a rectangle that is added to the square of side x (Fig. 1.12). Given this interpretation, the procedure amounts to cutting half of the rectangle off from one side of the square and moving it to the bottom. Adding a square of side $b/2$ "completes the square." It is then evident that the unknown length x is equal to the difference between the side of the new square and $b/2$, exactly as the formula implies.

For the analogous problem $x^2 - bx = c$, the Babylonian geometric procedure is equivalent to the formula $x = \sqrt{(b/2)^2 + c} + b/2$. This is illustrated by the example $x^2 - \frac{1}{3}x = \frac{1}{12}$. Keep in mind, however, that the "quadratic formula" as used by the Babylonian scribes was not the same as the modern version. First, the scribes gave different procedures for solving the two types, $x^2 + bx = c$ and $x^2 - bx = c$, because the two problems were different; they had different geometric meanings. To a modern mathematician, on the other

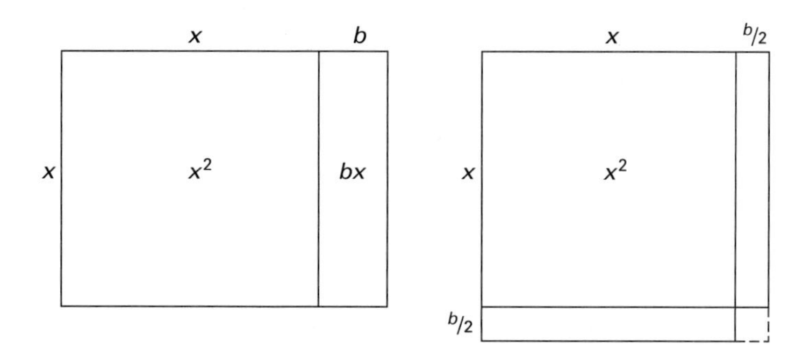

FIGURE 1.12 Geometric version of the quadratic formula for solving $x^2 + bx = c$

hand, these problems are the same, because the coefficient of x can be taken as positive or negative. Second, the modern quadratic formula in these two cases gives a positive and a negative solution to each equation. The negative solution, however, makes no geometrical sense and was completely ignored by the Babylonians.

Note that in the two quadratic equation problems mentioned from BM 13901, the coefficients of x, $\frac{4}{3}$ and $\frac{1}{3}$, respectively, are not "natural"; that is, they are arbitrary fractions chosen to give scribes practice in using a particular method. Yet there is another problem on this tablet of the form $x^2 + bx = c$, which is solved by a somewhat different procedure. Modern scholars believe that this problem is an example of an original problem coming directly from surveyors, a problem that later turns up both in Islamic mathematics and in medieval European mathematics. The problem states that the sum of four sides and the (square) surface is $\frac{25}{36}$. Here, of course, the "four" is "natural," that is, essential to the problem. And as the first step reveals, the method depends directly on the "four." For in that first step, the scribe takes $\frac{1}{4}$ of the $\frac{25}{36}$ to get $\frac{25}{144}$. To this, he adds 1, giving $\frac{169}{144}$. The square root of this value is $\frac{13}{12}$. Subtracting 1 gives $\frac{1}{12}$. Thus, the length of the side is twice that value, namely, $\frac{1}{6}$. The procedure, seemingly different from the one in the previous problem, is best illustrated by a diagram (Fig. 1.13). What the scribe intends is that the four

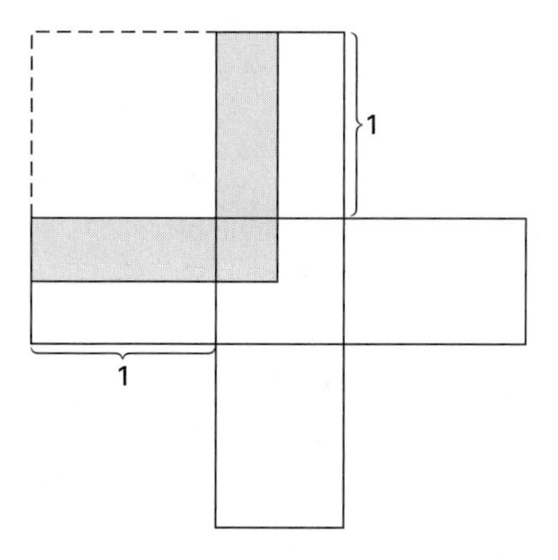

FIGURE 1.13 The sum of four sides and the square surface

"sides" are really projections of the actual sides of the square into rectangles of length 1. Taking $\frac{1}{4}$ of the entire sum means that we are only considering the shaded gnomon, which is one-fourth of the original figure. When we add a square of side 1 to that figure, we get a square whose side we can then find. Subtracting 1 from the side then gives us half of the original side of the square.

Other problems on BM 13901 deal with various other situations involving squares and sides, with each of the solution procedures having a geometric interpretation. As a final example, consider the problem $x^2 + y^2 = \frac{13}{36}$, $x - y = \frac{1}{6}$. The solution to this system, which can be generalized into the system $x^2 + y^2 = c$, $x - y = b$, was found by a procedure describable by the modern formula

$$x = \sqrt{\frac{c}{2} - \left(\frac{b}{2}\right)^2} + \frac{b}{2} \qquad y = \sqrt{\frac{c}{2} - \left(\frac{b}{2}\right)^2} - \frac{b}{2}.$$

It appears that the Babylonians developed the solution by using the geometric idea expressed in Figure 1.14. This figure shows that

$$x^2 + y^2 = 2\left(\frac{x+y}{2}\right)^2 + 2\left(\frac{x-y}{2}\right)^2.$$

It follows that

$$c = 2\left(\frac{x+y}{2}\right)^2 + 2\left(\frac{b}{2}\right)^2$$

and therefore that

$$\frac{x+y}{2} = \sqrt{\frac{c}{2} - \left(\frac{b}{2}\right)^2}.$$

Because $x = (x + y)/2 + (x - y)/2$ and $y = (x + y)/2 - (x - y)/2$, the result follows.

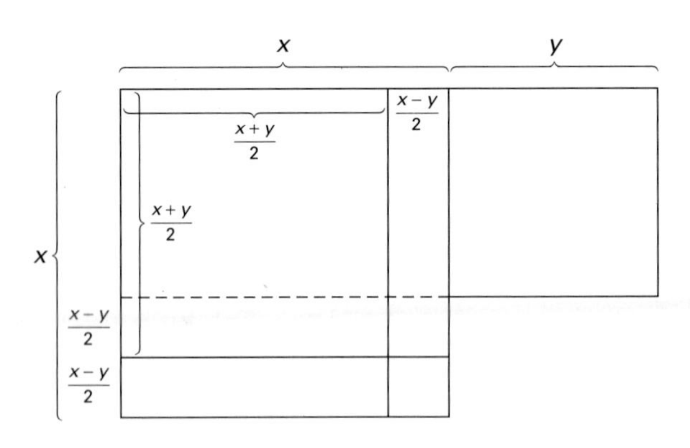

FIGURE 1.14 Geometric procedure for solving the system $x - y = b$, $x^2 + y^2 = c$

1.3 Conclusion

The extant papyri and tablets containing Egyptian and Babylonian mathematics were generally teaching documents, used to transmit knowledge from one scribe to another. Their function was to provide "trainee" scribes with a set of example-types, problems whose solutions could be applied in other situations. Learning mathematics for these trainees was learning how to select and perhaps modify an appropriate algorithm and then mastering the arithmetic techniques necessary to carry out the algorithm to solve any new problem. The reasoning behind the Egyptian and Babylonian algorithms was evidently transmitted orally, so that mathematicians today are forced to speculate as to their origins.

We should note that although the long lists of quadratic problems on some of the Babylonian tablets were generally given as "real-world" problems, the problems are in fact just as contrived as the ones found in most current algebra texts. That the authors knew they were contrived is shown by the fact that, typically, all problems of a given set have the same answer. But since the problems often grew in complexity, it appears that the tablets were used to develop techniques of solution. It is plausible, therefore, that the study of mathematical problem solving, especially problems involving quadratic equations, was a method for training the minds of future leaders of the country. In other words, it was not really that important to solve quadratic equations—there were few real situations that required them. What was important was that students develop skills in solving problems in general, skills that could be used in dealing with the everyday problems a nation's leaders need to solve. These skills included not only the ability to follow well-established procedures—algorithms—but also the knowledge of how and when one could modify the methods and how one could reduce more complicated problems to ones already solved. Today's students are often told that one reason to study mathematics is to "train the mind." It seems that teachers have been telling their students the same thing for the past 4000 years.

Exercises

1. Represent 275 in Egyptian hieroglyphics and in Babylonian cuneiform.

2. Use Egyptian techniques to divide 93 by 5.

3. Multiply $7\ \overline{2}\ \overline{4}\ \overline{8}$ by $12\ \overline{\overline{3}}$ using the Egyptian multiplication technique. Note that it is necessary to multiply each term of the multiplicand separately by $\overline{\overline{3}}$.

4. The *Rhind Mathematical Papyrus* table of division by 2 includes $2 \div 11 = \overline{6}\ \overline{66}$, $2 \div 13 = \overline{8}\ \overline{52}\ \overline{104}$, and $2 \div 23 = \overline{12}\ \overline{276}$. The calculation of $2 \div 13$ is given as follows:

1	13
$\overline{2}$	$6\ \overline{2}$
$\overline{4}$	$3\ \overline{4}$
$\overline{8}$	$1\ \overline{2}\ \overline{8}'$
$\overline{52}$	$\overline{4}'$
$\overline{104}$	$\overline{8}'$
$\overline{8}\ \overline{52}\ \overline{104}$	$1\ \overline{2}\ \overline{4}\ \overline{8}\ \overline{8}$
	2

Perform similar calculations for $2 \div 11$ and $2 \div 23$ to check the results.

5. Given the value $2 \div 13 = \overline{8}\ \overline{52}\ \overline{104}$ from Exercise 4, you can easily find the unit fraction values of $3, 4, 5, \ldots, 12$ divided by 13. For example, $3 \div 13 = \overline{8}\ \overline{13}\ \overline{52}\ \overline{104}$ (since $3 = 1 + 2$) and $4 \div 13 = \overline{4}\ \overline{26}\ \overline{52}$ (since $4 = 2 \times 2$). Find $5 \div 13$, $6 \div 13$, and $8 \div 13$.

6. Solve by the method of false position: A quantity and its $\frac{1}{7}$ added together become 19. What is the quantity? (Problem 24 of the *Rhind Mathematical Papyrus*)

7. Solve by the method of false position: A quantity and its $\frac{2}{3}$ are added together and from the sum $\frac{1}{3}$ of the sum is subtracted, and 10 remains. What is the quantity? (Problem 28 of the *Rhind Mathematical Papyrus*)

8. A quantity, its $\frac{1}{3}$, and its $\frac{1}{4}$, added together, become 2. What is the quantity? (Problem 32 of the *Rhind Mathematical Papyrus*)

9. Problem 72 of the *Rhind Mathematical Papyrus* reads: "100 loaves of *pesu* 10 are exchanged for loaves of *pesu* 45. How many of these loaves are there?" The solution given is, "Find the excess of 45 over 10. It is 35. Divide this 35 by 10. You get $3\ \overline{2}$. Multiply $3\ \overline{2}$ by 100. Result 350. Add 100 to this 350. You get 450. Say then that the exchange is 100 loaves of *pesu* 10 for 450 loaves of *pesu* 45." Translate this solution into modern terminology. How does this solution demonstrate proportionality?

10. Solve problem 11 of the *Moscow Mathematical Papyrus*: The work of a man in logs; the amount of his work is 100 logs of 5 handbreadths diameter; but he has brought them in logs of 4 handbreadths diameter. How many logs of 4 handbreadths diameter are there?

11. Various conjectures have been made for the derivation of the Egyptian formula for the area A of a circle of diameter d: $A = (\frac{8}{9}d)^2$. One of these hypotheses is based on circular counters, known to have been used in ancient Egypt. Show by experiment (using pennies, for example, whose diameter can be taken as 1) that a circle of diameter 9 can essentially be filled by 64 circles of diameter 1. (Begin with one penny in the center; surround it with a circle of 6 pennies, and so on.) Use the obvious fact that 64 circles of diameter 1 also fill a square of side 8 to show how the Egyptians may have derived their formula.

12. Show that $1 \div 7$ gives the periodic sexagesimal fraction $0;8,34,17,8,34,17\ldots$ by dividing in base 60.

13. Find the reciprocals in base 60 of 18, 32, 54, and 64 (= 1,04). What condition on the integer n insures that it is a regular sexagesimal—that is, that its reciprocal is a finite sexagesimal fraction?

14. In the Babylonian system, multiply 25 by 1,04 and 18 by 1,21. Divide 50 by 18 and 1,21 by 32 (using reciprocals). Use the standard modern multiplication algorithm modified for base 60.

15. Show that the area of the Babylonian "barge" is given by $A = \frac{2}{9}a^2$, where a is the length of the arc (one-quarter of the circumference). Also show that the length of the long transversal of the barge is $\frac{17}{18}a$ and the length of the short transversal is $\frac{7}{18}a$. (Use the Babylonian values of $C^2/12$ for the area of a circle and $\frac{17}{12}$ for $\sqrt{2}$.)

16. Show that the area of the Babylonian "bull's eye" is given by $A = \frac{9}{32}a^2$, where a is the length of the arc (one-third of the circumference). Also show that the length of the long transversal of the bull's eye is $\frac{7}{8}a$ and the length of the short transversal is $\frac{1}{2}a$. (Use the Babylonian values of $C^2/12$ for the area of a circle and $\frac{7}{4}$ for $\sqrt{3}$.)

17. For the truncated pyramid from the *Moscow Papyrus*, compare the correct volume given in the text with the volume calculated by means of the incorrect Babylonian formula $V = \frac{1}{2}(a^2 + b^2)h$. Find the percentage error. Do the same for a truncated pyramid of lower base 10, upper base 8, and height 2.

18. Convert 1;24,51,10, the Babylonian approximation to $\sqrt{2}$, to decimals and determine the accuracy of the approximation.

19. Use the assumed Babylonian square root algorithm a second time to get a better approximation to $\sqrt{2}$, starting with the value $1;25 = \frac{17}{12}$ already found. Since this number is not regular, you will first need to find an approximation to its reciprocal. Your answer should be the value 1;24,51,10.

20. Use the assumed Babylonian square root algorithm discussed in the text to show that $\sqrt{3} \approx 1;45$ by beginning with the value 2. Find a three-place sexagesimal approximation to the reciprocal of 1;45, and use it to calculate a three-place sexagesimal approximation to $\sqrt{3}$.

21. Show that $12\ \overline{\overline{3}}\ \overline{15}\ \overline{24}\ \overline{32}$ is a good approximation to $\sqrt{164}$. (This value appears in a late Greek-Egyptian papyrus.)

22. Show that taking $v + u = 1;48$ $(= 1\frac{4}{5})$ leads to line 15 of Table 1.1 (from tablet Plimpton 322) and that taking $v + u = 2;05$ $(= 2\frac{1}{12})$ leads to line 9. Find the values for $v + u$ that lead to lines 6 and 13 of the table.

23. The scribe of Plimpton 322 did not use the value $v + u = 2;18,14,24$ with its associated reciprocal $v - u = 0;26,02,30$ in his work on the tablet. Find the smallest Pythagorean triple associated with those values.
24. Solve this problem from the Old Babylonian tablet BM 13901: The sum of the areas of two squares is 1525. The side of the second square is $\frac{2}{3}$ of that of the first plus 5. Find the sides of each square.
25. Solve the following Babylonian problem taken from a tablet found at Susa: Let the width of a rectangle measure a quarter less than the length. Let 40 be the length of the diagonal. What are the length and width? Use false position, beginning with the assumption that 1 (or 60) is the length of the rectangle.
26. Solve the following problem from VAT 8391: One of two fields yields $\frac{2}{3}$ *sila* per *sar*, the second yields $\frac{1}{2}$ *sila* per *sar*. The sum of the yields of the two fields is 1100 *sila*; the difference of the areas of the two fields is 600 *sar*. How large is each field?
27. Solve the following problem from YBC 4652: I found a stone, but did not weigh it; after I subtracted one-seventh and then one-thirteenth [of the difference], it weighed 1 *mina* [= 60 *gin*]. What was the original weight of the stone?
28. Solve the following problem from YBC 4652: I found a stone, but did not weigh it; after I subtracted one-seventh, added one-eleventh [of the difference], and then subtracted one thirteenth [of the previous total], it weighed 1 *mina* [= 60 *gin*]. What was the stone's weight?
29. Give a geometric argument to justify the Babylonian "quadratic formula" that solves the equation $x^2 - bx = c$.
30. Consider the system of equations taken from an Old Babylonian text:

$$x = 30 \qquad xy - (x - y)^2 = 500$$

Show that the substitution of the first equation into the second leads to a quadratic equation in y that has two positive roots, a type of equation the Babylonians did not deal with. Show that subtraction of the second equation from the square of the first gives the equation $(x - y)^2 + 30(x - y) = 400$, a quadratic in $x - y$ that has only one positive root.
31. Solve the following Babylonian problem by first multiplying the second equation by 7 and then subtracting the first equation from the result, thus reducing the system to a standard form:

$$x + y = 5\frac{5}{6} \qquad \frac{x}{7} + \frac{y}{7} + \frac{xy}{7} = 2$$

32. Construct two or three real-life division problems for which giving the answers in unit fractions, rather than other common fractions, makes sense.
33. Devise a lesson to teach ideas of proportionality by using the Egyptian method of false position.
34. Devise a lesson on place value using the Babylonian system, in particular, the table of multiplication by 9 given in Figure 1.3.
35. Devise a lesson to teach the quadratic formula using geometric arguments similar to the (assumed) Babylonian ones.

References

The best source on Egyptian mathematics is Richard J. Gillings, *Mathematics in the Time of the Pharaohs* (Cambridge: MIT Press, 1972). See also Gillings, "The Mathematics of Ancient Egypt," *Dictionary of Scientific Biography* (New York: Scribners, 1978), vol. 15, 681–705. A more recent survey is James Ritter, "Egyptian Mathematics," in H. Selin, ed., *Mathematics across Cultures: The History of Non-Western Mathematics* (Dordrecht: Kluwer Academic Publishers, 2000), 115–136. The *Rhind Mathematical Papyrus* is available in an edition by Arnold B. Chace (Reston, Va.: National Council of Teachers of Mathematics, 1967). (This work is an abridgement of the 1927 and 1929 publications by the Mathematical Association of America.) Another English version of that papyrus as well as the *Moscow Mathematical Papyrus* and other Egyptian mathematical fragments is available in Marshall Clagett, *Ancient Egyptian Science: A Source Book. Volume Three: Ancient Egyptian Mathematics* (Philadelphia: American Philosophical Society, 1999).

The standard accounts of Babylonian mathematics are Otto Neugebauer, *The Exact Sciences in Antiquity* (Princeton: Princeton University Press, 1951; New York: Dover,

1969) and B. L. Van der Waerden, *Science Awakening I* (New York: Oxford University Press, 1961). More recent surveys are by Jens Høyrup, "Mathematics, Algebra, and Geometry," in *The Anchor Bible Dictionary*, David N. Freedman, ed. (New York: Doubleday, 1992), vol. IV, 601–612, and by Jöran Friberg, "Mathematik," *Reallexikon der Assyriologie* 7 (1987–1990), 531–585 (in English). Translations and analyses of the Babylonian tablets themselves are found principally in Otto Neugebauer, *Mathematische Keilschrift-Texte* (New York: Springer, 1973, reprint of 1935 original); Otto Neugebauer and Abraham Sachs, *Mathematical Cuneiform Texts* (New Haven: American Oriental Society, 1945); and Evert Bruins and M. Rutten, *Textes mathématiques de Suse* (Paris: Paul Geuthner, 1961). Two technical works analyzing the Babylonian tablets are Jens Høyrup, *Lengths, Widths, Surfaces: A Portrait of Old Babylonian Algebra and Its Kin* (New York: Springer, 2002) and Eleanor Robson, *Mesopotamian Mathematics, 2100–1600 BC: Technical Constants in Bureaucracy and Education* (Oxford: Clarendon Press, 1999). Robson's analysis of Plimpton 322 is found in "Words and Pictures: New Light on Plimpton 322," *American Mathematical Monthly* 109 (2002), 105–120. More general surveys of Mesopotamian mathematics in the context of Mesopotamian society include chapter 3 of Jens Høyrup, *In Measure, Number, and Weight: Studies in Mathematics and Culture* (Albany: State University of New York Press, 1994); Eleanor Robson, "Mesopotamian Mathematics: Some Historical Background," in Victor Katz, ed., *Using History to Teach Mathematics* (Washington: Mathematical Association of America, 2000), 149–158; and Eleanor Robson, "The Uses of Mathematics in Ancient Iraq, 6000–600 BC," in H. Selin, ed., *Mathematics across Cultures*, 93–113. Finally, the early Babylonian tokens are discussed in great detail and with many illustrations in Denise Schmandt-Besserat, *Before Writing: From Counting to Cuneiform* (Austin: University of Texas Press, 1992).

A book that discusses the mathematics of Egypt and Babylonia (as well as of other ancient societies) comparatively and also deals with the questions of transmission and a possible single origin of mathematics is B. L. Van der Waerden, *Geometry and Algebra in Ancient Civilizations* (New York: Springer, 1983).

CHAPTER TWO

Greek Mathematics to the Time of Euclid

Thales was the first to go to Egypt and bring back to Greece this study [geometry]; he himself discovered many propositions, and disclosed the underlying principles of many others to his successors, in some cases his method being more general, in others more empirical.

—Proclus's *Summary* (c. 450 CE) of Eudemus's *History* (c. 320 BCE)

As the quotation indicates, a new attitude toward mathematics appeared in Greece beginning in the sixth century BCE. It was no longer sufficient merely to calculate numerical answers to problems. It had become necessary to prove that the results were correct. This change in the nature of mathematics, beginning around 600 BCE, was related to the great differences between the emerging civilization of Greece and those of Egypt and Babylonia, from whom the Greeks learned. The physical nature of Greece, with its many mountains and islands, is such that large-scale agriculture was not possible. Perhaps because of this, Greece did not develop a central government. The basic political organization was the *polis*, or city-state. The governments of the city-states were of every possible variety, but in general they controlled populations of only a few thousands. Whether the governments were democratic or monarchical, they were not arbitrary. Each government was ruled by law and therefore encouraged its citizens to argue and debate. It was perhaps out of this characteristic that there developed the necessity for proof in mathematics—that is, for argument aimed at convincing others of a particular truth.

Because virtually every city-state had access to the sea, there was constant trade, both within Greece itself and with other civilizations. As a result, the Greeks were exposed to many different peoples, and, in fact, they themselves settled in areas all around the eastern Mediterranean. In addition, a rising standard of living helped to attract able people from other parts of the world. Hence, the Greeks were able to study differing answers to fundamental questions about the world. They began to create their own answers. In many areas of thought, they learned not to accept what had been handed down from ancient times. Instead, they began to ask, and to try to answer, "Why?" Greek thinkers gradually came to the realization that the world around them was knowable, that they could discover its characteristics by rational inquiry. Hence, they were anxious to discover and expound theories in such fields as physics, biology, medicine, and politics. The Greeks believed, however, that mathematics was central to rational thought. Thus, although Western civilization owes a great debt to the Greeks in literature, art, and architecture, it is the Greek idea of mathematical proof that is at the basis of modern mathematics and, by extension, at the foundation of our modern technological civilization.

2.1 The Earliest Greek Mathematics

2.1.1 Thales, Pythagoras, and the Pythagoreans

The earliest Greek mathematician of whom we have any knowledge is Thales (c. 624–547 BCE), from Miletus in Asia Minor (modern-day Turkey), who was credited even in ancient times with beginning the Greek mathematical tradition. Many stories are recorded about him, most written down several hundred years after his death. These include his prediction of a solar eclipse that occurred in 585 BCE and his application of the angle-side-angle criterion of triangle congruence to the problem of measuring the distance to a ship at sea. Thales is also credited with formulating the theorems that the base angles of an isosceles triangle are equal and that vertical angles are equal and with proving that the diameter of a circle divides the circle into two equal parts. Although exactly how Thales accomplished these feats is not known, it does seem clear that he advanced some logical arguments.

Thales is also said to have impressed officials in Egypt by determining the height of a pyramid by comparing the length of its shadow to that of the length of the shadow of a stick of known height. Thus, he evidently knew something about the basic principles of similarity. It appears that this knowledge had spread by the middle of the sixth century to Samos, an island just off the coast of Asia Minor not far from Miletus. For on that island, the architect Eupalinos designed and constructed a tunnel to bring water from a spring outside the capital through a mountain to a point inside the city walls. Modern archaeological excavations of the tunnel have revealed that it was dug by two teams that met in the middle (Fig. 2.1). There are no records as to how the construction crews managed to keep digging in the correct direction, but studies of the site and a reference to digging a tunnel in a work by Heron in the first century CE lend support to the conclusion that Eupalinos used some principle of triangle similarity to line up the direction of the digging.

Samos was also the birthplace of Pythagoras (c. 572–497 BCE), the next mathematician about whom there are extensive stories. Around 530 BCE, after having been forced to leave his native island, Pythagoras settled in Crotona, a Greek town in southern Italy. There he gathered around him a group of disciples, later known as the Pythagoreans; the group was both a religious order and a philosophical school. But since there are no extant works ascribed to Pythagoras or the Pythagoreans, the mathematical doctrines of this group can only be surmised from the works of later writers, including the neo-Pythagoreans.

One important such mathematical doctrine was that "number was the substance of all things," that numbers—that is, positive integers—formed the basic organizing principle of the universe. What the Pythagoreans meant by this was not only that all known objects have a number, or can be ordered and counted, but also that numbers are at the basis of all

FIGURE 2.1 Water tunnel on the island of Samos

physical phenomena. For example, a constellation in the heavens could be characterized both by the number of stars that compose it and by its geometrical form, which itself could be represented by a number. The motions of the planets could be expressed in terms of ratios of numbers. Musical harmonies depend on numerical ratios: two strings with ratio of length 2 : 1 give an octave when plucked; two with ratio 3 : 2 give a fifth; and two with ratio 4 : 3 give a fourth. Out of these intervals an entire musical scale can be created. Finally, the fact that triangles whose sides are in the ratio of 3 : 4 : 5 are right-angled established a connection between number and angle. Given the Pythagoreans' interest in number as a fundamental principle of the cosmos, it is only natural that they studied the properties of positive integers, what mathematicians today call the elements of the theory of numbers.

The starting point of this theory was the dichotomy between the odd and the even. The Pythagoreans probably represented numbers by dots or, more concretely, by pebbles. Hence, an even number would be represented by a row of pebbles that could be divided into two equal parts. An odd number could not be so divided because there would always be a single pebble left over. It was easy enough using pebbles to verify some simple theorems. For example, the sum of any collection of even numbers is even; the sum of an even collection of odd numbers is even, and that of an odd collection is odd (Fig. 2.2).

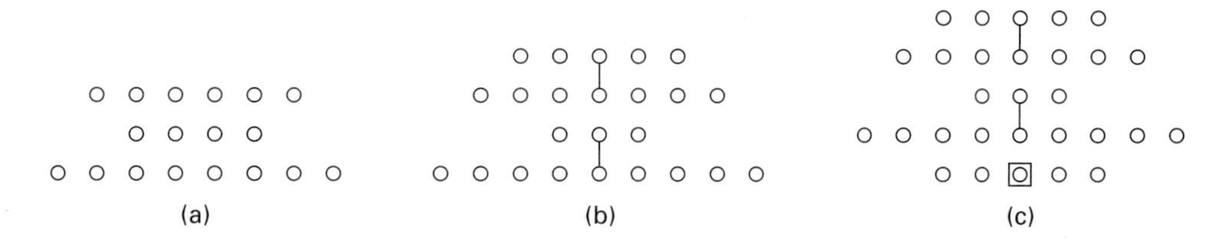

FIGURE 2.2 (a) The sum of even numbers is even. (b) An even sum of odd numbers is even. (c) An odd sum of odd numbers is odd.

Among other simple corollaries of the basic results mentioned above were the theorems that the square of an even number is even and the square of an odd number is odd. Squares themselves could also be represented using dots, providing simple examples of **figurate numbers**. If a given square—for example, the square of 4—is represented in this way, it is easy to see that the next higher square can be formed by adding a row of dots around two sides of the original figure. There are $2 \cdot 4 + 1 = 9$ of these additional dots. The Pythagoreans generalized this observation to show that one can form squares by adding the successive odd numbers to 1. For example, $1 + 3 = 2^2$, $1 + 3 + 5 = 3^2$, and $1 + 3 + 5 + 7 = 4^2$. The added odd numbers are in the shape of a gnomon (Fig. 2.3).

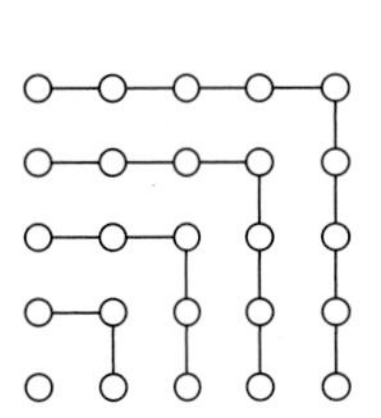

FIGURE 2.3 Square numbers

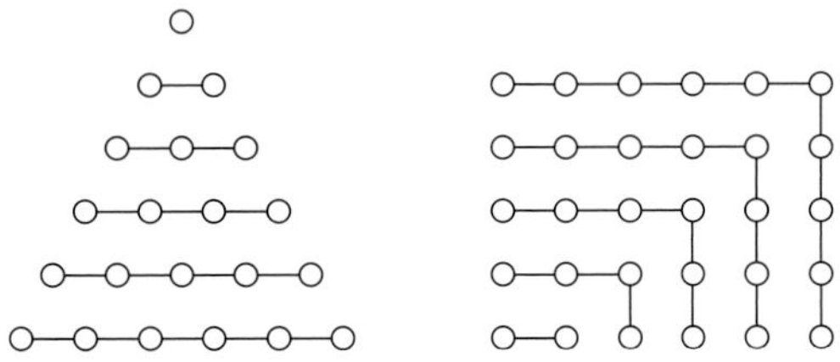

FIGURE 2.4 Triangular and oblong numbers

Other examples of figurate numbers include the triangular numbers, produced by successive additions of the natural numbers themselves (Fig. 2.4). Similarly, oblong numbers, numbers of the form $n(n + 1)$, are produced by beginning with 2 and adding the successive even numbers (Fig. 2.4). The first four of these oblong numbers are 2, 6, 12, and 20, that is, 1×2, 2×3, 3×4, and 4×5. Figure 2.5 provides demonstrations of the results that any square number is the sum of two consecutive triangular numbers and that any oblong number is the double of a triangular number.

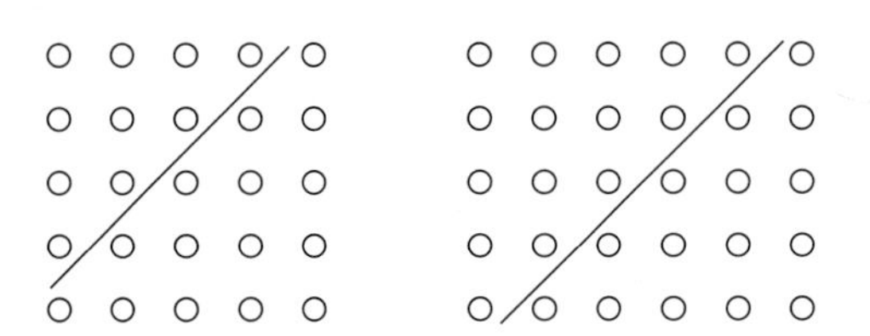

FIGURE 2.5 Two theorems on triangular numbers

Another number theory problem of particular interest to the Pythagoreans was the construction of Pythagorean triples. There is evidence that they saw that, for n an odd number, the triple $(n, (n^2 - 1)/2, (n^2 + 1)/2)$ is a Pythagorean triple, and, if m is even, $(m, (m/2)^2 - 1, (m/2)^2 + 1)$ is such a triple. The geometric theorem out of which the study of Pythagorean triples grew—namely, that in any right triangle the square on the hypotenuse is equal to the sum of the squares on the legs—has long been attributed to Pythagoras himself. There is no direct evidence that Pythagoras derived the theorem, and in fact, it was known in other cultures long before he lived. Nevertheless, by the fifth century BCE, knowledge of this theorem led to the first discovery of what is today called an **irrational number**.

Because the Pythagoreans considered number as the basis of the universe, everything could be counted, including lengths. In order to "count" a length, of course, one needs a measure. The Pythagoreans thus assumed that one could always find an appropriate measure. Once such a measure was found in a particular problem, it became the unit and thus could not be divided. In particular, the Pythagoreans assumed that one could find a measure by which both the side and the diagonal of a square could be counted. In other words, there should exist a length such that the side and the diagonal were integral multiples of that length. Unfortunately, this turned out not to be true. The side and diagonal of a square are **incommensurable**; there is no common measure. Whatever unit of measure is chosen such that an exact number will fit the length of one of these lines, the other line will require some number plus a portion of the unit—but the unit cannot be divided. The discovery of this result, in approximately 430 BCE, forced the Pythagoreans to give up their basic philosophy that all things were made up of numbers and enabled Greek mathematicians to develop some new theories. It also required an adjustment in mathematical philosophy, reflected in Aristotle's careful distinction between number and magnitude, made less than a century later. According to Aristotle (384–322 BCE), there were in fact two distinct types of "quantity": the discrete (number) and the continuous (magnitude). As examples of the latter, he cited lines, surfaces, bodies, and time. The primary distinction between these two classes is that a magnitude is "that which is divisible into divisibles that are infinitely divisible," whereas the basis of number is the indivisible unit. Thus, magnitudes cannot be composed of indivisible elements, while numbers inevitably are.

It is Aristotle who hints about the nature of the first proof of the incommensurability of the diagonal of a square and its side: He writes that "if it is assumed to be commensurable, then odd numbers will be equal to even." One possible form this proof could have taken is the following: Assume that the side BD and diagonal DH in Figure 2.6 are commensurable, that is, that each is represented by the number of times it is measured by their common measure. It may be assumed that at least one of these numbers is odd, for otherwise there would be a larger common measure. Then the squares on the side and diagonal, $DBHI$ and $AGFE$, respectively, represent square numbers. The latter square is clearly double the former, so it represents an even square number. Therefore, its side $AG = DH$ also represents an even number, and the square $AGFE$ is a multiple of 4. Since $DBHI$ is half of $AGFE$, it must be a multiple of 2; that is, it represents an even square. Hence, its side BD must also be even. But this contradicts the original assumption that either DH or BD must be odd. Therefore, the two lines are incommensurable.

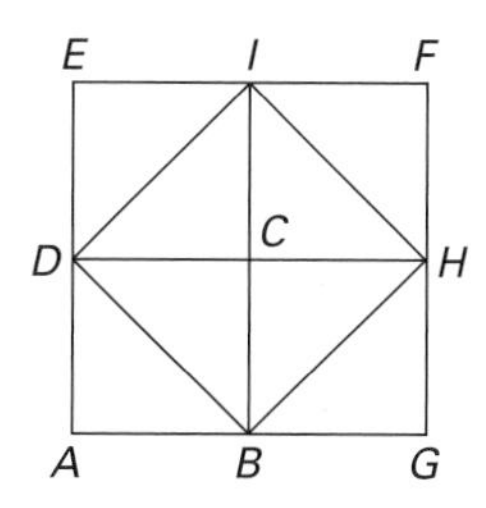

FIGURE 2.6 The incommensurability of the side and diagonal of a square (first possibility)

Note that such a proof presupposes that the notion of proof was already ingrained into the Greek conception of mathematics. Although there is no evidence that the Greeks of the fifth century BCE possessed the entire mechanism of an axiomatic system and had explicitly recognized that certain statements need to be accepted without proof, they certainly had decided that some form of logical argument was necessary for determining the truth of a particular result. Furthermore, this entire notion of incommensurability represents a break from the Babylonian and Egyptian concepts of calculation with numbers. There is naturally no question that one can assign a numerical value to the length of the diagonal of a square of side one unit, as the Babylonians did, but the notion that no "exact" value can be found was first formally recognized in Greek mathematics.

2.1.2 Geometric Problem Solving and the Need for Proof

The idea of proof and the change from numerical calculation are further exemplified in the mid–fifth-century attempts to solve two geometric problems, which were to occupy Greek mathematicians for centuries: the squaring of the circle (already attempted in Egypt), and the duplication of the cube, that is, finding the length of the side of a cube whose volume was twice that of another cube. The multitude of attacks on these particular problems and the slightly later one of trisecting an arbitrary angle serve to remind us that a central goal of Greek mathematics was geometrical problem solving and that, to a large extent, the great body of theorems found in the major extant works of Greek mathematics served as logical underpinnings for these solutions.

Hippocrates of Chios (c. 470–410 BCE) (no connection to the famous physician) was among the first to attack the cube and circle problems. Hippocrates perhaps realized that the first of these problems was analogous to the simpler problem of doubling a square of side a. As Socrates convinces the slave boy in Plato's dialogue *Meno*, the doubled square is found by using as its side the diagonal of the original square. That is, if b is the length of the diagonal of a square of side a, then $b^2 = 2a^2$. This statement implies that b is the mean proportional between a and $2a$, that is, $a : b = b : 2a$. Hippocrates then came up with the idea of reducing the problem of doubling the cube of side a to the problem of finding two mean proportionals b, c between a and $2a$. For, if $a : b = b : c = c : 2a$, then

$$a^3 : b^3 = (a : b)^3 = (a : b)(b : c)(c : 2a) = a : 2a = 1 : 2$$

and $b^3 = 2a^3$. Hippocrates was not, however, able to construct the two mean proportionals using the geometric tools at his disposal. It was left to some of his successors to find this construction.

Hippocrates similarly made progress in the squaring ot the circle, essentially by showing that certain **lunes**—figures bounded by arcs of two circles—could be "squared," that is, that their areas could be shown to be equal to certain regions bounded by straight lines. To do this, he first had to know the generalized Pythagorean theorem, that if similar figures are drawn on the legs and the hypotenuse of a right triangle, then the area of the region on the hypotenuse is equal to the sum of the areas of the regions on the legs. Assuming this, suppose that ABC is an isosceles right triangle. Circumscribe a semicircle with diameter AC about the triangle and let ADC be a segment of a circle similar to the two segments cut off by the sides of the right triangle, that is, circumscribed about an isosceles right triangle with leg AC (Fig. 2.7). By the generalized Pythagorean theorem, the sum of the areas of

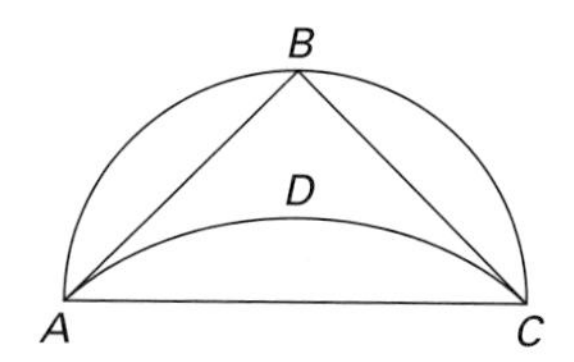

FIGURE 2.7 Hippocrates' lune

the two segments cut off by the sides is equal to the area of the segment cut off by the base. Therefore, the area of the lune between the semicircle and that segment is the same as the area of the triangle.

Hippocrates gave constructions for squaring other lunes or combinations of lunes, but he was unable actually to square a circle. Nevertheless, it is apparent that his attempts on the squaring problem and the doubling problem were based on a large collection of geometric theorems, which he organized into the first recorded book on the elements of geometry. Although the details of Hippocrates' book are unknown, it is assumed that he used certain logical principles in proving his theorems. There are, in fact, many examples of reasoning techniques in both political and philosophical works of his time. These include *reductio ad absurdum*, in which one assumes that a proposition to be proved is false and then derives a contradiction, and *modus tollens*, in which one shows first that if A is true, then B follows, shows next that B is not true, and concludes finally that A is not true. It was Aristotle, however, who took the ideas developed over the centuries and first codified the principles of logical reasoning, primarily in his treatises *Prior Analytics* and *Posterior Analytics*.

Aristotle believed that true proofs should be built out of **syllogisms**. He defined a *syllogism* as "discourse in which, certain things being stated, something other than what is stated follows of necessity from their being so." In other words, a syllogism consists of certain statements that are taken as true and certain other statements that are then necessarily true. For example, the argument "if all monkeys are primates, and all primates are mammals, then it follows that all monkeys are mammals" exemplifies one type of syllogism; the argument "if all Catholics are Christians and no Christians are Muslim, then it follows that no Catholic is Muslim" exemplifies a second type.

After clarifying the principles of dealing with syllogisms, Aristotle noted that syllogistic reasoning enables one to use "old knowledge" to impart new. If one accepts the premises of a syllogism as true, then one must also accept the conclusion. One cannot, however, obtain every piece of knowledge as the conclusion of a syllogism. One has to begin somewhere with truths that are accepted without argument. These truths are generally called **axioms** or **postulates**. Aristotle also discussed certain basic principles of argument, principles that earlier thinkers had used intuitively. One such principle is that a given assertion cannot be both true and false. A second principle is that an assertion must be either true or false; there is no other possibility.

For Aristotle, logical argument according to his methods was the only certain way of attaining scientific knowledge. There might be other ways of gaining knowledge, but demonstration via a series of syllogisms was the sole way by which one could be sure of the results. Because one cannot prove everything, however, one must always be careful that the premises, or axioms, are true and well-known. As he wrote, "syllogism there may indeed be without these conditions, but such syllogism, not being productive of scientific knowledge, will not be demonstration." In other words, one can choose any axioms one wants and draw conclusions from them, but if one wants to attain knowledge, one must start with "true" axioms. The question then becomes, how can one be sure that one's axioms are true? Aristotle answered that these primary premises are learned by induction, by drawing conclusions from one's sense perception of numerous examples. In other words, an axiom is "true" if it is virtually self-evident. Although the question of the "truth" of the basic axioms has always been a matter of debate, Aristotle's rules of attaining knowledge by beginning with axioms and using demonstrations to gain new results became the model for mathematicians to the present day.

Curiously, Greek mathematicians apparently did not use syllogisms explicitly to build their mathematical proofs. They used other basic forms of argument. These were analyzed in some detail in the third century BCE by the Stoics, of whom the most prominent was Chrysippus (280–206 BCE). The logic used in mathematical proofs was based on **propositions**, statements that can be either true or false, rather than on the Aristotelian syllogisms. The basic rules of inference dealt with by Chrysippus, with their traditional names, are the following, where p, q, and r stand for propositions:

(1) *Modus ponens*
If p, then q.
p.
Therefore q.

(2) *Modus tollens*
If p, then q.
Not q.
Therefore, not p.

(3) *Hypothetical syllogism*
If p, then q.
If q, then r.
Therefore, if p, then r.

(4) *Alternative syllogism*
p or q.
Not p.
Therefore q.

For example, from the statements "if it is daytime, then it is light" and "it is daytime," one concludes by *modus ponens* that "it is light." From "if it is daytime, then it is light" and "it is not light," one concludes by *modus tollens* that "it is not daytime." Adding to the first hypothesis the statement "if it is light, then I can see well," one concludes by the hypothetical syllogism that "if it is daytime, then I can see well." Finally, from "either it is daytime or it is nighttime" and "it is not daytime," the alternative syllogism allows one to conclude that "it is nighttime."

2.2 Euclid and His *Elements*

The logic of propositions is clearly used in the most important mathematical text of Greek times, and probably of all time, Euclid's *Elements*, written about 2300 years ago, a work that has appeared in more editions than any other except the Bible. It has been translated into countless languages and has been continuously in print in one country or another nearly since the beginning of printing. Yet to the modern reader the work is incredibly dull. There are no examples; there is no motivation; there are no witty remarks; there is no calculation. There are simply definitions, axioms, theorems, and proofs. Nevertheless, the *Elements* has been intensively studied. Biographies of many famous mathematicians indicate that Euclid's work provided their initial introduction to mathematics, that it in fact motivated them to become mathematicians. It provided them with a model of how "pure mathematics" should be written, with precise definitions, well-thought-out axioms, carefully stated theorems, and logically coherent proofs. Although there were earlier versions of *Elements* before that of Euclid, his is the only one to survive, perhaps because it was the first one written after the foundations of both proportion theory and the theory of irrationals had been developed and after Aristotle had propounded the necessary distinctions between number and magnitude. It was therefore both "complete" and well-organized. Since the mathematical community as a whole was of limited size, once Euclid's work was recognized for its general excellence, there was no reason to keep another inferior work in circulation.

Although essentially nothing is known about the life of the author of the *Elements*, it is generally assumed that Euclid taught and wrote at the Museum and Library at Alexandria. This complex was founded around 300 BCE by Ptolemy I Soter, the Macedonian general who served Alexander the Great and then became ruler of Egypt after the death of Alexander in 323 BCE. In this case, "Museum" meant a "Temple of the Muses," that is, a location where scholars meet and discuss philosophical and literary ideas. The Museum was to be, in effect, a government research establishment. The Fellows of the Museum received stipends and free board and were exempt from taxation. By offering these incentives, Ptolemy I and his successors hoped that men of eminence would be attracted to Alexandria from the entire Greek world. In fact, the Museum and Library soon became a focal point of the highest developments in Greek scholarship, in both the humanities and the sciences. The Fellows were initially appointed to carry on research, but since younger students gathered there as well, the Fellows soon turned to teaching. The aim of the Library was to collect the entire body of Greek literature in the best available copies and to organize it systematically. Ship captains who sailed from Alexandria were instructed to bring back scrolls from every port they stopped at. The Library ultimately contained over 500,000 volumes in every field of knowledge. Although parts of the collection were destroyed in various wars, some of it remained intact until the fourth century CE.

There are no copies of the *Elements* dating from Euclid's time. The earliest extant fragments include some potsherds discovered in Egypt and dating from about 225 BCE, on which are written what appear to be notes on two propositions from Book XIII, and pieces of papyrus containing parts of Book II dating from about 100 BCE. Copies of the work were, however, made regularly from Euclid's time onward. Various editors made emendations, added comments, or put in new lemmas. In particular, Theon of Alexandria (fourth century CE) was responsible for one important new edition. Most of the extant manuscripts of Euclid's *Elements* are copies of Theon's edition. The earliest such copy still in existence is in the Bodleian Library at Oxford University and dates from 888. There is, however, one manuscript in the Vatican Library, dating from the tenth century, that is not a copy of Theon's edition but of an earlier version. A detailed comparison of this manuscript with several old manuscript copies of Theon's version allowed the Danish scholar J. L. Heiberg to compile a definitive Greek version in the 1880s, as close to the Greek original as he believed was possible. The extracts to be discussed here are all adapted from Thomas Heath's 1908 English translation of Heiberg's Greek text. (It should be noted that some modern scholars believe that one can get closer to Euclid's original by taking more account of medieval Arabic translations than Heiberg was able to do.)

Euclid's *Elements* is a work in thirteen books. Books I through VI form a relatively complete treatment of two-dimensional geometric magnitudes, while Books VII through

IX deal with the theory of numbers, in keeping with Aristotle's instructions to separate the study of magnitude and number. In fact, Euclid includes two entirely separate treatments of proportion theory: in Book V for magnitudes and in Book VII for numbers. Book X provides the link between the two concepts; there Euclid introduces the notions of commensurability and incommensurability and shows that, in regard to proportions, commensurable magnitudes may be treated as if they were numbers. That book continues by presenting a classification of some incommensurable magnitudes. Euclid deals in Books XI and XII with three-dimensional geometric objects, and in Book XIII he constructs the five regular polyhedra and classifies some of the lines involved according to his scheme of Book X.

It is also useful to note that much of the ancient mathematics discussed in Chapter 1 is included in one form or another in Euclid's masterwork, with the exception of actual methods of arithmetic computation. The methodology in the *Elements*, however, is entirely different from that of the Egyptians and Babylonians. Mathematics in earlier cultures always involved numbers and measurement. Numerical algorithms for solving various problems are prominent. The mathematics of Euclid, however, is completely nonarithmetical. There are no numbers used in the entire work aside from a few small positive integers. There is also no measurement. Various geometrical objects are compared, but not by use of numerical measures. There are no cubits or acres or degrees. The only measurement standard—for angles—is the right angle. Nevertheless, the question of how much influence the mathematical cultures of Egypt and Mesopotamia had on Euclidean mathematics inevitably arises. Certain pieces of evidence that help answer this question are discussed in this chapter, but a complete answer cannot yet be given.

2.2.1 The Pythagorean Theorem and Its Proof

As Aristotle suggested, a scientific work needs to begin with definitions and axioms. Euclid therefore prefaced several of the thirteen books of *Elements* with definitions of the mathematical objects discussed, most of which are relatively standard. He also prefaced Book I with ten axioms; five of these are geometrical postulates and five are more general truths about mathematics called *common notions*. Euclid then proceeded to prove one result after another, each proof based on the previous results and/or the axioms. If one reads Book I from the beginning, one never has any idea what will come next. It is only when one gets to the end of the book, where Euclid proves the Pythagorean theorem, that one realizes that Book I's basic purpose is to lead to the proof of that result. Thus, in order to understand the reasons for various theorems, we will begin our discussion of Book I with the Pythagorean theorem and work backwards. This will also enable us to see why certain unproved results, namely, the axioms, must be assumed.

PROPOSITION I-47. *In right-angled triangles, the square on the hypotenuse is equal to the sum of the squares on the legs.*

Euclid proves the result by first constructing the line AL parallel to BD meeting the base DE of the square on the hypotenuse at L and then showing that rectangle $BNLD$ is equal to the square on AB and rectangle $CNLE$ is equal to the square on AC (Fig. 2.8). To accomplish the first equality, Euclid connects AD and CF to produce triangles ADB and CBF. He then shows that these two triangles are equal to each other, that rectangle $BNLD$ is double triangle ABD, and that the square on AB is double triangle CBF. His first equality then

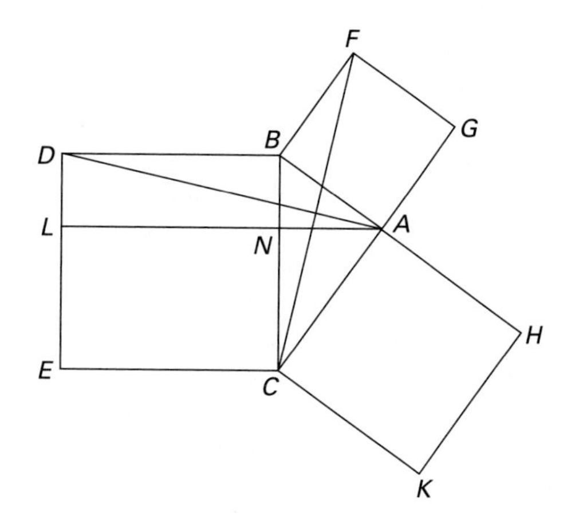

FIGURE 2.8 The Pythagorean theorem in Euclid's *Elements*

follows. The second one is proved similarly, and the sum of the two equalities proves the theorem.

So, what results are needed to make this proof work? First, of course, to make any sense of the theorem at all, one needs to know how to construct a square on a given straight line segment. After all, the theorem states a relationship between certain squares. We are therefore led to

PROPOSITION I-46. *On a given straight line to describe a square.*

There are many ways to accomplish this construction, so Euclid must make a choice. He begins by constructing a perpendicular AC to the given line AB and determining a point D so that $AD = AB$ (Fig. 2.9). He then constructs a line through D parallel to AB and a line through B parallel to AD, the two lines meeting at point E. His claim now is that quadrilateral $ADEB$ is the desired square. (Note that to get this far one needs to be able

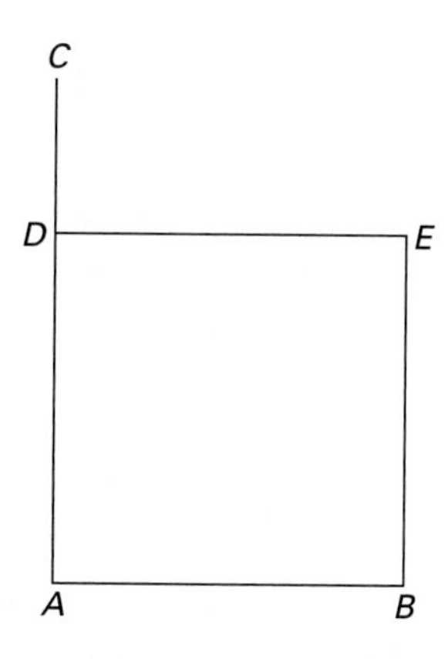

FIGURE 2.9 *Elements*, Proposition I-46

to construct lines perpendicular and parallel to given lines—these constructions are given in propositions I-11 and I-31—as well as to cut off on one line segment a line segment equal to another one—this is proposition I-3.) To prove that his construction is correct, Euclid begins by noting that quadrilateral $ADEB$ has two pairs of parallel sides, so it is a parallelogram. And by proposition I-34, the opposite sides are equal. It follows that all four sides of $ADEB$ are equal. To show that $ADEB$ is a square, it remains to show that all the angles are right angles. But line AD crosses the two parallel lines AB, DE. So, by proposition I-29, the two interior angles on the same side, namely, angles BAD and ADE, are equal to two right angles. But since we already know that angle BAD is a right angle, so is angle ADE. And since opposite angles in parallelograms are equal according to I-34, all four angles are right angles, and $ADEB$ is a square.

So although the actual construction of a square is fairly obvious, the proof that the construction is correct appears to require many other propositions. Before looking at some of those propositions, let us return to the main theorem and see what else we need.

The first result is the one that allows Euclid to conclude that triangles ADB and CBF (Fig. 2.8) are equal. That follows by the familiar side-angle-side theorem, proved by Euclid as

PROPOSITION I-4. *If two triangles have two sides equal to two sides respectively, and have the angles contained by the equal sides also equal, then the two triangles are congruent.*

The word *congruent* is used here as a modern shorthand for Euclid's conclusion that each part of one triangle is equal to the corresponding part of the other. Euclid proves this theorem by superposition. That is, he imagines the first triangle being moved from its original position and placed on the second triangle with one side placed on the corresponding equal side and the angles also matching. Euclid here tacitly assumes that such a motion is always possible without deformation. Rather than supply such a postulate, nineteenth-century mathematicians tended to assume this theorem itself as a postulate.

Euclid also needs the result that a rectangle is double a triangle with the same base and height. This follows from

PROPOSITION I-41. *If a parallelogram has the same base with a triangle and is in the same parallels, the parallelogram is double the triangle.*

Since "in the same parallels" means that the two figures have the same height, it would seem that this proposition follows from the formulas for the areas of a triangle and a parallelogram: $A = \frac{1}{2}bh$ and $A = bh$. But Euclid does not use formulas. The only way he could prove this proposition was by demonstrating that the parallelogram can be divided into two triangles, each equal to the given one. In Figure 2.10, the given parallelogram is $ABCD$ and the

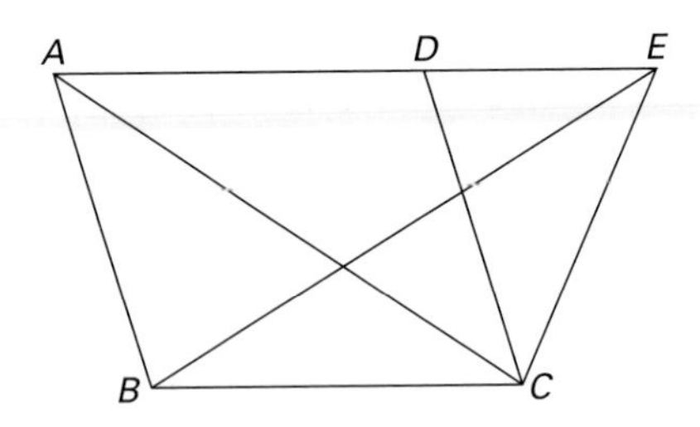

FIGURE 2.10 *Elements*, Proposition I-41

given triangle is BCE. Euclid draws AC, the diagonal of the parallelogram, then notes that triangle ABC is equal to triangle BCE because they have the same base and are in the same parallels (Proposition I-37). But parallelogram $ABCD$ is double the triangle ABC (by proposition I-34) and therefore is double the triangle BCE.

Recall that the construction of a square required the construction of both a perpendicular to a given line and a parallel to a given line. The first of these constructions requires the drawing of the equilateral triangle DFE whose base DE contains at its center the point C at which the perpendicular is drawn (Fig. 2.11). This construction is accomplished in proposition I-1 and requires the use of a compass and a straight edge. Thus, Euclid needs to postulate that a circle can be drawn with a given center and radius and that a line can be drawn connecting two points. These postulates are postulate 3 and postulate 1, respectively. Once the triangle is constructed, the line from the vertex not on the original line to the given point is the desired perpendicular. To prove this, Euclid notes that the two triangles DCF and ECF are congruent by side-side-side, a result proved as Proposition I-8. Since the sum of the equal angles DCF and ECF is two right angles, each of the angles DCF and ECF is right.

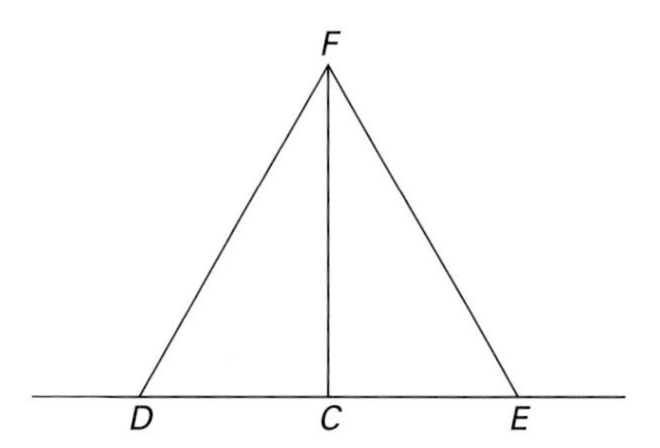

FIGURE 2.11 *Elements*, Proposition I-1

To construct a line through a given point A parallel to a given line BC, Euclid takes an arbitrary point D on BC and connects AD (Fig. 2.12). By proposition I-23, he then constructs the angle DAE equal to the angle ADC and extends AE into the straight line AF. That one can extend a straight line in a straight line is the substance of another construction postulate, postulate 2. To prove that EF is parallel to BC, Euclid notes that the alternate interior angles DAE and ADC are equal. By proposition I-27, the two lines are parallel.

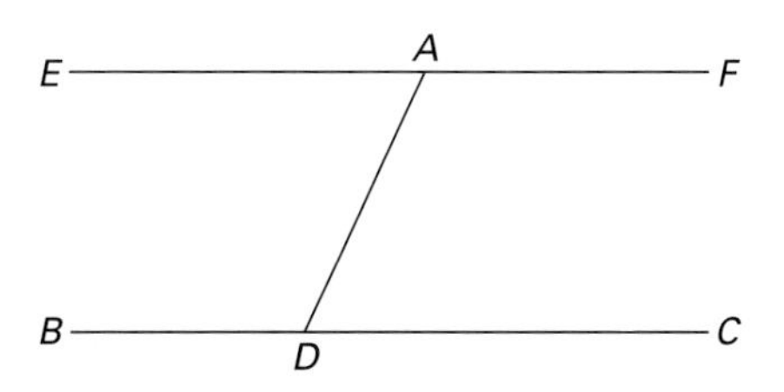

FIGURE 2.12 *Elements*, Proposition I-31

Rather than go through the proofs of these further results and then the proofs of the results that those depend on, which would force us to consider virtually every proposition in Book I, we will conclude this section by considering just two more important propositions that have already been referred to several times. First, we consider

PROPOSITION I-34. *In parallelograms, the opposite sides and angles are equal to one another and the diameter bisects the areas.*

Note that the proofs of propositions I-46 and I-41 used all three conclusions of this proposition. To prove it, one thinks of the diagonal as first cutting one pair of parallel sides and then cutting the other. In each case, one concludes from proposition I-29 that the alternate interior angles are equal. It then follows (by angle-side-angle) that the two triangles into which the parallelogram is cut by the diagonal are congruent. (The angle-side-angle triangle congruence theorem is proposition I-26.) The congruence of the two triangles then implies that each pair of opposite sides and each pair of opposite angles are equal. The third part of the proposition follows immediately.

The final proposition we consider is one on which both I-34 and I-46 depend:

PROPOSITION I-29. *A straight line falling on parallel straight lines makes the alternate angles equal to one another, the exterior angle equal to the interior and opposite angle, and the interior angles on the same side equal to two right angles.*

It is easy enough to see that any two of the statements are simple consequences of the third. So Euclid needed to decide which one to prove. From hints in various Greek texts, we know that before Euclid, the situation regarding this theorem was very unclear. How does one prove one of these results? What must one assume? It is in his answer to these questions that Euclid showed his genius. He had already proved the converse of this theorem in propositions I-27 and I-28. Evidently, however, he saw no way of proving any of the statements in this proposition directly. We can imagine that he struggled with this, but he eventually realized that he would have to take one of these results—or its equivalent—as a postulate. And so he decided, for reasons we can only guess, to take the contrapositive of the third statement in the proposition as a postulate. Thus, at the beginning of Book I, he placed

POSTULATE 5. *If a straight line intersecting two straight lines makes the interior angles on the same side less than two right angles, the two straight lines, if produced indefinitely, meet on that side on which the angles are less than two right angles.*

Given this postulate, the proof of proposition I-29 is straightforward by a *reductio* argument: Assume that angle $AGH >$ angle GHD (Fig. 2.13). Then the sum of angles AGH and BGH is greater than the sum of angles GHD and BGH. The first sum equals two right angles (by proposition I-13), so the second one is less than two right angles. Then, by the postulate, the lines AB and CD must meet. But this contradicts the hypothesis that those lines are parallel.

Thus, we see that the Pythagorean theorem, the culminating theorem of Book I, besides requiring very many of the earlier results in Book I, including all three triangle congruence theorems, rests on the critical parallel postulate. The parallel postulate, alone among Euclid's postulates, has caused immense controversy over the years, because many people claimed it was not "self-evident." And for Euclid, as for Aristotle, a postulate should be "self-evident." Thus, almost from the time the *Elements* appeared, people have attempted

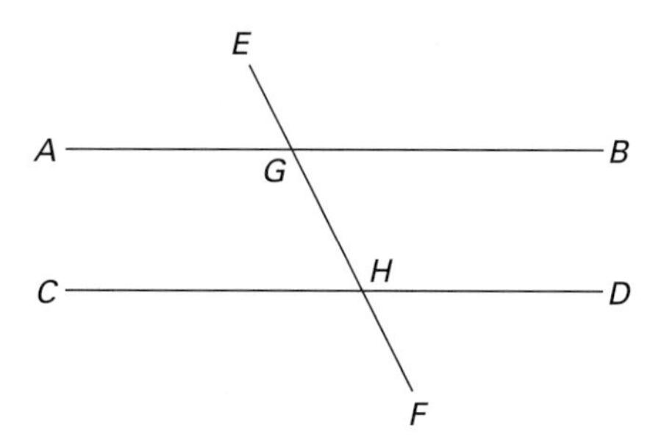

FIGURE 2.13 *Elements*, Proposition I-29

to prove this result as a theorem, using as a basis just the other axioms and postulates. Many people thought they had accomplished this task, but a close examination of every such proof always reveals either an error or, more likely, another assumption—one that perhaps is more "self-evident" than Euclid's postulate but that nevertheless cannot be proved from the other nine axioms. Probably the most familiar "other assumption" is what is now known as

PLAYFAIR'S AXIOM. *Through a given point outside a given line, exactly one line may be constructed parallel to the given line.*

Exercise 11 asks you to show that this result is entirely equivalent to Euclid's postulate.

2.2.2 Geometric Algebra

Book II of the *Elements* is quite different in tone from Book I. It deals with the relationships between various rectangles and squares, many of which are similar to those of Babylonian algebra. In fact, it is the propositions in Book II, together with propositions 42–45 of Book I and propositions 27–30 of Book VI, that form the content of what is called **geometric algebra**, the representation of algebraic concepts through geometric figures.

Euclid begins Book II with a definition:

Any rectangle is said to be **contained by** *the two straight lines forming the right angle.*

This definition shows Euclid's geometric usage. The statement does not mean that the area of a rectangle is the product of the length and the width. Euclid never multiplies two lengths together, because he has no way of defining such a process for arbitrary lengths. At various places, he multiplies lengths by numbers (that is, positive integers), but otherwise he only writes of rectangles "contained by" two lines. Although the phrase "contained by" is often translated as "product" in what follows, keep in mind that this is modern usage, not Euclid's own.

Propositions II-5 and II-6 illustrate Euclid's ideas in the manipulation of geometric figures "algebraically." In fact, centuries later, these propositions were used as the geometric justifications of the standard algebraic solutions of quadratic equations.

PROPOSITION II-5. *If a straight line is cut into equal and unequal segments, the rectangle contained by the unequal segments of the whole together with the square on the straight line between the points of section is equal to the square on the half.*

PROPOSITION II-6. *If a straight line is bisected and a straight line is added to it, the rectangle contained by the whole with the added straight line and the added straight line together with the square on the half is equal to the square on the straight line made up of the half and the added straight line.*

Figure 2.14 should help clarify these propositions. If AB is labeled in each diagram as b, AC and BC as $b/2$, and DB as x, proposition II-5 translates into $(b - x)x + (b/2 - x)^2 = (b/2)^2$, and proposition II-6 gives $(b + x)x + (b/2)^2 = (b/2 + x)^2$. The quadratic equation $bx - x^2 = c$ [or $(b - x)x = c$] can be solved using the first equality by writing $(b/2 - x)^2 = (b/2)^2 - c$ and then getting

$$x = \frac{b}{2} - \sqrt{\left(\frac{b}{2}\right)^2 - c}.$$

Similarly, the equation $bx + x^2 = c$ [or $(b + x)x = c$] can be solved from the second equality by using an analogous formula. Alternatively, one can label AD as y and DB as x in each diagram and translate the first result into the standard Babylonian system $x + y = b$, $xy = c$ and the second result into the system $y - x = b$, $yx = c$. In any case, note that Figure 2.14 is essentially the same as Figure 1.11, the figure representing the Babylonian scribes' probable method for solving the first of these systems.

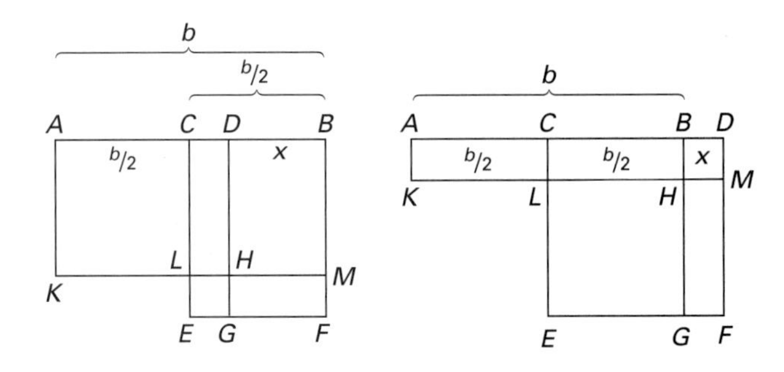

FIGURE 2.14 *Elements*, Propositions II-5 and II-6

Similarly, consider

PROPOSITION II-9. *If a straight line is cut into equal and unequal segments, the squares on the unequal segments of the whole are double the sum of the square on the half and the square on the straight line between the points of section.*

One possible algebraic translation of this proposition is

$$x^2 + y^2 = 2\left(\frac{x + y}{2}\right)^2 + 2\left(\frac{x - y}{2}\right)^2,$$

the result used by the Babylonians to solve the system $x - y = b$, $x^2 + y^2 = c$.

Euclid, of course, does not do any of the translations indicated. He just uses the constructions implied in Figure 2.14 to prove the equalities of the appropriate squares and rectangles in propositions II-5 and II-6 and uses a proof involving the Pythagorean theorem to demonstrate proposition II-9. He does not indicate anywhere that these propositions are of use in solving what are now called **quadratic equations**.

Yet Euclid does solve problems that can be translated into the form of quadratic equations. The first example is in

PROPOSITION II-11. *To cut a given straight line so that the rectangle contained by the whole and one of the segments is equal to the square on the remaining segment.*

The goal of the proposition is to find a point H on the line so that $AB \times HB$ equals the square on AH (Fig. 2.15). To translate this problem into algebra, one lets the line AB be a and lets AH be x. Then $HB = a - x$, and the problem amounts to solving the equation

$$a(a - x) = x^2 \quad \text{or} \quad x^2 + ax = a^2.$$

The Babylonian solution is

$$x = \sqrt{\left(\frac{a}{2}\right)^2 + a^2} - \frac{a}{2}.$$

Euclid's proof seemingly amounts to precisely this formula. To get the square root of the sum of two squares, the obvious method is to use the hypotenuse of a right triangle whose sides are the given roots, in this case, a and $a/2$. So Euclid draws the square on AB and then bisects AC at E. It follows that EB is the desired hypotenuse. To subtract $a/2$ from this length, he draws EF equal to EB and subtracts off AE to get AF; this is the needed value x. Since he wants the length marked off on AB, he simply chooses H so that $AH = AF$. To prove that this choice of H is correct, Euclid then appeals to proposition II-6, which provides the needed justification.

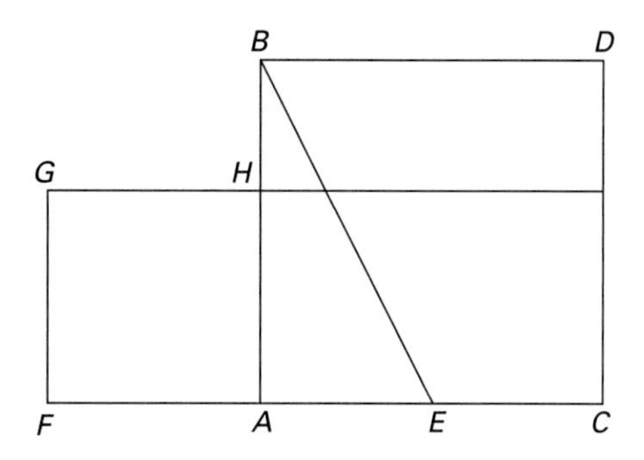

FIGURE 2.15 *Elements*, Proposition II-11

Before considering more complicated examples, it is necessary to make a slight digression back to Book I, where Euclid solves what amounts to a simple linear equation:

PROPOSITION I-44. *To a given straight line to apply, in a given rectilinear angle, a parallelogram equal to a given triangle.*

The aim of the construction is to find a parallelogram of given area with one angle given and one side equal to a given line segment. That is, the parallelogram is to be "applied" to the given line segment. That this too can be interpreted algebraically is easily seen if the given angle is a right angle. If the area of the triangle is taken to be c and the given line segment is assumed to have length a, the goal of the problem is to find a line segment b such that the rectangle with length a and width b has area c—that is, to solve the equation $ax = c$.

In Book VI, Euclid expands this notion of "application of areas," giving two important results that effectively provide solutions to certain types of quadratic equations.

PROPOSITION VI-28. *To a given straight line to apply a parallelogram equal to a given rectilinear figure and deficient by a parallelogram similar to a given one; thus the given rectilinear figure must not be greater than the parallelogram described on the half of the straight line and similar to the defect.*

PROPOSITION VI-29. *To a given straight line to apply a parallelogram equal to a given rectilinear figure and exceeding by a parallelogram similar to a given one.*

In these propositions, Euclid deals with application of areas that are "deficient" and "exceeding." In the first case, he proposes to construct a parallelogram of given area whose base is less than the given line segment AB (Fig. 2.16). The parallelogram on the deficiency, the line segment SB, is to be similar to a given one. In the second case, the constructed parallelogram of given area has base greater than the given line segment AB, whereas the parallelogram on the excess, the line segment BS, is again to be similar to a given one. The importance of these ideas of "deficient" and "exceeding" will be apparent in the discussion of conic sections later on. For the discussion of geometric algebra, it is simplest to assume that the given parallelogram in each case is a square. The propositions and their proofs will be translated into algebra under that assumption. Thus, the constructed parallelogram must in each case be a rectangle.

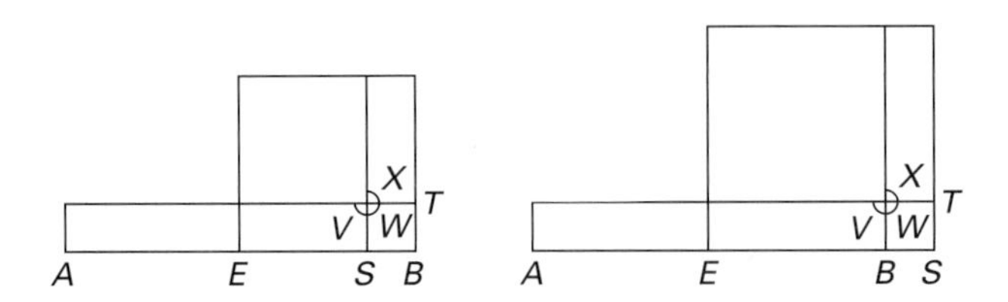

FIGURE 2.16 *Elements*, Propositions VI-28 and VI-29

Designate AB in both cases by b, and designate the area of the given rectilinear figure by c. The problems reduce to finding a point S on AB (proposition VI-28) or on AB extended (proposition VI-29) such that $x = BS$ satisfies $x(b - x) = c$ in the first case and $x(b + x) = c$ in the second. That is, it is necessary to solve the quadratic equations $bx - x^2 = c$ and $bx + x^2 = c$, respectively. In each case, Euclid finds the midpoint E of AB and constructs the square on BE whose area is $(b/2)^2$. In the first case, S is chosen so that ES is the side of a square whose area is $(b/2)^2 - c$. That is why the proposition states the condition that, in effect, c cannot be greater than $(b/2)^2$. This choice for ES implies that

$$x = BS = BE - ES = \frac{b}{2} - \sqrt{\left(\frac{b}{2}\right)^2 - c}.$$

In the second case, S is chosen so that ES is the side of a square whose area is $(b/2)^2 + c$. Then

$$x = BS = ES - BE = \sqrt{\left(\frac{b}{2}\right)^2 + c} - \frac{b}{2}.$$

In both cases, Euclid proves that his choice is correct by showing that the desired rectangle equals the gnomon XWV and that the gnomon is in turn equal to the given area c. Algebraically, that amounts in the first case to showing that

$$x(b - x) = \left(\frac{b}{2}\right)^2 - \left[\left(\frac{b}{2}\right)^2 - c\right] = c$$

and in the second case to showing that

$$x(b + x) = \left[\left(\frac{b}{2}\right)^2 + c\right] - \left(\frac{b}{2}\right)^2 = c.$$

Proposition VI-29 is actually applied immediately to the special case where the "given parallelogram" is a square:

PROPOSITION VI-30. *To cut a given finite straight line in extreme and mean ratio.*

Cutting a line AB in "extreme and mean ratio" means finding a point H on the line such that $AB : AH = AH : HB$. Euclid had already shown that this proportion is equivalent to the statement that the square on AH is equal to the rectangle contained by AB and AH, and so this is the same problem already solved in II-11. Here, Euclid begins by constructing a square $ABEC$ on AB (Fig. 2.17). He then uses VI-29 to apply to AC the rectangle $CFDG$ equal to BC and exceeding by the square $AHDG$. He then easily shows that the point H where the rectangle $CFDG$ cuts the line AB cuts that line in extreme and mean ratio.

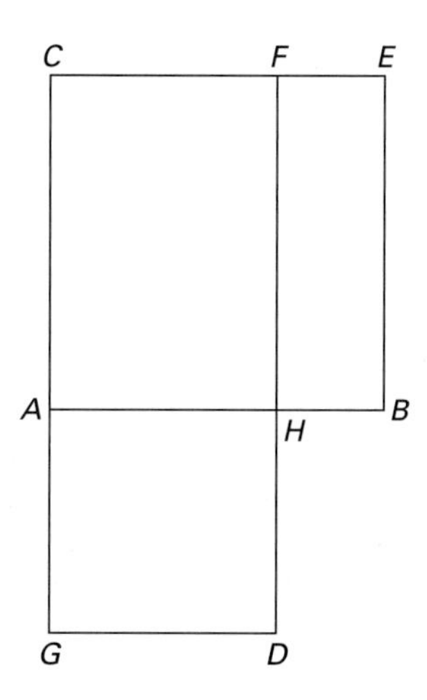

FIGURE 2.17 *Elements*, Proposition VI-30

There has long been a debate over whether the geometric algebra in Euclid stems from a deliberate transformation of the Babylonian quasi-algebraic results into formal geometry. As already noted, there is a strong similarity between the geometric procedures and the algebraic ones, at least in the special cases discussed. One can then argue that the Greek adaptation into a geometric viewpoint, given the necessity of proof, was related to the discovery that not every line segment could be represented by a number. One can further argue that, once one has translated the material into geometry, one might just as well state and prove certain results for parallelograms as for rectangles, since little extra effort is required. A further argument supporting the transformation is that the original Babylonian methodology itself may well have been couched in a "naive" geometric form, a form well suited to a translation into the more sophisticated Greek geometry. As has been noted, several of Euclid's propositions appear to be identical to the assumed Babylonian geometric bases for the algorithms they used to solve their equations. The chief difference is that in Euclid, the propositions are carefully proved.

Was there any opportunity for direct cultural contact between Babylonian mathematical scribes and Greek mathematicians? It used to be argued that this was virtually impossible, because there was no record of Babylonian mathematics at all between the sixth and the fourth centuries BCE, when this contact would have had to take place, and because those in the aristocracy to which the Greek mathematicians belonged would have been disdainful of the activities of the Old Babylonian scribes, who were not part of the elite. However, recent discoveries have indicated that mathematical activity did continue in Babylonia through the middle of the first millennium BCE. Furthermore, by this time, the Mesopotamian languages were often being written in ink on papyrus using a new alphabet. Cuneiform writing on clay tablets was restricted to important documents that needed to be preserved, and those who could perform this service had become members of the elite, experts in traditional wisdom who were central to the functioning of the state. Besides, from the sixth century BCE on, Mesopotamia was a province of the Persian empire, with whom the Greeks did maintain contact.

On the other hand, despite the possibilities for contact and the logic of the argument as to how Babylonian mathematics could have been translated into Greek geometry, there is no direct evidence of any transmission of Babylonian mathematics to Greece before or during the fourth century BCE. One could then argue that although the Greeks did employ what modern mathematicians think of as algebraic procedures, their mathematical thought was so geometrical that all such procedures were automatically expressed that way. The Greeks of the period up to 300 BCE had no algebraic notation and therefore no way of manipulating expressions that stood for magnitudes, except by thinking of them in geometric terms. In fact, Greek mathematicians became very proficient in manipulating geometric entities. Finally, there was no way the Greeks could express, other than geometrically, irrational solutions of quadratic equations.

There is as yet no clear answer to the related questions of whether Babylonian algebra was transmitted in some form to Greece by the fourth century BCE and whether the theorems discussed in this section should be considered as algebra. We need to await further discoveries. In the meantime, the interested reader should carefully read the original sources.

2.2.3 The Pentagon Construction

After dealing with rectilinear figures in Books I and II of the *Elements*, Euclid turns in Book III to the properties of the most fundamental curved figure, the circle. The Greeks were greatly impressed with the symmetry of the circle, the fact that no matter how one turned it, it always appeared the same. They thought of it as the most perfect of plane figures. Similarly, they believed the three-dimensional analog of the circle, the sphere, was the most perfect of solid figures. These philosophical ideas provided the basis for Greek ideas on astronomy, which will be discussed later.

If there is any organizing principle of Book III, it is to provide for the construction, in Book IV, of polygons inscribed in and circumscribed about circles. In particular, most of the propositions from the last half of Book III are used in the most difficult construction of Book IV, the construction of the regular pentagon. The constructions of the triangle, square, and hexagon are relatively intuitive, but the construction of the pentagon involves more advanced concepts. This construction is used in turn in Euclid's construction of some of the regular solids in Book XIII.

To get the flavor of Books III and IV, then, we will start with the construction of the pentagon and work backwards to see what is needed from Book III for its proof. The treatment of the pentagon begins in Book IV after Euclid has already shown the simpler techniques of inscribing triangles and squares in circles, inscribing circles in triangles and squares, circumscribing triangles and squares about circles, and circumscribing circles about triangles and squares. Euclid then divides his construction of a regular pentagon into two steps: first, the construction of an isosceles triangle with each of the base angles double the vertex (IV-10), and second, the actual inscribing of the pentagon in the circle (IV-11). As usual, Euclid does not show how he arrived at the construction, but a close reading gives a clue to his analysis of the problem. We will therefore assume the construction made and try to see where that assumption leads.

So, suppose $ABCDE$ is a regular pentagon inscribed in a circle (Fig. 2.18) and draw the diagonals AC and CE. Since angles CEA and CAE each subtend an arc double that subtended by angle ACE, it follows that triangle ACE is an isosceles triangle with base angles double the vertex angle. Thus, the pentagon construction has been reduced to the construction of that triangle. Assume then that ACE is such an isosceles triangle, and let AF bisect angle A. It follows that triangles AFE and CEA are similar, so $EF : AF =$

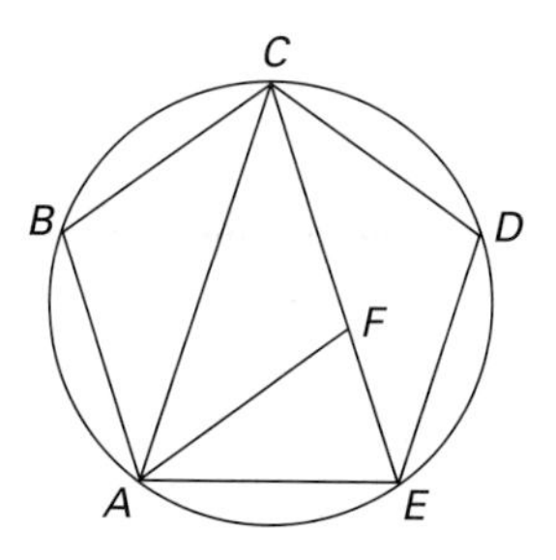

FIGURE 2.18 Construction of a regular pentagon

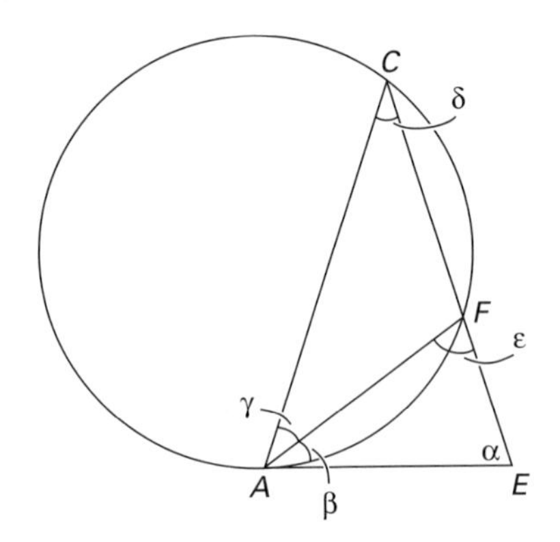

FIGURE 2.19 *Elements*, Proposition IV-10

$EA : CE$. But triangles AFE and AFC are both isosceles, so $EA = AF = FC$. Therefore, $EF : FC = FC : CE$, or, in modern notation, $FC^2 = EF \cdot CE$. The construction has therefore been reduced to finding a point F on a given a line segment CE such that the square on CF is equal to the rectangle contained by EF and CE. But this is precisely the construction of proposition II-11. Once F is found, the isosceles triangle with base angles double the vertex angle can be constructed by drawing a circle centered on C with radius CE and another circle centered on E with radius CF. The intersection A of the two circles is the third vertex of the desired triangle.

Euclid performs this construction in proposition IV-10 (Fig. 2.19), but, since he will not discuss similarity until Book VI, he needs to use alternative arguments to prove that proposition. The goal is to show that $\alpha = 2\delta$. If it can be shown that $\beta = \delta$, then $\beta + \gamma = \delta + \gamma = \epsilon$. Also, since $CA = CE$, we have $\alpha = \beta + \gamma$; thus, $\epsilon = \alpha$. But then $AE = AF$, and since by construction $AE = FC$, it follows that triangle AFC is isosceles and that $\delta = \gamma$. Finally, $\alpha = \beta + \gamma = \delta + \delta = 2\delta$, as desired. To show that $\beta = \delta$, circumscribe a circle around triangle AFC. In the diagram (Fig. 2.19), it appears that AE is tangent to the circle. Assuming this is the case, Euclid needs a theorem relating certain angles formed by tangents and secants in circles:

PROPOSITION III-32. *If a straight line is tangent to a circle, and from the point of tangency there is drawn a straight line cutting the circle, the angles which that line makes with the tangent will be equal to the angles in the alternate segments of the circle.*

In other words, this proposition asserts that one of the angles formed by the tangent EF and the secant BD (Fig. 2.20), say, angle DBF, is equal to any angle in the "alternate" segment BD of the circle, such as angle DAB. Similarly, the other angle made by the tangent, angle DBE, is equal to any angle in the remaining segment, such as angle DCB. (By proposition III-21, any two angles in the same segment are equal to one another.) To prove this result, draw a perpendicular AB to the tangent at the point B of tangency. Since a perpendicular to a tangent passes through the center of the circle (proposition III-19), the angle ADB, being an angle in a semicircle, is a right angle (proposition III-31). Therefore, angles DAB and ABD sum to a right angle. But angles DBF and ABD also sum to a right angle. It follows that angle DAB equals angle DBF, as claimed. The equality of the other two angles can then be easily established.

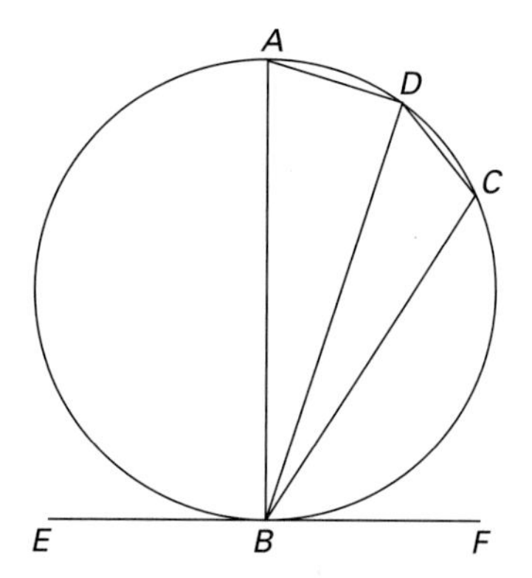

FIGURE 2.20 *Elements*, Proposition III-32

Returning to the proof of proposition IV-10, this result shows that $\beta = \delta$. It thus remains to show that in fact AE is tangent to the circle. For this, Euclid needs

PROPOSITION III-37. *If from a point outside a circle we draw two straight lines, one a secant and one meeting the circle, and if the rectangle contained by the whole secant and that segment which is outside the circle equals the square on the straight line which meets the circle, then that second straight line is tangent to the circle.*

In IV-10, the square on CF equals the rectangle contained by EF and CE (Fig. 2.19). But $AE = CF$. Thus, the conditions of proposition III-37 are met (where E is the "point outside the circle"), and AE is therefore tangent to the circle. The proof of IV-10 is complete, once III-37 is proved. Since this latter proposition is the converse of proposition III-36, we first look at that result:

PROPOSITION III-36. *If from a point outside a circle we draw a tangent and a secant to the circle, then the rectangle contained by the whole secant and that segment which is outside the circle equals the square on the tangent.*

The statement may remind the reader of proposition II-6. And, in fact, that proposition is used in the proof. We will consider only the easier case here, where the secant line $DCFA$ goes through the center F (Fig. 2.21). Join FB to form the right triangle FBD. Proposition

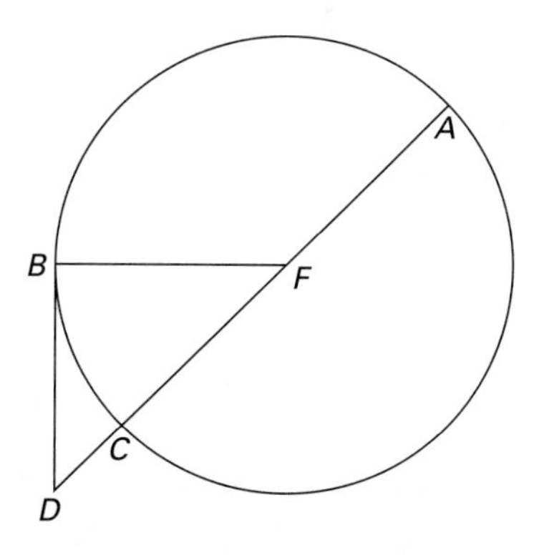

FIGURE 2.21 *Elements*, Proposition III-36

II-6 asserts that the rectangle contained by AD and CD, together with the square on FC, equals the square on FD. But $FC = FB$, and the sum of the squares on FB and BD equals the square on FD. Therefore, the rectangle contained by AD and CD equals the square on BD, as claimed. The case where the secant line does not pass through the center is slightly trickier.

To prove proposition III-37, Euclid first draws a tangent to the circle from the point outside, then concludes from III-36 that the length of that tangent is the same as the length of the line that "meets" the circle. It is then straightforward to show that this second line is also a tangent.

With the proof of IV-10, an isosceles triangle has been constructed with base angles double the vertex angle. To inscribe the pentagon in the circle, Euclid first inscribes that isosceles triangle ACE in the circle (Fig. 2.18). Next, he bisects the angles at A and E. The intersection of these bisectors with the circle are points D and B, respectively. Then A, B, C, D, E are the vertices of a regular pentagon.

2.2.4 Ratio, Proportion, and Incommensurability

The regular pentagon is part of the pentagram, evidently one of the symbols used by the Pythagoreans. Thus, it is believed that the Pythagoreans worked out a construction of the pentagon, although more likely their construction used similarity rather than the method described above. It is therefore plausible that the pentagram's property of reproducing itself when the diagonals of the inner pentagon are constructed (Fig. 2.22) could well have been an alternative path to the discovery of incommensurability (instead of the one described earlier). To explain this, we need to move to Book VII, the first of the three books of number theory in the *Elements*.

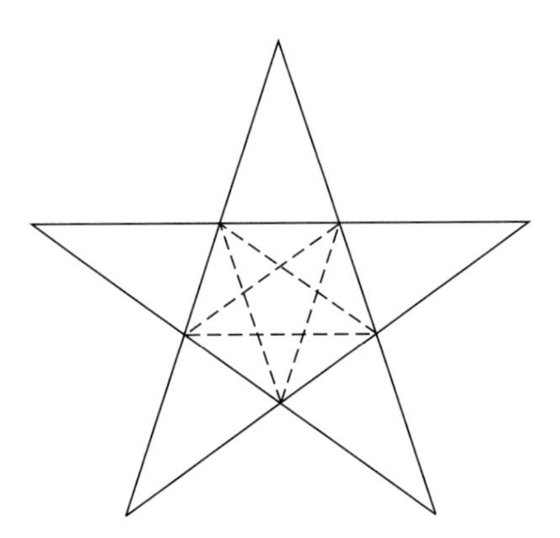

FIGURE 2.22 Diagonals of inner pentagon of a pentagram

Book VII, like all the number theory books, deals with the Greek concept of number, what modern mathematicians call the positive integers, in contrast to the geometrical magnitudes of the earlier books. And the first item of business for Euclid in this book is the familiar process for finding the greatest common divisor of two numbers. This algorithm, usually called the **Euclidean algorithm**, although it was certainly known long before Euclid, is presented in propositions VII-1 and VII-2. Given two numbers, a, b with $a > b$, one subtracts b from a as many times as possible; if there is a remainder c, which of course must be less than b, one then subtracts c from b as many times as possible. Continuing in this manner, one eventually comes either to a number m, which "measures" (divides) the one before (proposition VII-2), or to the unit (1) (proposition VII-1). In the first case, Euclid proves that m is the greatest common measure (divisor) of a and b. In the second case, he shows that a and b are prime to one another. For example, given the two numbers 18 and 80, first subtract 18 from 80. One can do this four times, with remainder 8. Next, subtract 8 from 18; this can be done twice, with remainder 2. Finally, one can subtract 2 exactly four times from 8. It then follows that 2 is the greatest common divisor of 18 and 80. In addition, this calculation shows that one can express the ratio of 80 to 18 in the form (4, 2, 4), in the sense that applying the algorithm to any other pair a, b such that $a : b = 80 : 18$ will also give (4, 2, 4). As another example, take the pair 7 and 32. One can subtract 7 four times from 32 with remainder 4. One can then subtract 4 once from 7 with remainder 3. Finally, one can subtract 3 once from 4 with remainder 1. Thus, 7 and 32 are prime to one another and their ratio can be expressed in the form (4, 1, 1).

It was probably Theaetetus (417–369 BCE) who first investigated the possibility of applying what is now known as the Euclidean algorithm to magnitudes. The results appear as propositions 2 and 3 of Book X, where Euclid shows how to determine whether two magnitudes A and B have a common measure (are commensurable) or do not (are incommensurable). The procedure is basically the same as for numbers. Thus, supposing that $A > B$, one first subtracts B from A as many times as possible, say n_0, getting a remainder b, which is less than B. One next subtracts b from B as many times as possible, say n_1, getting a remainder b_1 less than b. Euclid shows in proposition X-2 that if this process never ends, then the original two magnitudes are incommensurable. If, on the other hand, one of the magnitudes of this sequence measures the previous one, then that magnitude is the greatest common measure of the original two (proposition X-3). A natural question here is how one can tell whether or not the process ends. In general, that is difficult. But in certain cases, a repeated pattern becomes evident in the remainders, which shows that the process cannot end.

For example, let us consider the case of the diagonal and the side of the regular pentagon (Fig. 2.23). By the properties of the pentagon, $CG = KG$. Therefore, one can subtract the side $CG = KG$ once from the diagonal GD, leaving remainder KD. One next must

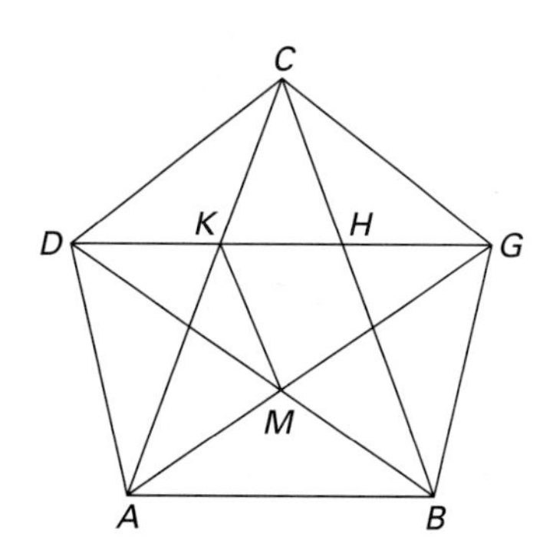

FIGURE 2.23 Incommensurability of diagonal and side of a regular pentagon

subtract KD from the side CG. But $CG = HD$, so KD can be subtracted once from $CG = HD$, with remainder KH. Note that KH is the side of another regular pentagon, whose diagonal is $KM = KD$. Therefore, at the next stage, one is again subtracting a side from a diagonal of a pentagon. Since one can continue getting new smaller and smaller pentagons by connecting diagonals of previous ones, it is clear that in this case the process never ends. Thus, the diagonal and the side of a regular pentagon are incommensurable. In fact, the ratio of the diagonal to the side may be written as $(1, 1, 1, \ldots)$.

Having recognized the existence of incommensurable quantities, the Greeks realized that they had to figure out a method of dealing with the ratios of such quantities. When they believed that any pair of quantities was commensurable, it was easy enough to see when two such pairs were proportional, or had equal ratio. Euclid, in fact, defines this concept in Book VII, when he is dealing with numbers:

Four numbers are **proportional** *when the first is the same multiple, or the same part, or the same parts, of the second that the third is of the fourth.*

As an example, $3 : 4 = 6 : 8$, because 3 is 3 "fourth" parts of 4 and 6 is 3 "fourth" parts of 8. But this definition cannot be used for general magnitudes. The side of a pentagon cannot be expressed either as a multiple or as a part or parts of the diagonal.

Thus, using the procedure of the Euclidean algorithm, Theaetetus gave a new definition of equal ratio, which applied to all magnitudes. Suppose there are two pairs of magnitudes A, B, and C, D. Applying the procedure to each pair gives two sequences of equalities:

$$\begin{array}{ll} A = n_0 B + b \ (b < B) & C = m_0 D + d \ (d < D) \\ B = n_1 b + b_1 \ (b_1 < b) & D = m_1 d + d_1 \ (d_1 < d) \\ b = n_2 b_1 + b_2 \ (b_2 < b_1) & d = m_2 d_1 + d_2 \ (d_2 < d_1) \\ \vdots & \vdots \end{array}$$

If the two sequences of numbers $(n_0, n_1, n_2, \ldots)$, $(m_0, m_1, m_2, \ldots)$ are equal term by term and both end at, say, $n_k = m_k$, then it is possible to check that the ratios $A : B$ and $C : D$ are both equal to the same ratio of integers. Hence, Theaetetus could give the general definition that $A : B = C : D$ if the (possibly neverending) sequences $(n_0, n_1, n_2, \ldots)$, $(m_0, m_1, m_2, \ldots)$ are equal term by term. Although in general it may be difficult to decide whether two ratios are equal, there are interesting cases in which the sequence $n_0, n_1, n_2, \ldots$ is relatively simple to determine. In any case, Aristotle notes that Theaetetus's definition of equal ratios was the one in use in his time.

Unfortunately, it turns out that Theaetetus's definition is very awkward to use in practice, so mathematicians continued to search for a better one. It was finally Eudoxus (408–355 BCE) who came up with the definition that Euclid adopted at the beginning of Book V:

Magnitudes are said to be **in the same ratio** *(alternatively,* **proportional***), the first to the second and the third to the fourth, when, if any equal multiples whatever are taken of the first and third, and any equal multiples whatever of the second and fourth, the former multiples alike exceed, are alike equal to, or alike fall short of, the latter multiples respectively taken in corresponding order.*

Translated into algebraic symbolism, this definition says that $a : b = c : d$ if, given any positive integers m, n, whenever $ma > nb$, also $mc > nd$, whenever $ma = nb$, also $mc = nd$, and whenever $ma < nb$, also $mc < nd$. In modern terms, this is equivalent to noting that for every fraction n/m, the quotients a/b and c/d are both greater than, equal to, or less than that fraction.

Although Book V of the *Elements* gives numerous properties of magnitudes in proportion, the main application of this theory for Euclid is in the treatment of similarity in Book VI. The foundation of the idea of similarity, the notion of equal ratio (or proportionality), was originally based on the idea that all quantities could be thought of as numbers. In fact, the notion of proportionality occurs in the definition of similarity:

Similar rectilinear figures *are such as have their angles respectively equal and the sides about the equal angles proportional.*

Once the basis for the idea of proportionality was destroyed, the foundation for these results no longer existed. That is not to say that mathematicians ceased to use them. Intuitively, they knew that the concept of equal ratios made perfectly good sense, even if they could not provide a formal definition. In Greek times, as in modern times, mathematicians often ignored foundational questions and proceeded to discover new results. The working mathematician knew that eventually the foundation would be strengthened. Once this occurred, around 360 BCE, the actual similarity results could be organized into a logically acceptable treatise. It is not known who provided this final organization. What is probably true is that there was actually very little to redo except for the proof of the first proposition of Book VI. That is the only one that depends directly on Eudoxus's definition.

PROPOSITION VI-1. *Triangles and parallelograms which have the same height are to one another as their bases.*

Given triangles ABC, ACD with the same height (Fig. 2.24), Euclid needs to show that as BC is to CD, so is the triangle ABC to the triangle ACD. Proceeding as required by Eudoxus's definition, he extends the base BD to both right and left so that he can take arbitrary multiples of both BC and CD along that line. Although Euclid writes about "any number" of line segments, he has no notation to express this and therefore just takes two segments. Presumably, Euclid felt that this was what may be called a **generalizable example**. Working with the two line segments on each side, Euclid notes that because triangles with equal heights and equal bases are equal, whatever multiple the base HC is of the base BC, the triangle AHC is the same multiple of triangle ABC. The same holds for triangle ALC with respect to triangle ACD. Since triangles AHC and ALC again have the same heights, the former is greater than, equal to, or less than the latter precisely when HC is greater than, equal to, or less than CL. Equal multiples having been taken of base BC and triangle ABC, other equal multiples having been taken of base CD and triangle ACD, and the results compared as required by Eudoxus's definition, it follows that

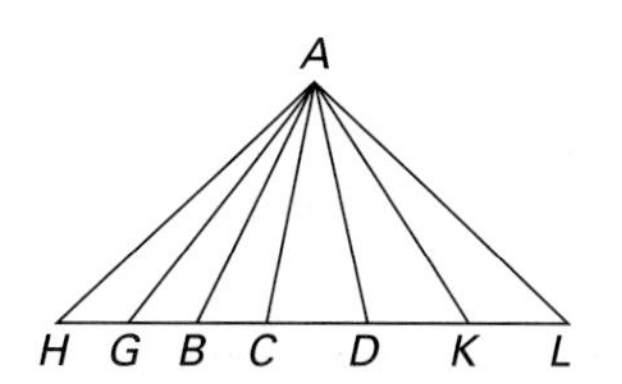

FIGURE 2.24 *Elements*, Proposition VI-1

$BC : CD = ABC : ACD$, as desired. The result for parallelograms is immediate, because each parallelogram is double the corresponding triangle.

The rest of Book VI contains many of the basic results on similarity. In particular, Euclid gives various sets of conditions under which two triangles are similar. Because the definition of similarity requires both that corresponding angles are equal and that corresponding sides are proportional, Euclid shows that one or the other of these two conditions is sufficient.

Proposition VI-16 is in essence the familiar one that in a proportion the product of the means is equal to the product of the extremes. But since Euclid never multiplies magnitudes, he could not have stated this result in terms of Book V. In the geometry of Book VI, however, he has the equivalent of multiplication, for line segments only:

PROPOSITION VI-16. *If four straight lines are proportional, the rectangle contained by the extremes is equal to the rectangle contained by the means; and if the rectangle contained by the extremes is equal to the rectangle contained by the means, the four straight lines will be proportional.*

An important concept of Book V, which is of importance in Book VI and later, is the notion of duplicate ratio:

When three magnitudes are proportional, the first is said to have to the third the **duplicate ratio** *of that which it has to the second.*

A modern statement of this definition would replace "duplicate ratio" by "square of the ratio." Explicitly, if $a : b = b : c$, we would write $a/c = (a/b)(b/c) = (a/b)(a/b) = (a/b)^2$; that is, the ratio $a : c$ is the square of the ratio $a : b$. But Euclid does not multiply ratios. They are not, in fact, quantities. Thus, Euclid needs the alternative indicated in this definition, an alternative he applies in

PROPOSITION VI-19. *Similar triangles are to one another in the duplicate ratio of the corresponding sides.*

To prove this proposition, Euclid constructs a point G on BC so that $BC : EF = EF : BG$ (Fig. 2.25). The ratio $BC : BG$ is then the duplicate of the ratio $BC : EF$ of the corresponding sides. Because triangles ABG, DEF are equal and because triangle ABC is to triangle ABG as BC is to BG, the conclusion of the proposition follows.

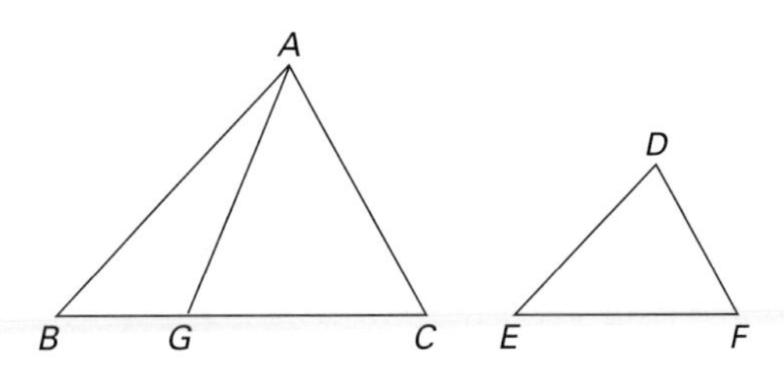

FIGURE 2.25 *Elements*, Proposition VI-19

Proposition VI-20 extends proposition VI-19 to the case of similar polygons. In particular, since all squares are similar, we conclude that a square has to another square a ratio duplicate of the ratio of the sides.

Two parallelograms, of course, can be equiangular without being similar. Euclid is also able to deal with the ratio of such figures, but only by using a concept not formally defined:

PROPOSITION VI-23. *Equiangular parallelograms have to one another the ratio compounded of the ratios of the sides.*

A "compounded" ratio is a generalization of the concept of a duplicate ratio. Euclid illustrates it here in the context of line segments. If the two ratios are $a : b$ and $c : d$, one first constructs a segment e such that $c : d = b : e$. The ratio **compounded** of $a : b$ and $c : d$ is then the ratio $a : e$. In modern terms, the fraction a/e is simply the product of the fractions a/b and $c/d = b/e$.

2.2.5 Number Theory

As already noted, Books VII through IX of Euclid's *Elements* deal with the elementary theory of numbers. There is no mention of the first six books in Books VII, VIII, and IX; these three books form an entirely independent unit. Here we will consider only the most important propositions from the number theory books, ones that have had far-reaching implications over the years. We begin with Euclid's version of the fundamental theorem of arithmetic—that any number can be expressed uniquely as a product of prime numbers. Euclid does not state this as a single proposition, but expresses its essence in several propositions. We first consider

PROPOSITION VII-31. *Any composite number is measured by some prime number.*

and

PROPOSITION VII-32. *Any number either is prime or is measured by some prime number.*

The latter proposition is a direct consequence of the former. Proposition VII-31 in turn is proved by a technique Euclid uses often in the arithmetic books—the least number principle. He begins with a composite number a, which is therefore measured (divided) by another number b. If b is prime, the result follows. If not, then b is in turn measured by c, which then measures a, and c is in turn either prime or composite. As Euclid then says, "if the investigation is continued in this way, some prime number will be found which will measure the number before it, which will also measure a. For, if it is not found, an infinite series of numbers will measure the number a, each of which is less than the other; which is impossible in numbers." Note here the distinction between number and magnitude. Any decreasing sequence of numbers has a least element, but the same is not true of magnitudes.

Although Euclid does not do so, it is straightforward to demonstrate from VII-32 that any number can be expressed as the product of prime numbers. To prove that this expression is unique, we first need

PROPOSITION VII-30. *If a prime number measures the product of two numbers, it will measure one of them.*

Suppose the prime number p divides ab but does not divide a. Then $ab = sp$, or $p : a = b : s$. But since p and a are relatively prime, they are the least numbers in that ratio. It follows that b is a multiple of p, or that p divides b. The actual uniqueness result is in

PROPOSITION IX-14. *If a number is the least of those that are measured by certain prime numbers, then no other prime number will measure it.*

Book IX contains a number of other important results, beginning with the result that mathematicians today express as "There are infinitely many prime numbers." But Euclid, and the Greek mathematicians in general, do not deal with the concept of infinite sets. Thus, Euclid simply offers

PROPOSITION IX-20. *Prime numbers are more than any assigned multitude of prime numbers.*

As with the proof of proposition VI-1, Euclid has no way of writing down an arbitrary "assigned multitude" of primes. Therefore, he again uses the method of generalizable example. He picks just three primes, A, B, C, and shows that one can always find an additional one. To do this, consider the number $N = ABC + 1$. If N is prime, a prime other than those given has been found. If N is composite, then it is divisible by a prime p. Euclid shows that p is distinct from the given primes A, B, C, because none of these divides N. It follows again that a new prime, p, has been found. Euclid presumably assumes that his readers are convinced that a similar proof will work, no matter how many primes are originally picked.

In proposition IX-35, Euclid essentially determines the sum of a geometric progression:

PROPOSITION IX-35. *If as many numbers as we please are in continued proportion, and there is subtracted from the second and the last numbers equal to the first, then, as the excess of the second is to the first, so will the excess of the last be to all those before it.*

Represent the sequence of numbers "in continued proportion" by $a, ar, ar^2, ar^3, \ldots, ar^n$ and the sum of "all those before [the last]" by S_n. Euclid's result states that

$$(ar^n - a) : S_n = (ar - a) : a.$$

The modern form for this sum is

$$S_n = \frac{a(r^n - 1)}{r - 1}.$$

The final proposition of Book IX, proposition IX-36, shows how to find perfect numbers, those which are equal to the sum of all their factors:

PROPOSITION IX-36. *If as many numbers as we please beginning from the unit are set out continuously in double proportion, until the sum of all becomes prime, then the product of the sum and the last number will be perfect.*

In modern symbols, this result states that if the sum of any number of terms of the sequence $1, 2, 2^2, \ldots, 2^n$ is prime, then the product of that sum and 2^n is perfect. For example, $1 + 2 + 2^2 = 7$ is prime; therefore, $7 \times 4 = 28$ is perfect. And, in fact, $28 = 1 + 2 + 4 + 7 + 14$. Other perfect numbers known to the Greeks were 6, corresponding to $1 + 2$, 496, corresponding to $1 + 2 + 4 + 8 + 16$; and 8128, corresponding to $1 + 2 + 4 + 8 + 16 + 32 + 64$. In the eighteenth century, Euler proved that any even perfect number is given by this criterion, but it is still not known whether there are any odd perfect numbers. It is curious, perhaps, that Euclid devoted the culminating theorem of the number theory books to the study of a class of numbers of which only four were known. Nevertheless, the theory of perfect numbers has always proved a fascinating one for mathematicians.

2.2.6 Incommensurability, Solid Geometry, and the Method of Exhaustion

Book X is the longest of the thirteen books of Euclid's *Elements* and probably the best organized. Its purpose is evidently the classification of certain incommensurable magnitudes. One of the motivations for the book was the desire to characterize the edge lengths of the regular polyhedra, whose construction in Book XIII provides a fitting climax to the *Elements*. Euclid needed a nonnumerical way of comparing the edges of the icosahedron and the dodecahedron to the diameter of the sphere in which they were inscribed. In a manner familiar in modern mathematics, this simple question was to lead to the elaborate classification scheme of Book X, far in excess of the direct answer. Because of the complications of this scheme, however, we will discuss only a few of the opening propositions of this book.

The first proposition of Book X is fundamental, not only in that book but also in Book XII.

PROPOSITION X-1. *Two unequal magnitudes being given, if from the greater there is subtracted a magnitude greater than its half, and from that which is left a magnitude greater than its half, and if this process is repeated continually, there will be left some magnitude less than the lesser of the given magnitudes.*

The result depends on definition 4 of Book V, the criterion that two given magnitudes have a ratio. That definition requires that some multiple n of the lesser magnitude exceed the greater. Then n subtractions of magnitudes greater than half of what is left at any stage give the desired result.

Propositions X-2 and X-3 are the results on the Euclidean algorithm discussed earlier. But since Euclid uses the same procedure for magnitudes as he did for numbers in Book VII, he can now connect these two distinct concepts:

PROPOSITION X-5. *Commensurable magnitudes have to one another the ratio which a number has to a number.*

PROPOSITION X-6. *If two magnitudes have to one another the ratio which a number has to a number, the magnitudes will be commensurable.*

Thus, even though number and magnitude are distinct notions, one can now apply the machinery of numerical proportion theory to commensurable magnitudes. The more complicated Eudoxian definition is then necessary only for incommensurable magnitudes.

In particular, Euclid next gives, in proposition X-9, a generalization of the result that presumably led to the study of incommensurables in the first place—that the side and diagonal of a square are incommensurable; that is, in modern terms, that $\sqrt{2}$ is irrational. Namely, Euclid shows, in effect, that the square root of every nonsquare integer is incommensurable with the unit. In Euclid's terminology, the theorem states that two sides of squares are commensurable in length if and only if the squares have the ratio of a square number to a square number. The more interesting part is the "only if" part. Suppose the two sides a, b are commensurable in length. Then $a : b = c : d$, where c, d are numbers. Hence, the duplicates of each ratio are equal. But Euclid has already shown (VI-20) that the square on a is to the square on b in the duplicate ratio of a to b as well as, in VIII-11, the analogous result for numbers that c^2 is to d^2 in the duplicate ratio of c to d. The result then follows.

Book XI is the first of three books dealing with solid geometry. This book contains the three-dimensional analogs of many of the two-dimensional results of Books I and VI, including some constructions. For example, proposition XI-11 shows how to draw a straight line perpendicular to a given plane from a point outside it, and proposition XI-12 shows how to draw such a line from a point in the plane. There is also a series of theorems on parallelepipeds. In particular, by analogy with proposition I-35, Euclid shows that parallelepipeds on the same base and with the same height are equal (XI-30 and XI-31), and then, in analogy with VI-1, that parallelepipeds of the same height are to one another as their bases (XI-32). Also, in analogy with VI-19 and VI-20, he shows in proposition XI-33 that similar parallelepipeds are to one another in the triplicate ratio of their sides. Hence, the volumes of two similar parallelepipeds are in the ratio of the cubes of any pair of corresponding sides. As before, Euclid computes no volumes. Nevertheless, one can easily derive from these theorems the basic results on volumes of parallelepipeds. The "formulas" for volumes of other solids are included in Book XII.

The central feature of Book XII, which distinguishes it from the other books of the *Elements*, is the use of a limiting process, generally known as the **method of exhaustion**. This process, developed by Eudoxus, is used to deal with the area of a circle, as well as the volumes of pyramids, cones, and spheres. "Formulas" giving some of these areas and volumes were known much earlier, but for the Greeks a proof was necessary, and Eudoxus's method provided a proof. What it did not provide was a way of discovering the formulas to begin with.

The main results of Book XII are the following:

PROPOSITION XII-2. *Circles are to one another as the squares on the diameters.*

PROPOSITION XII-7 (COROLLARY). *Any pyramid is a third part of the prism which has the same base with it and equal height.*

PROPOSITION XII-10. *Any cone is a third part of the cylinder which has the same base with it and equal height.*

PROPOSITION XII-18. *Spheres are to one another in the triplicate ratio of their respective diameters.*

The first of these results is Euclid's version of the ancient result concerning the area of a circle. In modern terms, it states that the area of a circle is proportional to the square on the diameter. It does not state what the constant of proportionality is, but the proof does provide a method for approximating this. Proposition XII-1, that similar polygons inscribed in circles are to one another as the squares on the diameters, serves as a lemma to this proof. This result in turn is a generalization of VI-20, that similar polygons are to one another in the duplicate ratio of the corresponding sides. It is not difficult to show, first, that one can take any corresponding lines in place of the "corresponding sides," even the diameter of the circle, and second, that one can replace "duplicate ratio" by "squares."

The main idea of the proof of XII-2 is to "exhaust" the area of a particular circle by inscribing in it polygons of increasingly many sides. In particular, Euclid shows that one can inscribe in the given circle a polygon whose area differs from that of the circle by less than any given area. His proof of the theorem begins by assuming that the result is not true. That is, if the two circles C_1, C_2 have areas A_1, A_2 and diameters d_1, d_2, respectively, he assumes that $A_1 : A_2 \neq d_1^2 : d_2^2$. Therefore, there is some area S, either greater than or less than A_2, such that $d_1^2 : d_2^2 = A_1 : S$. Suppose first that $S < A_2$ (Fig. 2.26). Then

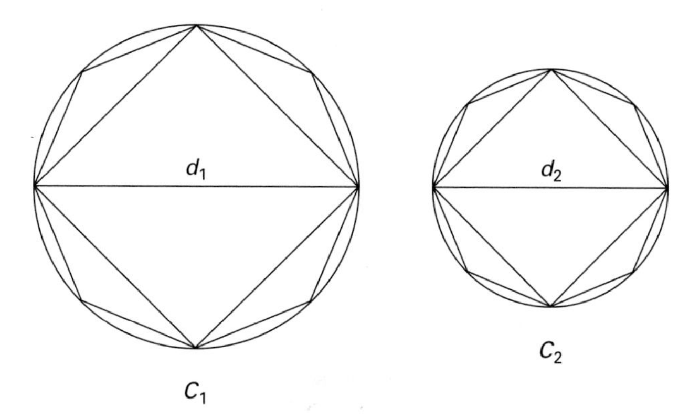

FIGURE 2.26 *Elements*, Proposition XII-2: The method of exhaustion

beginning with an inscribed square and continually bisecting the subtended arcs, inscribe in C_2 a polygon P_2 such that $A_2 > P_2 > S$. In other words, P_2 is to differ from A_2 by less than the difference between A_2 and S. This construction is possible by proposition X-1, since each bisection increases the area of the polygon by more than half of the difference between the circle and the polygon. Next, inscribe a polygon P_1 in C_1 similar to P_2. By proposition XII-1, $d_1^2 : d_2^2 = P_1 : P_2$. By assumption, this ratio is also equal to $A_1 : S$. Therefore, $P_1 : A_1 = P_2 : S$. But, clearly, $A_1 > P_1$. It follows that $S > P_2$, contradicting the assumption that $S < P_2$. Therefore, S cannot be less than A_2. Euclid proves that S also is not greater than A_2 by reducing it to the case already dealt with. It then follows that the ratio of the circles must be equal to the ratio of the squares on the diameters, as asserted.

It is virtually certain that the theorem giving the volume of the pyramid was known to both the Egyptians and the Babylonians. Archimedes, however, writes that although Eudoxus was the first to prove that theorem, the result was first discovered by Democritus (fifth century BCE). Unfortunately, we have no record of how the Egyptians, the Babylonians, or Democritus may have made this discovery. For the latter, we do have a hint in a report given by Chrysippus, in which Democritus discusses the problem of slicing a cone into "indivisible" sections by planes parallel to the base. He wonders whether these indivisible circles will be unequal or equal: "If they are unequal, they will make the cone irregular, as having many indentations, like steps, and unevennesses; but if they are equal, the sections will be equal, and the cone will appear to have the property of the cylinder, and to be made up of equal, not unequal, circles, which is very absurd."

Although we don't know what Democritus's final conclusion was, he evidently did think that the cone and, analogously, the pyramid are "made up" of indivisibles. If so, he could have derived Euclid's proposition XII-5, that pyramids of the same height and with triangular bases are to one another as their bases. For if one imagines the two pyramids cut by planes parallel to and at equal distances from the bases, then the corresponding sections of the two pyramids will be in the ratio of the bases. Since Democritus conceived of each pyramid as being "made up" of these infinitely many indivisible sections, the pyramids themselves would be in this same ratio. He could then have completed the demonstration by noting, as in *Elements* XII-7, that a prism with a triangular base can be divided into three pyramids, all of equal height and equal bases.

Euclid, of course, proves XII-5 as well as XII-10 and XII-18 by using *reductio* arguments, along the lines of his proof of XII-2. But the quotation from Democritus shows us that from the earliest period of Greek mathematics, there were attempts to discover certain results by use of indivisibles. On the other hand, given that the Egyptians knew the formula for the volume of a pyramid and given that Thales, among other early Greek mathematicians, is said to have studied in Egypt, it seems reasonable to assume that the formula itself, if not the derivation, was brought to Greece from Egypt. Other results discussed by Euclid, including the general idea of proportional thinking, were so highly developed in Egypt that it seems reasonable to believe that these were also brought back by Thales and others.

The final book of the *Elements*, Book XIII, is devoted to the construction of the five regular polyhedra—the cube, tetrahedron, octahedron, dodecahedron, and icosahedron—their "comprehension" (inscription) in a sphere, and a comparison of their edge lengths to the sphere's diameter. For the tetrahedron, Euclid shows that the square on the diameter is $1\frac{1}{2}$ times the square on the edge. In the cube, the square on the diameter is triple the square on the edge, while in the octahedron the square on the diameter is double that on the edge. The other two cases are somewhat trickier. In the case of the dodecahedron, Euclid shows, in effect, that if the diameter of the sphere is equal to 1, then the edge length is $\frac{1}{6}(\sqrt{15} - \sqrt{3})$. Similarly, the edge length of an icosahedron inscribed in a sphere of diameter 1 is $\frac{1}{10}\sqrt{50 - 10\sqrt{5}}$.

In a fitting conclusion to Book XIII and the *Elements*, Euclid constructs the edges of the five regular solids in one plane figure, thereby comparing them to each other and the diameter of the given sphere. He then demonstrates that there are no regular polyhedra other than these five.

It appears from the texts attributed to Euclid, including works in such fields as optics, music, and the conic sections, that he saw himself as a compiler of the Greek mathematical tradition up to his time. This would certainly have been appropriate if he was the first mathematician called to the Museum at Alexandria. It would have been his aim to demonstrate to his students not only the basic results known to that time but also some of the methods by which new problems could be approached. In fact, he demonstrated such methods in his work, the *Data*, which applied the results of Books I–VI of the *Elements* to the solving of geometric problems, including, for example, the solution of the standard Babylonian problems, such as the system $xy = c$, $x - y = b$. The two mathematicians in the third century BCE who advanced the field of mathematics most, Archimedes and Apollonius, probably received their earliest mathematical training from students of Euclid, training that enabled them to solve many problems left unsolved by Euclid and his predecessors.

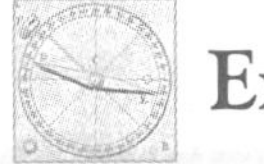

Exercises

1. Thales is said to have invented a method of finding distances of ships from shore by use of the angle-side-angle theorem. Here is a possible method: Suppose A is a point on shore and S is a ship (Fig. 2.27). Measure the distance AC along a perpendicular to AS and bisect it at B. Draw CE at right angles to AC and pick point E on it in a straight line with B and S. Show that $\triangle EBC \cong \triangle SBA$ and therefore that $SA = EC$.

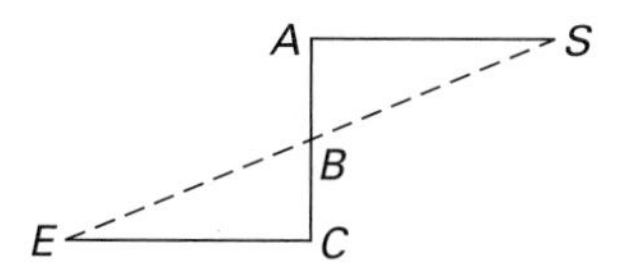

FIGURE 2.27 One method Thales could have used to determine the distance to a ship at sea

2. A second possibility for Thales' method is the following: Suppose Thales was atop a tower on the shore with an instrument made of a straight stick and a crosspiece AC that could be rotated to any desired angle and then would remain where it was put (Fig. 2.28). Thales rotates AC until he sights the ship S, then turns and sights an object T on shore without moving the crosspiece. Show that $\triangle AET \cong \triangle AES$ and therefore that $SE = ET$.

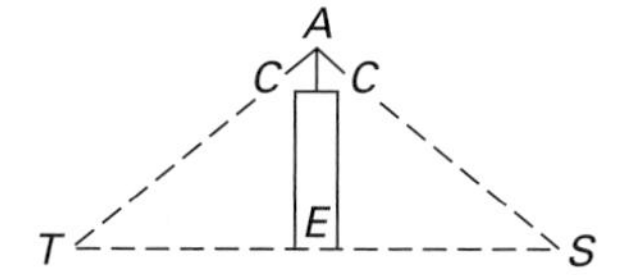

FIGURE 2.28 Second method Thales could have used to determine the distance to a ship at sea

3. Show that the nth triangular number is represented algebraically as $T_n = n(n+1)/2$ and therefore that an oblong number is double a triangular number.
4. Show algebraically that any square number is the sum of two consecutive triangular numbers.
5. Show using dots that eight times any triangular number plus 1 makes a square. Conversely, show that any odd square diminished by 1 becomes eight times a triangular number. Show these results algebraically as well.
6. Construct five Pythagorean triples using the formula $(n, (n^2-1)/2, (n^2+1)/2)$, where n is odd. Construct five different ones using the formula $(m, (m/2)^2 - 1, (m/2)^2 + 1)$, where m is even.
7. Show that $\sqrt{3}$ is incommensurable with 1 by an argument similar to the proposed Pythagorean argument that $\sqrt{2}$ is incommensurable with 1.
8. Read Plato's *Meno* and write a short essay discussing Socrates' method of convincing the slave boy that he knows how to construct a square double a given square. Consider both the "Socratic method" that Socrates uses and the mathematics.
9. Prove proposition I-32, that the three interior angles of any triangle are equal to two right angles. Show that the proof depends on I-29 and therefore on postulate 5.
10. Solve the (modified) problem of proposition I-44, to apply to a given straight line AB a rectangle equal to a given rectangle c. Use Figure 2.29, where $BEFG$ is the given rectangle, D is the intersection of the extension of the diagonal HB and the extension of the line FE, and $ABML$ is the rectangle to be constructed.

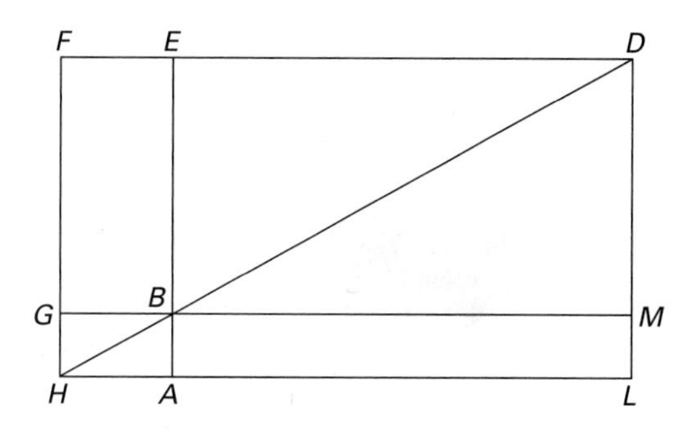

FIGURE 2.29 *Elements*, Proposition I-44

11. Show that Playfair's Axiom—Through a given point outside a given line, exactly one line may be constructed parallel to the given line—is equivalent to Euclid's postulate 5.
12. Proposition II-8 says: If a straight line is cut at random, four times the rectangle contained by the whole and one of the segments together with the square on the remaining segment is equal to the square on the whole and the former segment taken together. Translate this proposition into an algebraic result, and show that it is valid.
13. Provide the details of the proof of proposition III-20: In a circle, the angle at the center is double the angle at the circumference, when the angles cut off the same arc.
14. Prove proposition III-31, that the angle in a semicircle is a right angle.

15. Given that a pentagon and an equilateral triangle can be inscribed in a circle, show how to inscribe a regular 15-gon in a circle.

16. Prove proposition VI-2: If a straight line is drawn parallel to one of the sides of a triangle, it will cut the other two sides proportionally.

17. Prove proposition VI-3: The angle bisector of one angle of a triangle cuts the opposite side into two segments whose ratio is the same as that of the remaining two sides of the triangle. (Hint: Draw a line through one vertex of the triangle parallel to the angle bisector and extend the opposite side to meet it. Then use VI-2.)

18. Prove that the last nonzero remainder in the Euclidean algorithm applied to the numbers a, b is in fact the greatest common divisor of a and b.

19. Use the Euclidean algorithm to find the greatest common divisor of 963 and 657.

20. Suppose that a line of length 1 is divided in extreme and mean ratio; that is, the line is divided at x so that $1/x = x/(1-x)$. Show by the method of the Euclidean algorithm that 1 and x are incommensurable. In fact, show that $1 : x$ can be expressed using Theaetetus's definition as $(1, 1, 1, \ldots)$.

21. Use Theaetetus's definition of equal ratio to show that $46 : 6 = 23 : 3$. Show that each can be represented by $(7, 1, 2)$.

22. Solve the quadratic equation $x^2 + 10 = 7x$ geometrically (using VI-28). There are two positive solutions to this equation. Modify your diagram so that both solutions are evident.

23. Solve the quadratic equation $x^2 + 10x = 39$ geometrically (using VI-29).

24. Use proposition VII-30 to prove the uniqueness (up to order) of the prime decomposition of any positive integer. (This is essentially proposition IX-14.)

25. Give a modern proof of the result that there are infinitely many prime numbers. Compare your proof to Euclid's and comment on the differences.

26. Turn Euclid's proof of XII-2 into a recursive algorithm for calculating the area of a circle. Use the algorithm several times to approximate the area of a circle of radius 1.

27. Prove XIII-9: If the side of the hexagon and the side of the decagon inscribed in the same circle are placed together in a single straight line, then the meeting point divides the entire line segment in extreme and mean ratio, with the greater segment being the side of the hexagon. In Figure 2.30, BC is the side of a decagon and CD the side of a hexagon inscribed in the same circle. Show that $\triangle EBD$ is similar to $\triangle EBC$.

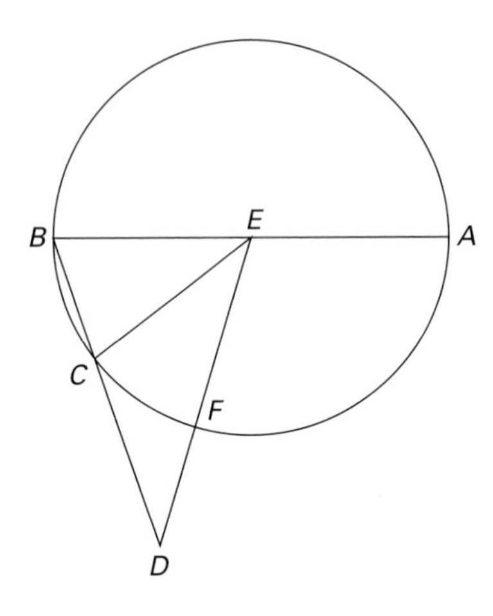

FIGURE 2.30 *Elements*, Proposition XIII-9

28. Prove XIII-10: If an equilateral pentagon, hexagon, and decagon are each inscribed in a given circle, then the square on the side of the pentagon equals the sum of the squares on the sides of the hexagon and the decagon.

29. Eratosthenes of Cyrene (276–194 BCE) is credited with measuring the earth by an argument from parallel lines: He found that at noon on the summer solstice the sun was directly overhead at Syene, a place on the Tropic of Cancer, while at the same time at Alexandria, approximately 5000 *stades* due north, the sun was at $7\frac{1}{5}^\circ$ from the zenith. Given that the rays from the sun to the earth are all parallel, he concluded that $\angle SOA = 7\frac{1}{5}^\circ$ (Fig. 2.31). Calculate Eratosthenes' value for the circumference of the earth in *stades*. If the length of a *stade* is taken to be 516.7 feet, calculate Eratosthenes' values for the circumference and diameter of the earth. How accurate is the method described? How would Eratosthenes know the distance from Alexandria to Syene?

30. Discuss the advantages and disadvantages of a geometric approach compared to a purely algebraic approach in the teaching of the quadratic equation in school.

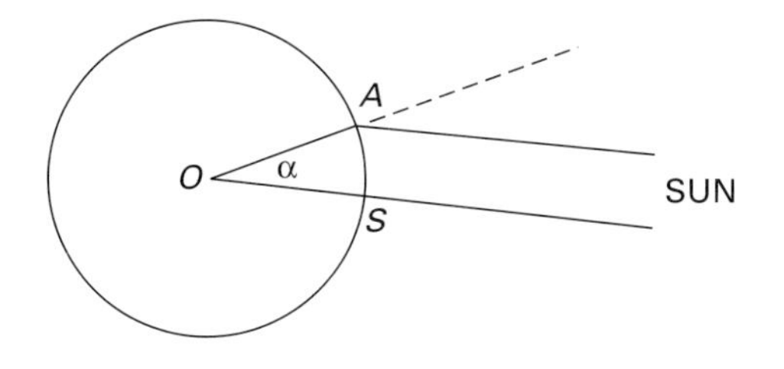

FIGURE 2.31 Eratosthenes' determination of the size of the earth

31. Compare the treatment of geometry in a typical high school geometry text with the treatment in Euclid's *Elements*. How are the two approaches similar? How are they different?
32. Should high school teachers base the study of geometry on Euclid's *Elements*, as was done for many years? Discuss the pros and cons of using Euclid versus a "modern" approach.

References

A good source of basic information on Greek civilization is H. D. F. Kitto, *The Greeks* (London: Penguin, 1951). Two excellent general works on early Greek science, by G. E. R. Lloyd, are *Early Greek Science: Thales to Aristotle* (New York: Norton, 1970) and *Magic, Reason and Experience* (Cambridge: Cambridge University Press, 1979). The latter work, in particular, deals with the beginnings of logical reasoning in Greece and the emergence of the idea of mathematical proof. The standard reference on Greek mathematics is Thomas Heath, *A History of Greek Mathematics* (New York: Dover, 1981, reprinted from the 1921 original). However, many of Heath's conclusions have been challenged in more recent works. The two best reevaluations of some central parts of the story of Greek mathematics are Wilbur Knorr, *The Ancient Tradition of Geometric Problems* (Boston: Birkhäuser, 1986), which argues that geometric problem solving was the motivating factor for much of Greek mathematics, and David Fowler, *The Mathematics of Plato's Academy: A New Reconstruction* (Oxford: Clarendon Press, 1987, 2nd ed., 1999), which claims that the idea of *anthyphairesis* (reciprocal subtraction) provides much of the impetus for the Greek development of the ideas of ratio and proportion. A newer work, Serafina Cuomo's *Ancient Mathematics* (London: Routledge, 2001) provides an excellent survey of Greek mathematics, while claiming that many of Heath's (and others') conclusions are based on very flimsy evidence. The emergence of the deductive method in Greek mathematics is discussed in Reviel Netz, *The Shaping of Deduction in Greek Mathematics: A Study in Cognitive History* (Cambridge: Cambridge University Press, 1999). An earlier, but still useful, work on the same topic is I. Mueller, *Philosophy and Deductive Structure in Euclid's Elements* (Cambridge: MIT Press, 1981).

The standard modern English version of Euclid's *Elements* is the three-volume set edited by Thomas Heath, which is still in print and available from Dover Publications, New York. This edition contains Heath's extensive notes, as well as a long introduction. Those who just want the text of Euclid in a convenient one-volume version should look for the edition prepared by Dana Densmore and published in 2002 by Green Lion Press, Santa Fe. A theorem-by-theorem analysis of *Elements*, which is very helpful to have alongside Euclid's text, is Benno Artmann, *Euclid: The Creation of Mathematics* (New York: Springer, 1999). Wilbur Knorr's *The Evolution of the Euclidean Elements* (Dordrecht: Reidel, 1975) is also useful for its analysis of theorems. Many of the available fragments from earlier Greek mathematics are collected in Ivor Thomas, *Selections Illustrating the History of Greek Mathematics* (Cambridge: Harvard University Press, 1941). Aristotle's works on logic, especially *Prior Analytics* and *Posterior Analytics*, are available in many modern editions.

Other useful works on Greek mathematics include B. L. Van der Waerden's *Science Awakening I* (New York: Oxford University Press, 1961); F. Lasserre, *The Birth of Mathematics in the Age of Plato* (Larchmont, N.Y.: American Research Council, 1964); J. Klein, *Greek Mathematical Thought and the Origin of Algebra* (Cambridge: MIT Press, 1968); and Asger Aaboe, *Episodes from the Early History of Mathematics* (Washington: Mathematical Association of America, 1964).

Although there have been many discussions of the algebraic nature of Euclid's Book II over the years, the debate was renewed with a vengeance in an article by Sabetai Unguru entitled "On the Need to Rewrite the History of Greek Mathematics," *Archive for History of Exact Sciences* 15 (1975), 67–114. He was answered by several other historians over the next two years. The most important responses were by B. L. Van der Waerden, "Defence of a Shocking Point of View," *Archive for History of Exact Sciences* 15

(1976), 199–210, and by Hans Freudenthal, "What is Algebra and What Has It Been in History?" *Archive for History of Exact Sciences* 16 (1977), 189–200. A reply to these was offered by Unguru and David Rowe, "Does the Quadratic Equation Have Greek Roots? A Study of Geometric Algebra, Application of Areas, and Related Problems," *Libertas Mathematica* 1 (1981) and 2 (1982). These articles are recommended as examples of the strong feelings historical controversy can bring out. A recent article by Jens Høyrup, "On a Collection of Geometrical Riddles and Their Role in the Shaping of Four to Six 'Algebras'," *Science in Context* 14 (2001), 85–131, applies the new discoveries about Babylonian algebra to this debate and tends to reject Unguru's arguments.

Finally, there has been much controversy over the relationship of Greek civilization to Egyptian civilization and, in particular, over the relationship of Greek mathematics to Egyptian mathematics. The opening shot in this battle was the publication of Martin Bernal's *Black Athena: The Afroasiatic Roots of Classical Civilization* (New Brunswick, N.J.: Rutgers University Press, 1987). This work asserted that classical Greek civilization had deep roots in Afroasiatic cultures, but that these influences have been systematically ignored or denied since the eighteenth century, chiefly for racist reasons. Bernal did not write much about science in this work, but summarized his views on the contributions of Egyptian science to Greek science in "Animadversions on the Origins of Western Science," *Isis* 83 (1992), 596–607. This article was answered by Robert Palter in his "*Black Athena*, Afro-Centrism, and the History of Science," *History of Science* 31 (1993), 227–287. Bernal responded in "Response to Robert Palter," *History of Science* 32 (1994), 445–464; and Palter answered Bernal in the same issue, 464–468. The last word on this issue remains to be uttered.

CHAPTER THREE

Greek Mathematics from Archimedes to Ptolemy

Plato . . . set the mathematicians the following problem: What circular motions, uniform and perfectly regular, are to be admitted as hypotheses so that it might be possible to save the appearances presented by the planets?

—Simplicius's *Commentary* on Aristotle's *On the Heavens*

Greek mathematics from the third century BCE through the second century CE was dominated by three major figures, each of whom contributed to both "theoretical" and "applied" mathematics. Archimedes of Syracuse (c. 287–212 BCE) pushed forward the study of the "limit" methods of Eudoxus and succeeded not only in applying them to determine areas and volumes of new figures, but also in developing new techniques to discover those results. Archimedes, unlike Euclid, was neither reluctant to share his methods of discovery nor afraid of performing numerical calculations and exhibiting numerical results. Furthermore, Archimedes wrote several treatises presenting mathematical models of certain aspects of what is now called *theoretical physics* and applied his physical principles to the invention of various mechanical devices.

Apollonius of Perge (c. 250–175 BCE), on the other hand, was instrumental in extending the domain of analysis to new and more difficult geometric construction problems. As a foundation for these new approaches, he created his magnum opus, the *Conics*, a work in eight books that synthetically developed the important properties of this class of curves, properties that were central in developing new solutions to such problems as the duplication of the cube and the trisection of the angle. But Apollonius was also a well-known astronomer who made extensive efforts to answer Plato's challenge of explaining the phenomena ("saving the appearances") in the heavens through the use of circular motions.

After three centuries of further work by Greek mathematicians and astronomers, Plato's challenge was definitively answered by Claudius Ptolemy (c. 100–178 CE). Around 150, he wrote the *Almagest*, a work studied, commented upon, and extensively criticized, yet never replaced for fourteen hundred years, a work in which he not only used ideas from plane and spherical geometry but also devised ways to perform the extensive numerical calculations necessary to make his book a useful one.

This chapter will survey some major contributions of these three mathematicians, as well as the work of certain others who considered similar problems.

3.1 Archimedes

Unlike Euclid, Archimedes did not write systematic treatises explaining all the details of a particular subject. Instead, he wrote what may be considered research monographs, treatises concentrating on the solution of a particular set of problems. These treatises were often sent as letters to mathematicians Archimedes knew, so many of them include prefaces describing the circumstances and purposes of their writing.

3.1.1 The Determination of π

One very brief treatise of Archimedes, *On the Measurement of the Circle*, does not contain a preface. It does, however, contain numerical results, unlike anything found in Euclid's work. Its first proposition, in addition, gives Archimedes' answer to the question of squaring the circle, by showing that the area of a circle of given radius can be found once the circumference is known.

PROPOSITION 1. *The area A of any circle is equal to the area of a right triangle in which one of the legs is equal to the radius and the other to the circumference.*

Archimedes' result is equivalent to the ancient result that $A = (C/2)(d/2)$, where d is the length of the diameter and C the length of the circumference. Archimedes, however, gives a rigorous proof, using a Eudoxian exhaustion argument: If K is the area of the given triangle, Archimedes first supposes that $A > K$. By inscribing in the circle regular polygons of successively more sides, he eventually determines a polygon of area P such that $A - P < A - K$. Thus, $P > K$. Now the perpendicular from the center of the circle to the midpoint of a side of the polygon is less than the radius, and the perimeter of the polygon is less than the circumference. It follows that $P < K$, a contradiction. Similarly, the assumption that $A < K$ also leads to a contradiction, and the result is proved.

The third proposition of *On the Measurement of the Circle* complements the first by giving a numerical approximation to the length of the circumference:

PROPOSITION 3. *The ratio of the circumference of any circle to its diameter is less than* $3\frac{1}{7}$ *but greater than* $3\frac{10}{71}$.

Archimedes' proof of this statement provides algorithms for determining the perimeter of certain regular polygons circumscribed about and inscribed in a circle. Archimedes begins with regular hexagons, the ratios of whose perimeters to the diameter of the circle are known from elementary geometry. He then in effect uses the following lemmas to calculate, in turn, the ratios of the perimeters of regular polygons with 12, 24, 48, and 96 sides, respectively, to the diameter of the circle.

LEMMA 1. *Suppose OA is the radius of a circle and CA is tangent to the circle at A. Let DO bisect $\angle COA$ and intersect the tangent at D. Then $DA/OA = CA/(CO + OA)$ and $DO^2 = OA^2 + DA^2$ (Fig. 3.1).*

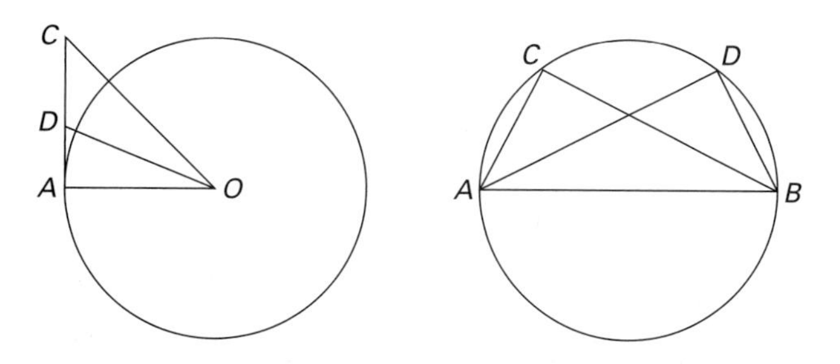

FIGURE 3.1 Lemmas 1 and 2 from *On the Measurement of the Circle*

LEMMA 2. *Let AB be the diameter of a circle and ACB a right triangle inscribed in the semicircle. Let AD bisect $\angle CAB$ and meet the circle at D. Connect DB. Then $AB^2/BD^2 = 1 + (AB + AC)^2/BC^2$ and $AD^2 = AB^2 - BD^2$ (Fig. 3.1).*

Archimedes uses Lemma 1 repeatedly to develop a recursive algorithm for determining the desired ratio using circumscribed polygons. He begins by assuming that $\angle COA$ is one-third of a right angle (30°), so CA is half of one side of a circumscribed regular hexagon. Therefore, CA and CO are known. Since $\angle DOA = 15°$, it follows that DA is half of one side of a regular 12-gon. DA and DO are then calculated by use of the lemma. Next, $\angle DOA$ is bisected to get an angle of $7\frac{1}{2}^\circ$. The piece of the tangent subtending that angle is then half of one side of a regular 24-gon. Its length can be calculated as well. If r is the radius of the circle, t_i half of one side of a regular (3×2^i)-gon $(i \geq 1)$, and u_i the length of the line from the center of the circle to a vertex of that polygon, the lemma can be translated into the following recursive formulas:

$$t_{i+1} = \frac{rt_i}{u_i + r} \qquad u_{i+1} = \sqrt{r^2 + t_{i+1}^2}.$$

The ratio of the perimeter of the ith circumscribed polygon to the diameter of the circle is then $6(2^i t_i) : 2r = 3(2^i t_i) : r$.

Archimedes developed a similar algorithm for inscribed polygons by use of Lemma 2, and in both cases he provided explicit numerical results at each stage. For example, in his calculations for both circumscribed and inscribed hexagons, he needed to evaluate the ratio $\sqrt{3} : 1$. What he wrote indicates that he knew that this ratio is greater than $265 : 163$ and less than $1351 : 780$. Although it is not known exactly how Archimedes found these results, it is certain that he, like many great mathematicians of later times, was a superb calculator. After four steps of both algorithms, in fact, he concluded that the ratio of the perimeter of the circumscribed 96-sided polygon to the diameter of the circle is less than $14{,}688 : 4673\frac{1}{2} = 3 + \frac{667\frac{1}{2}}{4673\frac{1}{2}} < 3\frac{1}{7}$, while the ratio of the inscribed 96-sided polygon to the diameter is greater than $6336 : 2077\frac{1}{4} > 3\frac{10}{71}$, thus proving the theorem.

Archimedes' proof is the first recorded method for actually computing π. Once the method was known, it was merely a matter of patience to calculate π to as great a degree of accuracy as desired. Archimedes does not tell us why he stopped at 96-sided polygons. But his value of $3\frac{1}{7}$ remains a standard approximation for π to the present day.

3.1.2 Archimedes' Method of Discovery

Another feature distinguishing Archimedes' work in geometry from that of Euclid is that Archimedes often presents his method of discovery of the theorem and/or his analysis of the situation before presenting a rigorous synthetic proof. The methods of discovery of several of his results are collected in a treatise called *The Method*, which was unexpectedly discovered in 1899. The manuscript dates from the tenth century, but the writing had been partially washed out in the thirteenth century so that the parchment could be reused for a religious work. (Parchment was a very valuable commodity in the Middle Ages; a reused parchment is called a **palimpsest**.) Fortunately, the old writing was in large part still readable. Heiberg deciphered it in Constantinople in 1906, using only a magnifying glass, and soon after published the Greek text. Interestingly, the original palimpsest disappeared during the First World War, only to reappear in an auction in 1998. Evidently, it had been owned for many years by a French family, who finally decided to sell. Although there were some legal challenges to the sale, the manuscript of Archimedes' *The Method* was sold for about $2,000,000 to an anonymous buyer, who then contracted with the Walters Art Gallery in Baltimore to preserve it and restore it where possible. At this writing, it is still at the gallery, although scholars have been permitted to inspect it using modern techniques. Whether anything new about Archimedes will be learned from the document is still unclear, but it is known that some diagrams that Heiberg had been unable to see are now visible.

In the introductory letter written to Eratosthenes, the chief librarian at the Library in Alexandria, Archimedes describes his purpose in writing *The Method*:

> Since, as I said, I know that you are diligent, an excellent teacher of philosophy, and greatly interested in any mathematical investigations that may come your way, I thought it might be appropriate to write down and set forth for you in this same book a certain special method, by means of which you will be enabled to recognize certain mathematical questions with the aid of mechanics. I am convinced that this is no less useful for finding the proofs of these same theorems. For some things, which first became clear to me by the mechanical method, were afterwards proved geometrically, because their investigation by the said method does not furnish an actual demonstration.... I now wish to describe the method in writing, partly because I have already spoken about it before, ... partly because I am convinced that it will prove very useful for mathematics; in fact, I presume there will be some among the present as well as future generations who by means of the method here explained will be enabled to find other theorems which have not yet fallen to our share.

The Method contains Archimedes' method of discovery by mechanics of many important results on areas and volumes, most of which are rigorously proved elsewhere. The essential features of *The Method* are, first, the assumption that figures are "composed" of their indivisible cross sections and, second, the balancing of cross sections against corresponding cross sections of a known figure, using the law of the lever. (The law of the lever states that weights m_1, m_2 balance at distances d_1, d_2 respectively from the fulcrum when $m_1 : m_2 = d_2 : d_1$.)

The first proposition of *The Method*, that a segment of a parabola is $\frac{4}{3}$ of the triangle inscribed in it, is a typical example of that work. By a segment ABC of a parabola, Archimedes means the region bounded by the curve and a line AC, where B is the point at which the line segment through the midpoint D of AC, drawn parallel to the axis of the parabola, meets the curve (Fig. 3.2). The point B is called the **vertex** of the parabolic segment. The vertex is also that point of the curve whose perpendicular distance to AC is the greatest. Given the parabolic segment ABC with vertex B, draw a tangent at C meeting the axis produced at E and a line through A parallel to the axis and meeting the tangent line at F. Produce CB to meet AF at K and extend CB to H so that $CK = KH$. Archimedes now considers CH as a lever with midpoint K. The idea of his demonstration is to show that triangle CFA, placed where it is in the figure, balances the parabolic segment ABC placed at H. He does this, line by line, by beginning with an arbitrary line segment MO of triangle CFA parallel to ED and showing that it balances the line PO of segment ABC placed at H. To show the balancing, two properties of the parabola are needed: first, that $EB = BD$, and second, that $MO : PO = CA : AO$. (It is evident that Archimedes was quite familiar with the elementary properties of parabolas.) From $EB = BD$, it follows from similarity that $FK = KA$ and $MN = NO$, and from the proportion and the fact that AF bisects CH, it also follows from proposition VI-2 of the *Elements* that $MO : PO = CA : AO = CK : KN = HK : KN$. If a line TG equal to PO is placed with its center at H, this proportion becomes $MO : TG = HK : KN$. Therefore, since N is the center of gravity of MO, by the law of the lever, MO and TG will be in equilibrium about K.

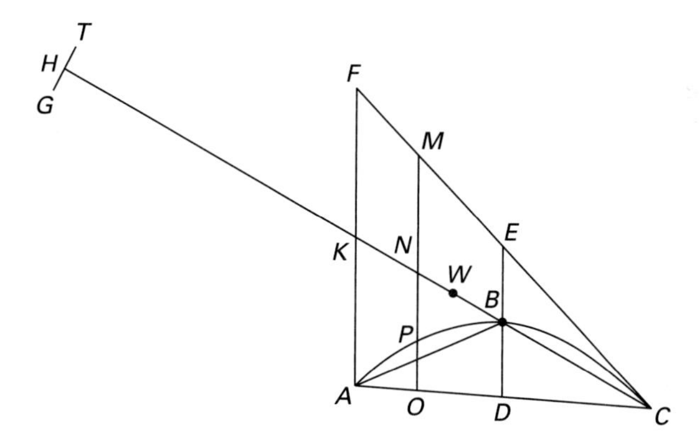

FIGURE 3.2 Balancing a parabolic segment

Archimedes continues, "since the triangle CFA is made up of all the parallel lines like MO, and the segment ABC is made up of all the straight lines like PO within the curve, it follows that the triangle, placed where it is in the figure, is in equilibrium about K with the segment ABC placed with its center of gravity at H." Because nothing is changed by considering the triangle as located at its center of gravity, the point W on CK two-thirds of the way from C to K, Archimedes derives the proportion $\triangle ACF$: segment $ABC = HK : KW = 3 : 1$. Therefore, segment $ABC = \frac{1}{3}\triangle ACF$. But $\triangle ACF = 4\triangle ABC$. Hence, segment $ABC = \frac{4}{3}\triangle ABC$, as asserted. Archimedes concludes this demonstration with a warning: "Now the fact here stated is not actually demonstrated by the argument used; but that argument has given a sort of indication that the conclusion is true. Seeing then that the theorem is not demonstrated, but at the same time suspecting that the conclusion is true, we shall have recourse to the geometrical demonstration which I myself discovered and have already published."

Interestingly, in this argument, Archimedes uses the same phraseology that Democritus used two centuries earlier, that indivisible sections of a plane area or a solid "make up" the entire figure. But although Archimedes used indivisibles throughout *The Method*, he does not explain how they are to be used, even heuristically. This leads us to believe that his contemporaries, and especially the mathematicians in Alexandria with whom he corresponded, understood the use of indivisibles and, perhaps, used them in similar arguments even though they knew that these arguments did not form a rigorous geometrical proof.

3.1.3 Sums of Series

The geometrical proof of the result for the segment of a parabola that Archimedes did consider valid appears in his treatise *Quadrature of the Parabola* and is based on Eudoxus's method of exhaustion. The idea as before is to construct, inside a parabolic segment, rectilinear figures whose total area differs from that of the segment by less than any given value. The figures Archimedes used for this purpose are triangles. Thus in each of the two parabolic segments PRQ, $PR'Q'$ left by the original triangle PQQ', he constructed a triangle PRQ, $PR'Q'$; in each of the four segments left by these triangles, he constructed new triangles, and so on (Fig. 3.3).

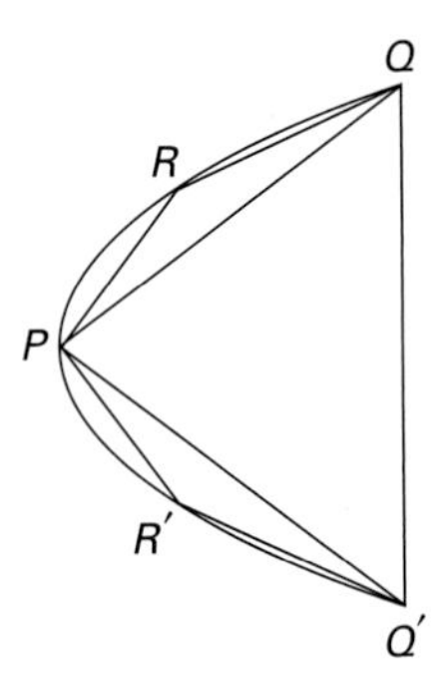

FIGURE 3.3 Determination of the area of a parabolic segment by summation of a geometric series

Archimedes next calculated that the total area of the triangles constructed at each stage is $\frac{1}{4}$ of the area of the triangles constructed in the previous stage. The more steps taken, the more closely the sum of the areas approaches the area of the parabolic segment. Therefore, to complete the proof, Archimedes in effect needed to find the sum of the geometric series $a + \frac{1}{4}a + (\frac{1}{4})^2a + \cdots + (\frac{1}{4})^na + \cdots$, where a is the area of triangle PQQ'. Archimedes did not use Euclid's formula for the sum of a geometric progression from *Elements* IX-35, but instead gave that sum in the form

$$a + \tfrac{1}{4}a + \left(\tfrac{1}{4}\right)^2 a + \cdots + \left(\tfrac{1}{4}\right)^n a + \tfrac{1}{3}\left(\tfrac{1}{4}\right)^n a = \tfrac{4}{3}a$$

and completed the argument through a double *reductio ad absurdum*, just as he had done in *On the Measurement of the Circle*. He assumed that $K = \frac{4}{3}a$ is not equal to the area B of the parabolic segment. If K is less than this area, then triangles can be inscribed as described above so that $B - T < B - K$, where T is the total area of the inscribed triangles. But then $T > K$, which is impossible because the summation formula shows that $T < \frac{4}{3}a = K$. On the other hand, if $K > B$, n is determined so that $(\frac{1}{4})^na < K - B$.

Because also $K - T = \frac{1}{3}(\frac{1}{4})^n a < (\frac{1}{4})^n a$, it follows that $B < T$, which is again impossible. Hence, $K = B$.

The important lemma to this proof shows how to find the sum of a geometric series. Archimedes' demonstration of this result was given for a series of five numbers, because, like Euclid, he had no notation to express a series with arbitrarily many numbers. But since his method generalizes easily, we can use modern notation, with n denoting an arbitrary positive integer. Archimedes began by noting that $(\frac{1}{4})^n a + \frac{1}{3}(\frac{1}{4})^n a = \frac{1}{3}(\frac{1}{4})^{n-1} a$. Then he calculated:

$$
\begin{aligned}
a + \tfrac{1}{4}a + \left(\tfrac{1}{4}\right)^2 a + \cdots + \left(\tfrac{1}{4}\right)^n a + \tfrac{1}{3}\left[\tfrac{1}{4}a + \left(\tfrac{1}{4}\right)^2 a + \cdots + \left(\tfrac{1}{4}\right)^n a\right] \\
= a + \left(\tfrac{1}{4}a + \tfrac{1}{3}\cdot\tfrac{1}{4}a\right) + \left[\left(\tfrac{1}{4}\right)^2 a + \tfrac{1}{3}\left(\tfrac{1}{4}\right)^2 a\right] + \cdots + \left[\left(\tfrac{1}{4}\right)^n a + \tfrac{1}{3}\left(\tfrac{1}{4}\right)^n a\right] \\
= a + \tfrac{1}{3}a + \tfrac{1}{3}\cdot\tfrac{1}{4}a + \cdots + \tfrac{1}{3}\left(\tfrac{1}{4}\right)^{n-1} a \\
= a + \tfrac{1}{3}a + \tfrac{1}{3}\left[\tfrac{1}{4}a + \cdots + \left(\tfrac{1}{4}\right)^{n-1} a\right].
\end{aligned}
$$

Subtracting equal quantities and rearranging gives the desired result:

$$a + \tfrac{1}{4}a + \left(\tfrac{1}{4}\right)^2 a + \cdots + \left(\tfrac{1}{4}\right)^n a + \tfrac{1}{3}\left(\tfrac{1}{4}\right)^n a = \tfrac{4}{3}a.$$

Another formula for a sum led to another result on area in *On Spirals*, a result again proved by Eudoxian methods. In proposition 10 of that book, Archimedes demonstrated a formula for determining the sum of the first n integral squares:

$$(n+1)n^2 + (1 + 2 + \cdots + n) = 3(1^2 + 2^2 + \cdots + n^2),$$

as a corollary to which he showed that

$$3[1^2 + 2^2 + \cdots + (n-1)^2] < n^3 < 3[1^2 + 2^2 + \cdots + n^2].$$

Archimedes needed the last inequality to determine the area bounded by one turn of the **Archimedean spiral**, the curve given in modern polar coordinates by the equation $r = a\theta$. In proposition 24 of *On Spirals*, he demonstrated that the area R bounded by one complete circuit of that curve and the radius line AL to its endpoint equals one-third of the area C of the circle with that line as radius. Archimedes first noted that one can inscribe and circumscribe about the region R figures whose areas differ by less than any assigned area ϵ (Fig. 3.4). By continued bisection (according to *Elements* X-1), one can determine an integer n such that the circular sector with radius AL and angle $(360/n)°$ has area less than

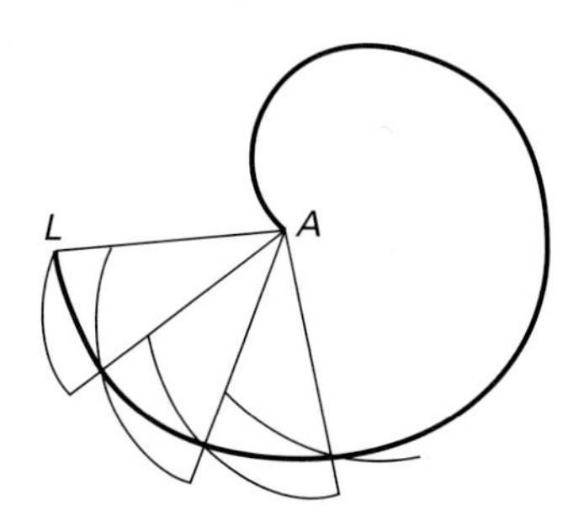

FIGURE 3.4 Area of the Archimedean spiral

ϵ. Then, inscribing a circular arc in and circumscribing a circular arc about the part of the spiral included in each of the n sectors with this angle, one notes that the difference between the complete circumscribed figure and the complete inscribed figure is equal to the area of the sector chosen initially and thus is less than ϵ.

The proof of the area result by a double *reductio* argument is now straightforward. Suppose that $R \neq \frac{1}{3}C$. Then either $R < \frac{1}{3}C$ or $R > \frac{1}{3}C$. In the first case, circumscribe a figure F about R as described above so that $F - R < \frac{1}{3}C - R$. Therefore, $F < \frac{1}{3}C$. From the defining equation of the curve, it follows that the radii of the sectors making up F are in arithmetic progression, which can be considered as $1, 2, \ldots, n$. Because $n \cdot n^2 < 3(1^2 + 2^2 + \cdots + n^2)$ and because the areas of the sectors (and the circle itself) are proportional to the squares on their radii, it follows that $C < 3F$ or $\frac{1}{3}C < F$, a contradiction. A similar argument using an inscribed figure shows that $R > \frac{1}{3}C$ also leads to a contradiction, and the proposition is proved.

3.1.4 Analysis

Two final examples of Archimedes' work show again his concern that his readers learn not only the solution to a geometric problem but also how the solution was found. In the case of proposition 3 of *On the Sphere and Cylinder II*, he provides his procedure in the context of a formal proof.

PROBLEM. *To cut a given sphere by a plane so that the surfaces of the segments may have to one another a given ratio.*

Archimedes' procedure, the method of analysis, is to assume the problem solved and then deduce consequences until he reaches a result already known. Thus, he assumes that the plane BB' cuts the sphere so that the surface of BAB' is to the surface of $BA'B'$ as H is to K (Fig. 3.5). He had already shown in *On the Sphere and Cylinder I* that the areas of such segments equal the areas of the circles on the radii AB, $A'B$. Hence, he concludes that $AB^2 : A'B^2 = H : K$ and therefore that $AM : A'M = H : K$ (since the areas of the triangles are as the bases). But the dividing of a line segment in a given ratio is a known procedure. Archimedes can therefore solve the original problem by beginning with that step and proceeding in reverse—namely, by taking M so that $AM : MA' = H : K$. The theorem already quoted then shows that $AM : MA' = AB^2 : A'B^2 =$ (circle with

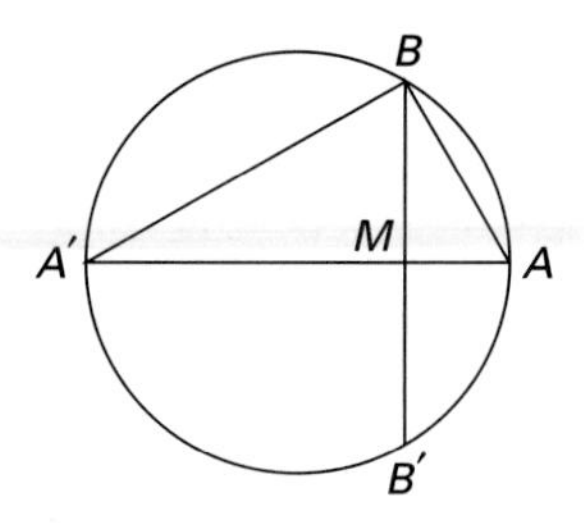

FIGURE 3.5 *On the Sphere and Cylinder II*, Proposition 3

radius AB) : (circle with radius $A'B$) = (surface of segment BAB') : (surface of segment $BA'B'$). The problem is solved.

Archimedes presented the analysis of a more complex problem in proposition 4 of the same book, where he proposed to cut a given sphere by a plane so that the volumes of the segments are in a given ratio. In this case, his analysis reduced the problem to the following: Given a straight line ABC with $AB = 2BC$ and given a point E on BC, to cut AB at a point M such that $AB^2 : AM^2 = MC : EC$ (Fig. 3.6). If one sets $AB = 2a$, $BC = a$, $EC = b$, and $AM = x$, the problem can be translated algebraically into $(2a)^2 : x^2 = 3a - x : b$, or $3ax^2 - x^3 = 4a^2b$. Hence, Archimedes needed to solve a cubic equation. He proceeded to do so by finding the desired point M as the intersection of a parabola and a hyperbola.

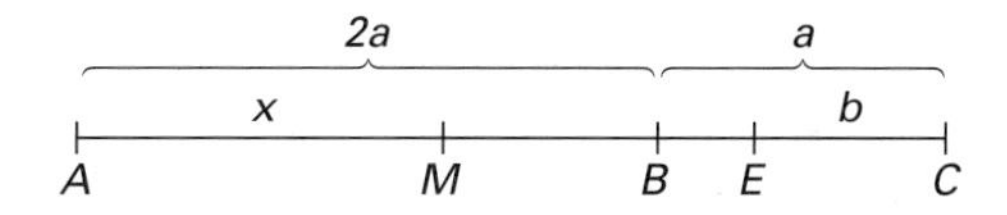

FIGURE 3.6 *On the Sphere and Cylinder II*, Proposition 4

Archimedes' mathematical genius was far-reaching. Only a few items from some of the fourteen extant treatises have been discussed here. Among other results, Archimedes proved that the volume of a sphere is four times that of the cone with base equal to a great circle of the sphere and height equal to its radius, that the volume of a segment of a paraboloid of revolution is $\frac{3}{2}$ that of the cone with the same base and axis, that the surface of a sphere is four times the greatest circle in it, and, what he evidently considered his most important result, that a cylinder whose base is a great circle in the sphere and whose height is equal to the diameter has volume $\frac{3}{2}$ of that of the sphere and also has surface area $\frac{3}{2}$ of the surface area of the sphere. At Archimedes' request, a sphere inscribed in a cylinder was engraved on his tombstone.

3.2 Apollonius and the Conic Sections

3.2.1 Conic Sections before Apollonius

The exact origins of the theory of conic sections are somewhat hazy, but they may well be connected to the problem of doubling the cube. Recall that Hippocrates in the fifth century BCE reduced the problem of constructing a cube with double the volume of a given cube of side a to the finding of two mean proportionals x, y between the lengths a and $2a$, that is, of determining x, y such that $a : x = x : y = y : 2a$. In modern terms, this is equivalent to solving simultaneously any two of the three equations $x^2 = ay$, $y^2 = 2ax$, and $xy = 2a^2$, the first two of which represent parabolas and the third a hyperbola.

It was Menaechmus (fourth century BCE) who first constructed curves that satisfy these algebraic properties and thus showed that the point of intersection of these curves gave the desired two means and solved the problem of doubling the cube. It is not known how he produced these curves, but it does appear that, soon thereafter, the conic sections were introduced as tools for the solution of certain geometric problems. We have already seen that Archimedes used them in solving a problem on spheres.

Modern mathematicians can only speculate as to how the Greeks realized that curves useful in solving the cube doubling problem could be generated as sections of a cone. One possibility is that these curves appeared as the path of the moving shadow of the gnomon on a sundial as the sun traveled through its circular daily path, which in turn was one base of a double cone whose vertex was the tip of the gnomon. In this suggestion, the plane in which the shadow falls would be the cutting plane. The Greeks might also have noted that the apparent shape of a circle viewed from a point outside its plane was an ellipse and determined that this shape results when a plane cuts the cone of vision. In any case, by the end of the fourth century BCE, there existed two extensive treatises on the properties of the curves obtained as sections of cones, and these properties were frequently referred to by Archimedes.

3.2.2 Definitions and Basic Properties of the Conics

Apollonius, in his *Conics*, evidently decided to look at the subject from the beginning, even giving a slightly different definition of a cone than had been given earlier. In fact, he defined what is today called a *double oblique cone*:

> If from a point a straight line is joined to the circumference of a circle which is not in the same plane as the point, and the line extended in both directions, and if, with the point remaining fixed, the straight line is rotated about the circumference of the circle . . . , then the generated surface composed of the two surfaces lying vertically opposite one another . . . [is] a **conic surface**. The fixed point [is] the **vertex** and the straight line drawn from the vertex to the center of the circle [is] the **axis**. . . . The circle [is] the **base** of the cone.

Note that, in general, the axis of the cone is not perpendicular to the base circle, but in what follows, for simplicity, we will take the axis perpendicular to the base.

To define the three conic sections, Apollonius first cut the cone by a plane through the axis. The intersection of this plane with the base circle is a diameter BC of that circle. The resulting triangle ABC is called the **axial triangle**. The parabola, ellipse, and hyperbola are then defined as sections of this cone produced by certain planes that cut the plane of the base circle in the straight line ST perpendicular to BC or BC extended (Figs. 3.7, 3.8, 3.9). The straight line EG is the intersection of the cutting plane with the axial triangle. If EG is parallel to one side of the axial triangle, the section is a parabola. If EG intersects both sides of the axial triangle, the section is an ellipse. Finally, if EG intersects one side of the axial triangle and the other side produced beyond A, the section is a hyperbola. In this situation, there are two branches of the curve.

For each case, Apollonius derived the **symptom** of the curve, the characteristic relation between the ordinate and the abscissa of an arbitrary point on the curve, which can easily be translated into an algebraic equation. He began by picking an arbitrary point L on the section and passing a plane through L parallel to the base circle. The section of the cone produced by the plane is a circle with diameter PR. Let M be the intersection of this plane with the line EG. Then LM is perpendicular to PR, and therefore $LM^2 = PM \cdot MR$.

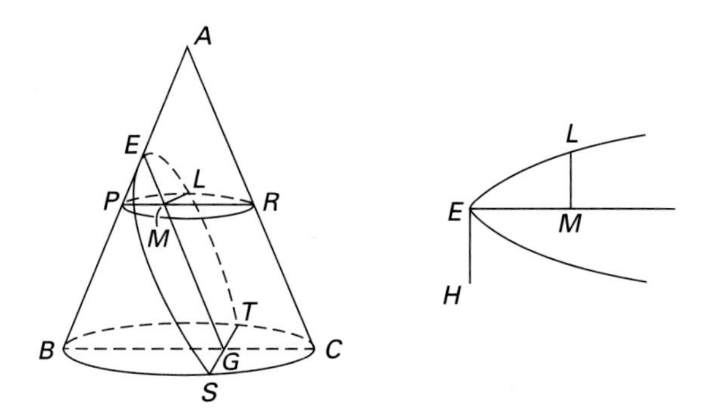

FIGURE 3.7 Derivation of the symptoms of a parabola

When EG is parallel to AC, a side of the axial triangle, Apollonius derived the standard symptom of a parabola, the relation between EM and LM, abscissa and ordinate, respectively, of the point L on the curve (proposition I-11). To do this, he drew EH perpendicular to EM (Fig. 3.7) such that

$$\frac{EH}{EA} = \frac{BC^2}{BA \cdot AC}.$$

The right-hand side of this equation can be written as the product of BC/BA and BC/AC. But, by similarity,

$$\frac{BC}{BA} = \frac{PR}{PA} = \frac{PM}{EP} = \frac{MR}{EA} \quad \text{and} \quad \frac{BC}{AC} = \frac{PR}{AR} = \frac{PM}{EM}.$$

It follows that

$$\frac{EH}{EA} = \frac{MR \cdot PM}{EA \cdot EM}.$$

But also

$$\frac{EH}{EA} = \frac{EH \cdot EM}{EA \cdot EM}.$$

Therefore, $MR \cdot PM = EH \cdot EM$ and $LM^2 = EH \cdot EM$. If we set $LM = y$, $EM = x$, and $EH = p$, we have derived the standard equation of the parabola: $y^2 = px$. The name **parabola** comes from the Greek word *paraboli* (meaning "applied"), because the square on the ordinate y is equal to the rectangle *applied* to the abscissa x. The constant p, which depends only on the cutting plane that determines the curve, is called the **parameter** of the parabola.

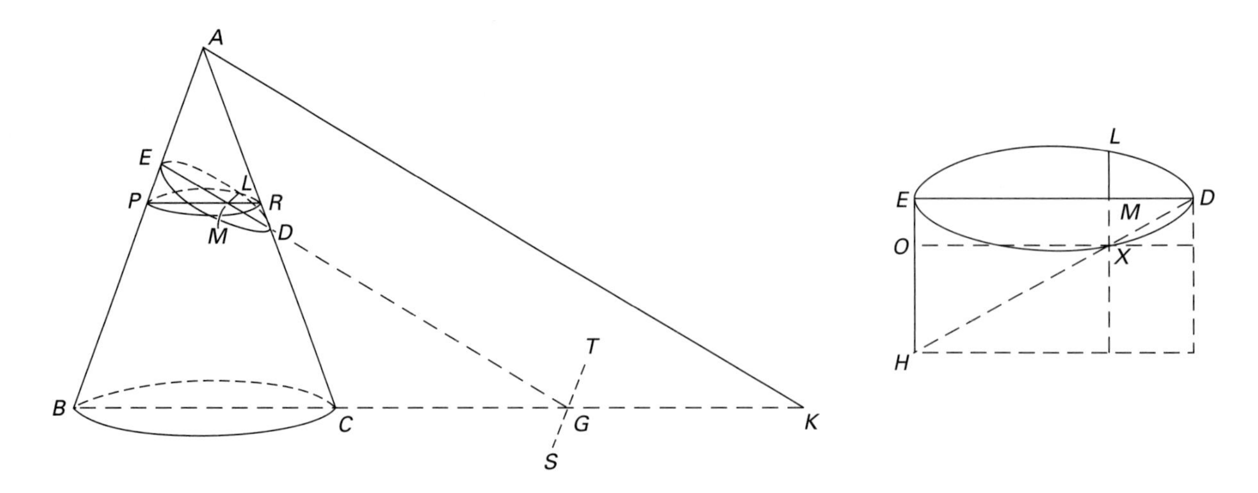

FIGURE 3.8 Derivation of the symptoms of an ellipse

In the other two cases, let D be the intersection of EG with the second side of the axial triangle (ellipse, Fig. 3.8) or with the second side produced (hyperbola, Fig. 3.9). Apollonius proved in these cases (propositions I-12 and I-13) that the square on LM is equal to a rectangle applied to a line EH with width equal to EM and exceeding (*yperboli*) or deficient (*ellipsis*) by a rectangle similar to the one contained by DE and EH, thus indicating the reason for the curves' names. He first chose EH, drawn perpendicular to

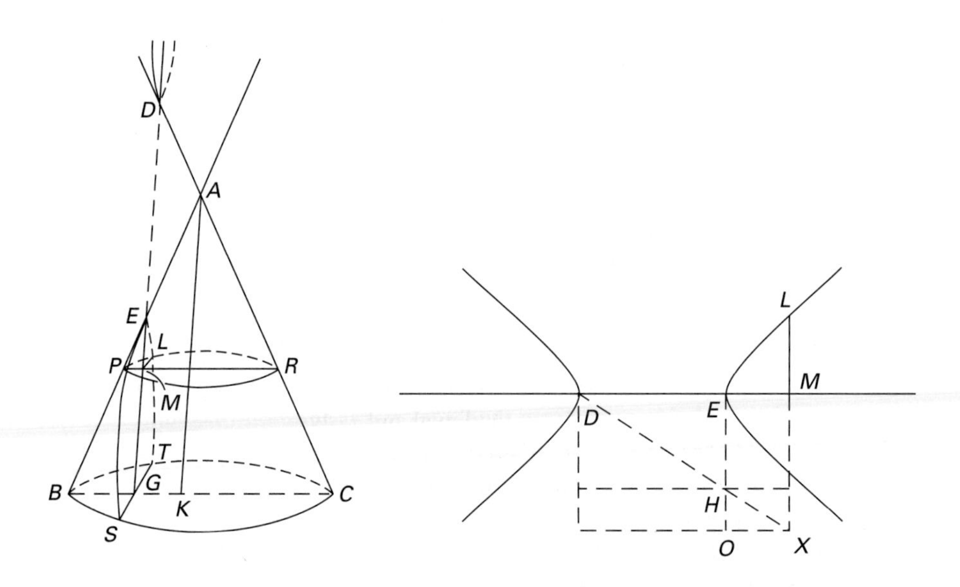

FIGURE 3.9 Derivation of the symptoms of a hyperbola

DE, so that

$$\frac{DE}{EH} = \frac{AK^2}{BK \cdot KC}.$$

As before, the right side of this equation can be written as a product: $(AK/BK) \cdot (AK/KC)$, where AK is parallel to DE. By similarity,

$$\frac{AK}{BK} = \frac{EG}{BG} = \frac{EM}{MP} \quad \text{and} \quad \frac{AK}{KC} = \frac{DG}{GC} = \frac{DM}{MR}.$$

Therefore,

$$\frac{DE}{EH} = \frac{EM \cdot DM}{MP \cdot MR}.$$

But also

$$\frac{DE}{EH} = \frac{DM}{MX} = \frac{DM}{EO} = \frac{EM \cdot DM}{EM \cdot EO}.$$

It follows that $MP \cdot MR = EM \cdot EO$ and therefore that $LM^2 = EM \cdot EO$. In the case of the hyperbola, $EO = EH + HO$, while for the ellipse, $EO = EH - HO$. In either case, because the rectangle contained by EM $(= OX)$ and HO is similar to the one contained by DE and EH, Apollonius has proved his result. In modern terms, because $EM/HO = DE/EH$, we have $HO = EM \cdot EH/DE$, and therefore, setting $LM = y$, $EM = x$, $EH = p$, and $DE = 2a$, Apollonius's symptoms become the modern equations for the hyperbola and the ellipse, respectively:

$$y^2 = x\left(p + \frac{p}{2a}x\right) \quad \text{and} \quad y^2 = x\left(p - \frac{p}{2a}x\right).$$

As before, the parameter p depends only on the cutting plane determining the curve.

After giving the symptoms in this form, Apollonius, in proposition I-21, proved that for both ellipse and hyperbola, the equation can be written in the form $y^2 = (p/2a)x_1x_2$, where x_1 and x_2 are distances of the point M from the two ends, E and D, of the axis of the curve. Note, in the case of the ellipse, that if the point (x, y) is the endpoint of the minor axis (of length $2b$), then this equation shows that $b^2 = (p/2a)a^2$, or $b^2 = pa/2$, which is the basic relationship between the parameter and the lengths of the two axes. For the hyperbola, as we will see below, the same equation holds, with b being the perpendicular distance from a vertex to an asymptote.

Furthermore, having derived the symptoms of the curves from their definitions as sections of a cone, Apollonius showed conversely in the final propositions of Book I that given a vertex (or a pair of vertices) at the end(s) of a given line (line segment) and a parameter, a cone and a cutting plane can be found such that the resulting section is a parabola (ellipse or hyperbola) with the given vertex (vertices), axis, and parameter. Henceforth, in Greek geometry as well as in medieval and early modern geometry, a mathematician could assert the "construction" of a conic section with given vertices, axes, and parameter in the same manner as the construction of a circle with given center and radius. New construction postulates had thus been added to the basic ones of Euclid's *Elements*.

Although Apollonius always used geometric language, much of his work can be characterized as geometric algebra. For example, the symptom of a curve can be thought of as an algebraic characterization of Apollonius's geometric derivation. Therefore, in our brief survey of highlights of the *Conics*, algebra will be used to simplify some of the statements and proofs.

3.2.3 Asymptotes, Tangents, and Foci

In Book II of the *Conics*, Apollonius dealt with the asymptotes to a hyperbola. These are constructed in proposition II-1 (Fig. 3.10). Drawing a tangent to the vertex A of the hyperbola and laying off on this tangent two segments AL, AL' (in opposite directions from the vertex) such that $AL^2 = AL'^2 = pa/2$ ($= b^2$), Apollonius showed that the lines CL, CL' drawn to L, L' from the center of the hyperbola do not meet either branch of the curve. (The Greek word *asymptotos* means "not capable of meeting.") Furthermore, in proposition II-14, Apollonius showed that the distance between the curve and these asymptotes, if both are extended indefinitely, becomes less than any given distance.

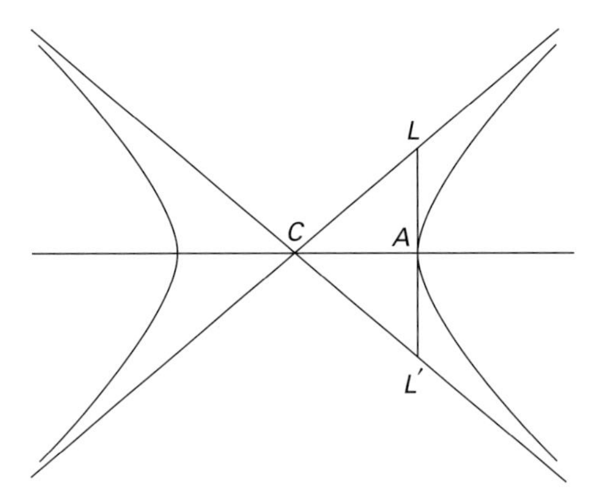

FIGURE 3.10 Construction of asymptotes to a hyperbola

In proposition II-4, Apollonius showed how to construct a hyperbola given a point on the hyperbola and its asymptotes, thus providing a further construction postulate. In proposition II-8, he then established the fact that segments cut off by a secant of a hyperbola between the hyperbola and the two asymptotes are equal. Then, in propositions II-10 and II-12, he showed that the symptom of a hyperbola can be expressed in terms of its asymptotes instead of its parameter and axis. In particular, he showed that the product of the lengths of the two lines drawn from any point of the hyperbola in given directions to the asymptotes is a constant. In modern notation, this result shows that a hyperbola can be defined by the equation $xy = k$.

Apollonius discussed the problem of drawing tangents to the conic sections in Book I. We will consider only the tangent to the parabola (Fig. 3.11):

PROPOSITION I-33. *Let C be a point on the parabola CET with CD perpendicular to the diameter EB. If the diameter is extended to A with $AE = ED$, then line AC will be tangent to the parabola at C.*

Set $DC = y$, $DE = x$, and $AE = t$. The theorem says that if $t = x$, then line AC is tangent to the curve at C. In other words, the tangent can be found by simply extending

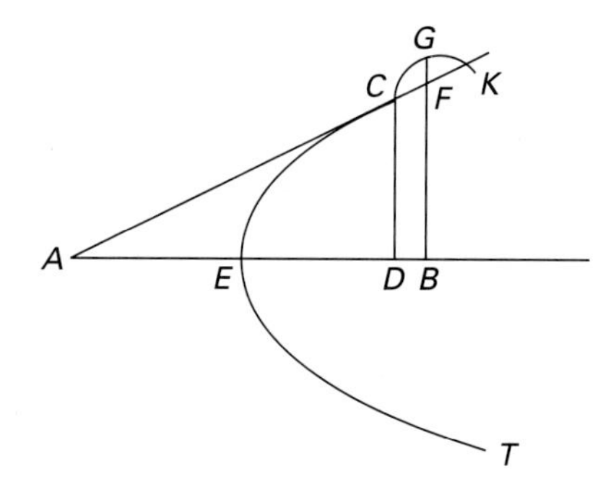

FIGURE 3.11 *Conics*, Proposition I-33

the diameter past E a distance equal to x and connecting the point so determined with C. Since a tangent line is a line that touches but does not cut the curve, Apollonius used a *reductio* argument and assumed that the line through A and C *does* cut the curve again, say, at K. Then the line segment from C to K lies within the parabola. Pick F on that segment and drop a perpendicular from F so that it meets the axis at B and the curve at G. Then $BG^2 : CD^2 > BF^2 : CD^2 = AB^2 : AD^2$. Also, since G and C lie on the curve, the symptom shows that $BG^2 = p \cdot EB$ and $CD^2 = p \cdot ED$, so $BG^2 : CD^2 = BE : DE$. Therefore, $BE : DE > AB^2 : AD^2$. Also, $4BE \cdot EA : 4DE \cdot EA > AB^2 : AD^2$, and therefore

$$4BE \cdot EA : AB^2 > 4DE \cdot EA : AD^2.$$

Now note that for any lengths a, b, proposition II-5 of *Elements* implies that $ab \leq [(a+b)/2]^2$, or $4ab \leq (a+b)^2$, with equality if and only if $a = b$. In this case, since $AE = DE$, we have $4DE \cdot EA = AD^2$, while since $AE < BE$, we have $4BE \cdot EA < AB^2$. Thus, the left side of the displayed inequality is less than 1, while the right side equals 1, a contradiction.

In Book III of *Conics*, Apollonius dealt with the focal properties of the ellipse and the hyperbola. In proposition III-45, for example, he found two points F, G on the axis AB of an ellipse such that the rectangle on AF, FB equals one-fourth of the rectangle on the parameter p and the axis AB, and similarly for the rectangle AG, GB (Fig. 3.12). These points are called the **foci**. In algebraic terms, if the distances from F and G to the center O

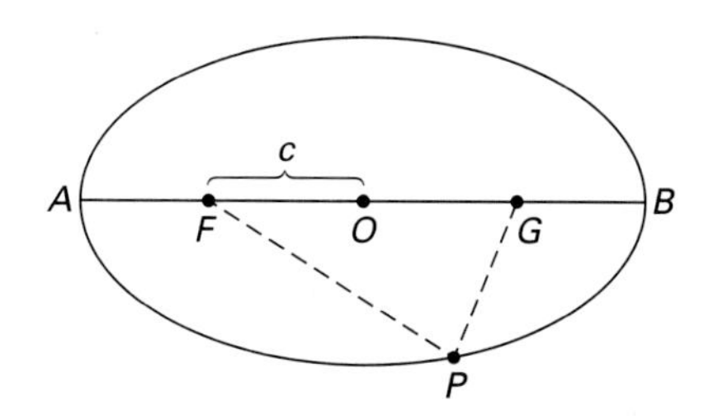

FIGURE 3.12 *Conics*, Proposition III-45

are equal to c, Apollonius's condition may be translated as the equation

$$(a-c)(a+c) = \frac{1}{4}\cdot 2ap \quad \text{or} \quad a^2 - c^2 = \frac{pa}{2} \quad \text{or} \quad a^2 - c^2 = b^2.$$

Given this definition, Apollonius presented a series of propositions that culminate in the well-known result that the lines from the two foci to any point on the ellipse make equal angles with the tangent to the ellipse at that point. Later in Book III, he also showed that if one connects an arbitrary point on the curve to each focus, then the sum of these two straight lines equals the axis. That is, if P is a point on the curve and F, G are the two foci, then $PF + PG = 2a$. This property is, in fact, the standard defining property for the ellipse in current textbooks.

Although Apollonius presented similar results for the hyperbola, he did not deal with the focal property of the parabola. That property—that any line from the focus to a point on the parabola makes an angle with the tangent at that point equal to the one made by a line parallel to the axis—was probably first proved by Diocles (early second century BCE), a contemporary of Apollonius. We will consider Diocles' proof.

Given a parabola LBM with axis BW, lay off BE along the axis equal to half the parameter and bisect BE at D (Fig. 3.13). This point D, whose distance from the vertex is $p/4$, is today called the **focus**. Pick an arbitrary point K on the parabola, draw a tangent line AKC through K meeting the axis extended at A, draw KS parallel to the axis, and connect DK. Diocles now asserted that $\angle AKD = \angle SKC$.

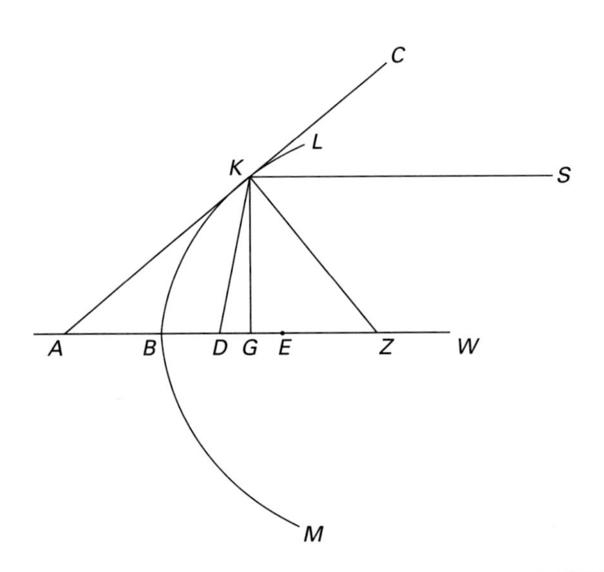

FIGURE 3.13 Focal property of a parabola, from Diocles' *On Burning Mirrors*

To prove this, first drop a perpendicular from K to the axis, meeting it at G. By *Conics* I-33, $AB = BG$. Next draw a line from K perpendicular to AK and meeting the axis at Z. Because $KG^2 = AG \cdot GZ$ and also $KG^2 = p \cdot BG$, it follows that $GZ = p/2$. Then $GZ = BE$, so $GB = EZ$, $AB = EZ$, and finally, $AD = DZ$. Because triangle AKZ is a right triangle whose hypotenuse is bisected at D, we have $AD = DK = DZ$. Therefore, $\angle DZK = \angle DKZ$. Since KS is parallel to AZ, it also follows that $\angle ZKS = \angle DKZ$. Subtracting these equal angles from the right angles ZKC and ZKA, we obtain the desired result.

3.2.4 Problem Solving Using Conics

Apollonius's aim in the *Conics* was not so much to develop the properties of the conic sections for their own intrinsic beauty, but to develop the theorems necessary for the application of these curves to the solution of geometric problems. We have already discussed the use of conics in the cube doubling problem and will conclude this section with two other examples of Greek use of these curves.

We first consider the angle trisection problem. Let angle ABC be the angle to be trisected (Fig. 3.14). Draw AC perpendicular to BC and complete the rectangle $ADBC$. Extend DA to the point E, which has the property that if BE meets AC at F, then the segment FE is equal to twice AB. It then follows that $\angle FBC = \frac{1}{3}\angle ABC$. For if FE is bisected at G, then $FG = GE = AG = AB$. Therefore, $\angle ABG = \angle AGB = 2\angle AEG = 2\angle FBC$, and the trisection is demonstrated. To complete the proof, however, it is necessary to show how to construct BE satisfying the given condition. Again, an analysis will help. Assuming $FE = 2AB$, draw CH and EH parallel to FE and AC, respectively. It follows that H lies on the circle of center C and radius FE $(= 2AB)$. Moreover, since $DE : DB = BC : CF$, or $DE : AC = DA : EH$, we have $DA \cdot AC = DE \cdot EH$, so H also lies on the hyperbola with asymptotes DB, DE and passing through C. Therefore, if the hyperbola and the circle are constructed and a perpendicular is dropped from the intersection point H to DA extended, the foot E of the perpendicular is the point needed to complete the solution.

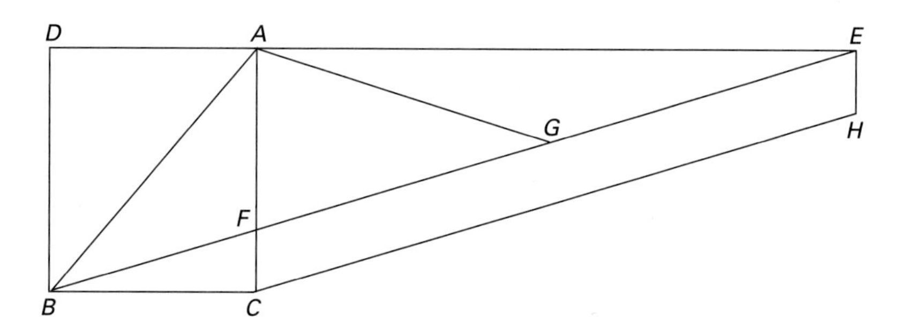

FIGURE 3.14 Angle trisection by way of conic sections

The second problem, one which had reverberations up to the seventeenth century, is the three- and four-line locus problem. The problem in its most elementary form can be stated as follows: Given three fixed straight lines, find the locus of a point moving so that the square of its distance to one line is in a constant ratio to the product of its distances to the other two lines. (Here distance is to be measured at a fixed angle to each line.) For the special case where two of the lines are parallel and the third perpendicular to the first two, it is easy to see analytically that the given locus is a conic section. Recall that one version of the equation of the ellipse and the hyperbola was $y^2 = (p/2a)x_1x_2$, where y is the distance of a given point from the diameter of the conic, and x_1, x_2 are the distances of the abscissa of that point from the endpoints of the diameter. If tangents are drawn to the conic at those two endpoints, this curve then provides a solution to the three-line locus problem with respect to the diameter and the two tangents.

The problem for the Greek mathematicians was to generalize this solution, that is, to show that the locus was a conic whatever the position of the three lines. Apollonius wrote that the three-line locus problem had been only partially worked out by Euclid, but that the new results in Book III of *Conics* would enable the problem to be completely solved. The text of Book III does not mention the problem as such, but in fact theorems in that book allow the derivation of the result that a conic has the property of the three-line locus relative to two tangents to the curve from a given point and the secant joining the two points of tangency. Other theorems there can be used to show that a conic also solves the four-line locus problem, which is to find the locus of a point such that the product of its distances to one pair of lines is in a constant ratio to the product of its distances to the other pair. In later Greek times, an attempt was made, without great success, to find the locus with regard to greater numbers of lines. It was this problem that both Descartes and Fermat demonstrated they could solve through their new method of analytic geometry in the seventeenth century, a method whose germ came from a careful reading of Apollonius's work. As should be evident from the foregoing description of many of the Greek problems in modern notation, the Greek tradition of geometric problem solving, which was carried on in the Islamic world long after its demise in the Hellenic world, ultimately led to new advances in mathematical technique, advances that finally reduced much of this kind of Greek mathematics to mere textbook exercises.

3.3 Ptolemy and Greek Astronomy

3.3.1 Astronomy before Ptolemy

Centuries of observation of the heavens had enabled the Babylonians to make relatively accurate predictions of the recurrence of various celestial phenomena, from such simple ones as the time of sunrise and sunset to such complicated ones as the times of lunar eclipses. But they never apparently applied more than arithmetic and simple algebra to this study, nor did they develop a model to connect the various celestial phenomena. The initial creation of such a model was a product of Greece in the fourth century BCE, the time of Plato's Academy.

The basic model developed at that time was one of two concentric spheres, the sphere of the earth and the sphere of the stars. The immediate evidence of our senses indicates that the earth is flat, but more sophisticated observations convinced the Greeks of the earth's sphericity. Their sense of esthetics—that a sphere was the most perfect solid shape—added to this conviction. That the shape of the heavens should mirror the shape of the earth was also only natural.

The evidence of the senses, and some logical argument as well, further convinced the Greeks that the earth was stationary in the middle of the celestial sphere. The second part of this conclusion came from the general symmetry of the major celestial phenomena; the first part came from the lack of any sensation of motion of the earth. The Greeks noted that if the earth did rotate on its axis once a day, its motion would of necessity be so swift that, in Ptolemy's words, "objects not actually standing on the earth would appear to have the same motion, opposite to that of the earth; neither clouds nor other flying or thrown objects would ever be seen moving toward the east, since the earth's motion toward the east would always outrun and overtake them, so that all other objects would seem to move in the direction of the west and the rear." With the earth considered immovable, the observed daily motion in the sky must be due to the rotation of the celestial sphere, to which were firmly attached the so-called **fixed stars**, grouped into patterns called **constellations**. These never change their positions with respect to each other and form the fixed background for the **wandering stars**, or planets.

The seven wanderers—the sun, the moon, Mercury, Venus, Mars, Jupiter, and Saturn—were more loosely attached to the celestial sphere. That they were attached was obvious; in general, they participated in the daily east to west rotation of the celestial sphere. But they also had their own motion, usually in the opposite direction (west to east) at much slower speeds. It is these motions that the Greek astronomers (and indeed all earlier astronomers) attempted to make sense of. The Greeks were limited in their attempts at explanation, however, by an overriding philosophical consideration: Since the universe beyond the earth was thought to be unchanging and perfect, according to Aristotle, the only movements in the heavens were the "natural" movements of these perfect bodies. Because the bodies were spherical, the natural movements were circular. Thus, Greek astronomers and mathematicians (usually the same people) attempted to solve Plato's problem, cited at the opening of the chapter—that is, to develop a model that would save the appearances—through a combination of geometrical constructs using circular and uniform motion. It was not the business of the astronomer-mathematicians to decide if or how such motions were physically possible, for celestial physics as we know it was never dealt with in ancient Greece. But they did, in fact succeed in finding several different systems that met Plato's challenge.

Because the basic Greek model of the heavens consisted of spheres, the first element of the study of celestial motion was the study of the properties of the sphere. Evidently, several texts were written in the fourth century BCE on the general subject of spherics, which did cover the basics, mostly in the context of results immediately useful in astronomy. These books contained such definitions as that of a **great circle** (a section of a sphere by a plane through its center) and its **poles** (the extremities of the diameter of the sphere perpendicular to this plane). The texts also included three important theorems very useful in what follows: First, any two points on the sphere that are not diametrically opposite determine a unique great circle. Second, any great circle through the poles of a second great circle is perpendicular to the original one, and, in this case, the second circle also contains the poles of the first. Third, any two great circles bisect one another.

There are several great circles on the celestial sphere that are important for astronomy. For example, the sun's path in its west to east movement through the stars is a great circle. This great circle, called the **ecliptic**, passes through the twelve constellations of the zodiac. (These constellations were first mentioned in Babylonian astronomy and appeared in Greek sources as early as 300 BCE.) The diameter of the earth through the north and south poles, extended to the heavens, is the axis around which the daily rotation of the celestial sphere takes place. The great circle corresponding to the poles of that axis is called the **celestial equator**. The equator and the ecliptic intersect at two diametrically opposite points, the **vernal** and **autumnal equinoxes**, for on those dates the sun is located on those intersections

(Fig. 3.15). The points on the ecliptic at the maximal distance north and south of the equator are the **summer** and **winter solstices**, respectively.

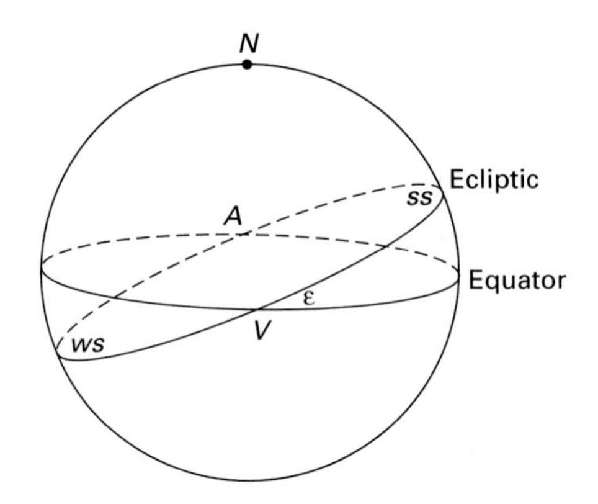

FIGURE 3.15 The ecliptic and the equator. (SS indicates the summer solstice; WS is the winter solstice.)

Since the Greeks knew that the earth is so small that it can in effect be considered as a point with respect to the sphere of the stars, they assumed that the horizon plane passed through the center of the celestial sphere and hence that the horizon itself was also a great circle. The horizon intersects the equator at the east and west points. Finally, the **local meridian** is the great circle that passes through the north and south points of the horizon and the point directly overhead, the **local zenith**. Because the meridian circle is perpendicular to both the horizon and the celestial equator, it also passes through the north and south poles of the latter. The angle ϵ between the equator and the ecliptic can be determined by taking half the distance (in degrees) between the noon altitudes of the sun at the summer and winter solstices. This value had been measured as 24° by the time of Euclid and was taken to be 23°51′20″ by Ptolemy. (In fact, this value is slowly decreasing and is now about $23\frac{1}{2}^\circ$.) The angle between the horizon and the equator is $90^\circ - \phi$, where ϕ is the geographical latitude of the observer (Fig. 3.16). The measure of the arc between the north celestial pole and the horizon is also given by ϕ.

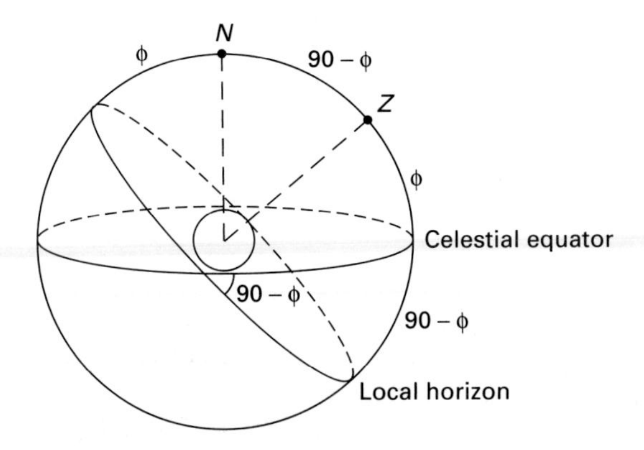

FIGURE 3.16 The horizon and the equator

3.3.2 Apollonius and Hipparchus

It was long known that the velocity of the sun around the ecliptic was not constant. The Babylonians had already discovered this in connection with their attempts to determine, for example, the time of first visibility of the moon each month. The Greeks discovered this by determining that the seasons of the year were not equal in length; for example, the time from the vernal equinox to the summer solstice is two days longer than the time from the summer solstice to the autumnal equinox. A simple model of the sun revolving in a circle centered on the earth at constant speed could not account for this phenomenon. Because nonuniform motion would not satisfy Plato's rules, Apollonius proposed the following solution: Place the center of the sun's orbit at a point (called the **eccenter**) displaced away from the earth. Then, if the sun moves uniformly around the new circle (called the **deferent circle**), an observer on earth will see more than a quarter of the circle against the spring quadrant (the upper right) than against the summer quadrant (the upper left) (Fig. 3.17a). The distance ED, or better, the ratio of ED to DS, is known as the **eccentricity** of the deferent. If line ED is extended to the deferent circle, the intersection point closest to the earth is called the **perigee** of the deferent, and the one farthest from the earth is called the **apogee**. Assuming that one can determine the correct parameters in this model (the length and direction of ED) so that the seasonal lengths come out right, the question in using the model is where the sun will be seen on a particular day. To answer this question, it is necessary to find angle DES. This requires solving triangle DES, which in turn requires trigonometry. In fact, it was the necessity for introducing numerical parameters into these geometric models that led to the invention of trigonometry.

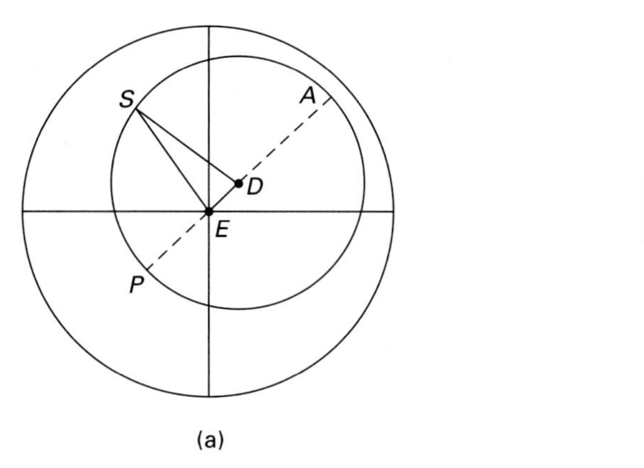

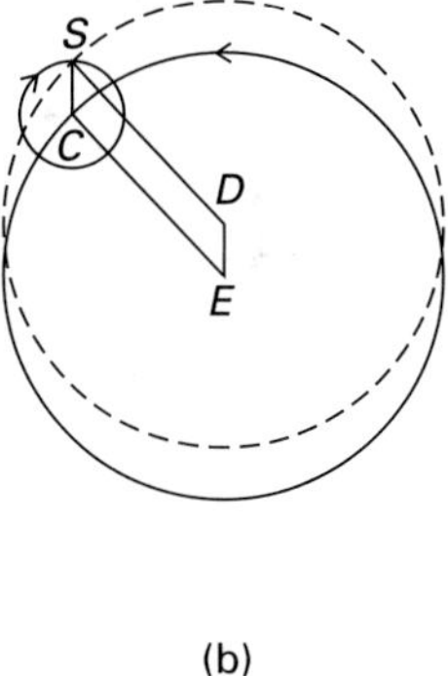

FIGURE 3.17 (a) Apollonius's eccenter model for the sun. (b) Apollonius's epicycle model for the sun

Apollonius also noticed that this eccentric model can be replaced by another geometric model, the epicyclic one. That is, instead of considering the sun as traveling on the eccentric circle, it may be imagined as traveling on a small circle, the **epicycle**, whose center travels on the original earth-centered circle (Fig. 3.17b). If the epicycle rotates once clockwise in the same time as its center rotates around the earth—that is, if the two motions always keep $DECS$ a parallelogram—the actual path of the sun will be the same as it was using the deferent circle. It then turns out that combining epicycles and eccentric circles produces the more complicated motions of the planets, including their occasional westward (retrograde) motion. Again, of course, there was the necessity for numerical parameters from which other calculations could be made.

Apollonius himself did not possess the trigonometric machinery necessary to complete the solution of these problems of celestial motion. It was Hipparchus of Bithynia (190–120 BCE) who systematically carried out numerous observations of planetary positions, introduced a coordinate system for the stellar sphere, and began the tabulation of trigonometric tables necessary to enable mathematicians to solve triangles easily and to attack Apollonius's questions successfully.

To deal quantitatively with the positions of the stars and planets, astronomers needed both a unit of measure for arcs and angles and a method of specifying where a particular body is located on the celestial sphere—that is, a system of coordinates. The Babylonians, sometime before 300 BCE, initiated the division of the circumference of the circle into 360 parts, called degrees, and within the next two centuries this measure, along with the sexagesimal division of degrees into minutes and seconds, was adopted in the Greek world. Why the Babylonians divided the circle into 360 parts is not known. Perhaps it was because 360 is easily divisible by many small integers or because it is the closest "round" number to the number of days in the year. The latter reason gives the convenient approximation that the sun travels 1° along the ecliptic each day.

It was also the Babylonians who first introduced coordinates into the sky. The system they used, later taken over by Ptolemy, is known as the **ecliptic system**. Positions of stars are measured both along and perpendicular to the ecliptic. The coordinate along the ecliptic (measured in degrees counterclockwise from the vernal point as seen from the north pole) is called the **longitude**, λ; the perpendicular coordinate, measured in degrees north or south of the ecliptic, is called the **latitude**, β (Fig. 3.18a). This coordinate system is particularly useful when dealing with the sun, moon, and planets. The sun, since it travels along the ecliptic, always has latitude 0°. Its longitude increases daily by approximately 1°, from 0° at the vernal equinox to 90° at the summer solstice, 180° at the autumnal equinox, and 270° at the winter solstice.

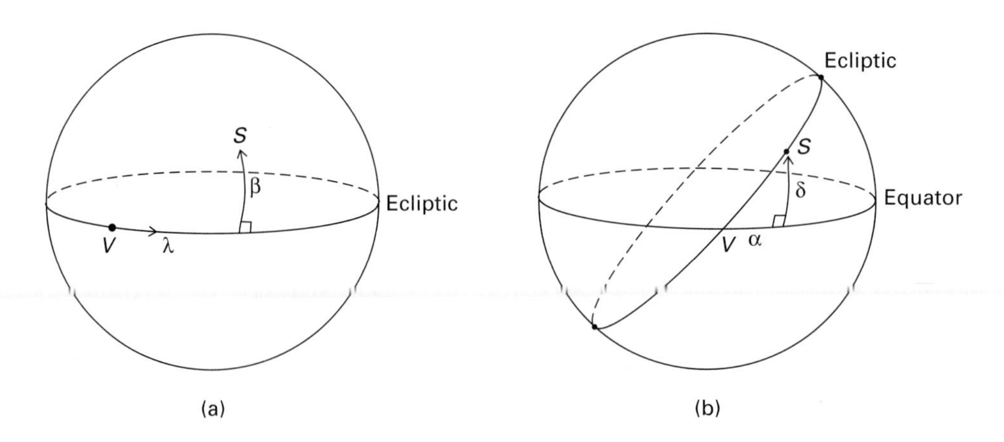

FIGURE 3.18 (a) Ecliptic coordinate system on the celestial sphere. (b) Equatorial coordinate system on the celestial sphere

In place of this ecliptic coordinate system, Hipparchus used a system based on the celestial equator. The coordinate along the equator, also measured counterclockwise from the vernal point, is called the **right ascension**, α; the perpendicular coordinate, measured north and south from the equator, is called the **declination**, δ (Fig. 3.18b). Hipparchus drew up a catalogue of fixed stars, in which he described some of their positions in terms of this coordinate system.

To be able to relate the coordinates of a point in one coordinate system to its coordinates in another—and this is necessary to solve astronomical problems—one needs spherical trigonometry. But before this could be developed, it was necessary to understand plane trigonometry. Hipparchus was evidently the first to attempt the detailed calculation of lengths that would enable plane triangles to be solved. But since his original work is lost, we will turn to the work of Ptolemy, who improved Hipparchus's methods.

3.3.3 Ptolemy and His Chord Table

The basic element in Ptolemy's trigonometry, detailed in the first book of his *Mathematiki Syntaxis* (*Mathematical Collection*), generally known today as the *Almagest* (a variant of its Islamic name), was the chord subtending a given arc (or central angle) in a circle of fixed radius. Ptolemy produced a table listing α and chord α for integral and half-integral values of the arc α, from 0° to 180°. Note that chord α, henceforth abbreviated $\operatorname{crd}\alpha$, is simply a length (Fig. 3.19). If the radius of the circle is denoted by R, then the chord is related to the sine by the equations

$$\frac{\frac{1}{2}\operatorname{crd}\alpha}{R} = \sin\frac{\alpha}{2} \qquad \text{or} \qquad \operatorname{crd}\alpha = 2R\sin\frac{\alpha}{2}.$$

Similarly,

$$\operatorname{crd}(180-\alpha) = 2R\cos\frac{\alpha}{2}.$$

The identity $\sin^2\alpha + \cos^2\alpha = 1$ is then equivalent to the chord result $\operatorname{crd}(180-\alpha) = \sqrt{(2R)^2 - \operatorname{crd}^2\alpha}$.

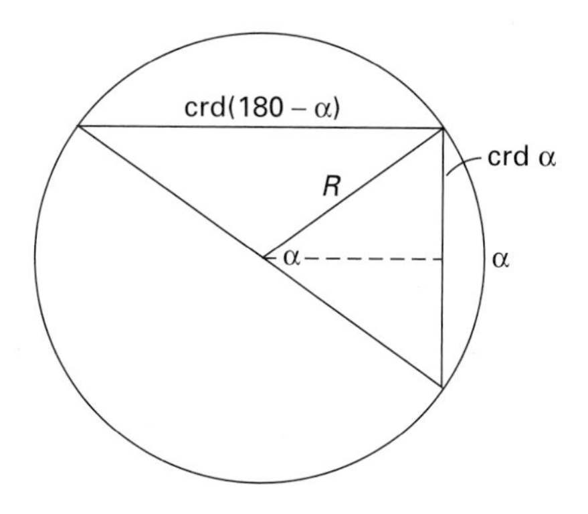

FIGURE 3.19 Chord α and chord$(180-\alpha)$

Because he was doing calculations in the sexagesimal system, it was natural for Ptolemy to choose as his circle radius $R = 60$. His first calculation established the chord of 36°, namely, the length of a side of a regular decagon inscribed in a circle. In Figure 3.20, ADC is the diameter of the circle with center D, BD is perpendicular to ADC, E bisects DC, and F is chosen so that $EF = EB$. By *Elements* II-6, we have $CF \cdot FD + ED^2 = BE^2$. Therefore, $CF \cdot FD = BE^2 - ED^2 = BD^2 = CD^2$, and the line CF has been divided at D in extreme and mean ratio. By *Elements* XIII-9, if the side of a hexagon and a decagon inscribed in the same circle are placed together in a straight line, then the meeting point divides the entire line segment in extreme and mean ratio. Because CD, the radius, equals the side of a hexagon inscribed in the circle, Ptolemy has shown that DF is the side of a decagon, that is, $DF = \text{crd}(36°)$. To calculate its length, he noted that

$$DF = EF - ED = EB - ED = \sqrt{BD^2 + ED^2} - ED$$
$$= \sqrt{3600 + 900} - 30 = 37;4,55.$$

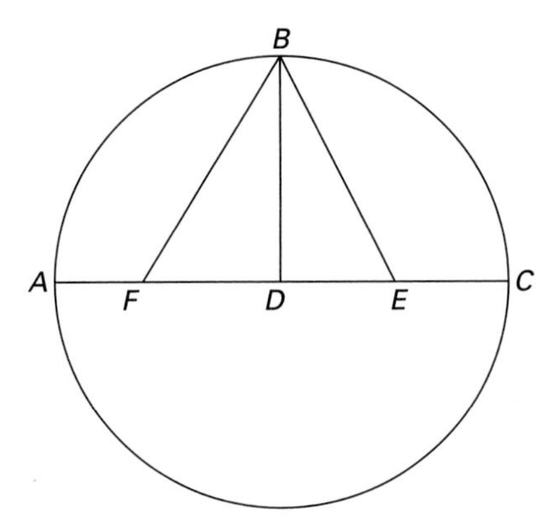

FIGURE 3.20 Ptolemy's calculation of crd(36°)

Ptolemy next noted that since the square on the side of a regular pentagon [= crd(72°)] equals the sum of the squares on the side of a regular decagon and the side of a regular hexagon (*Elements* XIII-10), it followed that $\text{crd}(72°) = \sqrt{R^2 + \text{crd}^2(36°)} = 70;32,3$, with, of course, $\text{crd}(60°) = R = 60$. Furthermore, $\text{crd}(90°) = \sqrt{2R^2} = \sqrt{7200} = 84;51,10$ and $\text{crd}(120°) = \sqrt{(2R)^2 - \text{crd}^2\, 60} = \sqrt{3R^2} = 103;55,23$. Finally, because $\text{crd}^2(180 - \alpha) = (2R)^2 - \text{crd}^2\alpha$, Ptolemy could also calculate the chord of the supplement to any arc whose chord was known. For example, $\text{crd}(144°) = 114;7,37$. He was therefore well started on a chord table simply from propositions of Euclidean geometry and the ability to calculate square roots.

To go further, Ptolemy needed a way to calculate the chords of sums and differences of angles, given the chords of the angles, and a way to calculate the chord of half an angle, given the chord of the angle. To accomplish the first, he proposed

PTOLEMY'S THEOREM. *Given any quadrilateral inscribed in a circle, the product of the diagonals equals the sum of the products of the opposite sides.*

To prove that $AC \cdot BD = AB \cdot CD + AD \cdot BC$ in quadrilateral $ABCD$ choose E on AC so that $\angle ABE = \angle DBC$ (Fig. 3.21). Then $\angle ABD = \angle EBC$. Also, $\angle BDA = \angle BCA$ since they both subtend the same arc. Therefore, $\triangle ABD$ is similar to $\triangle EBC$. Hence, $BD : AD = BC : EC$ or $AD \cdot BC = BD \cdot EC$. Similarly, since $\angle BAC = \angle BDC$, $\triangle ABE$ is similar to $\triangle DBC$. Hence, $AB : AE = BD : CD$ or $AB \cdot CD = BD \cdot AE$. Adding equals to equals gives $AB \cdot CD + AD \cdot BC = BD \cdot AE + BD \cdot EC = BD(AE + EC) = BD \cdot AC$, and the theorem is proved.

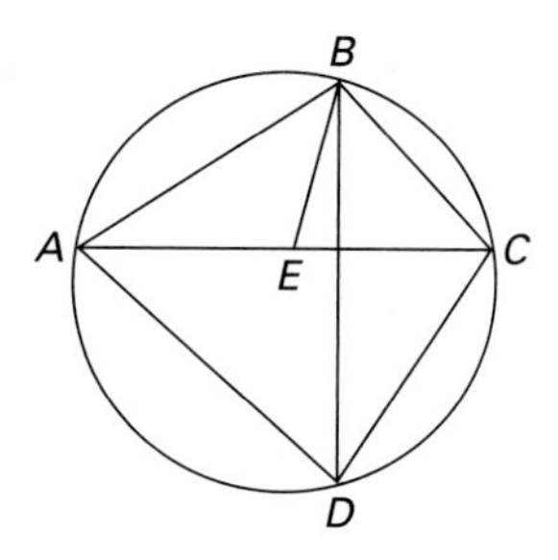

FIGURE 3.21 Ptolemy's theorem

To derive a formula for the chord of a difference of two arcs α, β, Ptolemy used the theorem with $AC = \text{crd } \alpha$ and $AB = \text{crd } \beta$ given. Applying the result to quadrilateral $ABCD$ gives $AB \cdot CD + AD \cdot BC = AC \cdot BD$ (Fig. 3.22). Because $BC = \text{crd}(\alpha - \beta)$,

$$120 \text{ crd}(\alpha - \beta) = \text{crd } \alpha \cdot \text{crd}(180 - \beta) - \text{crd } \beta \cdot \text{crd}(180 - \alpha).$$

This is easily translated into the modern difference formula for the sine:

$$\sin(\alpha - \beta) = \sin\alpha \cos\beta - \cos\alpha \sin\beta.$$

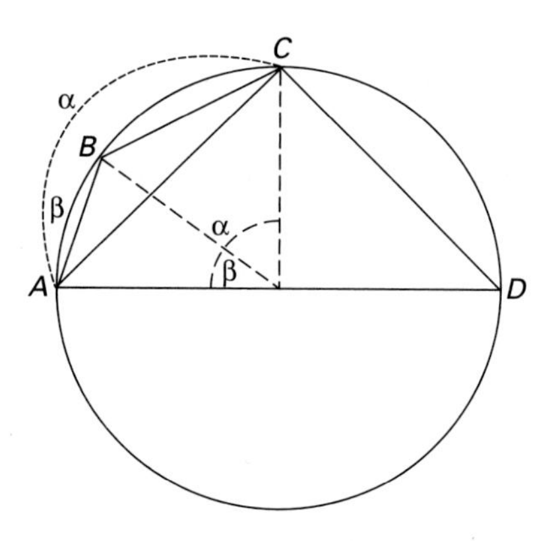

FIGURE 3.22 The difference formula for chords

A similar argument shows that

$$120 \text{ crd}[180 - (\alpha + \beta)] = \text{crd}(180 - \alpha) \text{ crd}(180 - \beta) - \text{crd } \beta \cdot \text{crd } \alpha,$$

a formula equivalent to the sum formula for the cosine:

$$\cos(\alpha + \beta) = \cos\alpha \cos\beta - \sin\alpha \sin\beta.$$

Ptolemy then proved a version of the half-angle formula. Suppose $\alpha = \angle BOC$ is bisected by OD (Fig. 3.23). To express $\text{crd}(\alpha/2) = DC$ in terms of $\text{crd } \alpha = BC$, choose E on AC so that $AE = AB$. Then $\triangle ABD$ is congruent to $\triangle AED$ and $BD = DE$. Since $BD = DC$, also $DC = DE$. If DF is drawn perpendicular to EC, then $CF = \frac{1}{2}CE = \frac{1}{2}(AC - AE) = \frac{1}{2}(AC - AB) = \frac{1}{2}[2R - \text{crd}(180 - \alpha)]$. But also, triangles ACD and DCF are similar, so $AC : CD = CD : CF$. Therefore,

$$\text{crd}^2 \frac{\alpha}{2} = CD^2 = AC \cdot CF = R[2R - \text{crd}(180 - \alpha)] = 60[120 - \text{crd}(180 - \alpha)].$$

Putting this into modern notation gives

$$\left(2R \sin\frac{\alpha}{4}\right)^2 = R\left(2R - 2R\cos\frac{\alpha}{2}\right),$$

or, replacing α by 2α,

$$\sin^2\frac{\alpha}{2} = \frac{1 - \cos\alpha}{2},$$

the standard half-angle formula.

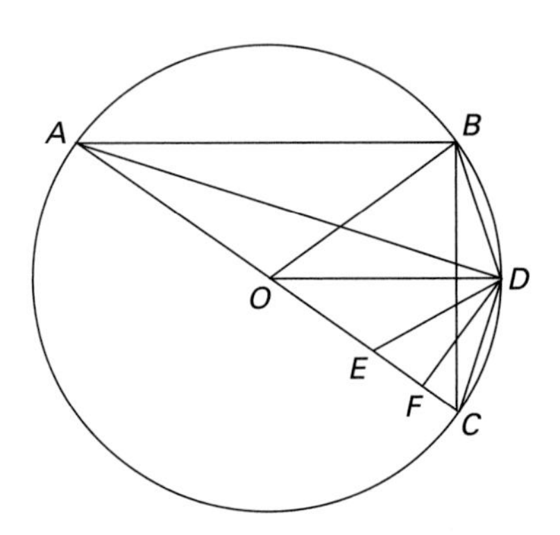

FIGURE 3.23 Ptolemy's half-angle formula

Using the difference formula and the half-angle formula, Ptolemy could now calculate $\text{crd}(12°) = \text{crd}(72° - 60°)$, $\text{crd}(6°) = \text{crd}(\frac{1}{2} \cdot 12°)$, $\text{crd}(3°)$, $\text{crd}(1\frac{1}{2}°)$, and $\text{crd}(\frac{3}{4}°)$. His last two results are $\text{crd}(1\frac{1}{2}°) = 1;34,15$ and $\text{crd}(\frac{3}{4}°) = 0;47,8$. Using the addition formula, he could have built up the table in intervals of $1\frac{1}{2}°$ or even $\frac{3}{4}°$. However, he wanted his table to be in intervals of $\frac{1}{2}°$. He realized that determining $\text{crd}(\frac{1}{2}°)$ from the already calculated

$\text{crd}(1\frac{1}{2}^\circ)$ by Euclidean tools would be equivalent to trisecting an angle using these tools. That, he believed, was impossible. He therefore needed an alternative method.

This alternative, an **approximation procedure**, is based on the lemma that if $\alpha < \beta$, then $\text{crd } \beta : \text{crd } \alpha < \beta : \alpha$, or, in modern notation, that $(\sin x)/x$ increases as x approaches 0. Applying this lemma first to $\alpha = \frac{3}{4}^\circ$ and $\beta = 1^\circ$, Ptolemy found $\text{crd}(1^\circ) < \frac{4}{3}\ \text{crd}(\frac{3}{4}^\circ) = \frac{4}{3}(0;47,8) = 1;2,50,40$. Applying it next to $\alpha = 1^\circ$ and $\beta = 1\frac{1}{2}^\circ$, he found $\text{crd}(1^\circ) > \frac{2}{3}\ \text{crd}(1\frac{1}{2}^\circ) = \frac{2}{3}(1;34,15) = 1;2,50$. Since all calculated values were rounded off to two sexagesimal places, it appears that, to that number of places, $\text{crd}(1^\circ) = 1;2,50$, and therefore $\text{crd}(\frac{1}{2}^\circ) = 0;31,25$. The addition formula now enabled Ptolemy to build up his table in steps of $\frac{1}{2}^\circ$ from $\text{crd}(\frac{1}{2}^\circ)$ to $\text{crd}(180^\circ)$. To aid in interpolation for calculating chords of any number of minutes, he added to his table a third column containing one-thirtieth of the increase from $\text{crd } \alpha$ to $\text{crd}(\alpha + \frac{1}{2}^\circ)$. A small portion of the table, whose accuracy is roughly equivalent to that of a modern five-decimal-place table, is shown in Table 3.1.

TABLE 3.1

Portion of Ptolemy's Chord Table

Arcs	Chords	Sixtieths	Arcs	Chords	Sixtieths
$\frac{1}{2}$	0;31,25	0;1,2,50	6	6;16,49	0;1,2,44
1	1;2,50	0;1,2,50	47	47;51,0	0;0,57,34
$1\frac{1}{2}$	1;34,15	0;1,2,50	49	49;45,48	0;0,57,7
2	2;5,40	0;1,2,50	72	70;32,3	0;0,50,45
$2\frac{1}{2}$	2;37,4	0;1,2,48	80	77;8,5	0;0,48,3
3	3;8,28	0;1,2,48	108	97;4,56	0;0,36,50
4	4;11,16	0;1,2,47	120	103;55,23	0;0,31,18
$4\frac{1}{2}$	4;42,40	0;1,2,47	133	110;2,50	0;0,24,56

3.3.4 Solving Plane Triangles

Given his chord table, Ptolemy could solve plane triangles. Although he never stated a systematic procedure for doing so, he does seem to have applied fixed rules. One difference to keep in mind when comparing Ptolemy's method to a modern one is that Ptolemy's table contains lengths of chords when the radius is 60 rather than ratios. Therefore, he always had to adjust his tabular values in a given problem to the actual length of the radius. We consider here two examples of his procedures.

First, consider how Ptolemy calculated the parameters for the eccentric model of the sun. The calculation amounts to solving the right triangle LDE, where D represents the center of the sun's orbit and E represents the earth (Fig. 3.24; compare with Fig. 3.17a). Divide the ecliptic into four quadrants by perpendicular lines through E and similarly divide the eccentric circle. To find LD and LE, one must first calculate the arcs $\theta = \frac{1}{2}\widehat{VV'}$ and

$\tau = \frac{1}{2}\widehat{WW'}$ using the known inequalities of the seasons. Given that the spring path of the sun is 94.5 days and the summer path is 92.5 days, and supposing that v is the mean daily angular velocity of the sun, Fig. 3.24 shows that $90 + \theta + \tau = 94.5v$ for the spring and $90 + \theta - \tau = 92.5v$ for the summer. Because v equals the length of the year (observed to be 365;14,48 days) divided by 360°, or $0°59'8''$ per day, it follows that $90° + \theta + \tau = 93°9'$ and $90° + \theta - \tau = 91°11'$. A simple calculation then shows that $\theta = 2°10'$ and $\tau = 0°59'$.

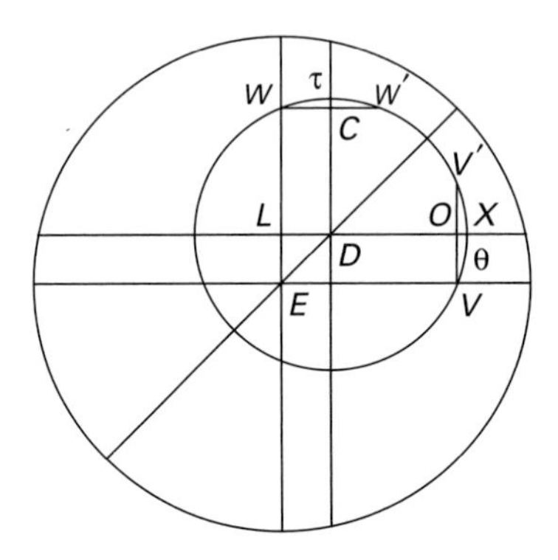

FIGURE 3.24 Calculation of the parameters in the eccentric model of the sun

The sides of the triangle DLE can now be determined under the assumption that the radius DX of the deferent is 60. Since DX bisects arc $\widehat{VV'}$, it is evident that $LE = OV = \frac{1}{2}VV' = \frac{1}{2}$ crd$(2\theta) = \frac{1}{2}$ crd$(4°20') = 2;16$. Similarly, $DL = \frac{1}{2}$ crd$(2\tau) = \frac{1}{2}$ crd$(1°58') = 1;2$. By the Pythagorean theorem, $DE^2 = LE^2 + DL^2 = 6;12,20$, and $DE = 2;29,30$ or, approximately, $2;30 = 2\frac{1}{2}$. In modern terminology, Ptolemy has simply calculated $LE = OV = R \sin\theta$ and $DL = CW = R \sin\tau$.

To complete the solution of the triangle, Ptolemy calculated $\angle LED$ by circumscribing a circle around ΔLDE. Since $LD = 1;2$ when $DE = 2;29,30$, LD would be 49;46 if DE were 120. Using the table of chords in reverse, Ptolemy found that the corresponding arc is about 49°; hence, $\angle LED$ is half of that, or $24°30'$. Then $\angle LDE = 65°30'$, and the triangle is solved. Again, in modern terminology, Ptolemy first calculated $120a/c = 2R\sin\alpha$ or $\sin\alpha = a/c$ and then used the inverse sine relation to determine α.

As a second example, consider Ptolemy's solution of an oblique triangle. The problem here is to find the direction $\angle DES$ of the sun, from the eccentric model, given that $DE = 2;30$ if DS is arbitrarily picked to be 60 (Fig. 3.25). For a given day, the angle PDS is known from the speed of the sun in its orbit, and hence the angle EDS is known. Ptolemy made the calculation for $\angle PDS = 30°$ and $\angle EDS = 150°$. He first constructed EK perpendicular to SD extended. Considering as before the circle about triangle DKE, he concluded that arc $DK = 120°$. From his chord table, he noted that if the radius were 60 (or $DE = 120$), then DK would be crd$(120°) = 103;55$. Since, however, $DE = 2;30$, by proportionality $DK = 2;10$. Then $SK = SD + DK = 62;10$. Since $\angle KDE = 30°$, also $EK = \frac{1}{2}DE = 1;15$. Applying the Pythagorean theorem to ΔSKE gives $SE = 62;11$. Next, consider the circle circumscribing ΔSKE. Because $KE = 1;15$ when $SE = 62;11$, it would be 2;25 if SE were 120. The chord table can be used in reverse to find that 2;25 corresponds to an arc of $2°18'$. It follows that $\angle KSE = 1°9'$ and therefore that $\angle DES$ is $180° - 150° - 1°9' = 28°51'$.

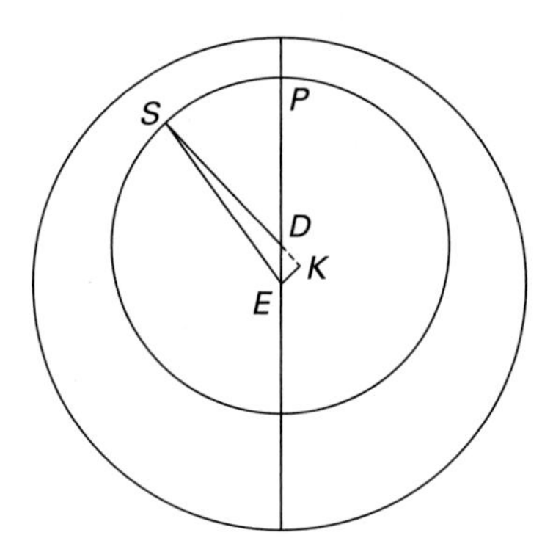

FIGURE 3.25 Finding the position of the sun

Ptolemy's procedure can be translated as follows. Given $\triangle ABC$ with a, b, and $\gamma > 90°$ known, drop AD perpendicular to BC extended (Fig. 3.26). If $AD = h$ and $CD = p$, then $p = [\text{crd}(2\gamma - 180) \cdot b]/2R$ and $h = [\text{crd}(360 - 2\gamma) \cdot b]/2R$. It follows that

$$\begin{aligned} c^2 &= h^2 + (a + p)^2 \\ &= a^2 + \left(\frac{\text{crd}^2(360 - 2\gamma)}{4R^2} + \frac{\text{crd}^2(2\gamma - 180)}{4R^2} \right) b^2 + \frac{2ab\ \text{crd}(2\gamma - 180)}{2R} \\ &= a^2 + b^2 + 2ab\frac{\text{crd}(2\gamma - 180)}{2R}, \end{aligned}$$

or

$$c^2 = a^2 + b^2 - 2ab\cos\gamma,$$

precisely the law of cosines for the case where two sides and the included angle are known. To find the angles, Ptolemy then noted that $\text{crd}(2\beta) = (h \cdot 2R)/c$ and found β from the

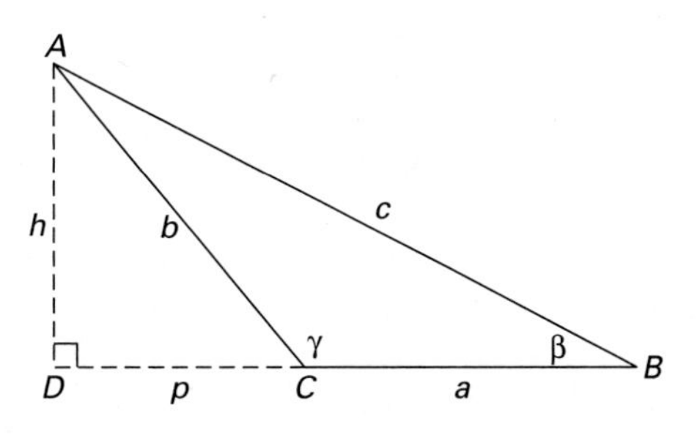

FIGURE 3.26 Ptolemy's law of cosines

table. This translates as $\sin\beta = h/c = (b\sin\gamma)/c$. Hence, Ptolemy has also used the equivalent of the law of sines.

Note that in giving the above example, Ptolemy explicitly provided an algorithm for calculating c and β given values of a, b, and γ. In fact, such algorithms are common in the *Almagest*. These algorithms of plane trigonometry can be translated into modern formulas without doing injustice to Ptolemy's own procedure.

3.3.5 Solving Spherical Triangles

Ptolemy's algorithms for solving right spherical triangles can also be translated into modern formulas. These formulas are based on the earliest known work on spherical trigonometry, the *Spherica*, by Menelaus (c. 100 CE). The formulas that Ptolemy used are equivalent to the following, all dealing with a right spherical triangle with sides a, b, c, and opposite angles A, B, and $C = 90°$, respectively (Fig. 3.27):

$$\tan A = \frac{\tan a}{\sin b} \tag{3.1}$$

$$\sin A = \frac{\sin a}{\sin c} \tag{3.2}$$

$$\cos B = \frac{\tan a}{\tan c} \tag{3.3}$$

$$\cos c = \cos a \cdot \cos b \tag{3.4}$$

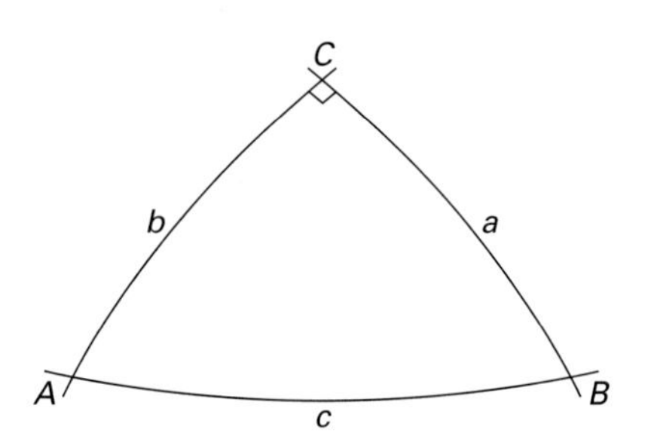

FIGURE 3.27 Right spherical triangle

Ptolemy's first application of these results was to find the declination δ and right ascension α of the sun, given its longitude λ (Fig. 3.28). Here, VA is the equator, VB the ecliptic, and V the vernal point. The angle ϵ between the equator and the ecliptic, according to Ptolemy, is $23°51'20''$. Suppose the sun is at H, a point with longitude λ. To determine $HC = \delta$ and $VC = \alpha$, the right triangle VHC must be solved. From equation 3.2, $\sin\epsilon = \sin\delta/\sin\lambda$ or $\sin\delta = \sin\epsilon\sin\lambda$. Ptolemy performed this calculation with both $\lambda = 30°$ and $\lambda = 60°$ to get (in the first case) $\delta = 11°40'$ and (in the second case) $\delta = 20°30'9''$. Having thus demonstrated the algorithm, he was able to produce a table for δ given each integral value of λ from $1°$ to $90°$. Similarly, from equation 3.3, $\cos\epsilon = \tan\alpha/\tan\lambda$ or $\tan\alpha = \cos\epsilon\tan\lambda$. Again, Ptolemy calculated the value of α corresponding to $\lambda = 30°$ to be $27°50'$ and that corresponding to $\lambda = 60°$ to be $57°44'$. He then listed the values of α corresponding

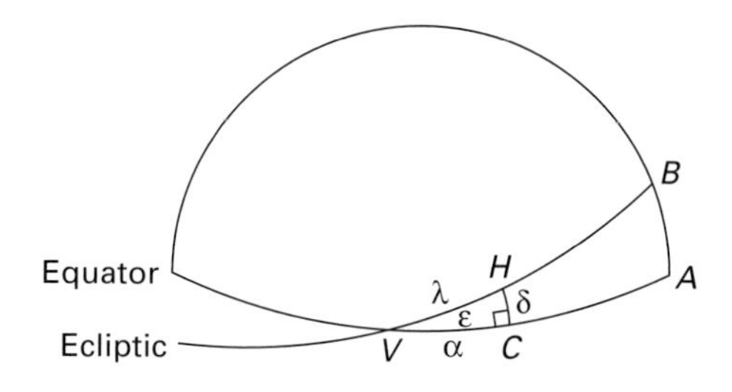

FIGURE 3.28 Method for determining the declination and right ascension of the sun, given its longitude

to other values of λ. Note further that, by symmetry, $\alpha(\lambda + 180) = \alpha(\lambda) + 180°$ and $\delta(\lambda + 180) = -\delta(\lambda)$.

Many of the other problems that Ptolemy solved are closely related to the determination of the **rising time** of an arc of the ecliptic. Namely, Ptolemy wanted to determine for a given geographical latitude the arc of the celestial equator that crosses the horizon at the same time as a given arc of the ecliptic. Since it is sufficient to determine this for an arc one endpoint of which is the vernal point, it is only necessary to determine the length VE of the equator that crosses the horizon simultaneously with the given arc VH of the ecliptic (Fig. 3.29). This arc length is called the rising time because time is measured by the uniform motion of the equator around its axis. One complete revolution takes 24 hours, so 15° along the equator corresponds to 1 hour, and 1° corresponds to 4 minutes. In any case, to solve Ptolemy's problem, it is sufficient to solve the triangle HCE for $EC = \sigma(\lambda, \phi)$ and then subtract that value from $VC = \alpha(\lambda)$ already determined. For example, suppose that the latitude $\phi = 36°$ and that $\lambda = 30°$. By the calculation above, $\delta = 11°40'$. Equation 3.1 then gives $\sin\sigma = \tan\delta/\tan(90 - \phi) = \tan\delta\tan\phi$ and, therefore, $\sigma = 8°38'$. Since $\alpha = 27°50'$, the rising time $VE = 27°50' - 8°38' = 19°12'$. Ptolemy calculated the rising time $\rho(\lambda, \phi)$ for values of λ in 10-degree intervals from 10° to 360° at eleven different latitudes ϕ and presented the results in an extensive table. This table can be used to calculate the length of daylight $L(\lambda, \phi)$ on any date at any given latitude. If the sun is at longitude λ, the point at longitude $\lambda + 180$ is rising when the sun is setting. Hence, we simply need to subtract the rising time of λ from that of $\lambda + 180$. We can simplify matters somewhat by noting that

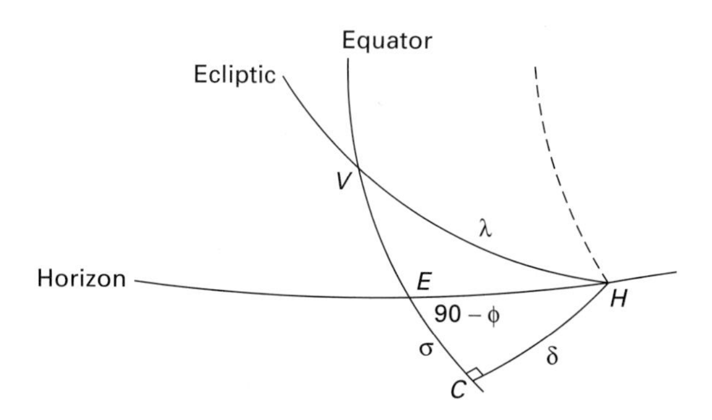

FIGURE 3.29 Calculation of the rising time

since $\sigma(\lambda + 180, \phi) = -\sigma(\lambda, \phi)$, we have

$$L(\lambda, \phi) = \rho(\lambda + 180, \phi) - \rho(\lambda) = \alpha(\lambda + 180) - \sigma(\lambda + 180, \phi) - \alpha(\lambda) + \sigma(\lambda, \phi)$$
$$= 180^\circ + 2\sigma(\lambda, \phi).$$

For example, when $\phi = 36^\circ$ and $\lambda = 30^\circ$, $L(30, 36) = 180^\circ + 2\sigma(30, 36) = 180^\circ + 17^\circ16' = 197^\circ16'$, which corresponds to approximately 13 hours, 9 minutes.

As a final application of spherical trigonometry, we calculate the distance of the sun from the zenith at noon. The sun on any given day is always at a distance δ from the equator. Hence, at noon, when it crosses the meridian, it is (assuming $\delta > 0$) between the north pole N and the intersection T of the meridian with the equator at a distance δ from that intersection (Fig. 3.30). Because arc $NT = 90^\circ$ and arc $NY = \phi$, it follows that arc $SZ = 90^\circ - (90^\circ - \phi) - \delta = \phi - \delta$. Note that if $\phi - \delta > 0$, or $\phi > \delta$, the sun will be in the south at noon and shadows will point north. Because the maximum value of δ is $23^\circ51'20''$, this will always be the case at latitudes greater than that value. On the other hand, when $\phi = \delta$, the sun is directly overhead at noon. The dates on which that occurs and also the dates when the sun is in the north at noon can easily be calculated for a given latitude. In any case, given the angular distance of the sun from the zenith, Ptolemy was able to use plane trigonometry to calculate shadow lengths of a pole of given height. He presented his results in a long table in which he gave the length of the longest day for 39 different parallels of latitude as well as the shadow lengths of a pole of height 60 at noon on the summer solstice, the equinoxes, and the winter solstice.

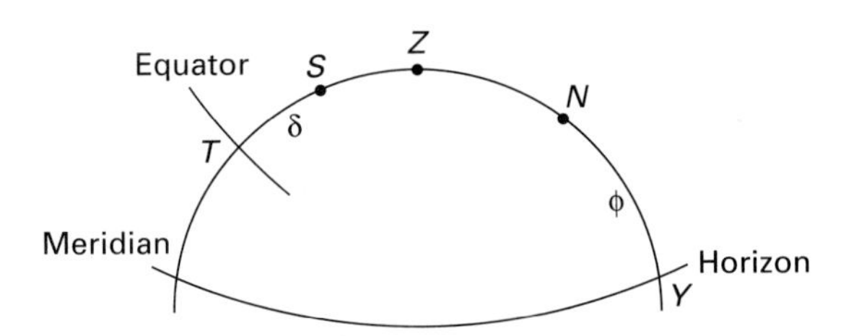

FIGURE 3.30 Calculation of the distance of the sun from the zenith

The examples above deal only with the sun and are taken from the first three books of the *Almagest*. In the remainder of the work, Ptolemy discussed the moon and the planets. For each heavenly body, he gave first a brief qualitative sketch of the phenomena to be explained, then a description of the postulated geometrical model, which combined epicycles and eccenters, and finally a detailed deduction of the parameters of the model from certain observations that he had personally made or of which he had records—records often based on Babylonian observations as far back as the eighth century BCE. He generally concluded by showing that his model with the calculated parameters in fact predicted a new planetary position, which was verified by observation. Ptolemy is thus the first mathematical scientist for whom there is documented evidence of the use of mathematical models in actually "doing" science. He began with a model and then used observations to improve it to the point that it predicted observed phenomena to within the limits of his observational accuracy.

Ptolemy was proud of his accomplishments in "saving the appearances"—that is, in showing that for all seven of the wandering heavenly bodies "their apparent anomalies can be represented by uniform circular motions, since these are proper to the nature of divine beings. ... Then it is right that we should think success in such a purpose a great thing and truly the proper end of the mathematical part of theoretical philosophy. But, on many grounds, we must think that it is difficult and that there is good reason why no one before us has yet succeeded in it." Ptolemy, however, overcame the difficulties and gave to posterity a masterful mathematical work that did predict the celestial phenomena, a work not superseded for 1400 years.

Exercises

1. Prove the two lemmas (see pages 68 and 69) that Archimedes used to derive his algorithms for calculating π.
2. Use a calculator (or program a computer) to calculate π by iterating the algorithm of Archimedes given by lemma 1. How many iterations are necessary to get five-decimal-place accuracy?
3. Translate lemma 2 into a recursive algorithm for calculating π. Iterate this algorithm to calculate π to five-decimal-place accuracy. How many iterations are necessary?
4. Show that if a is the nearest positive integer to the square root of $a^2 \pm b$, then

$$a \pm \frac{b}{2a} > \sqrt{a^2 \pm b} > a \pm \frac{b}{2a \pm 1}.$$

Beginning with $2^2 - 1 = 3$, and therefore, as a first approximation, $2 - \frac{1}{4} > \sqrt{3}$, show first that $\sqrt{3} > \frac{5}{3}$, second that $\sqrt{3} < \frac{26}{15}$, third that $\sqrt{3} < \frac{1351}{780}$, and fourth that $\sqrt{3} > \frac{265}{153}$. Note that these last two approximations are the values Archimedes used in his *Measurement of a Circle*.

5. Given the parabolic segment with MO parallel to the axis of the segment and MC tangent to the parabola (Fig. 3.2), show analytically that $MO : OP = CA : AO$.
6. The proof of proposition 2 of Archimedes' *The Method* is outlined here:

PROPOSITION 2. *Any sphere is (in respect of solid content) four times the cone with base equal to a great circle of the sphere and height equal to its radius.*

Let $ABCD$ be a great circle of a sphere with perpendicular diameters AC, BD. Describe a cone with vertex A and axis AC, and extend its surface to the circle with diameter EF. On the latter circle erect a cylinder with height and axis AC. Finally, extend AC to H such that $HA = AC$. Certain pieces of the figures described are to be balanced using CH as the lever (Fig. 3.31).

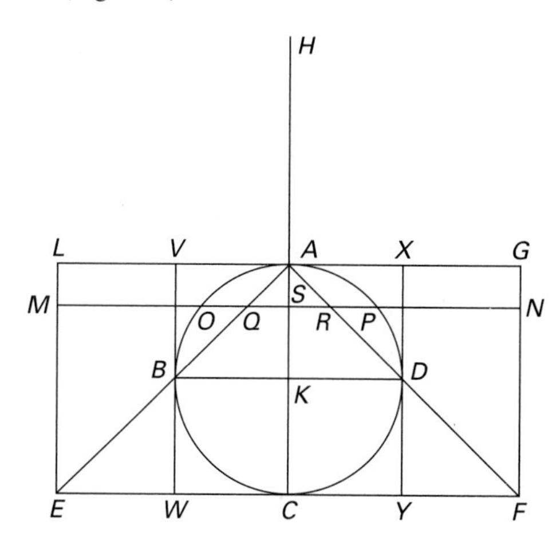

FIGURE 3.31 Archimedes' *Method*, Proposition 2

Let MN be an arbitrary line in the plane of the circle $ABCD$ and parallel to BD, with its various intersections marked as in the figure. Through MN, draw a plane at right angles to AC. This plane will cut the cylinder in a circle with diameter MN, the sphere in a circle with diameter OP, and the cone in a circle with diameter QR.

(a) Show that $MS \cdot SQ = OS^2 + SQ^2$.

(b) Show that $HA : AS = MS : SQ$. Then, multiplying both parts of the last ratio by MS, show that $HA : AS = MS^2 : (OS^2 + SQ^2) = MN^2 : (OP^2 + QR^2)$. Show that this last ratio equals the ratio of the circle with diameter MN to the sum of the circle with diameter OP and the circle with diameter QR.

(c) Conclude that the circle in the cylinder, placed where it is, is in equilibrium about A with the circle in the sphere together with the circle in the cone, if both the latter circles are placed with their centers of gravity at H.

(d) Archimedes concluded from the above that the cylinder, placed where it is, is in equilibrium about A with the sphere and the cone together, when both are placed with their center of gravity at H. Show therefore that

$$HA : AK = (\text{cylinder}) : (\text{sphere} + \text{cone AEF}).$$

(e) From the fact that the cylinder is three times the cone AEF and the cone AEF is eight times the cone ABD, conclude that the sphere is equal to four times the cone ABD.

7. Use the general technique of *The Method* to demonstrate proposition 4, that the volume of a segment of paraboloid of revolution cut off by a plane at right angles to the axis is $\frac{3}{2}$ the volume of a cone with the same base and the same height. Begin with triangle ABC inscribed in the segment $BOAPC$ of a parabola with both inscribed in rectangle $EFCB$ (Fig. 3.32). Rotate the entire figure around the axis AD to get a cone inside of a paraboloid of revolution, which is in turn inside of a cylinder. Extend DA to H so that $AD = AH$; draw MN parallel to BC; and imagine the plane through MN producing sections of the cone, the paraboloid, and the cylinder. Finally, imagine that HD is a lever with midpoint A and use Archimedes' balancing techniques to show that the circle in the cylinder of radius MS, placed where it is, balances the circle in the paraboloid of radius OS, if the latter is placed at H. Use the result that the volume of a cone is $\frac{1}{3}$ that of the inscribing cylinder to conclude the proof of the theorem.

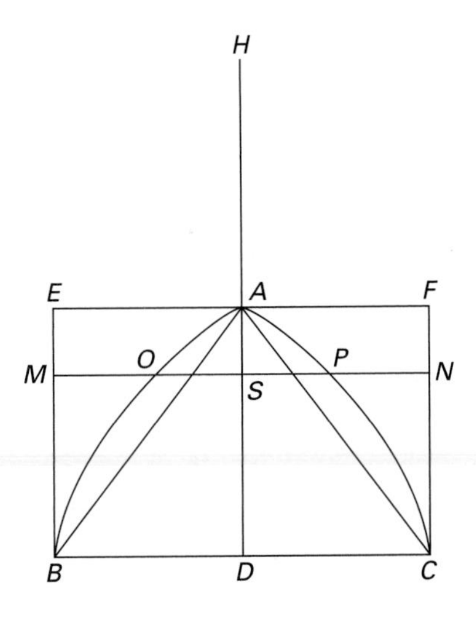

FIGURE 3.32 Archimedes' *Method*, Proposition 4

8. Use calculus to prove Archimedes' result that the area of a parabolic segment is $\frac{4}{3}$ the area of the inscribed triangle.

9. Show analytically that the vertex of a parabolic segment (see the definition on p. 70) is that point on the curve whose perpendicular distance to the base of the segment is greatest.

10. Use calculus to prove Archimedes' result that a cylinder whose base is a great circle in the sphere and whose height is equal to the diameter has volume $\frac{3}{2}$ of that of the sphere and also has surface area $\frac{3}{2}$ of the surface area of the sphere.

11. Consider proposition 1 of *On the Sphere and Cylinder II*: Given a cylinder, to find a sphere equal to the cylinder. Provide the analysis of this problem. That is, assume that V is the given cylinder and that a new cylinder P has been constructed with volume $\frac{3}{2}V$. Assume further that another cylinder Q has been constructed equal to P but with height equal to its diameter. The sphere whose diameter equals the height of Q would then solve the problem, because the volume of the sphere is $\frac{2}{3}$ that of the cylinder. So, given the cylinder P of given diameter and height, determine how to construct a cylinder Q of the same volume but with height and diameter equal.

12. Determine the equations of the parabola and the hyperbola whose intersection provides the solution x to the cubic equation $3ax^2 - x^3 = 4a^2b$ needed by Archimedes to solve proposition 4 of *On the Sphere and Cylinder II*. Sketch the two curves on the same pair of axes.

13. Show that in the curve $y^2 = px$, the value p represents the length of the **latus rectum**, the straight line through the focus perpendicular to the axis.

14. Rewrite the equations $y^2 = x[p + (p/2a)x]$ and $y^2 = x[p - (p/2a)x]$ for the hyperbola and the ellipse, respectively, in the current standard forms for those

equations. What point is the center of the curve? Show in the case of the ellipse, where $2b$ is the length of the minor axis, that $b^2 = pa/2$.

15. Use calculus to prove *Conics* I-33.

16. Apollonius states and proves many of the properties of conics in a more general form than we have considered. Instead of restricting himself to the principal diameters of the conic, such as the major and minor axes of the ellipse, he deals with any pair of conjugate diameters. For an ellipse, given a tangent at any point, the parallel to this tangent passing through the center of the ellipse is **conjugate** to the straight line passing through the point of tangency and the center (Fig. 3.33). (A similar definition can be given for a hyperbola, but this problem is restricted to the case of the ellipse.)

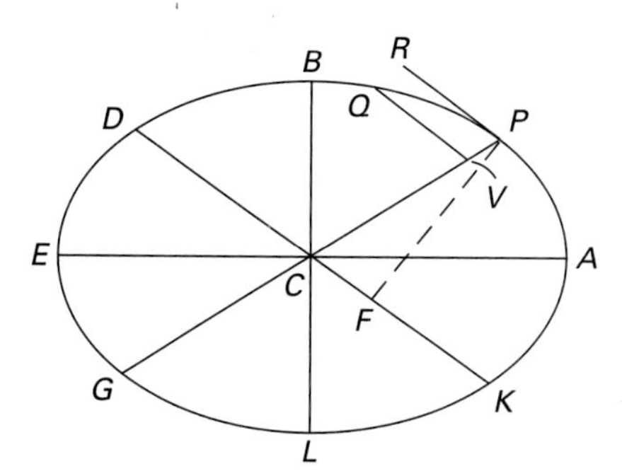

FIGURE 3.33 Conjugate diameters in an ellipse

(a) Show that if DK is conjugate to PG, then PG is also conjugate to DK.

(b) Assume the equation of the ellipse is given by $b^2x^2 + a^2y^2 = a^2b^2$ in rectangular coordinates x, y. Denote angle PCA by θ and angle DCA by α. Given that diameter DK is parallel to the tangent to the ellipse at $P = (x_0, y_0)$, show that $\tan\theta = y_0/x_0$ and that $\tan\alpha = -b^2x_0/a^2y_0$.

(c) Transform the equation of the ellipse to new oblique coordinates x', y' based on the conjugate diameters PG and DK. Show that the transformation is given by

$$x = x'\cos\theta + y'\cos\alpha$$
$$y = x'\sin\theta + y'\sin\alpha$$

and that the new equation for the ellipse is $Ax'^2 + Cy'^2 = a^2b^2$, where $A = b^2\cos^2\theta + a^2\sin^2\theta$ and $C = b^2\cos^2\alpha + a^2\sin^2\alpha$.

(d) Let $Q = (x', y')$ be a point on the ellipse with coordinates relative to the conjugate diameters PG and DK, and let QV be drawn to diameter PG with QV parallel to DK. Then set $QV = y'$, $PV = x'_1$, $GV = x'_2$, $PC = a'$, and $DC = b'$ and show that the equation of the ellipse can be written in the form

$$y'^2 = \frac{b'^2}{a'^2}x'_1x'_2,$$

thus generalizing the version of Apollonius's proposition I-21 given in this chapter.

(e) Show that the rectangles constructed on any pair of conjugate diameters are equal (proposition VII-31). In other words, if PF is drawn perpendicular to DK, show that $PF \times CD = AC \times BC = ab$.

17. Use *Conics* II-8 to show that the two line segments of a tangent to a hyperbola between the point of tangency and the asymptotes are equal. Then show, without calculus, that the slope of the tangent line to the curve $y = 1/x$ at $(x_0, 1/x_0)$ equals $-1/x_0^2$.

18. Use calculus to give a proof that the line from the focus to a point on a parabola makes an angle with the tangent at that point equal to the angle made by a line parallel to the axis.

19. Show analytically that the solution to the three-line locus problem is a conic section in the case where two of the lines are parallel and the third is perpendicular to the other two. Characterize the curve in reference to the distance between the two parallel lines and the given ratio.

20. Show analytically that the solution to the general three-line locus problem is always a conic section.

21. Calculate crd(30°), crd(15°), and crd($7\frac{1}{2}$°) using Ptolemy's half-angle formula, beginning with the fact that crd(60°) $= R = 60$.

22. Prove the sum formula

$$120\ \text{crd}[180 - (\alpha + \beta)]$$
$$= \text{crd}(180 - \alpha)\ \text{crd}(180 - \beta) - \text{crd}\ \alpha\ \text{crd}\ \beta,$$

using Ptolemy's theorem on quadrilaterals inscribed in a circle.

23. Use Ptolemy's difference formula to calculate crd(12°) and then apply the half-angle formula to calculate crd(6°), crd(3°), crd($1\frac{1}{2}$°), and crd($\frac{3}{4}$°).

24. Calculate the declination and right ascension of the sun when it is at longitude 90° (summer solstice) and longitude 45°. By symmetry, find the declination at longitudes 270° and 315°.

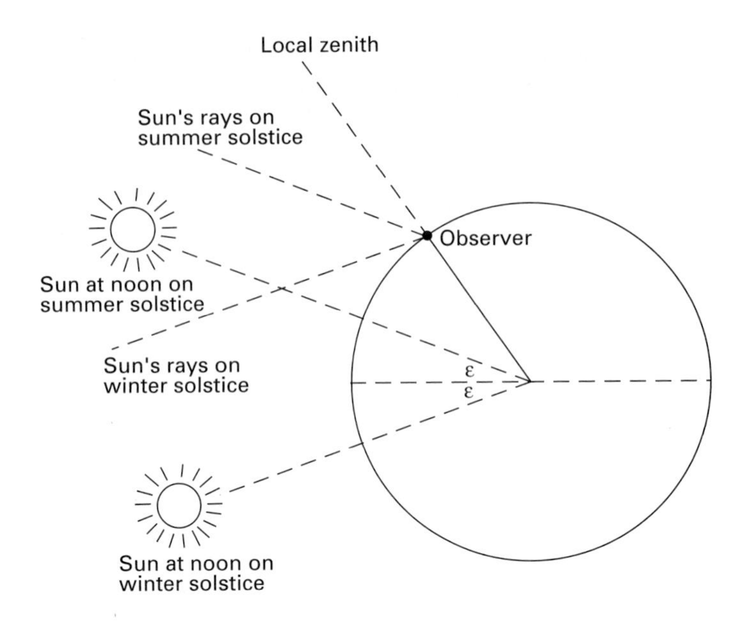

FIGURE 3.34 Calculation of the inclination of the ecliptic

25. Calculate the rising times $\rho(\lambda, \phi)$ for $\phi = 45°$ and $\lambda = 60°$ and $90°$.

26. Write a program for your graphing calculator to calculate the length of daylight $L(\lambda, \phi)$ for any value of longitude λ and geographic latitude ϕ.

27. Explain why the angle ϵ between the equator and the ecliptic can be determined by taking half of the angular distance between the noon altitudes of the sun at the summer and winter solstices (Fig. 3.34).

28. Calculate the length of daylight on a day when $\lambda = 60°$ at latitude 36°. Calculate the local time of sunrise and sunset.

29. Suppose that the maximum length of day at a particular location is known to be 15 hours. Calculate the latitude of that location.

30. The formula $\sin\sigma = \tan\delta\tan\phi$ makes sense only if the right-hand side is less than or equal to 1. Since the maximum value of δ is $23\frac{1}{2}°$, show that the right-hand side will be greater than 1 whenever $\phi > 66\frac{1}{2}°$. Interpret the formula in this case in terms of the length of daylight.

31. Calculate the angular distance of the sun from the zenith at noon at latitude 45° when $\lambda = 45°$ and $90°$.

32. On approximately what dates is the sun directly overhead at noon at a place whose geographical latitude is 20°?

33. Calculate the sun's maximal northerly sunrise point for latitude 36°.

34. On approximately what date is the "midnight sun" first observed at latitude 75°?

35. Devise a lesson for a precalculus course that will demonstrate the formula for the sum of a geometric series as in Archimedes' work.

36. Discuss whether Archimedes' procedure for determining the area of a parabolic segment and/or the area bounded by one turn of the spiral can be adapted to introduce students in a precalculus class (or even a calculus class) to the calculation of areas bounded by curves.

37. Prepare lessons for a precalculus course deriving the equations of the conic sections from their definitions as sections of a cone, as in the work of Apollonius. How does this method compare to the use of the standard modern textbook definitions?

38. Why did the Greeks continue to search for solutions to the problems of squaring the circle, trisecting an angle, and doubling the cube, when solutions had already been found?

39. Outline a trigonometry course following Ptolemy's order of presentation; that is, derive the major formulas as tools for producing a sine table. Discuss the advantages and disadvantages of this approach compared to the approach used in most modern textbooks.

40. What observations would have convinced the Greeks that the radius of the celestial sphere is so large that the earth can in effect be considered a point with respect to that sphere?

41. List evidence that convinces you that the earth (a) rotates on its axis once a day and (b) revolves around the sun once a year. Would this evidence have convinced the Greeks? How would you refute the reasons the Greeks gave for the earth's immovability?

42. Read the article "Why December 21 Is the Longest Day of the Year," in *Mathematics Magazine 63* (1990), 307–311, to learn why the extrema of sunrise and sunset times do not occur on the solstices, even though those days are in fact the longest and shortest of the year. Prepare a brief report to explain this surprising phenomenon.

References

Many of the books on Greek mathematics referred to in chapter 2 have sections on Archimedes, Apollonius, and Ptolemy. In particular, Thomas Heath's *A History of Greek Mathematics*, B. L. Van der Waerden's *Science Awakening*, and Wilbur Knorr's *The Ancient Tradition of Geometric Problems* are good sources of further reading on the material in this chapter. Selections from the works of Archimedes, Apollonius, and Ptolemy, as well as of others discussed in this chapter, can be found in Ivor Thomas, *Selections Illustrating the History of Greek Mathematics*.

A complete translation of the extant works of Archimedes, slightly edited for modern readers, is found in Thomas Heath, *The Works of Archimedes* (New York: Dover, 1953). The most detailed discussion of Archimedes' works, however, is in E. J. Dijksterhuis, *Archimedes* (Princeton: Princeton University Press, 1987). This edition of Dijksterhuis's work has a bibliographic essay by Wilbur Knorr, which gives the latest information on research on the work of Archimedes. A discussion of the use of indivisibles in Greek mathematics in Archimedes' work and elsewhere is found in Wilbur Knorr, "The Method of Indivisibles in Ancient Geometry," in Ronald Calinger, ed., *Vita Mathematica* (Washington: Mathematical Association of America, 1996), pp. 67–86. A valuable discussion of *The Method* as well as of Ptolemy's trigonometry can be found in Asger Aaboe, *Episodes from the Early History of Mathematics* (Washington: MAA, 1964). A detailed analysis of the manuscript of *The Method* can be found in *The Archimedes Palimpsest*, published by Christie's Auction House in New York City in 1998, prior to the auction.

An English translation of the first three books of Apollonius's *Conics*, edited by Dana Densmore, has recently been published by Green Lion Press (Santa Fe, 2000). Thomas Heath's older work, *Apollonius of Perga* (Cambridge: W. Heffer and Sons, 1961) contains all seven extant books of the *Conics*. But since Heath modifies the order of the books and often combines several theorems, this cannot be considered a translation. A new analysis of Apollonius's work is Michael Fried and Sabetai Unguru, *Apollonius of Perga's Conics: Text, Context, Subtext* (Leiden: Brill Academic Publishers, 2001). This work contains a translation of Book 4 of the *Conics*. Diocles' work is discussed by Gerald Toomer in *Diocles On Burning Mirrors* (New York: Springer, 1976). Toomer's book provides a complete translation of Diocles' treatise as well as a discussion of its importance.

The best available English translation of Ptolemy's *Almagest* is by Gerald Toomer: *Ptolemy's Almagest* (New York: Springer, 1984). As a companion to the translation, consult Olaf Pedersen, *A Survey of the Almagest* (Odense: University Press, 1974), which provides background and commentary on all of Ptolemy's mathematical and astronomical material.

The standard reference on the applications of mathematics to astronomy from Babylonian times to the sixth century CE is Otto Neugebauer, *A History of Ancient Mathematical Astronomy* (New York: Springer, 1975). This work provides a detailed study of the mathematical techniques used by Ptolemy and other astronomers as they worked out their versions of the system of the heavens. A more popular work is James Evans, *The History and Practice of Ancient Astronomy* (New York: Oxford University Press, 1998). Thomas Kuhn's *The Copernican Revolution* (Cambridge: Harvard University Press, 1957) provides excellent background reading on the nature of astronomy in Greek times. A more recent work, especially designed for undergraduates, is Michael J. Crowe, *Theories of the World from Antiquity to the Copernican Revolution* (New York: Dover Publications, 1990). Len Berggren provides more mathematical details on Ptolemy's applications of trigonometry to astronomical problems in "Mathematical Methods in Ancient Science: Astronomy," in I. Grattan-Guinness, ed., *History in Mathematics Education* (Paris: Belin, 1987).

CHAPTER FOUR

Greek Mathematics from Diophantus to Hypatia

This tomb holds Diophantus . . . [and] tells scientifically the measure of his life. God granted him to be a boy for the sixth part of his life, and adding a twelfth part to this, he clothed his cheeks with down. He lit him the light of wedlock after a seventh part, and five years after his marriage He granted him a son. Alas! late-born wretched child; after attaining the measure of half his father's life, chill Fate took him. After consoling his grief by this science of numbers for four years, he ended his life.

—Epigram 126 of Book XIV of *Greek Anthology* (c. 500 CE)

Although the Egyptian-Greek empire of the Ptolemies, under whose rule the Museum and Library flourished, collapsed under the onslaught of Roman might in 31 BCE, the intellectual-scientific tradition of Alexandria continued for many centuries. In chapter 3, we discussed the work of Claudius Ptolemy, who flourished under Roman rule in Egypt. In this chapter, we will deal briefly with three other mathematicians of Alexandria, all of whom had influence stretching into the Renaissance: Diophantus, Pappus, and Hypatia.

Diophantus, who lived in the mid–third century, is best known for his *Arithmetica*, a work in what we call algebra, consisting mostly of an organized collection of problems translatable into what are today called **indeterminate equations**, all to be solved in rational numbers. The geometer Pappus lived in the early fourth century. He is best known not for his original work, but for his commentaries on various aspects of Greek mathematics, and in particular for his discussion of the Greek method of geometric analysis. The chapter will conclude with a brief discussion of the work of Hypatia, the first woman mathematician of whom any details are known. It was her death at the hands of an enraged mob that effectively marked the end of the Greek mathematical tradition in Alexandria.

4.1 Diophantus and the *Arithmetica*

Little is known about Diophantus's life, other than what is found in the epigram at the beginning of the chapter, except that he lived in Alexandria. It is through his major work, the *Arithmetica*, that his influence has reached modern times. Diophantus tells us in his introduction that the *Arithmetica* is divided into thirteen books. Only six have survived in Greek. Four others were recently discovered in an Arabic version; from internal references, it appears that these form books 4 through 7 of the complete work, while the final three Greek books come later. We will refer to the Greek books as I through VI and to the Arabic ones as A, B, C, and D. The style of the Arabic books is somewhat different from that of the Greek ones in that each step in the solution of a problem is explained more fully. It is quite possible, therefore, that the Arabic books are translations not of Diophantus's original work, but of a commentary on *Arithmetica*, written by Hypatia around 400.

Before dealing with the problems of the *Arithmetica*, it is worthwhile to discuss Diophantus's major advance in the solution of equations—his introduction of symbolism. The Egyptians and Babylonians wrote out equations and solutions in words. Diophantus, on the other hand, introduced symbolic abbreviations for the various terms involved in equations. And in a clear break with traditional Greek usage, he dealt with powers higher than the third.

Diophantus's symbols are abbreviations. For example, the *square* [of the unknown quantity] is called *dynamis* and its symbol is Δ^{Υ}; the cube is called *kubos* and has for its symbol K^{Υ}; the square multiplied by itself is called *dynamo-dynamis* and its symbol is $\Delta^{\Upsilon}\Delta$; the square multiplied by the cube formed from the same root is called *dynamo-kubos*, and its [symbol is] ΔK^{Υ}; the cube multiplied by itself is called *kubo-kubos*, and its symbol is $K^{\Upsilon}K$.

The number which has none of these characteristics, but merely has in it an undetermined multitude of units, is called *arithmos*; its symbol is ς. This symbol is a contraction of the first two letters of $\alpha\rho\iota\theta\mu o\varsigma$ (*arithmos*, or "number"). Finally, there is a symbol denoting the unit, namely, $\overset{\circ}{M}$. This symbol stands for $\mu o\nu\alpha\varsigma$ (*monas*, or "unit"). These symbols were combined in various ways with the Greek symbols for numbers, namely, the letters of the Greek alphabet. Thus, the manuscripts contain expressions such as $\Delta^{\Upsilon}\gamma\varsigma\iota\beta\overset{\circ}{M}\theta$, which stands for 3 ($\gamma$) squares, 12 ($\iota\beta$) numbers, and 9 ($\theta$) units, or, as we write it, $3x^2 + 12x + 9$. Diophantus also used the above symbols with the mark χ to designate reciprocals. For example, $\Delta^{\Upsilon\chi}$ represented $1/x^2$. In addition, the symbol ⋔, perhaps coming from an abbreviation for $\lambda\epsilon\iota\psi\iota\varsigma$ (*lepsis*, or "wanting" or "negation"), is used for minus, as in $K^{\Upsilon}\alpha\varsigma\gamma$⋔$\Delta^{\Upsilon}\gamma\overset{\circ}{M}\alpha$ for $x^3 - 3x^2 + 3x - 1$. (Negative terms are always collected, so a single ⋔ refers to all terms following it.) In the discussion of Diophantus's problems, however, we will use modern notation.

Although Diophantus did not deal with negative numbers as entities, he was certainly aware of the rules for multiplying with the minus: "A minus multiplied by a minus makes a plus, a minus multiplied by a plus makes a minus." These rules, then, are those necessary for multiplying algebraic expressions involving subtractions. He did not explicitly state the rules for adding and subtracting with positive and negative terms, but simply assumed they are known. And then, near the conclusion of his introduction, he stated the basic rules for solving equations:

> If a problem leads to an equation in which certain terms are equal to terms of the same species but with different coefficients, it will be necessary to subtract like from like on both sides, until one term is found equal to one term. If by chance there are on either side or on both sides any negative terms, it will be necessary to add the negative terms on both sides, until the terms on both sides are positive, and then again to subtract like from like until one term only is left on each side. This should be the object aimed at in framing the hypotheses of propositions, that is to say, to reduce the equations, if possible, until one term is left equal to one term; but I will show you later how, in the case also where two terms are left equal to one term, such a problem is solved.

In other words, Diophantus's general method of solving equations is designed to lead to an equation of the form $ax^n = bx^m$, where, in the first three books at least, m and n are no greater than 2. On the other hand, he does know how to solve quadratic equations of, for example, the form $ax^2 + c = bx$.

4.1.1 Linear and Quadratic Equations

Most of Diophantus's problems are indeterminate; that is, they can be written as a set of k equations in more than k unknowns. Often there are infinitely many solutions. For these problems, Diophantus generally gave only one solution explicitly, but his method can easily be extended to give other solutions. For determinate problems, once certain quantities are made explicit, there is only one solution. Examples of both of these types will be described in what follows.

PROBLEM I-1. *To divide a given number into two having a given difference.*

Diophantus presented the solution for the case where the given number is 100 and the given difference is 40. If x is the smaller of the two numbers of the solution, then $2x + 40 = 100$, so $x = 30$, and the required numbers are 30 and 70. This problem is determinate, once the "given" numbers are specified, but Diophantus's method works for any pair. If a is the given number and $b < a$ the given difference, then the equation will be $2x + b = a$, and the required numbers will be $\frac{1}{2}(a - b)$ and $\frac{1}{2}(a + b)$.

In this problem, as in most of the problems in Book I, the given values are picked to ensure that the answers are integers. In the other books, however, the only general condition on solutions is that they be positive rational numbers. Evidently Diophantus began with integers merely to make these introductory problems easier. In what follows, then, the word "number" should always be interpreted as "rational number."

PROBLEM I-28. *To find two numbers such that their sum and the sum of their squares are given numbers.*

It is a necessary condition that double the sum of the squares exceeds the square of the sum by a square number. In the problem presented, the given sum is 20 and the sum of the squares is 208.

This problem is of the general form $x + y = b$, $x^2 + y^2 = c$, a type solved by the Babylonians. Three other Babylonian types appear in problems I-27, I-29, and I-30: $x + y = b$, $xy = c$; $x + y = b$, $x^2 - y^2 = c$; and $x - y = b$, $xy = c$, respectively. As we have seen, the methods of solutions of these problems are also found in Book II of Euclid's *Elements*. Although he presented his solution to the present problem strictly algebraically, Diophantus used the same basic procedure as the Babylonians. That is, he took as his "unknown" z half the difference between the two desired numbers. Therefore, since 10 is half the sum of the two numbers, the two numbers themselves are $x = 10 + z$ and $y = 10 - z$. The Babylonian result and the Euclidean one tell us that the sum of the squares, here 208, is twice the sum of the squares on half the sum and half the difference. In this case, then, we get $200 + 2z^2 = 208$. It follows that $z = 2$, and the required two numbers are 12 and 8. Diophantus's method, applicable to any system of the given form, can be translated into the modern formula

$$x = \frac{b}{2} + \sqrt{\frac{c}{2} - \left(\frac{b}{2}\right)^2} \qquad y = \frac{b}{2} - \sqrt{\frac{c}{2} - \left(\frac{b}{2}\right)^2},$$

from which his condition ensuring a rational solution is easily derived. Interestingly, the answers to problems I-27, I-29, and I-30 are also 12 and 8—a reminder of the common Babylonian practice of having the same answers to a series of related problems.

Did Diophantus have access to the Babylonian materials? Or did he learn his methods from a careful study of Euclid's *Elements*? These questions cannot be answered. It is, however, apparent that there is no geometric methodology in Diophantus's procedures. Perhaps by this time the Babylonian algebraic methods, stripped of their geometric origins, were known in the Greek world.

PROBLEM II-8. *To divide a given square number into two squares.*

> Let it be required to divide 16 into two squares. And let the first square $= x^2$; then the other will be $16 - x^2$; it shall be required therefore to make $16 - x^2 =$ a square. I take a square of the form $(ax - 4)^2$, a being any integer and 4 the root of 16; for example, let the side be $2x - 4$, and the square itself $4x^2 + 16 - 16x$. Then $4x^2 + 16 - 16x = 16 - x^2$. Add to both sides the negative terms and take like from like. Then $5x^2 = 16x$, and $x = \frac{16}{5}$. One number will therefore be $\frac{256}{25}$, the other $\frac{144}{25}$, and their sum is $\frac{400}{25}$, or 216, and each is a square.

This is an example of an indeterminate problem. It translates into one equation in two unknowns: $x^2 + y^2 = 16$. This problem also demonstrates one of Diophantus's most common methods. In many problems from Book II onward, Diophantus required a solution, expressed in the form of a quadratic polynomial, which must be a square. To ensure a rational solution, he chose his square in the form $(ax \pm b)^2$, with a and b selected so that either the quadratic term or the constant term is eliminated from the equation. In this case, where the quadratic polynomial is $16 - x^2$, he used $b = 4$ and the negative sign, so that the constant term will be eliminated and the resulting solution will be positive. The rest of the solution is then obvious. The method can be used to generate as many solutions as desired to $x^2 + y^2 = 16$, or, in general, to $x^2 + y^2 = b^2$. Take any value for a and set $y = ax - b$. Then $b^2 - x^2 = a^2x^2 - 2abx + b^2$ or $2abx = (a^2 + 1)x^2$, so $x = 2ab/(a^2 + 1)$.

As another example where Diophantus needed a square, consider

PROBLEM II-19. *To find three squares such that the difference between the greatest and the middle has a given ratio to the difference between the middle and the least.*

Diophantus assumed that the given ratio is 3 : 1. If the least square is x^2, then he took $(x + 1)^2 = x^2 + 2x + 1$ as the middle square. Because the difference between these two squares is $2x + 1$, the largest square must be $x^2 + 2x + 1 + 3(2x + 1) = x^2 + 8x + 4$. To make that quantity a square, Diophantus set it equal to $(x + 3)^2$, choosing the coefficient of

x so that the x^2 terms cancel. Then $8x + 4 = 6x + 9$, so $x = 2\frac{1}{2}$ and the desired squares are $6\frac{1}{4}$, $12\frac{1}{4}$, and $30\frac{1}{4}$. Notice, however, that given his initial choice of $(x + 1)^2$ as the middle square, 3 is the only integer b Diophantus could use in $(x + b)^2$ that would give him a solution. Of course, with other values of the initial ratio, there would be more possibilities, as there would with a different choice for the second square. In any case, in this problem, as in all of Diophantus's problems, only one solution is required.

4.1.2 Higher-Degree Equations

Since the problems in Book A involve cubes and even higher powers, Diophantus began with a new introduction in which he described the rules for multiplying and dividing such powers. For example, since x^2, x^3, x^4, x^5, and x^6 are represented by Δ^Υ, K^Υ, $\Delta^\Upsilon\Delta$, ΔK^Υ, and $K^\Upsilon K$, respectively, Diophantus wrote, for example, that ΔK^Υ multiplied by ς equals K^Υ multiplied by itself, which equals Δ^Υ multiplied by $\Delta^\Upsilon\Delta$, and all equal $K^\Upsilon K$. Similarly, if $K^\Upsilon K$ is divided by $\Delta^\Upsilon\Delta$, the result is Δ^Υ. Thus, although Diophantus's results are equivalent to modern laws of exponents, his notation did not allow him to express them in the modern way of adding the exponents when multiplying powers.

Diophantus did, however, explain that as before, his equations will end up with a term in one power equaling a term in another, that is, $ax^n = bx^m$ $(n < m)$, where m may be any number up to 6. To solve, one must use the rules to divide both sides by the lesser power and end up with one "species" equal to a number—that is, in modern notation, $a = bx^{m-n}$. The latter equation is easily solved.

As an example of Diophantus's use of higher powers of x, consider

PROBLEM A-25. *To find two numbers, one a square and the other a cube, such that the sum of their squares is a square.*

The goal is to find x, y, and z such that $(x^2)^2 + (y^3)^2 = z^2$. Thus, this is an indeterminate problem with one equation in three unknowns. Diophantus set x equal to $2y$ (the 2 is arbitrary) and performs the exponentiation to conclude that $16y^4 + y^6$ must be a square, which he takes to be the square of ky^2. So $16y^4 + y^6 = k^2y^4$, $y^6 = (k^2 - 16)y^4$, and $y^2 = k^2 - 16$. It follows that $k^2 - 16$ must be a square. Diophantus chose the easiest value, namely, $k^2 = 25$, so that $y = 3$. It then follows that the desired numbers are $y^3 = 27$ and $(2y)^2 = 36$. This solution is easily generalized. Take $x = ay$ for any positive a. Then k and y must be found so that $k^2 - a^4 = y^2$ or so that $k^2 - y^2 = a^4$. Diophantus had, however, already demonstrated in problem II-10 that one can always find two squares whose difference is given.

Problem B-7 shows that Diophantus knew the expansion of $(x + y)^3$. As he put it, "whenever we wish to form a cube from some side made up of the sum of, say, two different terms—so that a multitude of terms does not make us commit a mistake—we have to take the cubes of the two different terms, and add to them three times the results of the multiplication of the square of each term by the other."

PROBLEM B-7. *To find two numbers such that their sum and the sum of their cubes are equal to two given numbers.*

The problem is to solve $x + y = b$, $x^3 + y^3 = c$. This system of two equations in two unknowns is determinate. It is a generalization of the Babylonian-style problem I-28,

$x + y = b$, $x^2 + y^2 = c$. Diophantus's method of solution for problem B-7 generalized his method for problem I-28. Letting $b = 20$ and $c = 2240$, he began as before by setting $x = 10 + z$ and $y = 10 - z$. The second equation then becomes $(10 + z)^3 + (10 - z)^3 = 2240$, or, using the expansion already discussed, $2000 + 60z^2 = 2240$, or $60z^2 = 240$, $z^2 = 4$, and $z = 2$. Diophantus gave, of course, a condition for a rational solution, namely, that $(4c - b^3)/3b$ is a square (equivalent to the more natural condition that $[c - 2(b/2)^3]/3b$ is a square). It is interesting that the answers here are the same as in problem I-28, namely, 12 and 8.

4.1.3 The Method of False Position

In Book IV, Diophantus began to use a new technique, one reminiscent of the Egyptian false position. Among many problems he solved using this technique, the following one will be important in our later discussion of elliptic curves.

PROBLEM IV-24. *To divide a given number into two parts such that their product is a cube minus its side.*

If a is the given number, the problem is to find x and y such that $y(a - y) = x^3 - x$. This is an indeterminate problem. As usual, Diophantus began by choosing a particular value for a, here $a = 6$. So $6y - y^2$ must equal a cube minus its side. He chose the side x to be of the form $x = my - 1$. The question is what value should he choose for m. Diophantus picked $m = 2$ and calculated: $6y - y^2 = (2y - 1)^3 - (2y - 1)$, or $6y - y^2 = 8y^3 - 12y^2 + 4y$. We note immediately that the 1 in $x = my - 1$ was chosen so that there would be no constant term in this equation. Nevertheless, this is still an equation with three separate species, not the type Diophantus could solve most easily. So he noted that if the coefficients of y on each side were the same, then the solution would be simple. The 6 on the left is the "given number," so that cannot be changed. But the 4 on the right comes from the calculation $3 \cdot 2 - 2$, which in turn depends on the choice $m = 2$ in $x = my - 1$. Therefore, Diophantus needed to find m so that $3 \cdot m - m = 6$. Therefore, $m = 3$. He could then begin again: $x = 3y - 1$ and $6y - y^2 = (3y - 1)^3 - (3y - 1)$, or $6y - y^2 = 27y^3 - 27y^2 + 6y$. Therefore, $27y^3 = 26y^2$ and $y = \frac{26}{27}$. The two parts of 6, therefore, are $\frac{26}{27}$ and $\frac{136}{27}$, while the product of those two numbers is $(\frac{17}{9})^3 - \frac{17}{9}$. The general solution to this problem, for arbitrary a, is given by $y = (6a^2 - 8)/a^3$, $x = [(3a^2 - 4)/a^2] - 1$.

In problem IV-31, Diophantus found again that his original assumption did not work. But here he had no choice but to get a mixed quadratic equation. His problem is that, with his original assumption, the quadratic equation fails to have a rational solution.

PROBLEM IV-31. *To divide unity into two parts so that, if given numbers are added to them respectively, the product of the two sums is a square.*

Diophantus set the given numbers at 3, 5 and the parts of unity as x, $1 - x$. Therefore, $(x + 3)(6 - x) = 18 + 3x - x^2$ must be a square. Since neither of his usual techniques for determining a square will work here (neither 18 nor -1 is a square), he tried $(2x)^2 = 4x^2$ as the desired square. But the resulting quadratic equation, $18 + 3x = 5x^2$ "does not give a rational result." He needs to replace $4x^2$ by a square of the form $(mx)^2$ that does give a rational solution. Thus, since $5 = 2^2 + 1$, he noted that the quadratic equation will be solvable if $(m^2 + 1) \cdot 18 + (\frac{3}{2})^2$ is a square. This implies that $72m^2 + 81$ is a square, say, $(8m + 9)^2$. (Here, his usual technique succeeds.) Then $m = 18$, and, returning to the beginning, he sets $18 + 3x - x^2 = 324x^2$. He then simply presented the solution: $x = \frac{78}{325} = \frac{6}{25}$, and the desired numbers are $\frac{6}{25}, \frac{19}{25}$.

Although Diophantus did not give details in problem IV-31 on the solution of the quadratic equation, he did give them in problem IV-39. His words in that problem are easily translated into the formula for solving the equation $c + bx = ax^2$:

$$x = \frac{(b/2) + \sqrt{ac + (b/2)^2}}{a}.$$

This formula is the same as the Babylonian formula, assuming that one first multiplies the equation through by a and solves for ax. That technique is in fact used in Babylonian problems. Diophantus was sufficiently familiar with this formula and its variants that he used it in various later problems to solve not only quadratic equations but also quadratic inequalities.

Diophantus's work, the only example of a genuinely algebraic work surviving from ancient Greece, was highly influential. Not only was it commented on in late antiquity, but it was also studied by Islamic authors. Many of its problems were taken over by Rafael Bombelli and published in his *Algebra* of 1572; Pierre Fermat carefully studied Bachet's Greek edition, published in 1621, which led him to numerous general results in number theory, at which Diophantus himself only hinted. Perhaps more important, however, is the fact that the *Arithmetica*, as a work of algebra, was in effect a treatise on the analysis of problems. The solution of each problem began with the assumption that the answer x, for example, had been found. The consequences of this fact were then followed to the point where a numerical value of x could be determined by solving a simple equation. Diophantus never gave the synthesis, which in this case is the proof that the answer satisfies the desired conditions, because it only amounted to an arithmetic computation. Thus, Diophantus's work is at the opposite end of the spectrum from the purely synthetic work of Euclid.

4.2 Pappus and Analysis

Although analysis and synthesis had been used by all of the major Greek mathematicians, there was no systematic study of the methodology published, as far as is known, until the work of Pappus, who lived in Alexandria early in the fourth century. Pappus was one of the last mathematicians in the Greek tradition. He was familiar with the major and minor works of the men already discussed, and even extended some of their work in certain ways. He is best known for his *Collection*, a group of eight separate works on various topics in mathematics, probably put together shortly after his death by an editor attempting to preserve Pappus's papers. The books of the collection vary greatly in quality, but most of the material consists of surveys of certain mathematical topics collected from the works of Pappus's predecessors.

The most influential book of the *Collection* is Book 7, *On the Domain of Analysis*, which contains the most explicit discussion from Greek times of the method of analysis, the methodology used by Greek mathematicians to solve problems. The central ideas are spelled out in Pappus's introduction to Book 7:

> Now analysis is the path from what one is seeking, as if it were established, by way of its consequences, to something that is established by synthesis. . . . There are two kinds of analysis;

> one of them seeks after truth and is called "theorematic," while the other tries to find what was demanded and is called "problematic." In the case of the theorematic kind, we assume what is sought as a fact and true, then advance through its consequences, as if they are true facts according to the hypothesis, to something established; if this thing that has been established is a truth, then that which was sought will also be true, and its proof the reverse of the analysis; but if we should meet with something established to be false, then the thing that was sought too will be false. In the case of the problematic kind, we assume the proposition as something we know, then proceed through its consequences, as if true, to something established; if the established thing is possible and obtainable, which is what mathematicians call "given," the required thing will also be possible, and again the proof will be the reverse of the analysis; but should we meet with something established to be impossible, then the problem too will be impossible.

According to Pappus, then, to solve a problem or prove a theorem by analysis, begin by assuming what is required, then consider the consequences flowing from it until a result is reached that is known to be true or "given." That is, begin by assuming that which is required, p, for example, and then prove that p implies q_1, q_1 implies $q_2, \ldots, q_n$ implies q, where q is something known to be true. To give the formal synthetic proof of the theorem, or to solve the problem, reverse the process beginning with q implies q_n. This method of reversal has always been controversial; after all, not all theorems have valid converses. In fact, however, most important theorems from Euclid and Apollonius do have at least partial converses. Thus, the method does often provide the desired proof or solution, or at least demonstrates, when there are only partial converses, the conditions under which a problem can be solved.

There are few examples in the extant literature of theorematic analysis, because Euclid, for example, never shared his method of discovery of his proofs. But some of the manuscripts of Book XIII of the *Elements*, contain, evidently as an interpolation made in the early years of the common era, an analysis of each of the first five propositions. Consider

PROPOSITION XIII-1. *If a straight line is cut in extreme and mean ratio, the square on the sum of the greater segment and half of the whole is five times the square on the half.*

Let AB be divided in extreme and mean ratio at C, with AC being the greater segment, and let $AD = \frac{1}{2}AB$ (Fig. 4.1). To perform the analysis, assume the truth of the conclusion, namely, $CD^2 = 5AD^2$, and determine its consequences. Since also $CD^2 = AC^2 + AD^2 + 2AC \cdot AD$, therefore $AC^2 + 2AC \cdot AD = 4AD^2$. But $AB \cdot AC = 2AC \cdot AD$, and since $AB : AC = AC : BC$, also $AC^2 = AB \cdot BC$. Therefore $AB \cdot BC + AB \cdot AC = 4AD^2$, or $AB^2 = 4AD^2$, or, finally, $AB = 2AD$, a result known to be true. The synthesis can then proceed by reversing each step: Since $AB = 2AD$, we have $AB^2 = 4AD^2$. Since also $AB^2 = AB \cdot AC + AB \cdot BC$, it follows that $4AD^2 = 2AD \cdot AC + AC^2$. Adding to each side the square on AD gives the result, $CD^2 = 5AD^2$.

FIGURE 4.1 *Elements*, Proposition XIII-1

More important for Greek mathematics than theorematic analysis is problematic analysis. We have already discussed several examples of this type of analysis, including the problems of angle trisection and cube duplication and Archimedes' problems on the division of a sphere by a plane. And although Euclid did not present the analysis as such, one can carry out the procedure in solving his VI-28, the geometric algebra problem leading to the solution of the quadratic equation $x^2 + c = bx$. The analysis would show that an additional condition is required for the solution, namely, that $c \leq (b/2)^2$.

Pappus's Book 7 is a companion to the *Domain of Analysis*, a collection of several geometric treatises, all written many centuries before Pappus. These works—Apollonius's *Conics* as well as six other books (all but one lost), Euclid's *Data* and two other lost works, and single works by Aristaeus and Eratosthenes (both lost)—provided the Greek mathematician with the tools necessary to solve problems by analysis. For example, to deal with problems that result in conic sections, one had to be familiar with Apollonius's work. To deal with problems solvable by Euclidean methods, the material in the *Data* was essential.

Pappus's work does not include the *Domain of Analysis* itself. It is designed to be read along with those treatises. Therefore, it includes a general introduction to most of the individual books as well as a large collection of lemmas that are intended to help the reader work through the actual texts. Pappus evidently decided that the texts themselves were too difficult for most readers of his day to understand without guidance. The teaching tradition had been weakened through the centuries, and there were few, like Pappus, who could appreciate these several-hundred-year-old works. Pappus's goal was to increase the number of students who could understand the mathematics in these classical works by helping his readers through the steps where the authors wrote "clearly... !" He also included various supplementary results as well as additional cases and alternative proofs.

Among Pappus's additional remarks is the generalization of the three- and four-line locus problems discussed by Apollonius. Pappus noted that in those problems the locus is a conic section. But, he wrote, if there are more than four lines, the loci are as yet unknown; that is, "their origins and properties are not yet known." He was disappointed that no one had given the construction of these curves that satisfy the five- and six-line locus. The problem in these cases is, given five (six) straight lines, to find the locus of a point such that the rectangular parallelepiped contained by the lines drawn at given angles to three of these lines has a given ratio to the rectangular parallelepiped contained by the remaining two lines and some given line (remaining three lines). Pappus noted that one can generalize the problem even further, to more than six lines, but in that case, "one can no longer say 'the ratio is given between some figure contained by four of them to some figure contained by the remainder' since no figure can be contained in more than three dimensions." Nevertheless, according to Pappus, one can express this ratio of products by compounding the ratios of individual lines to one another, so that one can in fact consider the problem for any number of lines. But, Pappus complained, "[geometers] have by no means solved [the multiline locus problem] to the extent that the curve can be recognized. . . . The men who study these matters are not of the same quality as the ancients and the best writers."

Much of the explicit analysis in Greek mathematics has to do with material now generally regarded as algebraic. The examples from *Elements* XIII-1 and VI-28 are clearly such. The examples using the conic sections are ones that today would be solved using analytic geometry, a familiar application of algebra. It is somewhat surprising, then, that Pappus does not mention the strictly algebraic *Arithmetica* of Diophantus as a prime example of analysis, because, in effect, every problem in Diophantus's work is solved according to

Pappus's model. Possibly Pappus did not include this work because it was not on the level of the classic geometric works. In any case, it was the algebraic analysis of Diophantus and the "quasi-algebraic" analysis of many of the other works mentioned, rather than pure geometric analysis, that provided the major impetus for sixteenth- and seventeenth-century European mathematicians to expand on the notion of algebra and develop it into a major tool to solve even purely geometric problems.

4.3 Hypatia

Pappus's aim of reviving Greek mathematics was unsuccessful, probably in part because the increasingly confused political and religious situation affected the stability of the Alexandrian Museum and Library. In Pappus's time, Christianity was changing from the beliefs of a persecuted sect into the official religion of the Roman Empire. In 313, Emperor Galerius issued an edict of toleration in the eastern empire, and two years later Constantine did the same in the west. Constantine was in fact converted to Christianity before his death in 337. Within sixty years, Christianity became the state religion of the empire and the ancient worship of the Roman gods was banned. Of course, the banning of paganism did not cause everyone to adopt Christianity. In fact, in the late fourth and early fifth centuries, Hypatia (c. 355–415), the daughter of Theon of Alexandria, was a respected and eminent teacher in that city, one who taught not only mathematics but also some of the philosophic doctrines dating back to Plato's Academy. And although she maintained her non-Christian religious beliefs, she enjoyed intellectual independence and even had eminent Christians among her students, including Synesius of Cyrene (in Libya), who later became a bishop.

Although there is some evidence of earlier women being involved in Greek mathematics, only for Hypatia is there substantial indication of mathematical accomplishments. Hypatia received a very thorough education in mathematics and philosophy from her father. Although the only surviving documents with a clear reference to Hypatia are Synesius's letters to her requesting scientific advice, detailed textual studies of Greek, Arabic, and medieval Latin manuscripts have led to the conclusion that she was responsible for many mathematical works. These include several parts of her father's commentary on Ptolemy's *Almagest*, the edition of Archimedes' *Measurement of the Circle* from which most later Arabic and Latin translations stem, a work on areas and volumes that reworked Archimedean material, and a text on isoperimetric figures related to Pappus's Book 5. She is also responsible for commentaries on Apollonius's *Conics* and, as noted earlier, on Diophantus's *Arithmetica*.

Unfortunately, although Hypatia had many influential friends in Alexandria, including the Roman prefect Orestes, they were primarily from the upper classes. The general populace supported the patriarch Cyril in his struggle with Orestes for control of the city. So when Cyril spread rumors that the famous woman philosopher practiced sorcery as part of her philosophical, mathematical, and astronomical work, a group became intent on eliminating this "satanic" figure. Hypatia's life was then cut short by a fanatical mob. Her death effectively ended the Greek mathematical tradition of Alexandria.

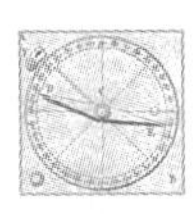

Exercises

1. Determine Diophantus's age at his death from the epigram at the opening of the chapter.
2. Solve Diophantus's problem I-27 by the method of problem I-28: To find two numbers such that their sum and product are given. Diophantus gives the sum as 20 and the product as 96.
3. Solve Diophantus's problem II-10: To find two square numbers having a given difference. Diophantus puts the given difference as 60. Also, give a general rule for solving this problem given any difference.
4. Generalize Diophantus's solution to problem II-19 by choosing an arbitrary ratio $n:1$ and the value $(x+m)^2$ for the second square.
5. Solve Diophantus's problem B-8: To find two numbers such that their difference and the difference of their cubes are equal to two given numbers. (Write the equations as $x-y=a$, $x^3-y^3=b$. Diophantus takes $a=10, b=2120$.) Derive necessary conditions on a and b that ensure a rational solution.
6. Solve Diophantus's problem B-9: To divide a given number into two parts such that the sum of their cubes is a given multiple of the square of their difference. [The equations become $x+y=a$, $x^3+y^3=b(x-y)^2$. Diophantus takes $a=20$ and $b=140$ and notes that the necessary condition for a solution is that $a^3(b-\frac{3}{4}a)$ is a square.]
7. Solve Diophantus's problem D-12: To divide a given square into two parts such that when each is subtracted from the given square, the remainder (in both cases) is a square. Note that the solution follows immediately from problem II-8.
8. Solve Diophantus's problem IV-9: To add the same number to a cube and its side and make the second sum the cube of the first. (The equation is $x+y=(x^3+y)^3$. Diophantus begins by assuming that $x=2z$ and $y=27z^3-2z$.)
9. Book VI of the *Arithmetica* deals with Pythagorean triples. For example, solve problem VI-16: To find a right triangle with integral sides such that the length of the bisector of an acute angle is also an integer. Hint: Use proposition VI-3 of the *Elements*: The bisector of an angle of a triangle cuts the opposite side into segments in the same ratio as that of the remaining sides.
10. Carry out the analysis of proposition VI-28 of the *Elements*: To a given straight line, to apply a parallelogram equal to a given rectilinear figure and deficient by a parallelogram similar to a given one. Consider only the case where the parallelograms are rectangles. Begin with the assumption that such a rectangle has been constructed and derive the condition that "the given rectilinear figure must not be greater than the rectangle described on the half of the straight line and similar to the defect."
11. Provide the analysis for proposition XIII-4 of the *Elements*: If a straight line is cut in extreme and mean ratio, the sum of the squares on the whole and on the lesser segment is triple the square on the greater segment.
12. Write an equation for the locus described by the problem of five lines. Assume for simplicity that all the lines are either parallel or perpendicular to one of them and that all the given angles are right.

Exercises 13–15 are from the *Greek Anthology* (c. 500 CE), the source of the chapter-opening epigram about Diophantus.

13. Solve Epigram 116: Mother, why do you pursue me with blows on account of the walnuts? Pretty girls divided them all among themselves. For Melission took two-sevenths of them from me, and Titane took the twelfth. Playful Astyoche and Philinna have the sixth and third. Thetis seized and carried off twenty, and Thisbe twelve, and look there at Glauce smiling sweetly with eleven in her hand. This one nut is all that is left to me. How many nuts were there originally?
14. Solve Epigram 130: Of the four spouts, one filled the whole tank in a day, the second in two days, the third in three days, and the fourth in four days. What time will all four take to fill it?
15. Solve Epigram 145: *A*: Give me ten coins and I have three times as many as you. *B*: And if I get the same from you, I have five times as much as you. How many coins does each have?
16. Devise a lesson teaching the method of problematic analysis. Use problems from ancient times and more recent problems.

17. Compare Diophantus's use of false position with that of the Egyptians and the Babylonians. Does it appear that these methods are in any way historically related?

18. Why were so few women involved in mathematics in Greek times?

References

Thomas Heath's *A History of Greek Mathematics* and B. L. Van der Waerden's *Science Awakening*, referred to in chapter 2, have sections on the material discussed in this chapter. The six books of Diophantus still extant in Greek are found in Thomas L. Heath, *Diophantus of Alexandria: A Study in the History of Greek Algebra* (New York: Dover, 1964). Heath does not, however, translate Diophantus literally but generally simply outlines Diophantus's arguments. More literal translations of certain of the problems are found in Thomas, *Selections Illustrating the History of Greek Mathematics*. A translation of and commentary on the four Arabic books of Diophantus's *Arithmetica* is J. Sesiano, *Books IV to VII of Diophantos' Arithmetica in the Arabic Translation of Qusṭā ibn Lūqā* (New York: Springer, 1982). A brief survey of Diophantus's work is in J. D. Swift, "Diophantus of Alexandria," *American Mathematical Monthly* 63 (1956), 163–170. An extensive discussion of the algebraic analysis of Diophantus and its effects on the development of algebra is found in J. Klein, *Greek Mathematical Thought and the Origin of Algebra* (Cambridge: MIT Press, 1968). The relationship of Diophantus's work to later work in what is now called *Diophantine analysis* is spelled out in Isabella Grigoryevna Bashmakova, *Diophantus and Diophantine Equations*, trans. Abe Shenitzer (Washington: MAA, 1997).

The entire extant text of Pappus's *Collection* is translated into French in Paul Ver Eecke, *Pappus d'Alexandrie, La Collection Mathematique* (Paris: Desclée, De Brouwer et Cie., 1933). A more recent English translation of Book 7, with commentary, is provided by Alexander Jones, *Pappus of Alexandria: Book 7 of the Collection* (New York: Springer, 1986). For more on Greek analysis, see Michael Mahoney, "Another Look at Greek Geometrical Analysis," *Archive for History of Exact Sciences* 5 (1968), 318–348, and J. Hintikka and U. Remes, *The Method of Analysis: Its Geometrical Origin and Its General Significance* (Boston: Reidel, 1974).

For a recent biography of Hypatia, see Maria Dzielska, *Hypatia of Alexandria*, trans. F. Lyra (Cambridge: Harvard University Press, 1995). Although the book has little discussion of her mathematics, that gap is filled by Michael A. B. Deakin in "Hypatia and her Mathematics," *American Mathematical Monthly* 101 (1994), 234–243. Details on the attribution of various mathematical works to Hypatia are found in Wilbur Knorr, *Textual Studies in Ancient and Medieval Geometry* (Boston: Birkhäuser, 1989).

CHAPTER FIVE

Ancient and Medieval China

Now the science of mathematics is considered very important. This book [Precious Mirror of the Four Elements, by Zhu Shijie] . . . therefore will be of great benefit to the people of the world. The knowledge for investigation, the development of intellectual power, the way of controlling the kingdom and of ruling even the whole world, can be obtained by those who are able to make good use of the book. Ought not those who have great desire to be learned take this with them and study it with great care?
—Introduction to *Precious Mirror of the Four Elements*, 1303

In the first four chapters, we discussed the mathematics of Greece as well as the mathematics of two civilizations known to have influenced Greek mathematics—Mesopotamia and Egypt. But mathematics was done in other parts of the world, even in ancient times. In this chapter, we will look at some mathematical ideas from ancient and medieval China, some of which may have, through paths so far undiscovered, reached Europe.

Although there are legends that date Chinese civilization back 5000 or more years, the earliest solid evidence of such a civilization is provided by the excavation of ruins at Anyang, near the Huang River, which are dated to about 1600 BCE. It is to the society centered there, the Shang dynasty, that the "oracle bones" belong, curious pieces of bone inscribed with very ancient writing, which were used for divination by the priests of the period and are the source of current knowledge of early Chinese number systems. Around the beginning of the first millennium BCE, the Shang dynasty was replaced by the Zhou dynasty, which in turn dissolved into numerous warring feudal states. In the sixth century BCE, there was a great period of intellectual flowering, in which the most famous philosopher was Confucius. Academies of scholars were founded in several of the feudal states. Other feudal lords hired scholars to advise them in a time of technological growth caused by the development of iron.

The feudal period ended as the weaker states were gradually absorbed by the stronger, until ultimately China was unified under the Emperor Qin Shi Huangdi in 221 BCE. Under his leadership, China was transformed into a highly centralized bureaucratic state. He enforced a severe legal code, levied taxes evenly, and demanded the standardization of weights, measures, money, and especially the written script. Legend holds that this emperor ordered the burning of all books from earlier periods to suppress dissent, but there is some reason to doubt that the book burning was actually carried out. The emperor died in 210 BCE, and his dynasty was soon overthrown and replaced by that of the Han, which was to last about 400 years. The Han completed the establishment of a trained civil service, for which a system of education was necessary. Among the texts that began to be used for this purpose were two mathematical works, probably compiled early in the Han dynasty but containing material from several hundred years earlier. These are the *Zhoubi suanjing* (*Arithmetical Classic of the Gnomon and the Circular Paths of Heaven*) and the *Jiuzhang suanshu* (*Nine Chapters on the Mathematical Art*). The latter work, in particular, became central to Chinese mathematical practice over the centuries. As originally written, it was a compilation of problems with answers and rules for determining the answers. Over the centuries, however, commentaries were written to explain or derive the rules.

The Han dynasty in China disintegrated early in the third century CE, and China broke up into several warring kingdoms. The period of disunity lasted until 581, when the Sui dynasty was established, followed 37 years later by the Tang dynasty, which was to last nearly 300 years. Although another brief period of disunity followed, much of China was again united under the Song dynasty (960–1279), a dynasty itself overthrown by the Mongols under Ghengis Khan. Despite the numerous wars and dynastic conflicts, a true Chinese culture was developing throughout most of east Asia, with a common language and common values. The system of imperial examinations for entrance into the civil service, instituted during the Han dynasty, lasted—with various short periods of disruption—into the twentieth century. Although the examination was chiefly based on Chinese literary classics, the demands of the empire for administrative services, including surveying, taxation, and calendar making, required that many civil servants be competent in various areas of mathematics. The Chinese imperial government therefore encouraged the study of applicable mathematics, as indicated in this chapter's opening quotation. In fact, at various times, there was an imperial Institute of Mathematics, where officials were trained in practical mathematics. At other times, mathematics was studied at the Institute of Astronomy or the Institute of Records. In general, the mathematical texts studied by candidates for such institutes were collections of problems with methods of solution, including the *Nine Chapters on the Mathematical Art*. New methods were rarely introduced. The examination system often required recitation of relevant passages from the mathematics texts, as well as the solving of problems in the manner described in those texts. Thus, there was no particular incentive for mathematical creativity.

Nevertheless, creative mathematicians did appear in China, and they applied their talents not only to improving old methods of solving practical problems, but also to extending those methods far beyond the requirements of practical necessity. We will look at developments in four major areas: numerical calculations, geometry, equation solving, and the solution of linear congruences. The first two areas are, in general, the product of the period prior to about 500 CE, while new discoveries in the latter two areas were being made into the thirteenth century.

5.1 Calculating with Numbers

The Chinese from earliest recorded times (the Shang dynasty) used a multiplicative system of writing numbers, based on powers of 10. That is, they developed symbols for the numbers 1 through 9 as well as for each of the powers of 10. For example, the number 659 was written using the symbol for 6 ([symbol]) attached to that for 100 ([symbol]), followed by the symbol for 5 ([symbol]) attached to the symbol for 10 (/), and finally the symbol [symbol] for 9:

[symbol]

Existing records indicate that by the fourth century BCE, mathematicians were using a physical system of representing numbers with counting rods, small bamboo rods about 10 centimeters long. These were manipulated on a counting board, where they were arranged in vertical columns standing for the various powers of 10. There were two possible arrangements of the rods to represent integers less than 10:

1	2	3	4	5	6	7	8	9
𝍠	𝍡	𝍢	𝍣	𝍤	𝍥	𝍦	𝍧	𝍨
𝍩	𝍪	𝍫	𝍬	𝍭	𝍮	𝍯	𝍰	𝍱

To represent numbers greater than 10, the rods were set up in columns, with the rightmost column holding the units, the next the tens, the next the hundreds, and so on. A blank column in a given arrangement represented a zero. So that the numbers could be read easily, the two arrangements of rods were alternated. The vertical arrangement was used in the units column, the hundreds column, and the ten thousands column, while the horizontal arrangement was used in the other columns. Thus, 1156 was represented by 𝍩 𝍠 𝍭 𝍥 and 6083 by 𝍮   𝍰 𝍢. These representations also occur in written records of counting board computations. There is some evidence that a dot was used in this situation to represent an empty column (intermediate zero) as early as the eighth century CE, but unambiguous evidence of the use of a small circle to represent zero in these situations dates from the twelfth century. Thus, by that time, Chinese number notation was clearly in the form of a decimal place value system. The earliest extant records of fractions in China are of common fractions, designated by symbols representing the words *fen zhi*. Thus, $\frac{2}{3}$ would be written 3 *fen zhi* 2, and could be translated as "2 parts from a whole broken into 3 equal parts." By medieval times, however, the Chinese were also using decimal fractions in many contexts.

Negative numbers, which were in use in China from at least the beginning of the common era, were represented on the counting board by using some feature to distinguish "negative" rods from "positive." One way was to use red rods for positive numbers and black ones for negative numbers. A negative number was represented in written records by drawing an oblique bar across one of the digits in rod numeral notation.

Calculating with fractions is recorded in the first chapter of the *Nine Chapters on the Mathematical Art*. The first rule for such calculation is the rule for reducing to lower terms:

> If the denominator and numerator can be halved, halve them. If not, lay down the denominator and numerator, subtract the smaller number from the greater. Repeat the process to obtain the greatest common divisor. Reduce them by the greatest common divisor.

For example, problem 6 of chapter 1 requires the reduction of $\frac{49}{91}$. Since the two numbers are not even, we cannot halve them. Therefore, we follow the subtraction process, the same process as the Euclidean algorithm. We subtract 49 from 91, giving 42. Then we subtract

42 from 49, giving 7. We next subtract 7 from 42 as many times as we can, eventually noting that we can do it six times with no remainder. Thus, 7 is the greatest common divisor. We "reduce" the numerator and denominator by 7, giving $\frac{7}{13}$ as the reduced fraction.

The rule for addition of fractions reads

> Each numerator is multiplied by the denominators of the other fractions. Add them as the dividend; multiply the denominators as the divisor. Divide. If there is a remainder, let it be the numerator and the divisor be the denominator. In the case of equal denominators, the numerators are to be added directly.

The basic idea, illustrated in several problems, is to use as a common divisor the product of the original divisors. Rules are also given for the other arithmetic operations on fractions, each preceded by several problems. For example, problem 19 is to find the area of a field that is $\frac{4}{7}$ *bu* in width and $\frac{3}{5}$ *bu* in length. The answer is given as $\frac{12}{35}$ square *bu*.

Another type of calculation, discussed in detail in chapter 4 of the *Nine Chapters*, is the determination of a square root. The algorithm is based on the algebraic formula $(x + y)^2 = x^2 + 2xy + y^2$, but most probably the author had in mind a diagram like that in Figure 5.1. This algorithm can be illustrated using problem 12 of that chapter: to determine the side of a square of area 55,225. The idea is to find digits a, b, c so that the answer can be written as $100a + 10b + c$. First, find the largest digit a so that $(100a)^2 < 55{,}225$. In this case, $a = 2$. The difference between the large square (55,225) and the square on $100a$ (40,000) is the large gnomon in the figure. If the outer thin gnomon is neglected, it is clear that b must satisfy $55{,}225 - 40{,}000 > 2(100a)(10b)$, or $15{,}225 > 4000b$. So certainly $b < 4$. To check that $b = 3$ is correct—that is, that with the square on $10b$ included, the area of the large gnomon is still less than 15,225—it is necessary to check that $2(100a)(10b) + (10b)^2 < 15{,}225$. Because this is in fact true, the same procedure can be repeated to find c: $55{,}225 - 40{,}000 - 30(2 \times 200 + 30) > 2 \times 230c$, or $2325 > 460c$. Evidently, $c < 6$. An easy check shows that $c = 5$ gives the correct square root: $\sqrt{55{,}225} = 235$.

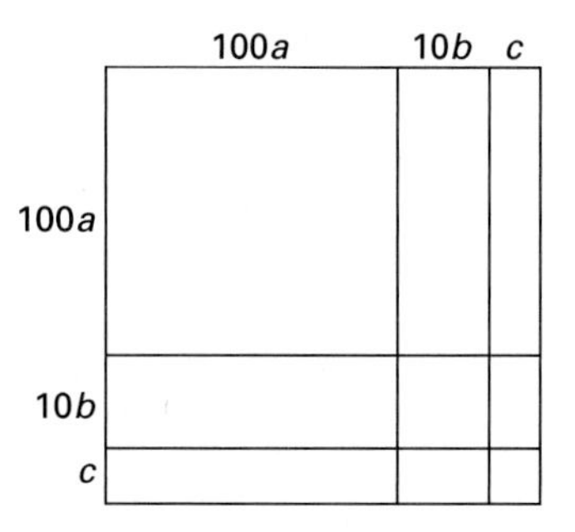

FIGURE 5.1 Algorithm for determination of the square root

The Chinese algorithm for calculating square roots is similar to one that was taught in Western schools in recent years. This method gives a series of answers, in this case, 200, 230, 235, each a better approximation to the true result than the one before. Although it appears clear to a modern reader that, if the answer is not a whole number, the procedure could continue indefinitely using decimal fractions, the Chinese author used common fractions as remainders in the cases where there was no integral square root.

Two further notes are in order here. First, a close examination of the algorithm shows that the solution of a quadratic equation (or, at least, a quadratic inequality) is part of the process. Second, *Nine Chapters* also contains an analogous algorithm for finding cube roots, perhaps derived from a consideration of actual cubes, as the one just discussed was derived from squares. The Chinese ultimately developed these ideas into a detailed procedure for solving polynomial equations of any degree, a procedure to be discussed below.

5.2 Geometry

5.2.1 The Pythagorean Theorem and Surveying

The *Nine Chapters* and other ancient Chinese documents contain many geometrical calculations. In particular, they all assume the result now known as the Pythagorean theorem. And the *Arithmetical Classic of the Gnomon*, probably written somewhat earlier than the *Nine Chapters*, contains an argument for the theorem. It presents a diagram like the one in Figure 5.2, along with the following commentary:

> Thus let us cut a rectangle [diagonally] and make the width 3 [units] and length 4 [units]. The diagonal between the [opposite] corners will then be 5 [units]. Now after drawing a square on this diagonal, circumscribe it by half-rectangles like that which has been left outside so as to form a [square] plate. Thus the [four] outer half-rectangles of width 3, length 4, and diagonal 5, together make two rectangles [of area 24]. Then [when this is subtracted from the square plate of area 49], the remainder is of area 25. This [process] is called "piling up the rectangles."

Although the commentary and diagram are given for the specific case of a (3, 4, 5) triangle, the proof (given in the last two lines) is quite general. Denote the width by a, the length by b, and the diagonal by c. The argument is then as follows: $(a+b)^2 - 2ab = c^2$; since $(a+b)^2 = a^2 + b^2 + 2ab$, the Pythagorean result $a^2 + b^2 = c^2$ is immediate. Looking at it geometrically, the argument depends simply on dissecting the large square in two ways: first as the square on a plus the square on b plus twice the rectangle ab; then as the square on c plus twice the rectangle ab.

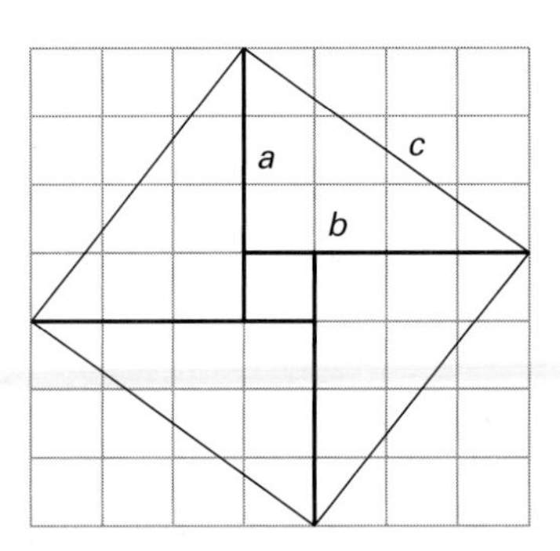

FIGURE 5.2 Diagram for demonstration of the Pythagorean theorem

Is this argument a proof? To meet modern standards, it would be necessary to show either that the inscribed figure (the square on c) or the circumscribed figure (the square on $a + b$) is in fact a square. To the ancients, however, this was obvious (as it probably is to most students today). The Chinese had no notion of an axiomatic system from which theorems could be derived. "Proof" in this context means simply a convincing argument. In fact, the Greek word *theorem* is derived from *theorein*, "to look at." If one looks at the diagram, one sees the theorem at once.

Again, assuming knowledge of the Pythagorean theorem, chapter 9 of the *Nine Chapters* contains many problems involving right triangles. For example, problem 6 concerns a square pond with side 10 feet, with a reed growing in the center whose top is 1 foot out of the water. If the reed is pulled to the shore, the top just reaches the shore. The problem is to find the depth of the water and the length of the reed. In Figure 5.3, $y = 5$ and $x + a = d$, where, in this case, $a = 1$. A modern solution might begin by setting $d^2 = x^2 + y^2$ and substituting for d. A brief algebraic calculation gives $x = (y^2 - a^2)/2a$. With the given numerical values, $x = 12$, and therefore $d = 13$. The Chinese rule states: "Multiply half of the side of the pond by itself; decrease this by the product of the length of the reed above the water with itself; divide the difference by twice the length of the reed above the water. This gives the depth. Add this to the length of the reed above the water. This gives the length of the reed." A translation of this rule into a formula gives $x = (y^2 - a^2)/2a$, the same result already derived. It is not clear, however, whether the Chinese author found the solution algebraically as above or by the equivalent geometric method, where $y^2 = AC^2 = AB^2 - BC^2 = BD^2 - EG^2 = DE^2 + 2 \times CE \times BC = a^2 + 2ax$. But what is certain is that the author was fluent in the use of the Pythagorean theorem.

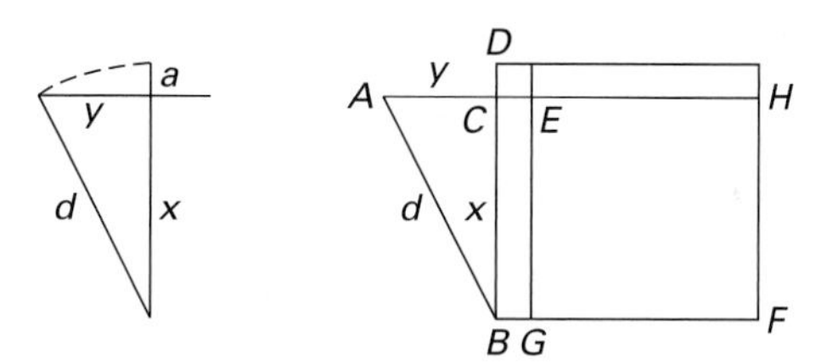

FIGURE 5.3 Problem 6 of chapter 9 of the *Nine Chapters*

Most of the final problems in chapter 9 of the *Nine Chapters* deal with surveying questions. When writing his commentary on the *Nine Chapters*, Liu Hui (third century CE) decided to add an addendum on more complicated problems of that type. This addendum ultimately became a separate mathematical work, the *Haidao suanjing* (*Sea Island Mathematical Manual*).

In the continuing tradition of problem texts, the *Sea Island Mathematical Manual* was simply a collection of nine problems with solutions, derivations, illustrations, and commentary. Unfortunately, all that remains today are the problems themselves with the computational directions for finding the solutions. No reasons are given for why these particular computations are to be performed, so the following discussion presents some possible methods by which Liu Hui worked out his rules.

The first of the nine problems, for which the text is named, shows how to find the distance and the height of a sea island. The other problems demonstrate how to determine such items as the height of a tree, the depth of a valley, and the width of a river. The sea island problem reads, "For the purpose of looking at a sea island, erect two poles of the same height, 5 feet, the distance between the front and rear pole being 1000 feet. Assume that the rear pole is aligned with the front pole. Move away 123 feet from the front pole and observe the peak of the island from ground level. Move backward 127 feet from the rear pole and observe the peak of the island from ground level again; the tip of the back pole also coincides with the peak. What is the height of the island and how far is it from the front pole?"

Liu Hui's answer is that the height of the island is 1255 feet and its distance from the pole is 30,750 feet. He also presents the rule for the solution (Fig. 5.4):

> Multiply the distance between poles by the height of the pole, giving the *shi*. Take the difference in distances from the points of observations as the *fa* to divide the *shi*. Add what is thus obtained to the height of the pole; the result is the height of the island. [Thus, the height h is given by the formula $h = a + [ab/(c - d)]$, where a is the height of the pole, b the distance between the poles, and c and d the respective distances from the poles to the observations points.] To find the distance of the island from the front pole, multiply the distance of the backward movement from the front pole by the distance between the poles, giving the *shi*. Take the difference in distance at the points of observation as the *fa* to divide the *shi*. The result is the distance of the island from the pole. [The distance s is given by $s = bd/(c - d)$.]

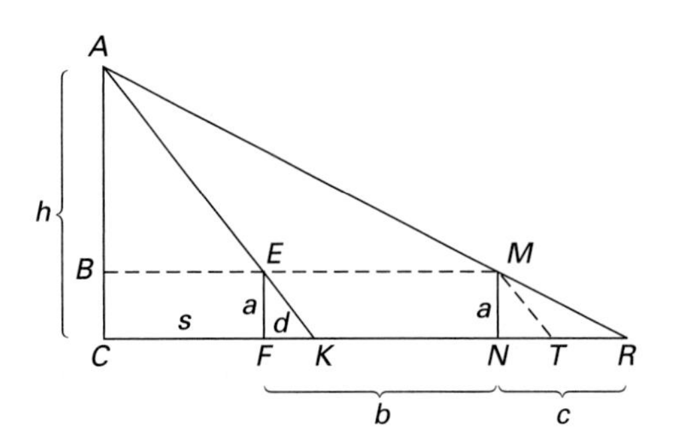

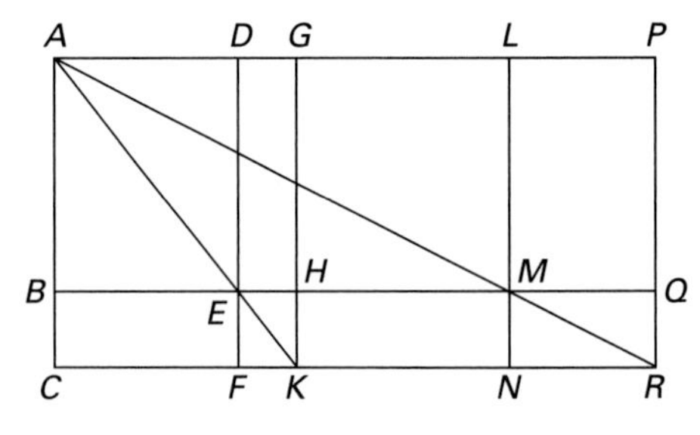

FIGURE 5.4 Problem 1 of the *Sea Island Mathematical Manual*

Liu Hui called his method the **method of double differences**, because two differences are used in the solution procedure. A modern derivation of the method would use similar triangles: Construct MT parallel to EK. Then $\triangle AEM$ is similar to $\triangle MTR$ and $\triangle ABM$ is similar to $\triangle MNR$. Therefore, $ME : TR = AM : MR = AB : MN$, so

$$AB = \frac{ME \cdot MN}{TR} = \frac{FN \cdot EF}{TR},$$

and the height h ($= AB + BC$) of the island is

$$h = \frac{FN \cdot EF}{TR} + EF = \frac{ab}{c - d} + a,$$

as noted above. A similar argument gives Liu Hui's result for the distance s of the island from the front pole.

However, there are other ways of deriving Liu Hui's formula. In the mid–thirteenth century, Yang Hui commented on this particular problem and gave a justification using only congruent triangles and area relationships, a justification more in keeping with what is known about early Chinese mathematical techniques. Since triangles APR and ACR are congruent, as are triangles ALM and ABM, trapezoid $LPRM$ has the same area as trapezoid $BMRC$. Subtracting off the congruent triangles MQR and MNR shows that rectangles $LPQM$ and $BMNC$ are also equal in area. By a similar argument, rectangle $DGHE$ equals rectangle $BECF$. It follows that rectangle $EMNF$ (= rectangle $BMNC-$ rectangle $BECF$) equals rectangle $LPQM-$ rectangle $DGHE$. Writing each of the areas of the rectangles as products gives

$$FN \cdot EF = PQ \cdot QM - GH \cdot HE = PQ \cdot RN - PQ \cdot FK$$
$$= PQ(RN - FK) = AB(RN - FK).$$

Therefore, $AB = (FN \cdot EF)/(RN - FK)$, and the height $h = AC$ is given by

$$h = AC = AB + BC = \frac{FN \cdot EF}{RN - FK} + EF,$$

as desired. The distance $s = CF$ can then be determined by beginning with the equality of the areas of rectangles $DGHE$ and $BCFE$, that is, with $CF \cdot BC = DE \cdot EH$, and replacing $DE = AB$ by the value already found.

5.2.2 Areas and Volumes

The Chinese developed numerous formulas for calculating the areas and volumes of geometrical figures. Many of them are standard formulas, such as those for the areas of rectangles and triangles or for the volume of a parallelepiped. The *Nine Chapters* also gives the correct formula for the volume of a pyramid. Here we will consider the formulas for the area of a circle and the volume of a sphere.

The Chinese presented several versions of the formula for the area of a circle. For example, consider problem 32 from the first chapter of the *Nine Chapters*:

> There is a round field whose circumference is 181 yards and whose diameter is $60\frac{1}{3}$ yards. What is the area of the field? Answer: $2730\frac{1}{12}$ square yards.

The first thing to notice is that the stated diameter of the field is $\frac{1}{3}$ of the circumference. In other words, at the time the *Nine Chapters* was written, the number used for the ratio of circumference to diameter of a circle was taken as 3, the same value used by the Babylonians. Second, the Chinese scribe stated not one but four separate formulas by which the calculation of area could be made:

1. The rule is: Half of the circumference and half of the diameter are multiplied together to give the area.
2. Another rule is: The circumference and the diameter are multiplied together, then the result is divided by 4.
3. Another rule is: The diameter is multiplied by itself. Multiply the result by 3 and then divide by 4.

4. Another rule is: The circumference is multiplied by itself. Then divide the result by 12.

Of course, given that π is taken to be 3, all of these formulas are equivalent. Note also that it is the fourth rule that is the same as the usual Babylonian rule, but, like the Babylonians, the author of the *Nine Chapters* does not tell why these formulas work.

On the other hand, Liu Hui, in his own commentary, noted that the value 3 for the ratio of circumference to diameter must be incorrect. He did so in the context of the area situation, where the Chinese formula for the area of a circle of radius 1 is 3, but where he could easily calculate that the area of a regular dodecagon inscribed in that circle is also 3. Thus, he concluded, the area of the circle must be larger. In fact, Liu then proceeded to approximate this area by an argument involving the construction of inscribed polygons with more and more sides, an argument that is reminiscent of Archimedes' determination of π by using perimeters of polygons. As Liu Hui wrote, "the larger the number of sides, the smaller the difference between the area of the circle and that of its inscribed polygons. Dividing again and again until it cannot be divided further yields a regular polygon coinciding with the circle, with no portion whatever left out." That is, although he does not use a formal *reductio ad absurdum* argument as in the Eudoxian method of exhaustion, Liu assumes that eventually, the polygons will in fact "exhaust" the circle.

We can describe Lui's argument by looking at a regular n-gon inscribed in a circle of radius r. Let c_n be the length of the side of the inscribed n-gon, a_n the length of the perpendicular from the center of the circle to the side, and S_n the total area of the n-gon (Fig. 5.5). We start with $c_6 = r$. In general,

$$a_n = \sqrt{r^2 - \left(\frac{c_n}{2}\right)^2} \quad \text{and} \quad c_{2n} = \sqrt{\left(\frac{c_n}{2}\right)^2 + (r - a_n)^2}.$$

Then

$$S_{2n} = 2n\frac{1}{2}\frac{c_n}{2}r = \frac{1}{2}nrc_n.$$

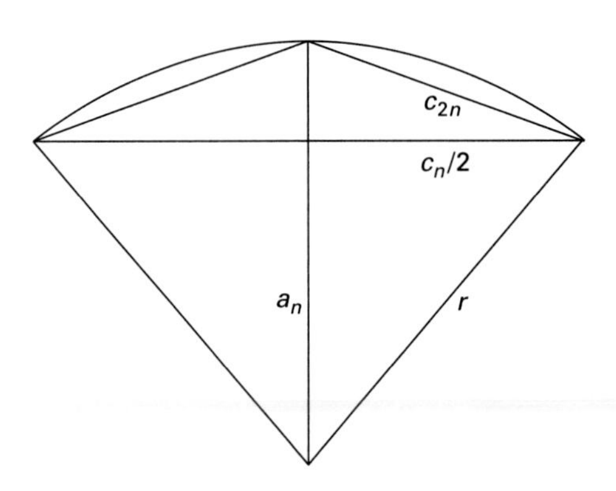

FIGURE 5.5 Inscribed regular n-gon in a circle of radius r

Liu calculated S_{2n} for $n = 96$ in the case of $r = 10$ to be $314\frac{64}{625}$, equivalent to a value for π of 3.141024, and then noted that it is "convenient" to take 3.14 as an approximation to π and neglect the fractional part. Two centuries later, however, Zu Chongzhi (c. 429–500) decided to carry out the calculations further. He found by use of S_{24576} that a better approximation to π was 3.1415926.

Chapter 4 of the *Nine Chapters* gives a rule for determining the diameter d of a sphere of given volume V, which is equivalent to proposing a formula for the volume of a sphere: "Lay down the given number (V). Multiply it by 16; divide it by 9; extract the cube root of the result." In other words, the rule is that $d = \sqrt[3]{\frac{16}{9}V}$, equivalent to the volume formula $V = \frac{9}{16}d^3$, or $V = \frac{9}{2}r^3$, where r is the radius. Even taking the usual approximation that $\pi = 3$, this result is incorrect—and Liu Hui described in his commentary how he knew that.

Consider a cylinder inscribed in a cube of side d, and consider the cross section of this figure by a plane perpendicular to the axis of the cylinder (Fig. 5.6). The plane cuts the cylinder in a circle of diameter d and the cube in a square of side d. The ratio of the areas of these two plane figures is $\pi : 4$. Since this is true for each cross section, the ratio of the volumes must be the same, so the volume of the cylinder is $(\pi/4)d^3$. (This principle, similar to Archimedes' procedure in *The Method*, is what is now known as **Cavalieri's principle**.) Now, consider a sphere inscribed in the cylinder. If the ratio of the volume of the sphere to that of the cylinder were also $\pi : 4$, then the volume of the sphere would be $(\pi^2/16)d^3$, which, with π taken equal to 3, is exactly the value given in the *Nine Chapters*. But Liu knew that this was incorrect, that in fact the ratio of the volume of the sphere to that of the cylinder was not $\pi : 4$. His argument was as follows: Inscribe a second cylinder in the cube, whose axis is perpendicular to that of the first cylinder, and consider the intersection of the two cylinders (Fig. 5.7). He called this intersection the "double box-lid." Since the sphere is contained in each cylinder, it is contained in their intersection. Now any cross section of the box-lid perpendicular to its axis is a square, so the ratio of the volume of the sphere to that of the box-lid is $\pi : 4$. But the box-lid is smaller than the cylinder, so the ratio of the volume of the sphere to the cylinder must be less than $\pi : 4$. So to find a correct formula for the volume of a sphere, it was necessary to find the volume of the box-lid.

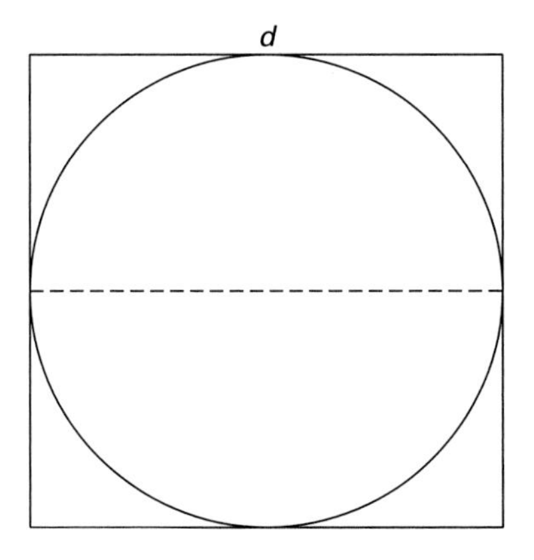

FIGURE 5.6 Cross section of a cylinder inscribed in a cube

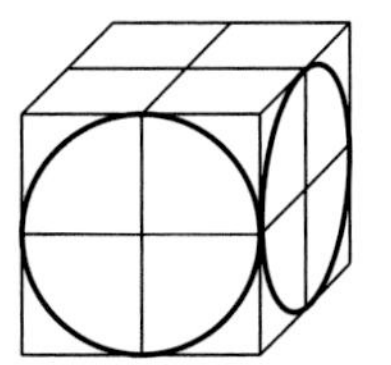
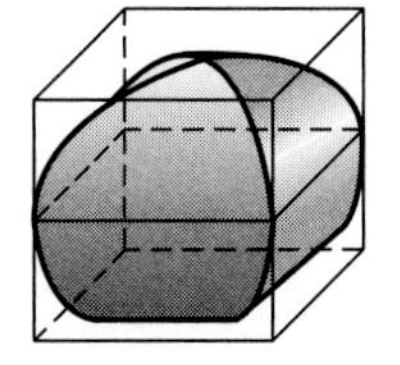

FIGURE 5.7 Intersection of two cylinders inscribed in the same cube

Liu Hui could not find this volume; as he wrote, "Let us leave the problem to whomever can tell the truth." That person was Zu Geng (late fifth to early sixth century), the son of Zu Chongzhi. He formalized Cavalieri's principle as "If the corresponding section areas of two solids are equal everywhere, then their volumes cannot be unequal." In the case of the double box-lid, his argument is as follows: Consider $\frac{1}{8}$ of the box-lid and inscribe it in a cube of side $r = d/2$ (Fig. 5.8). If we pass a plane through the box-lid at height h, the cross section is a square of side s, where $s^2 = r^2 - h^2$. Therefore, since the plane intersects the circumscribing cube in a square of area r^2, the difference between the two cross sections is h^2. But we know that if we take an inverted pyramid of height r and square base of side r and pass through it a plane at height h (from the vertex), the cross section is also a square of area h^2. It follows that the volume of that part of the inscribing cube outside of the box-lid is equal to the volume of the pyramid, namely, $\frac{1}{3}r^3$. Subtracting this from the volume of the cube itself, we find that the volume of $\frac{1}{8}$ of the box-lid is $\frac{2}{3}r^3$, and therefore the volume of the entire box-lid is $\frac{16}{3}r^3$. But the ratio of the volume of the sphere to that of the box-lid is $\pi : 4$. Thus, the volume of the sphere is $\frac{4}{3}\pi r^3$.

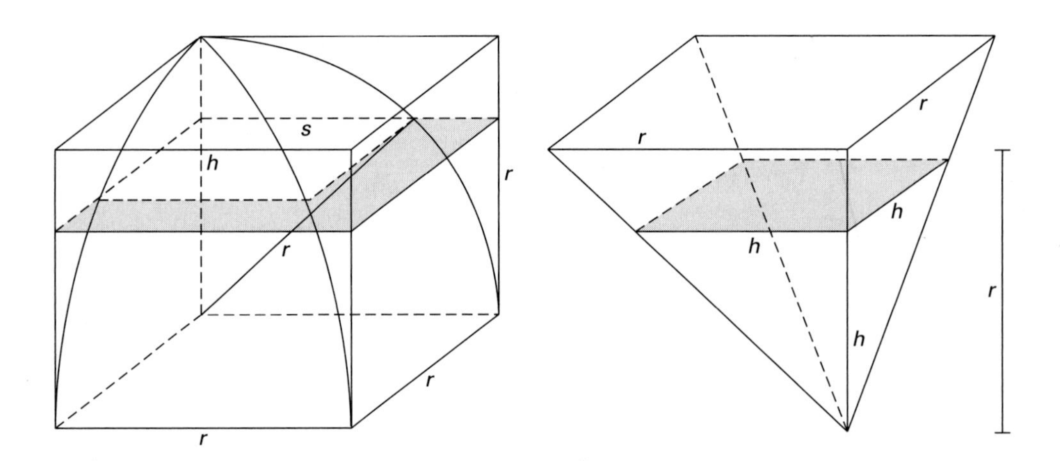

FIGURE 5.8 One-eighth of the box-lid inscribed in a cube of side r

5.3 Solving Equations

5.3.1 Systems of Linear Equations

The *Nine Chapters* includes techniques for solving systems of linear equations. In fact, chapter 8 describes a method virtually identical to the method of Gaussian elimination. Because the method is supposed to be performed on a counting board, it is presented essentially in matrix form. As an example, consider problem 1 of that chapter: "There are three classes of grain, of which three bundles of the first class, two of the second, and one of the third make 39 measures. Two of the first, three of the second, and one of the third make 34 measures. And one of the first, two of the second and three of the third make 26 measures. How many measures of grain are contained in one bundle of each class?" The problem can be translated into modern terms as the following system:

$$\begin{aligned} 3x + 2y + z &= 39 \\ 2x + 3y + z &= 34 \\ x + 2y + 3z &= 26 \end{aligned}$$

The algorithm for the solution is then stated: "Arrange the 3, 2, and 1 bundles of the three classes and the 39 measures of their grains at the right. Arrange other conditions at the middle and at the left." This arrangement is shown below:

$$\begin{array}{ccc} 1 & 2 & 3 \\ 2 & 3 & 2 \\ 3 & 1 & 1 \\ 26 & 34 & 39 \end{array}$$

The text continues: "With the first class on the right column multiply currently the middle column and directly leave out." This means to multiply the middle column by 3 (the first class on the right) and then subtract off a multiple (in this case, 2) of the right-hand column so that the first number in the middle column becomes 0. The same operation is then performed with respect to the left column. The results are

$$\begin{array}{ccc} 1 & 0 & 3 \\ 2 & 5 & 2 \\ 3 & 1 & 1 \\ 26 & 24 & 39 \end{array} \qquad \begin{array}{ccc} 0 & 0 & 3 \\ 4 & 5 & 2 \\ 8 & 1 & 1 \\ 39 & 24 & 39 \end{array}$$

"Then with what remains of the second class in the middle column, directly leave out." That is, perform the same operations using the middle column and the left column. The result is

$$\begin{array}{ccc} 0 & 0 & 3 \\ 0 & 5 & 2 \\ 36 & 1 & 1 \\ 99 & 24 & 39 \end{array}$$

Because this last diagram is equivalent to the triangular system

$$\begin{aligned} 3x + 2y + z &= 39 \\ 5y + z &= 24, \\ 36z &= 99 \end{aligned}$$

the author explains how to solve that system by what is today called **back substitution** beginning with $z = \frac{99}{36} = 2\frac{3}{4}$.

One might wonder what happened when these matrix manipulations led to a negative quantity. A glance at problem 3 of the same chapter shows that this was not a limitation. The method was carried through perfectly correctly for the system

$$\begin{aligned} 2x + \ y \qquad &= 1 \\ 3y + \ z &= 1, \\ x \qquad + 4z &= 1 \end{aligned}$$

a system in which negative quantities appear in the process of completing the algorithm. In fact, the author gave the rules for adding and subtracting with positive and negative quantities: "For subtraction—with the same signs, take away one from the other; with different signs, add one to the other; positive taken from nothing makes negative, negative from nothing makes positive. For addition—with different signs subtract one from the other; with the same signs add one to the other; positive and nothing makes positive; negative and nothing makes negative." Thus, interestingly, rules for dealing with negative numbers arose in China not in the context of solving equations that have no positive solution, but as an intermediate step in the use of a known algorithm designed to solve a problem that does have positive solutions.

5.3.2 Polynomial Equations

As noted earlier, the Chinese of the Han period had to solve certain quadratic equations in the context of using their square root algorithm. Over the centuries, the method developed into a procedure for solving polynomial equations of any order. This was first explicated by Jia Hian (c. 1050) in a work whose contents were described in the thirteenth century, although the work itself has not survived. Jia's basic idea stems from the original square and cube root algorithms, which made use of the binomial expansions $(r+s)^2 = r^2 + 2rs + s^2$ and $(r+s)^3 = r^3 + 3r^2s + 3rs^2 + s^3$, respectively. For example, consider the solution of the equation $x^3 = 12{,}812{,}904$, which a reasonable guess shows is a three-digit number starting with 2. In other words, the closest integer solution can be written as $x = 200 + 10b + c$. Ignoring the c temporarily, we need to find the largest b so that $(200 + 10b)^3 = 200^3 + 3 \cdot 200^2 \cdot 10b + 3 \cdot 200 \cdot (10b)^2 + (10b)^3 \leq 12{,}812{,}904$, or so that $3 \cdot 200^2 \cdot 10b + 3 \cdot 200 \cdot 100b^2 + 1000b^3 = b(1{,}200{,}000 + 60{,}000b + 1000b^2) \leq 4{,}812{,}904$. By trying in turn $b = 1, 2, 3, \ldots$, we discover that $b = 3$ is the largest value satisfying the inequality. Since $3(1{,}200{,}000 + 60{,}000 \cdot 3 + 1000 \cdot 3^2) = 4{,}167{,}000$, we next subtract 4,167,000 from 4,812,904 and derive a similar inequality for c: $c(3 \cdot 230^2 + 3 \cdot 230c + c^2) \leq 645{,}904$. In this case, it turns out that $c = 4$ satisfies this as an equality, so the solution to the original equation is $x = 234$.

Jia realized that this solution process could be generalized to nth-order roots for $n > 3$ by determining the binomial expansion $(r+s)^n$. In fact, not only did he write out the Pascal triangle of binomial coefficients through the sixth row (Fig. 5.9), but he also developed the usual method of generating the triangle: "Add the numbers in the two places above in order to find the number in the place below." He then used the binomial coefficients to find higher-order roots by a method analogous to that just described.

Evidently, Jia went even further. He saw that his method could be used to solve arbitrary polynomial equations, especially since these appeared as part of the root extraction process, but that it would be simpler on the counting board to generate step by step the various multiples by binomial coefficients rather than from the triangle itself.

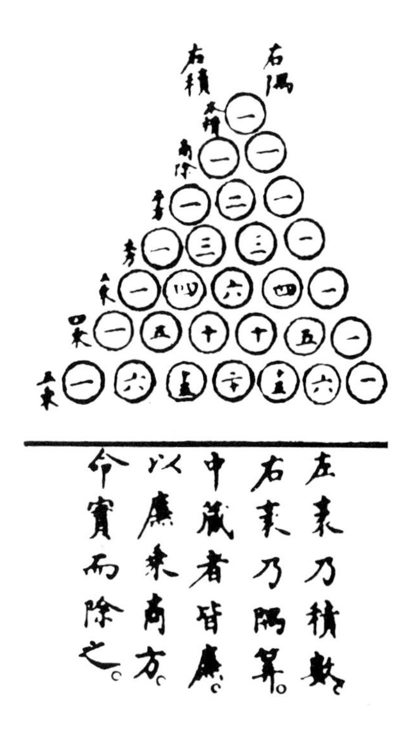

FIGURE 5.9 Jia Hian's diagram of the Pascal triangle, as recorded by a thirteenth-century author. (Source: From "The Chinese Connection between the Pascal Triangle and the Solution of Numerical Equations of Any Degree" by Lam Lay-Yong, *Historia Mathematica*, Vol. 7, No. 4. November 1980. Copyright © 1980 by Academic Press, Inc. Reprinted by permission of Academic Press, Inc. and Dr. Lam Lay-Yong.)

The first detailed account of Jia's method for solving equations, probably somewhat improved, appeared in the 1247 work *Shushu jiuzhang* (*Mathematical Treatise in Nine Sections*) of Qin Jiushao (1202–1261). We consider his method in the context of a particular equation, $-x^4 + 763{,}200x^2 - 40{,}642{,}560{,}000 = 0$, which arises from a geometrical problem of finding the area of a pointed field (see exercise 14). The initial steps in solving such an equation are the same as those in solving the pure equation: First, determine the number of decimal digits in the answer, and second, guess the appropriate first digit. In this case, the answer is found (by experience or by trial and error) to be a three-digit number beginning with 8. Qin's approach, like that of the old cube root algorithm, is, in effect, to set $x = 800 + y$, substitute this value into the equation, and then derive a new equation in y whose solution will be only a two-digit number. One can then guess the first digit of y and repeat the process.

The Chinese did not, of course, use modern algebra techniques to "substitute" $x = 800 + y$ into the original equation, as William Horner did in his essentially similar method of 1819. The problem was set up on a counting board with each row standing for a particular power of the unknown (Fig. 5.10). For reasons of space, however, we will write the coefficients horizontally. Thus, for the problem at hand, the opening configuration is

$$-1 \qquad 0 \qquad 763{,}200 \qquad 0 \qquad -40{,}642{,}560{,}000.$$

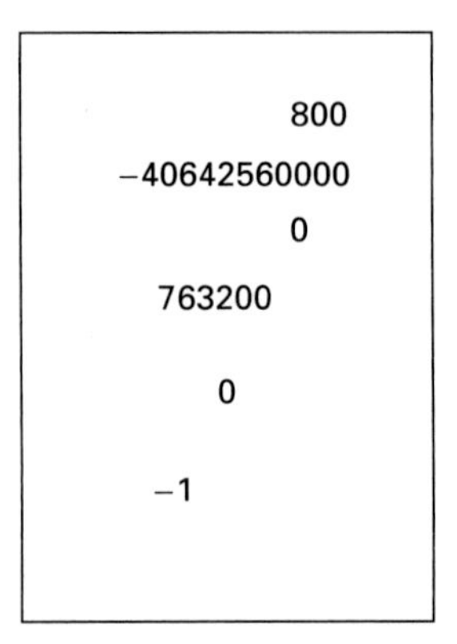

FIGURE 5.10 Initial counting board configuration for solution of $-x^4 + 763{,}200x^2 - 40{,}642{,}560{,}000 = 0$

Given that the initial approximation to the root was 800, Qin described what is now called the repeated (synthetic) division of the original polynomial by $x - 800$ $(= y)$. The first step gives

800					
	−1	0	763200	0	−40642560000
		−800	−640000	98560000	78848000000
	−1	−800	123200	98560000	38205440000

Qin's description of the counting board process tells exactly what numbers to multiply and add (or subtract) to give the arrangement on the third line. For example, −1 is multiplied by 800, and the result is added to 0. That result (−800) is then multiplied by 800, and the product is subtracted from the 763,200. In algebraic symbolism, this first step shows that the original polynomial has been replaced by

$$\begin{aligned}(x - 800)(-x^3 - 800x^2 + 123200x + 98560000) + 38205440000 \\ = y(-x^3 - 800x^2 + 123200x - 98560000) + 38205440000.\end{aligned}$$

Qin repeated the procedure three more times, dividing each quotient polynomial by the same $y = x - 800$. The result is finally

$$\begin{aligned}0 &= -x^4 + 763200x^2 - 40642560000 \\ &= y\{y[y(-y - 3200) - 3076800] - 826880000\} + 38205440000\end{aligned}$$

or

$$-y^4 - 3200y^3 - 3076800y^2 - 826880000y + 38205440000 = 0.$$

At this point, the procedure begins again with a guess that $y = 40 + z$. It turns out, in this instance, that $z = 0$, or, alternatively, that the polynomial in y is exactly divisible by $y - 40$. The solution to the original equation is then $x = 840$.

It is not difficult to see the relationship of Qin's description to Jia's method by the Pascal triangle by using Qin's method explicitly to solve the cube root problem already considered,

$x^3 = 12{,}812{,}904$. But Qin himself gave no theoretical justification of his algorithm, nor did he mention the Pascal triangle. This algorithm, since it was rediscovered in Europe more than five centuries after Qin's time, deserves a few additional comments.

First, the texts only briefly state how the guessed values for the digits of the root are found. In some cases, it is clear that the solver simply made a trial division of the constant term by the coefficient of the first power of the unknown, as is generally done in the square root algorithm itself. Sometimes several trials are indicated, and the solver picks one that works. But, in general, one can only surmise that the Chinese mathematicians possessed extensive tables of powers, which could be used to make the various guesses. Second, there is no mention in the texts of multiple roots. Qin's fourth-degree equation above, in fact, has another positive root, 240, as well as two negative ones. The root 240 could easily have been found by the same method, provided one had guessed 2 for the initial digit. But, in this case, the geometric problem from which the equation was derived had only one solution, 840, and Qin did not deal with equations in the abstract. Third, operations with negative numbers were performed as easily as those with positive ones. On the other hand, negative roots do not appear, again because the problems from which the equations arise have positive solutions. Fourth, the Chinese generally represented equations in a form equivalent to $f(x) = 0$. This represents an approach that is fundamentally different from the ancient Babylonian method or the medieval Islamic one. Finally, it appears that the Chinese method of solving quadratic equations is completely different from that of the Babylonians. The latter essentially developed a formula that could only be applied to such equations. The Chinese developed a numerical algorithm that they ultimately generalized to equations of any degree.

5.4 The Chinese Remainder Theorem

Probably the most famous mathematical technique developed in China is the technique long known as the **Chinese remainder theorem**, the procedure for solving systems of linear congruences. The earliest example of such a problem in Chinese mathematics is in the *Sunzi suanjing* (*Master Sun's Mathematical Manual*), a work probably written late in the third century: "We have things of which we do not know the number; if we count them by threes, the remainder is 2; if we count them by fives, the remainder is 3; if we count them by sevens, the remainder is 2. How many things are there?" In modern notation, the problem is to find N that simultaneously satisfies

$$N = 3x + 2 \qquad N = 5y + 3 \qquad N = 7z + 2$$

for integral values x, y, z, or, what amounts to the same thing, that satisfies the congruences

$$N \equiv 2 (\text{mod } 3) \qquad N \equiv 3 (\text{mod } 5) \qquad N \equiv 2 (\text{mod } 7).$$

Sun Zi gives the answer, 23, as well as his method of solution: "If you count by threes and have the remainder 2, put 140. If you count by fives and have the remainder 3, put 63. If you count by sevens and have the remainder 2, put 30. Add these numbers and you get 233. From this subtract 210 and you get 23." Sun Zi explains further: "For each unity as remainder when counting by threes, put 70. For each unity as remainder when counting by fives, put 21. For each unity as remainder when counting by sevens, put 15. If the sum is 106 or more, subtract 105 from this and you get the result."

In modern notation, Sun Zi has apparently noted that

$$70 \equiv 1 \pmod 3 \equiv 0 \pmod 5 \equiv 0 \pmod 7,$$
$$21 \equiv 1 \pmod 5 \equiv 0 \pmod 3 \equiv 0 \pmod 7,$$

and

$$15 \equiv 1 \pmod 7 \equiv 0 \pmod 3 \equiv 0 \pmod 5.$$

Hence, $2 \times 70 + 3 \times 21 + 2 \times 15 = 233$ satisfies the desired congruences. Since any multiple of 105 is divisible by 3, 5, and 7, one subtracts 105 twice to get the smallest positive value. Because this problem is the only one of its type that Sun Zi presented, it is not known whether he had developed a general method of finding integers congruent to 1 modulo m_i but congruent to 0 modulo m_j, $j \neq i$, for given integers $m_1, m_2, m_3, \ldots, m_k$, the most difficult part of the complete solution. The numbers in this particular problem are easy enough to find by inspection, but note for future reference that

$$70 = \frac{3 \times 5 \times 7}{3} \times 2, \qquad 21 = \frac{3 \times 5 \times 7}{5} \times 1, \qquad \text{and} \qquad 15 = \frac{3 \times 5 \times 7}{7} \times 1.$$

Although other congruence problems were given in Chinese texts over the centuries, particularly problems dealing with finding a date with a given relationship to several different calendrical cycles, a general method for solving such problems only first appeared in Qin Jiushao's *Mathematical Treatise in Nine Sections*. There, Qin described what he called the *ta-yen* rule for solving simultaneous linear congruences, which in modern notation are written $N \equiv r_i \pmod{m_i}$ for $i = 1, 2, \ldots, n$. For simplicity, we will assume that the m_i are relatively prime in pairs, although Qin himself dealt with the more general case. In particular, we will follow Qin's method to solve the following congruence:

$$N \equiv 32 \pmod{83} \qquad N \equiv 70 \pmod{110} \qquad N \equiv 3 \pmod{27}$$

The first step is to determine M, the product of the moduli. Here, $M = 83 \times 110 \times 27 = 246{,}510$. Since any two solutions to the system will be congruent modulo M, once Qin found one solution, he generally found the smallest positive solution by subtracting off sufficient copies of this value.

For the second step, Qin divided M by each of the moduli m_i in turn to get values we designate by M_i. Here $M_1 = M \div m_1 = 246{,}510 \div 83 = 2970$, $M_2 = 246{,}510 \div 110 = 2241$, and $M_3 = 246{,}510 \div 27 = 9130$. Each M_i satisfies $M_i \equiv 0 \pmod{m_j}$ for $j \neq i$.

In the third step, Qin subtracted from each of the M_i as many copies of the corresponding m_i as possible; that is, he found the remainders of M_i modulo m_i. These remainders, P_i, are $P_1 = 2970 - 35 \times 83 = 65$, $P_2 = 2241 - 20 \times 110 = 41$, and $P_3 = 9130 - 338 \times 27 = 4$. Of course, $P_i \equiv M_i \pmod{m_i}$ for each i, so P_i and m_i are relatively prime.

It is finally time to solve congruences, in particular, the congruences $P_i x_i \equiv 1 \pmod{m_i}$. Once this is done, one answer to the problem is easily seen to be

$$N = \sum_{i=1}^{n} r_i M_i x_i,$$

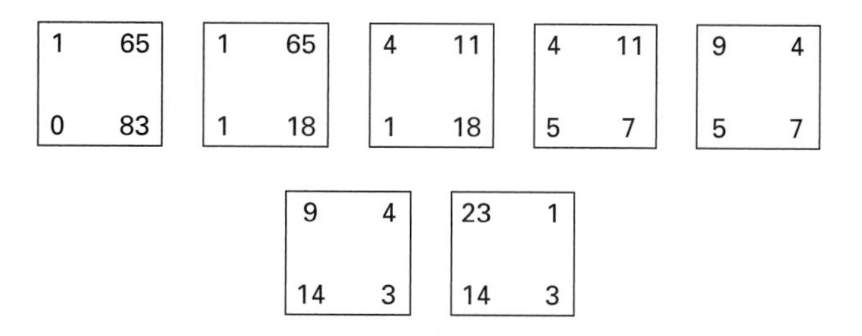

FIGURE 5.11 Counting board diagrams for solving $65x \equiv 1 \pmod{83}$ by the method of Qin Jiushao

in analogy with the solution to Sun Zi's problem. Because each m_i divides M, any multiple of M can be subtracted from N to get other solutions.

To solve $P_i x_i \equiv 1 \pmod{m_i}$ with P_i and m_i relatively prime, Qin used what he called the "technique of finding one," essentially the Euclidean algorithm. Qin described it using diagrams of the counting board. We can demonstrate the technique by solving $P_1 x_1 \equiv 1 \pmod{m_1}$, that is, $65x_1 \equiv 1 \pmod{83}$. Qin began by placing 65 in the upper right of a counting board with four squares, 83 in the lower right, 1 at the upper left, and nothing in the lower left. As he wrote, "first divide right bottom by right top, multiply the quotient obtained by the top left and [add it to] the bottom left, [at the same time replacing the bottom right by the remainder of the division]. And then use the right column top and bottom; using the smaller to divide the greater, dividing alternately, immediately multiply by the quotient obtained [and add it] successively ... into the left column top or bottom until finally the top right is just 1, then stop. Then take the top left result [as the solution]." The diagrams in Figure 5.11 represent the following computations:

$$
\begin{aligned}
83 &= 1 \cdot 65 + 18 & \qquad 1 \cdot 1 + 0 &= 1 \\
65 &= 3 \cdot 18 + 11 & 3 \cdot 1 + 1 &= 4 \\
18 &= 1 \cdot 11 + 7 & 1 \cdot 4 + 1 &= 5 \\
11 &= 1 \cdot 7 + 4 & 1 \cdot 5 + 4 &= 9 \\
7 &= 1 \cdot 4 + 3 & 1 \cdot 9 + 5 &= 14 \\
4 &= 1 \cdot 3 + 1 & 1 \cdot 14 + 9 &= 23
\end{aligned}
$$

The last numbers in the second column can be thought of as representing the absolute values of the successive coefficients of 65 obtained by substitution. That is, begin with $18 = 83 - 1 \cdot 65$ and substitute this into $11 = 65 - 3 \cdot 18$ to get $11 = 65 - 3 \cdot (83 - 1 \cdot 65) = 4 \cdot 65 - 3 \cdot 83$, where the 4 is the result of the second calculation in the second column. Similarly, $7 = 18 - 1 \cdot 11 = (83 - 1 \cdot 65) - 1 \cdot (4 \cdot 65 - 3 \cdot 83) = 4 \cdot 83 - 5 \cdot 65$. The final result is that $1 = 23 \cdot 65 - 18 \cdot 83$, and $x_1 = 23$ is a solution to the congruence. (Qin always adjusted matters so that the final coefficient was positive.)

To complete the original problem, we note that $x_2 = 51$ and $x_3 = 7$. It follows that

$$
\begin{aligned}
N = \sum_{i=1}^{3} r_i M_i x_i &= 32 \cdot 2970 \cdot 23 + 70 \cdot 2241 \cdot 51 + 3 \cdot 9130 \cdot 7 \\
&= 2{,}185{,}920 + 8{,}000{,}370 + 191{,}730 = 10{,}378{,}020.
\end{aligned}
$$

We then determine the smallest positive solution by subtracting off $42M = 42 \cdot 246{,}510 = 10{,}353{,}420$ to get the final answer: $N = 24{,}600$.

Two of the Chinese remainder problems in Qin's text are calendrical problems in which the question is to find the common cycle of several different cyclical events. The search for solutions to these types of problems may well have been the starting point for Qin's method, but there are no records of any development of the method over the years. And for various reasons, it is unlikely that the method was transmitted from other cultures. So at the moment, the only conclusion to be drawn is that Qin developed the methods himself.

5.5 Transmission to and from China

Not much is known about the possible transmission of mathematical ideas between China and other cultures before the sixteenth century. All that is known is that there are certain similarities in techniques in the mathematics of China, India, Europe, and the Islamic world. For example, the Chinese essentially used a decimal place value system on their counting board and even represented an empty place by a dot as early as the eighth century. But whether the Chinese system influenced the Indian development of the modern decimal place value system is not known. Similarly, Indian mathematicians used a technique involving the Euclidean algorithm to solve simultaneous congruences, while Islamic mathematicians used a technique related to Horner's method to solve polynomial equations numerically. However, in both cases, there are sufficient differences in detail to rule out direct copying from one civilization to the other. Whether the ideas traveled, however, is much more difficult to answer.

At the end of the sixteenth century, the Jesuit priest Mateo Ricci (1552–1610) came to China. Ricci and one of his Chinese students, Xu Guangqi (1562–1633), translated the first six books of Euclid's *Elements* into Chinese in 1607. And although it took many years for the Chinese to understand that the form and content of Euclidean geometry were inseparable (to Western minds, at least), at this time, Western mathematics began to enter China and the indigenous mathematics began to disappear.

Exercises

1. The basic Chinese symbols for numbers from the Shang period are

1	2	3	4	5	6	7	8	9	10	100	1000
—	=	≡	≣	☒	介	+	)(	⺌	|	ⓐ	千

There were compound symbols for 20, 30, and 40 (namely, U Ш Ш), but in general notation followed the system described in the text. Hence, 88 is)( )(and 162 is ⓐ介=. Write the Chinese forms of 56, 554, 63, and 3282.

2. Use the Chinese square root algorithm to find the square root of 142,884.

3. Solve problem 3 of chapter 3 of the *Nine Chapters*: Three people, who have 560, 350, and 180 coins, respectively, are required to pay a total tax of 100 coins in proportion to their wealth. How much does each pay?

4. Solve problem 26 of chapter 6 of the *Nine Chapters*: There is a reservoir with five channels bringing in water. If only the first channel is open, the reservoir can be filled in $\frac{1}{3}$ of a day. The second channel by itself will

fill the reservoir in 1 day, the third channel in $2\frac{1}{2}$ days, the fourth one in 3 days, and the fifth one in 5 days. If all the channels are open together, how long will it take to fill the reservoir? (This problem is the earliest known of this type. Similar problems appear in later Greek, Indian, and Western mathematics texts.)

5. Solve problem 28 of chapter 6 of the *Nine Chapters*: A man is carrying rice on a journey. He passes through three customs stations. At the first, he gives up $\frac{1}{3}$ of his rice, at the second, $\frac{1}{5}$ of what was left, and at the third, $\frac{1}{7}$ of what remains. After passing through all three customs stations, he has left 5 pounds of rice. How much did he have when he started? (Versions of this problem occur in later sources from various civilizations.)

6. Solve problem 8 of chapter 9 of the *Nine Chapters*: The height of a wall is 10 *ch'ih*. A pole of unknown length leans against the wall so that its top is even with the top of the wall. If the bottom of the pole is moved 1 *ch'ih* further from the wall, the pole will fall to the ground. What is the length of the pole?

7. Solve problem 20 of chapter 9 of the *Nine Chapters*: A square-walled city of unknown dimensions has four gates, one at the center of each side. A tree stands 20 *pu* from the north gate. One must walk 14 *pu* southward from the south gate and then turn west and walk 1775 *pu* before one can see the tree. What are the dimensions of the city?

8. Solve problem 24 of chapter 9 of the *Nine Chapters*: A deep well 5 feet in diameter is of unknown depth (to the water level). If a 5-foot post is erected at the edge of the well, the line of sight from the top of the post to the edge of the water surface below will pass through a point 0.4 foot from the lip of the well below the post. What is the depth of the well? (This is an example of the type of elementary surveying problem that stimulated Liu Hui to write his *Sea Island Mathematical Manual*.)

9. Solve problem 3 of the *Sea Island Mathematical Manual*: To measure the size of a square-walled city $ABCD$, we erect two poles 10 feet apart at F and E (Fig. 5.12). By moving northward 5 feet from E to G and sighting on D, the line of observation intersects the line EF at a point H such that $HE = 3\frac{93}{120}$ feet. Moving to point K such that $KE = 13\frac{1}{3}$ feet, the line of sight to D passes through F. Find DC and EC. (Liu Hui calculated $DC = 943\frac{3}{4}$ feet and $EC = 1245$ feet.)

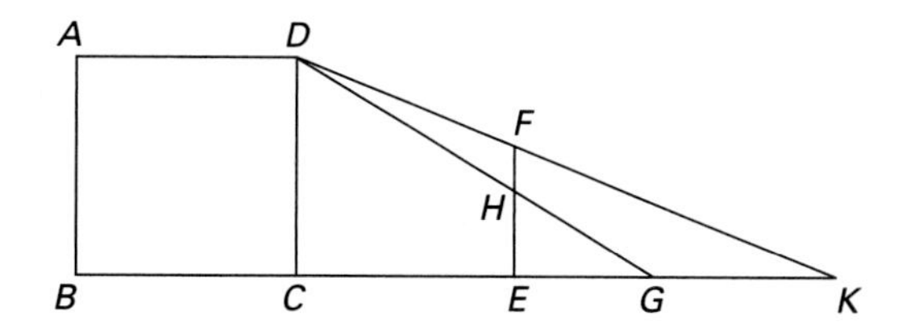

FIGURE 5.12 Problem 3 of the *Sea Island Mathematical Manual*

10. Perform the calculations of Liu Hui's algorithm for determining π to determine S_{2n} for $n = 6, 12, 24, 48,$ and 96.

11. Program your calculator to use Liu Hui's algorithm for determining π. Check Zu Chongzhi's value of S_{24576}. How many further iterations of the algorithm are necessary to achieve an accuracy of ten decimal places?

12. Use calculus to confirm that the volume of the "double box-lid" for a sphere of radius r is $\frac{16}{3}r^3$.

13. Find the solution to problem 3 of chapter 8 of the *Nine Chapters* using the Chinese method: The yields of 2 bundles of the best grain, 3 bundles of ordinary grain, and 4 bundles of the worst grain are neither sufficient to make a whole measure. If we add to the good grain 1 bundle of the ordinary, to the ordinary 1 bundle of the worst, and to the worst 1 bundle of the best, then each yield is exactly one measure. How many measures does 1 bundle of each of the three types of grain contain? Show that the solution according to the Chinese method involves the use of negative numbers.

14. Solve the following two equations numerically using Qin Jiushao's procedure. Both are taken from his text.
 (a) $16x^2 + 192x - 1863.2 = 0$
 (b) $-x^4 + 15{,}245x^2 - 6{,}262{,}506.25 = 0$

15. Use Qin's method to solve the pure quadratic equation $x^2 = 55{,}225$. Compare this method to the earlier square root procedure.

16. Use Qin's method to solve the pure cubic equation $x^3 = 12{,}812{,}904$. Show where the third-order coefficients of the Pascal triangle 3 3 1 appear in the solution procedures.

17. Solve the pure fourth-degree equation $y^4 = 279{,}841$ using Qin's procedure. Show how the fourth-order coefficients of the Pascal triangle 4 6 4 1 appear in the solution procedure.

18. The numerical equation from Qin Jiushao's text that is analyzed in section 5.3.2 came from the geometrical problem of finding the area of a pointed field. If the sides and one diagonal are labeled as in Figure 5.13, show that the area of the lower triangle is given by $B = (c/2)\sqrt{b^2 - (c/2)^2}$ and that of the upper triangle by $A = (c/2)\sqrt{a^2 - (c/2)^2}$. Then the area x of the entire field is given by $x = A + B$. Show that x satisfies the fourth-degree polynomial equation $-x^4 + 2(A^2 + B^2)x^2 - (A^2 - B^2)^2 = 0$. If $a = 39$, $b = 20$, and $c = 30$, show that this equation becomes the one solved by Qin and discussed in the text.

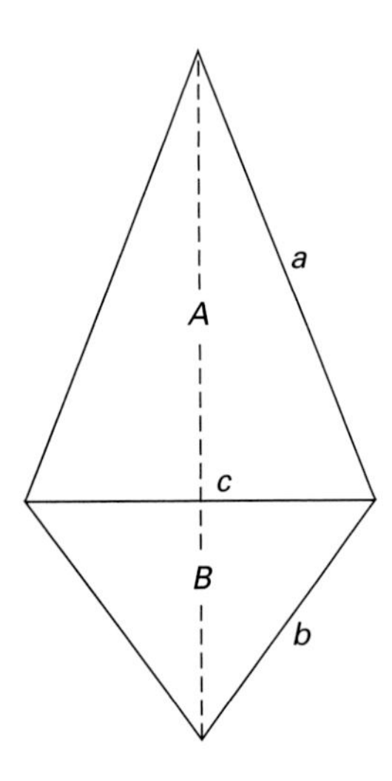

FIGURE 5.13 A pointed field from the *Mathematical Treatise in Nine Sections*

19. Solve problem 1 from chapter 1 of the *Mathematical Treatise in Nine Sections*, which is equivalent to solving the system $N \equiv 0 \pmod 3$, $N \equiv 1 \pmod 4$.
20. Solve problem 4 from chapter 1 of the *Mathematical Treatise*, which is equivalent to solving the system $N \equiv 0 \pmod{11}$, $N \equiv 0 \pmod 5$, $N \equiv 4 \pmod 9$, $N \equiv 6 \pmod 8$, $N \equiv 0 \pmod 7$.
21. Devise a lesson to teach the Pythagorean theorem using material from Chinese sources.
22. Explain why Qin's method of solving the congruence $P_i x_i \equiv 1 \pmod{m_i}$ works.
23. Imbed your explanation from exercise 22 into a lesson to teach the Chinese remainder theorem in a number theory course, based on the methods of Qin Jiushao.
24. Devise a lesson to teach a method of solving simultaneous linear equations using Chinese sources. In the lesson, you should explain why the method works.
25. Liu Hui's method for finding the height of a distant object was used in many cultures around the world until the seventeenth century. Curiously, this method was even used in cultures whose mathematicians understood methods of solving triangles using trigonometry. Discuss why this method would continue to be used, even in those circumstances.

References

Two relatively recent surveys of Chinese mathematics are available in English: Li Yan and Du Shiran, *Chinese Mathematics—A Concise History*, trans. John N. Crossley and Anthony W. C. Lun (Oxford: Clarendon Press, 1987) and Jean-Claude Martzloff, *A History of Chinese Mathematics*, trans. Stephen S. Wilson (Springer: Berlin, 1997). Two older surveys, which are still useful, are J. Needham, *Science and Civilization in China* (Cambridge: Cambridge University Press, 1959), vol. 3, and Yoshio Mikami, *The Development of Mathematics in China and Japan* (New York: Chelsea, 1974). Briefer surveys include Frank Swetz, "The Evolution of Mathematics in Ancient China," *Mathematics Magazine* 52 (1979), 10–19; Philip D. Straffin, Jr., "Liu Hui and the First Golden Age of Chinese Mathematics," *Mathematics Magazine* 71 (1998), 163–181; and Man-Keung Siu, "An Excursion in Ancient Chinese Mathematics," in Victor J. Katz, ed., *Using History to Teach Mathematics: An International Perspective* (Washington: MAA, 2000).

An English translation of chapter 9 of *Jiuzhang suanshu* with commentary was published by Frank Swetz and T. I. Kao as *Was Pythagoras Chinese?* (Reston, Va.: N.C.T.M, 1977). The entire work has been translated into English, with much commentary, by Shen Kangshen, John N. Crossley, and Anthony W.-C. Lun as *The Nine Chapters on the Mathematical Art: Companion and Commentary* (Oxford: Oxford University Press, 1999). The *Sea Island Mathematical Manual* is also available in English in Frank J. Swetz, *The Sea Island Mathematical Manual: Surveying and Mathematics in Ancient China* (University Park, Pa.: Pennsylvania State University Press, 1992). A detailed work on aspects of Chinese mathematics in the thirteenth century and its relationship to mathematics at other times and in other countries is Ulrich Libbrecht, *Chinese Mathematics in the Thirteenth Century: The Shu-shu chiu-chang of Ch'in Chiu-shao* (Cambridge: MIT Press, 1973). Finally, there is a guide to the literature on Chinese mathematics in Frank Swetz and Ang Tian Se, "A Brief Chronological and Bibliographic Guide to the History of Chinese Mathematics," *Historia Mathematica* 11 (1984), 39–56.

CHAPTER SIX

Ancient and Medieval India

In all those transactions which relate to worldly, Vedic, or . . . religious affairs, calculation is of use. In the science of love, in the science of wealth, in music and in the drama, in the art of cooking, and similarly in medicine and in things like the knowledge of architecture; in prosody, in poetics and poetry, in logic and grammar and such other things, . . . the science of computation is held in high esteem. In relation to movements of the sun and other heavenly bodies, in connection with eclipses and the conjunction of planets . . . it is utilized. The number, the diameter and the perimeter of islands, oceans and mountains, the extensive dimensions of the rows of habitations and halls belonging to the inhabitants of the world, . . . all of these are made out by means of computation.

—Mahāvīra's *Gaṇitāsarasaṅgraha* (9th century)

A civilization called the Harappan arose in India on the banks of the Indus River in the third millennium BCE, but there is no direct evidence of its mathematics. The earliest Indian civilization for which there is such evidence was formed along the Ganges River by Aryan tribes migrating from the Asian steppes late in the second millennium BCE. From about the eighth century BCE there were monarchical states in the area, which had to deal with such activities as fortifications, administrative centralization, and large-scale irrigation works. These states had a highly stratified social system headed by the king and the priests (brahmins). The literature of the brahmins was oral for many generations, expressed in lengthy verses called *Vedas*. Although these verses probably achieved their current form by 600 BCE, there are no written records dating from before the common era.

Some of the material from the *Vedas* describes the intricate sacrificial system of the priests, the bearers of the religious traditions that grew into Hinduism. It is in these works, the *Śulbasūtras*, that mathematical ideas are found. Curiously, however, although this mathematics deals with the theoretical requirements for building altars out of bricks, as far as is known, the early Vedic civilization did not have a tradition of brick technology, while the Harappan culture did. Thus, there is a possibility that the mathematics in the *Śulbasūtras* was created in the Harappan period, although the mechanism of its transmission to the later period is currently unknown. In any case, it is the *Śulbasūtras* that are the sources for current knowledge of ancient Indian mathematics.

In 327 BCE, Alexander the Great crossed the Hindu Kush mountains into northwestern India and, during the following two years, conquered the small Indian kingdoms of the area. Greek influence began to spread into India. Alexander came with scientists and historians in his entourage, not just as a conqueror interested in plunder but on a mission to "civilize" the east. (Naturally, the Indians believed they were already "civilized." Each people considered the other "barbarians.") Alexander's grand designs ended with his premature death in 323 BCE. His Indian provinces were soon reconquered by Chandragupta Maurya, who had earlier become king of Magadha, the major northern Indian kingdom of the time. Chandragupta established friendly relations with Seleucus, Alexander's successor in western Asia, and through this relationship there was evidently some interchange of ideas. Shortly after Chandragupta's death, Ashoka succeeded to the throne. He proceeded to conquer most of India but then converted to Buddhism; instead of continuing to acquire greater power, he sent missionaries both east and west to convert the neighboring kingdoms. Ashoka left records of his reign in edicts carved on pillars throughout his kingdom. These pillars contain some of the earliest written evidence of Indian numerals.

During the first century CE, northern India was conquered by Kushan invaders. The Kushan empire soon became the center of a flourishing trade between the Roman world and the east. Early in the fourth century, northern India was again united under a native dynasty, that of the Guptas. Under their rule, which only lasted about a century and a half, India reached a high point of culture with the flowering of art and medicine and the opening of universities. It was also during this period that Indian colonists spread Hindu culture to various areas of southeast Asia, including Burma, Malaya, and Indochina.

A northern Indian kingdom was revived in 606 by Harsha, a remarkably tolerant and just ruler. After his death in 647, however, his empire collapsed, and northern India broke up into many small states. Southern India at the time also consisted of numerous small kingdoms. Nevertheless, there was still cultural unity in the Indian subcontinent, primarily based on the use of Sanskrit as a common language of scholars, and so, even after the seventh century, it is possible to speak of Indian mathematics. Beginning in the eighth century, Muslim Arabs made periodic incursions into the north, and major battles between Muslims and Hindus occurred. Finally, toward the end of the twelfth century, northern India was conquered by a Muslim army under Mohammed Ghori, and in 1206 the Muslim Sultanate of Delhi was established, an empire that was to last over 300 years. The Sultanate even succeeded in conquering parts of the Hindu kingdoms in the south of India, which had generally been independent of the earlier northern kingdoms.

Throughout the various invasions and kingdoms, it does appear that the study of astronomy was always encouraged. Whoever ruled the country seemed to need astronomers to help with calendrical questions and, of course, to give astrological advice. Thus, much of Indian mathematics is recorded in astronomical works. Nevertheless, here, as elsewhere, creative mathematicians went beyond the strict requirements of practical problem solving to develop new areas of mathematics which they found of interest. We will consider in this chapter the Indian number systems and methods of calculation; the geometry of the *Śulbasūtras* and later works; the algebraic methods developed in the medieval period to solve equations, including the so-called Pell equation; the beginning of combinatorics;

and the development of trigonometry and associated techniques. We will conclude with a study of the development of power series in south India during the fourteenth and fifteenth centuries.

6.1 Indian Number Systems and Calculations

There is no record of the written number system of ancient India, but there is literary evidence that numerical symbols did exist. Examples of written numbers are available only from about the time of Ashoka in the third century BCE. The numbers appear in various decrees of the king inscribed on pillars throughout India. The system used at this time was mixed. There was a ciphered base-10 system with separate symbols for the numbers 1 through 9 and 10 through 90. For larger numbers, the system was a multiplicative one similar to that of the Chinese. For example, the symbol for 200 was a combination of the symbol for 2 and that for 100, while the symbol for 70,000 combined the symbols for 70 and 1000.

The details of the shift from this system to the modern Hindu-Arabic decimal place value system are still somewhat sketchy. By the time of Āryabhaṭa (late fifth century CE), special names were assigned to the various powers of ten: *dasa* (ten), *sata* (hundred), *sahasra* (thousand), *ayuta* (ten thousand), *niyuta* (hundred thousand). Then, sometime around 600 CE, the Indians dropped their old symbols for numbers higher than 9 and began to use their symbols for 1 through 9 in a place value arrangement. The earliest dated reference to this use, however, does not come from India itself. A fragment of a work by Severus Sebokht, a Syrian priest, dated 662, includes the remark that the Hindus have a valuable method of calculation "done by means of nine signs." Severus only speaks of nine signs, and there is no mention of a sign for zero. However, in the Bakhshālī manuscript, a mathematical manuscript in poor condition discovered in 1881 in the village of Bakhshālī in northwestern India, numbers are written using the place value system and a dot is used to represent zero. The best evidence is that this manuscript also dates from the seventh century. Perhaps Severus did not consider the dot as a "sign." In other Indian works from about the same period, numbers are generally written in a quasi–place value system to accommodate the poetic nature of the documents. For example, in the work of Mahāvīra in the ninth century, certain words stand for numbers: *moon* for 1, *eye* for 2, *fire* for 3, and *sky* for 0. So, the sequence *fire-sky-moon-eye* stands for 2103, and *moon-eye-sky-fire* stands for 3021. Note that the place values begin on the left, with the units.

Curiously, the earliest dated inscriptions using the modern decimal place value system including the zero are found in Cambodia and date from around 683. One inscription has 605 represented by three digits with a dot in the middle; another has 608 written in three digits with a modern zero in the middle. The dot as symbol for zero as part of a decimal place value system also appears in the *Chiu-chih li*, a Chinese astronomical work of 718 compiled by Indian scholars in the employ of the Chinese emperor. Although the actual symbols for the other Indian digits are not known, the author does give details of how the place value system works: "Using the [Indian] numerals, multiplication and division are carried out. Each numeral is written in one stroke. When a number is counted to ten, it is advanced into the higher place. In each vacant place a dot is always put. Thus, the numeral is always denoted in each place. Accordingly there can be no error in determining the place. With the numerals, calculation is easy. . . ."

The question remains as to why the Indians early in the seventh century dropped their own multiplicative system and introduced the place value system, including a symbol for zero. We cannot answer that definitively. It has been suggested that the origins of the system in India come from the Chinese counting board. Since the counting boards were portable, Chinese traders who visited India certainly carried them along. In fact, since southeast Asia is the border between Hindu culture and Chinese influence, it may have well been that the interchange took place in that area. What may have happened is that the Indians were impressed with the idea of using only nine symbols. But they naturally took for their symbols the ones they had already been using. They then improved the Chinese system of counting rods by using exactly the same symbols for each place value rather than alternating two types of symbols in the various places. And because they needed to be able to write numbers in some form, rather than just put them on a counting board, they were forced to use a symbol, the dot (and later the circle), to represent the blank column of the counting board. If this theory is correct, it is somewhat ironic that Indian scientists then returned the favor and brought this new system back to China early in the eighth century.

In any case, a fully developed decimal place value system for integers was most certainly in use in India by the eighth century, even though the earliest definitively dated decimal place value inscription there dates to 870. Well before then, though, this system had been transmitted not only to China but also west to Baghdad, the center of the developing Islamic culture. Note, however, that there is no early evidence in India of the decimal place value system being extended to decimal fractions. It was the Muslims who completed the Indian written decimal place value system by introducing these decimal fractions.

Even before the decimal place value system was fully developed, the Indians were adept at calculations. Brahmagupta, born in 598 in what is now Bhinmal in Rajasthan, gave many details of arithmetic calculation in his major work, *Brāhmasphuṭasiddhānta* (*Correct Astronomical System of Brahma*). Not only did he present the standard arithmetical rules for calculating with fractions, but in chapter 18 he gave the rules for dealing with negatives:

> The sum of two positives is positive, of two negatives negative; of a positive and a negative the sum is their difference; if they are equal it is zero. The sum of a negative and zero is negative, that of a positive and zero positive, and that of two zeros, zero. If a smaller positive is to be subtracted from a larger positive, the result is positive; if a smaller negative from a larger negative, the result is negative; if a larger negative or positive is to be subtracted from a smaller negative or positive, the sign of their difference is reversed—negative becomes positive and positive negative. A negative minus zero is negative, a positive minus zero positive; zero minus zero is zero. When a positive is to be subtracted from a negative or a negative from a positive, then it is to be added.
>
> The product of a negative and a positive is negative, of two negatives positive, and of two positives positive; the product of zero and a negative, of zero and a positive, or of two zeros is zero. A positive divided by a positive or a negative divided by a negative is positive; a zero divided by a zero is zero; a positive divided by a negative is negative; a negative divided by a positive is also negative. A negative or a positive divided by zero has that zero as its divisor, or zero divided by a negative or a positive has that negative or positive as its divisor.

Brahmagupta does not indicate how these rules were derived or exactly what he means by the rules dealing with division by zero. Still, the basic rules of operation were available and, as we will see below, were used effectively in the solution of equations.

6.2 Geometry

Many important geometric ideas are expressed in the *Śulbasūtras* as part of their treatment of the construction of altars. But since these literary pieces were not designed to teach mathematics as such, there are no derivations, just assertions. On the other hand, later commentators sometimes did give demonstrations. We will look at several results from the *Baudhāyana Śulbasūtra*, which probably dates from around 600 BCE. The first is the Pythagorean theorem:

> The areas of the squares produced separately by the length and the breadth of a rectangle together equal the area of the square produced by the diagonal. This is observed in rectangles having sides 3 and 4, 12 and 5, 15 and 8, 7 and 24, 12 and 35, 15 and 36.

A proof of this result is given in the *Yuktibhāsā*, written by Jyesthadeva in the mid–sixteenth century. The idea is to put two right triangles together, then draw the square on each of the two sides and on the hypotenuse (Fig. 6.1). If one cuts along each of the two lines indicated, then rotates each of the triangles outside the large square, the two pieces together will fill up the square on the hypotenuse. Again, as in the Chinese proof, there is no principle of beginning with axioms. One just studies the diagram, rotates the pieces, and understands that the theorem is true. This procedure could be thought of as an empirical proof.

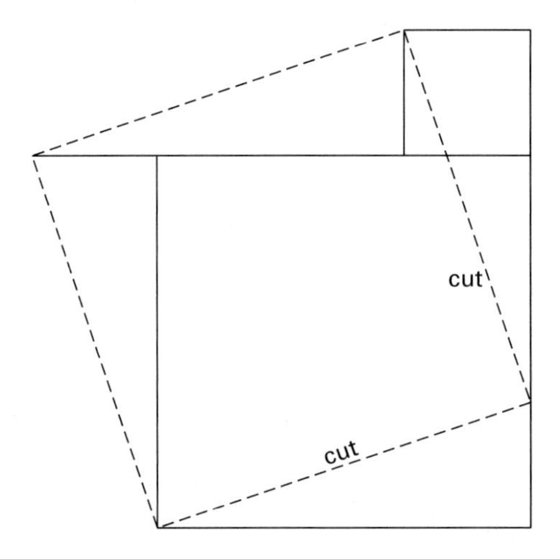

FIGURE 6.1 Proof of Pythagorean theorem, as given in the *Yuktibhāsā*

The Pythagorean theorem is then used implicitly to justify each of the following constructions:

> If it is desired to remove a square from another, a rectangular part is cut off from the larger square with the side of the smaller one to be removed; the longer side of the cut-off rectangular part is placed across so as to touch the opposite side; by this contact, the side is cut off. With the cut-off part, the difference of the two squares is obtained. (Fig. 6.2)
>
> If it is desired to transform a rectangle into a square, its breadth is taken as the side of a square, and this square on the breadth is cut off from the rectangle. The remainder of the rectangle is

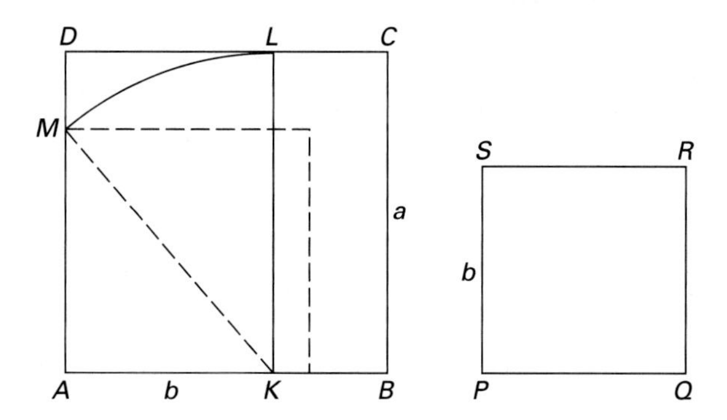

FIGURE 6.2 Procedure for determining a square equal to the difference of two squares, from the *Baudhāyana Śulbasūtra*

divided into two equal parts and placed on two sides (one part on each). The empty space in the corner is filled up with a square piece. The removal of it has been stated [in the previous construction]. (Fig. 6.3)

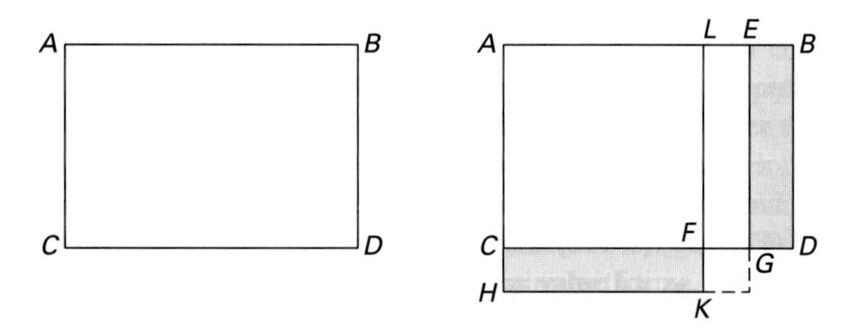

FIGURE 6.3 Procedure for transforming a rectangle into a square, from the *Baudhāyana Śulbasūtra*

Note that this second construction uses the "completing the square" technique, which is familiar from Babylonian mathematics. It is quite different, however, from Euclid's construction of the same problem found in *Elements* II-14. Later in the *Śulbasūtra* are two results involving circles:

> If it is desired to transform a square into a circle, a cord of length half the diagonal of the square is stretched from the center to the east, a part of it lying outside the eastern side of the square. With one-third of the part lying outside added to the remainder of the half diagonal, the requisite circle is drawn. (Fig. 6.4)
>
> To transform a circle into a square, the diameter is divided into eight parts; one such part, after being divided into twenty-nine parts is reduced by twenty-eight of them and further by the sixth of the part left less the eighth of the sixth part. [The remainder is then the side of the required square.]

In the first of these constructions, MN is the radius r of the desired circle. It is straightforward to show that if the side of the original square is s, then $r = [(2 + \sqrt{2})/6]s$. This implies a value for π of 3.088311755. In the second construction, the side of the required

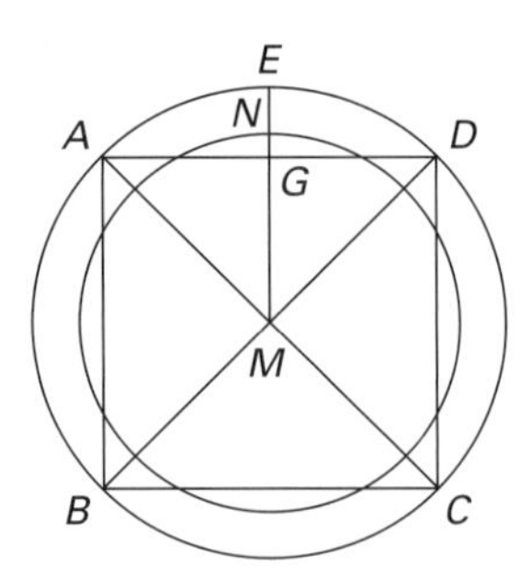

FIGURE 6.4 Indian procedure for "circling" the square

square is to be taken equal to

$$\frac{7}{8} + \frac{1}{8 \times 29} - \frac{1}{8 \times 29 \times 6} + \frac{1}{8 \times 29 \times 6 \times 8}$$

of the diameter of the circle. This is equivalent to taking 3.088326491 for π. The Indian work does not mention in either case that these constructions are approximations. What seems remarkable is that the two constructions imply values for π that are equal to four decimal places. Yet there is no indication of whether one of these constructions was derived from the other. On the other hand, Jyesthadeva does give an alternative construction for squaring the circle, one that he explicitly indicates is only approximate:

> Divide the diameter into fifteen parts and reduce it by two of them. This gives the approximate side of the desired square.

In other words, the side is here given as $\frac{13}{15}$ of the diameter. It is easy to see that this results in a value for π of $4(\frac{13}{15})^2 = 3.00444444$.

Many other geometric formulas, some exact, some stated as if exact but actually only approximate, and some explicitly described as approximate, occur in various Indian mathematical texts. But we will conclude this section with two remarkable results of Brahmagupta dealing with cyclic quadrilaterals (quadrilaterals inscribed in circles), given in chapter 12 of *Brāhmasphuṭasiddhānta*. The first is

> The accurate area [of a cyclic quadrilateral] is the square root of the product of the halves of the sums of the sides diminished by the square of its segments.

This result says that if $s = \frac{1}{2}(a + b + c + d)$, where a, b, c, d are the sides of the quadrilateral in cyclic order (Fig. 6.5), then the area S is given by $S = \sqrt{(s-a)(s-b)(s-c)(s-d)}$. A special case of this formula, when $d = 0$, gives the area of a triangle in terms of the lengths of its three sides as $S = \sqrt{s(s-a)(s-b)(s-c)}$, where again s is half the perimeter. This formula, usually known as **Heron's formula**, was probably known to Archimedes and was stated and proved by Heron, a Greek mathematician who lived in Alexandria in the first century CE. We have no idea how Brahmagupta discovered his formula, or whether he was aware of Heron's result. A complete proof first appeared in the *Yuktibhāsā*, based on another of Brahmagupta's results:

> One should multiply the sum of the products of the arms adjacent to the diagonals, after it has been mutually divided on either side, by the sum of the two products of the arms and the counter-arms. In an unequal cyclic quadrilateral, the two square roots are the two diagonals.

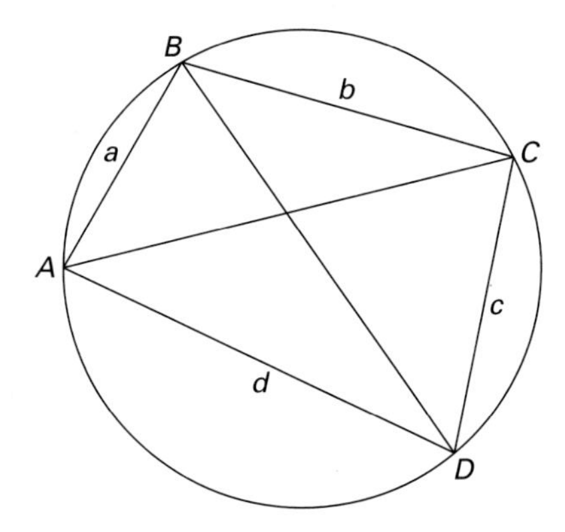

FIGURE 6.5 Area of a cyclic quadrilateral, from the *Brahmasphuṭasiddhānta*

This statement translates into formulas for the lengths of the diagonals AC and BD of the quadrilateral. Since the "sum of the products of the arms adjacent to" diagonal AC, or $ad + bc$, is "mutually divided" (that is, divided by the corresponding sum in respect to the second diagonal, or $ab + cd$) and then multiplied by the "sum of the two products of the arms and counter-arms" (or $ac + bd$), the result is

$$AC = \sqrt{\frac{(ac + bd)(ad + bc)}{ab + cd}} \quad \text{and, similarly,} \quad BD = \sqrt{\frac{(ac + bd)(ab + cd)}{ad + bc}}.$$

Again, a proof appears in the *Yuktibhāsā*.

6.3 Algebra

The rule for solving quadratic equations seems to have been known in India from at least the end of the fifth century. Āryabhaṭa, in dealing with arithmetic progressions in two stanzas of his *Āryabhaṭīya*, provides what amounts to the quadratic formula in a special case. Brahmagupta, a century and a quarter later, however, wrote out the algorithm explicitly for the equation $ax^2 + bx = c$:

> Take absolute number on the side opposite to that on which the square and simple unknown are. To the absolute number multiplied by four times the [coefficient] of the square, add the square of the [coefficient of the] unknown; the square root of the same, less the [coefficient of the] unknown, being divided by twice the [coefficient of the] square is the [value of the] unknown.

Brahmagupta's words can easily be translated into the formula

$$x = \frac{\sqrt{4ac + b^2} - b}{2a}.$$

As an example, Brahmagupta presented the solution to the equation $x^2 - 10x = -9$:

> Now to the absolute number [−9] multiplied by four times the [coefficient of the] square [−36], and added to the square [100] of the [coefficient of the] unknown, (making 64), the square root being extracted [8], and lessened by the [coefficient of the] unknown [−10], the remainder 18 divided by twice the [coefficient of the] square [2] yields the value of the unknown 9.

Note that although the given equation actually has a second positive solution, corresponding to the negative of the square root, Brahmagupta does not mention it. Several hundred years later, Bhāskara (1114–1185) (sometimes called Bhāskara II), the most famous of the medieval Indian mathematicians, did deal with multiple roots, at least when both are positive. Unfortunately, there is no record as to whether any of the Indian mathematicians learned the quadratic formula from the Babylonians or from Diophantus or discovered it independently.

On the other hand, Indian mathematicians very probably originated a method for solving linear congruences, because there is no comparable method described anywhere else. As in China, it seems the Indians' motivation for solving linear congruences was the desire to determine times when certain astronomical phenomena would recur. However, in contrast to the Chinese, the Indians generally only considered the case of two simultaneous congruences. For example, Brahmagupta described a method for finding an integer which when divided by each of two positive integers, leaves a given remainder. In modern notation, his goal is to find N satisfying $N \equiv a \pmod{r}$ and $N \equiv b \pmod{s}$, or to find x and y such that $N = a + rx = b + sy$, or so that $a + rx = b + sy$, or finally, setting $c = a - b$, so that $rx + c = sy$.

We will follow Brahmagupta's method by using an example from his text: $N \equiv 10 \pmod{137}$ and $N \equiv 0 \pmod{60}$. The problem can be written as the single equation $137x + 10 = 60y$. The basic procedure is to use the Euclidean algorithm to determine the greatest common divisor of 137 and 60, while at the same time, to make substitutions in accordance with the successive quotients coming from the algorithm.

$$
\begin{array}{lll}
137x + 10 = 60y & 60y - 10 = 137x & 137 = 2 \cdot 60 + 17 \\
60y - 10 = (2 \cdot 60 + 17)x & 60(y - 2x) - 10 = 17x & t = y - 2x \\
60t - 10 = 17x & 17x + 10 = 60t & 60 = 3 \cdot 17 + 9 \\
17x + 10 = (3 \cdot 17 + 9)t & 17(x - 3t) + 10 = 9t & u = x - 3t \\
17u + 10 = 9t & 9t - 10 = 17u & 17 = 1 \cdot 9 + 8 \\
9t - 10 = (1 \cdot 9 + 8)u & 9(t - 1u) - 10 = 8u & v = t - 1u \\
9v - 10 = 8u & 8u + 10 = 9v & 9 = 1 \cdot 8 + 1 \\
8u + 10 = (1 \cdot 8 + 1)v & 8(u - 1v) + 10 = 1v & w = u - 1v \\
8w + 10 = 1v & 1v - 10 = 8w &
\end{array}
$$

Brahmagupta solved the last equation, $1v - 10 = 8w$, by inspection: $v = 18$, $w = 1$. The remaining variables were then found by substitution, working up the column of variables.

$$
\begin{array}{ll}
u = 1v + w = 1 \cdot 18 + 1 = 19 & t = 1u + v = 1 \cdot 19 + 18 = 37 \\
x = 3t + u = 3 \cdot 37 + 19 = 130 & y = 2x + t = 2 \cdot 130 + 37 = 297
\end{array}
$$

Although $x = 130$, $y = 297$ is a solution to the original equation, Brahmagupta wanted a smaller solution, and, of course, he wanted to find N. He realized that given any solution x and y, he could get a new solution by adding or subtracting any multiple of 60 to x and the same multiple of 137 to y. That is, the solutions can be written as $x = 130 + 60z$, $y = 297 + 137z$. In this case, to determine the smallest value of x, he subtracts 2×60 from 130 to get $x = 10$. The corresponding value for y is then $297 - 2 \cdot 137 = 23$. Since $N = 60y$, Brahmagupta found that $N = 1380$. In fact, since N is only determined modulo $137 \cdot 60 = 8220$, we can write $N \equiv 1380 \pmod{8220}$.

The ability to solve systems of pairs of linear congruences turned out to be important in the solution of another type of indeterminate equation, the quadratic equation of the form $Dx^2 \pm b = y^2$. Today, the special case where $b = 1$ is usually referred to as the *Pell equation* (mistakenly named after the seventeenth-century Englishman John Pell). But although there are indications that the Greeks could solve a few of these equations, the general case, first developed in India, was undoubtedly the high point of medieval Indian algebra.

We will outline Brahmagupta's method of solving certain special cases of this quadratic equation, then continue with the general solution worked out over the next several hundred years. As usual, we will consider the rule in connection with an example: $92x^2 + 1 = y^2$. Brahmagupta began by picking an arbitrary value x_0 for x and another arbitrary value b_0 such that $92x_0^2 + b_0$ is a square, say, y_0^2. In this case, we take $x_0 = 1$ and choose $b_0 = 8$. Then $y_0 = 10$, and three numbers x_0, b_0, y_0 have been found satisfying the equation $Dx_0^2 + b_0 = y_0^2$. For convenience, we will say that (x_0, y_0) is a solution for additive b_0. In this case, (1, 10) is a solution for additive 8. Brahmagupta next noted that, in general, if (u_0, v_0) is a solution for additive c_0 and (u_1, v_1) is a solution for additive c_1, then $(u_0v_1 + u_1v_0, Du_0u_1 + v_0v_1)$ is a solution for additive c_0c_1. To check this result, we only need to verify the identity

$$D(u_0v_1 + u_1v_0)^2 + c_0c_1 = (Du_0u_1 + v_0v_1)^2,$$

given that $Du_0^2 + c_0 = v_0^2$ and $Du_1^2 + c_1 = v_1^2$. We will call this new solution the **composition** of the solutions (u_0, v_0) and (u_1, v_1). In the present example, Brahmagupta composed the solution (1, 10) with itself to get $(x_1, y_1) = (20, 192)$ as a solution for additive $b_1 = 64$. Since 64 is the square of 8, Brahmagupta divided that solution by 8. That is, $(\frac{5}{2}, 24)$ is a solution for additive 1. Because one of these roots is not an integer, this is not a satisfactory answer. So Brahmagupta composed this solution with itself to get an integral solution for additive 1: (120, 1151). In other words, $92 \cdot 120^2 + 1 = 1151^2$.

This example, as well as illustrating Brahmagupta's method, shows its limitations. The solution for additive 1 in the general case is the pair $(x_1/b_0, y_1/b_0)$. There is no guarantee that these will be integers or even that integers will be generated if this solution is combined with itself. But Brahmagupta simply gave several more rules and examples.

For example, Brahmagupta noted that composition allows him to get other solutions for any additive, provided he knows one solution for this additive as well as a solution for additive 1. Next, he showed how to find a solution for additive 1 if he has found a solution (u, v) for additive 4: If v is odd or u is even, then

$$(u_1, v_1) = \left(u\left(\frac{v^2 - 1}{2}\right), \; v\left(\frac{v^2 - 3}{2}\right)\right)$$

is the desired solution. In the case where v is even and u is odd,

$$(u_1, v_1) = \left(\frac{2uv}{4}, \; \frac{Du^2 + v^2}{4} = \frac{2v^2 - 4}{4}\right)$$

is an integral solution. Of course, if both u and v are even, then $(u/2, v/2)$ is already an integral solution. Brahmagupta gave a similar rule for subtractive 4, as well as rules for solving the Pell equation in other special circumstances.

Although his methods are always correct, the text has no proofs, nor do we learn how Brahmagupta discovered the method. In any case, the Pell equation became a tradition in Indian mathematics. It was studied through the next several centuries and was solved completely by the otherwise unknown Acarya Jayadeva (c. 1000). The solution given by Bhāskara is more easily followed, however.

Bhāskara's goal in his algebra text, *Līlāvatī*, was to show how any equation of the form $Dx^2 + 1 = y^2$ can be solved in integers. He began by recapitulating Brahmagupta's procedure. In particular, he emphasized that once one has found one solution pair, indefinitely many others can be found by composition. More importantly, however, he showed how to choose appropriate solution pairs for various additives so that one eventually reaches a pair that has the desired additive 1. We will follow Bhāskara's rule for the equation $Dx^2 + 1 = y^2$ using one of his examples: $67x^2 + 1 = y^2$.

As before, we begin by choosing a solution pair (u_0, v_0) for any additive b_0. For this example, we take $(1, 8)$ as a solution for additive -3. Next, we solve the indeterminate equation $u_0m + v_0 = b_0n$ for m (here, $1m + 8 = -3n$) by the method discussed above. The general result is $m = 1 + 3t$, $n = -3 - t$ for any integer t. We then choose t so that the square of m is as close to D as possible and take $b_1 = (m^2 - D)/b_0$ for the new additive (which may be negative). The new solution is $(u_1, v_1) = \left(|(u_0m + v_0)/b_0|, \sqrt{Du_1^2 + b_1}\right)$. In the given example, Bhāskara wanted m^2 close to 67, so he chose $t = 2$ and $m = 7$. Then $(m^2 - D)/b_0 = (49 - 67)/(-3) = 6$, so the new additive is 6. We calculate that $u_1 = |(1 \cdot 7 + 8)/(-3)| = 5$ and $v_1 = \sqrt{67 \cdot 25 + 6} = \sqrt{1681} = 41$. So $(5, 41)$ is a solution for additive 6.

Bhāskara claimed that if the above operation is repeated, eventually a solution for additive or subtractive 4, 2, or 1 will be reached. As already noted, a solution for additive 1 can be found from a solution with additive or subtractive 4. This is also easy to do with additive or subtractive 2 and with subtractive 1. Before continuing with the example, however, we need to discuss two questions, neither of which Bhāskara addressed. First, why does the method always give integral values at each stage? Second, why does the repetition of the method eventually give a solution pair for additives ± 4, ± 2, or ± 1?

To answer the first question, note that Bhāskara's method can be derived by composing a solution (u, v) for additive b with the obvious solution $(1, m)$ for additive $m^2 - D$. It follows that $(u', v') = (um + v, Du + mv)$ is a solution for additive $b(m^2 - D)$. Dividing the resulting equation by b^2 gives the solution $(u_1, v_1) = ((um + v)/b, (Du + mv)/b)$ for additive $(m^2 - D)/b$. It is then clear why m must be found so that $um + v$ is a multiple of b. It is not difficult to prove (although as usual the text does not have a proof) that if $(um + v)/b$ is integral, so are $(m^2 - D)/b$ and $(Du + mv)/b = \pm\sqrt{Du_1^2 + b_1}$.

The reason that $m^2 - D$ is chosen to be small is so that the second question can be answered. Unfortunately, the proof that the process eventually reaches additive 1 is quite difficult; the first published version did not appear until 1929. It may well be that neither Bhāskara nor Jayadeva proved the result. They may simply have done enough examples to convince themselves of its truth.

In any case, let us continue with Bhāskara's example. Beginning with $67 \cdot 1^2 - 3 = 8^2$, we have derived $67 \cdot 5^2 + 6 = 41^2$. The next step is to solve $5m + 41 = 6n$, with $|m^2 - 67|$ small. The appropriate choice is $m = 5$. Then $(u_2, v_2) = (11, 90)$ is a solution for additive -7, or $67 \cdot 11^2 - 7 = 90^2$. Again, solve $11m + 90 = -7n$. The value $m = 9$ works, and $(u_3, v_3) = (27, 221)$ is a solution for additive -2, or $67 \cdot 27^2 - 2 = 221^2$. At this point, since additive -2 has been reached, it is only necessary to compose $(27, 221)$ with itself.

This gives $(u_4, v_4) = (11{,}934, 97{,}684)$ as a solution for additive 4. Dividing by 2, Bhāskara finally found the desired solution to the original equation $67x^2 + 1 = y^2$: $x = 5967$, $y = 48{,}842$.

6.4 Combinatorics

The earliest recorded statements of combinatorial rules appear in India, although again without any proofs or justifications. For example, the medical treatise of Susruta, perhaps written in the sixth century BCE, states that sixty-three combinations can be made out of six different tastes—bitter, sour, salty, astringent, sweet, and hot—by taking them one at a time, two at a time, three at a time, ... In other words, there are six single tastes, fifteen combinations of two, twenty combinations of three, and so on. Other works from the same general time period include similar calculations dealing with such topics as philosophical categories and senses. In all these examples, however, the numbers are small enough that simple enumeration is sufficient to produce the answers. We do not know whether relevant formulas had been developed.

On the other hand, a sixth-century work by Varāhamihira dealt with a larger value. It plainly states that "if a quantity of 16 substances is varied in four different ways, the result will be 1820." In other words, since Varāhamihira was trying to create perfumes using four ingredients out of a total of 16, he had calculated that there were precisely $1820 (= C_4^{16})$ different ways of choosing the ingredients. It is unlikely that the author actually enumerated these 1820 combinations, and so we assume that he knew a method to calculate that number.

In the ninth century, Mahāvīra gave an explicit algorithm for calculating the number of combinations:

> The rule regarding the possible varieties of combinations among given things: Beginning with one and increasing by one, let the numbers going up to the given number of things be written down in regular order and in the inverse order (respectively) in an upper and a lower horizontal row. If the product of one, two, three, or more of the numbers in the upper row taken from right to left be divided by the corresponding product of one, two, three, or more of the numbers in the lower row, also taken from right to left, the quantity required in each such case of combination is obtained as the result.

Mahāvīra did not, however, give any proof of this algorithm, which can be translated into the modern formula

$$C_r^n = \frac{n(n-1)(n-2)\cdots(n-r+1)}{r!}.$$

He simply applied the rule to two problems, one about combinations of the tastes—as his predecessor did—and another about combinations of diamonds, sapphires, emeralds, corals, and pearls on a necklace.

Bhāskara gave many other calculations using this basic formula and also calculated that the number of permutations of a set of order n was $n!$ He was therefore able to ask and answer the question

> How many are the variations of form of the god Sambhu by the exchange of his ten attributes held reciprocally in his several hands: namely, the rope, the elephant's hook, the serpent, the tabor, the skull, the trident, the bedstead, the dagger, the arrow, and the bow?

6.5 Trigonometry

During the first centuries of the common era, in the period of the Kushan empire and that of the Guptas, Greek astronomical knowledge was likely transmitted to India, probably along the Roman trade routes. Just as the needs of Greek astronomy led to the development of trigonometry, the needs of Indian astronomy led to Indian improvements in this field.

Recall that Ptolemy, in order to solve triangles using a table of chords, often had to deal with half the chord of double the angle. It was probably an unknown Indian mathematician who decided that it would be much simpler to tabulate the half-chords of double the angle rather than the chords themselves. Thus, Indian trigonometry works use this half-chord "function." Of course, the length of the half-chord depends on the radius of the circle. In what follows, we will use the word *sine* rather than *half-chord*, always bearing in mind that until the eighteenth century, the word *sine* meant the length of a line in a circle of a particular radius.

Ptolemy tabulated his chords in a circle of radius 60, while Hipparchus, several centuries earlier, had used a radius of 3438. Since there are $360 \cdot 60 = 21{,}600$ minutes of arc in the circumference C, and since $C = 2\pi r$, Hipparchus took the radius as $21{,}600 \div 2\pi \approx 3438$, so that the radius would also be measured in minutes. Because the Indians used 3438 for the radius, it seems likely that Hipparchus's trigonometry rather than Ptolemy's first reached India.

The early Indian sine tables contained only values for arcs that were multiples of $3\frac{3}{4}^\circ$, along with the differences between adjacent values. The first part of such a table is shown in Table 6.1. (The arc numbers are simply for identification; for example, the sine of $11\frac{1}{4}^\circ$ is called the *third sine*, and 215 is called the *fifth sine difference*.) The Indians presumably calculated these values just as Ptolemy had. In other words, they began by noting that the sine of 90° is equal to the radius 3438′; the sine of 30° is half the radius, 1719′; and the sine of 45° is $3438/\sqrt{2} = 2431'$. They then calculated the sines of the other multiples of $3\frac{3}{4}^\circ$ using the Pythagorean theorem and the half-angle formula. They made use of the sine differences for approximating other sine values.

TABLE 6.1
Portion of an Early Indian Sine Table

Arc Number	Arc	Sine	Sine Difference
1	$3\frac{3}{4}^\circ$	225	225
2	$7\frac{1}{2}^\circ$	449	224
3	$11\frac{1}{4}^\circ$	671	222
4	15°	890	219
5	$18\frac{3}{4}^\circ$	1105	215
6	$22\frac{1}{2}^\circ$	1315	210
7	$26\frac{1}{4}^\circ$	1520	205
8	30°	1719	199

In fact, as early as the seventh century, Brahmagupta had developed an interpolation scheme similar to Newton's interpolation formula. In modern notation, if Δ_i represents the ith sine difference, α_i the ith arc, and $h = 3\frac{3}{4}^\circ$ the interval between these arcs, then Brahmagupta's result is that

$$\sin(\alpha_i + \theta) \approx \sin(\alpha_i) + \frac{\theta}{2h}(\Delta_i + \Delta_{i+1}) - \frac{\theta^2}{2h^2}(\Delta_i - \Delta_{i+1}).$$

For example, to calculate $\sin(20^\circ)$, note that $20 = 18\frac{3}{4} + 1\frac{1}{4}$, where $18\frac{3}{4} = \alpha_5$. The formula then gives

$$\begin{aligned}\sin(20) \approx \sin\left(18\tfrac{3}{4} + 1\tfrac{1}{4}\right) &= \sin\left(18\tfrac{3}{4}\right) + \frac{1\frac{1}{4}}{2\left(3\frac{3}{4}\right)}(215 + 210) - \frac{\left(1\frac{1}{4}\right)^2}{2\left(3\frac{3}{4}\right)^2}(215 - 210)\\ &= 1105 + \tfrac{1}{6}(425) - \tfrac{1}{18}(5) = 1176\end{aligned}$$

to the nearest integer.

Brahmagupta gave no justification for this interpolation formula, but we note that the right side of the formula is the unique quadratic polynomial in θ that agrees with the left side for $\theta = -3\frac{3}{4}^\circ$, $\theta = 0^\circ$, and $\theta = 3\frac{3}{4}^\circ$. Curiously, Brahmagupta himself often used an alternate procedure to approximate sines—namely, an algebraic formula that seems to have been first given by his older contemporary Bhāskara I (early seventh century).

Bhāskara's procedure can be translated into the formula

$$\sin\theta = \frac{4R\theta(180 - \theta)}{40{,}500 - \theta(180 - \theta)}.$$

If we use the formula to calculate the sine of $\theta = 20^\circ$, we get

$$\sin 20 = 3438 \cdot \frac{4 \cdot 20 \cdot 160}{40{,}500 - 20 \cdot 160} = 1180$$

to the nearest integer, a value with error of approximately 0.3%.

Again, we would like to know how this formula was derived. One modern suggestion is that Bhāskara started with the best quadratic approximation to the sine function. That would be the function $P(\theta) = R\theta(180 - \theta)/8100$, since this agrees with the sine function at $\theta = 0$, $\theta = 90$, and $\theta = 180$. He then noted that the same is true for the function $F(\theta) = \theta(180 - \theta)\sin\theta/8100$. Because $P(30) = \frac{5}{9}R$ and $F(30) = \frac{5}{18}R$, he proceeded to get a formula giving the correct value $R/2$ for $\theta = 30$ by the use of simple proportions:

$$\frac{P(\theta) - \sin\theta}{F(\theta) - \sin\theta} = \frac{\frac{5}{9}R - \frac{1}{2}R}{\frac{5}{18}R - \frac{1}{2}R}.$$

This reduces to the equation

$$\frac{R\theta(180 - \theta) - 8100\sin\theta}{\theta(180 - \theta)\sin\theta - 8100\sin\theta} = -\frac{1}{4},$$

which in turn gives Bhāskara's formula.

The apparent method of producing an approximation formula by beginning with a good guess and then tinkering with it to make it agree with the correct result on a few selected values appears in other parts of Indian mathematics. But since no author says that he is just tinkering, it is difficult to know not only how the results were obtained but also why. It may simply be that, as usual, mathematicians exercised their creative faculties to produce clever and beautiful results. And because the sine function was necessary in so many calculations for astronomical purposes, it was convenient for astronomers to have a sufficiently accurate rational approximation to the sine, which saved them the labor of constantly doing interpolations in the published sine tables. As we have already noted, there is little evidence that Indian mathematicians restricted themselves to methods based on a particular formal proof structure. Thus, even though it is certain that they knew how to prove mathematical results, the extant texts often demonstrate that once a result seemed sufficiently plausible, it was just passed down through the generations.

Now, in the time of Bhāskara I and Brahmagupta, algebraic approximations or interpolation schemes using differences were sufficient for the use to which these sine values were put in astronomy. But over the next several hundred years, it became increasingly important to have more accurate sine tables, particularly for navigation, as sailors in the Indian Ocean needed to be able to determine precisely their latitude and longitude. Since observation of the pole star was difficult in the tropics, one had to determine latitude by observation of the solar altitude at noon, μ. A standard formula for determining the latitude ϕ, given in an astronomical work of Bhāskara I, was $\sin\delta = \sin\phi \sin\mu$, where δ is the sun's declination (known from tables or calculations). Determination of longitude was somewhat more difficult, but this could also be accomplished using trigonometry if one knew the distance on the earth's surface of one degree along a great circle. In any case, the more accurate the sine values, the more accurately one could determine one's location. Thus, beginning late in the fourteenth century, mathematicians in south India, in what is now the state of Kerala, developed power series for the sine, cosine, and arctangent. These series appeared in written form in about 1530, in the *Tantrasaṃgraha-vyākhyā*, a commentary on a work by Nīlakaṇṭha (late fifteenth century). Derivations appear in the *Yuktibhāsā*, whose author, Jyesthadeva, credits these series to Madhava (1349–1425).

The Indian derivations of these results begin with the obvious approximations to the cosine and sine for small arcs and then use a "pull yourself up by your own bootstraps" approach to improve the approximation step by step. The derivations all make use of the notion of sine differences, an idea already used much earlier. In our discussion of the Indian method, we will use modern notation.

We first consider the circle of radius R with a small arc $\alpha = \widehat{AC} \approx AC$ (Fig. 6.6). From the similarity of triangles AGC and OEB, we get

$$\frac{x_1 - x_2}{\alpha} = \frac{y}{R} \quad \text{and} \quad \frac{y_2 - y_1}{\alpha} = \frac{x}{R}$$

or

$$\frac{\alpha}{R} = \frac{x_1 - x_2}{y} = \frac{y_2 - y_1}{x}.$$

In modern terms, if $\angle BOF = \theta$ and $\angle BOC = \angle AOB = d\theta$, these equations amount to

$$\sin(\theta + d\theta) - \sin(\theta - d\theta) = \frac{y_2 - y_1}{R} = \frac{\alpha x}{R^2} = \frac{2R\, d\theta}{R}\cos\theta = 2\cos\theta\, d\theta$$

and

$$\cos(\theta + d\theta) - \cos(\theta - d\theta) = \frac{x_2 - x_1}{R} = -\frac{\alpha y}{R^2} = -\frac{2R\, d\theta}{R}\sin\theta = -2\sin\theta\, d\theta.$$

(These results, of course, almost give the derivative of the sine and cosine.)

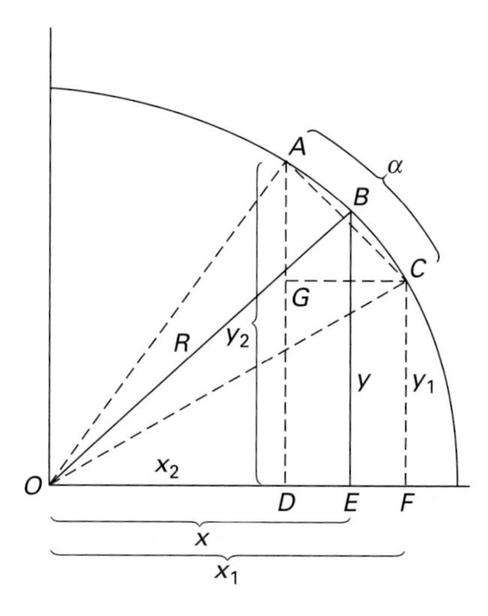

FIGURE 6.6 Derivation of power series for sine and cosine

Now, suppose we have a small arc s divided into n equal subarcs, with $\alpha = s/n$. For simplicity, we take $R = 1$, although the Indian mathematicians did not. By applying the previous results repeatedly, we get the following sets of differences for the y's (where $y_n = y = \sin s$) (Fig. 6.7):

$$\Delta_n y = y_n - y_{n-1} = \alpha x_n$$
$$\Delta_{n-1} y = y_{n-1} - y_{n-2} = \alpha x_{n-1}$$
$$\vdots$$
$$\Delta_2 y = y_2 - y_1 = \alpha x_2$$
$$\Delta_1 y = y_1 - y_0 = \alpha x_1.$$

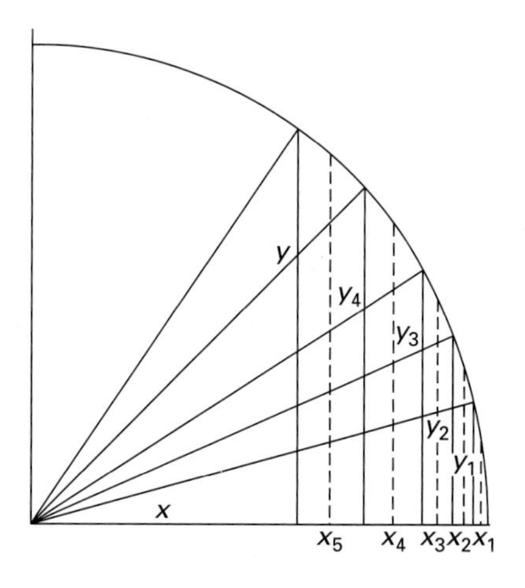

FIGURE 6.7 Differences of y's

Similarly, the differences for the x's can be written

$$\begin{aligned}\Delta_{n-1}x &= x_n - x_{n-1} = -\alpha y_{n-1}\\ &\vdots\\ \Delta_2 x &= x_3 - x_2 = -\alpha y_2\\ \Delta_1 x &= x_2 - x_1 = -\alpha y_1.\end{aligned}$$

We next consider the second differences for the y's:

$$\Delta_2 y - \Delta_1 y = y_2 - y_1 - y_1 + y_0 = \alpha(x_2 - x_1) = -\alpha^2 y_1.$$

In other words, the second difference of the sines is proportional to the negative of the sine. But since $\Delta_1 y = y_1$, we can write this result as

$$\Delta_2 y = y_1 - \alpha^2 y_1.$$

In general, we find that

$$\Delta_k y = y_1 - \alpha^2 y_1 - \alpha^2 y_2 - \cdots - \alpha^2 y_{k-1}.$$

But the sine equals the sum of its differences:

$$\begin{aligned}y = y_n &= \Delta_1 y + \Delta_2 y + \cdots + \Delta_n y\\ &= ny_1 - [y_1 + (y_1 + y_2) + (y_1 + y_2 + y_3) + \cdots + (y_1 + y_2 + \cdots + y_{n-1})]\alpha^2.\end{aligned}$$

Also, $s/n \approx y_1 \approx \alpha$, or $ny_1 \approx s$. Naturally, the larger the value of n, the better each of these approximations is. Therefore,

$$y \approx s - \lim_{n\to\infty}\left(\frac{s}{n}\right)^2 [y_1 + (y_1 + y_2) + \cdots + (y_1 + y_2 + \cdots + y_{n-1})].$$

Next, we add the differences for the x's. We get

$$x_n - x_1 = -\alpha(y_1 + y_2 + \cdots + y_{n-1}).$$

But $x_n \approx x = \cos s$ and $x_1 \approx 1$. It then follows that

$$x \approx 1 - \lim_{n\to\infty}\left(\frac{s}{n}\right)(y_1 + y_2 + \cdots + y_{n-1}).$$

To continue the calculation, the Indian mathematicians needed to approximate each y_i and use these approximations to get approximations for $x = \cos s$ and $y = \sin s$. Each new approximation in turn is placed back in the expressions for x and y and leads to a better approximation. Note first that if y is small, y_i can be approximated by is/n. It follows that

$$\begin{aligned}
x &\approx 1 - \lim_{n\to\infty}\left(\frac{s}{n}\right)\left[\frac{s}{n} + \frac{2s}{n} + \cdots + \frac{(n-1)s}{n}\right] \\
&= 1 - \lim_{n\to\infty}\left(\frac{s}{n}\right)^2 [1 + 2 + \cdots + (n-1)] \\
&= 1 - \lim_{n\to\infty}\frac{s^2}{n^2}\left[\frac{n^2}{2} - \frac{n}{2}\right] \\
&= 1 - \frac{s^2}{2}.
\end{aligned}$$

Note that in this calculation, we replaced the sum of the first $n-1$ integers by a simple expression. To go further, Jyesthadeva needed to know similar formulas for the sums of integral squares, integral cubes, and so on. In particular, he needed to know that

$$\sum_{i=0}^{n-1} i^k = \frac{n^{k+1}}{k+1} \pm \text{ lower-order terms.}$$

The preceding result was known in India, as was the result

$$\sum_{p=1}^{n-1}\left(\sum_{i=1}^{p} i^k\right) = n\sum_{i=1}^{n-1} i^k - \sum_{i=1}^{n-1} i^{k+1},$$

from which the earlier result was proved. Since both of these results were discovered several hundred years earlier in the Islamic world, we will postpone discussion of them until the next chapter. But we will use them in what follows. In particular, the first result will be used in the form

$$\lim_{n\to\infty}\frac{\sum_{i=1}^{n-1} i^k}{n^{k+1}} = \frac{1}{k+1}.$$

To get our new approximation for y, we proceed as follows:

$$\begin{aligned}
y &\approx s - \lim_{n\to\infty}\left(\frac{s}{n}\right)^2\left[\frac{s}{n} + \left(\frac{s}{n} + \frac{2s}{n}\right) + \cdots + \left(\frac{s}{n} + \frac{2s}{n} + \cdots + \frac{(n-1)s}{n}\right)\right] \\
&= s - \lim_{n\to\infty}\frac{s^3}{n^3}\{1 + (1+2) + (1+2+3) + \cdots + [1 + 2 + \cdots + (n-1)]\} \\
&= s - \lim_{n\to\infty}\frac{s^3}{n^3}\{n[1 + 2 + \cdots (n-1)] - [1^2 + 2^2 + \cdots + (n-1)^2]\} \\
&= s - s^3\lim_{n\to\infty}\left[\frac{\sum_{i=1}^{n-1} i}{n^2} - \frac{\sum_{i=1}^{n-1} i^2}{n^3}\right] \\
&= s - s^3\left(\frac{1}{2} - \frac{1}{3}\right) \\
&= s - \frac{s^3}{6}.
\end{aligned}$$

We thus have a new approximation for y and therefore for each y_i.

To improve the approximation for sine and cosine, we assume that $y_i \approx (is/n) - (is)^3/(6n^3)$ in the expression for $x = \cos s$ and proceed as before. We use the two sum formulas in the case $k = 3$ to get

$$x \approx 1 - \frac{s^2}{2} + \frac{s^4}{24}.$$

Similarly, we get a new approximation for $y = \sin s$:

$$y \approx s - \frac{s^3}{6} + \frac{s^5}{120}.$$

Because Jyesthadeva considered each new term in these polynomials as a correction of the previous value, he understood that the more terms taken, the more closely the polynomials approach the true values for sine and cosine. The polynomial approximations can thus be continued as far as necessary to achieve any desired approximation. The Indian mathematicians had therefore discovered the sine and cosine power series.

6.6 Transmission to and from India

We are much better informed about Indian mathematics throughout history than we are about the mathematics of China. We know, for example, that Indian mathematicians learned trigonometry (and also some astronomy) from Greek sources. We also know that Islamic scholars learned Indian trigonometry when Indian works were brought to Baghdad in the eighth century. And, of course, the decimal place value system traveled from India through the Middle East and Africa to western Europe over a period of several hundreds of years. On the other hand, there is no record of the Indian solution of the Pell equation being known in Europe before European scholars solved it themselves (and in a way different from that of the Indians).

The most interesting question in this context, however, relates to the transmission of the power series for the sine and cosine. There is certainly no available documentation showing that any Europeans knew of Indian developments in this area before they themselves worked out the power series in the mid–seventeenth century. However, there is some circumstantial evidence. First, the Europeans, just like the Indians, needed precise trigonometric values for navigation. Second, the texts in which these power series were described were easily available in south India. Third, the Jesuits, in their quest to proselytize in Asia, established a center in south India in the late sixteenth century. In general, wherever the Jesuits went, they learned the local languages, collected and translated local texts, and then set up educational institutions to train disciples. But the question remains as to whether, in fact, the Jesuits did find the texts describing power series and bring them back in some form to Europe. As we will discuss in the chapter on calculus, in the years from 1630 to 1680, some of the basic ideas present in these Indian texts began to appear in European works. Newton kept extensive records of his thinking in his notebooks, which reveal no evidence that he was aware of Indian material. But many of the other European mathematicians left little documentary evidence of how they discovered and elaborated on their ideas. So, we can only speculate as to whether Indian trigonometric series were transmitted in some form to Europe by the early seventeenth century.

Exercises

1. Show that the construction given in the text for constructing a square equal to the difference of two squares is correct (Fig. 6.2). Here, $ABCD$ is the larger square with side equal to a, and $PQRS$ is the smaller square with side equal to b. Cut off $AK = b$ from AB, and draw KL perpendicular to AK and intersecting DC at L. With K as center and radius KL, draw an arc meeting AD at M. Show that the square on AM is the required square.

2. Show that the construction given in the text for transforming a rectangle into a square is correct (Fig. 6.3). The rectangle is $ABCD$. Find L on AB so that $AL = AC$. Then find the midpoint E of LB, and draw EG parallel to LF. Move the rectangle $EBGD$ from where it is to the bottom of the diagram, forming the rectangle $CFHK$. Complete the square by adding the square on FG. Show that using the result of exercise 1 gives the result.

3. This is the method presented in the text for finding a circle whose area is equal to a given square: In square $ABCD$, let M be the intersection of the diagonals (Fig. 6.4). Draw the circle with M as center and MA as radius; let ME be the radius of the circle perpendicular to the side AD and cutting AD at G. Let $GN = \frac{1}{3}GE$. Then MN is the radius of the desired circle. Show that if $AB = s$ and $MN = r$, then $r/s = (2 + \sqrt{2})/6$. Show that this implies a value for π equal to 3.088311755.

4. The *Śulbasūtra* method of squaring a circle of diameter d takes the side of the desired square to be $7/8 + 1/(8 \times 29) - 1/(8 \times 29 \times 6) + 1/(8 \times 29 \times 6 \times 8)$ times d. Show that this is equivalent to using a value for π equal to 3.088326491.

5. Brahmagupta asserts that if $ABCD$ is a quadrilateral inscribed in a circle, with side lengths a, b, c, d (in cyclic order) (Fig. 6.5), then the lengths of the diagonals AC and BD are given by

$$AC = \sqrt{\frac{(ac + bd)(ad + bc)}{ab + cd}}$$

and

$$BD = \sqrt{\frac{(ac + bd)(ab + cd)}{ad + bc}}.$$

Prove this result, as follows:

(a) Let $\angle ABC = \theta$. Then $\angle ADC = \pi - \theta$. Let $x = AC$. Use the law of cosines on each of triangles ABC and ADC to express x^2 two different ways. Then, since $\cos(\pi - \theta) = -\cos\theta$, use these two formulas for x^2 to determine $\cos\theta$ as a function of a, b, c, and d.

(b) Replace $\cos\theta$ in your expression for x^2 in terms of a and b with the value for the cosine determined in part a.

(c) Show that $cd(a^2 + b^2) + ab(c^2 + d^2) = (ac + bd)(ad + bc)$.

(d) Simplify the expression for x^2 found in part b by using the algebraic identity found in part c. By then taking square roots, you should get the desired expression for $x = AC$. (Of course, a similar argument will then give you the expression for $y = BD$.)

6. Brahmagupta asserts that if $ABCD$ is a quadrilateral inscribed in a circle, as in exercise 5, and if $s = \frac{1}{2}(a + b + c + d)$, then the area of the quadrilateral is given by $S = \sqrt{(s - a)(s - b)(s - c)(s - d)}$ (Fig. 6.8). Prove this result, as follows:

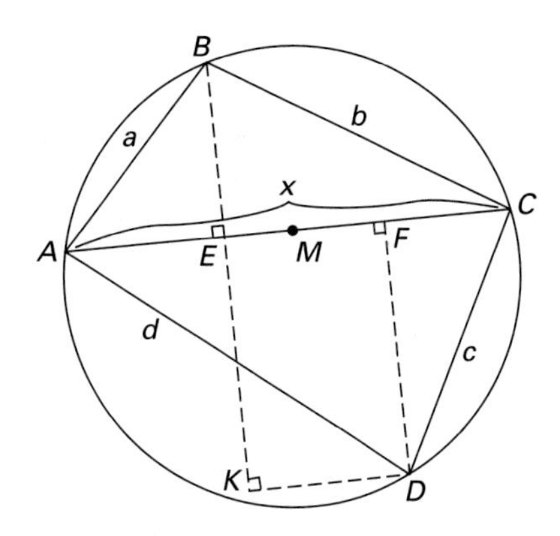

FIGURE 6.8 Area of a quadrilateral inscribed in a circle

(a) In triangle ABC, drop a perpendicular from B to point E on AC. Apply the law of cosines to that triangle to show that $b^2 - a^2 = x(x - 2AE)$.

(b) Let M be the midpoint of AC, so $x = 2AM$. Use the result of part a to show that $EM = (b^2 - a^2)/2x$.

(c) In triangle ADC, drop a perpendicular from D to point F on AC. Use arguments similar to those in parts a and b to show that $FM = (d^2 - c^2)/2x$.

(d) Denote the area of quadrilateral $ABCD$ by P. Show that $P = \frac{1}{2}x(BE + DF)$ and therefore that $P^2 = \frac{1}{4}x^2(BE + DF)^2$.

(e) Extend BE to K such that $\angle BKD$ is a right angle, and complete the right triangle BKD. Then $BE + DF = BK$. Substitute this value in your expression from part d; then use the Pythagorean theorem to conclude that $P^2 = \frac{1}{4}x^2(y^2 - EF^2)$.

(f) Since $EF = EM + FM$, conclude that $EF = [(b^2 + d^2) - (a^2 + c^2)]/2x$. Substitute this value into the expression for P^2 found in part e, along with the values for x^2 and y^2 found in exercise 5. Conclude that $P^2 = \frac{1}{4}(ac + bd)^2 - \frac{1}{16}[(b^2 + d^2) - (a^2 + c^2)]^2 = \frac{1}{16}\{4(ac + bd)^2 - [(b^2 + d^2) - (a^2 + c^2)]^2\}$.

(g) Since $s = \frac{1}{2}(a + b + c + d)$, show that

$$s - a = \tfrac{1}{2}(b + c + d - a),$$
$$s - b = \tfrac{1}{2}(a + c + d - b),$$
$$s - c = \tfrac{1}{2}(a + b + d - c),$$

and

$$s - d = \tfrac{1}{2}(a + b + c - d).$$

(h) To prove the theorem, it is necessary to show that the expression for P^2 given in part f is equal to the product of the four expressions in part g. It is clear that the denominators are both equal to 16. To prove that the numerators are equal involves a lot of algebraic manipulation. Work carefully and show that the two numerators are in fact equal.

7. This problem is from Brahmagupta's work on congruences: Given that the sun makes 30 revolutions through the ecliptic in 10,960 days, how many days have elapsed (since the sun was at a given starting point) if the sun has made an integral number of revolutions plus 8080/10,960 of a revolution, that is, "when the remainder of solar revolutions is 8080"? If y is the number of days sought and x is the number of revolutions, then, because 30 revolutions take 10,960 days, x revolutions take $(1096/3)x$ days. Therefore, $y = (x + 808/1096)(1096/3)$, or $1096x + 808 = 3y$. Solve $N \equiv 808 \pmod{1096}$ and $N \equiv 0 \pmod{3}$.

8. Solve the congruence $N \equiv 23 \pmod{137}$, $N \equiv 0 \pmod{60}$ using Brahmagupta's procedure.

9. Solve $1096x + 1 = 3y$ using Brahmagupta's procedure. Given a solution to this equation (with additive 1), it is easy to find solutions to equations with other additives by simply multiplying. For example, solve $1096x + 10 = 3y$.

10. Prove that Brahmagupta's procedure does give a solution to the simultaneous congruences. Begin by noting that the Euclidean algorithm allows one to express the greatest common divisor of two positive integers as a linear combination of these integers. Note further that in order for there to be a solution, this greatest common divisor must divide the "additive." Brahmagupta does not mention this, but Bhāskara and others do.

11. Solve the problem $N \equiv 5 \pmod 6 \equiv 4 \pmod 5 \equiv 3 \pmod 4 \equiv 2 \pmod 3$ by the Indian procedure (applied three times) and by the Chinese procedure. Compare the methods.

12. Solve the congruence $N \equiv 10 \pmod{137} \equiv 0 \pmod{60}$ by the Chinese procedure and compare your solution step by step with the solution by Brahmagupta's procedure. How do the two methods compare?

13. Solve the indeterminate equation $17n - 1 = 75m$ by both the Indian and Chinese procedures using the Euclidean algorithm explicitly. Compare the solutions.

14. Prove $D(u_0v_1 + u_1v_0)^2 + c_0c_1 = (Du_0u_1 + v_0v_1)^2$ given that $Du_0^2 + c_0 = v_0^2$ and $Du_1^2 + c_1 = v_1^2$.

15. Solve $83x^2 + 1 = y^2$ by Brahmagupta's procedure. Begin by noting that $(1, 9)$ is a solution for subtractive 2.

16. Show that if (u, v) is a solution to $Dx^2 - 4 = y^2$, then

$$(u_1, v_1) =$$
$$(\tfrac{1}{2}uv(v^2 + 1)(v^2 + 3), (v^2 + 2)[\tfrac{1}{2}(v^2 + 1)(v^2 + 3) - 1])$$

is a solution to $Dx^2 + 1 = y^2$ and that both u_1 and v_1 are integers regardless of the parity of u or v.

17. Solve $13x^2 + 1 = y^2$ by noting that $(1, 3)$ is a solution for subtractive 4 and applying the method of exercise 16.

18. Show that if (u, v) is a solution to $Dx^2 + 2 = y^2$, then $(u_1, v_1) = (uv, v^2 - 1)$ is a solution to $Dx^2 + 1 = y^2$. Deduce a similar rule if (u, v) is a solution to $Dx^2 - 2 = y^2$.

19. Solve $61x^2 + 1 = y^2$ by Bhāskara's process. The solution is $x = 226{,}153{,}980$, $y = 1{,}766{,}319{,}049$.

20. Solve the following problem from Mahāvīra: "One night, in a month of the spring season, a certain young lady ... was lovingly happy along with her husband on ... the floor of a big mansion, white like the moon, and situated in a pleasure garden with trees bent down with the load of the bunches of flowers and fruits, and resonant with the sweet sounds of parrots, cuckoos and bees, which were all intoxicated with the honey obtained from the flowers therein. Then on a love quarrel arising between the husband and the wife, that lady's necklace made up of pearls became sundered and fell on the floor. One-third of that necklace of pearls reached the maid-servant there; one-sixth fell on the bed; then one-half of what remained (and one-half of what remained thereafter and again one-half of what remained thereafter and so on, counting six times in all) fell all of them everywhere; and there were found to remain (unscattered) 1,161 pearls.... Give out the (numerical) measure of the pearls (in that necklace)."
21. Solve the following problem from Mahāvīra: There are 4 pipes leading into a well. Among these, each fills the well (in order) in $\frac{1}{2}$, $\frac{1}{3}$, $\frac{1}{4}$, and $\frac{1}{5}$ of a day. In how much of a day will all of them together fill the well and each of them to what extent?
22. Solve the following problem from the Bakhshālī manuscript: One person goes 5 *yojanas* a day. When he has proceeded for 7 days, the second person, whose speed is 9 *yojanas* a day, departs. In how many days will the second person overtake the first?
23. Here is another problem from Mahāvīra: Two travelers found a purse containing money. The first said to the second, "By securing half of the money in this purse, I shall become twice as rich as you." The second said to the first, "By securing two-thirds of the money in the purse, I shall, with the money I have on hand, have three times as much money as what you have on hand." How much did each have and how much was in the purse?
24. Use a graphing calculator and/or calculus techniques to show that the algebraic formula of Bhāskara I approximates the sine between 0 and 180° with an error of no more than 1%. Find the values that are most in error.
25. Use both the interpolation scheme of Brahmagupta and the algebraic formula of Bhāskara I to approximate $\sin(16°)$. Compare the two values to each other and to the exact value. What are the respective errors?
26. Continue the process described in the text for determining the power series for the sine and cosine for two more steps in each case. That is, beginning with $y_i \approx (is/n) - (is)^3/(6n^3)$, show that $x = \cos s \approx 1 - s^2/2 + s^4/24 - s^6/720$ and $y = \sin s \approx s - s^3/6 + s^5/120 - s^7/5040$.
27. For a number theory course, devise a lesson on solving indeterminate equations of the form $rx + c = sy$, using the methods of Brahmagupta.
28. Why would the Indians have thought it better to use an algebraic approximation to the sine function rather than calculating values using geometric methods and methods of interpolation?

References

Among the surveys of the general history of Indian mathematics are B. Datta and A. N. Singh, *History of Hindu Mathematics* (Bombay: Asia Publishing House, 1961) (reprint of 1935–1938 original); A. K. Bag, *Mathematics in Ancient and Medieval India* (Varanasi: Chaukhambha Orientalia, 1979); and C. N. Srinivasiengar, *The History of Ancient Indian Mathematics* (Calcutta: The World Press Private Ltd., 1967). Other surveys include volume 2 of D. Chattopadhyaya, ed., *Studies in the History of Science in India* (New Delhi: Editorial Enterprises, 1992); S. N. Sen, "Mathematics," in D. M. Bose, S. N. Sen, and B. V. Subbarayappa, *A Concise History of Science in India* (New Delhi: Indian National Science Academy, 1971), 136–212; chapters 8 and 9 in G. G. Joseph, *The Crest of the Peacock: Non-European Roots of Mathematics* (Princeton: Princeton University Press, 2000); and T. K. Puttaswamy, "The Mathematical Accomplishments of Ancient Indian Mathematicians," in Helaine Selin, ed., *Mathematics across Cultures: The History of Non-Western Mathematics* (Dordrecht: Kluwer Academic Publishers, 2000).

A basic study of the mathematical contents of the *Sulbasutras* is B. Datta, *The Science of the Sulba* (Calcutta: University of Calcutta, 1932); a more general work on Indian geometry is T. A. Sarasvatī, *Geometry in Ancient and Medieval India* (New Delhi: Motilal Banarsidass, 1979). A detailed discussion of the entire process of solving the

Pell equation is found in C. O. Selenius, "Rationale of the Chakravāla Process of Jayadeva and Bhāskara II," *Historia Mathematica* 2 (1975), 167–184. More on approximation methods in India can be found in the doctoral dissertation of Kim Plofker: *Mathematical Approximation by Transformation of Sine Functions in Medieval Sanskrit Astronomical Texts* (Brown University, 1995). R. H. Gupta has a survey of South Indian mathematics in "South Indian Achievements in Medieval Mathematics," *Gaṇita Bhārati* 9 (1987), 15–40. The power series methods are discussed in three articles: Ranjan Roy, "The Discovery of the Series Formula for π by Leibniz, Gregory and Nilakantha," *Mathematics Magazine* 63 (1990), 291–306; Victor J. Katz, "Ideas of Calculus in Islam and India," *Mathematics Magazine* 68 (1995), 163–174; and David Bressoud, "Was Calculus Invented in India?" *College Mathematics Journal* 33 (2002), 2–13.

The *Śulbasūtras* are available in English in S. N. Sen and A. K. Bag, eds., *The Śulbasūtras of Baudhāyana, Āpastamba, Kātyāyana and Mānava with Text, English Translation and Commentary* (New Delhi: Indian National Science Academy, 1983). The text of Āryabhaṭa is available in an English translation by Walter E. Clark: *The Āryabhaṭīya of Āryabhaṭa* (Chicago: University of Chicago Press, 1930). The major mathematical texts of Bhāskara and Brahmagupta were translated by H. T. Colebrooke in *Algebra with Arithmetic and Mensuration from the Sanskrit of Brahmegupta and Bhāscara* (London: John Murray, 1817). More modern translations should be available in the near future. Mahāvīra's *Gaṇitāsarasaṅgraha* was edited and translated by M. Rangācārya and published in 1912 by Government Press in Madras.

CHAPTER SEVEN

Mathematics in the Islamic World

You know well . . . for which reason I began searching for a number of demonstrations proving a statement due to the ancient Greeks . . . and which passion I felt for the subject . . . so that you reproached me my preoccupation with these chapters of geometry, not knowing the true essence of these subjects, which consists precisely in going in each matter beyond what is necessary . . . . Whatever way he [the geometer] may go, through exercise will he be lifted from the physical to the divine teachings, which are little accessible because of the difficulty to understand their meaning . . . and because of the circumstance that not everybody is able to have a conception of them, especially not the one who turns away from the art of demonstration.

—Preface to the *Book on Finding the Chords in the Circle* by al-Bīrūnī, c. 1030

In the first half of the seventh century CE, a new civilization came out of Arabia. Under the inspiration of the prophet Muḥammad, the new monotheistic religion of Islam quickly attracted the allegiance of the inhabitants of the Arabian peninsula. In less than a century after Muḥammad's capture of Mecca in 630, the Islamic armies conquered an immense territory as they propagated the new religion, first among the previously polytheistic tribes of the Middle East and then among the adherents of other faiths. Syria and then Egypt were wrested from the Byzantine empire. Persia was conquered by 642, and soon the victorious armies had reached as far as India and parts of central Asia. In the west, North Africa was quickly overrun, and in 711 Islamic forces entered Spain. Their forward progress was eventually halted at Tours by the army of Charles Martel in 732. Already, however, the problems of conquest were being replaced by the new problems of governing the immense new empire. Muḥammad's successors, the caliphs, originally set up their capital in Damascus, but after about a hundred years of wars, including great victories but also some substantial defeats, the caliphate split up into several parts. In the eastern segment, under the Abbasid caliphs, the spread of wealth and the cessation of wars of conquest created favorable conditions for the development of a new culture.

In 766, the caliph al-Manṣūr founded his new capital of Baghdad, which soon became a flourishing commercial and intellectual center. The initial impulses of Islamic orthodoxy were soon replaced by a more tolerant atmosphere, and the intellectual accomplishments of all residents of the caliphate were welcomed. The caliph Hārūn al-Rashīd, who ruled from 786 to 809, established a library in Baghdad. Manuscripts were collected from various academies in the Near East, which had been established by those fleeing from the persecutions inflicted on scholars in the ancient academies in Athens and Alexandria. These manuscripts included many of the classic Greek mathematical and scientific texts. A program of translating these manuscripts into Arabic was soon begun. Hārūn's successor, the caliph al-Ma'mūn (813–833), established a research institute, the *Bayt al-Ḥikma* (House of Wisdom), which was to last over 200 years. To this institute were invited scholars from all parts of the caliphate to translate Greek and Indian works as well as to conduct original research. By the end of the ninth century, many of the principal works of Euclid, Archimedes, Apollonius, Diophantus, Ptolemy, and other Greek mathematicians had been translated into Arabic and were available to the scholars gathered in Baghdad. Islamic scholars also absorbed the ancient mathematical traditions of the Babylonian scribes, still evidently available in the Tigris-Euphrates valley, and learned the mathematics of the Hindus.

The Islamic scholars during the first few hundred years of Islamic rule did more than just bring these sources together. They amalgamated them into a new whole and, in particular, as the opening quotation indicates, infused their mathematics with what they felt was divine inspiration. Creative mathematicians of the past had always carried investigations well beyond the dictates of immediate necessity, but many Muslims believed that this was a requirement of God. Islamic culture in general regarded secular knowledge not as in conflict with holy knowledge but as a way to it. Learning was therefore encouraged, and those who had demonstrated sparks of creativity were often supported by the rulers (usually both secular and religious authorities) so that they could pursue their ideas as far as possible. Mathematicians responded by always invoking the name of God at the beginning and end of their works and even occasionally referring to Divine assistance throughout the texts. Furthermore, since the rulers were naturally interested in the needs of daily life, Islamic mathematicians, unlike their Greek predecessors, nearly all contributed not only to theory but also to practical applications.

By the eleventh century, however, the status of mathematical thought in Islamic society was beginning to change. It appears that, even when mathematics was being highly developed, the areas of mathematics more advanced than basic arithmetic were classified as "foreign sciences"—in contrast to the "religious sciences," including religious law and speculative theology. To many Islamic religious leaders, the foreign sciences were potentially subversive to the faith and certainly superfluous to the needs of life, either here on earth or in the hereafter. And although the earliest Islamic leaders encouraged the study of the foreign sciences, the support for such study lessened over the centuries, as more orthodox religious leaders came to power. More and more, the institutions of higher learning throughout the Islamic world, the *madrasas*, tended to concentrate on the teaching of Islamic law. A scholar in charge of one of these schools could, of course, teach the foreign sciences, but if he did, he could be the subject of a legal ruling from traditionalists—a ruling that would be based on the law establishing the school and specifying that nothing inimical to the tenets of Islam could be taught. Thus, although there were significant mathematical achievements in Islamic cultures through the fifteenth century, gradually science became less important.

In this chapter, we will consider some of the highlights of Islamic mathematics. In particular, we will see that Islamic mathematicians fully developed the decimal place value number system to include decimal fractions; systematized the study of algebra and began to consider the relationship between algebra and geometry; brought the rules of combinatorics from India and reworked them into an abstract system; studied and made advances on the major Greek geometrical treatises of Euclid, Archimedes, and Apollonius; and made significant improvements in plane and spherical trigonometry.

7.1 Arithmetic

The decimal place value system had spread from India at least as far as Syria by the middle of the seventh century and was certainly known to Islamic mathematicians by the time of the founding of the House of Wisdom. The earliest arithmetic text that deals with the Hindu numbers is the *Book on Addition and Subtraction after the Method of the Indians* by Muḥammad ibn-Mūsā al-Khwārizmī (c. 780–850), an early member of the House of Wisdom. Unfortunately, there is no extant Arabic manuscript of this work, only several different Latin versions made in Europe in the twelfth century. In his text, al-Khwārizmī introduced nine characters to designate the first nine numbers and, according to the Latin versions, a circle to designate zero. He demonstrated how to use these characters to write any number in place value notation. He then described the algorithms of addition, subtraction, multiplication, division, halving, doubling, and determining square roots and gave examples of their use. The algorithms were usually set up to be performed on a dust board, a writing surface on which sand was spread. Thus, the steps in calculations were generally erased as one proceeded to the final answer. Although al-Khwārizmī was not consistent in his style, he most often expressed fractions in the Egyptian mode as sums of unit fractions. It is important to note that one of the most important features of the modern place value system, decimal fractions, was still missing. Nevertheless, al-Khwārizmī's work introduced many Europeans to the basics of the decimal place value system.

Over the next several centuries, Islamic mathematicians gradually improved the Indian system, developing new algorithms for calculations on paper and eventually extending the system to decimal fractions. As far as we know, the first Islamic scholar not only to use decimal fractions but also to grasp their significance with regard to approximation was al-Samaw'al ibn Yaḥyā ibn Yahūda al-Maghribī (c. 1125–1174), born a Jew in Baghdad but who converted to Islam when he was about 40. He described the basic idea of these fractions in his *Treatise on Arithmetic* of 1172:

> Given that proportional places, starting with the place of the units, follow one another indefinitely according to the tenth proportion, we therefore suppose that on the other side [of the units] the place of the parts [of ten follow one another] according to the same proportion, and the place of units lies half-way between the place of the integers whose units are transferred in the same way indefinitely, and the place of indefinitely divisible parts.

As an example, al-Samaw'al divides 210 by 13, and notes that the division does not come out even, but can be carried as far as desired. He wrote the result to five places as "16 plus 1 part of 10 plus 5 parts of 100 plus 3 parts of 1000 plus 8 parts of 10,000 plus 4 parts of 100,000." Similarly, he expressed the square root of 10 in words as "3 plus 1 part of 10 plus 6 parts of 100 plus 2 parts of 1000 plus 2 parts of 10,000 plus 7 parts of 100,000 plus 7 parts of 1,000,000" (3.162277). Although he still used words to describe the various places, it is clear that he understood the value of using decimal fractions for approximating rational numbers or irrational numbers. In fact, when al-Samaw'al calculated higher roots by a method similar to that used in China, he explicitly noted the purpose of the successive steps of the algorithm: "And thus we operate to determine the side of a cube, of a square-square, a square-cube and other [powers]. This method enables us . . . to obtain an infinite number of answers, each one being more precise and closer to the truth than the preceding one." Al-Samaw'al evidently realized that one can potentially calculate an infinite decimal expansion of a number and that the finite decimals of this expansion "converge" to the exact value, a value not expressible in any finite form.

But even with this important work, the development of the place value system was not complete. It was the work of Ghiyāth al-Dīn Jamshīd al-Kāshī (d. 1429), who lived in Samarkand, that first demonstrated both a total command of the idea of decimal fractions and a convenient notation for them—namely, a vertical line to separate the integer part of a number from the decimal fraction part. With this work, the Hindu-Arabic place value system was complete.

7.2 Algebra

The most important contributions of the Islamic mathematicians lie in the area of algebra. They took the material already developed by the Babylonians, combined it with the classical Greek heritage of geometry, and produced a new algebra, which they proceeded to extend. By the end of the ninth century, the chief Greek mathematical classics were well known in the Islamic world. Islamic scholars studied them and wrote commentaries on them. The most important idea they learned from their study of these Greek works was the notion of proof. They absorbed the idea that one could not consider a mathematical problem solved unless one could demonstrate that the solution was valid. How does one demonstrate this, particularly for an algebra problem? The answer seemed clear. The only real proofs were geometric. After all, geometry was what was found in Greek texts, not algebra. Hence, Islamic scholars generally set themselves the task of justifying algebraic rules (either the ancient Babylonian ones or new ones they themselves discovered) through geometry.

7.2.1 The Algebra of al-Khwārizmī

One of the earliest Islamic algebra texts, written in about 825 by al-Khwārizmī, was titled *The Condensed Book on the Calculation of al-Jabr and al-Muqābala*, a work that ultimately had even more influence than his arithmetical work. The term *al-jabr* can be translated as "restoring" and refers to the operation of transposing a subtracted quantity from one side of an equation to the other side, where it becomes an added quantity. The word *al-muqābala* can be translated as "comparing" and refers to the reduction of a positive term by subtracting equal amounts from both sides of the equation. Thus, the conversion of $3x + 2 = 4 - 2x$ to $5x + 2 = 4$ is an example of *al-jabr*, and the conversion of the latter to $5x = 2$ is an example of *al-muqābala*. The word *algebra* is a corrupted form of the Arabic *al-jabr*.

When al-Khwārizmī's work and other similar treatises were translated into Latin, no translation was made of the word *al-jabr*, which thus came to be taken for the name of this science.

Al-Khwārizmī wrote in his introduction that he was interested in writing a practical manual, not a theoretical one. Nevertheless, he had been sufficiently influenced by the introduction of Greek mathematics into the House of Wisdom that he felt constrained to give geometric proofs of his algebraic procedures. The geometric proofs, however, are not Greek proofs. They appear to be, in fact, very similar to the Babylonian geometric arguments out of which the algebraic algorithms grew. Again, like his Eastern predecessors, al-Khwārizmī gave numerous examples and problems, but the Greek influence showed through in his systematic classification of the problems he intended to solve, as well as in the very detailed explanations of his methods.

Al-Khwārizmī's text was to be a manual for solving equations. The quantities he dealt with were generally of three kinds, the square (of the unknown), the root of the square (the unknown itself), and the absolute numbers (the constants in the equation). He noted that six types of equations can be written using these three kinds:

1. Squares are equal to roots. ($ax^2 = bx$)
2. Squares are equal to numbers. ($ax^2 = c$)
3. Roots are equal to numbers. ($bx = c$)
4. Squares and roots are equal to numbers. ($ax^2 + bx = c$)
5. Squares and numbers are equal to roots. ($ax^2 + c = bx$)
6. Roots and numbers are equal to squares. ($bx + c = ax^2$)

The reason for the six-fold classification is that Islamic mathematicians, unlike the Hindus, did not deal with negative numbers at all. Coefficients, as well as the roots of the equations, had to be positive. The types listed are the only types that have positive solutions. The standard modern form $ax^2 + bx + c = 0$ would make no sense to al-Khwārizmī, because if the coefficients are all positive, the roots cannot be.

Al-Khwārizmī's solutions to the first three types of equations are straightforward. We need only note that zero is not considered as a solution to the first type. His rules for the compound types of equations are more interesting. We will present his solution to type 4. Because al-Khwārizmī used no symbols, we will express everything in his example in words, including the numbers: "What must be the square which, when increased by ten of its own roots, amounts to thirty-nine? The solution is this: You halve the number of roots, which in the present instance yields five. This you multiply by itself; the product is twenty-five. Add this to thirty-nine; the sum is sixty-four. Now take the root of this, which is eight, and subtract from it half the number of the roots, which is five; the remainder is three. This is the root of the square which you sought for."

Al-Khwārizmī's verbal description of his procedure was essentially the same as that of the Babylonian scribes. Namely, in modern notation, the solution of $x^2 + bx = c$ is

$$x = \sqrt{\left(\frac{b}{2}\right)^2 + c} - \frac{b}{2}.$$

Al-Khwārizmī's geometric justification of this procedure also demonstrates his Babylonian heritage. Beginning with a square representing x^2, he adds two rectangles, each of width

5 ("half the number of roots") (Fig. 7.1). The sum of the area of the square and the two rectangles is then $x^2 + 10x = 39$. One now completes the square with a single square of area 25 to make the total area 64. The solution $x = 3$ is then easily found. This geometric description corresponds to the Babylonian description of the solution of $x^2 + \frac{4}{3}x = \frac{11}{12}$. (See chapter 1, p. 22 and Fig. 1.12, p. 23.)

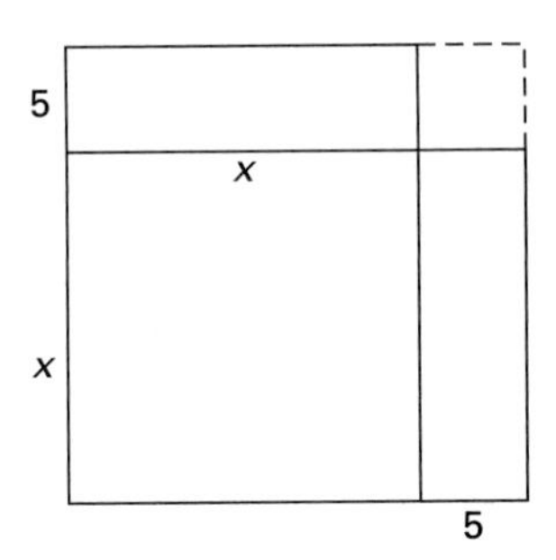

FIGURE 7.1 Al-Khwārizmī's geometric justification for the solution of $x^2 + 10x = 39$

Although al-Khwārizmī's geometric justification of his method appears to have been taken over from Babylonian sources, he succeeded in directing the focus of quadratic equation solving away from the actual finding of sides of squares to finding numbers satisfying certain conditions. For example, he explained the term *root* not as a side of a square but as "anything composed of units which can be multiplied by itself." Also, his procedure for solving quadratic equations of type 4, when the coefficient of the square term is other than 1, was the arithmetical method of first multiplying or dividing appropriately to make the initial coefficient 1, and then proceeding as before.

Finally, al-Khwārizmī's presentation of the method and geometric description for equations of type 5, squares and numbers equal to roots, showed that, unlike the Babylonians, he could deal with an equation with two positive roots, at least numerically. In this case, $x^2 + c = bx$, his verbal description of the solution procedure easily translates into the modern formula

$$x = \frac{b}{2} \pm \sqrt{\left(\frac{b}{2}\right)^2 - c}. \tag{7.1}$$

In fact, he stated that one could employ either addition or subtraction to get a root and also noted the condition on the solution: "If the product [of half the number of roots with itself] is less than the number connected with the square, then the instance is impossible; but if the product is equal to the number itself, then the root of the square is equal to half of the number of roots alone, without either addition or subtraction." The geometric demonstration in this case only deals with the subtraction in formula 7.1. In Figure 7.2, square $ABCD$ represents x^2, and rectangle $ABNH$ represents c. Therefore, HC represents b. Bisect HC at G, extend TG to K so that $GK = GA$, and complete the rectangle $GKMH$. Finally, choose L on KM so that $KL = GK$, and complete the square $KLRG$. It is then clear that rectangle $MLRH$ equals rectangle $GATB$. Since the area of square $KMNT$

is $(b/2)^2$, and that square less square $KLRG$ equals rectangle $ABNH$, or c, it follows that square $KLRG$ equals $(b/2)^2 - c$. Since the side of that square is equal to AG, it follows that $x = AC = CG - AG$ is given by formula 7.1 using the minus sign. Although al-Khwārizmī briefly noted that CR could also represent a solution, he did not demonstrate this by a diagram, nor did he deal in his diagram with the special conditions mentioned in his verbal description.

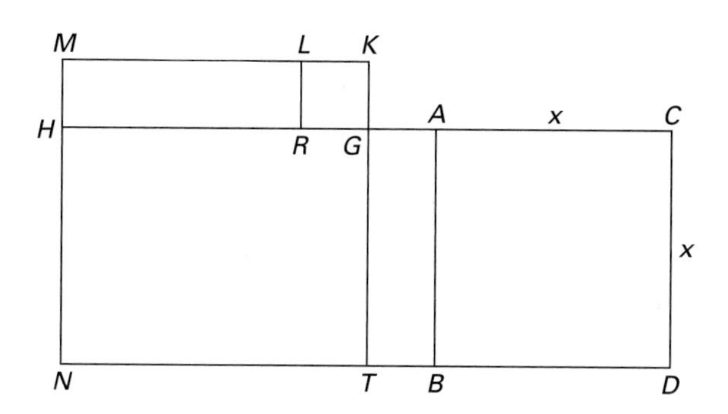

FIGURE 7.2 Al-Khwārizmī's geometric justification for the solution of $x^2 + c = bx$

Al-Khwārizmī's text contains much else of interest besides the rules for solving quadratic equations, including an introduction to manipulation with algebraic expressions, explained by reference to similar manipulations with numbers. For example, he noted that if $a \pm b$ is multiplied by $c \pm d$, then four multiplications are necessary. Although none of his numbers are negative, he certainly knew the rules for dealing with multiplication and signs. As he stated, "If the units [b and d in our notation] ... are positive, then the last multiplication is positive; if they are both negative, then the fourth multiplication is likewise positive. But if one of them is positive and one negative, then the fourth multiplication is negative."

The text continues with a large collection of problems, many of which involve these algebraic manipulations and most of which result in a quadratic equation. For example, one problem states, "I have divided ten into two parts, and having multiplied each part by itself, I have put them together, and have added to them the difference of the two parts previous to their multiplication, and the amount of all this is fifty-four." It is not difficult to translate this problem into the equation $(10 - x)^2 + x^2 + (10 - x) - x = 54$. The author reduced this to the equation $x^2 + 28 = 11x$ and then used his rule for an equation of type 5 to get $x = 4$. He ignored here the second root, $x = 7$, for then the sum of the two squares would be 58 and the conditions of the problem could not be met. In another example, al-Khwārizmī dealt with a nonrational root: "I have divided ten into two parts; I have multiplied the one by ten and the other by itself, and the products were the same." The equation here is $10x = (10 - x)^2$, and the solution is $x = 15 - \sqrt{125}$. He again ignored the root with the positive sign, because $15 + \sqrt{125}$ could not be a "part" of 10.

Although al-Khwārizmī promised in his preface that he would write about what is "useful," very few of his problems leading to quadratic equations deal with any practical ideas. Many of them are similar to the previous examples and begin with "I have divided ten into two parts." Among the few problems written in real-world terms is the following:

"You divide one *dirhem* among a certain number of men. Now you add one man more to them, and divide again one *dirhem* among them. The quota of each is then one-sixth of a *dirhem* less than at the first time." If x represents the number of men, the equation becomes $1/x - 1/(x+1) = 1/6$, which reduces to $x^2 + x = 6$, for which the solution is $x = 2$.

One can only conclude that although al-Khwārizmī was interested in teaching his readers how to solve mathematical problems, and especially how to deal with quadratic equations, he could think of few real-life situations that required these equations. Things apparently had not changed in this regard since the time of the Babylonians.

7.2.2 The Algebra of Abū Kāmil

Within 50 years of al-Khwārizmī's text, Islamic mathematicians had decided that the necessary geometric foundations of the algebraic solution of quadratic equations should be based on the work of Euclid rather than on the ancient traditions. Among these authors was the Egyptian mathematician Abū Kāmil ibn Aslam (c. 850–930), who wrote in his own algebra text, "I shall explain their rule using geometric figures clarified by wise men of geometry and which are explained in the Book of Euclid." For example, Abū Kāmil proved al-Khwārizmī's rule for equations of type 4 using Figure 7.3. Here, $ABGD$ is the original square, $ABHW$ represents bx, and L is the midpoint of BH. He then noted that, according to Euclid (II–6), the square on LG is equal to the rectangle formed by HG and BG plus the square on BL. The desired rule then follows easily.

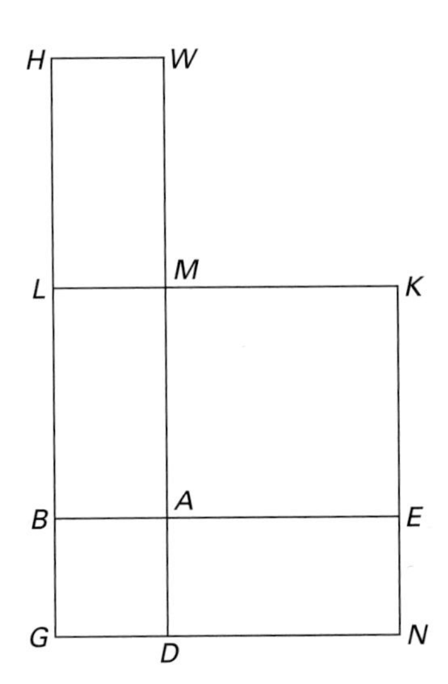

FIGURE 7.3 Abū Kāmil's geometric justification for the solution of Al-Khwārizmī's equation of type 4 ($ax^2 + bx = c$)

Like his predecessor, Abū Kāmil followed his discussion of the various forms of quadratic equations by a treatment of various algebraic rules and then a large selection of problems. But he made some advances over al-Khwārizmī by considering many more complicated identities and more complex problems, including in particular manipulations with *surds*. Abū Kāmil was not at all worried about dealing with these irrational quantities. He used them freely in his problems, many of which, like those of al-Khwārizmī, start with "divide 10 into two parts." For example, consider problem 37: "If one says that ten is divided into two parts, and one part is multiplied by itself and the other by the root of eight, and subtract the quantity of the product of one part times the root of eight from ... the product of the other part multiplied by itself, it gives forty." The equation in this case is $(10 - x)(10 - x) - x\sqrt{8} = 40$. After rewriting it in the form $x^2 + 60 = 20x + \sqrt{8x^2}$ $[= (20 + \sqrt{8})x]$, Abū Kāmil carried out the algorithm for the case where squares and numbers are equal to roots to conclude that $x = 10 + \sqrt{2} - \sqrt{42 + \sqrt{800}}$ and that $10 - x$, the "other part," is equal to $\sqrt{42 + \sqrt{800}} - \sqrt{2}$.

Thus, it appears that Abū Kāmil was willing to use the algebraic algorithms with any type of positive number. He made no distinction between operating with 2 or with $\sqrt{8}$ or even with $\sqrt{42 + \sqrt{800}}$. Since these algorithms came from geometry, on one level that is not surprising. After all, the Greek failure to find a numerical representation of the diagonal of a square was one of the reasons for their use of the geometric algebra of line segments and areas. But in dealing with these quantities, Abū Kāmil interpreted all of them in the same way. It did not matter whether a magnitude was technically a square or a fourth power or a root or a root of a root. For Abū Kāmil, the solution of a quadratic equation was not a line segment, as it would be in the interpretation of the appropriate propositions of the *Elements*. It was a number, even though Abū Kāmil may not have been able to give a proper definition of that term. He therefore had no compunction about combining the various quantities that appeared in the solutions, using general rules. Abū Kāmil's willingness to handle all of these quantities by the same techniques helped pave the way toward a new understanding of the concept of number, fully as important as al-Samaw'al's use of decimal approximations.

7.2.3 The Algebra of Polynomials

The process of relating arithmetic to algebra, begun by al-Khwārizmī and Abū Kāmil, continued in the Islamic world over the next two centuries. In particular, al-Samaw'al helped to show that the techniques of arithmetic could be fruitfully applied in algebra and, reciprocally, that ideas originally developed in algebra could be important in dealing with numbers.

Al-Samaw'al, basing his work on that of some of his predecessors, established in full the algebra of polynomials. For example, he introduced negative coefficients and the rules for dealing with them when adding and subtracting polynomials:

> If we subtract an additive number from an empty power $[0x^n - ax^n]$, the same subtractive number remains; if we subtract the subtractive number from an empty power $[0x^n - (-ax^n)]$, the same additive number remains. If we subtract an additive number from a subtractive number, the remainder is their subtractive sum; if we subtract a subtractive number from a greater subtractive number, the result is their subtractive difference; if the number from which one subtracts is smaller than the number subtracted, the result is their additive difference.

To multiply polynomials, al-Samaw'al needed the law of exponents. He decided that this law could best be expressed by using a table consisting of columns, each column representing a different power of either a number or an unknown. In fact, he also saw that he could deal with powers of $1/x$ as easily as with powers of x. In his work, the columns were headed by the Arabic letters standing for the numerals, reading both ways from the central column labeled 0. Each column had the name of the particular power or reciprocal power. For example, the column headed by a 2 on the left was named "square," that headed by a 5 on the left was named "square-cube," that headed by a 3 on the right was named "part of cube," and so on. To simplify matters, we will just use the Arabic numerals themselves and the powers of x.

7	6	5	4	3	2	1	0	1	2	3	4	5	6	7
x^7	x^6	x^5	x^4	x^3	x^2	x	1	x^{-1}	x^{-2}	x^{-3}	x^{-4}	x^{-5}	x^{-6}	x^{-7}

Al-Samaw'al used his table to explain how to multiply monomials: "The distance of the order of the product of the two factors from the order of one of the two factors is equal to the distance of the order of the other factor from the unit. If the factors are in different directions then we count [the distance] from the order of the first factor towards the unit; but, if they are in the same direction, we count away from the unit." So, for example, to multiply x^3 by x^4, count four orders to the left of column 3 and get the result as x^7. To multiply x^3 by x^{-2}, count two orders to the right from column 3 and get the answer x^1. Using these rules, al-Samaw'al could easily multiply polynomials in x and $1/x$ as well as divide such polynomials by monomials.

Al-Samaw'al was also able to divide polynomials by polynomials, although he did not hesitate to use negative powers of x as well as positive ones. For example, to divide $20x^2 + 30x$ by $6x^2 + 12$, he began by dividing $20x^2$ by $6x^2$, getting $3\frac{1}{3}$. The product of $3\frac{1}{3}$ and $6x^2 + 12$ is $20x^2 + 40$. He then subtracted this result from $20x^2 + 30x$ to get $30x - 40$. He repeated the process, dividing that expression by $6x^2 + 12$. The initial quotient is $5 \cdot 1/x$, which is in turn multiplied by $6x^2 + 12$, and the algorithm thus continues.

In this particular example, the division is not exact. Al-Samaw'al continued the process through eight steps to get

$$3\frac{1}{3} + 5\left(\frac{1}{x}\right) - 6\frac{2}{3}\left(\frac{1}{x^2}\right) - 10\left(\frac{1}{x^3}\right) + 13\frac{1}{3}\left(\frac{1}{x^4}\right) + 20\left(\frac{1}{x^5}\right) - 26\frac{2}{3}\left(\frac{1}{x^6}\right) - 40\left(\frac{1}{x^7}\right).$$

To show his fluency with the multiplication procedure, he checked the answer by multiplying it by the divisor. Because the product differs from the dividend only by terms in $1/x^6$ and $1/x^7$, he called the result given "the answer approximately." Nevertheless, he also noted that there is a pattern to the coefficients of the quotient. In fact, if a_n represents the coefficient of $1/x^n$, the pattern is given by $a_{n+2} = -2a_n$. He then proudly wrote out the next 21 terms of the quotient, ending with $54{,}613\frac{1}{3}(1/x^{28})$.

Given that al-Samaw'al thought of extending division of polynomials into polynomials in $1/x$ and thought of partial results as approximations, it is not surprising that he would divide whole numbers by simply replacing x by 10. As we have seen, al-Samaw'al was the first to recognize that one could approximate fractions more and more closely by calculating more and more decimal places. His work was thus extremely important in developing the idea that algebraic manipulations and manipulations with numbers are parallel. Virtually any technique that applies to one can be adapted to apply to the other.

7.2.4 Induction, Sums of Powers, and the Pascal Triangle

Another important idea used by Islamic mathematicians was that of an inductive argument for dealing with certain formulas about integers. For example, the formulas for the sums of the integers and their squares had long been known, while the formula for the sum of cubes is easy to discover if one considers a few examples. To give an argument for their validity that generalizes to enable one to find a formula for the sum of fourth powers, however, is more difficult. This, however, was accomplished early in the eleventh century by the Egyptian mathematician Abū' Alī al-Ḥasan ibn al-Ḥasan ibn al-Haytham (965–1039). That he did not generalize his result to find the sums of higher powers is probably due to his needing only the formulas for the second and fourth powers in his computation of the volume of a paraboloid, to be discussed in section 7.4.2.

The central idea in ibn al-Haytham's proof of the sum formulas was the derivation of the equation

$$(n+1)\sum_{i=1}^{n} i^k = \sum_{i=1}^{n} i^{k+1} + \sum_{p=1}^{n}\left(\sum_{i=1}^{p} i^k\right). \qquad \textbf{(7.2)}$$

Ibn al-Haytham did not state this result in general form but only for particular integers: $n = 4$ and $k = 1,\ 2,\ 3$. His proof, however, uses inductive reasoning and is immediately generalizable to any values of n and k. We consider his proof for $k = 3$ and $n = 4$:

$$\begin{aligned}(4+1)(1^3+2^3+3^3+4^3) &= 4(1^3+2^3+3^3+4^3) + 1^3+2^3+3^3+4^3 \\ &= 4\cdot 4^3 + 4(1^3+2^3+3^3) + 1^3+2^3+3^3+4^3 \\ &= 4^4 + (3+1)(1^3+2^3+3^3) + 1^3+2^3+3^3+4^3\end{aligned}$$

But, because formula 7.2 is assumed true for $n = 3$, we have

$$(3+1)(1^3+2^3+3^3) = 1^4+2^4+3^4+(1^3+2^3+3^3)+(1^3+2^3)+1^3.$$

Thus, formula 7.2 is proved for $n = 4$.

Ibn al-Haytham used formula 7.2 to derive formulas for the sums of second, third, and fourth powers. Thus, for $k = 2$ and $k = 3$, he showed that

$$\begin{aligned}\sum_{i=1}^{n} i^2 &= \left(\frac{n}{3}+\frac{1}{3}\right)n\left(n+\frac{1}{2}\right) = \frac{n^3}{3}+\frac{n^2}{2}+\frac{n}{6} \\ \sum_{i=1}^{n} i^3 &= \left(\frac{n}{4}+\frac{1}{4}\right)n(n+1)n = \frac{n^4}{4}+\frac{n^3}{2}+\frac{n^2}{4}\end{aligned}$$

We consider here the proof of the analogous result for fourth powers. This result, although stated (at the end) in all generality, is only proved for the case $n = 4$. But we can consider this proof to represent the method of generalizable example, a method which, as we saw earlier, was used by Euclid. We begin with formula 7.2, with $k = 3$ and $n = 4$:

$$\begin{aligned}(1^3+2^3+3^3+4^3)5 = {}&1^4+2^4+3^4+4^4+(1^3+2^3+3^3+4^3) \\ &+(1^3+2^3+3^3)+(1^3+2^3)+1^3\end{aligned}$$

$$= 1^4 + 2^4 + 3^4 + 4^4 + \left(\frac{4^4}{4} + \frac{4^3}{2} + \frac{4^2}{4}\right) + \left(\frac{3^4}{4} + \frac{3^3}{2} + \frac{3^2}{4}\right)$$

$$+ \left(\frac{2^4}{4} + \frac{2^3}{2} + \frac{2^2}{4}\right) + \left(\frac{1^4}{4} + \frac{1^3}{2} + \frac{1^2}{4}\right)$$

$$= 1^4 + 2^4 + 3^4 + 4^4 + \frac{1}{4}(1^4 + 2^4 + 3^4 + 4^4)$$

$$+ \frac{1}{2}(1^3 + 2^3 + 3^3 + 4^3) + \frac{1}{4}(1^2 + 2^2 + 3^2 + 4^2)$$

$$= \frac{5}{4}(1^4 + 2^4 + 3^4 + 4^4) + \frac{1}{2}(1^3 + 2^3 + 3^3 + 4^3)$$

$$+ \frac{1}{4}(1^2 + 2^2 + 3^2 + 4^2).$$

Therefore,

$$1^4 + 2^4 + 3^4 + 4^4 = \frac{4}{5}(1^3 + 2^3 + 3^3 + 4^3)\left(4 + \frac{1}{2}\right) - \frac{1}{5}(1^2 + 2^2 + 3^2 + 4^2)$$

$$= \frac{4}{5}\left(4 + \frac{1}{2}\right)\left(\frac{4}{4} + \frac{1}{4}\right)4(4+1)4 - \frac{1}{5}\left(\frac{4}{3} + \frac{1}{3}\right)4\left(4 + \frac{1}{2}\right)$$

and finally

$$1^4 + 2^4 + 3^4 + 4^4 = \left(\frac{4}{5} + \frac{1}{5}\right)4\left(4 + \frac{1}{2}\right)\left[(4+1)4 - \frac{1}{3}\right].$$

From this result for the case $n = 4$, ibn al-Haytham simply stated his general result in words, which can be translated into the modern formula

$$\sum_{i=1}^{n} i^4 = \left(\frac{n}{5} + \frac{1}{5}\right)n\left(n + \frac{1}{2}\right)\left[(n+1)n - \frac{1}{3}\right].$$

Al-Samaw'al also used an inductive argument, this time in relation to the binomial theorem and the Pascal triangle. The binomial theorem is the result

$$(a+b)^n = \sum_{k=0}^{n} C_k^n a^{n-k} b^k,$$

where n is a positive integer and the values C_k^n are the binomial coefficients, the entries in the Pascal triangle. Naturally, al-Samaw'al, having no symbolism, wrote this formula in words in each individual instance. For example, for the case $n = 4$, he wrote, "For a number divided into two parts, its square-square [fourth power] is equal to the square-square of each part, four times the product of each by the cube of the other, and six times the product of the squares of each part." He then provided a table of binomial coefficients up to the twelfth power, with the coefficients for the nth power written in the nth column. His procedure for constructing this table is the familiar one, that any entry comes from adding the entry to the left of it to the entry just above that one. He noted that one can use the table to read off the expansion of any power up to the twelfth of "a number divided into two parts."

With this table in mind, let us see how al-Samaw'al demonstrated the quoted result for $n = 4$. Assume the number c is equal to $a + b$. Since $c^4 = cc^3$ and c^3 is already known to be given by $c^3 = (a+b)^3 = a^3 + b^3 + 3ab^2 + 3a^2b$, it follows that $(a+b)^4 = (a+b)(a+b)^3 = (a+b)(a^3 + b^3 + 3ab^2 + 3a^2b)$. By using repeatedly the result $(r+s)t = rs + rt$, which he quoted from Euclid's *Elements* II, al-Samaw'al found that this last quantity equals $(a+b)a^3 + (a+b)b^3 + (a+b)3ab^2 + (a+b)3a^2b = a^4 + a^3b + ab^3 + b^4 + 3a^2b^2 + 3ab^3 + 3a^3b + 3a^2b^2 = a^4 + b^4 + 4ab^3 + 4a^3b + 6a^2b^2$. The coefficients here are the appropriate ones from the fourth column of his table, and the expansion shows that the new coefficients are formed from the old ones exactly as stated in the table construction. Al-Samaw'al next quoted the result for $n = 5$ and asserted his general result: "He who has understood what we have just said can prove that for any number divided into two parts, its quadrato-cube [fifth power] is equal to the sum of the quadrato-cubes of each of its parts, five times the product of each of its parts by the square-square of the other, and ten times the product of the square of each of them by the cube of the other. And so on in ascending order." Like ibn al-Haytham's proof, al-Samaw'al's argument contained the two basic elements of an inductive proof: He began with a value for which the result is known, here $n = 2$, and then used the result for a given integer to derive the result for the next. Although al-Samaw'al did not have any way of stating, and therefore proving, the general binomial theorem, for modern readers there is only a short step from al-Samaw'al's argument to a full inductive proof of the binomial theorem, provided that in the statement of that theorem the coefficients themselves are defined inductively, essentially as al-Samaw'al did define them, as $C_m^n = C_{m-1}^{n-1} + C_m^{n-1}$. In any case, the Pascal triangle was used by both Islamic and, as we have noted, Chinese mathematicians to develop an algorithm to calculate roots of numbers. In the Islamic case, this algorithm is documented from the time of al-Samaw'al, although there are strong indications that it was known at least a century earlier.

7.2.5 The Solution of Cubic Equations

Another strand of development in algebra in the Islamic world that proceeded alongside of arithmetization and inductive ideas was the application of geometry. By the end of the ninth century, Islamic mathematicians, having read the major Greek texts, had noticed that certain geometric problems led to cubic equations, which could be solved by finding the intersection of two conic sections. Such problems included the doubling of the cube and Archimedes' splitting of a sphere into two parts whose volumes are in a given ratio. Several Islamic mathematicians during the tenth and eleventh centuries also solved other cubic equations by taking over this Greek idea of intersecting conics. But it was the mathematician and poet 'Umar ibn Ibrāhīm al-Khayyāmī (1048–1131) (usually known in the West as Omar Khayyam) who first systematically classified and then proceeded to solve all types of cubic equations by this general method.

Al-Khayyāmī's major mathematics text, *Treatise on Demonstrations of Problems of Al-jabr and Al-muqābala*, is primarily devoted to the solution of cubic equations. Although the author suggests that the reader should be familiar with Euclid's *Elements* and Apollonius's *Conics*, nevertheless, the text addresses algebraic, not geometric, problems. In fact, al-Khayyāmī would have liked to have provided algebraic algorithms for solving cubic equations, analogous to al-Khwārizmī's three algorithms for solving mixed quadratic equations. As he wrote, "When, however, the object of the problem is an absolute number, neither we, nor any of those who are concerned with algebra, have been able to solve this equation—perhaps others who follow us will be able to fill the gap." Although there were later attempts by Islamic mathematicians to fill the gap, it was not until the sixteenth century in Italy that al-Khayyāmī's hope was realized.

Al-Khayyāmī began his work, in the style of al-Khwārizmī, by giving a complete classification of equations of degree up to three. Since for al-Khayyāmī, as for his predecessors, all numbers were positive, he had to list separately the various forms that might possess positive roots. Among these were fourteen not reducible to quadratic or linear equations. These were in three groups: one binomial equation, $x^3 = d$, six equations with three terms, and seven with four. For each of these forms, the author described the conic sections necessary for a solution and proved that his solution was correct. We will discuss al-Khayyāmī's solution of $x^3 + cx = d$ or, as he puts it, the case where "a cube and sides are equal to a number."

To construct the solution, al-Khayyāmī set AB equal in length to $\sqrt{c}$ (Fig. 7.4). He then constructed BC perpendicular to AB so that $BC \cdot AB^2 = d$, or $BC = d/c$. Next, he extended AB in the direction of Z and constructed a parabola with vertex B, axis BZ, and parameter AB. In modern notation, this parabola has the equation $x^2 = \sqrt{c}y$. Similarly, he constructed a semicircle on the line BC, whose equation is

$$\left(x - \frac{d}{2c}\right)^2 + y^2 = \left(\frac{d}{2c}\right)^2 \quad \text{or} \quad x\left(\frac{d}{c} - x\right) = y^2.$$

The circle and the parabola intersect at a point D. It is the x coordinate of this point, here represented by the line segment BE, that provides the solution to the equation.

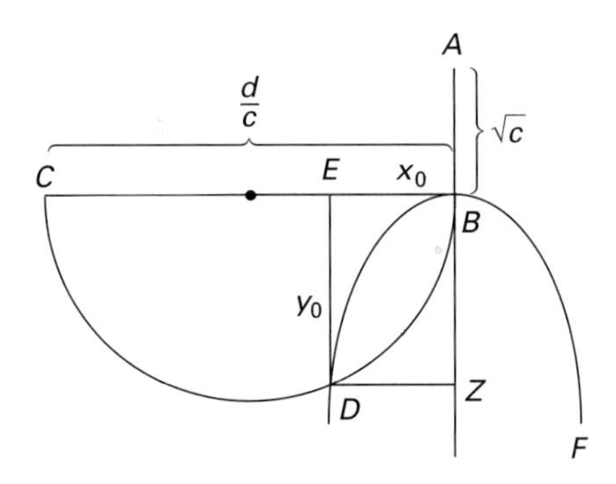

FIGURE 7.4 Al-Khayyāmī's construction for the solution of $x^3 + cx = d$

Al-Khayyāmī proved that his solution is correct by using the basic properties of the parabola and the circle. If $BE = DZ = x_0$ and $BZ = ED = y_0$, then, first, $x_0^2 = \sqrt{c}y_0$, or $\sqrt{c}/x_0 = x_0/y_0$, because D is on the parabola, and, second, $x_0(d/c - x_0) = y_0^2$, or $x_0/y_0 = y_0/(d/c - x_0)$, because D is on the semicircle. It follows that

$$\frac{c}{x_0^2} = \frac{x_0^2}{y_0^2} = \frac{y_0^2}{(d/c - x_0)^2} = \frac{y_0}{d/c - x_0} \cdot \frac{x_0}{y_0} = \frac{x_0}{d/c - x_0}$$

and then that $x_0^3 = d - cx_0$, so x_0 is the desired solution. Al-Khayyāmī noted, without any indication of a proof, that this class of equations always has a single solution. In other words, the parabola and the circle always intersect in one point other than the origin. The origin, though, does not provide a solution to the problem. Al-Khayyāmī's remark reflects the modern statement that the equation $x^3 + cx = d$ always has exactly one positive solution.

Al-Khayyāmī treated each of his fourteen cases in the same manner. For those for which a positive solution does not always exist, he noted that there are zero, one, or two solutions, depending on whether the conic sections involved do not intersect or intersect at one or two points. In general, however, he did not relate the existence of one or two solutions to conditions on the coefficients.

This gap was filled to some extent by Sharaf al-Dīn al-Ṭūsī (d. 1213), a mathematician born in Tus, Persia. Like his predecessor, he began by classifying the cubic equations into several groups. His groups differed from those of al-Khayyāmī, because he was interested in determining conditions on the coefficients that determine the number of solutions. Therefore, his first group consisted of those equations that could be reduced to quadratic ones, plus the equation $x^3 = d$. The second group consisted of the eight cubic equations that always have at least one (positive) solution. The third group consisted of those types that may or may not have (positive) solutions, depending on the particular values of the coefficients. These include $x^3 + d = bx^2$, $x^3 + d = cx$, $x^3 + bx^2 + d = cx$, $x^3 + cx + d = bx^2$, and $x^3 + d = bx^2 + cx$. It is with his study of this group that Sharaf al Dīn made his most original contribution.

Consider Sharaf al-Dīn's analysis of $x^3 + d = bx^2$. He began by putting the equation in the form $x^2(b - x) = d$. He then noted that the question of whether the equation has a solution depends on whether the "function" $f(x) = x^2(b - x)$ reaches the value d or not (Fig. 7.5). He therefore carefully proved that the value $x_0 = 2b/3$ provides the maximum value for $f(x)$; that is, for any x between 0 and b, $x^2(b - x) \leq (2b/3)^2(b/3) = 4b^3/27$. Although he did not say why he chose this particular value for x_0, he was able to complete a similar analysis for each of the five equations of his third group.

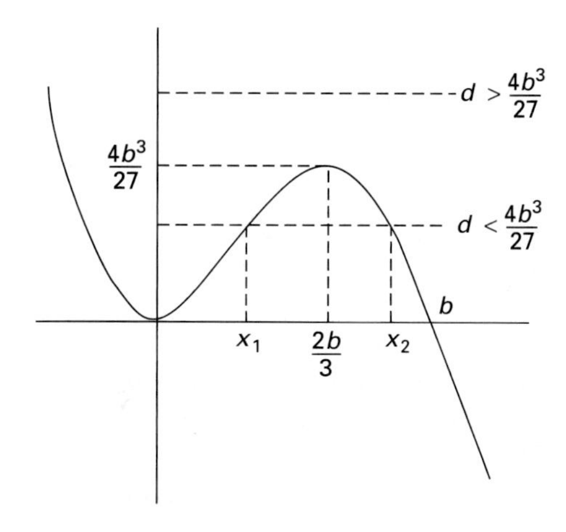

FIGURE 7.5 Modern graphic interpretation of Sharaf al-Dīn's analysis of the cubic equation $x^3 + d = bx^2$

Given that $2b/3$ provides the maximum, Sharaf al-Dīn then noted that if the maximum value $4b^3/27$ is less than the given d, there can be no solutions to the equation. If $4b^3/27$ equals d, there is only one solution, $x = 2b/3$. Finally, if $4b^3/27$ is greater than d, there are two solutions, x_1 and x_2, where $0 < x_1 < 2b/3$ and $2b/3 < x_2 < b$. Once Sharaf al-Dīn knew something about the solutions, he proceeded to find them numerically. He often did so by a method related to the Chinese method discussed in chapter 5.

Islamic algebraists evidently made great strides in developing the algebra they had received from the Babylonians. Not only did they develop new techniques, but they incorporated into their work the notion of proof derived from the Greeks and thereby put their methods on a firm footing.

7.3 Combinatorics

As we have seen, the basic formulas for combinations and permutations were known in India by the ninth century and probably even earlier. Islamic mathematicians too were interested in such questions from the eighth century onward, but it is only from the thirteenth century that we have evidence of the derivation of the basic combinatorial formulas. We will consider the contributions of two Islamic mathematicians to this work.

7.3.1 Counting Combinations

Early in the thirteenth century Aḥmad al-Ab'dari ibn Mun'im discussed the calculation of the number of combinations of r things from a set of n by looking at this number in terms of combinations of $r - 1$ things. Little is known about ibn Mun'im, but he probably lived at the court of the Almohade caliph in Marrakech (now in Morocco) during the reign of Mohammed ibn Ya'kub al-Nasir (1199–1213).

Ibn Mun'im basically wanted to examine the question of the number of possible words that could be formed out of the letters of the Arabic alphabet. But before dealing with that question, he considered a different problem: How many different bundles of colors can one make out of ten different colors of silk? He calculated these carefully. First, he noted that with bundles of only one color, there are ten possibilities, that is, $C_1^{10} = 10$. To calculate the possibilities for bundles of two colors, ibn Mun'im listed the pairs in order (where c_i represents the ith color):

$$(c_2, c_1);\ (c_3, c_1),\ (c_3, c_2);\ \ldots;\ (c_{10}, c_1),\ (c_{10}, c_2), \ldots;\ (c_{10}, c_9)$$

He then noted that

$$C_2^{10} = C_1^1 + C_1^2 + \cdots + C_1^9 = 1 + 2 + \cdots + 9 = 45.$$

C_2^k can be calculated similarly for values of k less than 10.

To calculate C_3^{10}, ibn Mun'im proceeded analogously. That is, for each c_k with $k = 3, 4, \ldots, 10$, he considered the pairs from the previous calculation having all indices less than k, for example,

$$(c_3, (c_2, c_1));\ (c_4, (c_2, c_1)),\ (c_4, (c_3, c_1)),\ (c_4, (c_3, c_2));\ (c_5, (c_2, c_1)), \ldots.$$

Thus, C_3^{10} is the sum $1 + 3 + 6 + \cdots + 36 = C_2^2 + C_2^3 + C_2^4 + \cdots + C_2^9$, each of which numbers had already been calculated. At each stage, he wrote down the numbers calculated, thus generating again the Pascal triangle. In fact, he essentially showed that for $n \leq 10$ and $k \leq n$,

$$C_k^n = C_{k-1}^{k-1} + C_{k-1}^{k} + C_{k-1}^{k+1} + \cdots + C_{k-1}^{n-1}.$$

Since he wanted to address the question of the number of words, he next considered permutations. He showed that in a word of n letters, the number of permutations of the letters is found by multiplying 1 by 2 by 3 by 4 by 5, and so on, up to n, the number of letters of the word. Ibn Mun'im pursued several other problems, including permutations with repetitions, before dealing with matters of pronunciations and vowel signs. His aim was to determine the number of possible Arabic words and, after some discussion of exactly what this meant, he used some of the above ideas to calculate explicitly the number of words of nine letters, each word having two nonrepeated letters, two letters repeated twice, and one letter repeated three times. The number turns out to have sixteen decimal digits.

7.3.2 Deriving the Combinatorial Formulas

It was a direct successor of ibn Mun'im in Morocco, Abu-l-'Abbas Aḥmad al-Marrakushi ibn al-Bannā (1256–1321), also of Marrakech, who was able to derive the standard multiplicative formula for finding combinations. In addition, he dealt with combinatorics in the abstract, without being concerned with what kinds of objects were being combined.

Ibn al-Bannā began by using a counting argument to show that $C_2^n = n(n-1)/2$: An element a_1 is associated with each of $n-1$ elements, a_2 is associated with each of $n-2$ elements, and so on, so C_2^n is the sum of $n-1, n-2, n-3, \ldots, 2, 1$. He then showed that to find the value C_k^n, "we always multiply the combination that precedes the combination sought by the number that precedes the given number, and whose distance to it is equal to the number of combinations sought. From the product, we take the part that names the number of combinations." These words correspond to the modern formula

$$C_k^n = \frac{n-(k-1)}{k} C_{k-1}^n.$$

To prove this result, ibn al-Bannā began with C_3^n. To each set of 2 elements from the n elements, one associates one of the $n-2$ remaining elements. One obtains $(n-2)C_2^n$ different sets. But, because $C_2^3 = 3$, each of these sets is repeated three times. For example, $\{a, b, c\}$ occurs as $\{\{a, b\}, c\}$, as $\{\{a, c\}, b\}$, and as $\{\{b, c\}, a\}$. Therefore,

$$C_3^n = \frac{n-2}{3} C_2^n,$$

as claimed. For the next step, we know that $C_3^4 = 4$. It follows that if we associate to each set of three elements one of the $n-3$ remaining elements, the total $(n-3)C_3^n$ is four times larger than C_4^n, or

$$C_4^n = \frac{n-3}{4} C_3^n.$$

A similar argument holds for other values of k. Putting these results together, it follows that

$$C_k^n = \frac{n(n-1)(n-2)\cdots[n-(k-1)]}{1 \cdot 2 \cdot 3 \cdots k},$$

which is the standard formula for the number of ways to pick k elements out of a set of n. Using this result and the result of ibn Mun'im—that the number of permutations of a set of n objects was $n!$—ibn al-Bannā showed by multiplication that the number P_k^n of permutations of k objects from a set of n is

$$P_k^n = n(n-1)(n-2)\cdots[n-(k-1)].$$

Ibn al-Bannā's proof of the formula for C_k^n as well as ibn Mun'im's proof of the permutation rule are, like earlier proofs we have seen, in inductive style. That is, the mathematician begins with a known result for a small value and uses it to build up step by step to higher values. But neither ibn al-Bannā nor any of his predecessors explicitly stated an induction principle to be used as a basis for proofs. Such a statement was first made by Levi ben Gerson, a younger contemporary of ibn al-Bannā, and will be considered in chapter 8.

7.4 Geometry

Islamic mathematicians dealt at an early stage with practical geometry. In fact, a section of al-Khwārizmī's text was devoted to elementary methods of determining areas and volumes. But later on, many mathematicians worked on various theoretical aspects of geometry, including the parallel postulate of Euclid and the exhaustion principle for determining volumes of solids.

7.4.1 The Parallel Postulate

Islamic mathematicians were greatly influenced by their reading of the Greek theoretical works and soon became interested in pure geometrical questions stemming from their knowledge of Euclid and other Greek authors. One of the ideas that recurs in Islamic geometry is that of parallel lines and the provability of Euclid's fifth postulate. Even in Greek times, mathematicians were disturbed by this postulate. Many attempts were made, both in Greece and in the Islamic world to prove it from the other postulates or to replace it by a more "self-evident" postulate.

Among those who proposed a new postulate was al-Khayyāmī, who began with the principle that two convergent straight lines intersect, and it is impossible for them to diverge in the direction of convergence. By convergent lines, he meant lines that approached one another. Given this postulate, al-Khayyāmī proceeded to prove a series of eight propositions, culminating in Euclid's fifth postulate. He began by constructing a quadrilateral with two perpendiculars of equal length, AC and BD, at the two ends of a given line segment AB and then connecting the points C and D (Fig. 7.6). He proceeded to prove that the two angles at C and D were right angles by showing that the two other possibilities—that they were both acute or both obtuse—led to contradictions. If they were acute, CD would be longer than AB; if they were obtuse, CD would be shorter than AB. In each case, he showed that the lines AC and BD would diverge or converge on both sides of AB, and this would contradict his original postulate. Al-Khayyāmī was now able to demonstrate Euclid's fifth postulate.

About a century after al-Khayyāmī, another mathematician, Naṣīr al-Dīn al-Ṭūsī (1201–1274), subjected the works of his predecessor to detailed criticism and then attempted his own proof of the fifth postulate. He considered the same quadrilateral as al-Khayyāmī and

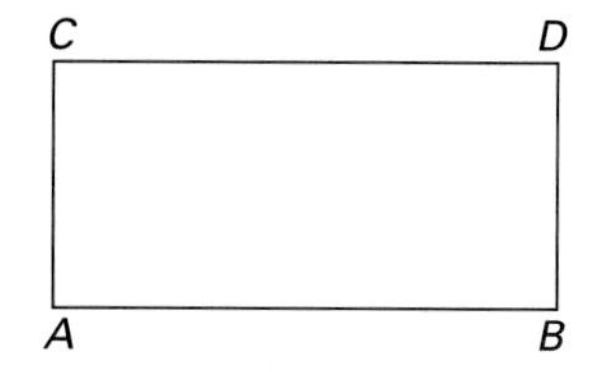

FIGURE 7.6 Al-Khayyāmī's quadrilateral: $AC = BD$, $AC \perp AB$, and $BD \perp AB$. Are the angles at C and D acute, obtuse, or right?

also tried to derive a contradiction from the hypotheses of the acute and obtuse angles. But in a manuscript probably written by his son Ṣadr al-Dīn in 1298, based on Naṣīr al-Dīn's later thoughts on the subject, there is a new argument based on another hypothesis, also equivalent to Euclid's: If a line GH is perpendicular to CD at H and oblique to AB at G, then the perpendiculars drawn from AB to CD are greater than GH on the side on which GH makes an obtuse angle with AB and less on the other side (Fig. 7.7). The importance of this latter work is that it was published in Rome in 1594 and was studied by European geometers.

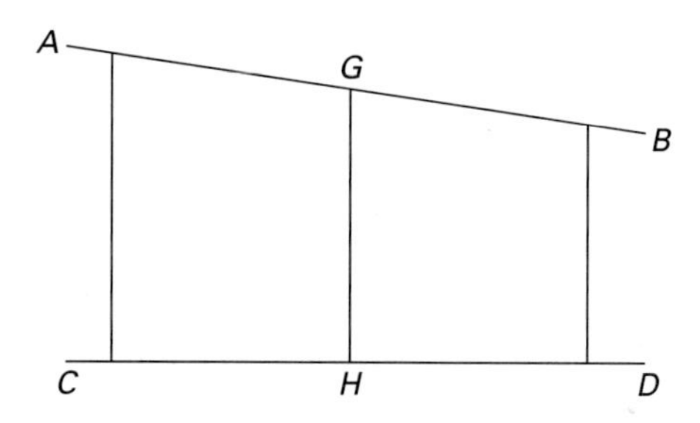

FIGURE 7.7 Naṣīr al-Dīn al-Ṭūsī's hypothesis on parallels and perpendiculars

7.4.2 Volumes and the Method of Exhaustion

Islamic mathematicians understood Eudoxus's method of exhaustion for calculating volumes of solids and read some, but not all, of Archimedes' works applying this method. They went on both to recalculate some of the same volumes as Archimedes and to calculate new volumes. Here we will look at ibn al-Haytham's method for determining the volume of the solid formed by revolving a segment of a parabola about a line perpendicular to its axis.

In modern terminology, ibn al-Haytham proved that the volume of the solid formed by rotating the parabola $x = ky^2$ around the line $x = kb^2$ (which is perpendicular to the axis of the parabola) is $\frac{8}{15}$ of the volume of the cylinder of radius kb^2 and height b. His formal argument was a typical exhaustion argument: He assumed that the desired volume was greater than $\frac{8}{15}$ of that of the cylinder and derived a contradiction, then assumed that it was less and derived another contradiction. But the essence of ibn al-Haytham's argument involved slicing the cylinder into n disks, each of thickness $h = b/n$, the intersection of each

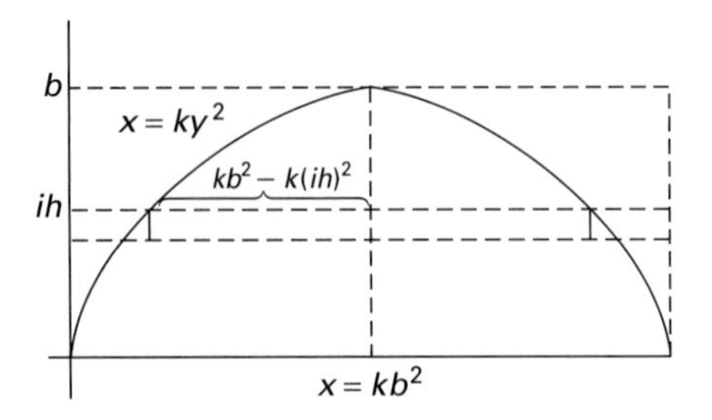

FIGURE 7.8 Revolving a segment of a parabola around a line perpendicular to its axis

with the paraboloid providing an approximation to the volume of a slice of the paraboloid (Fig. 7.8). The ith disk in the paraboloid has radius $kb^2 - k(ih)^2$ and therefore has volume $\pi h(kh^2n^2 - ki^2h^2)^2 = \pi k^2h^5(n^2 - i^2)^2$. The total volume of the paraboloid is therefore approximated by

$$\pi k^2h^5 \sum_{i=1}^{n-1}(n^2 - i^2)^2 = \pi k^2h^5 \sum_{i=1}^{n-1}(n^4 - 2n^2i^2 + i^4).$$

But ibn al-Haytham already knew formulas for the sums of integral squares and integral fourth powers. Using these, he could calculate that

$$\sum_{i=1}^{n-1}(n^4 - 2n^2i^2 + i^4) = \tfrac{8}{15}(n-1)n^4 + \tfrac{1}{30}n^4 - \tfrac{1}{30}n = \tfrac{8}{15}n \cdot n^4 - \tfrac{1}{2}n^4 - \tfrac{1}{30}n$$

and therefore that

$$\tfrac{8}{15}(n-1)n^4 < \sum_{i=1}^{n-1}(n^2 - i^2)^2 < \tfrac{8}{15}n \cdot n^4.$$

But the volume of a typical slice of the circumscribing cylinder is $\pi h(kb^2)^2 = \pi k^2h^5n^4$, and therefore the total volume of the cylinder is $\pi k^2h^5n \cdot n^4$, while the volume of the cylinder less its top slice is $\pi k^2h^5(n-1)n^4$. Therefore, the inequality shows that the volume of the paraboloid is bounded between $\frac{8}{15}$ of the cylinder less its top slice and $\frac{8}{15}$ of the entire cylinder. Since the top slice can be made as small as desired by taking n sufficiently large, it follows that the volume of the paraboloid is exactly $\frac{8}{15}$ of that of the cylinder, as asserted.

7.5 Trigonometry

An Indian astronomical work, with its accompanying trigonometry, was brought to Baghdad late in the eighth century and translated into Arabic. Thus, Islamic scholars were made aware of the trigonometric knowledge of the Hindus, which had earlier been adapted from the Greek version of Hipparchus. They also became aware of Ptolemy's trigonometry, as detailed in his *Almagest*, when that work was translated into Arabic as well. As in other areas of mathematics, Islamic mathematicians absorbed knowledge of trigonometry from other cultures and gradually infused the subject with new ideas.

As in both Greece and India, trigonometry in the Islamic world was intimately tied to astronomy, so in general, mathematical texts on trigonometry were written as chapters of more extensive astronomical works. The mathematicians were particularly interested in using trigonometry to solve spherical triangles, because Islamic law required that Muslims face the direction of Mecca when they prayed. To determine the appropriate direction at one's own location required an extensive knowledge of the solution of such triangles on the sphere of the earth. The solution of both plane and spherical triangles was also important in the determination of the correct time for prayers. These times were generally defined in relation to the onset of dawn and the end of twilight as well as the length of daylight and the altitude of the sun on a given day, values whose accurate determination also required spherical trigonometry.

7.5.1 The Trigonometric Functions

Recall that Ptolemy used only one trigonometric "function," the chord, in his trigonometric work, and the Hindus modified that into the more convenient sine. Early in Islamic trigonometry, both the chord and the sine were used concurrently, but eventually the sine won out. (The sine of an arc, for both Islamic and Hindu mathematicians, was the length of a particular line in a circle of given radius R.) It is not entirely clear who introduced the other functions, but Abū 'Abdallāh Muḥammad ibn Jābir al-Battānī (c. 855–929) used the "sine of the complement to 90°" (the modern cosine) in his astronomical work, which was designed to be an improvement on Ptolemy's *Almagest*. Because he did not use negative numbers, he defined the cosine only for arcs up to 90°. For arcs between 90° and 180°, he used the *versine*, defined as $\text{versin}\,\alpha = R + R\,\sin(\alpha - 90°)$. Even with the cosine, his methods for solving plane triangles were somewhat clumsy. For example, if sides b, c and angle C were given (Fig. 7.9), he would calculate side a as follows: Drop a perpendicular AH from A to side BC. Then find AH by multiplying the sine value for C (in his sine table for radius R) by b/R. Find CH similarly by using the cosine. BH is then found from the Pythagorean theorem applied to triangle AHB, so $a = CH + BH$ can be determined.

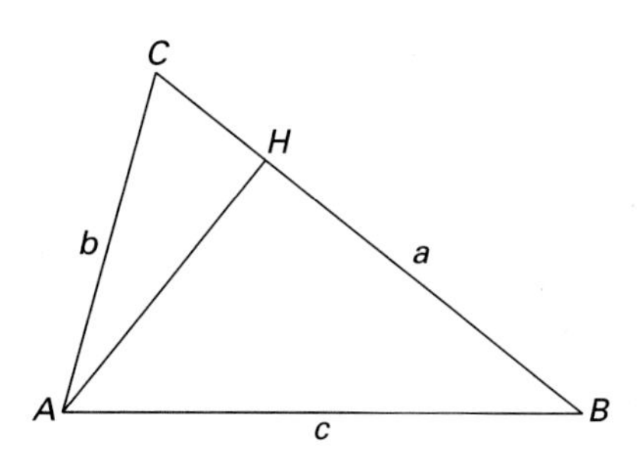

FIGURE 7.9 Al-Battānī's use of the sine and cosine to solve a plane triangle

The tangent, cotangent, secant, and cosecant functions made their appearance in Islamic works in the ninth century. We consider here, however, the discussion of these functions by Abu l-Rāyḥan Muḥammad ibn Aḥmad al-Bīrūnī (973–1055) in his *Exhaustive Treatise on Shadows*. "An example of the direct shadow [cotangent] is: Let A be the body of the sun and BG the gnomon perpendicular to EG, which is parallel to the horizon plane, and ABE the sun's ray passing through the head of the gnomon BG [Fig. 7.10a].... EG is that which is called the direct shadow such that its base is G and its end E. And EB, the line joining the two ends of the shadow and the gnomon, is the hypotenuse of the shadow [cosecant]." The tangent and secant are defined similarly by using a gnomon parallel to the horizon plane. GE in Fig. 7.10b is called the "reversed shadow" (tangent), and BE is called the "hypotenuse of the reversed shadow" (secant).

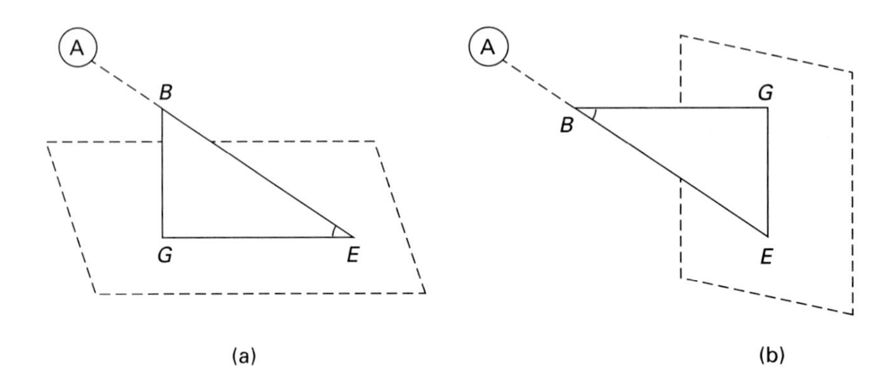

FIGURE 7.10 Al-Bīrūnī's definition of tangent, cotangent, secant, and cosecant. In (a), GE is the cotangent of angle E and EB is the cosecant. In (b), GE is the tangent of angle B and BE is the secant

Al-Bīrūnī used the various relationships among the trigonometric functions to solve problems. For example, to find the altitude α of the sun from its shadow, he wrote, "we multiply the shadow by its equal and the gnomon [g] by its equal and we take [the square root] of the sum, and it will be the cosecant. Then we divide by it the product of the gnomon by the total sine, and there comes out the sine of the altitude. We find its corresponding arc in the sine table and there comes out the altitude of the sun at the time of that shadow." In modern notation, the relationships al-Bīrūnī used are

$$\sqrt{g^2 \cot^2 \alpha + g^2} = g \csc \alpha \quad (\text{or } \cot^2 \alpha + 1 = \csc^2 \alpha) \qquad \text{and} \qquad \frac{gR}{g \csc \alpha} = R \sin \alpha.$$

He now had the value of the sine function of α based on the particular radius R used in his sine tables. He consulted those tables in reverse to determine α. Al-Bīrūnī similarly gave rules equivalent to $\tan^2 \alpha + 1 = \sec^2 \alpha$ and $\tan \alpha = \sin \alpha / \cos \alpha$ and presented a table for the tangent and cotangent in which he used the relationship $\cot \alpha = \tan(90° - \alpha)$.

It is perhaps surprising that al-Bīrūnī used the wealth of trigonometric knowledge collected in his text only for dealing with astronomical problems. For determining terrestrial heights and distances, he described nontrigonometric methods. For example, to determine the height of a minaret whose base is accessible, he suggested, "if surveyed at a time when the altitude of the sun equals an eighth of a revolution [45°], there will be between the end of the shadow and the foot of the vertical a distance equal to [its height]." If the base is not accessible, however, al-Bīrūnī described a procedure similar to the Chinese procedure discussed in chapter 6, a procedure also used in India. Unlike his Indian and Chinese predecessors, however, he gave a description in his text of his reasoning, using the idea of similar triangles.

7.5.2 Spherical Trigonometry

Although there were a few instances of the use of trigonometry to solve earthbound problems, the major use of the trigonometric functions was to solve spherical triangles arising from astronomical problems. Islamic mathematicians were able to derive simpler methods than those of Ptolemy for dealing with these problems. It appears that the basic results were discovered independently by two contemporaries of al-Bīrūnī: Abū Naṣr Manṣūr ibn 'Iraq (d. 1030), one of al-Bīrūnī's teachers; and Muḥammad Abu'l-Wafā' al-Būzjānī (940–997), an important astronomer of Baghdad. We will follow the work of the latter.

The first result is what has become known as the **rule of four quantities**:

THEOREM. *If ABC and ADE are two spherical triangles with right angles at C and E, respectively, and a common acute angle at A, then* $\sin BC : \sin BA = \sin DE : \sin DA$ *(Fig. 7.11).*

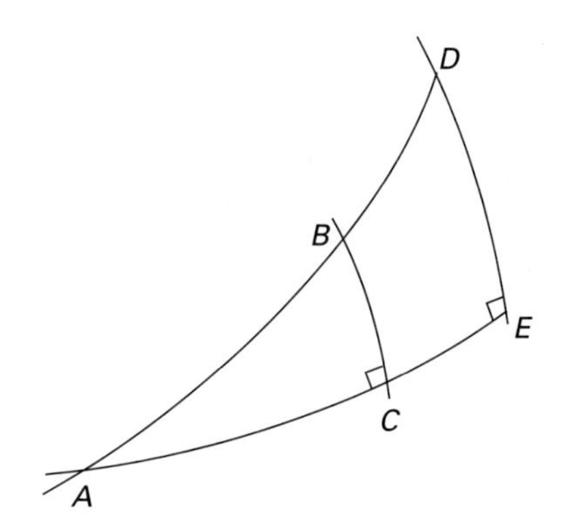

FIGURE 7.11 The rule of four quantities

An immediate corollary of this theorem is formula 3.2: If ABC is a right spherical triangle with right angle at C, then $\sin A = \sin a / \sin c$. To prove this, extend the hypotenuse AB and the base AC to points D and E, respectively, such that both AD and AE are quadrants of a great circle. Then the great circle arc from E to D is perpendicular to AE, and we can apply the theorem. Because $\sin DE = \sin A$, the result is proved. This corollary was in essence used by Ptolemy in many of his calculations. Abu'l-Wafā also gave proofs of some of the other results used by Ptolemy. In addition, he gave a proof of the sine theorem for arbitrary spherical triangles.

THEOREM. *In any spherical triangle ABC,* $\sin a/\sin A = \sin b/\sin B = \sin c/\sin C$ *(Fig. 7.12).*

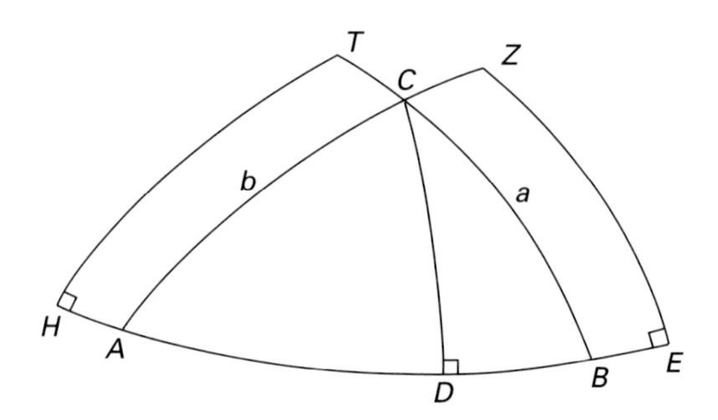

FIGURE 7.12 Abu'l-Wafā's proof of the sine theorem

Given the spherical triangle ABC, let CD be an arc of a great circle perpendicular to AB. Extend AB and AC to AE and AZ, both quadrants, and extend BA and BC to BH and BT, also both quadrants. Then A is a pole for the great circle EZ, and B is a pole for the great circle TH. Because the angles at E and H are right angles, it follows that the triangles ADC and AEZ are spherical right triangles with a common angle at A, while the triangles BDC and BHT are spherical right triangles with a common angle at B. By the rule of four quantities, we have

$$\frac{\sin DC}{\sin b} = \frac{\sin ZE}{\sin ZA} \quad \text{and} \quad \frac{\sin DC}{\sin a} = \frac{\sin TH}{\sin TB}.$$

But because A and B are poles of ZE and TH, respectively, arc ZE equals $\angle A$ and arc TH equals $\angle B$. Thus, the equations can be rewritten as

$$\frac{\sin DC}{\sin b} = \frac{\sin A}{R} \quad \text{and} \quad \frac{\sin DC}{\sin a} = \frac{\sin B}{R}.$$

Thus, $\sin A \sin b = \sin B \sin a$, and the sine theorem is proved.

Given the sine theorem and the other results, Islamic mathematicians worked out various methods of determining the *qibla*, the direction of Mecca, which a Muslim must face during prayer. The basic problem is the following: Assume that M is the position of Mecca and that P is one's current location (Fig. 7.13). Let arc AB represent the equator and T the north pole, and draw meridians from T through P and M. The *qibla* is then $\angle TPM$ on the earth's sphere. Assuming that the latitudes α, β and the longitudes γ, δ of P and M, respectively, are known, then arcs TP and TM are known ($90° - \alpha$, $90° - \beta$, respectively), and also $\angle PTM$ ($= \delta - \gamma$) is known. Thus, the problem is to solve a spherical triangle given two sides and the included angle. The sine theorem does not suffice to do this, but al-Bīrūnī and others worked out various procedures to solve the problem using the sine theorem applied to a series of triangles constructed from the original one.

It was not until the thirteenth century, however, that systematic methods for solving all types of plane and spherical triangles were completely worked out. These were published

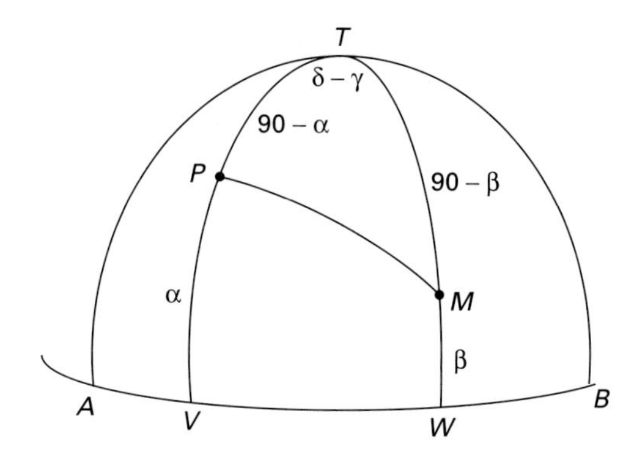

FIGURE 7.13 The problem of the *qibla*

in the first comprehensive work on spherical and plane trigonometry, independent of astronomy, written in the Islamic world: the *Treatise on the Transversal Figure*, usually known as the *Treatise on the Complete Quadrilateral*, by Naṣīr al-Dīn al-Ṭūsī. In it, for example, the author proved the theorem of sines for plane triangles, then used it systematically to solve plane triangles. For spherical triangles, he dealt with all the possible cases. In particular, he showed how to solve the case where the three sides are known and provided the first discussion of the solution of the case in which the three angles are known.

We consider here the first of these procedures. Let ABC be the triangle with given arcs AB, AC, and BC (Fig. 7.14). Extend AB and AC to quadrants AD and AE, respectively. Then draw the great circle through DE and extend it to meet BC extended at F. Because the angles at D and E are right, the rule of four quantities implies that $\sin CF : \sin BF = \sin CE : \sin BD$. Because $CE = 90 - CA$ and $BD = 90 - BA$ are known, the ratio of the sine of CF to the sine of BF is known. In addition, $BF - CF = BC$ is known. Therefore, both arcs BF and CF can be found. Then arcs DF and EF can be found by using results on right spherical triangles. It follows that $DE = DF - EF$ can be found and therefore that angle A (= arc DE) can be found. The remaining angles can then be found by the law of sines.

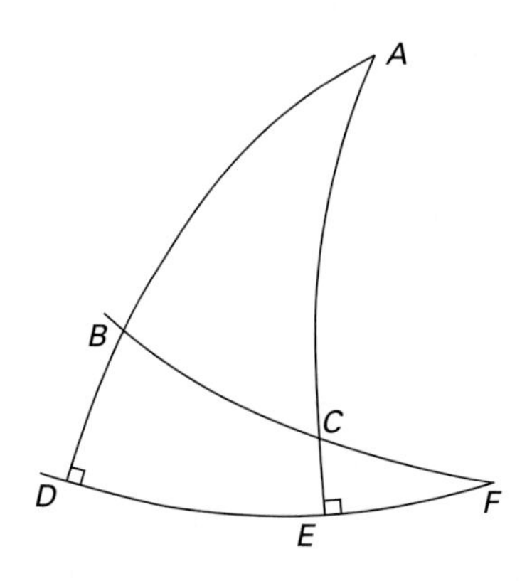

FIGURE 7.14 Solving a spherical triangle with all sides known

7.5.3 Values of Trigonometric Functions

Solving problems in astronomy and geography required not only formulas for solving triangles but also highly accurate tables. These were gradually developed. As for the calculations of Ptolemy, the accuracy of such a table depended primarily on the accuracy of the calculation of $\sin 1°$. Various methods were used to do this calculation, the most impressive being the method of al-Kāshī in the early fifteenth century. He started with the triple angle formula $\sin 3\theta = 3\sin\theta - 4\sin^3\theta$. Setting $\theta = 1°$ gives a cubic equation for $x = \sin 1°$: $3x - 4x^3 = \sin 3°$. Because al-Kāshī calculated his sine table based on a circle of radius 60, he needed to calculate $y = 60\sin 1° = 60x$. His equation was therefore $3y - 4y^3/60^2 = 60\sin 3°$, or

$$y = \frac{900(60\sin 3°) + y^3}{45 \cdot 60}.$$

Recall that $\sin 3°$ can be calculated to whatever degree of accuracy is needed by use of the difference and half-angle formulas. Al-Kāshī in fact used as his value for $60\sin 3°$ the sexagesimal 3;8,24,33,59,34,28,15. His equation, written in sexagesimal notation, was therefore

$$y = \frac{47,6;8,29,53,37,3,45 + y^3}{45,0}.$$

He proceeded to solve this equation by an iterative procedure, given his knowledge that the solution was a value close to 1. Writing the equation symbolically as $y = (q + y^3)/p$, and assuming that the solution is given as $y = a + b + c + \cdots$, where the various letters represent successive sexagesimal places, begin with the first approximation, $y_1 = q/p \approx a\ (= 1)$. To find the next approximation, $y_2 = a + b$, solve for b by setting

$$y_2 = \frac{q + y_1^3}{p} \qquad \text{or} \qquad a + b = \frac{q + a^3}{p}.$$

Then

$$b \approx \frac{q - ap + a^3}{p}\ (= 2).$$

Similarly, if $y_3 = a + b + c$, set

$$y_3 = \frac{q + y_2^3}{p} \qquad \text{or} \qquad a + b + c = \frac{q + (a + b)^3}{p}$$

and find

$$c \approx \frac{q - (a + b)p + (a + b)^3}{p}\ (= 49).$$

Al-Kāshī did not justify this iterative approximation procedure, but he evidently knew it converged more rapidly than the solution procedures for cubic equations used by his predecessors. In this case, he calculated $y = 1;2,49,43,11,14,44,16,26,17$, a result equivalent to a decimal value for $\sin 1°$ of 0.017452406437283571, quite a feat for the days before electronic calculators. Al-Kāshī's patron, Ulūgh Beg, an astronomer who ruled a domain in central Asia from his capital of Samarkand, used this work to calculate sine and tangent tables for every minute of arc to five sexagesimal places, a total of 5400 entries in each table!

7.6 Transmission of Islamic Mathematics

By the time of al-Kāshī, Islamic scientific civilization was in a state of decline. There were few other scientists of consequence in the years following. Even before the fifteenth century, though, mathematical activity had resumed in Europe. A central factor of this revival was the work of the translators of the twelfth century who made available to Europeans a portion of the Islamic mathematical corpus, most importantly al-Khwārizmī's arithmetic and algebra works. The work of Abū Kāmil also became available in Europe, chiefly through the inclusion of numerous problems from it in Leonardo of Pisa's *Liber abbaci* (1202) and the fifteenth-century translation of this work into Hebrew in Italy. As far as is known, the more advanced algebraic materials of al-Samaw'al and others did not reach Europe before or during the Renaissance.

We have already noted that both the idea and the notation for decimal fractions were present in the work of al-Kāshī. The system, including decimal fractions, also appeared around this time in a Byzantine textbook, with the method described as "Turkish," i.e., Islamic. This textbook was brought to Venice in 1562, but even before that the same notation appeared occasionally in European works. So, although the complete decimal system in Europe is traditionally ascribed to Simon Stevin in the late sixteenth century, it does appear that at least some aspects of it traveled to Europe from the Islamic world before that time.

In combinatorics, there is no known Renaissance translation into a European language of the work of ibn Mun'im or ibn al-Bannā. On the other hand, as we will see in the next chapter, ideas very closely related to theirs were developed in southern France in the fourteenth century by Levi ben Gerson, who in all probability was aware of Islamic advances in this area. However, it is not known whether the work of Levi had any influence on combinatorics in Europe later on. The Pascal triangle itself first appeared in Europe in the thirteenth century. Whether this idea traveled to Europe from the Islamic world (or from China) or was discovered in Europe independently is a matter for speculation.

As already noted, an important Islamic work dealing with the parallel postulate appeared in Rome in the late sixteenth century, in Arabic, but with a Latin title page. The work was not, as far as is known, formally translated into Latin. However, John Wallis in England was certainly aware of its contents and wrote about its ideas in developing his own ideas on the parallel postulate. Gerolamo Saccheri also knew of this work and used some of its ideas in his own work, which ultimately led to the development of non-Euclidean geometry.

The only known manuscript containing ibn al-Haytham's work on the volume of a paraboloid of revolution was acquired by the library of the India Office in England in the nineteenth century. Thus, although results similar to ibn al-Haytham's on the sum of integral powers began to appear in Europe in the seventeenth century, we have no way of knowing whether anyone in Europe was aware, either directly or indirectly, of that particular treatise of the Egyptian mathematician. Certainly, however, Europeans were aware of ibn al-Haytham's major work on optics, a work that was translated into Latin early and had a major influence on European work on that subject.

In trigonometry, our knowledge is more certain. Although Naṣīr al-Dīn al-Ṭūsī's trigonometric work did not reach Europe during the Renaissance, some of the earlier Islamic work did. In particular, the sine theorem and the rule of four quantities, along with some of their corollaries, appeared in Spain, in the work of Abū Muḥammad Jābir ibn Aflaḥ al-Ishbīlī (early twelfth century). Virtually nothing is known of Jābir's life except that he was from Seville, but his major book, a critique of Ptolemy's *Almagest*, was translated into Latin late in the twelfth century and provided Europeans with one of the earliest chronicles of the Islamic advances on the trigonometry of Ptolemy. In fact, Regiomontanus, the first European author to write a work on pure trigonometry, clearly took much of his material on spherical trigonometry directly from the book of Jābir.

Exercises

1. Al-Khwārizmī gives the following rule for his sixth case, $bx + c = x^2$: Halve the number of roots. Multiply this by itself. Add this square to the number. Extract the square root. Add this to the half of the number of roots. That is the solution. Translate this rule into a formula. Give a geometric argument for its validity using Figure 7.15, where $x = AB = BD$, $b = HC$, c is represented by rectangle $ABRH$, G is the midpoint of HC, and $HK = HG$.

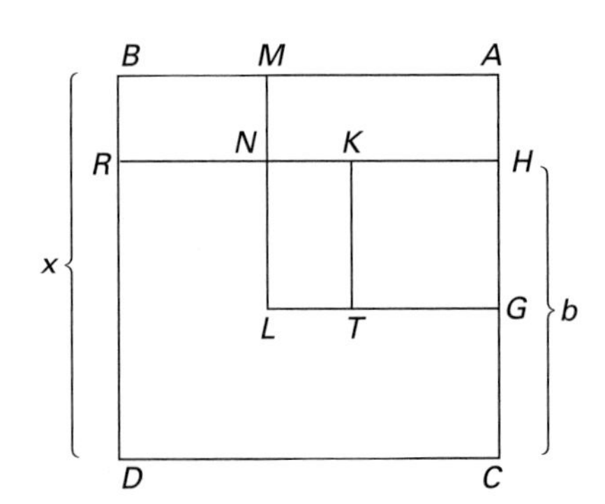

FIGURE 7.15 Al-Khwārizmī's justification for the solution rule for $bx + c = x^2$

2. Solve the following problems of al-Khwārizmī by applying the appropriate formula:
   (a) $(\frac{1}{3}x + 1)(\frac{1}{4}x + 1) = 20$
   (b) $x^2 + (10 - x)^2 = 58$

3. Solve $\frac{1}{2}x^2 + 5x = 28$ by first multiplying by 2 and then using al-Khwārizmī's procedure. Similarly, solve $2x^2 + 10x = 48$ by first dividing by 2.

4. Solve this problem of al-Khwārizmī: I have divided 10 into two parts, and have divided the first by the second, and the second by the first and the sum of the quotients is $2\frac{1}{6}$. Find the parts.

5. Solve the following problems of Abū Kāmil:
   (a) Suppose 10 is divided into two parts and the product of one part by itself equals the product of the other part by the square root of 10. Find the parts.
   (b) Suppose 10 is divided into two parts, each one of which is divided by the other, and when each of the quotients is multiplied by itself and the smaller is subtracted from the larger, then there remains 2. Find the parts.

6. Solve the following problems of Abū Kāmil:
   (a) $[x - (2\sqrt{x} + 10)^2]^2 = 8x$ (First substitute $x = y^2$.)
   (b) $\left(x + \sqrt{\frac{1}{2}x}\right)^2 = 4x$

7. Complete al-Samaw'al's procedure of dividing $20x^2 + 30x$ by $6x^2 + 12$ to get the result stated in the text. Prove that the coefficients of the quotient satisfy the rule $a_{n+2} = -2a_n$, where a_n is the coefficient of $1/x^n$.

8. Use al-Samaw'al's procedure to divide $20x^6 + 2x^5 + 58x^4 + 75x^3 + 125x^2 + 96x + 94 + 140(1/x) + 50(1/x^2) + 90(1/x^3) + 20(1/x^4)$ by $2x^3 + 5x + 5 + 10(1/x)$. (His answer is $10x^3 + x^2 + 4x + 10 + 8(1/x^2) + 2(1/x^3)$.)

9. Use ibn al-Haytham's procedure to derive the formula for the sum of the fifth powers of the integers:
$$1^5 + 2^5 + \cdots + n^5 = \tfrac{1}{6}n^6 + \tfrac{1}{2}n^5 + \tfrac{5}{12}n^4 - \tfrac{1}{12}n^2.$$

10. Show, using the formulas for sums of fourth powers and squares, that

$$\sum_{i=1}^{n-1}(n^4 - 2n^2i^2 + i^4) = \tfrac{8}{15}(n-1)n^4 + \tfrac{1}{30}n^4 - \tfrac{1}{30}n$$
$$= \tfrac{8}{15}n \cdot n^4 - \tfrac{1}{2}n^4 - \tfrac{1}{30}n.$$

11. Give a formal proof of formula 7.2 by induction on k.

12. Show that one can solve $x^3 + d = cx$ by intersecting the hyperbola $y^2 - x^2 + (d/c)x = 0$ with the parabola $x^2 = \sqrt{c}y$. Sketch the two conics. Find sets of values for c and d for which these conics do not intersect, intersect once, and intersect twice.

13. Show that one can solve $x^3 + d = bx^2$ by intersecting the hyperbola $xy = d$ and the parabola $y^2 + dx - db = 0$. Assuming that $\sqrt[3]{d} < b$, determine the conditions on b and d that give zero, one, or two intersections of these two conics. Compare your answer with Sharaf al-Dīn's analysis of the same problem.

14. Show that $x^3 + cx = bx^2 + d$ is the only one of al-Khayyāmī's cubics that could have three positive solutions. Under what conditions do these three positive solutions exist?

15. Show using calculus that $x_0 = 2b/3$ does maximize the function $x^2(b - x)$. Then use calculus to analyze the graph of $y = x^3 - bx^2 + d$ and confirm Sharaf al-Dīn's conclusion on the number of positive solutions to $x^3 + d = bx^2$.

16. Analyze the possibilities of positive solutions to $x^3 + d = cx$ by first showing that the maximum of the function $x(c - x^2)$ occurs at $x_0 = \sqrt{c/3}$. Use calculus to consider the graph of $y = x^3 - cx + d$ and determine the conditions on the coefficients giving it zero, one, or two positive solutions.

17. Give a modern proof by induction on n of ibn al-Bannā's result that

$$C_k^n = \frac{n - (k-1)}{k} C_{k-1}^n.$$

18. Al-Battānī developed a formula equivalent to the following relating three sides and one angle of a spherical triangle:

$$\cos a = \cos b \cos c + \sin b \sin c \cos A.$$

Although al-Battānī did not do so, the formula can be used to determine the third side of a triangle of which two sides and the included angle are given. Use this result, along with the law of sines, to determine the *qibla* for Rome (latitude 41°53′ N, longitude 12°30′ E), given that Mecca has latitude 21°45′ N and longitude 39°49′ E.

19. Use al-Battānī's formula to determine the distance on the earth between New York (latitude 41° N, longitude 74° W) and London (latitude 52° N, longitude 0°). Use the fact that the earth's circumference is approximately 25,000 miles.

20. Al-Bīrūnī devised a method for determining the radius r of the earth by sighting the horizon from the top of a mountain of known height h. That is, he assumed that one could measure α, the angle of depression from the horizontal at which one sights the apparent horizon (Fig. 7.16). Show that r is determined by the formula

$$r = \frac{h \cos\alpha}{1 - \cos\alpha}.$$

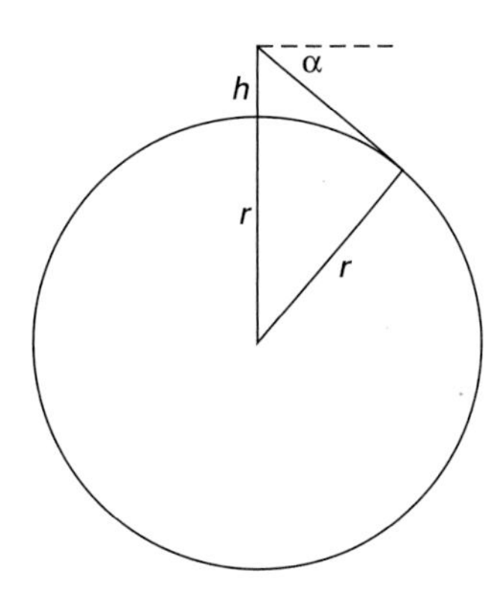

FIGURE 7.16 Al-Bīrūnī's method for calculating the earth's radius

Al-Bīrūnī performed this measurement in a particular case, determining that $\alpha = 0°34'$ measured from the summit of a mountain whose height was 652; 3,18 cubits. Calculate the radius of the earth in cubits. Assuming that a cubit equals 18 inches, convert your answer to miles, and compare to a modern value. Comment on the efficacy of al-Bīrūnī's procedure.

21. Naṣir al-Dīn al-Ṭūsī proved the following result known to Ptolemy: In a right spherical triangle with right angle at C,

$$\frac{\tan b}{\tan B} = \sin a.$$

Use this result to demonstrate the following equalities in an arbitrary spherical triangle ABC, where CD is a perpendicular from C to AB (Fig. 7.17):

$$\frac{\tan A}{\tan B} = \frac{\sin BD}{\sin AD} \quad \text{and} \quad \frac{\tan C_1}{\tan C_2} = \frac{\tan AD}{\tan BD}$$

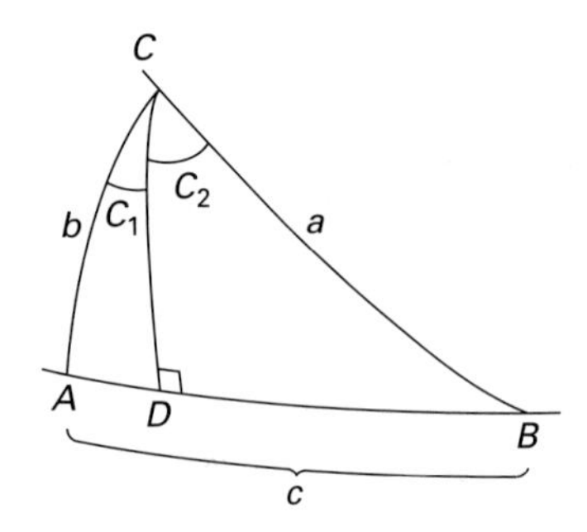

FIGURE 7.17 Determining tangent relationships in a spherical triangle

22. Use Naṣir al-Dīn al-Ṭūsī's procedure described in the text to solve a spherical triangle of sides 60°, 75°, and 31°.

23. Use al-Battānī's formula from exercise 18, along with the law of sines, to solve the triangle from exercise 22. Which method is easier?

24. Naṣir al-Dīn al-Ṭūsī demonstrated a method to solve a spherical triangle if all three angles are known. Suppose the three angles of triangle ABC are given (Fig. 7.18), and assume that all three sides of the triangle are less than a quadrant. Extend each side of the triangle two different ways to form a quadrant. That is, extend AB to AD and BH, AC to AE and CG, and BC to BK and FC, where all of the six new arcs are quadrants. Next, draw great circle arcs through D and E, F and G, and H and K to form the new spherical triangle, LMN. The vertices of the original triangle are now the poles of the three sides of the new triangle. Then, for example, $MD = EN = 90° - DE = 90° - A$, or $MN = 180° - A$. Thus, the three sides of triangle LMN are known, and therefore the triangle can be solved by the procedure discussed in the text. But the vertices of triangle LMN are the poles of the original triangle. So, for example, $BF = CK = 90° - BC$, and $L = FK = 180° - BC$. The sides of the original triangle can therefore be determined. Use this procedure to solve the triangle ABC, where $A = 75°$, $B = 80°$, and $C = 85°$.

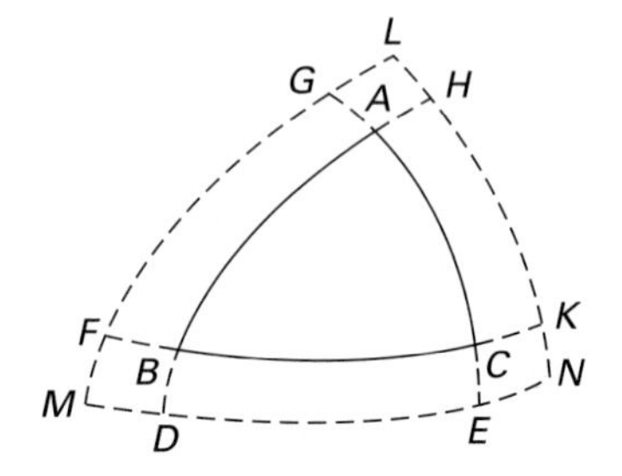

FIGURE 7.18 Solving a spherical triangle when the three angles are given

25. Calculate the first four sexagesimal places of the approximation to $y = 60 \sin 1°$ following the method indicated in the text. Your calculation should show why the iteration method works.

26. Why did it take many centuries after its introduction for the decimal place value system to become the system of numeration universally used in the Islamic world?

27. Outline a lesson to teach the quadratic formula using geometric arguments in the style of al-Khwārizmī.

28. Design a lesson for a calculus class that uses the methods of Sharaf al-Dīn al-Ṭūsī to demonstrate the number of solutions to various cubic equations.

29. Design a lesson deriving the multiplicative formula for C_k^n based on the work of ibn al-Bannā.

30. Design a lesson for a trigonometry class showing the application of the rules for solving spherical triangles to various interesting problems.

31. Should spherical trigonometry be taught in schools today? Give arguments for and against, but first determine in which countries this subject is in fact taught.

References

The best general work on Islamic mathematics is still Adolf P. Youschkevitch, *Les Mathématiques Arabes (VIIIe–XVe siècles)* (Paris: J. Vrin, 1976). This is one part of a more general work on medieval mathematics originally published in 1961 in Russian and translated into French by M. Cazenave and K. Jaouiche. Another excellent work, in English, that treats various important ideas of Islamic mathematics is J. Lennart Berggren, *Episodes in the Mathematics of Medieval Islam* (New York: Springer Verlag, 1986). This work, although not a history of Islamic mathematics as such, treats certain important mathematical ideas considered by Islamic mathematicians at a level accessible to university mathematics students. A somewhat more comprehensive survey is Jacques Sesiano, "Islamic Mathematics," in Helaine Selin, ed., *Mathematics across Cultures: The History of Non-Western Mathematics* (Dordrecht: Kluwer, 2000). A collection of papers by Roshdi Rashed has been edited and translated into English as *The Development of Arabic Mathematics: Between Arithmetic and Algebra* (Dordrecht: Kluwer, 1994). This collection presents much of Rashed's recent research and provides a new look at the arithmetical and algebraic aspects of Islamic mathematics. Rashed has also written three major articles for the *Encyclopedia of the History of Arabic Science* (London: Routledge, 1996), which together provide an excellent summary of the history of much of Islamic mathematics: "Algebra" (Vol. 2, 349–375); "Combinatorial analysis, numerical analysis, Diophantine analysis and number theory" (Vol. 2, 376–417); and "Infinitesimal determinations, quadrature of lunules and isoperimetric problems" (Vol. 2, 418–446). These articles are complemented by additional articles on other topics in the same *Encyclopedia*, including Marie-Thérèse Debarnot's article "Trigonometry" (Vol. 2, 495–538).

Many of the major works of Islamic mathematics have been translated into English. Al-Khwārizmī's *Algebra* was edited and translated by Frederic Rosen as *The Algebra of Muhammed ben Musa* (London: Oriental Translation Fund, 1831) (reprinted Hildesheim: Olms, 1986). A more recent English translation, from a twelfth-century Latin translation, is Louis Karpinski, *Robert of Chester's Latin Translation of the Algebra of al-Khowarizmi* (Ann Arbor: University of Michigan Press, 1930). Abu Kāmil's *Algebra* is available in Martin Levey, *The Algebra of Abū Kāmil, Kitāb fī'l-muqābala, in a Commentary by Mordecai Finzi* (Madison: University of Wisconsin Press, 1966). This English translation of the fifteenth-century Hebrew translation of Abū Kāmil's work provides a detailed discussion of the relation of his algebra to previous Greek and Islamic work. For Omar Khayyam, see Daoud S. Kasir, *The Algebra of Omar Khayyam* (New York: Columbia Teachers College, 1931), an English translation of al-Khayyāmī's work on algebra that includes a detailed discussion of his contributions to the subject. Sharaf al-Dīn's work containing his contribution to the solution of cubic equations has been translated into French by Roshdi Rashed as *Sharaf al-Dīn al-Ṭūsī, oeuvres mathématiques: Algèbre et géométrie au XIIe siècle* (Paris: Société d'Édition Les Belles Lettres, 1986). Rashed gives not only an edited Arabic text and a translation, but also an extensive commentary on the mathematics involved, relating it in particular to various modern concepts. Ibn Mun'im's work is available in Ahmed Djebbar, *L'Analyse Combinatoire au Maghreb: L'Exemple d'Ibn Mun'im (XIIe–XIIIe s.)* (Orsay: Université de Paris Sud, Publications Mathématiques D'Orsay, 1985). Al-Bīrūnī's trigonometry is covered in E. S. Kennedy, *The Exhaustive Treatise on Shadows by Abū al-Rayhān al-Bīrūnī* (Aleppo: University of Aleppo, 1976).

There is a large and growing literature dealing with various aspects of Islamic mathematics. A summary and bibliography are found in J. Lennart Berggren, "Mathematics and Her Sisters in Medieval Islam: A Selective Review of Work Done from 1985 to 1995," *Historia Mathematica* 24 (1997), 407–440. Some useful works on topics discussed in this chapter include J. Lennart Berggren, "Innovation and Tradition in Sharaf al-Dīn al-Ṭūsī's *al Mu'ādalāt*," *Journal of the American Oriental Society* 110 (1990), 304–309; Jan P. Hogendijk, "Sharaf al-Dīn al-Ṭūsī on the Number of Positive Roots of Cubic Equations," *Historia Mathematica* 16 (1989), 69–85; chapter 2 of B. A. Rosenfeld, *A History of Non-Euclidean Geometry: Evolution of the Concept of a Geometric Space*, translated by Abe Shenitzer (New York: Springer Verlag, 1988); and chapter 3 of Jeremy Gray, *Ideas of Space: Euclidean, Non-Euclidean, and Relativistic*, second edition (Oxford: Clarendon Press, 1989). An excellent introduction to the setting of Islamic algebra can be found in Jens Høyrup, "The Formation of 'Islamic Mathematics': Sources and Conditions," in *In Measure, Number, and Weight: Studies in Mathematics and Culture* (Albany: State University of New York Press, 1994), 89–122. This article contains a detailed development of these ideas. The author contends that there was an Islamic "miracle," comparable to the Greek one, involving the integration of mathematical theory and practice and crucial for the creation of modern science.

CHAPTER EIGHT

Mathematics in Medieval Europe

Who wishes correctly to learn the ways to measure surfaces and to divide them, must necessarily thoroughly understand the general theorems of geometry and arithmetic, on which the teaching of measurement . . . rests. If he has completely mastered these ideas, he . . . can never deviate from the truth.

—Introduction to Plato of Tivoli's Latin translation of the Hebrew *Treatise on Mensuration and Calculation* by Abraham bar Ḥiyya, 1116

The Roman Empire in the West collapsed in 476 under the onslaught of various "barbarian" tribes. Feudal societies were soon organized in parts of the old empire, and the long process of the development of the European nation-states began. For the next five centuries, however, the general level of culture in Europe was very low. Serfs worked the land, and few of the barons could read or write, let alone understand mathematics. In fact, there was little practical need for the subject, because the feudal estates were relatively self-sufficient and trade was almost nonexistent, especially after the Muslim conquest of the Mediterranean sea routes.

Despite the lack of mathematical activity, the early Middle Ages had inherited from antiquity the notion that the *quadrivium*—the subjects of arithmetic, geometry, music, and astronomy—was part of the requirements for an educated man, even in the evolving Roman Catholic culture. Yet the only texts available for study of these subjects were brief introductions, especially those by the Roman scholar Boethius (480–524) and the seventh-century bishop Isidore of Seville (560–636), making the quadrivium only a shell, nearly devoid of substance.

Virtually the only centers of learning in early medieval Europe were the monasteries, where the monks copied Greek and Latin manuscripts, thus preserving to some extent the classical heritage. The monks did have to consider one significant mathematical problem—the determination of the calendar. In particular, there was a debate in the Church as to whether Easter should be determined using the Roman solar calendar or the Jewish lunar calendar. The two reckonings could be reconciled, but only by those with some mathematical knowledge. Charlemagne, in fact, even before his coronation in 800 as Holy Roman Emperor, formally recommended that the mathematics necessary for Easter computations be part of the curriculum in church schools.

To assist him in establishing more schools, Charlemagne brought in Alcuin of York (735–804) as his educational adviser. There is no direct evidence as to Alcuin's knowledge of mathematics, but a collection of fifty-three arithmetical problems from his time, entitled *Propositions for Sharpening Youths*, is generally attributed to him. Solving the problems in the collection often requires some ingenuity, but the solutions do not depend on any particular mathematical theory or rules of procedure.

In the tenth century, a revival of interest in mathematics began with the work of Gerbert d'Aurillac (945–1003), who became Pope Sylvester II in 999. In his youth, Gerbert studied in Spain, where he probably learned some of the mathematics of the Muslims. Later, he reorganized the cathedral school at Rheims and successfully reintroduced the study of mathematics. Besides teaching basic geometry and astronomy, he also taught the use of a counting board, divided into columns representing the (positive) powers of ten, in each of which he would place a single counter marked with the western Arabic form of one of the numbers $1, 2, 3, \ldots, 9$. Zero was represented by an empty column. Gerbert's work represents the first appearance in the Christian West of the Hindu-Arabic numerals, although the absence of the zero and the lack of suitable algorithms for calculating with these counters showed that Gerbert did not understand the full significance of the Hindu-Arabic system.

Despite the limited mathematical sources available to Europeans at the turn of the millennium, scholars did know that there was an ancient tradition in mathematics that reached back to the Greeks. At the time, however, that tradition was virtually entirely inaccessible. It was only with the work of the translators that the Greek mathematical corpus, as well as a portion of the mathematics developed in the Islamic world, was brought to Western Europe to begin to fill in the shell of the quadrivium. European scholars discovered the major Greek scientific works (primarily in Arabic translation) beginning in the twelfth century and started the process of translating these into Latin. Much of this work was accomplished at Toledo in Spain, which at the time had only recently been retaken by the Christians from the former Muslim rulers. Here could be found repositories of Islamic scientific manuscripts as well as people at home in two cultures. In particular, there was a flourishing Jewish community, many of whose members were fluent in Arabic. The translations thus were often made in two stages: first by a Spanish Jew from Arabic into Spanish, and then by a Christian scholar from Spanish into Latin.

Among the Greek works translated into Latin from Arabic in the twelfth century were Aristotle's *Prior Analytics* and *Posterior Analytics*, Euclid's *Elements*, Ptolemy's *Almagest*, and Archimedes' *Measurement of the Circle*. The Arabic works translated included both the *Arithmetic* and the *Algebra* of al-Khwārizmī and trigonometric works written by al-Battānī and Jābir ibn Aflaḥ. In the thirteenth century, most of Archimedes' works were translated as well as parts of Apollonius's *Conics* and Abū Kāmil's *Algebra*. Other works reached Europe indirectly through this period as well, including some written by Spanish Jewish scholars based on Arabic works. European scholars absorbed the new (and old) ideas included in these materials and soon began to create new mathematics.

This chapter will discuss both the Jewish and the Christian contributions of the twelfth through the fourteenth centuries. We will begin by looking at developments in geometry, then move on to combinatorics and algebra. We will conclude with a consideration of the new mathematics of kinematics that stemmed from the study of Aristotle's works in the medieval universities.

8.1 Geometry

8.1.1 Abraham bar Ḥiyya's *Treatise on Mensuration*

Euclid's *Elements* was translated into Latin early in the twelfth century. Before then, of course, Arabic versions were available in Spain. And so when Abraham bar Ḥiyya (d. 1136) of Barcelona wrote his *Treatise on Mensuration and Calculation* in 1116 to help French and Spanish Jews with the measurement of their fields, he began the work with a summary of some important definitions, axioms, and theorems from Euclid, as noted in this chapter's opening quote. Not much is known of the life of Abraham bar Ḥiyya, but from his Latin title of Savasorda, a corruption of the Arabic words meaning "captain of the bodyguard," it is likely that he had a court position, probably one that required him to give mathematical and astronomical advice to the Christian monarch. The *Treatise on Mensuration* was translated into Latin in 1145 under the title *Liber embadorum* (*Book of Areas*).

Like most of those who dealt with geometry over the next few centuries, Abraham was not as interested in the theoretical aspects of Euclid's *Elements* as he was in the practical application of geometric methods to measurement. Thus, he gave rules for finding areas and volumes of numerous geometric figures. His most original contribution is found in his section on measurement in circles. He began by giving the standard rules for finding the circumference and area of a circle, first using $3\frac{1}{7}$ for π but then noting that if one wants a more exact value, as in dealing with the stars, one should use $3\frac{8\frac{1}{2}}{60}$ $(= 3\frac{17}{120})$. Curiously, in the Hebrew version of the text, but not in the Latin, there is a justification of the area formula $A = (C/2)(d/2)$ based on indivisibles. Namely, one thinks of the circle as made up of concentric circles of indivisible threads (Fig. 8.1). If one then slices this circle from

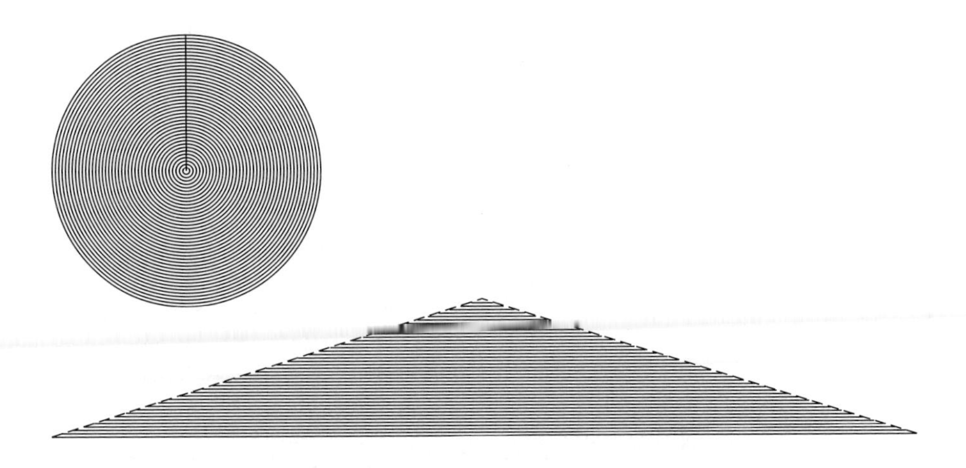

FIGURE 8.1 Circle unfolded into a triangle

Partes Cordatum	Arcus		
	Partes	Min.	Sec.
1	1	0	2
2	2	0	8
3	3	0	26
4	4	0	55
5	5	1	44
6	6	2	54
7	7	4	42
8	8	7	11
9	9	9	56
10	10	13	42
11	11	18	54
12	12	24	38
13	13	31	9
14	14	40	0
15	15	50	10
16	17	2	16
17	18	16	36
18	19	33	27
19	20	53	26
20	22	17	10
21	23	45	6
22	25	19	24
23	27	0	0
24	28	49	56
25	31	26	37
26	33	20	52
27	36	27	32
28	44	0	0

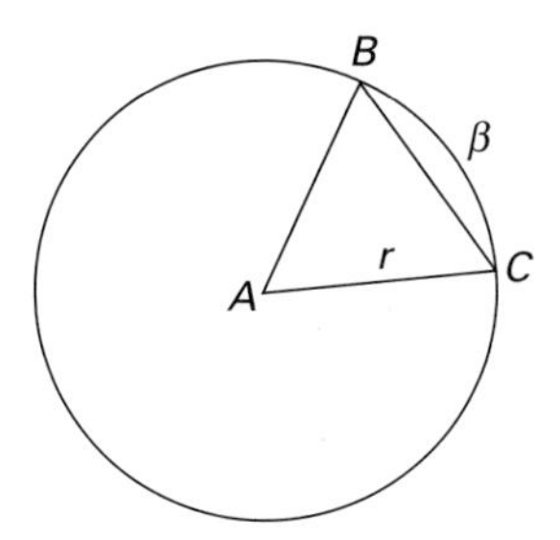

FIGURE 8.2 Area of segment $B\beta C$ = area of sector $AB\beta C$ − area of triangle ABC; area of sector $= r\beta/2$

FIGURE 8.3 Arcchord table of Abraham bar Ḥiyya

the center to the circumference and unfolds it into a triangle, the base of the triangle is the original circumference and the height is the radius. The area formula follows immediately.

To measure areas of segments of circles, Abraham noted that one must first find the area of the corresponding sector by multiplying the radius by half the length of the arc (Fig. 8.2). One then subtracts the area of the triangle formed by the chord of the segment and the two radii at its ends. But how does one calculate the length of the arc, assuming one knows the length of the chord? Abraham's answer is, by the use of a table relating chords and arcs. And so for the first time in Europe there appeared what one can call a trigonometric table (Fig. 8.3). But instead of being a table of chords for given arcs in a circle of radius 60, Abraham's table was a table of arcs for given chords, using what seemed to Abraham a more convenient measure. That is, he used a radius of 14 parts, so the semicircumference would be integral (44), and then gave the arc (in parts, minutes, and seconds) corresponding

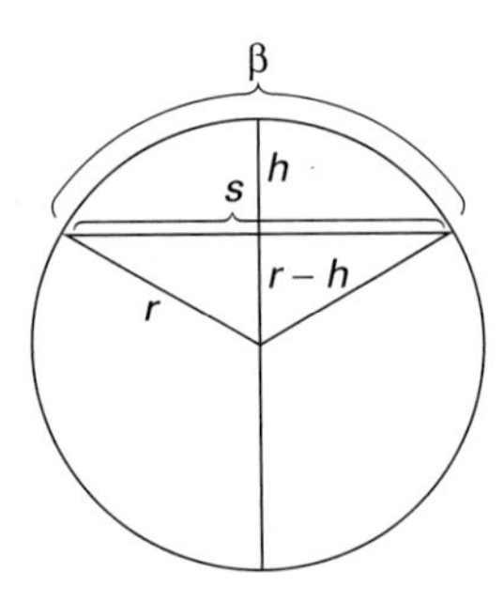

FIGURE 8.4 Length of arc $\beta = \frac{d}{28}$ arcchord$\left(\frac{28}{d}s\right)$, where $d = 2r = s^2/4h + h$

to each integral value of the chord from 1 to 28. So to determine the length of the arc of a segment of a circle, given the chord s and the distance h from the center of the chord to the circumference, Abraham first determined the diameter d of the circle by the formula $d = s^2/4h + h$ (Fig. 8.4). Then he multiplied the given chord by $28/d$ (to convert to a circle of diameter 28), consulted his table to determine the corresponding arc α, and multiplied α by $d/28$ to find the actual arc length.

8.1.2 Leonardo of Pisa's *Practica geometriae*

Abraham's Hebrew text was one of the earliest of many practical geometrical works to appear in medieval Europe. Probably the most important is the *Practica geometriae* by one of the first Italian mathematicians, Leonardo of Pisa (c. 1170–1240), usually known as Fibonacci since the nineteenth century. It is clear that Leonardo's work was heavily influenced by that of Abraham; in fact, some of the sections appear to be taken almost directly from Abraham's book. However, Leonardo's text is more extensive.

Leonardo, like Abraham, wrote a section on circles in which he quoted the standard $\frac{22}{7}$ for π. But Leonardo, in addition, showed how to calculate this value by the procedure of Archimedes. He found that the ratio of the perimeter of a 96-sided polygon circumscribed about a circle to the diameter of the circle is 1440 to $458\frac{1}{5}$, and the ratio of the perimeter of an inscribed 96-sided polygon to the diameter is 1440 to $458\frac{4}{9}$. Noting that $458\frac{1}{3}$ is approximately halfway between $458\frac{1}{5}$ and $458\frac{4}{9}$, he asserted that the ratio of circumference to diameter is close to $1440 : 458\frac{1}{3} = 864 : 275$. Because $864 : 274\frac{10}{11} = 3\frac{1}{7} : 1$, Leonardo had rederived the Archimedean value.

Leonardo also calculated areas of segments and sectors of circles. To do this, he too needed a table of arcs and chords. Strangely enough, although he defined the sine of an arc in the standard way, he did not give a table of sines, but one of chords, and in fact reproduced the Ptolemaic procedure for determining the chord of half an arc from that of the whole arc. His chord table, though, was not Ptolemaic. In fact, it may well be original to Leonardo, because it is based on a radius of 21. Like Abraham's value of 14, Leonardo's value of 21 was chosen so the semicircumference of the circle would be integral (66). But unlike Abraham's table, Leonardo's table is a direct chord table. For each integral arc from 1 to 66 rods (and also from 67 to 131), the table gives the corresponding chord, in the same measure, with fractions of the rods not in sixtieths, but in the Pisan measures of feet (6 to the rod), *unciae* (18 to the foot), and points (20 to the *uncia*). Leonardo then demonstrated how to use the chord table in circles of radius other than 21.

Like Abraham bar Ḥiyya, Leonardo used the table of chords only to calculate areas of circular sectors and segments. When, later in the same chapter of his book, he calculated the lengths of the sides and diagonals of a regular pentagon inscribed in a circle, he did not use what seems to us the obvious method of consulting his table of chords. He returned to Euclid and quoted appropriate theorems from Book XIII of the *Elements* relating the sides of a hexagon, pentagon, and decagon to enable him actually to perform the calculations. And toward the end of his book, when he wanted to calculate heights, again he did not use trigonometry. He used the old methods of similar triangles, starting with a pole of known height to help sight the top of the unknown object, then measuring the appropriate distances along the ground.

8.2 Combinatorics

We have already discussed the Indian and Islamic interest in combinatorics. In medieval Europe, there was also interest in such questions, primarily in the Jewish community. The earliest Jewish source on this topic seems to be the mystical work *Sefer yetsirah* (*Book of Creation*), written sometime before the eighth century and perhaps as early as the second century. In it, the unknown author calculated the various ways in which the 22 letters of the Hebrew alphabet can be arranged. He was interested in this calculation because the Jewish mystics believed that God had created the world and everything in it by naming these things (in Hebrew, of course): "God drew them, combined them, weighed them, interchanged them, and through them produced the whole creation and everything that is destined to be created. . . . Two stones [letters] build two houses [words], three build six houses, four build twenty-four houses, five build one hundred and twenty houses, six build seven hundred and twenty houses, seven build five thousand and forty houses." Evidently, the author understood that the number of possible arrangements of n letters was $n!$. An Italian rabbi, Shabbetai Donnolo (913–970), derived this factorial rule very explicitly in a commentary on the *Sefer yetsirah*:

> The first letter of a two-letter word can be interchanged twice, and for each initial letter of a three-letter word the other letters can be interchanged to form two two-letter words—for each of three times. And all the arrangements there are of three-letter words correspond to each one

of the four letters that can be placed first in a four-letter word: a three-letter word can be formed in six ways, and so for every initial letter of a four-letter word there are six ways—altogether making twenty-four words, and so on.

8.2.1 The Work of Abraham ibn Ezra

Although the author of the *Sefer yetsirah* briefly mentioned how to calculate the number of combinations of letters taken two at a time, a more detailed study of combinations was carried out by Rabbi Abraham ben Meir ibn Ezra (1090–1167), a Spanish-Jewish philosopher, astrologer, and Biblical commentator. It was in an astrological text that ibn Ezra discussed the number of possible conjunctions of the seven "planets" (including the sun and the moon). It was believed that these conjunctions would have a powerful influence on human life. Ibn Ezra thus calculated C_k^7 for each integer k from 2 to 7 and noted that the total was 120. He began with the simplest case, that the number of binary conjunctions was 21. This number was equal to the sum of the integers from one to six and could be calculated by a well-known formula: $C_2^n = n(n-1)/2$.

To calculate ternary combinations, ibn Ezra explained that there are five such combinations involving Jupiter and Saturn, four involving Jupiter and Mars, but not Saturn, and so on. Hence, there are $C_2^6 = 15$ ternary conjunctions involving Jupiter. Similarly, to find the ternary conjunctions involving Saturn but not Jupiter, ibn Ezra needed to calculate the number of choices of two planets from the remaining five: $C_2^5 = 10$. He then found the ternary conjunctions involving Mars, but neither Jupiter nor Saturn, and finally concluded with the result

$$C_3^7 = C_2^6 + C_2^5 + C_2^4 + C_2^3 + C_2^2 = 15 + 10 + 6 + 3 + 1 = 35.$$

Ibn Ezra next calculated the quaternary conjunctions by analogous methods. The conjunctions involving Jupiter require choosing three planets from the remaining six. Those involving Saturn but not Jupiter require choosing three from five. So $C_4^7 = C_3^6 + C_3^5 + C_3^4 + C_3^3 = 20 + 10 + 4 + 1 = 35$. Ibn Ezra then just stated the results for the conjunctions involving five, six, and seven planets. Essentially, he had given an argument for the case $n = 7$, easily generalizable to the general combinatorial rule:

$$C_k^n = \sum_{i=k-1}^{n-1} C_{k-1}^i.$$

Interestingly, in a somewhat later work, ibn Ezra essentially repeated the above calculations for C_2^7 and C_3^7 and then noted that by symmetry, $C_4^7 = C_3^7$ and $C_5^7 = C_2^7$. We note that ibn Ezra's calculations are similar to those of ibn Mun'im several decades later, but that neither ibn Mun'im nor ibn al-Bannā took explicit notice of the symmetry in their versions of the Pascal triangle.

8.2.2 Levi ben Gerson and Induction

Early in the fourteenth century, Levi ben Gerson (1288–1344), a French Jew from Orange, gave careful, rigorous proofs of various combinatorial formulas in a major work, the *Maasei ḥoshev* (*The Art of the Calculator*) (1321, with a second edition in 1322). Levi's text is divided into two parts, a first theoretical part, in which every theorem receives a detailed proof, and a second "applied" part, in which explicit instructions are given for performing various types of calculation. Levi's theoretical first section begins with a quite modern justification for considering theory at all:

> Because the true perfection of a practical occupation consists not only in knowing the actual performance of the occupation but also in its explanation, why the work is done in a particular way, and because the art of calculating is a practical occupation, it is clear that it is pertinent to concern oneself with its theory. There is also a second reason to inquire about the theory in this field. Namely, it is clear that this field contains many types of operations, and each type itself concerns so many different types of material that one could believe that they cannot all belong to the same subject. Therefore, it is only with the greatest difficulty that one can achieve understanding of the art of calculating, if one does not know the theory. With the knowledge of the theory, however, complete mastery is easy. One who knows it will understand how to apply it in the various cases which depend on the same foundation. If one is ignorant of the theory, one must learn each kind of calculation separately, even if two are really one and the same.

In his theoretical section, Levi assumed the knowledge of Books VII, VIII, and IX of Euclid's *Elements* and then gave careful, Euclidean-style proofs of all his results. The most important aspects of Levi's work are the combinatorial theorems. It is here that he used, somewhat more explicitly than his Islamic predecessors, the essentials of the method of mathematical induction, what he called the process of "rising step by step without end." In general, when Levi used such a proof, he first proved the **inductive step**, the step that allows one to move from k to $k+1$, next noted that the process begins at some small value of k, and then finally stated the complete result. Nowhere did he state the modern principle of induction, but it does appear that he knew how to use it.

For example, he used induction in his proof of the formula for the sum of the first n integral cubes. The basic inductive step is

PROPOSITION 41. *The square of the sum of the natural numbers from* 1 *up to a given number is equal to the cube of the given number added to the square of the sum of the natural numbers from* 1 *up to one less than the given number. [In modern notation, the theorem says that* $(1+2+\cdots+n)^2 = n^3 + (1+2+\cdots+(n-1))^2$.*]*

Here we present Levi's proof in modern notation. First, $n^3 = n \cdot n^2$. Also, $n^2 = (1+2+\cdots+n) + (1+2+\cdots+(n-1))$. (This result is Levi's proposition 30.) Then $n^3 = n[(1+2+\cdots+n) + (1+2+\cdots+(n-1))] = n^2 + n[2(1+2+\cdots+(n-1))]$. But $(1+2+\cdots+n)^2 = n^2 + 2n(1+2+\cdots+(n-1)) + (1+2+\cdots(n-1))^2$. It follows that $n^3 + (1+2+\cdots+(n-1))^2 = (1+2+\cdots+n)^2$.

Levi next noted that although 1 has no number preceding it, "its third power is the square of the sum of the natural numbers up to it." In other words, he gave the first step of a proof by induction for the result stated as

PROPOSITION 42. *The square of the sum of the natural numbers from* 1 *up to a given number is equal to the sum of the cubes of the numbers from* 1 *up to the given number.*

Levi's proof is not quite what we would expect of a proof by induction. Instead of arguing from n to $n+1$, he argued from n to $n-1$. He noted first, that $(1+2+\cdots+n)^2 = n^3 + (1+2+\cdots+(n-1))^2$. The final summand is, also by the previous proposition, equal to $(n-1)^3 + (1+2+\cdots+(n-2))^2$. Continuing in this way, Levi eventually reached $1^2 = 1^3$, and the result is proved. Note further that, although the proposition is stated in terms of an arbitrary natural number, in his proof, Levi wrote only a sum of five numbers in his first step rather than the n used in this adaptation. The five are represented by the five initial letters of the Hebrew alphabet. Like many of his predecessors, Levi had no way of writing the sum of arbitrarily many integers and so used the method of generalizable example. Nevertheless, the idea of a proof by induction is evident in Levi's demonstration.

As another example, we consider Levi's demonstration that the number of permutations of a given number n of elements is what we call $n!$:

PROPOSITION 63. *If the number of permutations of a given number of different elements is equal to a given number, then the number of permutations of a set of different elements containing one more number equals the product of the former number of permutations and the given next number.*

Symbolically, the proposition states that $P_{n+1} = (n+1)P_n$ (where P_k stands for the number of permutations of a set of k elements). This result provides the inductive step in the proof of the proposition $P_n = n!$. Levi's proof of proposition 63 is very detailed. Given a permutation, say $abcde$, of the original n elements and a new element f, he noted that $fabcde$ is a permutation of the new set. Because there are P_n such permutations of the original set, there are also P_n permutations of the new set beginning with f. Also, if one of the original elements, for example, e, is replaced by the new element f, there are P_n permutations of the set a, b, c, d, f and therefore also P_n permutations of the new set with e in the first place. Because any of the n elements of the original set, as well as the new element, can be put in the first place, it follows that the number of permutations of the new set is $(n+1)P_n$. Levi finished the proof of proposition 63 by showing that all of these $(n+1)P_n$ permutations are different. He then concluded, "Thus it is proved that the number of permutations of a given set of elements is equal to that number formed by multiplying together the natural numbers from 1 up to the number of given elements. For the number of permutations of 2 elements is 2, and that is equal to $1 \cdot 2$, the number of permutations of 3 elements is equal to the product $3 \cdot 2$, which is equal to $1 \cdot 2 \cdot 3$, and so one shows this result further without end." In other words, Levi mentioned the beginning step and then noted that with the inductive step already proved, the complete result is also proved.

After proving, using a counting argument, that $P_2^n = n(n-1)$ (where P_k^n is the number of permutations of k elements in a set of n), Levi proved that $P_k^n = n(n-1)(n-2) \cdots (n-k+1)$ by induction on k. As before, he stated the inductive step as a theorem:

PROPOSITION 65. *If a certain number of elements is given and the number of permutations of order a number different from and less than the given number of elements is a third number, then the number of permutations of order one more in this given set of elements is equal to the number which is the product of the third number and the difference between the first and the second numbers.*

Modern symbolism replaces Levi's convoluted wording with a brief equation: $P_{j+1}^n = (n-j)P_j^n$. Levi's proof is quite similar to that of proposition 63. At the end, he stated the complete result: "It has thus been proved that the permutations of a given order in a given number of elements are equal to that number formed by multiplying together the number of integers in their natural sequence equal to the given order and ending with the number of elements in the set." To clarify this statement, Levi first gave the initial step of the induction by quoting his previous result in the case $n = 7$, that is, the number of permutations of order 2 in a set of 7 is equal to $6 \cdot 7$. Then, the number of permutations of order 3 is equal to $5 \cdot 6 \cdot 7$ (since $5 = 7 - 2$). Similarly, the number of permutations of order 4 is equal to $4 \cdot 5 \cdot 6 \cdot 7$, "and so one proves this for any number."

In the final three propositions of the theoretical part of the *Maasei ḥoshev*, Levi completed his development of formulas for permutations and combinations. Proposition 66 shows that $P_k^n = C_k^n P_k^k$, while proposition 67 simply rewrites this as $C_k^n = P_k^n / P_k^k$. Since he has already given formulas for both the numerator and denominator of this quotient, Levi thus has demonstrated the standard formula for C_k^n:

$$C_k^n = \frac{n(n-1)\cdots(n-k+1)}{1 \cdot 2 \cdots k}.$$

And finally, proposition 68 demonstrates the result noted by ibn Ezra: $C_k^n = C_{n-k}^n$.

Levi gave examples of many of these results in the second section of his book. For example, he noted that to determine the sum of the cubes of the numbers from 1 to 6, one first calculates that the sum of the numbers themselves is 21 and therefore the sum of the cubes is the square of 21, namely, 441. Or, to find the number of permutations of five elements out of a set of eight, P_5^8, one multiplies $4 \cdot 5 \cdot 6 \cdot 7 \cdot 8$ to get 6720. Then the number of combinations of five elements out of eight, C_5^8, is that number divided by $1 \cdot 2 \cdot 3 \cdot 4 \cdot 5$, or 120. The result is 56.

Finally, at the end of the second section, Levi presented a number of interesting problems, most of which were seemingly practical and could be solved through a knowledge of ratio and proportion. These problems, including some familiar ones, become progressively harder, but Levi gave a detailed explanation of the solution to each. Some of these problems are included in the exercises to this chapter.

8.3 Medieval Algebra

Although the theory of combinatorics appears to have developed in Europe through the Jewish tradition, the writers on algebra in medieval Europe were direct heirs to Islamic work. Probably the earliest European work that included Islamic algebra was Abraham bar Ḥiyya's *Treatise on Mensuration*. In particular, Abraham took over the Islamic tradition of proof for algebraic procedures and therefore gave geometric justifications of methods for solving the algebraic problems he included as part of his geometric discussions.

For example, Abraham posed the question, "If from the area of a square one subtracts the sum of the (four) sides and there remains 21, what is the area of the square and what is the length of each of the equal sides?" We can translate Abraham's question into the quadratic equation $x^2 - 4x = 21$, an equation that he solves in the familiar way by halving 4 to get 2, squaring this result to get 4, adding this square to 21 to get 25, taking the square root to get 5, and then adding that to the half of 4 to get the answer 7 for the side and the answer 49 for the area. Abraham's statement of the problem was not geometric, but in his geometric justification, he restated the problem to mean the cutting off of a rectangle of sides 4 and x from the original square of unknown side x to leave a rectangle of area 21. He then bisected the side of length 4 and applied *Elements* II-6 to justify the algebraic procedure. Thus, Abraham evidently had learned his algebra not from al-Khwārizmī, but from an author such as Abū Kāmil, who used Euclidean justifications. Abraham similarly presented the method and a Euclidean proof for examples of the two other Islamic classes of mixed quadratic equations: $x^2 + 4x = 77$, and $4x - x^2 = 3$. In the latter case, he gave both positive solutions.

8.3.1 Leonardo of Pisa's *Liber abbaci*

Leonardo of Pisa also briefly treated quadratic equations in his *Practica geometriae*, with geometric justifications, but treated these more extensively in his magnum opus, the *Liber abbaci*, or *Book of Calculation*. [The word *abbaci* (from *abacus*) does not refer to a computing device, but simply to calculation in general.] The first edition of this work appeared in 1202, while a slightly revised one was published in 1228. The many surviving manuscripts testify to the wide readership the book enjoyed. The sources for the *Liber abbaci* were largely from the Islamic world, which Leonardo visited during many journeys, but he enlarged and arranged the material he collected through his own genius. The book contained not only the rules for computing with the new Hindu-Arabic numerals, but also numerous problems of various sorts in such practical topics as calculation of profits, currency conversions, and measurement, supplemented by topics that are now standard in algebra texts, such as mixture problems, motion problems, container problems, the Chinese remainder problem, and, as mentioned, various forms of problems solvable by use of quadratic equations. Interspersed among the problems is a limited amount of theory.

Leonardo used a great variety of methods in his solution of problems, including the old Egyptian method of false position in which a convenient, but wrong, answer is given first and then adjusted appropriately to get the correct result. For many of the problems, it is possible to cite Leonardo's sources. For example, he often took problems verbatim from al-Khwārizmī and Abū Kāmil; many of these he found in Arabic manuscripts discovered in his travels. Some of the problems seem ultimately to have come from China or India, but Leonardo probably learned these in Arabic translations. The majority of the problems, however, are of his own devising and show his creative abilities. A few of Leonardo's problems and solutions should give the flavor of this most influential mathematical work.

One classic problem is the following: Suppose a lion can eat a sheep in 4 hours, a leopard can eat a sheep in 5 hours, and a bear can eat a sheep in 6 hours. How long will it take the three animals to eat one sheep together? To solve this, Leonardo noted that in 60 hours, the lion can eat 15 sheep, the leopard 12 sheep, and the bear 10 sheep, for a total of 37 sheep. To find how long it takes to eat one sheep, therefore, he simply had to divide 60 by 37, giving $1\frac{23}{37}$ hours.

Leonardo's example of the Chinese remainder problem asked for a number that when divided by 2, had remainder 1; by 3, had remainder 2; by 4, had remainder 3; by 5, had remainder 4; by 6, had remainder 5; and by 7, had remainder 0. To solve this, he noted that 60 was evenly divisible by 2, 3, 4, 5, and 6. Therefore, $60 - 1 = 59$ satisfied the first five conditions, as did any multiple of 60, less 1. Thus, he had to find a multiple of 60 that had remainder 1 on division by 7. The smallest such number is 120, and therefore, 119 is the number sought.

Many of the later problems in the text are more explicitly algebraic. For example, consider the problem where two men each have some money. The first says to the second, "If you give me one *denarius*, we will each have the same amount." The second says to the first, "If you give me one *denarius*, I will have ten times as much as you." How much does each have? In modern notation, if x and y represent the amounts held by the first and second man, respectively, this problem can be translated into the system of equations $x + 1 = y - 1$, $y + 1 = 10(x - 1)$. Leonardo, however, looked at the problem somewhat differently by introducing the new unknown $z = x + y$ (the total sum of money). Then $x + 1 = \frac{1}{2}z$ and $y + 1 = \frac{10}{11}z$. Adding these two equations together gives $z + 2 = \frac{31}{22}z$, from which $z = \frac{44}{9}$, $x = 1\frac{4}{9}$, and $y = 3\frac{4}{9}$.

Leonardo also dealt comfortably with problems in more than two unknowns. For example, suppose there are four men such that the first, second, and third together have 27 *denarii*; the second, third, and fourth together have 31; the third, fourth, and first have 34; and the fourth, first, and second have 37. To determine how much each man has requires solving a system of four equations in four unknowns. Leonardo accomplished this expeditiously by adding the four equations together to determine that three times the total sum of money equals 129 *denarii*. The individual amounts are then easily calculated. On the other hand, in a similar question reducible to the four equations $x + y = 27$, $y + z = 31$, $z + w = 34$, $x + w = 37$, Leonardo first noted that this system is impossible since the two different ways of calculating the total sum of money give two different answers, 61 and 68. However, if one changes the fourth equation to $x + w = 30$, one can simply choose x arbitrarily ($x \leq 27$) and calculate y, z, and w by using the first, second, and third equations, respectively.

The most famous problem of the *Liber abbaci*, the rabbit problem, is tucked inconspicuously between a problem on perfect numbers and the problem with four unknowns just discussed: "How many pairs of rabbits can be bred in one year from one pair? A certain person places one pair of rabbits in a certain place surrounded on all sides by a wall. We want to know how many pairs can be bred from that pair in one year, assuming it is their nature that each month they give birth to another pair, and in the second month after birth, each new pair can also breed." Leonardo proceeded to calculate: After the first month there will be two pairs, after the second, three. In the third month, two pairs will produce, so at the end of that month there will be five pairs. In the fourth month, three pairs will produce, so there will be eight. Continuing in this fashion, he showed that there will be 377 pairs by the end of the twelfth month. Listing the sequence 1, 2, 3, 5, 8, 13, 21, 34, 55, 89, 144, 233, 377 in the margin, he noted that each number is found by adding the two previous numbers, and "thus you can do it in order for an infinite number of months." This sequence, calculated recursively, is known today as a **Fibonacci sequence**. It turns out that it has many interesting properties unsuspected by Leonardo, not the least of which is its connection with the Greek problem of dividing a line in extreme and mean ratio.

In his final chapter, Leonardo demonstrated his complete command of the algebra of his Islamic predecessors by showing how to solve equations that reduce ultimately to quadratic equations. He discussed in turn each of the six basic types of quadratic equation given by al-Khwārizmī, and then gave geometric proofs of the solution procedures for each of the three mixed cases. He followed the proofs with some fifty pages of examples, most taken from the works of al-Khwārizmī and Abū Kāmil, including the familiar ones beginning with "divide 10 into two parts."

The content of the *Liber abbaci* contained no particular advance over mathematical works then current in the Islamic world. In fact, as far as the algebra was concerned, Leonardo was presenting tenth-century Islamic mathematics and ignoring the advances of the eleventh and twelfth centuries. The chief value of the work, nevertheless, was that it did provide Europe's first comprehensive introduction to Islamic mathematics. Those reading it were afforded a wide variety of methods to solve mathematical problems, methods that provided the starting point from which further progress could ultimately be made.

8.3.2 The Work of Jordanus de Nemore

One of the first mathematicians who made some advance over the work of Leonardo was a contemporary, Jordanus de Nemore. About the author virtually nothing is known, although it is suspected that he taught in Paris around 1220. His writings include several works on arithmetic, geometry, astronomy, mechanics, and algebra, and it appears that he took upon himself the task of creating a Latin version of the quadrivium, but a version based solidly on theory. For example, his work on arithmetic, the *Arithmetica*, is far different from the demonstrationless arithmetic of Boethius then circulating widely in Europe. It is firmly based on a Euclidean model, with definitions, axioms, postulates, propositions, and careful proofs. And like Euclid, Jordanus did not give any numerical examples.

Among the interesting features of Jordanus's work is the first appearance in Europe of the Pascal triangle. The construction of the triangle is the standard one:

> Put 1 at the top and below two 1's. Then the row of two 1's is doubled so that the first 1 is in the first place and another 1 in the last place as in the second row; and 1, 2, 1 will be in the third row. The numbers are added two at a time, the first 1 to the 2 in the second place, and so on through the row until a final 1 is put at the end. Thus the fourth row has 1, 3, 3, 1. In this way subsequent numbers are made from pairs of preceding numbers.

Most of the medieval manuscripts of the *Arithmetica* display a version of the triangle at this point, some even up to the tenth line. Jordanus then used the triangle explicitly to construct series of terms in given ratios. For example, if $a = b = c = d = 1$, then the numbers $e = 1a = 1$, $f = 1a + 1b = 2$, $g = 1a + 2b + 1c = 4$, and $h = 1a + 3b + 3c + 1d = 8$ form a continued proportion with constant ratio 2. Similarly, $k = 1e = 1$, $\ell = 1e + 1f = 3$, $m = 1e + 2f + 1g = 9$, and $n = 1e + 3f + 3g + 1h = 27$ form a continued proportion with ratio 3.

Jordanus's *Arithmetica* was widely read, at least judging from the number of extant manuscripts. Similarly, his major work on algebra, *De numeris datis* (*On Given Numbers*), also had a large circulation in medieval Europe. *De numeris datis* is an analytic work on algebra, based on but differing in spirit from the Islamic algebras. It presents problems in which certain quantities are given and then shows that other quantities are therefore also determined. In fact, one of Jordanus's aims is apparently to base the new algebra on arithmetic, the most fundamental of the subjects of the quadrivium, rather than on geometry, and especially on his own work on the subject. He organized his book in a logical fashion, but, in a major departure from the Euclidean model and even from his own *Arithmetica*, provided numerical examples for most of his theoretical results.

Although many of the actual problems and the numerical examples were available in the Islamic algebras, Jordanus adapted them to his own purposes. In particular, he made the major change of using letters to stand for arbitrary numbers. Thus, Jordanus's algebra is not entirely rhetorical. That is not to say that his symbolism is modern-looking. He picked his letters in alphabetical order with no distinction between letters representing known quantities and those representing unknowns and used no symbols for operations. Sometimes a single number is represented by two letters. At other times the pair of letters *ab* represents the sum of the two numbers *a* and *b*. The basic arithmetic operations are always written in words. And Jordanus did not use the new Hindu-Arabic numerals. All of his numbers are written as Roman numerals. Nevertheless, the idea of symbolism, so crucial to any major advance in algebraic technique, is found, at least in embryonic form, in Jordanus's work.

To understand Jordanus's contribution, we will consider a few of the text's more than 100 propositions, which are organized into four books. Jordanus wrote each proposition in a standard form. The general enunciation is followed by a restatement in terms of letters. By use of general rules, the letters representing numbers are manipulated into a canonical form from which the general solution can easily be found. Finally, a numerical example is calculated, following the general outlines of the abstract solution. The canonical forms themselves are among the earliest of the propositions.

PROPOSITION I-1. *If a given number is divided into two parts whose difference is given, then each of the parts is determined.*

Jordanus's proof is straightforward: "Namely, the lesser part and the difference make the greater. Thus, the lesser part with itself and the difference make the whole. Subtract therefore the difference from the whole and there will remain double the lesser given number. When divided [by two], the lesser part will be determined; and therefore also the greater part. For example, let 10 be divided in two parts of which the difference is 2. When this is subtracted from 10 there remains 8, whose half is 4, which is thus the lesser part. The other is 6."

In modern symbolism, Jordanus's problem amounts to the solution of the two equations $x + y = a$, $x - y = b$. Jordanus noted first that $y + b = x$, so that $2y + b = a$ and therefore $2y = a - b$. Thus, $y = \frac{1}{2}(a - b)$ and $x = a - y$.

Jordanus used this initial proposition in many of the remaining problems of Book I. For example, consider

PROPOSITION I-3. *If a given number is divided into two parts, and the product of one by the other is given, then of necessity each of the two parts is determined.*

This proposition presents one of the standard Babylonian problems: $x + y = m$, $xy = n$. Jordanus's method of solution, however, is different from the classic Babylonian solution, or indeed the Islamic one. In addition, he used symbolism as indicated: Suppose the given number abc is divided into the parts ab and c. Suppose ab multiplied by c is d and abc multiplied by itself is e. Let f be the quadruple of d, and g be the difference of e and f. Then g is the square of the difference between ab and c. Its square root is then the difference between ab and c. Since now both the sum and difference of ab and c are given, both ab and c are determined according to the first proposition. Jordanus's numerical example has 10 as the sum of the two parts and 21 as the product. He noted that 84 is quadruple 21, that 100 is 10 squared, and that 16 is their difference. Then the square root of 16, namely, 4, is the difference of the two parts of 10. By the proof of the first proposition, 4 is subtracted from 10 to get 6. Then 3 is the desired smaller part and 7 is the larger.

Jordanus's solution, translated into modern symbolism, amounts to using the identity $(x - y)^2 = (x + y)^2 - 4xy = m^2 - 4n$ to determine $x - y$ and reduce the problem to proposition I-1. The solution is then $y = \frac{1}{2}(m - \sqrt{m^2 - 4n})$, $x = m - y$. Jordanus's method appears to be one that he himself devised, as are his solution methods to other quadratic problems in Book I. Nevertheless, the numerical examples themselves have a familiar look. In fact, every proposition in Book I deals with a number divided into two parts, and in every example but one the number to be divided is 10. The solution methods may differ somewhat from those in the Islamic texts, but it is clear that al-Khwārizmī's problems live on!

In Book IV, however, there are three consecutive problems representing al-Khwārizmī's three forms of mixed quadratics. Here Jordanus does use the Islamic algorithms, but he justifies these algebraically rather than geometrically. Consider

PROPOSITION IV-9. *If the square of a number added to a given number is equal to the number produced by multiplying the root and another given number, then two values are possible.*

Thus, Jordanus asserted that there are two solutions to the equation $x^2 + c = bx$. He then gave the procedure for solving the equation: Let d be half of b; square it to get f, and let g be the difference of x and d, that is, $g = \pm(x - \frac{1}{2}b)$. Then, since $bx = x^2 + c$, and $g^2 = x^2 - bx + d^2 = x^2 - bx + f$, we have $x^2 + f = x^2 + c + g^2$ and $f = c + g^2$. Then $g = \sqrt{f - c}$. Jordanus concluded by noting that x may be obtained either by subtracting g from $b/2$ or by adding g to $b/2$, that is,

$$x = \frac{b}{2} \pm \sqrt{\left(\frac{b}{2}\right)^2 - c}.$$

His example made his symbolic procedure clearer. To solve $x^2 + 8 = 6x$, he squared half of 6, giving 9, and then subtracted 8 from it, leaving 1. The square root of 1 is 1, and this is the difference between x and 3. Hence, x can be either 2 or 4.

Interestingly, the solution of every quadratic equation Jordanus solved is a positive integer. While he often used fractions as part of his solution, he carefully arranged matters so that final answers are always whole numbers. If, in fact, he had read Abū Kāmil's *Algebra*, which was available in Latin, Jordanus would have seen nonintegral, and even nonrational, solutions to this type of problem. He nevertheless rejected such solutions when he made up his examples. Given his very formal style, however, Jordanus may still have been under the influence of Euclid and have felt that irrational numbers simply did not belong in a work based on arithmetic. Hence, although *De numeris datis* represents an advance from the Islamic works in the constant striving toward generality and in the use of some symbolization, it returns to the strict Greek separation of number from magnitude, an idea from which Jordanus's Islamic predecessors had departed. Thus, it appears that although Jordanus certainly made use of the new Islamic material available in Europe, his goal was to provide his readers with a mathematics based as much as possible on Greek principles.

8.4 The Mathematics of Kinematics

The algebraic work of Jordanus de Nemore was not developed further in the thirteenth century, even though a group of followers had appeared in Paris by the middle of that century. Perhaps Europe was not then ready to resume the study of pure mathematics. By early in the fourteenth century, however, other aspects of mathematics began to develop in the universities of Oxford and Paris out of attempts to clarify certain remarks in Aristotle's physical treatises. In particular, ideas coming from Aristotle's ideas on motion were developed by Thomas Bradwardine (1295–1349) and William Heytesbury (early fourteenth century) at Merton College, Oxford.

Some Greek mathematicians had considered to some extent the notions of uniform velocity and accelerated motion, but they never considered velocity or acceleration as independent quantities that could be measured. Velocities were only dealt with by comparing distances and times, and therefore, in essence, only average velocities (over certain time periods) could be compared. The fourteenth century, however, saw the beginning of the notion that velocity, in particular instantaneous velocity, is a measurable entity. Thus, Bradwardine in his *Treatise on the Continuum* (c. 1330) saw how to compare velocities by the use of proportion. If two objects travel at (uniform) velocities v_1, v_2 in times t_1, t_2 and cover distances s_1, s_2, respectively, then (1) if $t_1 = t_2$, then $v_1 : v_2 = s_1 : s_2$, and (2) if $s_1 = s_2$, then $v_1 : v_2 = t_2 : t_1$. Bradwardine thus considered uniform velocity itself as a type of magnitude, capable of being compared with other velocities.

Heytesbury, only a few years later in his *Rules for Solving Sophisms* (1335), gave a careful definition of instantaneous velocity for a body whose motion is not uniform: "In nonuniform motion . . . the velocity at any given instant will be measured by the path which would be described by the . . . point if, in a period of time, it were moved uniformly at the same degree of velocity with which it is moved in that given instant, whatever [instant] be assigned." Having given this explicit definition, Heytesbury noted that if two bodies have the same instantaneous velocity at a particular instant, they do not necessarily travel equal distances in equal times, because their velocities may well differ at other instants.

Heytesbury also dealt with acceleration: "Any motion whatever is uniformly accelerated if, in each of any equal parts of the time whatsoever, it acquires an equal increment of velocity. . . . But a motion is nonuniformly accelerated . . . when it acquires . . . a greater increment of velocity in one part of the time than in another equal part. . . . And since any degree of velocity whatsoever differs by a finite amount from zero velocity . . . , therefore any mobile body may be uniformly accelerated from rest to any assigned degree of velocity." This statement provides not only a very clear definition of uniform acceleration, but also, in nascent form at least, the notion of velocity changing with time. In other words, velocity is being described by Heytesbury as a function of time.

How does one determine the distance traveled by a body being uniformly accelerated? The answer, generally known today as the **mean speed rule**, was first stated by Heytesbury: "When any mobile body is uniformly accelerated from rest to some given degree [of velocity], it will in that time traverse one-half the distance that it would traverse if, in that same time it were moved uniformly at the [final] degree [of velocity] . . . . For that motion, as a whole, will correspond to . . . precisely one-half that degree which is its terminal velocity." In modern notation, if a body is accelerated from rest in a time t with a uniform acceleration a, then its final velocity is $v_f = at$. What Heytesbury is saying is that the distance traveled by this body is $s = \frac{1}{2}v_f t$. Substituting the first formula in the second gives the standard modern formulation $s = \frac{1}{2}at^2$.

Heytesbury gave a proof of the mean speed theorem by an argument from symmetry, taking as his model a body d accelerating uniformly from rest to a velocity of 8 in one hour. (The number 8 does not represent any particular speed, but is just used as the basis for his example.) He then considered three other bodies, a moving uniformly at a speed of 4 throughout the hour, b accelerating uniformly from 4 to 8 in the first half hour, and c decelerating uniformly from 4 to 0 in that same half hour. First, he noted that body d goes as far in the first half hour as does c and as far in the second half hour as does b. Therefore, d travels as far in the whole hour as the total of b and c in the half hour. Second, he argued that since b increases precisely as much as c decreases, together they will traverse as much distance in the half hour as if they were both held at the speed of 4. This latter distance is the same that a travels in the whole hour. It follows that d goes exactly as far as does a in the hour, and the mean speed theorem is demonstrated. Heytesbury then proved the easy corollary, that the body d traverses in the second half hour exactly three times the distance it covered in the first half hour.

Other scholars at Merton College in the same time period began to explore the idea of representing velocity, as well as other varying quantities, by line segments. The basic idea seems to come, in effect, from Aristotle, because such notions as time, distance, and length (of line segments) were conceived of as magnitudes in the Greek philosopher's distinction between the two types of quantities. All were infinitely divisible. Hence, it was not unreasonable to attempt to represent the somewhat abstract idea of velocity, now itself being quantified, by the concrete geometric idea of a line segment. Velocities of different "degrees" would thus be represented by line segments of different lengths.

This idea was carried to its logical conclusion by Nicole Oresme (1320–1382), a French cleric and mathematician associated with the University of Paris. Oresme introduced a two-dimensional representation of velocity changing with respect to time. In fact, in his *Treatise on the Configuration of Qualities and Motions*, published about 1350, Oresme even generalized this idea to other cases where a given quantity varied in intensity over either distance or time.

Oresme explained that, since "every measurable thing except numbers is imagined in the manner of continuous quantity," therefore, to measure such a thing, "it is necessary that points, lines, and surfaces, or their properties, be imagined." Thus, he concluded, "every intensity which can be acquired successively ought to be imagined by a straight line perpendicularly erected on some point." From these straight lines, Oresme constructed what he called a **configuration**, a geometric figure consisting of all the perpendicular lines drawn over the base line. In the case of velocities, the base line represented time, while the perpendiculars represented the velocities at each instant. The entire figure represented the whole distribution of velocities, which Oresme interpreted as representing the total distance traveled by the moving object. Oresme did not use what we call coordinates. There was no particular fixed length by which a given degree of velocity was represented. The important idea was only that "equal intensities are designated by equal lines, a double intensity by a double line, and always in the same way if one proceeds proportionally."

For Oresme, then, a uniform quality—for example, a body moving with uniform velocity—is represented by a rectangle, for at each point the velocity is the same (Fig. 8.5). The area of the rectangle represents the total distance traveled. The distance traveled by a body beginning at rest and then moving with constant acceleration, representing what Oresme calls a "uniformly difform" quality, one whose intensity changes uniformly, is the

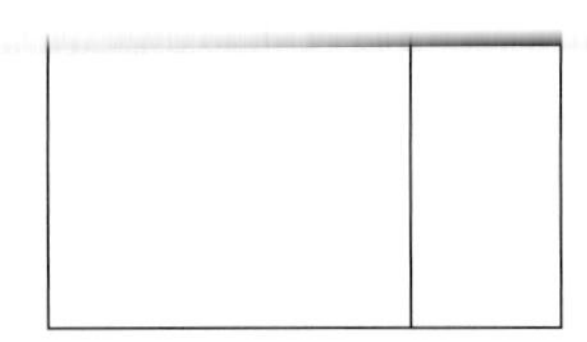

FIGURE 8.5 Uniform velocity

area of a right triangle (Fig. 8.6). As Oresme notes, "a quality uniformly difform is one in which if any three points [of the subject line] are taken, the ratio of the distance between the first and the second to the distance between the second and the third is as the ratio of the excess in intensity of the first point over that of the second point to the excess of that of the second point over that of the third point." This equality of ratios defines a straight line, the hypotenuse of the right triangle. Finally, a "difformly difform" quality, such as nonuniform acceleration, is represented by a figure whose "line of summit" is a curve that is not a straight line (Fig. 8.7). In other words, Oresme in essence developed the idea of using a curve to represent the functional relationship between velocity and time.

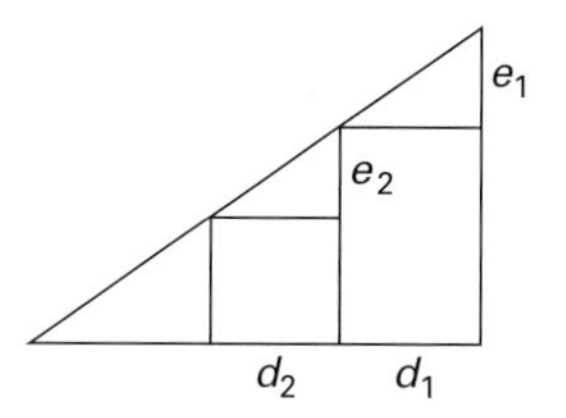

FIGURE 8.6 Uniformly difform velocity, where $d_1 : d_2 = e_1 : e_2$

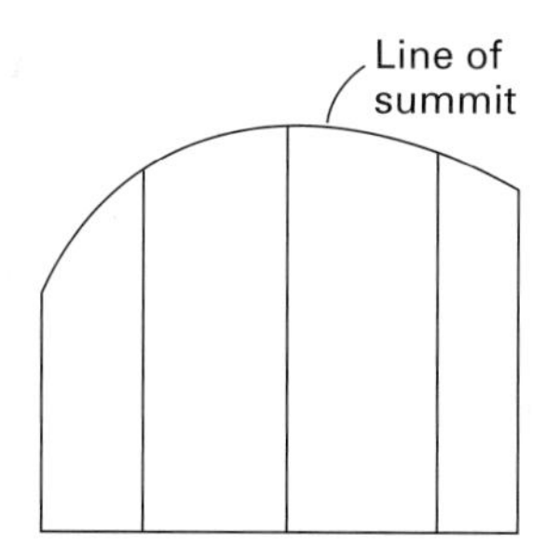

FIGURE 8.7 Difformly difform velocity, or nonuniform acceleration

Given this representation of the motion of bodies, it was easy for Oresme to give a geometric proof of the mean speed theorem. For if triangle ABC represents the configuration of a body moving with a uniformly accelerated motion from rest, and if D is the midpoint of the base AB, then the perpendicular DE represents the velocity at the midpoint of the journey and is half the final velocity (Fig. 8.8). The total distance traveled, represented by

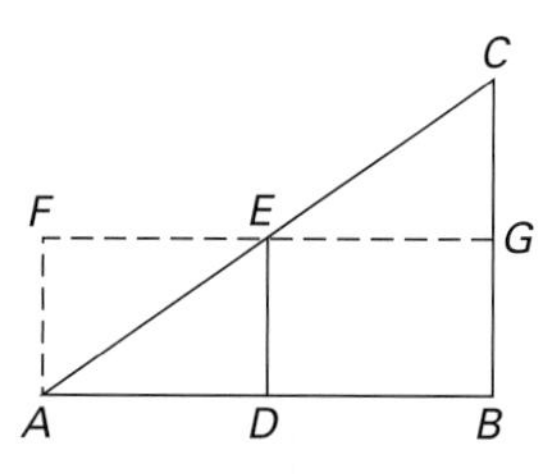

FIGURE 8.8 Oresme's proof of mean speed theorem

triangle ABC, is then equal to the area of the rectangle $ABGF$, precisely as stated by the Mertonians.

Oresme's idea of representing velocities, as well as other qualities, geometrically, was continued in various works by others over the next century. However, no one was able to extend the representation of distances to situations more complex than Oresme's uniformly difform qualities. Eventually, even this idea was lost. Much the same fate befell the ideas of the other major European mathematicians of the medieval period. Their works were not studied and their new ideas had to be rediscovered centuries later. This lack of progress is evident in the stagnant mathematical curricula at the first universities as well as at the many new ones founded in succeeding centuries. With the works of Aristotle as the basis of the curriculum, the only mathematics studied was whatever was useful for helping the student understand the works of the great philosopher. Although an Oresme might carry these ideas further, such men were rare. In addition, the ravages of the Black Death and the Hundred Years War caused a marked decline in learning in France and England. It was therefore mathematicians in Italy and Germany who studied a few of the ideas of the medieval French and English mathematicians, and it was in those countries that new ideas were generated in the Renaissance.

Exercises

1. This problem and the next are from Alcuin's *Propositions for Sharpening Youths*. A cask is filled to 100 *metreta* capacity through three pipes. One-third of its capacity plus 6 *modii* flows in through one pipe; one-third of its capacity flows in through another pipe; but only one-sixth of its capacity flows in through the third pipe. How many *sextarii* flow in through each pipe? (Here a *metreta* is 72 *sextarii*, and a *modius* is 200 *sextarii*.)
2. A man must ferry a wolf, a goat, and a head of cabbage across a river. The available boat, however, can only carry the man and one other thing. The goat cannot be left alone with the cabbage, nor the wolf with the goat. How should the man ferry his three items across the river?
3. Use Abraham bar Ḥiyya's table (Fig. 8.3) to find the length of the arc cut off by a chord of length 6 in a circle of diameter $10\frac{1}{2}$.
4. Find the area of the circle segment determined by the chord in exercise 3.
5. Find the length of the chord that cuts off an arc of length $5\frac{1}{2}$ in a circle of diameter 33.
6. If a chord of length 8 has distance 2 from the circumference, find the diameter of the circle.
7. Prove proposition 32 of the *Maasei ḥoshev*:
$$1 + (1 + 2) + (1 + 2 + 3) + \cdots + (1 + 2 + \cdots + n) = \begin{cases} 1^2 + 3^2 + \cdots + n^2 & n \text{ odd;} \\ 2^2 + 4^2 + \cdots + n^2 & n \text{ even.} \end{cases}$$
8. Prove proposition 33 of the *Maasei ḥoshev*:
$$(1 + 2 + 3 + \cdots + n) + (2 + 3 + \cdots + n) + (3 + \cdots + n) + \cdots + n = 1^2 + 2^2 + \cdots + n^2.$$
9. Prove proposition 34 of the *Maasei ḥoshev*:
$$[(1 + 2 + \cdots + n) + (2 + 3 + \cdots + n) + \cdots + n] + [1 + (1 + 2) + \cdots + (1 + 2 + \cdots + (n - 1))] = n(1 + 2 + \cdots + n).$$
10. Use the results of exercises 7, 8, and 9 to prove
$$1^2 + 2^2 + \cdots + n^2 = [n - \tfrac{1}{3}(n - 1)](1 + 2 + \cdots + n).$$
11. Solve one of the problems from the *Maasei ḥoshev*: A barrel has various holes. The first hole empties the full barrel in 3 days; the second hole empties the full barrel in 5 days; another hole empties the full barrel in 20 hours; and another hole empties the full barrel

in 12 hours. All the holes are opened together. How much time will it take to empty the barrel?

12. Solve another problem from the *Maasei ḥoshev*: A merchant sells four drugs. The cost of the first drug is 2 *dinars* per *litra*; the cost of the second is 3 *dinars* per *litra*; the cost of the third is 12 *dinars* per *litra*; the cost of the fourth is 20 *dinars* per *litra*. How many *litras* should one buy of each of the drugs so that the cost for each is the same?

13. Recall that Jordanus used the Pascal triangle in the *Arithmetica* to determine series of numbers in continued proportion. For example, beginning with the series $1, 1, 1, 1, \ldots$, he derived first the series $1, 2, 4, 8, \ldots$, and by using those terms, he derived the series $1, 3, 9, 27, \ldots$. Use this latter series in the same way to derive the series $1, 4, 16, 64, \ldots$. Formulate a generalization of this result.

14. If one uses the numbers 8, 4, 2, 1 in the Pascal triangle as Jordanus did (but in the reverse order of the way he used them originally), one gets the series 8, 12, 18, 27—a geometric series whose common ratio is $\frac{3}{2}$. Explore a generalization of this result.

15. Leonardo has the following problem in the *Liber abbaci*: There is a lion at the bottom of a pit 50 feet deep. The lion climbs up $\frac{1}{7}$ of a foot each day and then falls back $\frac{1}{9}$ of a foot each night. How long will it take him to climb out of the pit? (Leonardo claims that the answer is 1575 days, but this answer is incorrect.)

16. The Fibonacci sequence (the sequence of rabbit pairs) is determined by the recursive rule $F_0 = F_1 = 1$ and $F_n = F_{n-1} + F_{n-2}$. Show that

$$F_{n+1} \cdot F_{n-1} = F_n^2 - (-1)^n$$

and that

$$\lim_{n\to\infty} \frac{F_n}{F_{n-1}} = \frac{1+\sqrt{5}}{2}.$$

17. This problem and the next two are from Jordanus's *De numeris datis*. If the sum of the product of the two parts of a given number and of their difference is known, then each of them is determined. Namely, solve the system $x + y = a$, $xy + x - y = b$. Use Jordanus's example where $a = 9$ and $b = 21$.

18. If the sum of the two quotients formed by dividing the two parts of a given number by two different known numbers is given, then each of the parts is determined. Namely, solve the system $x + y = a, x/b + y/c = d$. Jordanus sets $a = 10$, $b = 3$, $c = 2$, and $d = 4$.

19. If the sum of two numbers is given together with the product of their squares, then each of them is determined. Jordanus's example is $x + y = 9$, $x^2y^2 = 324$.

20. Show that under the assumptions of the mean speed theorem, if one divides the time interval into four equal subintervals, the distances covered in each interval will be in the ratio $1 : 3 : 5 : 7$. Generalize this statement to a division of the time interval into n equal subintervals, and prove your result.

21. Determine what mathematics was necessary to solve the Easter problem. What was the result of the debate in the church? How is the date of Easter determined today? (Note that the procedure used in the Roman Catholic Church is different from that in the Eastern Orthodox Church.)

22. Compare Levi ben Gerson's use of what we call induction to that of al-Samaw'al and ibn al-Bannā. Should the methods of either be considered proof by induction? Discuss.

23. Write a lesson demonstrating proof by induction using some of Levi ben Gerson's examples.

24. Write a lesson developing some of the basic combinatorial rules using the methods of Abraham ibn Ezra and Levi ben Gerson.

25. Explain in detail why the area of one of Oresme's configurations should represent the total distance traveled by a moving object.

References

Among the best general sources on the mathematics of medieval Europe are Marshall Clagett, *Mathematics and Its Applications to Science and Natural Philosophy in the Middle Ages* (Cambridge: Cambridge University Press, 1987) and David C. Lindberg, ed., *Science in the Middle Ages* (Chicago: University of Chicago Press, 1978). In particular, in the latter work, chapter 5, "Mathematics," by Michael S. Mahoney and chapter 7, "The Science of Motion," by John E. Murdoch and Edith D. Sylla provide good surveys.

An excellent collection of original source materials translated into English is Edward Grant, ed., *A Source Book in Medieval Science* (Cambridge: Harvard University

Press, 1974). Another source book dealing mainly with aspects of mechanics is Marshall Clagett, *The Science of Mechanics in the Middle Ages* (Madison: University of Wisconsin Press, 1961). The Latin version of Abraham bar Ḥiyya's *Treatise on Mensuration* is available in a German translation in Maximilian Curtze, "Der Liber Embadorum des Abraham bar Chijja Savasorda in der Ubersetzung des Plato von Tivoli," *Abhandlungen zur Geschichte der mathematischen Wissenschaften* 12 (1902), 1–183. There is no complete modern translation directly from the Hebrew. There is also a German translation of Levi ben Gerson's *Maasei ḥoshev*: Gerson Lange, *Sefer Maasei Choscheb. Die Praxis der Rechners. Ein hebräisch-arithmetisches Werk des Levi ben Gerschom aus dem Jahre 1321* (Frankfurt: Louis Golde, 1909). And although there is no English translation of the entire work, there is a new English translation of the problems at the end: Shai Simonson, "The missing problems of Gersonides: A critical edition," *Historia Mathematica* 27 (2000), 243–302, 384–431. The *Sefer yetzirah* is available in English: Isidor Kalisch, ed. and trans., *The Sepher Yezirah (The Book of Formation)* (Gillette, N.J.: Heptangle Books, 1987). Leonardo of Pisa's *Liber abbaci* has recently been translated into English by L. E. Sigler as *Fibonacci's Liber Abaci: A Translation into Modern English of Leonardo Pisano's Book of Calculation* (New York: Springer, 2002). Both the *Arithmetica* and the *De numeris datis* of Jordanus are available in English. The first is in H. L. L. Busard, *Jordanus de Nemore, De Elementis Arithmetice Artis* (Stuttgart: Franz Steiner Verlag, 1991); the second is in Barnabas Hughes, *Jordanus de Nemore: De numeris datis* (Berkeley: University of California Press, 1981). Hughes also discusses the sources of the work and gives a modern symbolic translation of each proposition. Oresme's work is available in Marshall Clagett, *Nicole Oresme and the Medieval Geometry of Qualities and Motions. A Treatise on the Uniformity and Difformity of Intensities Known as Tractatus de configurationibus qualitatum* (Madison: University of Wisconsin Press, 1968). Besides the Latin text and a complete English translation, this edition contains a detailed commentary as well as translations of some related works.

Discussions of Jewish mathematics in the Middle Ages are in Nachum L. Rabinovitch, *Probability and Statistical Inference in Ancient and Medieval Jewish Literature* (Toronto: University of Toronto Press, 1973); in Y. Tzvi Langermann and Shai Simonson, "The Hebrew Mathematical Tradition," in Helaine Selin, ed., *Mathematics across Cultures: The History of Non-Western Mathematics* (Dordrecht: Kluwer, 2000), 167–188; and in Tony Lévy, "Hebrew Mathematics in the Middle Ages: An Assessment," in F. Jamil Ragep and Sally P. Ragep, eds., *Tradition, Transmission, Transformation* (Leiden: Brill, 1996), 71–88. An earlier work is M. Steinschneider, *Mathematik bei den Juden* (Hildesheim: Georg Olms, 1964), a reprint of original articles of 1893–1901. A discussion of ibn Ezra's work can be found in J. Ginsburg, "Rabbi ben Ezra on Permutations and Combinations," *The Mathematics Teacher* 15 (1922), 347–356. A broad overview of the work of Jordanus is Jens Høyrup, "Jordanus de Nemore, 13th Century Mathematical Innovator: An Essay on Intellectual Context, Achievement, and Failure," *Archive for History of Exact Sciences* 38 (1988), 307–363. The article by Wilbur Knorr, "On a Medieval Circle Quadrature: *De circulo quadrando*," *Historia Mathematica* 18 (1991), 107–128, makes a new attempt to place Jordanus in the context of Paris in the early thirteenth century. Finally, a study of the use of lines to express functions is in Edith Sylla, "Medieval Concepts of the Latitude of Forms: The Oxford Calculators," *Archives d'Histoire Doctrinal et Littérarie du Moyen Âge* 40 (1973), 223–283.

CHAPTER NINE

Mathematics in the Renaissance

But of number, cosa [unknown], and cubo [cube of the unknown], however they are compounded . . . , nobody until now has formed general rules, because they are not proportional among them. . . . And therefore, until now, for their equations, one cannot give general rules except that, sometimes, by trial, . . . in some particular cases. And therefore when in your equations you find terms with different intervals without proportion, you shall say that the art, until now, has not given the solution to this case, . . . even if the case may be possible.

—From the *Summa de arithmetica, geometrica, proportioni et proportionalita* of Luca Pacioli, 1494

Many changes began to take place in the European economy in the fourteenth century that eventually had an effect on mathematics. The general cultural movement of the next two centuries, known as the Renaissance, also had its impact, particularly in Italy, so it is in that country that we begin our discussion of Renaissance mathematics.

The Italian merchants of the Middle Ages generally were what today we might call venture capitalists. They traveled to distant places in the East, bought goods that were wanted back home, then returned to Italy to sell them in the hope of making a profit. These traveling merchants needed very little mathematics other than the ability to determine their costs and revenues for each voyage. By the early fourteenth century, a commercial revolution spurred originally by the demands of the Crusades had begun to change this system greatly. New technologies in shipbuilding and greater safety on the shipping lanes helped to replace the traveling merchants of the Middle Ages with the sedentary merchants of the Renaissance. These "new men" were able to remain at home in Italy and hire others to travel to the various ports, make the deals, act as agents, and arrange for shipping. Thus, international trading companies began to develop in the major Italian cities, companies that had a need for more sophisticated mathematics than did their predecessors. These new companies had to deal with letters of credit, bills of exchange, promissory notes, and interest calculations. Double-entry bookkeeping began as a way of keeping track of the various transactions. Business was no longer composed of single ventures but of a continuous flow of goods consisting of many shipments from many different ports en route simultaneously. The medieval economy, based in large part on barter, was gradually being replaced by a money economy.

The Italian merchants needed a new facility in mathematics to be able to deal with the new economic circumstances, but the mathematics they needed was not the mathematics of the quadrivium studied in the universities. They needed new tools for calculating and problem solving. To meet this need, a new class of professional mathematicians, the *maestri d'abbaco*, or abacists, appeared in early fourteenth-century Italy. These professionals wrote the texts from which they taught the necessary mathematics to the sons of the merchants in new schools created for this purpose.

The first section of this chapter will discuss the algebra of the abacists, followed by the new discoveries in algebra partly inspired by their work. These include the solution of the cubic and quartic equations and also the development of the use of symbolism in algebra.

Of course, there was much more going on in the Renaissance than just the revival of trade. There were striking new developments in art, particularly as artists strove to represent reality more accurately on their canvases. To do this, they needed to develop and understand the mathematics of perspective. Similarly, as ships began sailing beyond the confines of the Mediterranean, new methods of navigation became imperative; these often depended on a better knowledge of trigonometry, as well as on improved methods of calculation. And with these new methods, scientists were better able to interpret motion in the heavens and on the earth. Thus, this chapter will also deal with developments in the geometry of perspective, with improved methods of trigonometry, with the extension of the Hindu-Arabic decimal place value system to fractions, with the invention of logarithms, and with new developments in astronomy and physics that were influenced by and in turn had impacts on mathematics.

9.1 Algebra

9.1.1 The Abacists

The Italian abacists of the fourteenth century were instrumental in teaching the merchants the Hindu-Arabic decimal place value system and the algorithms for using it. As is usual when a new system replaces an old traditional one, there was great resistance to the change. For many years, account books were still kept in Roman numerals. It was believed that the Hindu-Arabic numerals could be altered too easily, and thus it was risky to depend on them alone in recording large commercial transactions. (The current system of writing out the amounts on checks in words dates from this time.) The advantages of the new system, however, eventually overcame the merchants' initial hesitation. The old counting board system required accountants to carry around not only a board but a bag of counters, while the new system required only pen and paper and could be used anywhere. In addition, using a counting board required that preliminary steps in the calculation be eliminated as one worked toward the final answer. With the new system, all the steps were available for checking when the calculation was finished. (Of course, these advantages would have meant nothing had not a steady supply of cheap paper been recently introduced.) The abacists instructed entire generations of middle-class Italian children in the new methods of calculation, and these methods soon spread throughout the continent.

In addition to the algorithms of the Hindu-Arabic number system, the abacists taught their students methods of problem solving using the tools of both arithmetic and Islamic algebra. The texts written by the abacists, of which several hundred different ones still exist, are generally large compilations of problems along with their solutions. These include not only genuine business problems of the type the students would have to solve when they joined their fathers' companies, but also plenty of recreational problems typical of the kind found in modern elementary algebra texts. The solutions in the texts were written in great detail with every step fully described, but, in general, no reasons were given for the various steps. Perhaps the teachers did not want to disclose their methods in written form, fearing that then there would no longer be any reason to hire them. In any case, it seems clear that these abacus texts were designed not only for classroom use, but also to serve as reference manuals for the merchants themselves. A merchant could easily find and readily follow the solution of a particular type of problem without the necessity of understanding the theory behind the solution.

Although these texts were strictly practical, they did have significant influence on the development of mathematics, because they instilled in the Italian merchant class a facility with numbers without which no future advances could be made. Furthermore, some of these texts also brought to this middle class the study of Islamic algebra as a basic part of the curriculum. During the fourteenth and fifteenth centuries, the abacists extended the Islamic methods in several directions. In particular, they introduced abbreviations and symbolism and expanded the rules of algebra into the domain of equations of degree higher than the second.

Recall that Islamic algebra was entirely rhetorical. There were no symbols for the unknown or its powers or for the operations performed on these quantities. Everything was written out in words. The same was generally true in the works of the early abacists and in the earlier Italian work of Leonardo of Pisa. Early in the fifteenth century, however, some of the abacists began to substitute abbreviations for unknowns. For example, in place of the standard words *cosa* ("thing"), *censo* ("square"), *cubo* ("cube"), and *radice* ("root"), some authors used the abbreviations *c*, *ce*, *cu*, and *R*. Combinations of these abbreviations were used for higher powers. Thus, *ce ce* stood for *censo di censo*, or fourth power; *ce cu*, an abbreviation for *censo di cubo*, stood for fifth power (x^2x^3); and *cu cu*, meaning *cubo di cubo*, stood for sixth power (x^3x^3). By the end of the fifteenth century, however, the naming scheme for higher powers had changed, and authors used *ce cu*, or *censo di cubo*, to designate the sixth power $[(x^3)^2]$ and *cu cu*, or *cubo di cubo*, to represent the ninth power $[(x^3)^3]$. The fifth power was then designated as *p.r.*, or *primo relato*, and the seventh power as *s.r.*, or *secondo relato*.

Near the end of the fifteenth century, the abbreviations $\overline{p}$ and $\overline{m}$ were introduced to represent plus and minus (*più* and *meno*). (These particular abbreviations probably came from a more general practice of using a bar over a letter to indicate that some other letters were missing.) As with other innovations, however, there was no great movement on the part of all the writers to use the same abbreviations. This change was a slow one. It took several centuries until modern algebraic symbolism was fully formed.

Even without much symbolism, the Italian abacists, like their Islamic predecessors, were competent in handling operations on algebraic expressions. And they used this competency to improve and extend Islamic methods for solving equations. For example, Maestro Dardi of Pisa, in a work of 1344, extended the normal Islamic list of six types of equations to 198 types of equations of degree up to four, some of which involved radicals. Most of the equations can be solved by a simple reduction to one of the standard forms, although in each case Dardi gave the solution anew. For example, he noted that the equation $ax^4 = bx^3 + cx^2$ has the solution given by

$$x = \sqrt{\left(\frac{b}{2a}\right)^2 + \frac{c}{a}} + \frac{b}{2a}.$$

That is, it has the same solution as the standard equation $ax^2 = bx + c$. (Note that zero is never considered as a solution.)

More interesting than quadratic equations, however, are four examples of irreducible cubic and quartic equations. Dardi's cubic equation is $x^3 + 60x^2 + 1200x = 4000$. His rule says to divide 1200 by 60 (giving 20), cube the result (which gives 8000), add 4000 (giving 12,000), take the cube root ($\sqrt[3]{12{,}000}$), and finally subtract the quotient of 1200 by 60. Dardi's answer, which is correct, is $x = \sqrt[3]{12{,}000} - 20$. If we write this equation using modern notation and then give Dardi's solution rule, we obtain the solution to the equation $x^3 + bx^2 + cx = d$ in the form

$$x = \sqrt[3]{\left(\frac{c}{b}\right)^3 + d} - \frac{c}{b}.$$

It is easy enough to see that this solution is wrong in general, and Dardi even admits as much. How then did Dardi figure out the correct solution to his particular case? We can answer this question by considering the problem that illustrates the rule, a problem in compound interest: A man lent 100 *lire* to another and after 3 years received back a total of 150 *lire* in principal and interest, where the interest was compounded annually. What was the interest rate? Dardi set the rate for 1 *lira* for 1 month at x *denarii*. Then the annual interest on 1 *lira* is $12x$ *denarii*, or $\frac{1}{20}x$ *lire*, since 1 *lira* is composed of 240 *denarii*. So the amount owed after 1 year is $100(1 + x/20)$ and the amount after 3 years is $100(1 + x/20)^3$. Dardi's equation therefore is

$$100\left(1 + \frac{x}{20}\right)^3 = 150 \qquad \text{or} \qquad 100 + 15x + \frac{3}{4}x^2 + \frac{1}{80}x^3 = 150$$

or finally

$$x^3 + 60x^2 + 1200x = 4000.$$

Because the left side of this equation comes from a cube, it can be completed to a cube once again by adding an appropriate constant. In general, because $(x + r)^3 = x^3 + 3rx^2 + 3r^2x + r^3$, to complete $x^3 + bx^2 + cx$ to a cube, we must find r satisfying two separate conditions, $3r = b$ and $3r^2 = c$, which can only be satisfied when $b^2 = 3c$. In Dardi's example, with $b = 60$ and $c = 1200$, the condition is satisfied and $r = c/b = 20$.

Dardi gave a similar rule for solving special quartic equations, while Piero della Francesca (c. 1420–1492), more famous as a painter than as an abacist, extended these rules to fifth- and sixth-degree equations in his own *Trattato d'abaco*. There is another (anonymous) manuscript of this period that suggests that the equation $x^3 + px^2 = q$ can be solved by setting $x = y - p/3$, where y is a solution of $y^3 = 3(p/3)^2y + [q - 2(p/3)^3]$. This is correct as far as it goes, but the author has only managed to replace one cubic equation by another. In the numerical example presented, he solved the new equation by trial, but this could also have been done with the original. Nevertheless, although the abacists did not manage to give a complete general solution to the cubic equation, they, like their Islamic predecessors, wrestled with the problem and arrived at partial results, as noted in this chapter's opening quotation from the work of Luca Pacioli (1445–1517).

Pacioli, one of the last of the abacists, was ordained as a Franciscan friar in the 1470s and taught mathematics at various places in Italy during the remainder of his career. Because he felt that there was a scarcity of available subject material, he gathered mathematical materials for some twenty years and in 1494 completed the most comprehensive mathematics text of the time and one of the earliest mathematics texts to be printed: the *Summa de arithmetica, geometrica, proportioni et proportionalita*. It contained not only practical arithmetic but also much of the algebra already discussed, the first published treatment of double-entry bookkeeping, and a section on practical geometry. There was little that was original in this work. In fact, a large number of the algebra problems are taken directly from della Francesca's treatise, while the practical geometry is very similar to that of Leonardo of Pisa. Nevertheless, its comprehensiveness and the fact that it was the first such work to be printed made it a widely circulated and influential text, extensively studied by sixteenth-century mathematicians. It became the common base from which they were able to extend the range of algebra.

9.1.2 Algebra in Northern Europe

The medieval economy was also changing in northern Europe during the fourteenth and fifteenth centuries, although developments were generally a bit behind those in Italy. And so mathematics texts began to appear there to meet the new needs of the society. There is much similarity among the new works in algebra and also between these works and the Italian algebra of the fifteenth century, so it is clear that all these authors had some knowledge of the contemporaneous work elsewhere in Europe. And certainly the spread of printing helped new ideas circulate more rapidly throughout the continent. The ideas generally perceived to be most important were absorbed into a new European algebra. We will consider a few of these new ideas.

Christoff Rudolff (sixteenth century) wrote his *Coss*, the first comprehensive German algebra, in Vienna in the early 1520s. [The word *Coss* is the German form of the Italian *cosa*, or "thing" (unknown); in fact, German algebraists were generally known as *Cossists*.] Among Rudolff's innovations was new material on what are now known as the laws of exponents. For example, Rudolff displayed a list of non-negative powers of 2 alongside their respective exponents and noted that multiplication in the powers corresponded to addition in the exponents. He then extended this idea to powers of the unknown. He did not explicitly use numerical values for these powers, however, but symbols and names adapted from the Italian ones. For example, *radix*, or "root," was symbolized by 𝔯; *zensus*, or "square," by 𝔷; *cubus*, or "cube," by 𝔠; and *zens de zens*, or "fourth power," by 𝔷𝔷.

To help the reader understand the use of these terms, Rudolff presented a multiplication table to use with them, which showed, for instance, that 𝔯 times 𝔷 was 𝔠. To simplify matters, he included numerical values for his symbols. Thus, *radix* was labeled as 1, *zensus* as 2, *cubus* as 3, and so on, and he noted that in multiplying expressions, one could simply add the corresponding numbers to find the correct symbol. Rudolff also dealt with binomials, terms connected by an operation sign, and included, for the first time in an algebra text, the symbols + and − used today to represent addition and subtraction. Rudolff also introduced in this work the modern symbol $\surd$ for square root and sometimes used a period in place of the German word *gleich* to represent "equals."

In 1544, Michael Stifel (1487–1567) prepared his own text, the *Arithmetica integra*. In this work, Stifel used the same symbols as Rudolff for the powers of the unknowns, but he was more consistent in using the correspondence between these letters and the integral exponents. He went further than Rudolff in writing out a table of powers of 2 along with their exponents, which included the negative values -1, -2, -3, corresponding to $\frac{1}{2}$, $\frac{1}{4}$, and $\frac{1}{8}$.

Although Stifel, like most of his contemporaries, did not accept negative roots to equations, he was the first to compress the three standard forms of the quadratic equation into the single form $x^2 = bx + c$, where b and c were either both positive or were of opposite parity. The solution, expressed in words, was then equivalent to

$$x = \frac{b}{2} \pm \sqrt{\left(\frac{b}{2}\right)^2 + c},$$

where the negative sign was only possible in the case where b was positive and c negative. In that case, as long as $(b/2)^2 + c > 0$, there were two positive solutions. Combining the three cases of the quadratic into one may not seem like a major advance, but in the context of the sixteenth century it was significant. It was another step toward the extension of the number concept, although two centuries were to pass before all algebra texts adopted Stifel's procedure.

Stifel's work was also the first European work both to present the Pascal triangle of binomial coefficients and to make use of the table for finding roots. Stifel noted that he had discovered these coefficients and the root-finding procedure only with great difficulty, as he had been unable to find any written accounts of them. Thus, although these coefficients had been used for that purpose in China and in Islamic countries several centuries earlier, the knowledge of this procedure evidently only reached Stifel indirectly.

The works of Rudolff and Stifel strongly influenced the first English algebra, *The Whetstone of Witte*, published in 1557 by Robert Recorde (1510–1558). Although there was little here that was original in technique, Recorde did create the modern symbol for equality: "To avoid the tedious repetition of these words—is equal to—I will set as I do often in work use, a pair of parallels, or *gemow* [twin] lines of one length, thus ═══, because no 2 things can be more equal." He also modified and extended the German symbolization of powers of the unknown to powers as high as the eightieth, setting the integer of the power next to each symbol and noting that multiplication of these symbols corresponds to addition of the corresponding integers. In fact, he showed how to build the symbol for any power out of the square ʒ, the cube ȼ, and various sursolids (prime powers higher than the third) * ʃʒ (where * stands for a letter designating the order of the prime). The fifth power is written ʃʒ, the seventh power as ^{b}ʃʒ (second sursolid), and the eleventh power as ^{c}ʃʒ (third sursolid). Then, for instance, the ninth power is written ȼȼ (cube of the cube), the twentieth power as ʒʒʃʒ (square of the square of the fifth power), and the twenty-first power as ȼ^{b}ʃʒ (cube of the seventh power).

Interestingly, a true exponential notation had been invented in 1484 by Nicolas Chuquet (d. 1487), a French physician from Lyon, in his *Triparty*, a work designed to meet the needs of that thriving commercial center. In his discussion of operations on polynomials, Chuquet introduced an exponential notation for the powers of the unknown. For example, he wrote 12^2 for what we write as $12x^2$ and, introducing actual negative numbers for the first time in a European work, wrote $\overline{m}12^{2\overline{m}}$ for $-12x^{-2}$. He even noted that the exponent 0 is to be used when one is dealing with numbers themselves. He then showed how to add, subtract, multiply, and divide these expressions using the standard modern rules, even when one of the exponents is negative. Thus, "whoever would multiply 8^3 by $7^{1\overline{m}}$ it is first necessary to multiply 8 by 7 coming to 56, then he must add the denominations, that is to say $3\overline{p}$ with $1\overline{m}$, coming to 2. Thus, this multiplication comes to 56^2, and so should others be understood." Unfortunately, the *Triparty* was never printed and exists today only in manuscript form. Thus, it evidently had little contemporary influence.

9.1.3 The Solution of the Cubic Equation

Fra Luca Pacioli noted in 1494 that there was not yet an algebraic solution to the general cubic equation, but throughout the fifteenth and early sixteenth centuries many mathematicians were working on this problem. Finally, sometime between 1500 and 1515, Scipione del Ferro (1465–1526), a professor at the University of Bologna, discovered an algebraic method of solving the cubic equation $x^3 + cx = d$. Recall that, since most mathematicians still did not deal with negative numbers even as coefficients of equations, there were thirteen different types of mixed irreducible cubic equations depending on the relative positions of the (positive) quadratic, linear, and constant terms. So del Ferro had only begun the process of "solving" the cubic equation with his solution of one of these cases.

In modern academia, professors announce and publish new results as quickly as possible to ensure priority, so it may be surprising to learn that del Ferro did not publish, or even publicly announce, his major breakthrough. But academic life in sixteenth-century Italy was far different from that of today. There was no tenure. University appointments were mostly temporary and subject to periodic renewal by the university senate. One of the ways a professor convinced the senate that he was worthy of continuing in his position was by winning public challenges. Two contenders for a given position would present each other with a list of problems, and, in a public forum some time later, each would present his solutions to the other's problems. Often, considerable amounts of money, aside from the university positions themselves, were dependent on the outcome of such a challenge. As a result, if a professor discovered a new method for solving certain problems, it was to his advantage to keep it secret. He could then pose these problems to his opponents, secure in the knowledge that he would prevail.

Before he died, del Ferro disclosed his solution of the cubic equation to his pupil, Antonio Maria Fiore (first half of the sixteenth century), and to his successor at Bologna, Annibale della Nave (1500–1558). Although neither of these men publicized the solution, word began to circulate among Italian mathematicians that this old problem had been, or soon would be, solved. Another mathematician, Niccolò Tartaglia of Brescia (1499–1557), in fact, boasted that he too had discovered the solution to a form of the cubic, $x^3 + bx^2 = d$. In 1535, Fiore challenged Tartaglia to a public contest, hoping to win on the strength of his knowledge of del Ferro's solution to the earlier case. Each of his thirty submitted problems dealt with that class of cubic equations. But Tartaglia, the better mathematician, worked long and hard on that case, and, as he later wrote, on the night of February 12, 1535, he discovered the solution. Since Fiore was unable to solve many of Tartaglia's questions covering other areas of mathematics besides the cubic, Tartaglia was declared the winner, in this case of thirty banquets prepared by the loser for the winner and his friends. (Tartaglia, probably wisely, declined the prize, accepting just the honor of the victory.)

Word of the contest and the new solutions of the cubic soon reached Milan, where Gerolamo Cardano (1501–1576) was giving public lectures in mathematics. Cardano wrote to Tartaglia, asking that Tartaglia show him the solution so it could be included, with full credit, in the arithmetic text Cardano was then writing. Tartaglia initially refused, but after many entreaties and a promise from Cardano to introduce him and his new inventions in artillery to the Milanese court, he finally came to Milan in early 1539. Tartaglia, after extracting an oath from Cardano that he would never publish Tartaglia's discoveries—Tartaglia planned to publish them himself at some later date—divulged his secrets to Cardano.

Cardano kept his promise not to publish Tartaglia's result in his arithmetic book, which soon appeared. In fact, he sent Tartaglia a copy off the press to show his good faith. Cardano then began to work on the problem himself, probably assisted by his servant and student, Lodovico Ferrari (1522–1565). Over the next several years, he worked out the solutions and their justifications to all of the various cases of the cubic. Ferrari managed to solve the fourth-degree equation as well. Meanwhile, Tartaglia still had not published anything on the cubic. Cardano did not want to break his solemn oath, but he was eager that the solutions should be made available. Acting on rumors of the original discovery by del Ferro, he and Ferrari journeyed to Bologna and called on della Nave. The latter graciously gave the two permission to inspect del Ferro's papers, and they were able to verify that del Ferro had discovered the solution first. Cardano no longer felt an obligation to Tartaglia. After all, he would not be publishing Tartaglia's solution, but one discovered some twenty years earlier by a man now deceased. So, in 1545, Cardano published his most important mathematical work, the *Ars magna, sive de regulis algebraicis* (*The Great Art, or On the Rules of Algebra*), chiefly devoted to the solution of cubic and quartic equations. Tartaglia, of course, was furious when Cardano's work appeared. He felt he had been cheated of the rewards of his labor, even though Cardano did mention that Tartaglia was one of the original discoverers of the method. Tartaglia's protests availed him nothing. In an attempt to recoup his prestige, he had another public contest, this time with Ferrari, but was defeated. To this day, the formula providing the solution to the cubic equation is known as **Cardano's formula**.

We consider here Cardano's treatment of one of the thirteen cases of the cubic, the case $x^3 = cx + d$ ("cube equal to the thing and number") presented in chapter 12 of the *Ars magna*. Cardano asserted that the solution is of the form $x = u + v$, where $uv = c/3$ and $u^3 + v^3 = d$. He demonstrated this geometrically, but his argument can be translated into algebra:

$$x^3 = (u + v)^3 = u^3 + 3u^2v + 3uv^2 + v^3 = 3uv(u + v) + u^3 + v^3 = cx + d.$$

Given this proof, Cardano then showed how to find u and v that satisfy the two relationships: "When the cube of one-third the coefficient of x is not greater than the square of one-half the constant of the equation, subtract the former from the latter and add the square root of the remainder to one-half the constant of the equation and, again, subtract it from the same half.... The sum of the cube roots of [these two quantities] constitutes the value of x." Cardano has therefore provided an algorithm for solving $x^3 = cx + d$. This algorithm can be translated into the formula

$$x = \sqrt[3]{\frac{d}{2} + \sqrt{\left(\frac{d}{2}\right)^2 - \left(\frac{c}{3}\right)^3}} + \sqrt[3]{\frac{d}{2} - \sqrt{\left(\frac{d}{2}\right)^2 - \left(\frac{c}{3}\right)^3}}.$$

Although Cardano did not describe how he found this formula, we note that the two conditions on u and v give us the sum and product of the two quantities u^3 and v^3. The solution of this problem dates back to the Babylonians and was one of the standard problems in Islamic algebra as well. So Cardano knew well how to solve the system for u^3 and v^3.

Following the rule, Cardano presented two examples. He showed that the solution to $x^3 = 6x + 40$ is $x = \sqrt[3]{20 + \sqrt{392}} + \sqrt[3]{20 - \sqrt{392}}$. (In this formula, the square and cube root symbols as well as the operation symbols are modern. Cardano himself wrote the solution as $\Re$ v : cu.20 p :$\Re$ 392 p : $\Re$ v : cu.20 m :$\Re$ 392.) Similarly, Cardano calculated that the solution to $x^3 = 6x + 6$ is $x = \sqrt[3]{4} + \sqrt[3]{2}$. He then noted the difficulty with his procedure if $(c/3)^3 > (d/2)^2$. In that case, one could not take the square root. To circumvent the difficulty, Cardano described other methods for special cases. As we will see below, it was Rafael Bombelli who showed how to deal with the square roots of negative numbers in the Cardano formula.

Cardano's discussion of the solutions of the thirteen cases of the cubic are in chapters 11–23 of the *Ars magna*. But the text opens with some general results, including a discussion of the number of roots a given equation could have, whether the roots are positive ("true") or negative ("fictitious"), and how the roots of one equation determine the roots of a related equation. For example, Cardano wrote that equations of the form $x^3 + cx = d$ always have one positive solution and no negative ones. Conversely, the number and sign of the roots of the equation $x^3 + d = cx$ depend on the coefficients. If $(2c/3)\sqrt{c/3} = d$, then this equation has one positive root, $r = \sqrt{c/3}$, and one negative one, $-s = -2\sqrt{c/3}$. If $(2c/3)\sqrt{c/3} > d$, then there are two positive roots, r and s, and one negative root, $-t$, where $t = r + s$ is the positive root of the equation $x^3 = cx + d$. Parenthetically, Cardano noted that in the first case, one could consider the positive root r as two separate roots, for the negative root equals $-2r$. Finally, if $(2c/3)\sqrt{c/3} < d$, then there are no positive roots. There is one negative root, $-s$, where s is the positive root of $x^3 = cx + d$. Sharaf al-Dīn al-Ṭūsī had given a similar discussion of the roots of this equation (and certain others) some three hundred years earlier and had arrived at the same criteria for the existence of positive roots, but whether Cardano used the same method of considering maximums is unknown. Cardano did provide more information than his Islamic predecessor, however, since he considered negative roots. Thus, he was also able to understand, if not to prove, that when there are three real roots to a cubic equation, their sum is equal to the coefficient of the x^2 term.

Cardano's pupil Lodovico Ferrari succeeded in finding the solution to the fourth-degree equation. Cardano presented this solution briefly near the end of the *Ars magna*, where he listed the twenty different types of quartic equations, outlined a basic procedure, and calculated a few examples. This basic procedure begins with a linear substitution, which eliminates the term in x^3, leaving an equation of the form $x^4 + cx^2 + e = dx$, for instance. To solve this equation, second-degree and constant terms are added to both sides to turn each side into a perfect square. One then takes square roots and calculates the answer. We illustrate the procedure with one of Cardano's examples: $x^4 + 3 = 12x$. If we add $2bx^2 + b^2 - 3$ to both sides (where b is to be determined), the left side becomes $x^4 + 2bx^2 + b^2$, a perfect square, while the right side becomes $2bx^2 + 12x + b^2 - 3$. For this latter to be a perfect square, we must have $2b(b^2 - 3) = (12/2)^2$, or $2b^3 = 6b + 36$. Therefore, we need to solve a cubic equation in b. Cardano, of course, had a rule for solving this equation, but in this case it is clear that $b = 3$ is a solution. Thus, the added polynomial is $6x^2 + 6$, and the original equation is transformed into $x^4 + 6x^2 + 9 = 6x^2 + 12x + 6$, or $(x^2 + 3)^2 = 6(x + 1)^2$. Taking the positive square root of each side gives $x^2 + 3 = \sqrt{6}(x + 1)$, the solutions to which are easily found to be

$$x = \sqrt{1\tfrac{1}{2}} \pm \sqrt{\sqrt{6} - 1\tfrac{1}{2}}.$$

Are these the only roots of the quartic? One can attempt to find others by taking a negative square root on the right in the equation $(x^2 + 3)^2 = 6(x + 1)^2$, but that leads to complex values for x, which Cardano ignores. In other examples, he did use both sets of roots. One could also look for other roots by using a second solution of the resolvent cubic. Cardano evidently considered this possibility but teases the reader about what happens: "I need not say whether having found another value for b ... we would come to two other solutions [for x]. If this operation delights you, you may go ahead and inquire into this for yourself."

9.1.4 Bombelli and Complex Numbers

Cardano's *Ars magna* was extremely influential, marking the first substantive advance over the Islamic algebra so long studied in Europe. Nevertheless, the book itself was difficult to read. Its arguments were often prolix and not easily followed, and its organization left much to be desired. To improve the teaching of the subject, and to clear up some of the difficulties still remaining, Rafael Bombelli (1526–1572) decided some fifteen years later to write a more systematic text in Italian to enable students to master the material on their own. Although only the first three of the five parts were published in Bombelli's lifetime, and although in the questions concerning multiple roots of cubic equations he did not achieve as much as Cardano, nevertheless, his *Algebra* marks the high point of the Italian algebra of the Renaissance.

Recall that algebraic symbolism was gradually replacing the strictly verbal accounts of the Muslims and of the earliest Italian algebraists. Cardano had used some symbolism, but Bombelli's was a bit different. For example, he used $R.q.$ to denote the square root, $R.c.$ to denote the cube root, and similar expressions to denote higher roots. He used $\lfloor\ \rfloor$ as parentheses to enclose long expressions, as in $R.c.\lfloor 2\ p\ R.q.21 \rfloor$, but kept the standard Italian abbreviations of p for plus and m for minus. His major notational innovation was the use of a semicircle around a number n to denote the nth power of the unknown. Thus, $x^3 + 6x^2 - 3x$ would be written as $1^{\underset{\smile}{3}}\,p\,6^{\underset{\smile}{2}}m\,3^{\underset{\smile}{1}}$. Writing powers numerically rather than in the German form of symbols allowed him easily to express the exponential laws for multiplying and dividing monomials.

Late in the first part of the *Algebra*, Bombelli introduced "another sort of cube root much different from the former, which comes from the chapter on the cube equal to the thing and number; ... this sort of root has its own algorithms for various operations and a new name." This root is the one that occurs in the cubic equations of the form $x^3 = cx + d$ when $(d/2)^2 - (c/3)^3$ is negative. Bombelli proposed a new name for these numbers, which are neither positive (*più*) nor negative (*meno*)—that is, the modern imaginary numbers. The numbers written today as bi and $-bi$, Bombelli called *più di meno* ("plus of minus") and *meno di meno* ("minus of minus"), respectively. For example, he wrote $2 + 3i$ as $2\ p\ di\ m\ 3$ and $2 - 3i$ as $2\ m\ di\ m\ 3$. Bombelli presented the various laws of multiplication for these new (complex) numbers, such as *più di meno* times *più di meno* gives *meno* and *più di meno* times *meno di meno* gives *più*, or $[(bi)(ci) = -bc, bi(-ci) = bc]$.

To illustrate his rules, Bombelli gave numerous examples of the four arithmetic operations on these new numbers. Thus, to find the product of $\sqrt[3]{2 + \sqrt{-3}}$ and $\sqrt[3]{2 + \sqrt{-3}}$, one first multiplies $\sqrt{-3}$ by itself to get -3, then 2 by itself to get 4, and then adds these two to get 1 for the "real" part. Next, one multiplies 2 by $\sqrt{-3}$ and doubles the result to get $\sqrt{-48}$. The answer is $\sqrt[3]{1 + \sqrt{-48}}$. To divide 1000 by $2 + 11i$, Bombelli multiplied both numbers by $2 - 11i$. He then divided the new denominator, 125, into 1000, giving 8, which in turn he multiplied by $2 - 11i$ to get $16 - 88i$ as the result. Bombelli, although he noted that "the whole matter seems to rest on sophistry rather than on truth," nevertheless gave here for the first time the rules of operation for complex numbers. It seems clear from his discussion that he developed these rules strictly by analogy to the known rules for dealing with real numbers. Arguing by analogy is a common method of making mathematical progress, even if one is not able to give rigorous proofs. Of course, because Bombelli did not know what these numbers "really" were, he could give no such proofs.

Proofs notwithstanding, with the rules for dealing with complex numbers now available, Bombelli could discuss how to use Cardano's formula for the case $x^3 = cx + d$, whether $(d/2)^2 - (c/3)^3$ is positive or negative. He first considered the example $x^3 = 6x + 40$. Cardano's procedure gives $x = \sqrt[3]{20 + \sqrt{392}} + \sqrt[3]{20 - \sqrt{392}}$, even though it is obvious that the answer is $x = 4$. Bombelli showed how one can see that the sum of the two cube roots is in fact 4. He assumed that $20 + \sqrt{392}$ equals the cube of a quantity of the form $a + \sqrt{b}$ for some numbers a and b, or $\sqrt[3]{20 + \sqrt{392}} = a + \sqrt{b}$. This implies that $\sqrt[3]{20 - \sqrt{392}} = a - \sqrt{b}$. Multiplying these two equations together gives $\sqrt[3]{8} = a^2 - b$, or $a^2 - b = 2$. Furthermore, cubing the first equation and equating the parts without square roots gives $a^3 + 3ab = 20$. Bombelli did not attempt to solve this system of two equations in two unknowns by a general argument. Rather, he noted that the only possible integral value for a is $a = 2$. Fortunately, $b = 2$ then provides the other value in each equation, so Bombelli had shown that $\sqrt[3]{20 + \sqrt{392}} = 2 + \sqrt{2}$. It follows that the solution to the cubic equation may be written as $x = (2 + \sqrt{2}) + (2 - \sqrt{2}) = 4$, as desired.

For the equation $x^3 = 15x + 4$, the Cardano formula gives $x = \sqrt[3]{2 + \sqrt{-121}} + \sqrt[3]{2 - \sqrt{-121}}$, although again it is clear that the answer is $x = 4$. Bombelli used his newfound knowledge of complex numbers to apply the same method as before. He first assumed that $\sqrt[3]{2 + \sqrt{-121}} = a + \sqrt{-b}$. Then $\sqrt[3]{2 - \sqrt{-121}} = a - \sqrt{-b}$, and a short calculation leads to the two equations $a^2 + b = 5$ and $a^3 - 3ab = 2$. Again, Bombelli carefully showed that $a = 2$ was the only possibility. Then $b = 1$ provides the other solution, and the desired cube root is $2 + \sqrt{-1}$. It followed that the solution to the cubic equation is $x = (2 + \sqrt{-1}) + (2 - \sqrt{-1})$, or $x = 4$.

Bombelli presented several more examples of the same type, where in each case he was able somehow to calculate the appropriate values of a and b. He did note, however, that this was not possible in general. Bombelli also showed that complex numbers could be used to solve quadratic equations that previously had been thought to have no solution. For example, he used the standard quadratic formula to show that $x^2 + 20 = 8x$ has the solutions $x = 4 + 2i$ and $x = 4 - 2i$. Although he could not answer all questions about the use of complex numbers, his ability to use them to solve certain problems provided mathematicians with the first hint that there was some sense to dealing with them. Since mathematicians were still not entirely happy with using negative numbers—Cardano called them "fictitious" and Bombelli did not consider them as roots at all—it is not surprising that it took many years before they were entirely comfortable with using complex numbers.

9.1.5 Viète, Algebraic Symbolism, and Analysis

The European algebraists of the sixteenth century had achieved about as much as possible in their continuation of the Islamic algebra of the Middle Ages. They were now expert in algebraic manipulations, and they knew how to solve any polynomial equation of degree up to four. The solutions, however, were given in the form of rules of procedure. Most of these authors used some symbols for the unknown and its powers, but there were no symbols for the coefficients. Thus, the best that could be done to illustrate a procedure was to use numerical examples. None of these algebra texts contain a written formula, like the quadratic formula found in every current elementary algebra textbook. To be able to write down such formulas required a new approach to symbols.

This new approach was first taken by François Viète (1540–1603). In the closing years of the sixteenth century, he composed the several treatises collectively known as *The Analytic Art*, in which he effectively reformulated the study of algebra, not only by introducing symbols for the constants of an equation, but also by using this new idea to replace the search for solutions to equations with the detailed study of the structure of these equations. Thus, Viète developed the earliest consciously articulated theory of equations.

But Viète also approached algebra differently from his predecessors. In fact, as the title of his treatises indicates, he was interested not just in solving equations—the main thrust of algebra before him—but also in using algebra as a form of the ancient Greek method of analysis, a method for solving problems. As he wrote, "the analytic art . . . claims for itself the greatest problem of all, which is TO LEAVE NO PROBLEM UNSOLVED." Thus, he began his program in the *Introduction to the Analytic Art* of 1591 with an announcement of what he wanted to accomplish:

> There is a certain way of searching for the truth in mathematics that Plato is said first to have discovered. Theon called it analysis, which he defined as assuming that which is sought as if it were admitted and working through the consequences of that assumption to what is admittedly true. . . . Therefore the whole analytic art . . . may be called the science of correct discovery in mathematics.

Viète's algebra, then, was to be a renewed version of this analysis. To solve a problem using this analysis meant to assume that the answer, say x, was known, then to work through a series of operations (consequences) until one arrived at an expression where x is written in terms of the known quantities. In essence, he was subsuming Diophantus's procedures under Pappus's geometric analysis.

To solve problems effectively, better symbolism was necessary, so Viète introduced this: "Numerical logistic is that which employs numbers; symbolic logistic that which uses symbols, as, say, the letters of the alphabet." Viète thus manipulated letters as well as numbers. "Given terms are distinguished from unknown by constant, general and easily recognized symbols, as (say) by designating unknown magnitudes by the letter A and the other vowels E, I, O, U, and Y and given terms by the letters B, G, D and the other consonants." While Viète's convention differed from the modern one in distinguishing knowns from unknowns, he was now able to manipulate completely with these letters. Furthermore, these letters did not need to stand for numbers only. They could stand for any quantity to which one can apply the basic operations of arithmetic. Viète had not entirely broken away from his predecessors, however. He continued to use words or abbreviations for powers rather than exponents, as suggested by Bombelli and Chuquet. Instead of using A^2, B^3, or C^4, Viète wrote A *quadratum*, B *cubus*, or C *quadrato-quadratum*, the first and third of which he sometimes abbreviated to A *quad* and C *quad-quad*. He therefore had to give verbal rules for multiplying and dividing powers—for example, *latus* (side) times *quadratum* equals *cubus*, and *quadratum* times itself equals *quadrato-quadratum*.

For operations, Viète generally adopted the German forms $+$ and $-$ for addition and subtraction, although sometimes he still used words. For multiplication, he used the word *in*, while for division, he used the fraction bar. Hence,

$$\frac{A\ in\ B}{C\ quadratum}$$

means, in modern notation, AB/C^2. Viète wrote square roots using the symbol $l.$ for *latus*: $l.64$ meant the square root of 64 and $l.c.64$ stood for the cube root of 64. Sometimes, however, he used R for *radix* to symbolize square root. Viète, like most of his predecessors, insisted on the law of homogeneity: that all terms in a given equation must be of the same degree. So to make sense of the modern equation $x^3 + cx = d$, Viète insisted that c be a plane (so that cx is a solid) and that d be a solid. He wrote this equation as $A\ cubus + C\ plano\ in\ A\ aequetus\ D\ solido$. Note that he did not use a symbol for equals, but a word (*aequetus*).

While Viète had come only partway toward modern symbolism, the crucial step of allowing letters to stand for numerical constants enabled him to break away from the style of examples and verbal algorithms of his predecessors. He could now treat general examples, rather than specific ones, and give formulas rather than rules. In addition, eliminating the possibility of actually carrying out numerical computations using the symbolic constants made it possible to focus on the procedures of the solution rather than the solution itself. Further, solving equations symbolically made the structure of the solution more evident. Instead of replacing $5 + 3$ by 8, for example, one kept the expression $B + D$ in the displayed formula so that at the end of the argument one could consider its relationship to the original constants. Viète was thus able in some circumstances to discover how the roots of an equation were related to the expressions from which the equation was constructed.

We will consider a few of Viète's problems and methods of solution in the various treatises that make up *The Analytic Art*. We begin with his rule, stated in the *Introduction*, that "an equation is not changed by *antithesis*":

> Let it be given that A square minus D plane is equal to G square minus B in A. I say that A square plus B in A is equal to G square plus D plane and that by this transposition under opposite signs of conjunction the equation is not changed. For since A square minus D plane is equal to G square minus B in A, add D plane plus B in A to both sides. Then by common agreement A square minus D plane plus D plane plus B in A is equal to G square minus B in A plus D plane $+ B$ in A. The negative affection on each side of this equation cancels the positive: on one side

> the affection D plane vanishes; on the other the affection B in A. This leaves A square plus B in A equal to G square plus D plane.

In modern notation, what Viète wrote is simply that if one adds $D + BA$ to each side of the equation $A^2 - D = G^2 - BA$, one gets the transposed version $A^2 + BA = G^2 + D$, and this new equation has the same meaning as the original one. Viète's expressions are still very wordy, but the basic new symbolism is present. We will see in the next chapter how Harriot rewrote Viète's rule.

In the *Prior Notes on Symbolic Logistic*, probably written at the same time as the *Introduction* but not published until 1631, Viète showed how to operate on symbolic quantities. He derived many of the standard algebraic identities, most of which were previously known in verbal form at least but were here written for the first time using symbols. For example, Viète noted that $A - B$ times $A + B$ equals $A^2 - B^2$, and he also wrote out the expansion of $(A + B)^n$ for each integer n from 2 to 6, as well as the products of $A - B$ with $A^2 + AB + B^2$, $A^3 + A^2B + AB^2 + B^2, \ldots$.

In the *Five Books of Zetetics* (1591), Viète used his symbolic methods of calculation to deal with a large number of algebraic problems drawn from a variety of sources, both ancient and contemporary. In each problem, he used "analysis"—representing the unknowns by letters, then operating on the unknowns and the knowns until it was clear how to express the former in terms of the latter. He began with the same problem with which both Diophantus and Jordanus de Nemore began their texts: Given the difference between two numbers and their sum, find the numbers. Viète's procedure was straightforward: Letting B be the difference, D the sum, and A the smaller of the two numbers, he noted that $A + B$ is the greater. Then the sum of the numbers is $2A + B$, which equals D. Hence, $2A = D - B$ and $A = \frac{1}{2}D - \frac{1}{2}B$. The other number is then $E = \frac{1}{2}D + \frac{1}{2}B$. Having written down the solution in symbols, Viète then restated it in words: "Half the sum of the numbers minus half the difference equals the lesser number, plus that difference, the greater." He concluded with an example: If B is 40 and D is 100, then A is 30 and E is 70. This format is typical of Viète's work. Although he introduced symbolic methods, he often restated his answers in words as if to convince skeptical readers that the new symbolic method can always be translated back into the more familiar verbal mode of expression.

It is enlightening to compare the same problem in Diophantus, Jordanus, and Viète to see the differences. Diophantus, although stating the problem generally, in fact solved it only for a particular numerical example, the same one that Viète used. Jordanus solved it generally but in words: "Subtract the difference from the whole and there will remain double the lesser given number." Viète solved it totally symbolically. This problem exemplifies the change in algebra over 1350 years.

The central work in Viète's theory of equations is found in the *Two Treatises on the Recognition and Emendation of Equations*. For example, we see how Viète solves the quadratic equation, which he writes as "A quad $+ B2$ in A equals Z plane" (or, in more modern notation, $A^2 + 2BA = Z$). Viète set $A + B$ to be E, or $E - B$ to be A. Then $(E - B)^2 + 2B(E - B) = Z$, which reduces to $E^2 = Z + B^2$. Therefore, A is equal to $\sqrt{Z + B^2} - B$. In Viète's notation, this is

$$A \text{ is } l.\overline{Z \text{ plane} + B \text{ quad}} - B,$$

the first occurrence of what can really be called the quadratic formula. However, we should also note that Viète gives two other versions of the formula, one for the case $A^2 - 2BA = Z$

and one for the case $2BA - A^2 = Z$, and, of course, only in the latter case is there the possibility of two (positive) solutions.

The treatment of cubics in *Two Treatises* is far more extensive, because there are many more types. We will consider the equation $x^3 = cx + d$, and in what follows, we will generally use modern notation. Recall that Cardano's formula gives complex numbers in this case when $(d/2)^2 < (c/3)^3$. Under those conditions, Viète decided to apply trigonometric reasoning. He began by rewriting the equation as $x^3 - 3b^2x = b^2d$, in keeping with homogeneity. The inequality then becomes $(b^2d/2)^2 < (b^2)^3$, which reduces to $2b > d$. Viète had earlier developed the triple angle formula $\cos 3\alpha = 4\cos^3\alpha - 3\cos\alpha$, or

$$\cos^3\alpha - \tfrac{3}{4}\cos\alpha = \tfrac{1}{4}\cos 3\alpha.$$

By setting $x = r\cos\alpha$ and substituting in his version of the cubic, Viète converted that equation to $r^3\cos^3\alpha - 3b^2r\cos\alpha = b^2d$, or

$$\cos^3\alpha - \frac{3b^2}{r^2}\cos\alpha = \frac{b^2d}{r^3}.$$

Comparing the two equations involving $\cos^3\alpha$ shows first that

$$\frac{3b^2}{r^2} = \frac{3}{4} \qquad \text{or} \qquad r = 2b$$

and second that

$$\frac{1}{4}\cos 3\alpha = \frac{b^2d}{r^3} = \frac{b^2d}{8b^3} \qquad \text{or} \qquad \cos 3\alpha = \frac{d}{2b}.$$

The inequality $2b > d$ for the coefficients ensures that this final equation makes sense. Thus, if α satisfies $\cos 3\alpha = d/2b$ and $r = 2b$, then $x = r\cos\alpha$ is a solution to the original cubic equation. For example, if $x^3 - 300x = 432$, then $b = 10$ and $d = \frac{432}{100}$. It follows that $\cos 3\alpha = \frac{432}{2000}$. By consulting tables, one determines that $\cos\alpha = 0.9$, and thus $x = 2b\cos\alpha = 18$.

For the same equation, $x^3 = cx + d$, when $(d/2)^3 > (c/3)^3$, Viète presented an algebraic solution. Rewriting the equation in the form $x^3 - 3bx = 2d$ (where b is plane and d is solid), Viète put the inequality in the form $b^3 < d^2$. Then, having earlier proved the identity $(r+s)^3 - 3rs(r+s) = r^3 + s^3$, he noted that x must be the sum of two numbers whose product is b. Therefore, $y(x-y) = b$, or $xy - y^2 = b$, or finally, $x = (b+y^2)/y$. Substituting this expression into $x^3 - 3bx = 2d$ and multiplying all terms by y^3 produces a quadratic equation in y^3: $2dy^3 - (y^3)^2 = b^3$. The solutions to this are $y^3 = d \pm \sqrt{d^2 - b^3}$, so the reason for the inequality condition is evident. Because the desired root x is the sum of the two values for y, the final result is the formula slightly modified from the description of Cardano,

$$x = \sqrt[3]{d + \sqrt{d^2 - b^3}} + \sqrt[3]{d - \sqrt{d^2 - b^3}}.$$

Of course, this formula is in modern notation. Viète's actual version of Cardano's formula stated that the solution to the equation

A cube $-$ *B* plane 3 in *A* equals *Z* solid 2

is given by

$$A \text{ is } \overline{\overline{l.c.Z \text{ solid} + \overline{l.Z \text{ solidsolid}}\ - B \text{ planeplaneplane}}}$$

$$+\ \overline{\overline{l.c.Z \text{ solid} - \overline{l.Z \text{ solidsolid}}\ - B \text{ planeplaneplane}}}.$$

Although Viète did not consider negative or complex roots to equations, he did deal to some extent with the relationship between the roots and the coefficients. For example, Viète was aware that the quadratic equation $bx - x^2 = c$ could have two positive roots. To discover the relationship between these two roots, x_1 and x_2, he equated the two expressions $bx_1 - x_1^2$ and $bx_2 - x_2^2$. Then $x_1^2 - x_2^2 = bx_1 - bx_2$, and, dividing through by $x_1 - x_2$, he found that $x_1 + x_2 = b$, that is, "b is the sum of the two roots being sought." Substituting $x_1 + x_2$ for b in the equation $bx_1 - x_1^2 = c$, he found the other relationship $x_1 x_2 = c$, or "c is the product of the two roots being sought."

At the very end of the second treatise on equations, Viète stated four propositions without proof, one for each degree of equation from two through five, generalizing the result for quadratics in terms of the elementary symmetric functions of the roots. Thus, for the third-degree equation, the proposition is "If $x^3 - x^2(b + d + g) + x(bd + bg + dg) = bdg$, x is explicable by any of the three, b, d or g"; for the fourth-degree equation, "If $x(bdg + bdh + bgh + dgh) - x^2(bd + bg + bh + dg + dh + gh) + x^3(b + d + g + h) - x^4 = bdgh$, x is explicable by any of the four, b, d, g, or h." Viète considered these theorems "elegant and beautiful" and a "crown" of his work.

9.2 Geometry and Trigonometry

9.2.1 Art and Perspective

Although the Renaissance developments in algebra provided the most striking change in mathematics, geometry was still an important topic of study. And the needs of artists—in particular their need to give visual depth to their works—required some new ideas in that subject. Although there was some use of perspective in ancient times, it was only in the Renaissance that painters began in earnest to try to figure out how to portray objects at a distance with fidelity. The earliest painters accomplished this through trial and error, but by the fifteenth century, artists were attempting to derive a mathematical basis for displaying three-dimensional objects on a two-dimensional surface. Clearly, objects that are farther away from the observer must be made smaller to give the picture realism. The question then becomes how small a given object should be. The answer to this question, painters ultimately realized, had to come from geometry. Filippo Brunelleschi (1377–1446) was the first Italian artist to make a serious study of the geometry of perspective, but Leon Battista Alberti (1404–1472) wrote the first text on the subject, the *Della pittura*, published in 1435. Alberti noted in this treatise that the first requirement of a painter is to know geometry. Thus, he presented a geometrical result showing how to represent a set of squares in the *ground plane* on the canvas, or the *picture plane*.

The picture plane may be thought of as pierced by rays of light extending from the various objects in the picture to the artist's eye, whose position is called the *station point*. Hence, the picture plane is a section of the projection from the eye (point A') to the scene to be pictured (Fig. 9.1). The perpendicular from the station point to the picture plane intersects the latter in a point V called the *center of vision*, or the *central vanishing point*. The horizontal line AV through the central vanishing point is called the *vanishing line*, or *horizon line*. All horizontal lines in the picture perpendicular to the picture plane must be drawn to intersect at the vanishing point. All other sets of parallel horizontal lines will intersect at some point on the vanishing line.

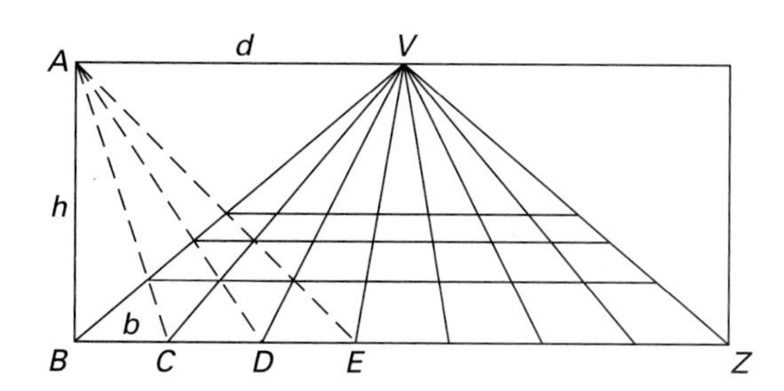

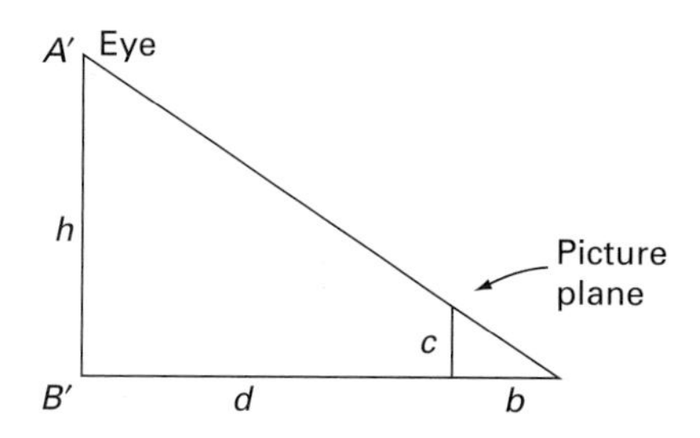

FIGURE 9.1 Alberti's rule for perspective drawing of a tiled floor

To represent a square-tiled pavement (a checkerboard) in the ground plane with sides parallel to and perpendicular to the picture plane, Alberti began by marking off a set of equally spaced points $B, C, D, E, \ldots$ on the line of intersection BZ of the picture plane and the ground plane, or the *ground line*. He connected these to the central vanishing point, thus giving one set of sides. To determine the set of sides parallel to the ground line, Alberti invented the following method: Mark the point A on the vanishing line at distance d from the vanishing point V, where d is the distance of the eye from the picture plane. Then draw lines connecting A to each of the points $B, C, D, E, \ldots$. The lines parallel to BZ through the intersections of BV with $AC, AD, AE, \ldots$ will then represent the lines of pavement parallel to the ground line.

To demonstrate that this construction is correct, we will use algebra, although Alberti himself did not give any demonstration. Suppose that the eye is situated at a height h above the ground line. If a line in the ground plane is parallel to the ground line and at a distance b behind it, then its position in the picture plane should be at a distance c above the ground line, where c is determined by the proportion $c : b = h : (d + b)$ derived from the similar triangles in Figure 9.1. Thus, $c = hb/(d + b)$. Now, if AB and the ground line BZ are taken as coordinate axes, the equation of the line connecting B and V is

$$y = \frac{h}{d}x,$$

while the equation of the line connecting $A = (0, h)$ and $C = (b, 0)$ is

$$y = -\frac{h}{b}x + h.$$

The y coordinate of the intersection of the two lines is then $hb/(d + b)$, as desired. One can easily demonstrate the correctness for as many parallel lines as desired.

This checkerboard construction is at the heart of the system of focused perspective used by artists from the fifteenth century to the present day. Alberti himself did not discuss any more advanced perspective constructions, but Piero della Francesca, in his work *On Perspective for Painting*, written sometime between 1470 and 1490, gave a detailed discussion of how to draw various two- and three-dimensional geometrical objects in focused perspective. Della Francesca, besides being an artist, was a competent mathematician. His text on perspective includes a drawing showing the calculations the artist would make in preparing a painting in focused perspective.

Another artist-mathematician of this same period was the German Albrecht Dürer (1471–1528), who spent several years in Italy studying works on perspective before finally writing his major treatise, the *Treatise on Mensuration with the Compass and Ruler in Lines, Planes, and Whole Bodies*. Published in 1525, this work was the first geometric text written in German. Dürer had to create a new German vocabulary for scientific terms, including abstract mathematical concepts. If possible, he used the expressions handed down from generation to generation by artisans. For example, *der neue Mondschein* ("crescent") denoted the intersection of two circles; *Gabellinie* ("fork line") meant hyperbola; and *Eierlinie* ("egg line") meant ellipse.

Dürer believed that he needed to instruct German artists in many of the preliminary geometrical ideas involved in drawing before they could approach perspective. Therefore, the work is eminently practical. Dürer showed how to apply geometric principles to the representation of objects on canvas.

The first of the four books of the *Treatise on Mensuration* deals with the representation of space curves. Dürer's idea is to project the curve onto both the yz plane and the xy plane in order to determine its nature. Unfortunately, this is not always a straightforward task. Consider his construction of an ellipse from its definition as a section of a right circular cone (Fig. 9.2). Dürer first projected the cone with its cutting plane onto the yz plane. The line segment fg representing the diameter of the ellipse is divided into 12 equal parts, and both vertical and horizontal lines are drawn through the division points. At each of the eleven points i, the horizontal line represents part of the diameter of the circular section C_i made by a horizontal cutting plane. The two points of intersection of this circle with the ellipse are symmetrically located on the ellipse with respect to its diameter and therefore determine the width w_i of the ellipse there. The projection of the cone onto the xy plane then consists of this series of concentric circles C_i. The continuation of each vertical line becomes a chord in the corresponding circle whose length is w_i. Dürer thus had a rough projection of the ellipse.

The outline of the ellipse is, however, not symmetric about its minor axis, since this projection is not taken from a direction perpendicular to the plane of the ellipse itself. But when Dürer attempted to draw the ellipse from its projection, he simply transferred the line segment representing the axis of the ellipse to a new vertical line fg, divided it at the same points i, drew horizontal line segments through each of width w_i, and then sketched the curve through the ends of these line segments. Dürer's drawing is therefore in error, because the curve is wider at the bottom than at the top. A possible reason Dürer did not realize that the ellipse should be symmetric about its minor axis is that the centerline of the cone, around the projection of which all the circles are drawn, does not pass through the center of the ellipse. Although one can prove that $w_i = w_{12-i}$ $(i = 1, 2, 3, 4, 5)$ by an analytic argument, Dürer probably believed that the ellipse was in fact egg-shaped—he does call it *Eierlinie*—because the cone itself widens toward the bottom.

After providing descriptions of the construction and representation by projections of other space curves, most of which were quite correct, Dürer continued in the remainder

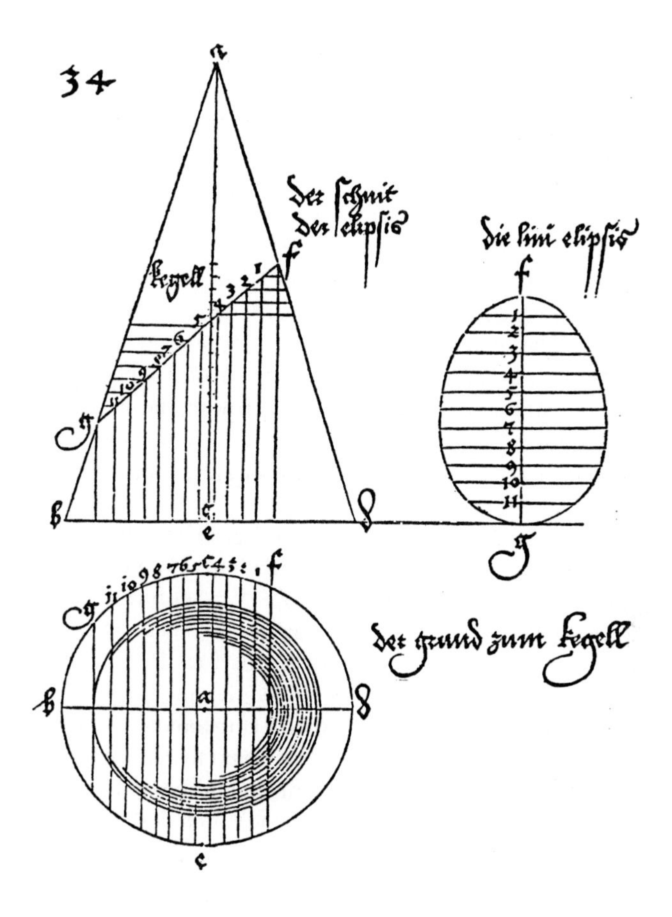

FIGURE 9.2 Dürer's construction of an ellipse by projection

of his *Treatise* to describe numerous other methods for making geometric and practical constructions. For example, he demonstrated methods for constructing various regular polygons, suggested new types of columns and roofs, showed methods of drawing both Roman and Gothic letters accurately, and, toward the end of his work, presented a method for constructing the five regular solids by folding paper. He concluded his work with the basic rules for the perspective drawing of solid figures.

9.2.2 The Conic Sections

One of the reasons that Dürer had such trouble drawing the ellipse was that he had probably never seen Apollonius's text on the conic sections. He knew that they were curves formed by cutting a cone with a plane, but exactly what they looked like was evidently not clear to him. Johannes Kepler (1571–1630), on the other hand, was very familiar with Apollonius's work. He realized that he needed a detailed understanding of conics as he was working on his treatise on optics as well as on astronomy (discussed in section 9.4). In thinking about the subject, Kepler was able to take more out of Apollonius's works than the author had put in. He described his thoughts on conic sections in his *Optics*.

For example, Kepler realized that all the conic sections were really part of the same family of curves. He wrote, "there exists among these curves the following order, by reason of their properties: It passes from the straight line through an infinity of hyperbolas to the parabola, and thence through an infinity of ellipses to the circle. For the most obtuse of all hyperbolas is a straight line; the most acute, a parabola. Likewise the most acute of all ellipses is a parabola; the most obtuse, a circle." In other words, the parabola is the boundary curve between the family of hyperbolas and the family of ellipses. Even though it is infinite, like the hyperbola, "the more it is extended, the more it becomes parallel to itself, and does not expand the arms like a hyperbola." The straight line is one of the outer limits of the family of curves. Therefore, the farther the hyperbola is extended, the more it becomes like a straight line. Similarly, since the circle is the other outer limit, the farther one goes from the center of an ellipse, the more it looks like a circle.

Kepler's use of analogy extended to his discussion of the foci of the conic sections, points to which Apollonius had referred but which Kepler named. The circle has one focus, at the center—thus, as far from the curve as possible. In an ellipse, there are two foci, and as the ellipses get more acute, the foci get further apart. Since the "most acute" ellipse is a parabola, the second focus of that curve is at an infinite distance from the first. The more obtuse the hyperbola, the closer the foci are to the curve, until in the case of the straight line, the foci merge again to lie on the line itself.

Finally, Kepler described constructions of the hyperbola and the ellipse via threads tied around pins at the foci. He then noted that since the parabola is in the middle, between the hyperbola and the ellipse, there should be an analogous construction of that curve. After searching for one, he finally found it and described it in his text as well.

9.2.3 Regiomontanus and Trigonometry

Astronomy was achieving even more importance in the Renaissance with the expansion of sea voyages beyond the confines of the Mediterranean. Sailors needed new methods to determine their location on the open seas, and astronomy and trigonometry provided these. For example, latitude could be determined at night by checking the altitude of the pole star, provided one was far enough north. Otherwise, one could find the latitude by determining the noon altitude of the sun, given a knowledge of its declination. The determination of longitude was more difficult. It required either a very accurate clock, which was not available until the eighteenth century, or detailed knowledge of the motion of the moon, which only gradually became available. In determining both latitude and longitude, however, good instruments and a solid knowledge of the motions of the heavenly bodies were essential, while trigonometry was certainly necessary for the latter.

Some works on trigonometry had been written during the medieval period in Europe, usually as parts of works on astronomy. The first "pure" trigonometry text written in Europe was *On Triangles of Every Kind* by Johannes Müller (1436–1476), who was generally known as Regiomontanus because he was born near Königsberg in Lower Franconia. The book was written about 1463 but not published until 70 years later.

Regiomontanus had made a new translation of Ptolemy's *Almagest* directly from the Greek and, after completing it, realized that a compact, systematic treatment of the rules governing the relationships of the sides and angles in both plane and spherical triangles would improve on Ptolemy's seemingly *ad hoc* approach. He considered such a treatment a necessary prerequisite to the study of the *Almagest*.

Regiomontanus presented his material in *On Triangles* in careful geometric fashion, beginning with definitions and axioms. He proved each theorem by using the axioms, results from Euclid's *Elements*, or earlier results in his own text. Most of the theorems are accompanied by diagrams, and many are followed by examples illustrating the material. Regiomontanus based his trigonometry on the sine of an arc, defined as the half chord of double the arc, but he did note that one can also consider the sine as depending on the corresponding central angle. Although he did use the cosine (written as sine of the complement) and the versine (= radius minus the cosine), Regiomontanus made no use of the tangent function, even though tangents were known in Europe in translations of Islamic astronomical works. In any case, the only table Regiomontanus needed in order to solve all of the standard problems of trigonometry was a sine table. He appended such a table, based on a radius of 60,000, to his text.

The first half of Regiomontanus's text deals with plane triangles, and the second half with spherical ones. Among his results are various methods for solving triangles. Conceptually, there is nothing particularly new in his methods, but unlike earlier authors, he often provided clear and explicit examples of his procedures. For example, theorem I-27 shows how to determine the angles of a right triangle if two sides are known, and theorem I-29 shows how to determine the unknown sides of a right triangle, given one of the two acute angles and one side. In both cases, Regiomontanus used his sine table. His example for the second of these theorems assumes that one acute angle is 36° and that the hypotenuse is 20. Thus, the other angle is 54°, and the two sides would be 35,267 and 48,541 respectively, if the hypotenuse were 60,000. Using proportions, Regiomontanus calculated that because the hypotenuse is 20, these sides are equal to $11\frac{3}{4}$ and $16\frac{11}{60}$, respectively.

In theorem I-49, Regiomontanus solved an arbitrary triangle when two sides and the included angle are known. Supposing AB and BC are known, together with the included angle ABC, Regiomontanus dropped a perpendicular AD to BC or BC extended (Fig. 9.3). In the right triangle ABD, one of the acute angles and a side are known. By theorem I-29, the remaining sides and angle can be calculated. Then, two sides of the right triangle ADC are known, and theorem I-27 and the Pythagorean theorem provide the missing side and angles.

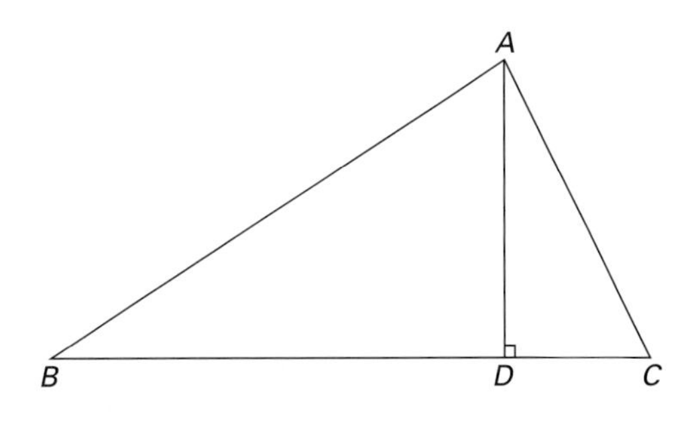

FIGURE 9.3 *On Triangles,* I-49

For the so-called ambiguous case, where two sides and an angle opposite one of them are known, Regiomontanus used a similar procedure. However, he was careful to deal with the various possibilities. He first dealt with the case where the given angle ACB opposite the known side AB is obtuse, AC also being known, by dropping a perpendicular AD to AC extended (Fig. 9.4). The triangle is then solved as in the proof of theorem I-49. In his treatment of the case where the given angle is acute, however, he noted that "there is not enough [information given] to find the [other] side and the remaining angles." For with an acute angle ABC given opposite side AC, two possible triangles can be constructed, one of which has an acute angle opposite AB, the other an obtuse angle. Regiomontanus showed how to find the unknown side and angles in each case, but failed to note the possibility that there might not be any solution, probably because he always assumed that the particular triangle whose unknown parts were being sought did exist.

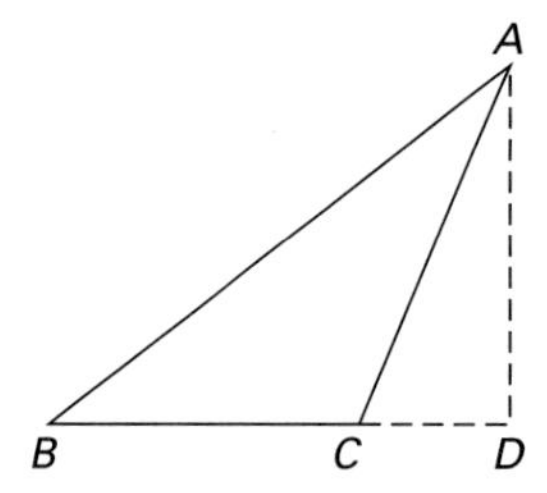

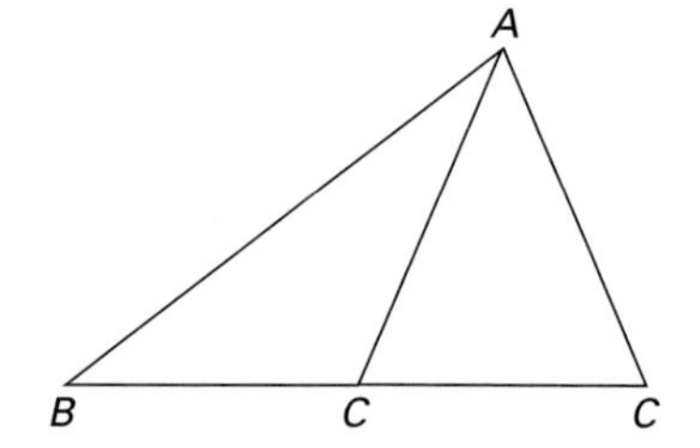

FIGURE 9.4 *On Triangles:* The ambigious case

In theorem II-1 Regiomontanus proved the law of sines: "In every rectilinear triangle the ratio of [one] side to [another] side is as that of the right sine of the angle opposite one of [the sides] to the right sine of the angle opposite the other side." Because Regiomontanus's sines are lines in a circle of given radius, his proof of the theorem for the triangle ABG requires circles with centers B and G having equal radii BD and GA, respectively (Fig. 9.5). Drawing perpendiculars to BG from A and D, intersecting that line at K and H, respectively, Regiomontanus then noted that DH is the sine of $\angle ABG$ and AK is the sine of $\angle AGB$, using circles of the same radius. Since triangle ABK is similar to triangle DBH,

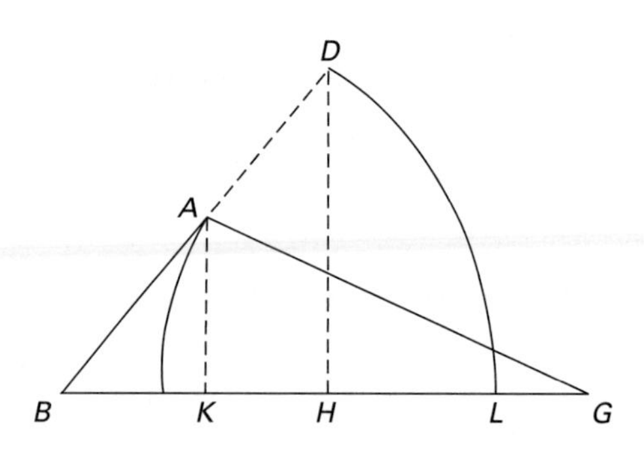

FIGURE 9.5 Proof of the law of sines

and since $BD = AG$, he had

$$\frac{AB}{AG} = \frac{AB}{BD} = \frac{AK}{DH} = \frac{\sin \angle AGB}{\sin \angle ABG},$$

as desired. Regiomontanus now used this result to solve anew the case of a triangle with two sides and the angle opposite one of them known.

In the remainder of Book II of *On Triangles*, Regiomontanus showed how to determine various parts of triangles if certain information is given, such as the ratio of the sides or the length of the perpendicular from a vertex to the opposite side. In two of these theorems, he even used arguments from algebra, because there was no geometric proof available.

Book III of *On Triangles* provides a basic introduction to spherical geometry, including many results on great circles. This discussion is preliminary to the standard material on spherical trigonometry contained in the final two books of the text. Regiomontanus included in those books the rule of four quantities, and then derived from it the law of sines for both right and arbitrary spherical triangles. He followed these with three other important results, the first two involving right triangles ABC, with the right angle at C, and the third about arbitrary spherical triangles. Theorem IV-18 is the result that $\cos B = \sin A \cos b$, evidently first proved by Jābir; theorem IV-19 is $\cos c = \cos a \cos b$, the spherical equivalent of the Pythagorean theorem, essentially known to Ptolemy; and theorem IV-20 shows that $\sin B_2 / \sin B_1 = \cos C / \cos A$, where the perpendicular BD from B to AC divides angle B into two angles B_1 and B_2 (Fig. 9.6).

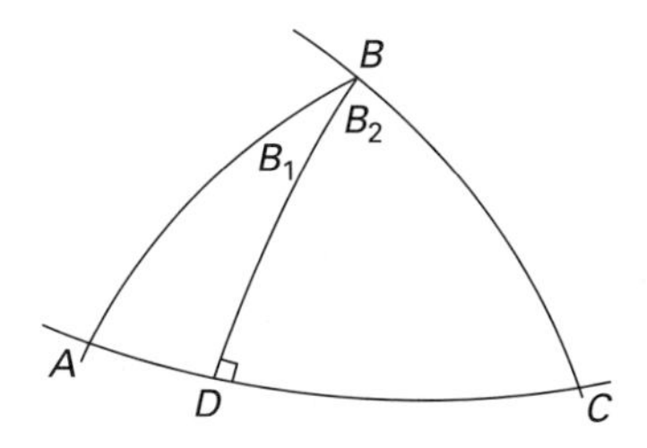

FIGURE 9.6 *On Triangles:* IV-20

Regiomontanus then gave a detailed discussion of how to solve spherical triangles, given various pieces of information, being very careful to distinguish cases where sides were less than or greater than quadrants and where certain angles were acute or obtuse. After showing how to use the law of sines as well as theorems IV-18 and IV-19 to solve right triangles, he proceeded through the various cases of arbitrary triangles. His usual technique is to drop a perpendicular from one vertex to the opposite side (or the opposite side extended) and then use results on right triangles. For example, theorem IV-20 gave him a simpler procedure than that of Naṣir al-Dīn for solving a spherical triangle all of whose angles are given. In that case, the ratio of $\sin B_1$ to $\sin B_2$ is known, as well as the sum of B_1 and B_2. It followed that both B_1 and B_2 can be found. Therefore, all angles in the right triangles ABD and BDC are known, so, by theorem IV-18 and the law of sines, the sides of both of these triangles can be found.

In Book V, Regiomontanus proved the result known to al-Battānī that gave him an alternate method for solving certain triangles:

$$\frac{\operatorname{versin} A}{\operatorname{versin} a - \operatorname{versin}(b - c)} = \frac{R^2}{\sin b \sin c}.$$

This result is equivalent to the spherical law of cosines: $\cos a = \cos b \cos c + \sin b \sin c \cos A$. In Book IV, he had provided two rather complicated methods for solving a spherical triangle with three sides given, one of them being the same as that of Naṣir al-Dīn, but the law of cosines gave him a much simpler method.

Even though Regiomontanus's book was not published until 1533, it was extremely influential in the development of European trigonometry and astronomy. In general, the other authors who wrote trigonometry texts in the last two-thirds of the sixteenth century modeled their texts on Regiomontanus's work, although they did improve his tables and introduce the other trigonometric functions—all, like the sine, defined as lengths of certain lines depending on a given arc in a circle of a fixed radius. George Joachim Rheticus (1514–1574), however, defined the trigonometric functions in his work directly in terms of angles of a right triangle, holding one of the sides fixed at a large numerical value. Rheticus thus called the sine the *perpendiculum* and the cosine the *basis* of the triangle with fixed hypotenuse. Other authors gave other names to the trigonometric functions. The first author to use the modern terms *tangent* and *secant* was Thomas Finck (1561–1656) in his *Geometria rotundi libra XIV* of 1583. He called the three co-functions *sine complement*, *tangent complement*, and *secant complement*.

Many of the trigonometry texts written during these years gave various numerical examples to illustrate methods of solving plane and spherical triangles, but it was not until the work of Bartholomew Pitiscus (1561–1613) in 1595 that such a text used a problem explicitly involving the solving of a real plane triangle on earth. Pitiscus, in fact, invented the term **trigonometry**. He titled his book *Trigonometriae sive, de dimensione triangulis, Liber* (*Book of Trigonometry, or the Measurement of Triangles*).

Pitiscus intended in the text to show how to measure triangles and, in appendix 2 on altimetry, he gave trigonometric methods for determining the height BC of a distant tower (Fig. 9.7). A quadrant is used to measure $\angle AKM = \angle ABC = 60°20'$. The distance AC from the observer to the tower is measured as 200 feet. Pitiscus set up the proportion

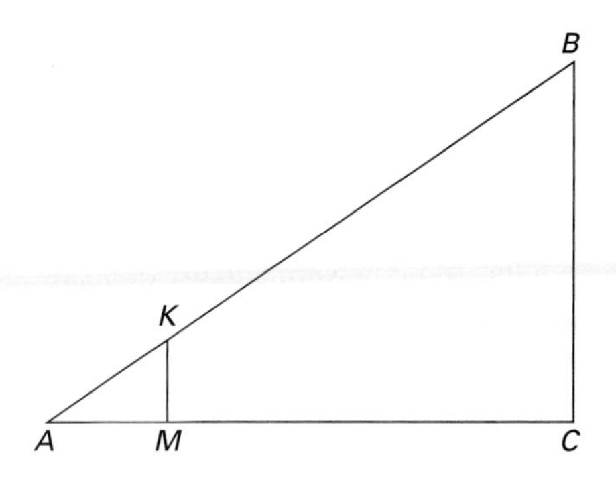

FIGURE 9.7 Measuring the height of a tower

$\sin 60°20' : AC = \sin 29°40' : BC$ and then calculated that $BC = 113\frac{80,204}{86,892}$, or approximately 114 feet. This calculation used Pitiscus's sine table, which was based on a radius of 100,000. He gave a second procedure using his tangent table, in which the required proportion is $AC : 100,000 = BC : \tan 29°40'$. The major difference between Pitiscus's methods and current ones is that he always adjusted for the fact that his trigonometric values are lengths of certain lines in a particular circle. The trigonometric ratios in use today had yet to appear.

9.3 Numerical Calculations

9.3.1 Simon Stevin and Decimal Fractions

Although the Hindu-Arabic decimal place value system was introduced into Europe in the twelfth century, it took many years before Europeans were totally comfortable with it. And even when they were regularly using this system to represent integers, Europeans generally used common fractions (or sexagesimal fractions) when fractions were necessary. One of the reasons that trigonometry tables were based on a large radius (such as 100,000 or 600,000) was that all the values could be expressed in integers, while still giving the astronomers the accuracy they needed. As we have noted, decimal fractions were known in Europe by late in the sixteenth century through Byzantine works, and occasionally appeared in works by Europeans, but the mathematician who strongly urged Europeans to adopt the decimal system in its entirety was Simon Stevin (1548–1620), a Dutch engineer. His well-thought-out notation for decimal fractions, plus his arguments for their adoption, were set forth in *De Thiende* (*The Art of Tenths*), known in its French version as *La Disme*, published in 1585.

In the preface to his book, Stevin made clear its purpose: "It teaches (to speak in a word) the easy performance of all reckonings, computations, and accounts, without broken numbers [common fractions], which can happen in man's business, in such sort as that the four principles of arithmetic, namely addition, subtraction, multiplication, and division, by whole numbers may satisfy these effects." Thus, Stevin promised to show that all operations using his new system can be performed exactly as if one were using whole numbers. That, of course, is the basic advantage of decimal fractions. Stevin began *De Thiende* by defining *thiende* as arithmetic based on geometric progression by tens, using the Hindu-Arabic numerals, and by calling a whole number a *commencement*, with the notation ⓪. Thus, 364 is to be thought of as 364 *commencements* and is written as 364⓪. He then continued: "And each tenth part of the unity of the commencement we call the *prime*, whose sign is ①, and each tenth part of the unity of the prime we call the *second*, whose sign is ② and so of the other; each tenth part of the unity of the precedent sign, always in order one further." To explain this, he gave examples: 3①7②5③9④ means 3 primes, 7 seconds, 5 thirds, and 9 fourths, or $\frac{3}{10}$, $\frac{7}{100}$, $\frac{5}{1000}$, and $\frac{9}{10,000}$, or, altogether $\frac{3759}{10,000}$. Similarly, $8\frac{937}{1000}$ is written as 8⓪9①3②7③. Stevin made the point that no fractions were used in his notation and that except in the case of ⓪, there were only single digits to the left of the signs (circled digits). The numbers written according to these rules Stevin named **decimal numbers**.

Stevin proceeded in the second part of his brief pamphlet to show how the basic operations are performed on decimal numbers. The important idea, naturally, is that operations are performed exactly as on whole numbers, with the proviso that one must take into account the respective signs. Thus, in addition and subtraction, the numbers must be lined up with all the ①s, for example, under one another. For multiplication, Stevin noted that once the multiplication in integers is performed, the sign of the rightmost digit is determined by adding the signs of the rightmost digits of the multiplicands. For division, one similarly subtracts the rightmost sign of the divisor from the rightmost sign of the dividend. Stevin also gave rules for determining the sign when finding square and cube roots. Thus, although his notation differs somewhat from what we use today, Stevin clearly set out the basic rules and rationale for using decimal fractions in calculations. The concluding section of *De Thiende* consisted of pleas to use his new decimal system for calculations in various trades. He suggested using a known basic unit in each case as the commencement and then applying his system for fractions of that unit. His suggestion, however, was not generally carried out until 200 years later, when the French revolutionary government introduced the metric system.

Stevin used his new system to argue that the old Euclidean distinction between (discrete) number and (continuous) magnitude was no longer valid. In particular, the unit itself could be divided "continuously." In fact, in some sense, this is the basis of the idea of a decimal fraction. One can continue the signs as far as one likes to determine any division of unity, however fine. Stevin's argument was certainly not new. Mathematicians over the centuries had gradually abandoned the Greek distinction, especially when they treated quantities of various sorts in the process of doing algebra. And from our current vantage point, because the discrete Euclidean "numbers" have long been incorporated into the continuous number line, it is somewhat difficult to appreciate Stevin's fundamental contribution. But Euclid had always been the center of the study of mathematics. His ideas always had to be confronted. If one wanted to change his notions, one needed to make strong and continued arguments. In fact, although most practicing mathematicians ignored Euclid's distinctions, there were some who were bothered by this generally cavalier attitude toward his work. These mathematicians needed to be convinced that there was no longer any mathematical necessity for the distinction. Naturally, Stevin alone did not do this. It was not until the nineteenth century that the work of embedding "discrete arithmetic" into "continuous magnitude" was completed. Nevertheless, Stevin stood at a watershed of mathematical thinking. Ultimately, he was so successful that it is difficult to understand how things were done before him.

9.3.2 Logarithms

Stevin's suggestions for using and writing decimal fractions were not accepted directly. It was only when it became necessary to prepare extensive tables using fractions that a workable notation became established. The initial appearance of modern decimal notation, in fact, was in the tables of logarithms worked out by John Napier (1550–1617), a Scottish nobleman. Napier noted that accuracy of computation required the use of large numbers such as 10,000,000 as the base for a table of sines, but then he wrote: "In computing tables, these large numbers may again be made still larger by placing a period after the number and adding ciphers .... In numbers distinguished thus by a period in their midst, whatever is written after the period is a fraction, the denominator of which is unity with as many ciphers after it as there are figures after the period." For example, 25.803 is the same as $25\frac{803}{1000}$ and 9,999,998.0005021 means $9{,}999{,}998\frac{5021}{10{,}000{,}000}$. The publication of Napier's tables, in which these decimal fractions appeared, soon led to their general use throughout Europe.

The idea of the logarithm itself probably had its source in the use of certain trigonometric formulas that transformed multiplication into addition or subtraction. Recall that if one needed to solve a triangle using the law of sines, a multiplication and a division were required. When sines were calculated to seven or eight digits (using a circle of radius 10,000,000 or 100,000,000), these calculations were long and errors were often made. Astronomers realized that calculations would be simpler and the number of errors would be reduced if one could replace the multiplications and divisions by additions and subtractions. To accomplish this task, sixteenth-century astronomers often used formulas such as $2 \sin \alpha \sin \beta = \cos(\alpha - \beta) - \cos(\alpha + \beta)$. Thus, if one wanted to multiply 4,378,218 by the sine of $27°15'22''$, one determined α such that $\sin \alpha = 2{,}189{,}109$, set $\beta = 27°15'22''$, and used a table to determine $\cos(\alpha - \beta)$ and $\cos(\alpha + \beta)$. The difference of these two latter values was then the desired product, found without any actual multiplying.

A second, more obvious, source of the idea of a logarithm was probably the work of algebraists such as Rudolff and Stifel, who had displayed tables relating the powers of 2 to the exponents and showed that multiplication in one table corresponded to addition in the other. But because these tables had increasingly large gaps, they could not be used for the necessary calculations. Around the turn of the seventeenth century, however, Napier came up with the idea of producing an extensive table that would allow one to multiply any desired numbers together (not just powers of 2) by performing additions.

Napier's logarithmic tables first appeared in 1614 in a book entitled *Description of the Wonderful Canon of Logarithms*. This work contained only a brief introduction, showing how the tables were to be used. His second work on logarithms, describing the theory behind the construction of the tables, *Construction of the Wonderful Canon of Logarithms*, appeared in 1619, two years after his death. In this latter work appears his imaginative idea of using geometry to construct a table for the improvement of arithmetic.

Realizing that astronomers' calculations involved primarily trigonometric functions, especially sines, Napier aimed to construct a table by which multiplications of these sines could be replaced by addition. For the definition of logarithms, Napier conceived of two number lines, on one of which an increasing arithmetic sequence, $0, b, 2b, 3b, \ldots$, was represented, and on the other a sequence whose distances from the right endpoint form a decreasing geometric sequence, $ar, a^2r, a^3r, \ldots$, where r is the length of the second line (Fig. 9.8). Napier chose r to be 10,000,000, because that was the radius for his table of sines, and a to be a number smaller than but very close to 1. The points on this second line can be marked $0, r - ar, r - a^2r, r - a^3r, \ldots$, with these values representing sines of certain angles.

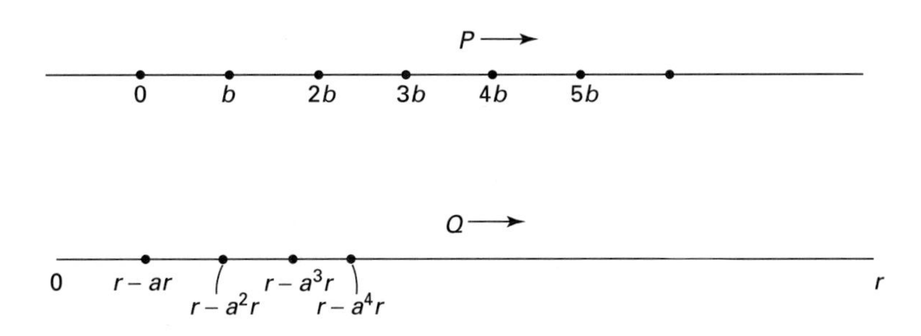

FIGURE 9.8 Napier's moving points

Napier now considered points P and Q moving to the right on each line, as follows: P moves on the upper line "arithmetically" (that is, with constant velocity). Thus, P covers each equal interval $[0, b]$, $[b, 2b]$, $[2b, 3b]$, ... in the same time. Q moves on the lower line "geometrically." Its velocity changes so that it too covers each (decreasing) interval $[0, r - ar]$, $[r - ar, r - a^2r]$, $[r - a^2r, r - a^3r]$, ... in the same time. The distances traveled in each interval form a decreasing geometric sequence, $r(1 - a)$, $ar(1 - a)$, $a^2r(1 - a)$, ..., each member of which is the same multiple of the distance from the left endpoint of the interval to the right end of the line. Because distances covered in equal times have the same ratios as the velocities, it follows that the point's velocity over each interval is proportional to the distance of the beginning of that interval from the right end of the line. It appears that Napier initially thought of the velocity of the lower point as changing abruptly when it passed each marked point, remaining constant in each of the given intervals. In his definition of logarithm, however, Napier smoothed out these changes by considering the second point's velocity as changing continuously (without, naturally, using that terminology). Thus, a point moves geometrically if its velocity is always proportional to its distance from the right end of the line. For Napier, "the logarithm of a given sine is that number which has increased arithmetically with the same velocity throughout as that with which radius began to decrease geometrically, and in the same time as radius has decreased to the given [number]." In other words, if the upper point P began to move from 0 with constant velocity equal to that with which the lower point Q also began to move (geometrically) from 0, and if P had reached y when Q had reached a point whose distance from the right endpoint (radius) is x, then Napier called y the **logarithm** of x.

In modern calculus notation, Napier's idea is reflected in the differential equations

$$\frac{dx}{dt} = -x, \qquad x(0) = r; \qquad \frac{dy}{dt} = r, \qquad y(0) = 0.$$

The solution to the first equation is $\ln x = -t + \ln r$, or $t = \ln(r/x)$. Combining this with the solution $y = rt$ of the second equation shows that Napier's logarithm y (here written as $y = \text{Nlog}\, x$) may be expressed in terms of the modern natural logarithm as $y = \text{Nlog}\, x = r \ln(r/x)$. Napier's logarithm is thus closely related to the natural logarithm. It does not, however, share the common properties of the natural logarithm, since, for example, its value decreases when the value of x increases.

Although Napier's definition is somewhat different from the modern one, he nevertheless was able to derive important properties of logarithms analogous to those of modern logarithms as well as to construct a table of logarithms of sines. He began by noting that the definition implied immediately that $\text{Nlog}\, r = 0$, for the upper point will not have moved at all. Also, if $\alpha/\beta = \gamma/\delta$, then $\text{Nlog}\, \alpha - \text{Nlog}\, \beta = \text{Nlog}\, \gamma - \text{Nlog}\, \delta$. This result follows from the definition, because the geometrical motion of the lower point implies that its time to travel from α to β equals its time to travel from γ to δ. From this result follow rules enabling one to use logarithms in calculation. For example, if $x : y = y : z$, then $2\text{Nlog}\, y = \text{Nlog}\, x + \text{Nlog}\, z$, and if $x : y = z : w$, then $\text{Nlog}\, x + \text{Nlog}\, w = \text{Nlog}\, y + \text{Nlog}\, z$. This law could immediately be applied to solve triangles.

Thus, consider the use of the law of sines

$$\frac{\sin\alpha}{a} = \frac{\sin\beta}{b}$$

to solve a plane triangle given two sides a, b and the angle α opposite side a. Napier applied his logarithm property for proportions to get $\text{Nlog}\, \sin\beta = \text{Nlog}\, \sin\alpha + \text{Nlog}\, a - \text{Nlog}\, b$.

Because he had found the logarithm of the sine of β, he could read his table in reverse to determine β itself. In this case, of course, he noted that there were two possible values for β, one less than a right angle and one greater. On the other hand, the standard method of solving a triangle given two sides a, b and the included angle γ required the dropping of a perpendicular, and was therefore not suited to logarithmic calculation. So Napier made use of the **law of tangents**:

$$\frac{a+b}{a-b} = \frac{\tan\frac{1}{2}(\alpha+\beta)}{\tan\frac{1}{2}(\alpha-\beta)}.$$

If γ is given, then $\alpha + \beta$ is known. Applying logarithms to this proportion allowed him to find $\tan\frac{1}{2}(\alpha-\beta)$, therefore $\frac{1}{2}(\alpha-\beta)$, and therefore both α and β.

How did Napier calculate the logarithm of a tangent from his table of logarithms of sines? To answer this question, consider the following line of Napier's actual table, which included seven columns for each minute of arc from 0° to 45°.

34°40′	5688011	5642242	3687872	1954370	8224751	55°20′

The first column gives the value of an arc (or angle), and the second gives the sine of that arc. The final column gives the arc that is complementary to the one in the first column, while the sixth column gives its sine. It follows that the sixth column gives the sine of the complement of the arc of the first column, that is, the cosine of that arc. The third and fifth columns give Napier's logarithms of the sines in the second and sixth columns, respectively, or, as Napier also notes, the logarithms of the sine complements of the sixth and second columns, respectively. Finally, the middle column represents the difference of the entries in the third and fifth columns. Since $\sin\alpha : \cos\alpha = \tan\alpha : r$ and $\text{Nlog}\, r = 0$, this difference is Napier's logarithm of the tangent of the arc of the first column. Because the logarithm of 10,000,000 is 0, logarithms of numbers greater than 10,000,000 must be negative and are defined by simply reversing the directions of the moving points in the original definition. These numbers, of course, cannot represent sines, but can represent tangents or secants. In this case, the negative of the logarithm in the middle column is the logarithm of the tangent of 55°20′, while the negative of the logarithm in the third column is the logarithm of the secant of that same angle.

Napier's actual construction of his table of logarithms took him 20 years. And even though this work was done in the era of hand calculations, there were remarkably few errors. Late in his life, however, Napier decided that it would be more convenient to have logarithms whose value was 0 at 1 rather than at 10,000,000. In that case the familiar properties of logarithms, $\log xy = \log x + \log y$ and $\log(x/y) = \log x - \log y$, would hold. Furthermore, if the logarithm of 10 were set at 1, the logarithm of $a \times 10^n$, where $1 \le a < 10$, would simply be n added to the logarithm of a. Napier died before he could construct a new table based on these principles, but Henry Briggs (1561–1631), who discussed this matter thoroughly with Napier in 1615, began the calculation of such a table. Rather than convert Napier's logarithms to these new "common" logarithms by simple arithmetic procedures, however, Briggs worked out the table from scratch. Starting with $\log 10 = 1$, he calculated successively $\sqrt{10}$, $\sqrt{\sqrt{10}}$, $\sqrt{\sqrt{\sqrt{10}}}$, ..., until after 54 such root extractions he reached a number very close to 1. All of these calculations were carried out to 30 decimal places. Since $\log\sqrt{10} = 0.5000$, $\log\sqrt{\sqrt{10}} = 0.2500, \ldots, \log(10^{1/2^{54}}) = 1/2^{54}$, he was able to build up a table of logarithms of closely spaced numbers using the laws of logarithms. Briggs's table, completed by Adrian Vlacq in 1628, became the basis for nearly all logarithm tables into the twentieth century. Astronomers very quickly discovered the great advantages of using logarithms for calculations. Logarithms became so important that the eighteenth-century French mathematician Pierre-Simon de Laplace was able to assert that the invention of logarithms, "by shortening the labors, doubled the life of the astronomer."

9.4 Astronomy and Physics

9.4.1 Copernicus and the Heliocentric Universe

Among the astronomers who found logarithms so useful was Johannes Kepler, who built on the work of Nicolaus Copernicus (1473–1543). It was Copernicus, who, having studied Ptolemy's system in great detail and having become aware of all its inaccuracies, came to the conclusion that it was impossible to continue to patch up the earth-centered approach. These inaccuracies had been building since Ptolemy's time. Although astronomers over the centuries had kept trying to fix Ptolemy's models by recalculating various parameters, by the fifteenth century, there was a growing consensus that new ideas were needed. As Copernicus wrote, "[The astronomers] have not been able to discover or deduce from [their hypotheses] the chief thing, that is the form of the universe, and the clear symmetry of its parts. They are just like someone including in a picture hands, feet, head, and other limbs from different places, well painted indeed, but not modeled from the same body, and not in the least matching each other, so that a monster would be produced from them rather than a man." To redo the painting and eliminate the monster, Copernicus decided to read all the opinions of the ancients to determine whether anyone had proposed a system of the universe different from the earth-centered one. Having discovered that some Greek philosophers had proposed a sun-centered (**heliocentric**) system in which the earth moves, Copernicus explored the consequences of reforming the system under that assumption, eventually coming to the conclusion that this arrangement gave better observational results than that of Ptolemy.

Copernicus's fundamental treatise in which he expounded his system of the universe was *De revolutionibus orbium coelestium* (*On the Revolutions of the Heavenly Spheres*), a book that represented the work of a lifetime but was only published in 1543, the year of his death. This book sets forth the first mathematical description of the motions of the heavens based on the assumption that the earth moves. *De revolutionibus*, interestingly enough, follows very closely the model of Ptolemy's *Almagest*. It is a very technical work in which the author used detailed mathematical calculation—based on the assumption that the sun is at the center of the universe and buttressed by the results of observations taken by Copernicus and his predecessors—to describe the orbits of the moon and the planets and to show how these orbits are reflected in the positions observed in the skies. Copernicus sketched his theory very briefly in the first book of *De revolutionibus* and presented a simplified diagram of the sun in the center of seven concentric spheres, one each for the six planets, including the earth, and one for the fixed sphere of the stars (Fig. 9.9). That is, like his predecessors, Copernicus conceived of the system of the universe as a series of nested spheres containing the planets, rather than as empty space through which the planets travel in circles.

Copernicus himself did not—and could not—present any real evidence for either the earth's daily rotation on its axis or its yearly revolution about the sun. For the first motion,

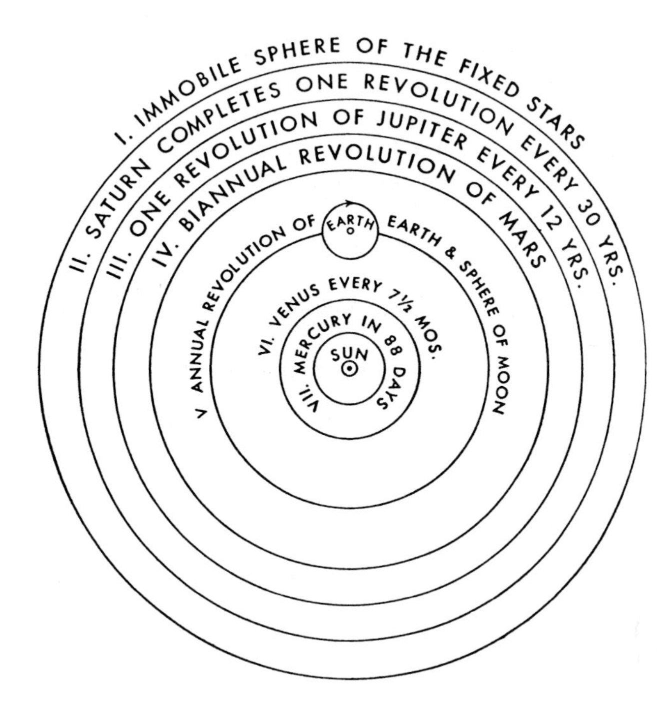

FIGURE 9.9 Copernicus's system of the universe

he simply argued that it was more reasonable to assume that the relatively small earth rotates rather than the immense sphere of the stars. For the second motion, his argument was in essence that the qualitative behavior of the planets could more easily be understood by attributing part of their motion to the earth's own yearly revolution. Thus, retrogression, the occasional east to west motion of the planets, could be explained in terms of the combined orbital motions of the earth and the planets rather than by an epicycle (Fig. 9.10). The observed variation in the planets' distances from the earth also is more easily understood in

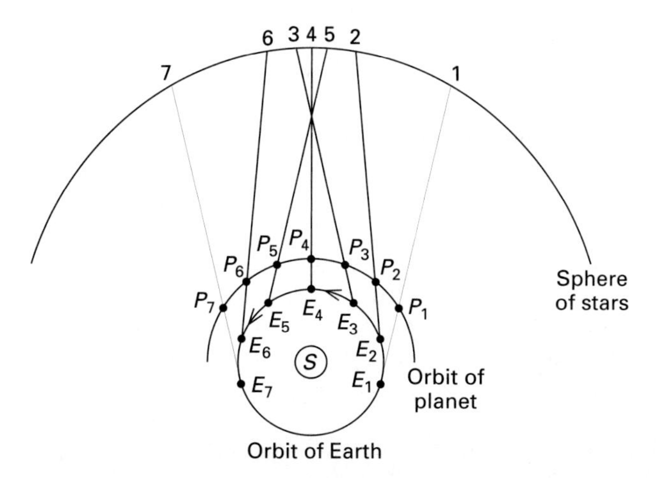

FIGURE 9.10 Retrograde motion for a planet outside the orbit of the earth. The observed positions of the planet against the sphere of the stars are marked, in order, 1, 2, 3, 4, 5, 6, 7. Retrogression takes place between 3 and 5.

terms of the two orbits. Copernicus answered the objection to the earth's motion around the sun, that it would cause the fixed stars to appear different at different times of the year (the so-called annual parallax), by assuming that the radius of the earth's orbit is so much smaller than the radius of the sphere of the stars that no such parallax could be observed. Thus, one of the effects of Copernicus's theory was vastly to increase the size of the perceived universe.

After his basic introduction to the new system, Copernicus followed his mentor Ptolemy by presenting an outline of the plane and spherical trigonometry necessary to solve the mathematical problems presented by the movements of the celestial bodies. Despite the advances in trigonometric technique available in Europe following the work of Regiomontanus, Copernicus's own treatment stayed very close to that of the second-century astronomer, even to the use of chords. He did, however, make some concessions to the 1400 years of work since the time of Ptolemy. First, he used 100,000 for his circle radius (now that Arabic numerals were in general use) rather than 60 as used by the ancients. Second, his table did not give the chords of the various arcs but, instead, half the chords of twice the arcs, although he did not use the term *sine*. Third, unlike Ptolemy, Copernicus did present specific methods for solving both plane and spherical triangles, rather than developing them *ad hoc*.

In the remaining books of his treatise, Copernicus used his new sun-centered model, along with both ancient and modern observations, to calculate the basic parameters of the orbits of the moon and the planets. Reading these later books reveals that moving the center of the universe away from the earth did not simplify Ptolemy's picture very much. Copernicus found that simply placing the planets on sun-centered spheres did not satisfy the requirements of observation. Thus, he, like Ptolemy, had to introduce various complications, including a return to epicycles. In the end, the full system as described in *De revolutionibus* was just as complex as Ptolemy's.

The mathematical details of Copernicus's work made it unreadable to all but the best astronomers of his day, its primary audience. Over the next several decades, these mathematicians found that calculations of astronomical phenomena were simplified by applying Copernicus's theory and techniques. It was unnecessary to believe in the movement of the earth to use these techniques. Therefore, many people, both astronomers and educated laymen, took Copernicus's work merely as a mathematical hypothesis and not as a physical theory.

During the latter half of the sixteenth century, however, various churchmen, particularly Protestant clerics deeply involved in the fierce conflict with the Roman Catholic Church, began to express severe opposition to Copernicus's ideas, because they explicitly contradicted certain Biblical passages asserting the earth's stability. These Protestant leaders believed that the Catholic Church had departed greatly from the views expressed in the Bible. They vehemently rejected any doctrines seen as deviating from the literal words of Scripture. During this same period, the Catholic Church itself had little to say about Copernicus's work. *De revolutionibus*, in fact, was taught at some Catholic universities, and the astronomical tables derived from it provided the basis for the reform of the calendar promulgated for the Catholic world by Pope Gregory XIII in 1582. Ironically, it was not until the seventeenth century, after most astronomers were convinced of the earth's movement by new evidence and a better heliocentric theory than that of Copernicus, that the Catholic Church brought its full power to bear against the heresy represented by the idea of a moving earth.

9.4.2 Johannes Kepler and Elliptical Orbits

The better heliocentric system was developed by Johannes Kepler, based on his analysis of the detailed and accurate observations of Tycho Brahe (1546–1601). It was perhaps his theological training, combined with a philosophical bent, that provided Kepler with the goal from which he never wavered—of discovering the mathematical rules God used for creating the universe. Throughout his life, Kepler attempted through both philosophical analysis and prodigious calculation to demonstrate the numerical relationships with which God had created the universe. His goal appeared to be nothing less than to reconfirm on a higher level the Pythagorean doctrine that the universe is made up of number. Taking as his starting point Copernicus's placing of the sun at the center of the universe, he was able to discover the three laws of planetary motion, today known as **Kepler's laws**, as well as many other relationships that today tend to be dismissed as mystical.

Kepler discussed one of these relationships in great detail in the *Mysterium cosmographicum* (*The Secret of the Universe*) of 1596: Why are there precisely six planets? Kepler's answer was that because "God is always a geometer," the Supreme Mathematician wanted to use the regular solids to separate the planets. Euclid had proved that there could be only five such solids, so Kepler took this as the reason that God chose to provide just six planets. He then worked out the idea that between each pair of spheres containing the orbits of adjacent planets there was inscribed one of the regular solids (Fig. 9.11). Thus, inside

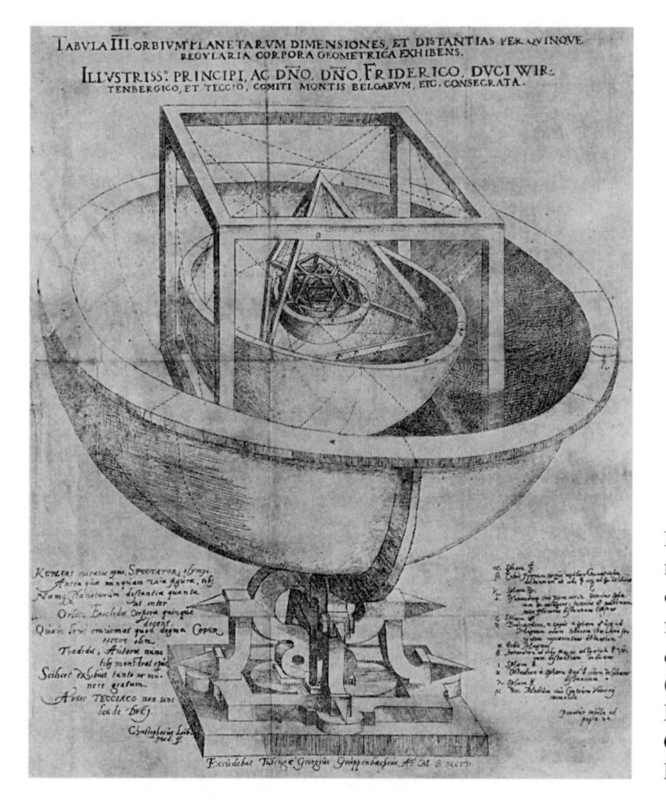

FIGURE 9.11 Kepler's regular solids representing the orbits of the planets, on a plate from the *Mysterium cosmographicum* of 1621. (Source: Courtesy of the Department of Special Collections, Stanford University Libraries)

the sphere of Saturn was to be inscribed a cube, which in turn circumscribed the orbit of Jupiter. Similarly, between the orbits of Jupiter and Mars was a tetrahedron, between Mars and Earth a dodecahedron, between Earth and Venus an icosahedron, and between Venus and Mercury an octahedron. These solids lay in the interspherical spaces, and their sizes provided a measure of the relationship between the sizes of the various planetary orbits. For example, Kepler noted that the diameter of Jupiter's orbit is triple that of Mars, while the ratio of the diameter of the sphere circumscribed about the tetrahedron is triple that of the sphere inscribed in the tetrahedron. Not all of the values came out exactly correct. There was still some discrepancy. But even this fact did not bother Kepler too much. He gave various reasons why the values could not be expected to be exact, including even the fact that the data from his tables were not entirely accurate. He was so convinced of the correctness of his basic idea that such discrepancies were of little moment. Kepler's views on this matter were not solely due to his youth (he was 25 when he wrote *Mysterium cosmographicum*). In fact, he returned time and again to this basic proposition, each time attempting to adduce new reasons for its correctness.

Kepler announced in his *Mysterium cosmographicum* that among his goals was to discover the "motion of the circles" of the planets, that is, to determine their orbits, for he had realized that the planets could not reside on spheres. He gave numerous arguments in that work for the basic correctness of the heliocentric system, but by the end of the century he realized that Copernicus's mathematical details did not give the complete solution to the problem. For example, Copernicus still treated the earth as special rather than as just another planet. To correct Copernicus's work, Kepler knew that he needed better observational data, data which only Tycho Brahe had collected. With these finally in hand by 1601, Kepler could proceed to determine the exact details of the planetary orbits. He began with Mars, because that planet's orbit had always been the most difficult to comprehend. If he could understand the orbit of Mars, Kepler believed, he could understand them all.

In his *Astronomia nova* (*New Astronomy*) of 1609, Kepler described his 8 years of detailed calculations, false starts, stupid mistakes, and continued perseverance to calculate the orbit of Mars. Unlike Ptolemy or Copernicus, however, Kepler was not only interested in the pure mathematics of the celestial motions (that is, in "saving the appearances") but in the physics as well. He was trying to describe the actual orbit of the earth through space and so wanted to know what caused the earth to move, what kept it in its orbit, and why the velocity changed with the distance to the sun. Having read the work of William Gilbert, *On Magnets* (1601), Kepler settled on the fact that some force emanating from the sun acts on the planet and sweeps it around in its orbit. He could understand this force acting on a planet as it moved around the sun much better than he could accept its acting on a planet moving on an epicycle. It also made sense that, like magnetic force, the sun's force weakened with distance, so that the planet's velocity was smaller at a greater distance.

After an immense amount of calculation and comparisons of various possible orbits with Brahe's data, Kepler eventually came to the realization that the orbit of Mars could not be a circle. It had to be an oval of some sort. It was somewhat strange to reject the comforting circularity of the Greeks and replace it with a rather vaguely shaped oval, because Kepler was expecting that God would insist on some regular figure. Nevertheless, Kepler began the long process of calculating the exact shape of the oval. After 2 years of calculation, the result came to Kepler virtually by accident: The oval must be an ellipse. Of course, since he had studied Apollonius's work extensively, he had enough understanding of the curve to work out all the details. Thus, he asserted what is now referred to as **Kepler's first law**: Every planet orbits the sun in an ellipse with the sun at one focus. As part of his study of the orbit, he also determined that the radius vector from the sun to a planet always swept out equal areas in equal times, giving a nice quantitative touch to the qualitative observation that the velocity of a planet was greater the closer it was to the sun (Fig. 9.12). This area rule is the content of **Kepler's second law**.

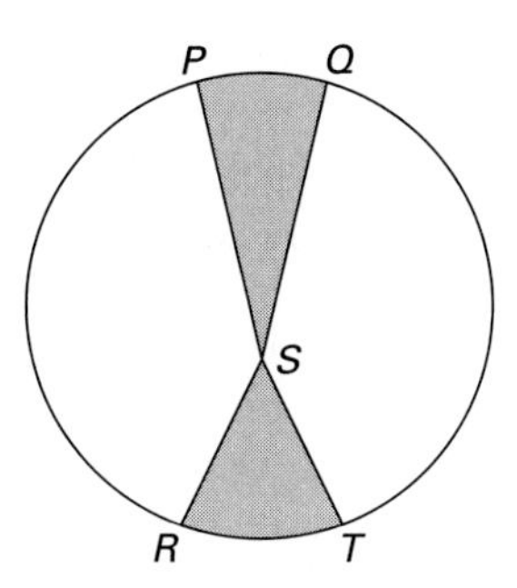

FIGURE 9.12 Kepler's second law: A planet sweeps out equal areas in equal times. The time of the planet's motion from Q to P is equal to that from R to T when area SPQ is equal to area SRT.

Kepler's third law appeared for the first time in the *Harmonice mundi* (*Harmonies of the World*) of 1618, stated as an empirical fact: "It is absolutely certain and exact that the ratio which exists between the periodic times of any two planets is precisely the ratio of the $\frac{3}{2}$th power of the mean distances [of the planet to the sun]." Kepler discovered the law by studying more of Brahe's measurements, but never gave any derivation of it from other principles.

Kepler's three laws of planetary motion had great consequences in the development of both astronomical and physical theory. Their discovery provides an excellent example of the procedures used by scientists. Scientists need some theory to begin with, but they must always compare the results of the theory with observations. If they have confidence in their observations, and these do not agree with the predictions of their theory, they must modify the theory. Kepler did this repeatedly until he finally reached theoretical results that agreed with his observations.

9.4.3 Galileo and Kinematics

Kepler believed that any mathematical theory of the universe had to be based on some physical principles. But he only vaguely understood these principles. The man generally considered to be the founder of modern physics, Galileo Galilei (1564–1642), understood these principles better, partly because he concentrated on applying mathematics to physical phenomena on earth. He did, however, believe, with Kepler, that the sun was the center of the universe. In fact, Galileo was eventually brought before the inquisition in Rome and convicted of teaching that the earth moved around the sun in his 1632 work *Dialogue Concerning the Two Chief World Systems*. But although he was sentenced to house arrest for the remainder of his life and forbidden to publish any more books, he was able to smuggle the manuscript of his most important work, the *Discourses and Mathematical Demonstrations Concerning Two New Sciences* of 1638, beyond the reach of the inquisition to Leiden in the Netherlands, where it was published by the Elseviers.

One of Galileo's two new sciences was the science of motion. To explicate this science, the *Two New Sciences* is written in the form of a dialogue among three people, whose discussion is carried out around the framework of a formal treatise on motion written in the Euclidean format, including a definition, a postulate, and many theorems and proofs. The definition reads, "Motion is equably or uniformly accelerated which, abandoning rest, adds on to itself equal momenta of swiftness in equal times." This definition is essentially the same as that put forth by Heytesbury 300 years earlier. Galileo, however, made two major advances on the work of his medieval predecessors. First, he had discovered by 1604 that uniformly accelerated motion is precisely that of a freely falling body; second, he worked out numerous mathematical consequences of this fact, some of which he could confirm by experiment.

At one time, Galileo believed that the velocity of a falling body increased in the ratio of the distances fallen rather than in the ratio of the times elapsed. In the *Two New Sciences*, he gave an argument showing that this first possibility is erroneous. First, he noted that if two different velocities of a given body are proportional to the distances covered while the body has each velocity in turn, then the times for the body to cover those distances are equal. This statement is virtually obvious for velocities constant over the given period of time. Galileo then assumed it to be true also for continuously changing velocities. Thus, "if the speeds with which the falling body passed the space of four *braccia* [an Italian measure of distance] were the doubles of the speeds with which it passed the first two *braccia*, as one space is double the other space, then the times of those passages are equal." Galileo was here comparing two (infinite) sets of velocities, those at each instant in which the falling body passed a point in the first two *braccia* with those at each instant it passed a point in the first four *braccia* twice as far from the point of origin. His statement that the total times are equal is the result of applying the argument for finite times to infinitesimal times and adding up the entire set of these infinitesimal times. Galileo concluded that it is ridiculous that a given falling body starting from rest could cover both two *braccia* and four *braccia* in the same time, and thus that it is false that speed increases as the distance traveled.

Galileo's argument by comparing two infinite sets of infinitesimals was one of the first such arguments in mathematical history, but one that he used in other contexts as well. In particular, he used it in his proof of the mean speed rule.

THEOREM. *The time in which a certain space is traversed by a moveable in uniformly accelerated movement from rest is equal to the time in which the same space would be traversed by the same moveable carried in uniform motion whose degree of speed is one-half the maximum and final degree of speed of the previous, uniformly accelerated, motion.*

With AB representing the time of travel, EB representing the maximum speed attained by the moveable, and F the midpoint of BE, Galileo constructed right triangle ABE and rectangle $ABFG$, whose areas are equal (Fig. 9.13). There are then one-to-one correspondences between the instants of time represented by points of the line AB and the parallels in the triangle representing the increasing degrees of speed, on the one hand, and those instants and the parallels in the rectangle representing the equal speeds at half the final speed, on the other. Galileo concluded that "there are just as many momenta of speed consumed in the accelerated motion according to the increasing parallels of triangle ABE as in the equable motion according to the parallels of the [rectangle $ABFG$]," because the deficit above the halfway point is made up by the surplus below it. Since these "momenta" of speed for each instant of time are proportional to the distances traveled in those instants, it follows that the total distance in each case is the same. As before, Galileo used an argument with infinitesimals. Although one may wonder why he believed in such arguments, given that they violated classic geometric concepts, use them he did.

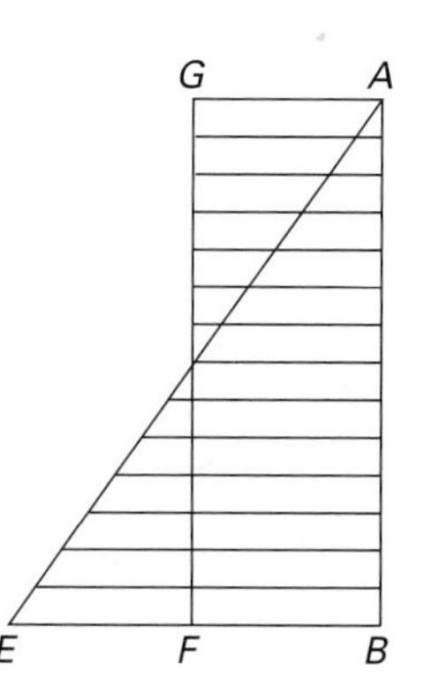

FIGURE 9.13 Galileo's proof of the mean-speed theorem

As a corollary to the theorem, Galileo proved that in the case of a moveable falling from rest, the distances traveled in any times are as the squares of those times. That is, he showed that if the body falls a distance d_1 in time t_1, and d_2 in time t_2, then $d_1 : d_2 = t_1^2 : t_2^2$. Galileo's result is written in modern notation as $d = kt^2$, but he always used Euclidean proportionality concepts instead of modern function concepts. To prove this corollary, Galileo first noted that for two bodies both traveling at constant but unequal velocities, the distances traveled are in the ratio compounded of the ratios of the speeds and the times, or $d_1 : d_2 = (v_1 : v_2)(t_1 : t_2)$. This result is derived from the facts that for equal times, distances are proportional to velocities, and for equal velocities, distances are proportional to times. By the theorem, the distances traveled by the falling body in the two times are the same as if in each case the body had a constant velocity equal to half its final velocity. These halves of the final velocities are also proportional to the times. It follows that in the compound ratio, the ratio of the velocities can be replaced by the ratio of the times, and the corollary is proved.

Galileo stated and proved some 38 propositions on naturally accelerated motion. He was interested in comparing velocities, times, and distances for motion along inclined planes as well as for the motion of free fall. Thus, he presented a postulate to the effect that the velocity acquired by an object sliding down an inclined plane (without friction) depends only on the height of the plane and not on the angle of inclination. Using this postulate, he deduced results such as that the times of descent for a given object along two different inclined planes of the same height are to one another as the lengths of the planes, and, conversely, the times of descent over planes of equal lengths are to one another inversely as the square roots of the heights of the planes. Galileo also made progress toward solving the **brachistochrone problem**, that is, discovering the path by which an object moves in shortest time from one point to another point at a lower level. He showed that for a given vertical circle, the time taken for a body to descend along a chord, say DC, from any point on the circle to the bottom of the circle is greater than its time to descend along the two chords DB, BC, the first beginning at the same point as the original chord, the second ending at the same bottom point (Fig. 9.14). (Here DC must subtend an arc no greater than 90°.) By extending this result to more and more chords, he concluded, erroneously, that the path of swiftest descent is a circular arc. It was not until the end of the century that several mathematicians deduced that this curve was in fact a cycloid.

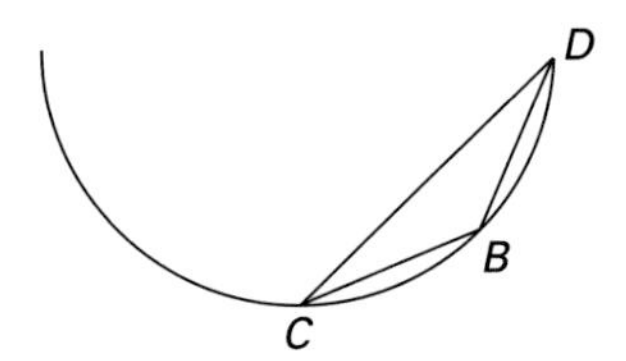

FIGURE 9.14 The brachistochrone problem

In the final part of the *Two New Sciences*, Galileo discussed the motion of projectiles. These motions are compounded from two movements: the horizontal one being of constant velocity, and the vertical one being naturally accelerated. In fact, he even stated part of the fundamental law of inertia, that a body moving on a frictionless horizontal plane at constant velocity will not change its motion, because "there is no cause of acceleration or retardation." He then proved

THEOREM. *When a projectile is carried in motion compounded from equable horizontal and from naturally accelerated downward* [*motions*], *it describes a semiparabolic line in its movement.*

Galileo formulated this theorem in 1608 in connection with an experiment involving rolling balls off tables, which convinced him that the horizontal motion was unaffected by the downward motion due to gravity. His proof in the *Two New Sciences* used this assumption. Galileo drew a careful graph of the path of the object, noting that the horizontal distances traveled in equal times are equal, while the vertical distances traveled in those same times increase in proportion to the squares of the times. Therefore, the curve has the property that for any two points on it, say F, H, the ratio of the squares of the horizontal distances, $FG^2 : HL^2$, is the same as that of the vertical distances (to the plane), $BG : BL$ (Fig. 9.15).

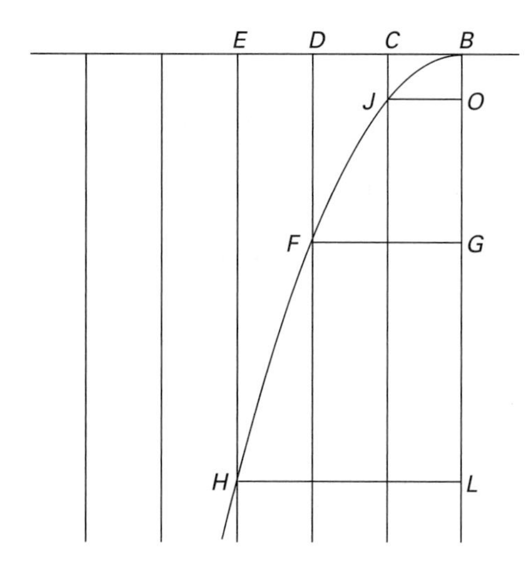

FIGURE 9.15 The parabolic motion of a projectile

Galileo then concluded from his familiarity with the work of Apollonius that the curve was a parabola.

Galileo continued this discussion of projectile motion by proving that for objects fired at an angle to the horizontal, as from a cannon, the path is also parabolic. In fact, he calculated several tables giving the height and distance traveled by such a projectile as functions of the initial angle of elevation, showing, for example, that the maximum range is achieved by an initial angle of 45°.

In developing his theories, Galileo displayed a solid understanding of the idea of a mathematical model. As he wrote, "No firm science can be given of such events of heaviness, speed, and shape which are variable in infinitely many ways. Hence to deal with such matters scientifically, it is necessary to abstract from them. We must find and demonstrate conclusions abstracted from the impediments, in order to make use of them in practice under those limitations that experience will teach us. . . . Indeed, in projectiles that we find practicable, . . . the deviations from exact parabolic paths will be quite insensible." Galileo thus stated his firm belief in the application of mathematics to physics. One must always form a mathematical model by considering only the most important ideas in a given situation. Only after deriving mathematically the consequences of one's model and comparing these to the results of experiment can one decide whether adjustments to the model are necessary. Galileo, like Kepler, followed these basic precepts of mathematical modeling of physical phenomena. Kepler, because he was dealing with astronomical phenomena, could only compare his theoretical results to observations. Galileo, on the other hand, conducted experiments to verify (or refute) the results of his reasoning. The detailed explication of this process of mathematical modeling, even more than his actual physical theorems, formed Galileo's most fundamental contribution to the mutual development of mathematics and physics. His ideas came into full flower as the scientific revolution of the seventeenth century reached its climax in the work of Isaac Newton.

Exercises

1. This problem is from the *Treviso Arithmetic*, the first printed arithmetic text, dated 1478: The Holy Father sent a courier from Rome to Venice, commanding him that he should reach Venice in 7 days. And the most illustrious Signoria of Venice also sent another courier to Rome, who should reach Rome in 9 days. And from Rome to Venice is 250 miles. It happened that by order of these lords the couriers started their journeys at the same time. It is required to find in how many days they will meet, and how many miles each will have traveled.

2. This problem and those in exercises 3 and 4 are from the work of Piero della Francesca: Three men enter into a partnership. The first puts in 58 ducats, the second 87; we do not know how much the third puts in. Their profit is 368, of which the first gets 86. What shares of profit do the second and third receive and how much did the third invest?

3. Of three workmen, the second and third can complete a job in 10 days. The first and third can do it in 12 days, while the first and second can do it in 15 days. In how many days can each of them do the job alone?

4. A fountain has two basins, one above and one below, each of which has three outlets. The first outlet of the top basin fills the lower basin in two hours, the second in three hours, and the third in four hours. When all three upper outlets are shut, the first outlet of the lower basin empties it in three hours, the second in four hours, and the third in five hours. If all the outlets are opened, how long will it take for the lower basin to fill?

5. Maestro Dardi gives a rule to solve the fourth-degree equation $x^4 + bx^3 + cx^2 + dx = e$ as $x = \sqrt[4]{(d/b)^2 + e} - \sqrt{d/b}$. His problem illustrating the rule is the following: A man lent 100 *lire* to another and after 4 years received back 160 *lire* for principal and (annually compounded) interest. What is the interest rate? As in the example in section 9.1.1, set x as the monthly interest rate in *denarii* per *lira*. Show that this problem leads to the equation $x^4 + 80x^3 + 2400x^2 + 32{,}000x = 96{,}000$ and that the solution found by "completing the fourth power" is given by the stated formula.

6. Piero della Francesa presented this problem: Divide 10 into two parts such that if their product is divided by their difference, the result is $\sqrt{18}$. To solve this, he used the rule for solving the fourth-degree equation $ax + bx^2 + cx^4 = d + ex^3$, which is $x = \sqrt[4]{(b/4c)^2 + (d/c)} + e/4c - \sqrt{a/2e}$. Show that this formula works in this case, but not in general. How did della Francesca derive the formula?

7. The equation $6x^3 = 43x^2 + 79x + 30$ is solved in Luca Pacioli's *Summa* as follows: "Add the number to the *cose* to form a number, and then you get one *cubo* equal to $7\frac{1}{6}$ *censi* plus $18\frac{1}{6}$, after you have reduced to one *cubo* [divided all the terms by 6]. Then divide the *censi* in half and multiply this half by itself, and add it onto the number. It will be $31\frac{1}{144}$ and the *cosa* is equal to the root of this plus $3\frac{7}{12}$, which is half of *censi*." Show that Pacioli's answer is incorrect. What was he thinking of in presenting his rule?

8. This problem and the one in exercise 9 are from Rudolff's *Coss*: I am owed 3240 *florins*. The debtor pays me 1 *florin* the first day, 2 the second day, 3 the third day, and so on. How many days does it take to pay off the debt?

9. Divide 10 into two parts such that their product is $13 + \sqrt{128}$.

10. This problem is from Stifel's *Arithmetica*: In the sequence of odd numbers, the first odd number equals 1^5. After skipping one number, the sum of the next four numbers $(5 + 7 + 9 + 11)$ equals 2^5. After skipping the next three numbers, the sum of the following nine numbers $(19 + 21 + 23 + 25 + 27 + 29 + 31 + 33 + 35)$ equals 3^5. At each successive stage, one skips the next triangular number of odd integers. Formulate this power rule of fifth powers in modern notation, and prove it.

11. The basis of Stifel's procedure for finding higher-order roots was the appropriate binomial expansion, or, more specifically, the entries in the appropriate row of the Pascal triangle. For example, to find the fourth root of 1,336,336, one first notes that the answer must be a two-digit number beginning with 3. One then subtracts $30^4 = 810{,}000$ from the original number to get remainder 526,336. Recalling that the entries in the fourth row of the triangle are 1, 4, 6, 4, 1, and guessing that the next digit is 4, one checks this by successively subtracting from that remainder $4 \times 30^3 \times 4 = 432{,}000$, $6 \times 30^2 \times 4^2 = 86{,}400$, $4 \times 30 \times 4^3 = 7680$, and $4^4 = 256$. In this

case, the result is 0, so the desired root is 34. Use this procedure to calculate the fourth root of 10,556,001.

12. This problem and the one in exercise 13 are from Recorde's *Whetstone of Witte*: There is a strange journey appointed to a man. The first day he must go $1\frac{1}{2}$ miles, and every day after the first he must increase his journey by $\frac{1}{6}$ of a mile, so that his journey shall proceed by an arithmetical progression. And he has to travel for his whole journey 2955 miles. In what number of days will he end his journey?

13. There is a certain army composed of dukes, earls, and soldiers. Each duke has under him twice as many earls as there are dukes. Each earl has under him four times as many soldiers as there are dukes. The two hundredth part of the number of soldiers is nine times as many as the number of dukes. How many of each are there?

14. Show that if r, s are two positive roots of $x^3 + d = cx$, then $t = r + s$ is a root of $x^3 = cx + d$.

15. Show that if t is a root of $x^3 = cx + d$, then $r = t/2 + \sqrt{c - 3(t/2)^2}$ and $s = t/2 - \sqrt{c - 3(t/2)^2}$ are both roots of $x^3 + d = cx$. Apply this rule to solve $x^3 + 3 = 8x$.

16. Prove that the equation $x^3 + cx = d$ always has one positive solution and no negative ones.

17. Use Cardano's formula to solve $x^3 + 3x = 10$.

18. Use Cardano's formula to solve $x^3 = 6x + 6$.

19. Consider the equation $x^3 = cx + d$. Show that if $(c/3)^3 > (d/2)^2$ (and thus that Cardano's formula involves imaginary quantities), then there are three real solutions.

20. Use Ferrari's method to solve the quartic equation $x^4 + 4x + 8 = 10x^2$. Begin by rewriting this as $x^4 = 10x^2 - 4x - 8$ and adding $-2bx + b^2$ to both sides. Determine the cubic equation that b must satisfy so that each side of the resulting equation is a perfect square. For each solution of that cubic, find all solutions for x. How many different solutions to the original equation are there?

21. It is obvious that 3 is a root of $x^3 + 3x = 36$. Show that Cardano's formula gives $x = \sqrt[3]{\sqrt{325} + 18} - \sqrt[3]{\sqrt{325} - 18}$. Using Bombelli's methods, show that this number is in fact equal to 3.

22. This problem and the one in exercise 22 are from Viète: Given the product of two numbers and their ratio, find the roots. Let A, E be the two roots, $AE = B$, $A : E = S : R$. Show that $R : S = B : A^2$ and $S : R = B : E^2$. Viète's example has $B = 20$, $R = 1$, $S = 5$. Show that, in this case, $A = 10$ and $E = 2$. (Jordanus has the same problem but with different numbers.)

23. Given the difference between two numbers and the difference between their cubes, find the numbers. Let E be the sum of the numbers, B the difference between them, and D the difference between the cubes. Show that $E^2 = (4D - B^3)/3B$. Once E^2 is known, so is E and then the numbers themselves. Find the solution when $B = 6$ and $D = 504$. (Diophantus has the same problem twice, once in Book IV with these numerical values and once in Book B.)

24. Make a perspective drawing of a checkerboard. First establish a reasonable distance for the vanishing line and vanishing point, and then construct the horizontal lines using the rules given in the text.

25. Suppose you are adding a row of telephone poles, all of the same actual height, to the picture of the tiled floor in Figure 9.1. The poles are equally distant from each other (each a distance of one square from the previous one) and are going off into the distance along the line from E to V in the figure. If the height on your canvas of the pole right at the picture plane is p, what should be the heights on your canvas of the remaining poles?

26. Kepler gave the following construction for a hyperbola with foci at A and B and with one vertex at C: Let pins be placed at A and B. To A let a thread with length AC be tied, and to B a thread with length BC. Let each thread be lengthened by an amount equal to itself. Then grasp the two threads together with one hand (starting at C) and little by little move away from C, paying out the two threads. With the other hand, draw the path of the join of the two threads at the fingers. Show that the path is a hyperbola.

27. Solve the problem from *On Triangles* of finding two sides AB, AG of triangle ABG given that $BG = 20$, the perpendicular $AD = 5$, and the ratio $AB : AG = 3 : 5$ (Fig. 9.16). Mark point E between D and G such that $DE = BD$ and set $EG = 2x$.

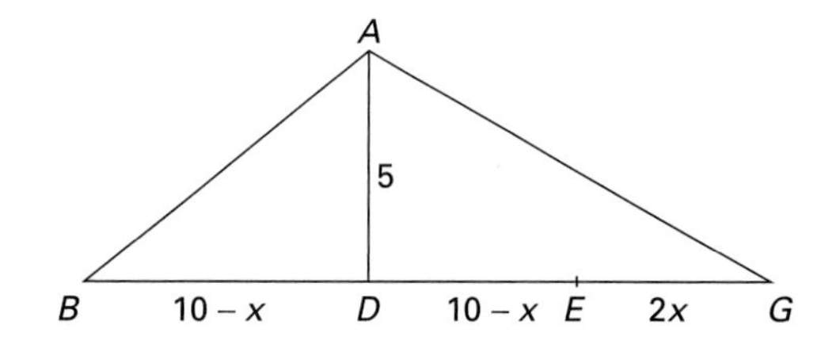

FIGURE 9.16 Regiomontanus's use of algebra

28. This problem and the one in exercise 29 are also from *On Triangles*: In triangle ABC, suppose the ratio $\angle A : \angle B = 10 : 7$ and the ratio $\angle B : \angle C = 7 : 3$. Find the three angles and the ratio of the sides.

29. Show that if the sum of two arcs is known and the ratio of their sines is known, then each arc may be found. In particular, suppose the sum of the two arcs is 40° and ratio of the sine of the larger part to that of the smaller is 7 : 4. Determine the two arcs. (Although Regiomontanus only used sines, it is probably easier to do this using cosines and tangents as well.)

30. Suppose the three angles of a spherical triangle are 90°, 70°, and 50°. Find the lengths of the sides.

31. Show that Regiomontanus's versine formula is equivalent to the spherical law of cosines: $\cos a = \cos b \cos c + \sin b \sin c \cos A$.

32. Prove that $2 \sin \alpha \sin \beta = \cos(\alpha - \beta) - \cos(\alpha + \beta)$.

33. Use the formula in exercise 32 to multiply 4,378,218 by the sine of $27°15'22''$. Check your answer using your calculator in the standard way.

34. Using the definition of the function Nlog presented in section 9.3.2, determine $\text{Nlog}(xy)$ and $\text{Nlog}(x/y)$ in terms of $\text{Nlog}\, x$ and $\text{Nlog}\, y$.

35. Given that the period of the earth is 1 year and that the mean distance of Mars from the sun is 1.524 times the mean distance of the earth from the sun, use Kepler's third law to determine the period of Mars.

36. According to Kepler's second law, at what point in a planet's orbit will the planet be moving the fastest?

37. Prove: If the same moveable is carried from rest on an inclined plane and also along a vertical of the same height, the times of movements will be to one another as the lengths of the plane and of the vertical. A corollary is that the times of descent along differently inclined planes of the same height are to one another as the lengths.

38. Prove: The times of motion of a moveable starting from rest over equal planes unequally inclined are to each other inversely as the square root of the ratio of the heights of the planes.

39. Show that a projectile fired at an angle α from the horizontal follows a parabolic path.

40. Galileo states that if a projectile fired at an angle α from the horizontal at a given initial speed reaches a distance of 20,000 if $\alpha = 45°$, then with the same initial speed it will reach a distance of 17,318 if $\alpha = 60°$ or $\alpha = 30°$. Check this statement.

41. Galileo states that if a projectile fired at a given initial speed at an angle α to the horizontal reaches a maximum height of 5000 if $\alpha = 45°$, then with the same initial speed, it will reach a height of 2499 when $\alpha = 30°$ and a height of 7502 when $\alpha = 60°$. Check this statement.

42. Given that the distances traveled in any times by a body falling from rest are as the squares of the times, show that the distances traveled in successive equal intervals are as the consecutive odd numbers $1, 3, 5, \ldots$.

43. Why is Cardano's formula no longer generally taught in college algebra courses? Should it be? What insights can it bring to the study of the theory of equations?

44. Outline a lesson that introduces the study of complex numbers by discussing the problems inherent in Cardano's formula giving a real root as the sum of two complex values. Discuss the merits of such an approach.

45. Compare the various notations for unknowns used by the mathematicians discussed in this chapter. Write a brief essay on the importance of good notation for increasing a student's understanding in algebra.

46. Why was the knowledge of mathematics necessary for the merchants of the Renaissance? Did they really need to know the solutions of cubic equations? What, then, was the purpose of the detailed study of these equations in the works of the late sixteenth century?

47. Compare the symbolism used by Jordanus and by Viète. In what way is Viète's work an advance on that of Jordanus?

48. Explain why mathematicians of the sixteenth century equated the new algebra with the Greek analysis as described by Pappus.

49. Compare Galileo's and Kepler's attitudes toward the interaction of experiment (or observation) and theory in developing a new body of knowledge.

50. Look up a treatment of geometrical perspective in a modern text on techniques of painting. How does it compare to Alberti's discussion?

References

General works on the history of algebra during the Renaissance include Paul Lawrence Rose, *The Italian Renaissance of Mathematics: Studies on Humanists and Mathematicians from Petrarch to Galileo* (Geneva: Droz, 1975); Warren van Egmond, "The Commercial Revolution and the Beginnings of Western Mathematics in Renaissance Florence, 1300–1500," Dissertation (Bloomington: Indiana University, 1976); and R. Franci and L. Toti Rigatelli, "Towards a History of Algebra from Leonardo of Pisa to Luca Pacioli," *Janus* 72 (1985), 17–82. Chapter 2 of B. L. Van der Waerden, *A History of Algebra from al-Khwarizmi to Emmy Noether* (New York: Springer, 1985) also provides a good introduction to the material. Art and mathematics during the Renaissance are discussed in J. V. Field, *The Invention of Infinity: Mathematics and Art in the Renaissance* (Oxford: Oxford University Press, 1997) and in Martin Kemp, *The Science of Art: Optical Themes in Western Art from Brunelleschi to Seurat* (New Haven: Yale University Press, 1990). There is also a discussion of perspective in three chapters of Julian Lowell Coolidge, *The Mathematics of Great Amateurs*, 2nd ed. (Oxford: Clarendon Press, 1990). Both Thomas S. Kuhn, *The Copernican Revolution* (Cambridge: Harvard University Press, 1957) and E. J. Dijksterhuis, *The Mechanization of the World Picture* (Princeton: Princeton University Press, 1986) contain sections on the developments in astronomy. The latter work, in particular, has mathematical details, including a study of Kepler's discovery of the elliptical orbit of the planets. In *The Sleepwalkers* (New York: Penguin, 1959), Arthur Koestler provides a lively treatment of the history of astronomy from Greek times to the time of Galileo, including biographies of Copernicus, Brahe, Kepler, and Galileo. Many of his interpretations are controversial. Finally, Ernan McMullin, ed., *Galileo, Man of Science* (Princeton Junction, N.J.: The Scholar's Bookshelf, 1988) is a collection of essays on various aspects of Galileo's scientific achievements.

Maestro Dardi's work on algebra has been translated by Warren Van Egmond in "The Algebra of Master Dardi of Pisa," *Historia Mathematica* 10 (1983), 399–421. There are no English versions available of Pacioli's *Summa* or the works of Rudolff or Stifel, but a photographic reprint of Robert Recorde, *The Whetstone of Witte*, was published by Da Capo Press, New York, in 1969. Chuquet's manuscript has been translated and edited by Graham Flegg, Cynthia Hay, and Barbara Moss in *Nicolas Chuquet, Renaissance Mathematician* (Boston: Reidel, 1985). In addition to Chuquet's text, this volume contains a discussion of the author's life and his place in the history of mathematics. Cardano's *Ars magna* is available in Girolamo Cardano, *The Great Art, or The Rules of Algebra*, translated and edited by T. Richard Witmer (Cambridge: MIT Press, 1968). Many aspects of the controversy surrounding the discovery of the solution to the cubic, together with many of the original documents, are presented in Martin A. Nordgaard, "Sidelights on the Cardan-Tartaglia Controversy," *Mathematics Magazine* 13 (1938), 327–346. Bombelli's algebra text is available only in an Italian reprint: Rafael Bombelli, *Algebra* (Milan: Feltrinelli, 1966). On the other hand, virtually all of Viète's works have been translated into English in *The Analytic Art*, translated and edited by T. Richard Witmer (Kent, Ohio: Kent State University Press, 1983). An English version of Dürer's *Treatise on Mensuration* is W. Strauss, trans., *The Painter's Manual* (New York: Abaris, 1977). Barnabas Hughes has prepared a modern edition of Regiomontanus's trigonometry: *Regiomontanus on Triangles* (Madison: University of Wisconsin Press, 1967). This work contains the original Latin of *De triangulis omnimodis* as well as an English translation, an introduction, and extensive notes. An English translation of Stevin's *De thiende* is available in Henrietta O. Midonick, ed., *The Treasury of Mathematics* (New York: Philosophical Library, 1965), 735–750. The translation was made by Robert Norton in 1608 and is also reprinted in *The Principal Works of Simon Stevin*, vol. II, edited by Dirk J. Struik (Amsterdam: Swets and Zeitlinger, 1958). Napier's two volumes on logarithms, the *Descriptio* and the *Constructio*, were translated in the seventeenth century. Photographic reprints are available as John Napier, *Descriptio*, translated by Edward Wright (New York: Da Capo Press, 1969) and John Napier, *Constructio*, translated by William R. MacDonald (London: Dawsons, 1966). The former includes Napier's original table; the latter shows in detail the ingenious interpolation schemes Napier used to guarantee eight-place accuracy. A translation of Copernicus's *De revolutionibus* is A. M. Duncan, trans., *Copernicus: On the Revolutions of the Heavenly Spheres* (New York: Barnes and Noble, 1976). Kepler's *Optics* is available in the translation of William H. Donahue (Santa Fe: Green Lion Press, 2000). The *Mysterium cosmographicum* is available as *The Secret of the Universe*, A. M. Duncan, trans. (New York: Abaris, 1981). This is a translation of the second edition, and so it includes not only the original text of 1596 but also Kepler's notes added in 1621. It also contains copies of the original diagrams. *New Astronomy* is also now available in a translation by William H. Donahue (Cambridge: Cambridge University Press, 1992), while a section of *Harmonice mundi*, translated by Charles

Glenn Wallis, is in volume 16 of the *Great Books* (Chicago: Encyclopaedia Britannica, 1952). Finally, Galileo's *Two New Sciences* was translated by Stillman Drake (Madison: University of Wisconsin Press, 1974). Drake also wrote an introduction, extensive notes, and a glossary of Galileo's technical terms.

There are numerous other works on various aspects of the mathematics discussed in this chapter. R. Franci and L. Toti Rigatelli, "Fourteenth-Century Italian Algebra," in Cynthia Hay, ed., *Mathematics from Manuscript to Print: 1300–1600* (Oxford: Clarendon Press, 1988), 11–29, summarizes the contents of several important fourteenth-century abacus manuscripts. Frank J. Swetz, *Capitalism and Arithmetic: The New Math of the 15th Century* (La Salle, Il.: Open Court, 1987) provides a complete translation of the *Treviso Arithmetic*, the first printed arithmetic book, as well as a detailed commentary on the mathematics and the social setting in which the book was produced. Various aspects of the development of logarithms are found in C. G. Knott, ed., *Napier Tercentenary Memorial Volume* (London: Longmans, Green, 1915). A general discussion of logarithms is found in E. M. Bruins, "On the History of Logarithms: Bürgi, Napier, Briggs, de Decker, Vlacq, Huygens," *Janus* 67 (1980), 241–260, and in F. Cajori, "History of the Exponential and Logarithmic Concepts," *American Mathematical Monthly* 20 (1913), 5–14, 35–47, 75–84, 107–117. For more information on Galileo's scientific life, see Stillman Drake, *Galileo Studies: Personality, Tradition and Revolution* (Ann Arbor: University of Michigan Press, 1970), and the classic work by Alexandre Koyré, *Galileo Studies*, translated by John Mepham (Atlantic Highlands, N.J.: Humanities Press, 1978). Drake has written numerous articles on various aspects of Galileo's scientific life. One of particular interest is "Galileo's Discovery of the Law of Free Fall," *Scientific American* 228 (May, 1973), 84–92, in which Drake analyzed certain of Galileo's manuscript notes, which provide details on the discovery.

CHAPTER TEN

Precalculus in the Seventeenth Century

Whenever two unknown magnitudes appear in a final equation, we have a locus, the extremity of one of the unknown magnitudes describing a straight line or a curve.
—Pierre de Fermat's *Introduction to Plane and Solid Loci*, 1637

It is in the early seventeenth century that the speeding up of mathematical developments becomes evident. Printing was well established, and communication, both through letter and through the printed word, was becoming much more rapid. The ideas of one mathematician were passed on to others, to be criticized, commented upon, and finally extended. In this chapter, we will survey some of the newly developing areas of mathematics.

Viète's ideas on the use of algebra in analysis were critical in the new developments. In the 1620s, William Oughtred, Thomas Harriot, Albert Girard, and others began to turn Viète's notation into recognizably modern notation, while at the same time further developing his theory of equations. In the next decade, Viète's analysis was applied to geometry and reformulated into the new subject of analytic geometry. The two central figures in the development of analytic geometry, which was to prove vital in the subsequent invention of the calculus, are Pierre de Fermat and René Descartes. Fermat was also involved, through his correspondence with Blaise Pascal, in the early development of probability theory, the first textbook on which was written by Christian Huygens in 1656; in addition, he was responsible for new work in number theory.

10.1 Algebraic Symbolism and the Theory of Equations

Algebraic methods for solving cubic and quartic equations were discovered in Italy in the sixteenth century and improved on somewhat by Viète near the turn of the seventeenth century. But Cardano was hampered by a lack of a convenient notation, and Viète always restricted himself to positive solutions. Thus, even though the former gave various examples of relationships among the roots of a single cubic equation and between roots of related equations and the latter was able to express algebraically the relationship between the coefficients and the solutions of equations of degree up to five, provided all values were positive, the general theory was still incomplete.

10.1.1 William Oughtred and Thomas Harriot

By the early seventeenth century, two English mathematicians, William Oughtred (1573–1660) and Thomas Harriot (1560–1621), had made careful studies of Viète's work and were converted to the method of symbolic reasoning he had introduced. Both of them attempted in their own work to go beyond Viète and to make algebraic arguments even more symbolic. Oughtred was a cleric, who evidently spent most of his time studying and teaching mathematics. His major work, the *Clavis mathematicae* (*Key of Mathematics*), first appeared in 1631 and had several subsequent editions, both in Latin and in English. The *Clavis* introduced English readers to Viète's symbolic algebra, and Oughtred attempted to show, as had his French predecessor, that algebra could really be considered the "analytical art, . . . in which by taking the thing sought as known, we find out that we seek." In other words, Oughtred felt that mathematical problems, including geometric ones, should be translated into symbolic equations and then solved by the methods of algebra.

Oughtred introduced many symbols, including $\times$ to represent multiplication; however, in part because there were so many and because they often confused his students and his typesetters, few of his symbols have lasted. For variables, constants, and their powers, however, Oughtred basically kept to Viète's plan, only making the notation a bit shorter by using abbreviations. For example, he used Aq for the square of A and Ac for its cube. But he did show how to use algebra to solve problems. For example, he rewrote Euclid's proposition II-11 as an equation. He let A stand for the greater of the two segments in which the line of length B was to be cut. Then $B - A$ was the lesser segment, so the rectangle contained by the whole and the lesser segment was $Bq - BA$. Since this was required to be equal to the square on A, he had $Aq = Bq - BA$, or $Aq + BA = Bq$. He then used his version of the quadratic formula to solve the equation. Note that in this case Oughtred used juxtaposition to indicate multiplication, although he did not do so consistently, sometimes reverting to Viète's "in."

Oughtred was hesitant to deal with negative numbers, but Harriot evidently understood the necessity of dealing with these quantities and even with complex roots of equations. He was therefore able to make some progress toward general theory. Harriot entered the service of Sir Walter Raleigh after finishing his undergraduate studies at Oxford and went on an expedition to Virginia in 1585 as an expert on cartography. Besides learning how to smoke tobacco while he was there, a habit that ultimately led to his death from cancer, Harriot also compiled a brief report on the colony and its inhabitants.

Much of Harriot's mathematical work exists today only in manuscript, although it appears that many of these manuscripts circulated in England during and after his lifetime. But because he never took the time to put his work into publishable form, many of his ideas are just sketched in his notes. Unfortunately, the editors of the published version, the *Artis analyticae praxis* (*Practice of the Analytic Art*), which appeared posthumously in 1631, did not fully understand some of Harriot's most important ideas.

Harriot took over from Viète the idea of using vowels for unknowns and consonants for knowns, although he used small letters instead of Viète's capital letters. He also improved the latter's notation for powers by using repeated copies of a single letter, writing $aaaa$ for a^4, rather than using an abbreviation for "square-square." He was thus able to simplify Viète's rules considerably. For example, he replaced Viète's justification of transposition (see section 9.1.5) by the following: "Let $aa - dc = gg - ba$. To be added to each $+ba + dc$. Whence $aa + ba = gg + dc$." Note that although Harriot replaced all words by symbols, he still felt constrained by the old notion of homogeneity.

Harriot also realized that equations could be generated from their roots $b, c, d, \ldots$ by multiplying together expressions of the form $b - a$, $c - a$, $d - a, \ldots$. Thus, he was led to the basic relationship between the roots and the coefficients of an equation, even in the case of negative and imaginary roots, although he never seems to have stated this explicitly as a theorem. For example, Harriot multiplied together $b - a$, $c + a$, and $df + aa$ to get the equation $bcdf + bdfa - dfaa + baaa - cdfa + bcaa - caaa - aaaa = 0$. He then indicated that the roots of the equation are $a = b$, $a = -c$, and $aa = -df$, or $a = \sqrt{-df}$. In this example, he did not seem to realize that there should be two square roots of $-df$, but in another example, he was able to come up with two complex roots to an equation. Taking the equation $12 = 8a - 13aa + 8aaa - aaaa$, he saw first that two of the roots are 2 and 6. Noting that the sum of these real roots is already equal to 8, the coefficient of a^3, he stated that there could not be any further real roots, because they would make the sum greater than 8. But he proceeded to solve the equation by substitution. Setting $a = 2 - e$, he found the new equation $-20e + 11ee - eeee = 0$, whose real roots are 0 and -4. The sum of the other roots must then be 4, while their product is 5. Harriot then wrote that these additional roots are $e = 2 + \sqrt{-1}$ and $e = 2 - \sqrt{-1}$, and therefore the complex roots of the original equation are $a = 2 - (2 + \sqrt{-1}) = -\sqrt{-1}$ and $a = 2 - (2 - \sqrt{-1}) = +\sqrt{-1}$.

10.1.2 Albert Girard and the Fundamental Theorem of Algebra

Albert Girard (1595–1632) was much clearer than Harriot about the relationship between the roots and coefficients of a polynomial in his 1629 work *A New Discovery in Algebra*, in which he also gave the first explicit statement of the fundamental theorem of algebra. Girard was probably born in St. Mihiel in the French province of Lorraine, but he spent much of his life in the Netherlands, where he studied at Leiden and served as a military engineer in the army of Frederick Henry of Nassau. Although he wrote a work on trigonometry and edited the works of Stevin, his most important contributions are to algebra. In *A New Discovery in Algebra*, Girard clearly introduced the notion of a fractional exponent ("the numerator is the power and the denominator the root") as well as the modern notation for higher roots (e.g., $\sqrt[3]{}$ for cube root as an alternative to exponent $\frac{1}{3}$.) However, the fractional exponent was not attached directly to an unknown. For example, Girard wrote $(\frac{3}{2})49$ to mean the cube of the square root of 49, or 343, and, taking his cue from Bombelli, $49(\frac{3}{2})$ to mean what is today written as $49x^{3/2}$.

Furthermore, Girard was among the first to note the geometric meaning of a negative solution to an equation. In an example of a geometric problem whose algebraic translation has two positive and two negative solutions, he noted on the relevant diagram that the negative solutions were to be interpreted as being laid off in the direction opposite that of the positive ones.

Not only did Girard understand the meaning of negative solutions to equations, he also systematized the work of Viète and Harriot and explicitly considered **factions**, today called the *elementary symmetric functions* of n variables: "When several numbers are proposed, the entire sum may be called the first *faction*; the sum of all the products taken two by two may be called the second *faction*; the sum of all the products taken three by three may be called the third *faction*; and always thus to the end, but the product of all the numbers is the last *faction*. Now, there are as many *factions* as proposed numbers. For example, for the numbers 2, 4, 5, the first *faction* is 11, their sum; the second is 38, the sum of all products of pairs; while the third is 40, the product of all three numbers." Girard also noted that the Pascal triangle of binomial coefficients tells how many terms each of the factions contains. In the case of four numbers, the first faction contains four terms, the second, six, the third, four, and the fourth and last, one. Girard's basic result in the theory of equations is the following theorem, for which he gave no proof:

THEOREM. *Every algebraic equation . . . admits of as many solutions as the denomination of the highest quantity indicates. And the first faction of the solutions is equal to the [coefficient of the second highest] quantity, the second faction of them is equal to the [coefficient of the third highest] quantity, the third to the [fourth], and so on, so that the last faction is equal to the [constant term]—all this according to the signs that can be noted in the alternating order.*

What Girard meant by the last statement about signs is that one must first arrange the equation so that the degrees alternate on each side of the equation. Thus, $x^4 = 4x^3 + 7x^2 - 34x - 24$ should be rewritten as $x^4 - 7x^2 - 24 = 4x^3 - 34x$. The roots of this equation being 1, 2, -3, and 4, the first faction is equal to 4, the coefficient of x^3; the second to -7, the coefficient of x^2; the third to -34, the coefficient of x; and the fourth to -24, the constant term.

In the first part of the theorem, Girard was asserting the truth of the **fundamental theorem of algebra**, that every polynomial equation has a number of solutions equal to its degree (denomination of the highest quantity). As his examples show, he acknowledged that a given solution could occur with multiplicity greater than 1. He also fully realized that in his count of solutions he would have to include imaginary ones (which he called "impossible"). So in his example $x^4 + 3 = 4x$, he noted that the four factions are 0, 0, 4, 3. Because 1 is a solution of multiplicity 2, the two remaining solutions have the property that their product is 3 and their sum is -2. It follows that these solutions are $-1 \pm \sqrt{-2}$.

Although Girard did not explain how he derived the theorem, given that he considered solutions with multiplicity greater than 1, he probably understood that equations of degree n come from multiplying together n expressions of the form $x - r_i$, where some of the r_i may be identical. It was Descartes, however, who made this procedure precise, as we will see.

10.2 Analytic Geometry

The new developments in algebra were immediately applied to geometry in 1637, when two mathematicians, René Descartes (1596–1650) and Pierre de Fermat (1601–1665), independently brought out works on what is now called *analytic geometry*. Early in that year, Fermat sent to his correspondents in Paris a manuscript entitled *Introduction to Plane and Solid Loci*. At about the same time, Descartes was readying for the printer the galley proofs of his *Discourse on the Method for Rightly Directing One's Reason and Searching for Truth in the Sciences* with its three accompanying essays, among which was the *Geometry*. Both Fermat's *Introduction* and Descartes's *Geometry* present the same basic techniques of relating algebra and geometry, the techniques whose further development culminated in the modern subject of analytic geometry. Both men came to the development of these techniques as part of the effort of rediscovering the "lost" Greek techniques of analysis. Both were intimately familiar with the Greek classics, in particular with the *Domain of Analysis* of Pappus, and both tested their new ideas against the four-line-locus problem of Apollonius and its generalizations. But Fermat and Descartes developed distinctly different approaches to their common subject, caused by their differing points of view toward mathematics.

10.2.1 Fermat and the *Introduction to Plane and Solid Loci*

Fermat began his study of mathematics with the normal university curriculum at Toulouse, which probably covered little more than an introduction to Euclid's *Elements*. Then, after completing his baccalaureate degree and before beginning his legal education, he spent several years in Bordeaux studying mathematics with former students of Viète, who during the late 1620s were engaged in the editing and publishing of their teacher's work. Fermat became familiar with both Viète's new ideas for symbolization in algebra and his program of using algebra as a form of analysis. In particular, Fermat began to work out an algebraic version of some of Apollonius's theorems. By some time in the 1630s, his ideas on relating geometry and algebra were sufficiently clear that he could organize them into a manuscript that was completed in 1637.

The central theme of Fermat's *Introduction to Plane and Solid Loci* was stated in the sentence quoted at the opening of this chapter. Fermat asserted that if, in solving a geometric problem algebraically, one ends up with an equation in two unknowns, the resulting solution is a locus, either a straight line or a curve, the points of which are determined by the motion of one endpoint of a variable line segment, the other endpoint of which moves along a fixed straight line.

Fermat's chief assertion in the *Introduction* was that if the moving line segment makes a fixed angle with the fixed line and if neither of the unknown quantities occurs to a power greater than the square, then the resulting locus will be a straight line, a circle, or one of the other conic sections. He proceeded to prove his result by treating each of the various possible cases. Let us consider first his treatment of the case of the straight line: "Let NZM be a straight line given in position, with point N fixed. Let NZ be equal to the unknown quantity A, and ZI, the line drawn to form the angle NZI, the other unknown quantity E. If D times A equals B times E, the point I will describe a straight line given in position." Fermat thus began with a single axis NZM and a linear equation (Fig. 10.1).

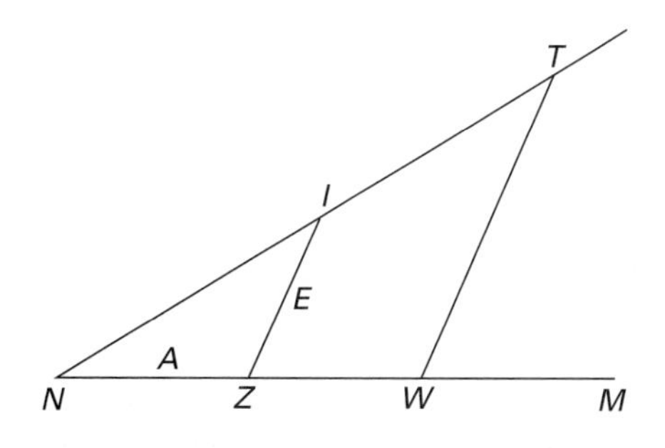

FIGURE 10.1 Fermat's analysis of the equation D times A equals B times E

(Fermat used Viète's convention of vowels for unknowns and consonants for knowns.) He wanted to show that this equation, which is written in modern notation as $dx = by$, represents a straight line. Because $D \cdot A = B \cdot E$, also $B : D = A : E$. Because $B : D$ is a known ratio, the ratio $A : E$ is determined, and so also is the triangle NZI. Thus, the line NI is, as Fermat wrote, "given in position." Fermat dismissed as "easy" the necessary completion of the argument, to show that any point T on NI determines a triangle TWN with $NW : TW = B : D$.

Although the basic notions of modern analytic geometry are apparent in Fermat's description, Fermat's ideas differed somewhat from current ones. First, Fermat used only one axis. The curve is thought of, not as made up of points plotted with respect to two axes, but as generated by the motion of the endpoint I of the variable line segment ZI as Z moves along the given axis. Fermat often took the angle between ZI and ZN as a right angle, although there was no particular necessity for so doing. Second, to Fermat, as to Viète and most others of the time, the only proper solutions to algebraic equations were positive. Thus, Fermat's "coordinates" ZN and ZI, solutions to his equation $D \cdot A = B \cdot E$, represented positive numbers. Hence, Fermat drew only the ray emanating from the origin into the first quadrant.

Fermat's restriction to the first quadrant was quite apparent also in his treatment of the parabola: "If *Aq* equals D times E, point I lies on a parabola." Fermat intended to show that the equation $x^2 = dy$ (in modern notation) determines a parabola. He began with the basic two line segments, NZ and ZI, in this case at right angles. Drawing NP parallel to ZI, he then asserted that the parabola with vertex N, axis NP, and latus rectum D is the parabola determined by the given equation (Fig. 10.2). Fermat was, of course, assuming that his readers were very familiar with Apollonius's *Conics*. For the parabola, Apollonius's construction showed that the rectangle contained by D and NP was equal to the square on PI (or NZ), a statement translated into algebra by the equation $dy = x^2$. Although Fermat knew what a parabola looked like, his diagram included only part of half of it. He did not deal with negative lengths along the axis.

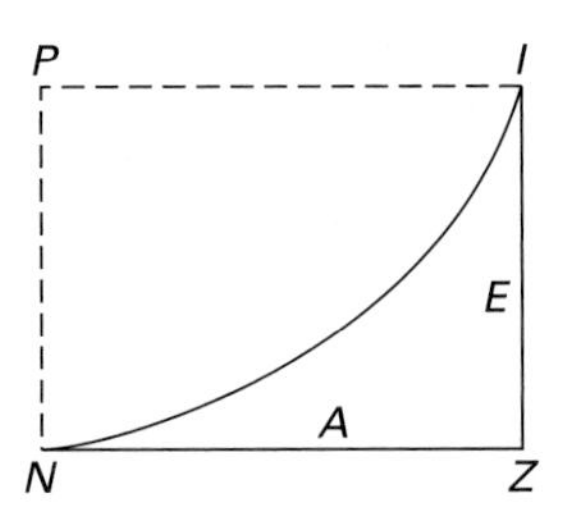

FIGURE 10.2 Fermat's analysis of the equation *Aq* equals *D* times *E*

Fermat proceeded to determine the curves represented by five other quadratic equations in two variables. In modern notation, $xy = b$ and $b^2 + x^2 = ay^2$ represent hyperbolas, $b^2 - x^2 = y^2$ represents a circle, $b^2 - x^2 = ay^2$ represents an ellipse, and $x^2 \pm xy = ay^2$ represents a straight line. In each case, his argument assumed the construction of a particular conic section according to Apollonius's procedures and showed that this conic had the desired equation. Finally, Fermat sketched a method of reducing any quadratic equation to one of his seven canonical forms, by showing how to change variables. For example, he asserted that any equation containing ax^2 and ay^2 along with bx and/or cy can be reduced to the canonical equation of a circle, provided that the angle between the axis and the line tracing out the curve is a right angle. For example, the equation $p^2 - 2hx - x^2 = y^2 + 2ky$ can be transformed by first adding k^2 to both sides so the right side becomes a square. Setting $r^2 = h^2 + k^2 + p^2$, or $r^2 - h^2 = k^2 + p^2$, Fermat could rewrite the original equation as

$$r^2 - (x + h)^2 = (y + k)^2,$$

the canonical equation of a circle if $x + h$ is replaced by x' and $y + k$ by y'. Fermat also dealt with equations containing an xy term by an appropriate change of variable.

Fermat was able to determine the locus corresponding to any quadratic equation in two variables and show that it had to be a straight line, a circle, or a conic section. To conclude his *Introduction*, Fermat noted that one could apply his methods to the following generalization of the four-line-locus problem: "If, given any number of lines in position, lines be drawn from one and the same point to each of them at given angles, and if [the sum of] the squares of all of the drawn lines is equal to a given area, the point will lie on a [conic section] given in position." Fermat, however, left the actual solution to the reader.

10.2.2 Descartes and the *Geometry*

Fermat's brief treatise created a stir when it reached Paris. A circle of mathematicians, centered on Marin Mersenne (1588–1648), had been gathering regularly to discuss new ideas in mathematics and physics. Acting as the recording and corresponding secretary of the group, Mersenne received material from various sources, copied it, and distributed it widely. He thus served as France's "walking scientific journal." Fermat had begun a regular correspondence with Mersenne in 1636, but because many of his manuscripts were brief and lacking in detail, Mersenne often forwarded to Fermat requests to amplify his work. Nevertheless, the *Introduction* was given a positive reception and established Fermat's reputation as a first-class mathematician. The manuscript had, however, reached Paris—and then Descartes—just prior to the publication of Descartes's own work on analytic geometry. One can only imagine Descartes's chagrin at seeing material similar to his own appearing before his own work reached its intended audience.

Descartes's analytic geometry was, nevertheless, somewhat different from that of Fermat. To understand it, one must realize that the *Geometry* was written to demonstrate the application to geometry of Descartes's methods of correct reasoning discussed in the *Discourse*—that is, reasoning based on self-evident principles. Like Fermat, Descartes had studied the works of Viète and saw in them the key to understanding the analysis of the Greeks. But rather than dealing with the relationship of algebra to geometry through the study of loci, Descartes was more concerned with demonstrating this relationship through the geometric construction of solutions to algebraic equations. In some sense, then, he was merely following in the ancient tradition, a tradition that had been continued by such Islamic mathematicians as al-Khayyāmī and Sharaf al-Dīn al-Ṭūsī. But Descartes did take the same crucial step as Fermat (a step his Islamic predecessors failed to take) of using coordinates to study this relationship between geometry and algebra.

The *Geometry* begins, "Any problem in geometry can easily be reduced to such terms that a knowledge of the lengths of certain straight lines is sufficient for its construction." Thus, such lengths must be found. In the first of the three books of this work, Descartes found these lengths by the use of lines and circles, the standard Euclidean curves. But Descartes made these Euclidean techniques appear modern in his clear use of algebraic techniques. For example, to find the solution of the quadratic equation $z^2 = az + b^2$, he constructed a right triangle NLM with $LM = b$ and $LN = \frac{1}{2}a$ (Fig. 10.3). Extending the hypotenuse to O, where $NO = NL$, and constructing the circle centered on N with radius NO, he concluded that OM is the required value z, because the value of z is given by the standard formula

$$z = \tfrac{1}{2}a + \sqrt{\tfrac{1}{4}a^2 + b^2}.$$

Under the same conditions, MP is the solution to $z^2 = -az + b^2$, and if MQR is drawn parallel to LN, then MQ and MR are the two solutions to $z^2 = az - b^2$.

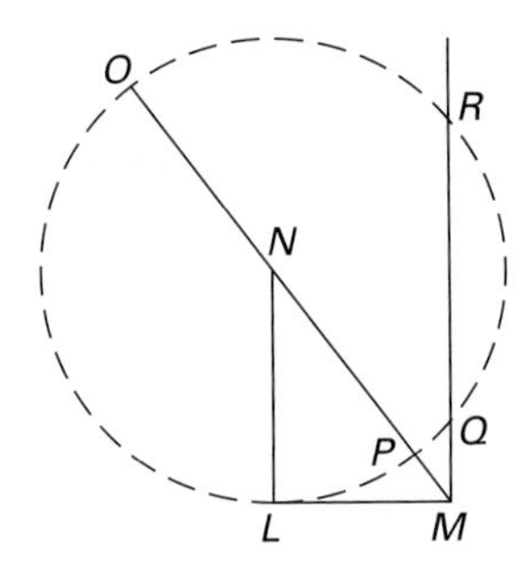

FIGURE 10.3 Descartes's construction of the solution to a quadratic equation

Descartes noted, however, that "often it is not necessary thus to draw the lines on paper, but it is sufficient to designate each by a single letter." As long as one knows what operations are possible geometrically, it is feasible just to perform the algebraic operations and state the result as a formula. In these algebraic operations, Descartes took another major step. He represented the terms aa (or a^2) and a^3 (he used modern exponential notation for powers three and higher and sometimes also for the second power) as line segments, rather than as geometric squares and cubes. Thus, he could also consider higher powers without worrying about their lack of geometric meaning. Descartes made a brief bow to homogeneity by noting that any algebraic expression could be considered to include as many powers of unity as necessary for this purpose, but in fact he freely added algebraic expressions, whatever the power of the terms. Furthermore, Descartes replaced Viète's vowel-consonant distinction of unknowns and knowns with the current system of using letters near the end of the alphabet for unknowns and those near the beginning for knowns.

Descartes concluded his first book with a detailed discussion of Apollonius's problem of the four lines. It is here that he introduced a coordinate axis to which all the lines as well as the locus of the solution are referred. The problem requires finding points from which lines drawn to four given lines at given angles satisfy the condition that the product of two of the line lengths bears a given ratio to the product of the other two. Descartes noted that "since there are always an infinite number of different points satisfying these requirements, it is also required to discover and trace the curve containing all such points."

Descartes noted that matters are simplified if all lines are referred to two principal ones. Thus, he set x as the length of segment AB along the given line EG and y as the length of segment BC along the line BC to be drawn, where C is one of the points satisfying the requirements of the problem (Fig. 10.4). The lengths of the required line segments CB, CH, CF, and CD (drawn to the given lines EG, TH, FS, and DR, respectively) can each be expressed as a linear function of x and y. For example, because all angles of the triangle ARB are known, the ratio $BR : AB = b$ is also known. It follows that $BR = bx$ and $CR = y + bx$. Because the three angles of triangle DRC are also known, so is the ratio $CD : CR = c$, and therefore $CD = cy + bcx$. Similarly, setting the fixed distances $AE = k$ and $AG = \ell$ and the known ratios $BS : BE = d$, $CF : CS = e$, $BT : BG = f$, and $CH : TC = g$, one shows in turn that $BE = k + x$, $BS = dk + dx$, $CS = y + dk + dx$, $CF = ey + dek + dex$, $BG = \ell - x$, $BT = f\ell - fx$, $CT = y + f\ell - fx$, and finally, $CH = gy + fg\ell - fgx$. Because the problem involves comparing the products of certain pairs of line lengths, it follows that the equation expressing the desired locus is a quadratic equation in x and y. Furthermore, as many points of the locus as desired can be constructed, because if any value of y is given, the value of x is expressed in the form of a determinate quadratic equation whose solution has already been provided. The required curve can then be drawn, point by point. In book two of the *Geometry*, Descartes returned to this problem and showed that the curve given by the quadratic equation in two variables is either a circle or one of the conic sections, depending on the values of the various constants involved.

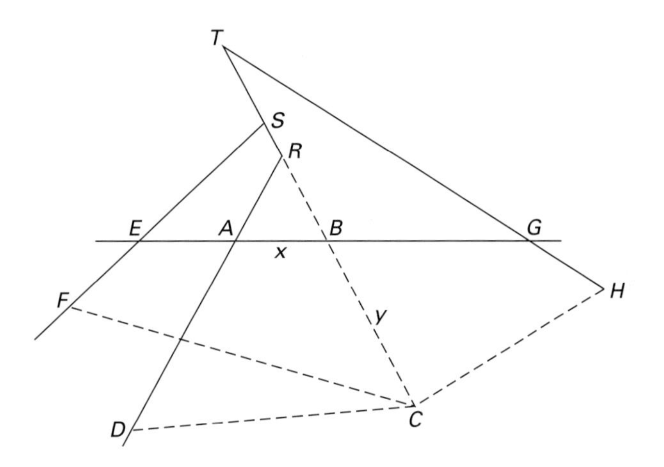

FIGURE 10.4 The problem of the four-line locus

Pappus (and Apollonius) already knew this result. What Descartes wanted to do was show that his new methods enabled him to go further and solve the problem for arbitrarily many lines. Thus, he presented a five-line problem: Suppose four of the lines are equally spaced parallel lines and the fifth is perpendicular to each of the others. Suppose that the lines drawn from the required point meet the given lines at right angles. Find the locus of all points such that the product of the lengths of the lines drawn to three of the parallel lines is equal to the constant spacing times the product of the lengths of the lines drawn to the remaining two lines. To solve this problem, let the four parallel lines be AB, IH, ED, and GF, the perpendicular line be GI, and the constant spacing be a. Let C be a point satisfying the problem, and let CB, CD, CF, CH, and CM be the line segments meeting the given lines at right angles (Fig. 10.5). Letting $CM = x$ and $CB = y$, we calculate that $CF = 2a - y$, $CD = a - y$, and $CH = y + a$. The conditions of the problem then give $(2a - y)(a - y)(y + a) = ayx$, or $y^3 - 2ay^2 - a^2y + 2a^3 = axy$.

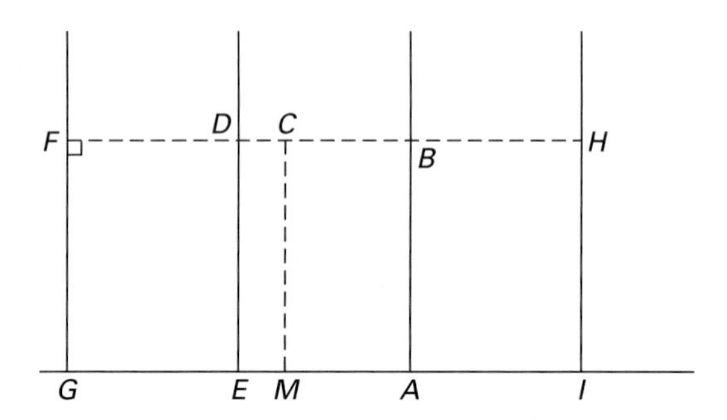

FIGURE 10.5 The problem of the five-line locus

With the equation known, Descartes wanted to construct the desired curve, both point by point and as a single unit. In fact, in order to establish his main thesis, he had to decide what curves were acceptable in geometry. He based his definition of such curves on Euclid's postulates 1 and 3 on drawing straight lines and circles and on the following new one: "Two or more lines can be moved, one upon the other, determining by their intersection other curves." He therefore accepted only curves traced by some continuous motion generated by certain machines. It is not entirely clear today how Descartes decided which curves fit his requirements, but he gave several examples of instruments designed to trace such curves. For example, in Figure 10.6, GL is a ruler that pivots at G. It is linked at L to a device $CNKL$ that allows L to be moved along AB, while keeping the line KN parallel to itself. The intersection C of the two moving lines GL and KN determines a curve. Descartes found the equation of this curve from simple geometric considerations. Setting $CB = y$, $BA = x$, and the constants $GA = a$, $KL = b$, and $NL = c$, Descartes calculated that $BK = (b/c)y$, $BL = (b/c)y - b$, and $AL = x + (b/c)y - b$. Since $CB : BL = GA : AL$, Descartes derived the equation

$$\frac{ab}{c}y - ab = xy + \frac{b}{c}y^2 - by,$$

or, after simplification,

$$y^2 = cy - \frac{c}{b}xy + ay - ac.$$

Descartes stated, without proof, that this curve is a hyperbola, and thus one of the curves that could be a solution to the four-line locus problem.

Because a similar machine could generate a parabola, Descartes could form a new machine by replacing the straight line CNK by a new "line"—namely, a parabola. That is, the point C in this case lies on the ruler but also on a parabola moving so that its axis remains along AB. With the parameter of the parabola chosen appropriately, Descartes showed that

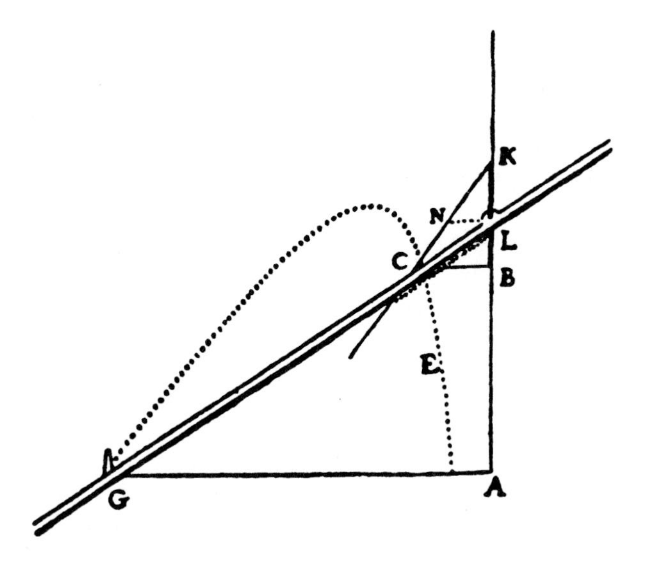

FIGURE 10.6 Descartes's curve-drawing instrument

this new machine generated the curve $y^3 - 2ay^2 - a^2y + 2a^3 = axy$, the solution to the five-line problem.

Besides constructing (or "tracing") the solution curve as a whole, Descartes also wanted to be able to construct points on it. To do so required a geometric solution of a cubic equation. To accomplish this, he needed to spend some time in his *Geometry* dealing with the solution of algebraic equations. Thus, in the third book, Descartes demonstrated how equations are built up from their solutions. Thus, if $x = 2$, or $x - 2 = 0$, and if also $x = 3$, or $x - 3 = 0$, Descartes noted that the product of the two equations is $x^2 - 5x + 6 = 0$, an equation of dimension 2 with the two roots 2 and 3. Again, if this latter equation is multiplied by $x - 4 = 0$, the result is an equation of dimension 3, $x^3 - 9x^2 + 26x - 24 = 0$, with the three roots 2, 3, and 4. Multiplying further by $x + 5 = 0$, an equation with a "false" root, -5, produces a fourth-degree equation with four roots, three "true" and one "false." Descartes concluded that "it is evident from the above that the sum of an equation having several roots [that is, the polynomial itself] is always divisible by a binomial consisting of the unknown quantity diminished by the value of one of the true roots, or plus the value of one of the false roots. In this way, the degree of an equation can be lowered. Conversely, if the sum of the terms of an equation is not divisible by a binomial consisting of the unknown quantity plus or minus some other quantity, then this latter quantity is not a root of the equation." This is the earliest statement of the modern **factor theorem**. In his usual fashion, Descartes did not give a complete proof. He just wrote that the result is "evident."

Given these basic results in the theory of equations, along with various methods of manipulating equations, Descartes was able to demonstrate explicitly some construction methods for solving equations of higher degree. In general, he accomplished such demonstrations by intersecting circles with curves (such as conic sections or curves of higher degree) that were constructed using one of his machines. Although he only briefly sketched his methods and applied them to a few examples, Descartes believed that it was "only necessary to follow the same general method to construct all problems, more and more complex, ad infinitum;

for in the case of a mathematical progression, whenever the first two or three terms are given, it is easy to find the rest."

How did Descartes decide on his defining characteristic of "geometric" curves? It appears that his basic reason for defining such curves was that "all points of those curves . . . must bear a definite relation to all points of a straight line, and that this relation must be expressed by means of a single equation." In other words, any such curve must be expressible as an algebraic equation. It is apparent that Descartes also believed the converse of this statement: Any algebraic equation in two variables determines a curve whose construction could be realized by an appropriate machine. He could not prove such a statement, but he was convinced that curves for which one could construct arbitrary points were amenable to a tracing by continuous motion via one of his machines.

There are probably several reasons why Descartes defined geometric curves by continuous motion rather than directly as curves having an algebraic equation. First, he was interested in reforming the study of geometry. Defining acceptable curves by a completely algebraic criterion would have reduced his work to algebra. Second, because he wanted to be able to construct the solution points of geometric problems, he needed to be able to determine intersections of algebraic curves. It was evident to him that defining curves by continuous motion would explicitly determine intersection points. It was not at all evident that curves defined by algebraic equations had intersection points. Because he was dealing with geometry, Descartes could not adopt the algebraic definition as an axiom. Finally, Descartes evidently was not convinced that an algebraic equation was the best way to define a curve. Nowhere in the *Geometry* did he begin with an equation. Unlike Fermat, Descartes always described a curve geometrically and then, if appropriate, derived its equation. An equation for Descartes was thus only a tool in the study of curves and not the defining criterion.

It is clear that both Fermat and Descartes understood the basic connection between a geometric curve and an algebraic equation in two unknowns. Both used as their basic tool a single axis along which one of the unknowns was measured, rather than the two axes used today, and neither insisted that the lines measuring the second unknown intersect the single axis at right angles. Both used as their chief examples the familiar conic sections, although both were also able to construct curves whose equations were of degree higher than two. And both recognized a new relationship of algebra to geometry. Recall that algebra grew out of some simple manipulation of geometric shapes. Then, during the medieval period and the Renaissance, algebra gradually freed itself from geometry. Now, algebra returned to the service of geometry. It became a much more flexible tool that could be used not only to determine solutions of equations but also to find entire curves. It therefore was available for use in the study of motion, a study central in the development of calculus.

As noted earlier, Fermat and Descartes came to the subject of analytic geometry from different viewpoints. Fermat gave a very clear statement that an equation in two variables determines a curve. He always started with the equation and then described the curve. Descartes, on the other hand, was more interested in geometry. For him, the curves were primary. Given a geometric description of a curve, he was able to come up with the equation. Thus, Descartes was forced to deal with algebraic equations that were considerably more complex than those addressed by Fermat.

Descartes and Fermat emphasized the two different aspects of the relationship between equations and curves. Unfortunately, Fermat never published his work. Although it was presented clearly and circulated through Europe in manuscript, it never had the influence of a published work. Descartes's work, on the other hand, though published, proved very difficult to read. It was published in French, rather than the customary Latin, and had so many gaps in arguments and such complicated equations that few mathematicians could fully understand it. A few years after the initial publication, however, Descartes agreed to have the work translated into Latin. This was done by Frans van Schooten (1615–1660), first in 1649 with commentary by van Schooten himself and by Florimond Debeaune (1601–1652), and then with even more extensive commentaries and additions in 1659–1661. It was only then that Descartes's work achieved the recognition he desired.

10.2.3 The Work of Jan de Witt

One of the additions to van Schooten's 1659–1661 edition of Descartes's *Geometry* was a treatise on conic sections by Jan de Witt (1623–1672). In his student days, de Witt had studied with van Schooten, who had known Descartes and had studied Fermat's works during a sojourn in Paris. Through van Schooten, de Witt became acquainted with the works of both of the inventors of analytic geometry. In 1646, at the age of 23, he composed the *Elements of Curves*, in which he treated the subject of conic sections from both a synthetic and an analytic point of view. The first of the two books of the *Elements of Curves* was devoted to developing the properties of the various conic sections using the traditional methods of synthetic geometry. In the second book, the first systematic treatise on conic sections using the new method, de Witt extended Fermat's ideas into a complete algebraic treatment of the conics, beginning with equations in two variables. Although the methodology was similar to that of Fermat, de Witt's notation was the modern one of Descartes.

For example, de Witt proceeded, as Fermat had, to show that $y^2 = ax$ represents a parabola. He also showed the graphs of parabolas determined by such equations as $y^2 = ax + b^2$, $y^2 = ax - b^2$, $y^2 = b^2 - ax$, and the equations formed from these by interchanging x and y. As before, de Witt showed only that part of the graph for which both x and y are positive. But he also considered in detail the more complicated equation

$$y^2 + \frac{2bxy}{a} + 2cy = bx - \frac{b^2x^2}{a^2} - c^2.$$

Setting $z = y + bx/a + c$ reduces this equation to

$$z^2 = \frac{2bc}{a}x + bx,$$

or, with $d = 2bc/a + b$, to $z^2 = dx$, an equation that de Witt knew represented a parabola. He then showed how to use this transformation to draw the locus. If the coordinates of D are (x, y), using AE as the x-axis and AF as the y-axis, set $BE = AG = c$ and extend DB to C such that $GB : BC = a : b$, or $BC = bx/a$ (Fig. 10.7). It follows that $DC = y + c + bx/a = z$. Also, setting $GB : GC = a : e$ gives $GC = ex/a$. In modern terminology, de Witt had used the transformation

$$x = \frac{a}{e}x' \qquad y = z - \frac{b}{e}x' - c$$

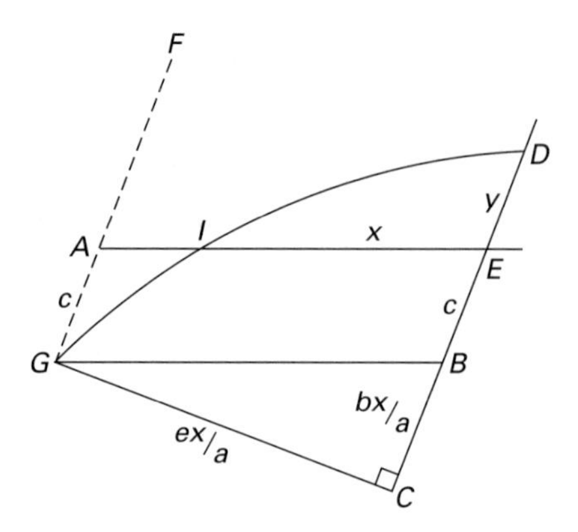

FIGURE 10.7 De Witt's construction of the parabola $y^2 + \frac{2bxy}{a} + 2cy = bx - \frac{b^2x^2}{a^2} - c^2$

to convert from the oblique axes AE and AF to the perpendicular axes GC and GF; thus, the point D with coordinates (x, y) has new coordinates (x', z) related to the original ones via this transformation. In the new coordinates, the equation of the curve is $z^2 = (da/e)x'$, a parabola with vertex G, axis GC, and latus rectum of length da/e. De Witt could thus draw this parabola, or, more particularly, the part ID above the original axis AE, because only that part can serve as the desired locus. De Witt completed the proof by noting that, given an arbitrary point D on this locus, the basic property of a parabola implies that the square on DC equals the rectangle on GC $(= ex/a)$ and the latus rectum da/e. Thus, $z^2 = dx$, and substituting $z = y + c + bx/a$ gives back the original equation.

De Witt similarly gave detailed treatments of both the ellipse and the hyperbola, presenting standard forms and then showing how other equations can be reduced to one of these by appropriate substitutions. He concluded his work by noting that any quadratic equation in two variables can be transformed into one of the standard forms and therefore represents a straight line, a circle, or a conic section. Although both Fermat and Descartes had sketched this same result, it was de Witt who provided all the details to solve the locus problem for quadratic equations.

10.3 Elementary Probability

10.3.1 Blaise Pascal and the Beginnings of the Theory of Probability

Around 1660, probability entered European thought, and in two senses: first, as a way of understanding stable frequencies in chance processes, and second, as a method of determining reasonable degrees of belief. The work of Blaise Pascal (1623–1662) exemplifies both of these senses. Pascal became interested in chance processes in attempting to respond to two gambling questions asked of him by Antoine Gombaud, the chevalier de Méré. One was on the equitable division of stakes in a set of games interrupted before its conclusion; the second was on the number of tosses of two dice necessary to have at least an even chance of getting a double six. On the other hand, in Pascal's decision-theoretic argument for belief in God there is no concept whatever of chance, just the idea of degrees of belief.

Because gambling is one of humanity's oldest leisure activities, it is likely that from earliest times people had considered the basic ideas of probability, at least on an empirical basis, and had formed some vague conception of how to calculate the odds of the occurrence of any given event in a gambling game. In fact, archaeologists have discovered dice from several ancient cultures, although there is no real evidence of how games were played. In the Renaissance, Cardano studied some mathematical ideas relating to gambling (he was an avid gambler himself), but we will begin our study of probability with Pascal's solution to the division problem, described in several letters to Fermat in 1654 and then in more detail a few years later at the end of his *Treatise on the Arithmetical Triangle*.

In the division problem, we assume that each player contributes a certain stake and that the rules state that the winner of a certain number of games is to win the entire sum of money. The question, then, is how to divide the money if the set of games is interrupted before either player has won the stated number. Pascal began his discussion with two basic principles to apply to the division. First, if the position of a given player is such that a certain sum belongs to him whether he wins or loses, he should receive that sum even if the set is halted. Second, if the position of the two players is such that if one wins, a certain sum belongs to him and if he loses, it belongs to the other, and if both players have equally good chances of winning, then they should divide the sum equally if they are unable to continue to play.

Pascal next noted that what determines the split of the stakes is the number of games remaining and the total number that the rules say either player must win to obtain the entire stake. Therefore, if they are playing to win 2 games and the score stands at 1 to 0, or to win 3 games and the score is 2 to 1, or to win 11 games and the score is 10 to 9, the results of the division of the stakes at the time of interruption should all be the same. In all these cases, the first player needs to win one more game, while the second player needs to win two.

As an example of Pascal's principles, suppose that the total stake in the contest is $80. First, if each player needs to win one game and the contest is stopped, simply divide the $80 in half, so each gets $40. Second, suppose that the first player needs to win one game to win the set and the second player needs to win two. If the first player wins the next game, he will receive $80. If he loses, then both players will need to win one game, so by the first case, the first player will receive $40. If they stop the contest now, the first player is therefore entitled to the $40 he would win in any case plus half of the remaining $40, that is, to $60, the mean of the two possible amounts he could win. Similarly, if the first player needs one game to win while the second player needs three, there are two possibilities for the next game. If the first player triumphs, he wins the $80; if he loses, the situation is the same as in the second case, in which he is entitled to $60. It follows that if that next game is not played, the first player should receive $60 plus half the remaining $20, that is, $70, the mean of his two possible winnings.

The general solution to the division problem, it turns out, requires some of the properties of Pascal's triangle. Before considering Pascal's solution, therefore, we must first look at his construction and use of what he called the **arithmetical triangle**, the triangle of numbers that had already been used in various parts of the world for more than 500 years. Pascal's *Treatise on the Arithmetical Triangle*, famous also for its explicit statement of the principle of mathematical induction, begins with his construction of the triangle starting with a 1 in the upper left-hand corner and then using the rule that each number is found by adding together the number above it and the number to its left (Fig. 10.8). In the discussion of Pascal's results, however, it will be clearer to use the modern table format (Table 10.1) and modern notation to identify the various entries in the triangle. The standard binomial

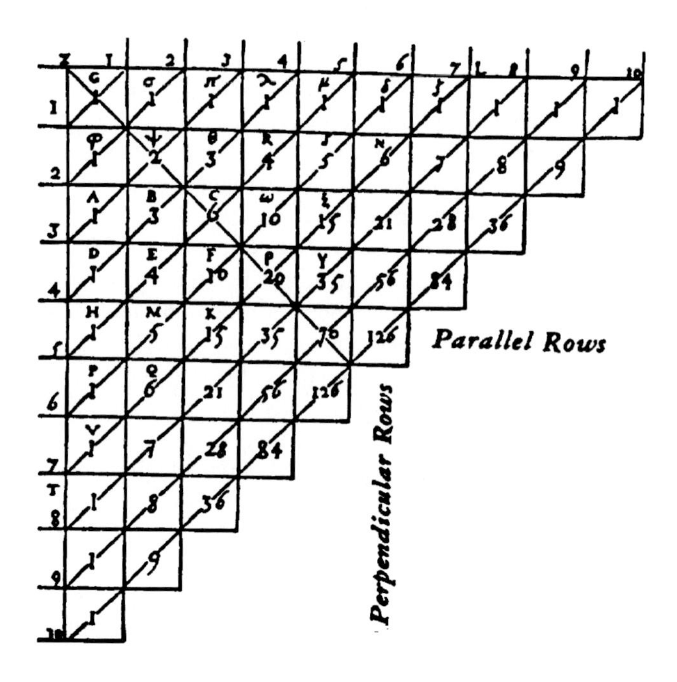

FIGURE 10.8 Pascal's version of the arithmetical triangle

TABLE 10.1
Tabular Arrangement of Pascal's Triangle

Row\Column	0	1	2	3	4	5	6	7	8
0	1								
1	1	1							
2	1	2	1						
3	1	3	3	1					
4	1	4	6	4	1				
5	1	5	10	10	5	1			
6	1	6	15	20	15	6	1		
7	1	7	21	35	35	21	7	1	
8	1	8	28	56	70	56	28	8	1

symbol $\binom{n}{k}$ will be used to name the kth entry in the nth row (where the initial column and initial row are each numbered 0). The basic construction principle is then that

$$\binom{n}{k} = \binom{n-1}{k} + \binom{n-1}{k-1}.$$

Pascal began his study by considering how various entries are related to sums of others. For example, Pascal's third consequence (of the definition of the triangle) states that any entry is the sum of all the elements in the preceding column up to the preceding row:

$$\binom{n}{k} = \sum_{j=k-1}^{n-1} \binom{j}{k-1}.$$

His eighth consequence is that the sum of the elements in the nth row is equal to 2^n.

Although his proof of the eighth consequence essentially used mathematical induction, it is only in the proof of the twelfth consequence that Pascal stated the principle explicitly, not in all generality but just in the context of the specific result to be proved:

$$\binom{n}{k} : \binom{n}{k+1} = (k+1) : (n-k).$$

Pascal noted that "although this proposition has an infinity of cases, I shall demonstrate it very briefly by supposing two lemmas"—namely, the two basic parts of an induction argument. "The first, which is self evident, [is] that this proportion is found in the [first row], for it is perfectly obvious that $\left[\binom{1}{0} : \binom{1}{1} =\right]$ 1 : 1. The second [is] that if this proportion is found in any [row], it will necessarily be found in the following [row]. Whence it is apparent that it is necessarily in all the [rows]. For it is in the [first row] by the first lemma; therefore by the second lemma it is in the [second row], therefore in the [third], and to infinity." Although this is a clear statement of the induction principle for the specific case at hand, and of the reason for its use in demonstrating a general result, Pascal did not prove the second lemma generally but used the method of generalizable example to show that the truth of the lemma in the third row implies its truth in the fourth. Thus, to demonstrate that $\binom{4}{1} : \binom{4}{2} = 2 : 3$, he first noted that $\binom{3}{0} : \binom{3}{1} = 1 : 3$ and therefore that $\binom{4}{1} : \binom{3}{1} = \left[\binom{3}{1} + \binom{3}{0}\right] : \binom{3}{1} = 4 : 3$. Next, since $\binom{3}{1} : \binom{3}{2} = 2 : 2$, it follows that $\binom{4}{2} : \binom{3}{1} = \left[\binom{3}{2} + \binom{3}{1}\right] : \binom{3}{1} = 4 : 2$. The desired result comes from dividing the first of these two proportions by the second. Pascal was aware that this proof is not general, for he completed it by noting that "the proof is the same for all other [rows], since it requires only that the proportion be found in the preceding [row], and that each [entry] be equal to the [entry above it and the entry to the left of that one], which is everywhere the case." In any case, this twelfth consequence enabled Pascal to demonstrate easily, by compounding ratios, that

$$\binom{n}{k} : \binom{n}{0} = (n-k+1)(n-k+2)\cdots n : k(k-1)\cdots 1,$$

or, since $\binom{n}{0} = 1$, that

$$\binom{n}{k} = \frac{n(n-1)\cdots(n-k+1)}{k!}.$$

Having set out the basic properties of the arithmetical triangle, Pascal applied it in several areas. He demonstrated that $\binom{n}{k}$ equals the number of combinations of k elements in a set of n elements. He showed that the row entries in the triangle are the binomial coefficients, that is, that the numbers in row n are the coefficients of the powers of a in the expansion

$(a+1)^n$. But one of the more important applications of the triangle, Pascal believed, was to the problem of the division of stakes, which he solved via the following

THEOREM. *Suppose that the first player lacks r games of winning the set while the second player lacks s games where both r and s are at least 1. If the set of games is interrupted at this point, the stakes should be divided so that the first player gets that proportion of the total as* $\sum_{k=0}^{s-1}\binom{n}{k}$ *is to* 2^n, *where* $n = r + s - 1$ *(the maximum number of games left).*

This theorem asserts that the probability of the first player's winning is the ratio of the sum of the first s terms of the binomial expansion of $(1+1)^n$ to the total 2^n. One can consider the first term of the expansion as giving the number of chances for the first player to win n games, the second the number of chances for that player to win $n-1$, and so on, while the sth term gives the number of chances to win $n-(s-1) = r$ games. Since one may as well assume that in fact exactly n more games must be played, these coefficients give all of the ways the first player can win.

Pascal had thus answered completely de Méré's problem of division. In his correspondence with Fermat, the two men discussed the same problem when there were more than two players and found themselves in agreement on the solution. Pascal also mentioned briefly the other problem: determining the number of throws of two dice for which there are even odds that a pair of sixes will occur. He noted that in the analogous problem for one die, the odds for throwing a six in four throws are 671 to 625, but he did not show his method of calculating the result. De Méré evidently believed that since four throws were sufficient to guarantee at least even odds in the case of one die (where there are six possible outcomes), the same ratio $4:6$ would hold no matter how many dice were thrown. Because there were 36 possibilities in tossing two dice, he thought that the correct value should be 24. He probably posed the question to Pascal because this value did not seem to be empirically correct. Pascal noted that the odds are against success in 24 throws, but did not detail in his letters or in any other work the theory behind this statement.

Pascal's decision-theoretic argument in favor of belief in God demonstrates the second side of probabilistic reasoning, as a method of coming to a "reasonable" decision. Either God is or God is not, according to Pascal. One has no choice but to "wager" on which of these statements is true, where the wager is in terms of one's actions. In other words, a person may act either with complete indifference to God or in a way compatible with the (Christian) notion of God. Which way should one act? If God is not, it does not matter much. If God is, however, wagering that there is no God will bring damnation, while wagering that God exists will bring salvation. Because the latter outcome is infinitely more desirable than the former, the outcome of the decision problem is clear, even if one believes that the probability of God's existence is small: The "reasonable" person will act as if God exists.

10.3.2 Christian Huygens and the Earliest Probability Text

Pascal's argument in favor of belief in God is certainly valid, given his premises. (Whether one accepts the premises is a different matter.) In fact, his notion of somehow calculating the "value" of a particular action became the basis for the first systematic treatise on probability, written in 1656 by Christian Huygens (1629–1695), a student of van Schooten. Huygens became interested in the question of probability during a visit to Paris in 1655 and wrote a brief book on the subject, *On the Calculations in Games of Chance*, which appeared in print in 1657.

Huygens's work contained only fourteen propositions and concluded with five exercises for the reader. The propositions included ones dealing with both of de Méré's problems, but Huygens also gave detailed discussions of the reasoning behind the solutions. In particular, he showed how to calculate the "value" of a particular outcome. Huygens's notion of value is similar to Pascal's notion of a wager on God's existence, but in the case of games of chance, Huygens could calculate it explicitly. In modern terms, the value of a chance is the **expectation**, the average amount that one would win if one played the game many times. It is this amount that a player would presumably pay to have the privilege of playing an equitable game. For example, this is Huygens's first proposition: "To have equal chances of winning [amounts] a or b is worth $(a + b)/2$ to me." This proposition is the same as one of the principles Pascal stated in solving the division problem. Huygens, however, gave a proof. He postulated two players, each putting in a stake of $(a + b)/2$ and each having the same chance of winning. If the first wins, he receives a and his opponent b. If the second wins, the payoffs are reversed. Because of the symmetry, Huygens considered this an equitable game. In modern terminology, since the probability of winning each of a or b is $\frac{1}{2}$, the expectation for each player is $\frac{1}{2}a + \frac{1}{2}b$, Huygens's "value" of the chance. Huygens then generalized this result in his third proposition: "To have p chances to win a and q chances to win b, the chances being equivalent, is worth $(pa + qb)/(p + q)$ to me."

Huygens took as an axiom that each player in an equitable game would be willing to risk the calculated fair stake and would not be willing to risk more. In fact, however, as the history of gambling shows, that assumption is, at the very least, debatable. It is not at all clear that the fair stake defined by Huygens is the most a given person is willing to pay for the chance to participate in a game. The success of state-run lotteries, not to mention the gambling palaces in Las Vegas and Atlantic City, testifies to precisely the opposite. Nevertheless, Huygens based the remainder of his treatise on the results of his third proposition, and even today the concept of expectation is considered a useful one.

Huygens discussed de Méré's problem of the dice in his eleventh proposition, where he showed how to determine the number of times two dice should be thrown, so that one would be willing to wager $\frac{1}{2}a$ in order to win a if two sixes appear in that many throws. Huygens proceeded in stages. Supposing that one wins a when two sixes turn up, he argued that on the first throw one has 1 chance of winning a and 35 chances of winning 0, so the value of a chance on one throw is $\frac{1}{36}a$. If the player fails on the first throw, he takes a second, whose value is naturally the same $\frac{1}{36}a$. Hence, for the first throw, the player has 1 chance of winning a and 35 chances of taking the second throw, which is worth $\frac{1}{36}a$. The value of his chance of throwing a double six on the two throws is, by the third proposition,

$$\frac{1a + 35\left(\frac{1}{36}\right)a}{1 + 35} \quad \text{or} \quad \frac{71}{1296}a.$$

Huygens next moved to the case of four throws. If the player gets a double six on one of the first two plays, he wins a; if not, he has a second pair of chances, the value of which is $\frac{71}{1296}a$. Since there are 71 chances of winning a on the first pair of plays and therefore 1225 chances of not winning (out of 1296), there are 1225 chances of reaching the second pair of plays, whose value is also $\frac{71}{1296}a$. Again by the third proposition, the value of the player's chance of throwing a double six in four throws is

$$\frac{71a + 1225\left(\frac{71}{1296}\right)a}{1296} \quad \text{or} \quad \frac{178{,}991}{1{,}679{,}616}a.$$

Because this value is still considerably less than the desired $\frac{1}{2}a$, Huygens had to continue the process. Although he did not present any further calculations, he noted that one next considers eight throws, then sixteen, and then twenty-four and twenty-five. The results show that in twenty-four throws the player is at a very slight disadvantage on the bet of $\frac{1}{2}a$, while for twenty-five throws he has a very slight advantage.

At the conclusion of his brief treatise, Huygens presented as exercises some problems of drawing different colored balls from a set of balls, problems of the type that today appear in every elementary probability text. These problems were discussed by many mathematicians over the next decades, especially since Huygens's text was the only introduction available to the theory of probability until the early eighteenth century. Even then, its influence continued, because James Bernoulli incorporated it into his own more extensive work on probability, the *Ars conjectandi* of 1713.

10.4 Number Theory

Fermat, involved in the beginnings of analytic geometry and probability, also made contributions to number theory, which were virtually ignored during his lifetime and indeed until the middle of the next century. One of the reasons for this was probably his deep secrecy about his methods. Thus, although many of his results are known, because he announced them proudly in letters to his various correspondents and presented them with challenges to solve similar problems, there is virtually no record of any of his proofs and only vague sketches of some of his methods.

Fermat's earliest interest in number theory grew out of the classical concept of a perfect number, one equal to the sum of all of its proper divisors. Book IX of Euclid's *Elements* contains a proof that if $2^n - 1$ is prime, then $2^{n-1}(2^n - 1)$ is perfect. The Greeks had, however, only been able to discover four perfect numbers—6, 28, 496, and 8128—because it was difficult to determine the values of n for which $2^n - 1$ is prime. Fermat discovered three propositions that could help in this regard, which he communicated to Mersenne in a letter in June, 1640. The first of these results was that if n is not itself prime, then $2^n - 1$ cannot be prime. The basic question therefore reduced to asking for which primes p is $2^p - 1$ prime. Such primes are today called **Mersenne primes** in honor of Fermat's favorite correspondent.

Fermat's second proposition was that if p is an odd prime, then $2p$ divides $2^p - 2$, or p divides $2^{p-1} - 1$. His third was that, with the same hypothesis, the only possible divisors of $2^p - 1$ are of the form $2pk + 1$. Fermat indicated no proofs of these results in his letter, but only gave a few numerical examples. He confirmed that $2^{37} - 1$ was composite by testing its divisibility by numbers of the form $74k + 1$ until he found the factor $223 = 74 \cdot 3 + 1$. But in a letter written a few months later to Bernard Frenicle de Bessy (1612–1675), he stated a more general theorem of which these two propositions are easy corollaries. This theorem, today known as **Fermat's Little Theorem**, is, in modern terminology, that if p is any prime and a any positive integer, then p divides $a^p - a$. Fermat gave no indication in any of his writings how he discovered or proved this result. In any case, the second of the propositions in the letter to Mersenne is simply the case $a = 2$ of the theorem (where $p > 2$). The third proposition requires only a bit more work. Suppose q is a prime divisor of $2^p - 1$. Since, according to the Little Theorem, q already divides $2^{q-1} - 1$, we know that p divides $q - 1$ or that $q - 1 = hp$ for some integer h. Because $q - 1$ is even, 2 must divide hp and therefore must divide h. It follows that $h = 2k$ and $q = 2kp + 1$, as asserted.

Fermat's Little Theorem turned out to be an extremely important result in number theory, with many applications. But his most famous contribution to number theory, the result known today as Fermat's Last Theorem, probably grew out of his study of a method of proof known as the **method of infinite descent**. This method demonstrates the nonexistence of positive integers having certain properties by showing that the assumption that one integer has such a property implies that a smaller one has the same property. By continuing the argument, one gets an infinite decreasing sequence of positive integers, an impossibility.

Fermat used the method of infinite descent in the only number theoretic proof that he actually wrote out in detail, the proof that it is impossible to find an integral right triangle whose area is a square. In other words, he showed that it is impossible to find integers x, y, z, w such that $x^2 + y^2 = z^2$ and $\frac{1}{2}xy = w^2$. Fermat knew that any Pythagorean triple (x, y, z) with the numbers relatively prime could be generated by a pair of relatively prime numbers p, q of opposite parity, with $p > q$, by setting $(x, y, z) = (2pq, p^2 - q^2, p^2 + q^2)$. Now suppose that there existed an integral right triangle whose area was a square. Then $\frac{1}{2}xy = pq(p^2 - q^2)$ would be a square. Because the factors of this product are relatively prime, each of them must also be a square. So $p = d^2$, $q = f^2$, and $p^2 - q^2 = d^4 - f^4 = c^2$. Next, Fermat noted that since $c^2 = (d^2 + f^2)(d^2 - f^2)$ and since d and f are relatively prime, $d^2 + f^2$ and $d^2 - f^2$ must both also be squares, say, $d^2 + f^2 = g^2$ and $d^2 - f^2 = h^2$. Subtracting the second of the two equations from the first gives $2f^2 = g^2 - h^2 = (g + h)(g - h)$. Because g^2 and h^2 are both odd and relatively prime, $g + h$ and $g - h$ are both even and can have no common factor other than 2. It follows that $g + h$ can be written as $2m^2$ and $g - h$ as n^2 (or vice versa), where n is even and m odd. So $g = m^2 + n^2/2$, $h = m^2 - n^2/2$, and $d^2 = \frac{1}{2}(g^2 + h^2) = (m^2)^2 + (n^2/2)^2$. But then m^2 and $n^2/2$ are sides of a new right triangle whose area $m^2n^2/4$ is also a square. Since the hypotenuse d of this new triangle is smaller than the hypotenuse of the original triangle, the method of infinite descent implies that the original assumption must be false.

One can pull out of this argument an argument by infinite descent showing that one cannot find three positive integers a, b, c such that $a^4 - b^4 = c^2$. It follows that one also cannot express a fourth power as a sum of two other fourth powers. Fermat wrote a generalization of this result—that "one cannot split a cube into two cubes, nor a fourth power into two fourth powers, nor in general any power beyond the square *in infinitum* into two powers of the same name"—as a marginal note to Diophantus's problem II-8 in his copy of the 1621 Latin edition of the *Arithmetica*. This generalization is the content of what has become known as **Fermat's Last Theorem**. In modern terms, the theorem asserts that there do not exist nonzero integers a, b, c, and $n > 2$, such that $a^n + b^n = c^n$. This result, of which Fermat claimed that he had "a truly marvelous demonstration ... which this margin is too narrow to contain," provided mathematicians since the seventeenth century with a major challenge, one that was finally met in 1995. We will discuss some of the mathematics that led to an eventual proof of the theorem in a later chapter. But since the techniques involved in the modern proof are well beyond anything Fermat knew, most historians believe that

Fermat erred in his own claim of a proof. Perhaps he erroneously assumed that the method of infinite descent, which works in the case $n = 4$, would generalize to larger values of n.

Fermat tried on many occasions to stimulate other European mathematicians to work on his various number theoretic problems, but his pleas fell on deaf ears. Not until the next century was there a successor to continue the work in number theory begun by the French lawyer.

Exercises

1. Solve $x^3 = 300x + 432$ using Girard's technique, given that $x = 18$ is one solution.
2. This problem illustrates Girard's geometric interpretation of negative solutions to polynomial equations: Let two straight lines DG, BC intersect at right angles at O (Fig. 10.9). Determine A on the line bisecting the right angle at O so that $ABOF$ is a square of side 4. Draw ANC as in the figure so that $NC = \sqrt{153}$. Find the length FN. (Girard notes that if $x = FN$, then $x^4 = 8x^3 + 121x^2 + 128x - 256$, and so there are four possible solutions, each of which can be calculated. The two positive solutions are represented by FN and FD, while the two negative ones are represented by FG and FH, the latter two taken in the opposite direction from the former two.)

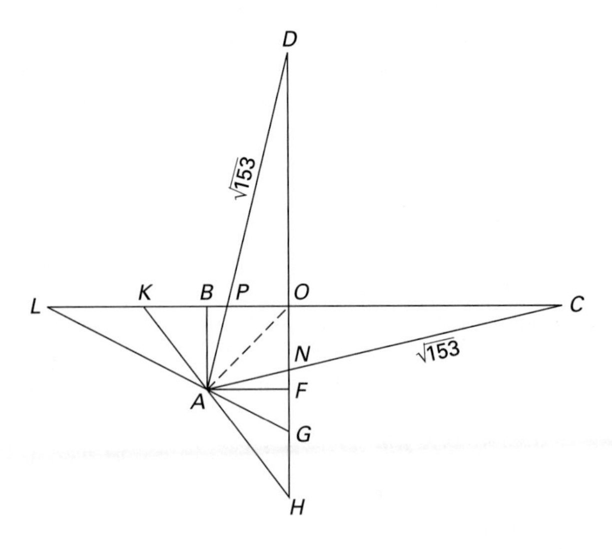

FIGURE 10.9 A problem from Girard

3. Assuming that $xy = c$ represents a hyperbola with the x and y axes as asymptotes, show that $xy + c = rx + sy$ also represents a hyperbola. Find its asymptotes.
4. Determine the locus of the equation $b^2 - 2x^2 = 2xy + y^2$. (Hint: Add x^2 to both sides.)
5. Descartes was able to construct the product and quotient of two quantities using similar triangles. Suppose AB is taken equal to 1, and one wants to multiply BD by BC (Fig. 10.10). Join AC and draw DE parallel to CA. Show that BE is the product of BD and BC. Similarly, given two lengths BE and BD, construct the quotient length.

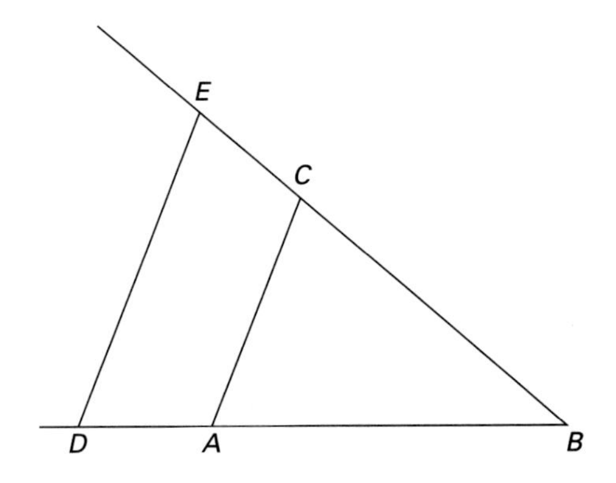

FIGURE 10.10 The product and quotient of two quantities

6. Show that MQ and MR in Fig. 10.3 represent the two solutions to the equation $z^2 = ax - b^2$.
7. Using the various constants mentioned in the text in the discussion of Descartes's *Geometry*, determine the equation of the locus that solves the four-line problem in the special case where the product of the first two lines equals the product of the second two. What type of curve is this?

8. Show that the equation Descartes found of the curve generated by the machine in Figure 10.6, $y^2 = cy - (c/b)xy + ay - ac$, is the equation of a hyperbola. In terms of Figure 10.6, determine the asymptotes of this hyperbola.

9. Using Figure 10.11, show that the curve CEG, which solves the five-line problem in section 10.2, comes from the intersection of the ruler with the parabola CKN. Let $KL = a$ and let a also be the parameter of the parabola. Let $GA = 2a$, $CB = MA = y$, and $CM = AB = x$. Use the similarity of triangles GMC and CBL to show that $GM : MC = CB : BL$. Translate this proportion into algebra to find BL and then BK. Then use the fact that BK is on the axis of a parabola with parameter a to show that the equation of curve CEG is $y^3 - 2ay^2 - a^2y + 2a^3 = axy$.

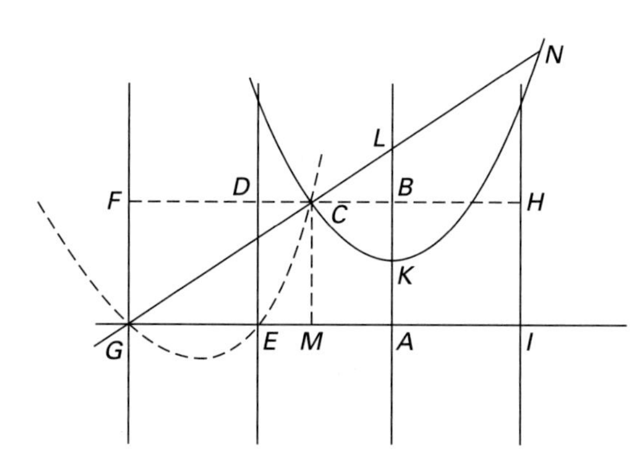

FIGURE 10.11 Curve CEG, a solution to the five-line problem

10. In de Witt's substitution $z = y + (b/a)x + c$, which simplifies the equation

$$y^2 + \frac{2bxy}{a} + 2cy = bx - \frac{b^2x^2}{a^2} - c^2,$$

he has rotated one of the axes through an angle α. Find the sine and cosine of α.

11. Show that de Witt's equation

$$y^2 + \frac{2bxy}{a} + 2cy = \frac{fx^2}{a} + ex + d$$

represents a hyperbola. (Use the substitution $z = y + (b/a)x + c$, and show that this substitution, when combined with a substitution of the form $x' = \beta x$, converts the original oblique x-y coordinate system into a new x'-z coordinate system based on perpendicular axes.) Sketch the curve.

12. Prove by induction on n that

$$\binom{n}{k} = \sum_{j=k-1}^{n-1} \binom{j}{k-1}$$

for all k less than n.

13. Prove that

$$\binom{n}{k} : \binom{n}{k+1} = (k+1) : (n-k).$$

14. Prove that

$$\binom{n}{k} : \binom{n-1}{k} = n : (n-k).$$

15. Pascal stated that the odds in favor of throwing a 6 in four throws of a single die are 671 to 625. Show why this is true.

16. Show that the odds against at least one 1 appearing in a throw of three dice are 125 : 91.

17. Determine the appropriate division of the stakes in a series between two players if the first player is lacking three games to win and the second four.

18. Suppose three players play a fair series of games under the condition that the first player to win three games wins the stakes. If they stop play when the first player needs one game while the second and third players each need two games, find the fair division of the stakes. (This problem was discussed in the correspondence between Pascal and Fermat.)

19. If two players play a game with two dice under the condition that the first player wins if the sum thrown is 7, the second wins if the sum is 6, and the stakes are split if there is any other sum, find the expectation (value of the chance) of each player.

20. If I and another player throw two dice alternately under the condition that I win when I have thrown a 7 and he wins when he throws a 6, and if he throws first, what is the ratio of my chance to his?

21. We have twelve balls, four of which are white and eight black. Three blindfolded players, A, B, C draw a ball in turn, first A, then B, then C. The winner is the one who first draws a white ball. Assuming that each (black) ball is replaced after being drawn, find the ratio of the chances of the three players.

22. Prove that if p is prime, then $2^p \equiv 2 \pmod{p}$ by writing $2^p = (1+1)^p$, expanding by the binomial theorem, and noting that all of the binomial coefficients $\binom{p}{k}$ for $1 \le k \le p-1$ are divisible by p. Prove $a^p \equiv a \pmod{p}$ by induction on a, using this result and the fact that $(a+1)^p \equiv a^p + 1 \pmod{p}$.

23. For a proof of Fermat's Little Theorem for the case where a and p are relatively prime, consider the remainders of the numbers $1, a, a^2, \ldots$ on division by p. These remainders must ultimately repeat (why?), and so $a^{n+r} \equiv a^r \pmod{p}$ or $a^r(a^n - 1) \equiv 0 \pmod{p}$ or $a^n \equiv 1 \pmod{p}$. (Justify each of these alternatives.) Take n as the smallest positive integer satisfying the last congruence. By applying the division algorithm, show that n divides $p - 1$.
24. The best-known of Descartes's statements is, "I think, therefore I am," from the *Discourse on Method.* The context is Descartes's resolve to accept only those ideas that are self-evidently true. There is a well-known joke based on this statement: Descartes goes into a restaurant. The waiter asks him, "Would you like tonight's special?" He replies, "I think not," and disappears. Comment on the logical validity of this joke.
25. Compare the analytic geometries of Descartes, Fermat, and De Witt. Adapt the formulation of one of these authors and develop a presentation of the subject to give to a precalculus class.
26. What advances in technique and/or understanding enabled Girard to state that every algebraic equation has as many solutions as its degree? Why were Cardano and Viète unable to make such an assertion?
27. Outline a lesson on the principle of mathematical induction using material from Pascal's *Treatise on the Arithmetical Triangle.*
28. Compare Pascal's use of mathematical induction to its use by ibn al-Haytham, al-Samaw'al, and Levi ben Gerson.

References

The first two chapters of Helena Pycior, *Symbols, Impossible Numbers, and Geometric Entanglements* (Cambridge: Cambridge University Press, 1997) discuss the contributions of Oughtred and Harriot to the development of symbolic algebra. A more recent book dealing with the same subject is Jacqueline A. Stedall, *A Discourse Concerning Algebra: English Algebra to 1685* (Oxford: Oxford University Press, 2002). Carl Boyer has written the only general history of analytic geometry: *History of Analytic Geometry* (New York: Scripta Mathematica, 1956). This work covers the subject from its beginnings in ancient times up through the nineteenth century. The best general histories of probability that deal to some extent with the seventeenth century and earlier are F. N. David, *Games, Gods, and Gambling* (New York: Hafner, 1962) and Ian Hacking, *The Emergence of Probability* (Cambridge: Cambridge University Press, 1975). The latter book is more philosophical, while the former discusses the relevant texts in more detail. A general history of number theory from a historical point of view is André Weil, *Number Theory: An Approach through History from Hammurapi to Legendre* (Boston: Birkhäuser, 1983).

Oughtred's *Clavis* was translated into English in 1647 by Robert Wood, under the title *The Key of the Mathematics, New Forged and Filed.* Although there is no recent reprint, there are studies of this work in Florian Cajori, *William Oughtred: A Great Seventeenth Century Teacher of Mathematics* (Chicago: Open Court, 1916) and in Jacqueline Stedall, "Ariadne's Thread: The Life and Times of Oughtred's *Clavis*," *Annals of Science* 51 (2000), 27–60. Harriot's *Praxis* was translated into English by Muriel Seltman as part of her MSc dissertation at the University of London. Although this translation has not been published, several recent articles discuss Harriot's work, including his manuscript notes. These include Jacqueline Stedall, "Rob'd of glories: The Posthumous Misfortunes of Thomas Harriot and His Algebra," *Archive for the History of Exact Sciences* 54 (2000), 455–497; Muriel Seltman, "Harriot's Algebra: Reputation and Reality," in Robert Fox, ed., *Thomas Harriot: An Elizabethan Man of Science* (Aldershot: Ashgate, 2000); and J. A. Lohne, "Essays on Thomas Harriot," *Archive for History of Exact Sciences* 20 (1979), 189–312. Girard's *A New Discovery in Algebra* is available in a translation by Ellen Black in *The Early Theory of Equations: On Their Nature and Constitution* (Annapolis: Golden Hind Press, 1986). Fermat's *Introduction to Plane and Solid Loci* has been translated by Joseph Seidlin and appears in David Eugene Smith, *A Source Book in Mathematics* (New York: Dover, 1959), vol. 2, 389–396. The standard modern version of Descartes's *Geometry* is David Eugene Smith and Marcia L. Latham, trans., *The Geometry of René Descartes* (New York: Dover, 1954). This book contains the original French and the English translation on facing pages, as well as Descartes's original notation and diagrams. The first book of Jan de Witt's *Elements of Curves* was recently translated by A. W. Grootendorst and Miente Bakker as *Jan de Witt's Elementa Curvarum linearum, Liber Primus* (New York: Springer, 2000). Pascal's *Treatise on the Arithmetical Triangle*, in a translation by Richard Scofield, is available in *Great Books of the Western World* (Chicago: Encyclopaedia Britannica, 1952), vol. 33. This edition also contains the letters between Pascal and Fermat on probability. Huygens's *On the Calculations in Games of Chance* is in his *Oeuvres completes* (The Hague: 1888–1950), vol. 14.

The best study of the life and mathematical work of Fermat is Michael S. Mahoney, *The Mathematical Career of Pierre de Fermat 1601–1665*, 2nd ed. (Princeton: Princeton University Press, 1994). This book contains a detailed analysis of Fermat's work, not only in analytic geometry and number theory but also in the various aspects of the calculus (which we will consider in chapter 11). A fascinating work on Descartes's discussion of curves is H. J. M. Bos, "On the Representation of Curves in Descartes' *Géométrie,*" *Archive for History of Exact Sciences* 24 (1981), 295–338. Bos discusses Descartes's general program for geometry, which is outlined in the *Geometry*. For more information on Descartes's methodology, see Judith Grabiner, "Descartes and Problem-Solving," *Mathematics Magazine* 68 (1995), 83–97. For more on his curve-drawing devices, see David Dennis, "René Descartes' Curve-Drawing Devices: Experiments in the Relations Between Mechanical Motion and Symbolic Language," *Mathematics Magazine* 70 (1997), 163–175. For a modern discussion of de Méré's dice problem, see Jane B. Pomeranz, "The Dice Problem—Then and Now," *College Mathematics Journal* 15 (1984), 229–237.

CHAPTER ELEVEN

Calculus in the Seventeenth Century

6accdæ13eff7i3l9n4o4qrr4s8t12ux. This is the anagram Newton wrote to conceal the basic goal of his version of the calculus in his second (and last) letter for Leibniz (the Epistola posterior), sent via Henry Oldenburg on October 24, 1676. It represented the Latin phrase Data æquatione quotcunque fluentes quantitates involvente, fluxiones invenire; et vice versa ("Given an equation involving any number of fluent quantities, to find the fluxions, and vice versa").

Building on the work of many mathematicians over the centuries who considered the problems of determining the areas of regions bounded by curves, the volumes of regions in space, the maximum or minimum values of certain functions, or the tangents to certain curves, two geniuses of the last half of the seventeenth century, Isaac Newton and Gottfried Leibniz, created the machinery of the calculus, the foundation of modern mathematical analysis and the source of application to an increasing number of other disciplines. These problems had been attacked and solved for various special cases over the years. But virtually every solution had required an ingenious construction. No one had developed any algorithm that would enable these problems to be solved easily in new situations.

New situations did not often occur in either the Greek or Islamic setting, since those mathematicians had few ways of describing new curves for which to calculate maximums or tangents or areas. But with the advent of analytic geometry in the first half of the seventeenth century, the possibility suddenly opened up of constructing all sorts of new curves and solids. After all, any algebraic equation determined a curve, and a new solid could be formed, for example, by rotating a curve around any line in its plane. With an infinity of new examples to deal with, mathematicians of the seventeenth century sought for and discovered new ways of finding maximums, constructing tangents, and calculating areas and volumes. These mathematicians were not, however, concerned with functions. They were concerned with curves, defined by some relation between two variables. And in the process of finding tangents, they often considered other geometric aspects of the curves. Figure 11.1 illustrates some of the quantities connected with a point p on a given curve: the abscissa x, the ordinate y, the arc length s, the subtangent t, the tangent τ, the normal n, and the subnormal ν.

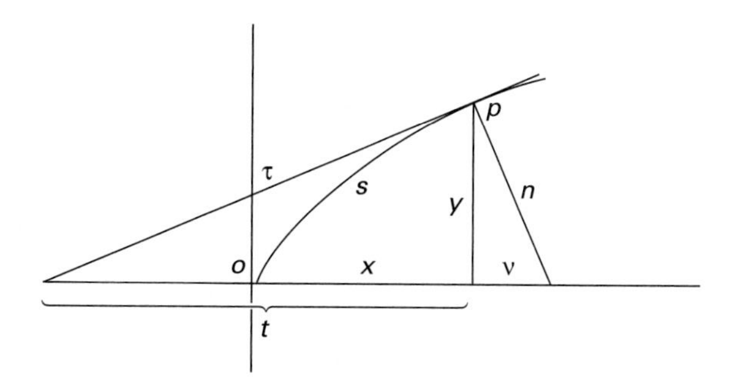

FIGURE 11.1 Quantities connected with a curve: x is the abscissa, y the ordinate, s the arc length, t the subtangent, τ the tangent, n the normal, and ν the subnormal

In this chapter, we will explore the various methods used to construct tangents and find extrema, the methods developed to determine areas and volumes, and the ways of determining the lengths of curves and their relationship to the discovery of the inverse nature of tangents and areas. We will then consider the work of Newton and Leibniz, the mathematicians who were first able to combine these various ideas into an organized whole.

11.1 Tangents and Extrema

11.1.1 Fermat's Method of Finding Extrema

Fermat, in the late 1620s, was stimulated to consider the question of extrema by a study of Viète's work relating the coefficients to the roots of a polynomial. Recall that Viète had shown that the sum of the two roots x_1, x_2 of $bx - x^2 = c$ was b by equating $bx_1 - x_1^2$ and $bx_2 - x_2^2$ and dividing through by $x_1 - x_2$. The equation $bx - x^2 = c$ comes from the geometric problem of dividing a line of length b into two parts whose product is c. Fermat knew from Euclid that the maximum possible value of c was $b^2/4$ and also that for any number less than the maximum, there were two possible values of x whose sum was b. But what happened as c approached its maximum value? The geometrical situation made it clear to Fermat that even for this maximum value, the equation had two solutions, each of the same value: $x_1 = b/2$ and $x_2 = b - x_1 = b - (b/2) = b/2$. This insight gave Fermat his method for maximizing a polynomial $p(x)$: Set $p(x_1) = p(x_2)$. Then divide through by $x_1 - x_2$ to find the relationship between the coefficients and any two roots of the polynomial. Finally, set the two roots equal to one another and solve.

From $bx_1 - x_1^2 = bx_2 - x_2^2$, Fermat derived the fact that $b = x_1 + x_2$, an equation holding for any two roots. Setting $x_1 = x_2$ $(= x)$ gives $b = 2x$. Thus, the maximum occurs when $x = b/2$. Similarly, to maximize $bx^2 - x^3$, Fermat set $bx_1^2 - x_1^3 = bx_2^2 - x_2^3$ and derived $b(x_1^2 - x_2^2) = x_1^3 - x_2^3$ and $bx_1 + bx_2 = x_1^2 + x_1x_2 + x_2^2$. He then set $x_1 = x_2$ $(= x)$ and determined that $2bx = 3x^2$, from which he concluded that $x = 2b/3$ provides the maximum value. He knew that this value was a maximum from the geometry of the situation. More generally, in other situations he used geometry to determine which answers gave maximums or minimums when there were two or more solutions to his final equation.

Fermat's method does raise a significant methodological question. How can one divide through by $x_1 - x_2$ and then set that value equal to 0? For Fermat, the geometric situation showed that the roots were distinguishable even when their difference was 0. Thus, he never felt he was dividing by 0. He simply assumed that the relationships worked out using Viète's methods were perfectly general (for example, $x_1 + x_2 = b$) and thus held for any particular values of the variables, even those at the maximum.

Fermat did realize, however, that if the polynomial $p(x)$ were somewhat complicated, the division by $x_1 - x_2$ might be rather difficult. Thus, he modified his method to avoid this. Instead of considering the two roots as x_1 and x_2, he wrote them as x and $x + e$. Then, after equating $p(x)$ with $p(x + e)$, he had only to divide by e or one of its powers. In the resulting expression, he removed any term that contained e to get an equation enabling the maximum to be found. Thus, using his original example of $p(x) = bx - x^2$, Fermat put $bx - x^2$ equal to $b(x + e) - (x + e)^2 = bx - x^2 + be - 2ex - e^2$. Canceling common terms gave him $be = 2ex + e^2$. After dividing by e, he found $b = 2x + e$. Removing the term containing e gave Fermat his known result: $x = b/2$.

By 1638, Fermat was able to adapt this method to determine the tangent to a curve. To draw a tangent line at B to a curve represented in modern notation by $y = f(x)$, pick an arbitrary point A on the tangent line and drop perpendiculars AI and BC to the axis (Fig. 11.2). Fermat's idea is then to set FI/BC equal to EI/CE, where F is the intersection of AI with the curve. If $CI = e$, $CD = x$, and $CE =$ the subtangent t, this can be written

$$\frac{f(x+e)}{f(x)} = \frac{t+e}{t},$$

or $tf(x + e) = (t + e)f(x)$. By applying his rules of canceling common terms, dividing through by e, and then removing any remaining terms containing e, Fermat could calculate the relation between t and x that determines the tangent line. For example, if the curve is the parabola $f(x) = \sqrt{x}$, then Fermat's method gives $t\sqrt{x+e} = (t + e)\sqrt{x}$. Squaring both

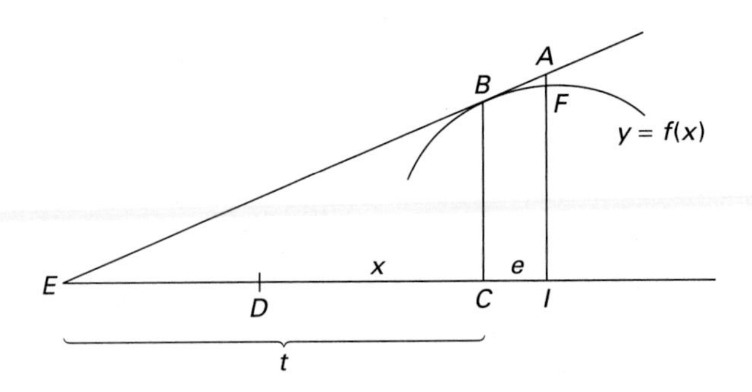

FIGURE 11.2 Fermat's method for determining subtangents

sides and simplifying gives $t^2e = 2etx + e^2x$. Dividing through by e and then removing the term still containing e produces the result $t = 2x$. This is, of course, Apollonius's result (proposition I–33) that the subtangent to the parabola at a point is double the abscissa.

11.1.2 Descartes and the Method of Normals

Descartes had also discovered a method of determining tangents, by beginning with the determination of a normal to a curve. He derived his idea for drawing a normal from the realization that a radius of a circle is always normal to the circumference. Thus, the radius of a circle that is tangent to a given curve at the given point will be normal to that curve as well. To construct a circle tangent to a curve required an idea similar to that of Fermat: that the two intersection points of a circle with the curve near the given point will become one if the circle is in fact tangent. To carry out this procedure at a point C of a curve given by $y = f(x)$, assume that P is the center of the required circle, take an arbitrary point A on the axis through P, and set $CP = n$ and $PA = v$ (Fig. 11.3). If $C = (x, y)$, then $PM = v - x$, and the equation of the circle is $n^2 = y^2 + (v - x)^2$. Now, if the circle is not tangent to the curve, it will cut the curve not only at C but also at another point E. Therefore, the equation giving the intersection of the curve $y = f(x)$ and the circle, $[f(x)]^2 + v^2 - 2vx + x^2 - n^2 = 0$, has two distinct roots. When the circle is tangent, however, these two roots coincide, and the equation has a double root. As Descartes knew from his study of roots of equations, this meant that the polynomial has a factor of $(x - x_0)^2$, where x_0 is the double root. Thus, setting $[f(x)]^2 + v^2 - 2vx + x^2 - n^2 = (x - x_0)^2q(x)$ and equating the coefficients of like powers of x, Descartes could solve for v in terms of x_0. Once he knew v, he could determine the point P.

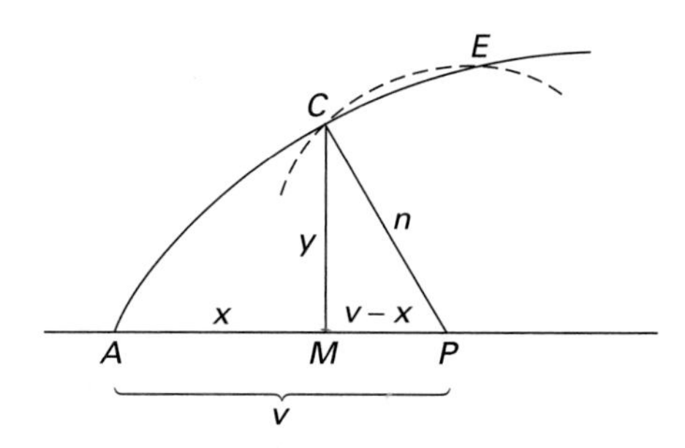

FIGURE 11.3 Descartes's method for finding normals

As an example, we determine the normal to the parabola $y = x^2$ at the point (x_0, x_0^2). In this case, the polynomial that has a double root is $(x^2)^2 + v^2 - 2vx + x^2 - n^2$. Because this is a fourth-degree polynomial, it must be equated to $(x - x_0)^2q(x)$, where $q(x)$ has degree two. Thus,

$$x^4 + x^2 - 2vx + v^2 - n^2 = (x - x_0)^2(x^2 + ax + b),$$

or

$$x^4 + x^2 - 2vx + v^2 - n^2$$
$$= x^4 + (a - 2x_0)x^3 + (b - 2x_0 a + x_0^2)x^2 + (ax_0^2 - 2bx_0)x + bx_0^2.$$

Equating coefficients gives

$$a - 2x_0 = 0$$
$$b - 2x_0 a + x_0^2 = 1$$
$$ax_0^2 - 2bx_0 = -2v$$
$$bx_0^2 = v^2 - n^2.$$

Solving the first three equations for v by setting $a = 2x_0$ and $b = 2ax_0 - x_0^2 + 1$ gives $v = 2x_0^3 + x_0$ as the (horizontal) coordinate of the desired point P. Thus, the normal is determined. We note further, however, that the slope of the normal line is

$$\frac{-y_0}{v - x_0} = \frac{-x_0^2}{2x_0^3} = \frac{-1}{2x_0},$$

and therefore the slope of the tangent line is $2x_0$, a familiar result.

11.1.3 Hudde's Algorithm

Both Fermat's and Descartes's procedures for finding tangents often led to very complicated algebra. A study of these methods, however, led Johann Hudde (1628–1704), a student of van Schooten, to discover a simpler procedure for determining a double root. This procedure was published in van Schooten's 1659 edition of Descartes's *Geometry*. **Hudde's rule**, for which he only sketched a proof, states that if a polynomial $f(x) = a_0 + a_1x + a_2x^2 + \cdots + a_nx^n$ has a double root $x = \alpha$, and if $p, p + b, p + 2b, \ldots p + nb$ is an arithmetic progression, then the polynomial $pa_0 + (p + b)a_1x + (p + 2b)a_2x^2 + \cdots + (p + nb)a_nx^n$ also has the root $x = \alpha$. In modern notation, the new polynomial can be expressed as $pf(x) + bxf'(x)$. Hudde's result follows immediately, because if $f(x)$ has a double root, then $f'(x)$ has the same root. Although his rule permitted the arbitrary choice of an arithmetic progression, Hudde most often used the progression with $p = 0, b = 1$. In this case, the new polynomial is $xf'(x)$, a result that helped to uncover the computational importance of what is now called the derivative.

As a first example of the rule, consider the problem of determining the normal to the parabola $y = x^2$, where it is necessary to find the relationship between the coefficient v and the double root x_0 of the polynomial $x^4 + x^2 - 2vx + v^2 - n^2$. Using Hudde's rule with $p = 0$ and $b = 1$ gives the new polynomial equation $4x^4 + 2x^2 - 2vx = 0$ or $4x^3 + 2x^2 - 2v = 0$. Because x_0 is a solution of this equation, it follows as before that $v = 2x_0^3 + x_0$ and therefore that the slope of the tangent line is $(v - x_0)/x_0^2 = 2x_0$. An easy generalization of this example makes it possible to show that the slope of the tangent line to $y = x^n$ at (x_0, x_0^n) is nx_0^{n-1}, a result extremely difficult to find using Descartes's procedure.

Hudde also applied his rule to the determination of extreme values, using Fermat's idea that if a polynomial $f(x)$ has an extreme value M, the polynomial $g(x) = f(x) - M$

has a double root. Thus, to maximize $x^2(b - x)$, use the rule with $p = 0, b = 1$ on the polynomial $-x^3 + bx^2 - M$. The new polynomial equation is $-3x^3 + 2bx^2 = 0$, the nonzero root of which, $x = 2b/3$, gives the desired maximum.

The importance of Hudde's rule was that it provided a general algorithm by which one could routinely construct tangents to curves given by polynomial equations. It was no longer necessary to develop a special technique for each particular curve. Anyone could now determine the tangent.

11.2 Areas and Volumes

Both Greek and Islamic mathematicians had been able to determine areas and volumes of certain regions bounded by curved lines or surfaces. The texts available, however, generally gave only the result with a proof based on the method of exhaustion. Such results gave seventeenth-century mathematicians few clues as to how to determine the areas bounded by the many new curves that had become available for study or the volumes of solid regions generated by revolving these curves around lines in the plane. The only clear idea passed down from Greek times was that somehow the given region needed to be broken up into very small regions, whose individual areas or volumes were known.

11.2.1 Infinitesimals and Indivisibles

In 1615, Kepler wrote *New Solid Geometry of Wine Bottles*, inspired by his inquiries into the method by which Austrian wine merchants determined how much wine remained in a given barrel. In this work, Kepler calculated numerous areas and volumes by procedures derived from his study of Archimedes. For example, he calculated the area of a circle of radius AB by first noting that the circle can be divided into infinitely many isosceles triangles, each with a vertex at the center A and base an infinitesimal part of the circumference. Kepler then stretched the circumference of the circle out into a straight line, and placed triangles on that line equal to the ones in the circle, arranged "one next to the other," all having the altitude AB (Fig. 11.4). It followed that the area of the triangle ABC was equal to the area of the circle, so the latter area must be one-half of the radius multiplied by the circumference. Similarly, Kepler calculated the volume of a ring (torus) by slicing it into an infinite number of very thin disks, each of which is thinner toward the center and thicker

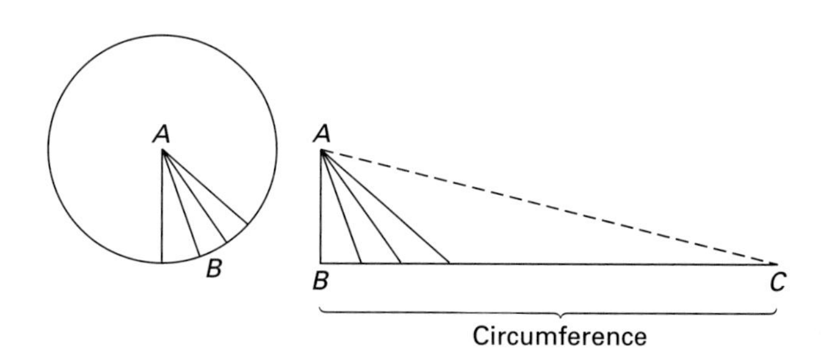

FIGURE 11.4 Kepler's method of determining the area of a circle

toward the outside. Kepler never claimed, however, that his method was rigorous, noting only that "we could obtain absolute and in all respects perfect demonstrations from these books of Archimedes themselves, were we not repelled by the thorny reading thereof."

Kepler's use of very thin disks or very small triangles are examples of what came to be called the **method of infinitesimals**. Archimedes himself, in his *Method*, had used the **method of indivisibles**, in which a given geometric object is considered to be made up of objects of dimension one lower.

It was Bonaventura Cavalieri (1598–1647), a disciple of Galileo, who first developed this idea of indivisibles into a complete theory, elaborated in his *Geometry, Advanced in a New Way by the Indivisibles of the Continua* of 1635 and his *Six Geometrical Exercises* of 1647. The central concept of Cavalieri's work was *omnes lineae*, or "all the lines" of a plane figure F, to be written as $\mathcal{O}_F(\ell)$. By this concept, Cavalieri meant the collection of intersections of the plane figure with a perpendicular plane moving parallel to itself from one side of the given figure to the other. These intersections are lines, and it is the collection of such lines, thought of as a single magnitude, that Cavalieri dealt with throughout his work. Cavalieri's lines in some sense made up the given figure, but he was careful to distinguish $\mathcal{O}_F(\ell)$ from F itself. He was also able to generalize the idea by considering objects of higher dimensions such as "all the squares" or "all the cubes" of a given figure. One can think of "all the squares" of a triangle, for example, as representing a pyramid, each of whose cross sections is a square of side the length of a particular line in the triangle.

The basis for Cavalieri's computations was a result to this day known as **Cavalieri's principle**, a principle that had already appeared in Chinese mathematics: "If two plane figures have equal altitudes and if sections made by lines parallel to the bases and at equal distances from them are always in the same ratio, then the plane figures are also in this ratio." Cavalieri proved this result by an argument using superposition. It followed that if there were a fixed ratio between corresponding lines of the two figures F and G, then $\mathcal{O}_F(\ell) : \mathcal{O}_G(\ell) = F : G$. For example, suppose the rectangle F of length a and width b is divided by its diagonal into two triangles T, S (Fig. 11.5). Since each line segment BM in triangle T corresponds to one and only one equal line segment HE in triangle S, then $\mathcal{O}_T(\ell) = \mathcal{O}_S(\ell)$. On the other hand, since every line segment BA of the rectangle is made up of one segment from triangle S and one from triangle T, $\mathcal{O}_F(\ell) = \mathcal{O}_T(\ell) + \mathcal{O}_S(\ell)$. It follows that $\mathcal{O}_F(\ell) = 2\mathcal{O}_T(\ell)$, or all the lines of the rectangle are double all the lines of the

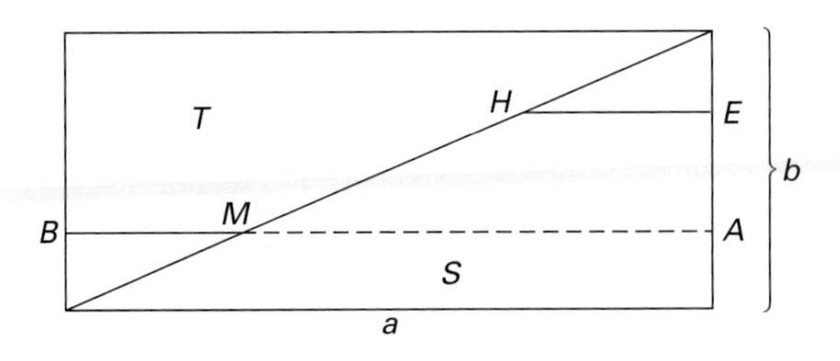

FIGURE 11.5 Cavalieri's method of "all the lines" in a triangle and rectangle

triangle. In modern notation, this result is equivalent to

$$ab = 2\int_0^b \frac{a}{b} t\, dt$$

or, more simply, to

$$b^2 = 2\int_0^b t\, dt.$$

Cavalieri was similarly able to demonstrate that "all the squares" of the rectangle F are triple of "all the squares" of each triangle, or, in modern notation, that

$$a^2 b = 3\int_0^b \frac{a^2}{b^2} t^2\, dt \qquad \text{or} \qquad b^3 = 3\int_0^b t^2\, dt.$$

By 1647, he had demonstrated analogous results for certain higher powers and was able to infer that the area under the "higher parabola" $y = x^k$ inscribed in a rectangle is $1/(k+1)$ times the area of the rectangle, or

$$\int_0^b x^k\, dx = \frac{1}{k+1} b^{k+1}.$$

This result was also discovered by others in the same time period.

11.2.2 Torricelli and the Infinitely Long Solid

Evangelista Torricelli (1608–1647), another disciple of Galileo, also worked with indivisibles. In particular, in 1643, he showed that the volume of the infinitely long solid formed by rotating the hyperbola $xy = k^2$ around the y-axis from $y = a$ to $y = \infty$ was finite and, in fact, that the sum of its volume and that of the cylinder of radius k^2/a and altitude a was equal to the volume of the cylinder of altitude k^2/a and radius equal to the semidiameter $AS = \sqrt{2}k$ of the hyperbola (Fig. 11.6). Torricelli used a method similar to the cylindrical shell method taught today, but expressed in terms of indivisibles, analogous to the lines of his friend Cavalieri. First, he showed that the lateral surface area of any cylinder, such as $POMN$, inscribed in his infinite hyperbolic solid was equal to the area of the circle of radius AS. [In modern terms, this is simply that $2\pi x(k^2/x) = \pi(\sqrt{2}k)^2$.] Next, he noted that the infinite solid (including its base cylinder) can be considered to be composed of all these cylindrical surfaces, to each of which there corresponds one of the circles making up the cylinder $ACHI$. It follows that the infinite solid is equal to the cylinder $ACHI$.

11.2.3 Fermat and the Area under Parabolas and Hyperbolas

In a letter to Gilles Persone de Roberval (1602–1675) of September 22, 1636, Fermat claimed that he had been able to calculate the area of a region under any higher parabola $y = px^k$. Roberval, writing back in October, claimed that he too had found the same result, using a formula for the sums of powers of the natural numbers: "The sum of the square numbers is always greater than the third part of the cube which has for its root the root of the greatest square, and the same sum of the squares with the greatest square removed is

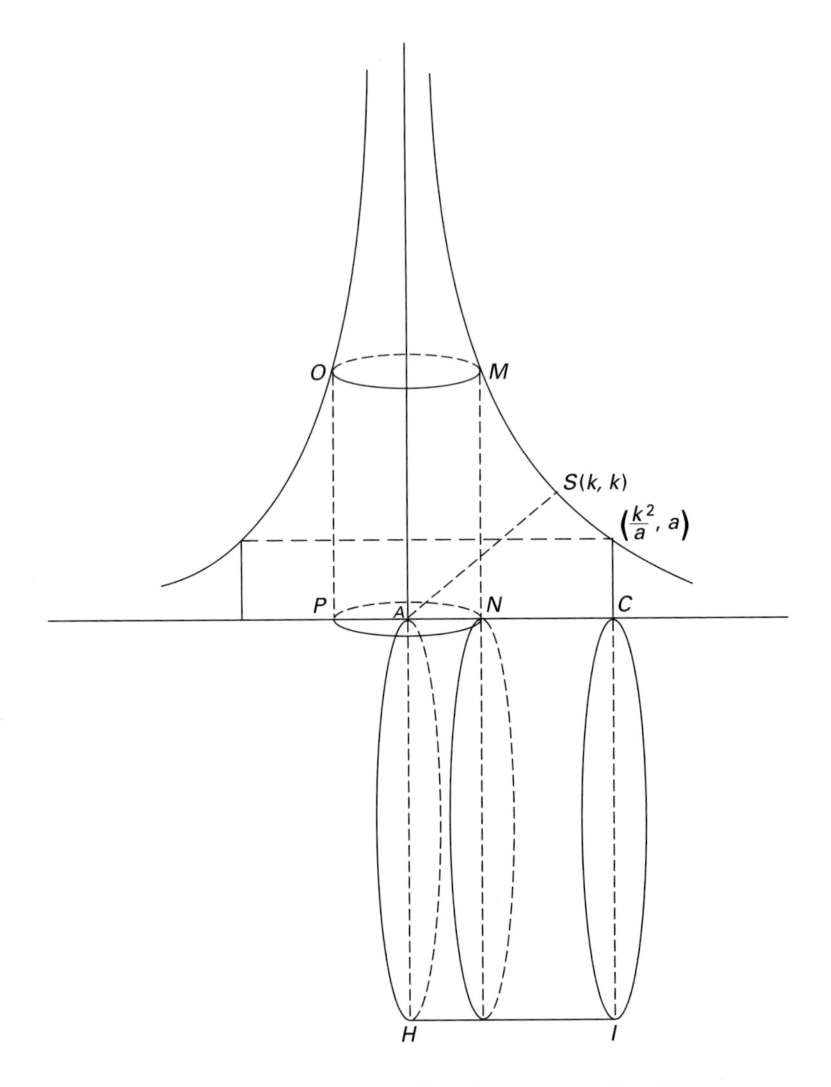

FIGURE 11.6 Torricelli's infinite hyperbolic solid

less than the third part of the same cube; the sum of the cubes is greater than the fourth part of the [fourth power] and with the greatest cube removed, less than the fourth part, etc." In other words, finding the area of the region bounded by the parabola $y = px^k$, the x-axis, and a given vertical line depends on the formula

$$\sum_{i=1}^{N-1} i^k < \frac{N^{k+1}}{k+1} < \sum_{i=1}^{N} i^k.$$

It is easy enough to see why this formula is fundamental, by considering the graph of $y = px^k$ over the interval $[0, x_0]$. Divide the base interval into N equal subintervals, each of length x_0/N, and erect over each subinterval a rectangle whose height is the y-coordinate

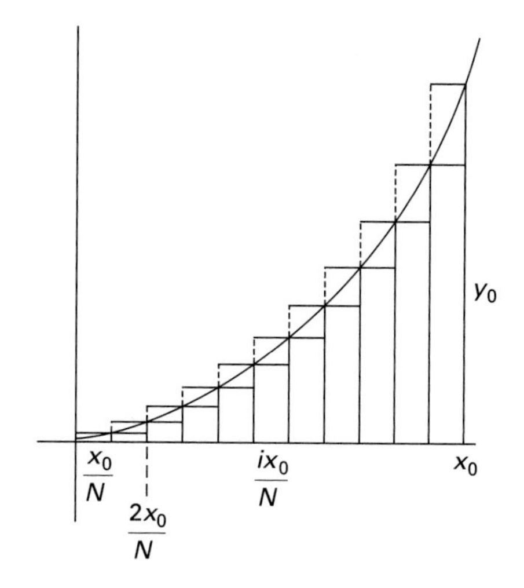

FIGURE 11.7 The area under $y = px^k$ according to Fermat and Roberval

of the right endpoint (Fig. 11.7). The sum of the areas of these N circumscribed rectangles is then

$$p\frac{x_0^k}{N^k}\frac{x_0}{N} + p\frac{(2x_0)^k}{N^k}\frac{x_0}{N} + \cdots + p\frac{(Nx_0)^k}{N^k}\frac{x_0}{N} = \frac{px_0^{k+1}}{N^{k+1}}\left(1^k + 2^k + \cdots + N^k\right).$$

Similarly, one can calculate the sum of the areas of the inscribed rectangles, those whose height is the y-coordinate of the left endpoint of the corresponding subinterval. If A is the area under the curve between 0 and x_0, then

$$\frac{px_0^{k+1}}{N^{k+1}}(1^k + 2^k + \cdots + (N-1)^k) < A < \frac{px_0^{k+1}}{N^{k+1}}(1^k + 2^k + \cdots + N^k).$$

The difference between the outer expressions of this inequality is simply the area of the rightmost circumscribed rectangle. Because x_0 and $y_0 = px_0^k$ are fixed, this difference may be made less than any assigned value simply by taking N sufficiently large. It follows from the inequality cited by Roberval that both the area A and the value

$$\frac{px_0^{k+1}}{k+1} = \frac{x_0 y_0}{k+1}$$

are "squeezed" between two values whose difference approaches 0. Thus, Fermat (and Roberval) found that

$$A = \frac{x_0 y_0}{k+1}.$$

Fermat then worked to extend the method from higher parabolas to higher hyperbolas, that is, curves of the form $y^m x^k = p$, as well as curves of the form $y^m = px^k$. In modern terms, the method described above for finding areas under $y = px^k$ only worked if k were a positive integer. Fermat wanted a method that would work if k were any rational number, positive or negative. He eventually discovered a new procedure in the 1640s, but did not announce it until writing his *Treatise on Quadrature* around 1658.

To apply his earlier method to the question of determining the area under $y = px^{-k}$ (k a positive integer) to the right of $x = x_0$ required dividing either the x-axis or the line segment $x = x_0$ from 0 to $y_0 = px_0^{-k}$ into finitely many intervals and summing the areas of the inscribed and circumscribed rectangles. Using the latter procedure, however, would give Fermat an infinite rectangle as the difference between his circumscribed and inscribed rectangles, one for which it was not at all clear that the area could be made as small as desired. On the other hand, there was no way of dividing the (infinite) x-axis into finitely many intervals ultimately to be made as small as one wished.

Fermat's solution to his dilemma was to partition the infinite interval to the right of x_0 into infinitely many intervals, whose length formed a geometric progression. Thus, he divided the interval as the points $a_0 = x_0, a_1 = (m/n)x_0, a_2 = (m/n)^2 x_0, \ldots, a_i = (m/n)^i x_0, \ldots,$ where m and n ($m > n$) are positive integers (Fig. 11.8). The intervals $[a_{i-1}, a_i]$ will ultimately be made as small as desired by taking m/n sufficiently close to 1. Fermat next circumscribed rectangles above the curve over each small interval. The first circumscribed rectangle has area

$$R_1 = \left(\frac{m}{n}x_0 - x_0\right)y_0 = \left(\frac{m}{n} - 1\right)x_0\frac{p}{x_0^k} = \left(\frac{m}{n} - 1\right)\frac{p}{x_0^{k-1}}.$$

The second rectangle has area

$$R_2 = \left[\left(\frac{m}{n}\right)^2 x_0 - \left(\frac{m}{n}\right)x_0\right]\frac{p}{\left(\frac{m}{n}x_0\right)^k} = \left(\frac{m}{n}\right)\left(\frac{m}{n} - 1\right)x_0\left(\frac{n}{m}\right)^k\frac{p}{x_0^k} = \left(\frac{n}{m}\right)^{k-1}R_1.$$

Similarly, the third rectangle has area

$$R_3 = \left(\frac{n}{m}\right)^{2(k-1)}R_1.$$

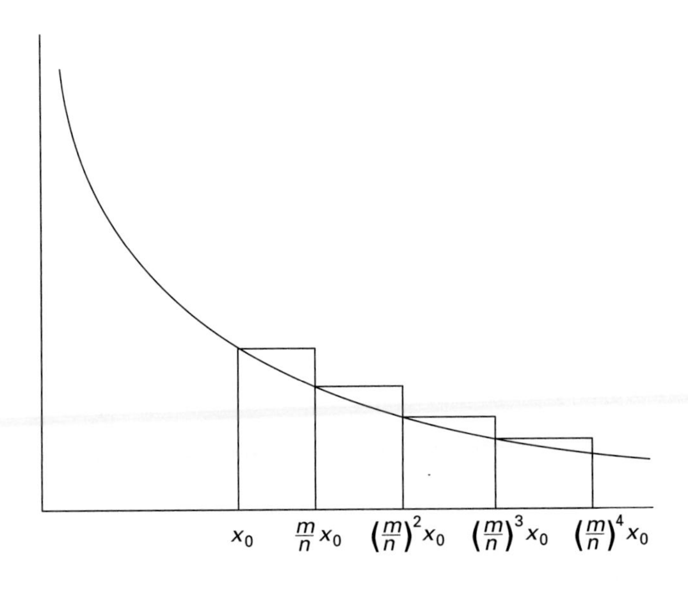

FIGURE 11.8 Fermat's procedure for determining the area under $y = px^{-k}$

It follows that the sum of the areas of all the circumscribed rectangles is

$$R = R_1 + \left(\frac{n}{m}\right)^{k-1} R_1 + \left(\frac{n}{m}\right)^{2(k-1)} R_1 + \cdots = R_1\left[1 + \left(\frac{n}{m}\right)^{k-1} + \left(\frac{n}{m}\right)^{2(k-1)} + \cdots\right]$$

or, using the formula for the sum of a geometric series,

$$R = \frac{1}{1-\left(\frac{n}{m}\right)^{k-1}} R_1 = \frac{1}{1-\left(\frac{n}{m}\right)^{k-1}}\left(\frac{m}{n}-1\right)\frac{p}{x_0^{k-1}} = \frac{1}{\frac{n}{m}+\left(\frac{n}{m}\right)^2+\cdots+\left(\frac{n}{m}\right)^{k-1}}\frac{p}{x_0^{k-1}}.$$

Fermat could have made a similar calculation for the inscribed rectangles, but decided it was not necessary. He let the area of the first rectangle "go to nothing," or, in modern terminology, he found the limiting value of his sum, by letting n/m approach 1. The value of R then approaches $[1/(k-1)](p/x_0^{k-1})$, and therefore, the desired area A is given by

$$A = \frac{1}{k-1}x_0 y_0.$$

Note that this method does not work for $k = 1$. Fermat was not able to solve the area problem in that case.

Fermat quickly noticed that this division of the axis into infinite intervals could also be applied to find the known area under the parabolas $y = px^k$ from $x = 0$ to $x = x_0$. He simply divided this finite interval $[0, x_0]$ into an infinite set of subintervals by beginning from the right: $a_0 = x_0, a_1 = (n/m)x_0, a_2 = (n/m)^2x_0, \ldots, a_i = (n/m)^i x_0, \ldots$, where $n < m$, and proceeded as above to show that this area is equal to $[1/(k+1)]x_0 y_0$. The other cases Fermat wanted to solve were the areas under the curves $x^k y^m = p$ and $y^m = px^k$, for which the method had to be modified slightly to avoid having the geometric series involve fractional powers. But Fermat did succeed in showing that the area under the "hyperbola" $x^k y^m = p$ to the right of $x = x_0$ is $m/(k-m)]x_0 y_0$, and that under the "parabola" $y^m = px^k$ from 0 to x_0 is $[m/(k+m)]x_0 y_0$.

11.2.4 Wallis and Fractional Exponents

Another mathematician who derived the same "integration" formulas as Fermat was John Wallis (1616–1703). Wallis, the first mathematician actually to explain fractional exponents and use them consistently, used indivisibles to solve the area problem for a curve of the form $y = x^{p/q}$ over the interval [0, 1] in his 1655 work *Arithmetica infinitorum*. That is, he found

$$\int_0^1 x^{p/q}\,dx = \frac{1}{\frac{p}{q}+1}.$$

He then tried to extend this result to curves of the form $y = x^{-k}$. His basic rule told him that for exponent -1, the area should be $1/(-1+1) = 1/0$, while for exponent -2, the area should be $1/-1$. It was reasonable to assume that the area under the hyperbola $y = 1/x$ was in some sense $1/0$, or infinity, but what did it mean that the area under the curve $y = 1/x^2$ was $1/(-1)$? Since for exponents 3, 2, 1, 0, the corresponding areas were $1/4$, $1/3$, $1/2$, and $1/1$, and these values formed an increasing sequence, he assumed that the area $1/(-1)$ for exponent -2 should be greater than the ratio $1/0$ for exponent -1. But what it meant for $1/(-1)$ to be greater than infinity, Wallis could never quite figure out.

Passing over this problem, but also realizing that his method could be applied to finding areas under curves given by sums of terms of the form $ax^{p/q}$, Wallis next attempted to generalize his methods to the more complicated problem of determining arithmetically the area of a circle of radius 1, namely, of finding the area under the curve $y = \sqrt{1-x^2} = (1-x^2)^{1/2}$. In fact, he tried to solve the more general problem, to find the ratio of the area of the unit square to the area enclosed in the first quadrant by the curve $y = (1 - x^{1/p})^n$. The case $p = \frac{1}{2}, n = \frac{1}{2}$ is the case of the circle, where the ratio is $4/\pi$. It was easy enough for Wallis to calculate by his known methods the ratios in the cases where p and n were integral. For example, if $p = 2$ and $n = 3$, the area under $y = (1 - x^{1/2})^3$ from 0 to 1 is that under $y = 1 - 3x^{1/2} + 3x - x^{3/2}$; that is, $1 - 2 + \frac{3}{2} - \frac{2}{5} = \frac{1}{10}$. Since the area of the unit square is 1, the ratio here is $1 : \frac{1}{10} = 10$. Wallis thus constructed the following table of these ratios, where, for $p = 0$, he simply used the area under $y = 1^n$:

$p\backslash n$	0	1	2	3	4	5	6	7	...
0	1	1	1	1	1	1	1	1	...
1	1	2	3	4	5	6	7	8	...
2	1	3	6	10	15	21	28	36	...
3	1	4	10	20	35	56	84	120	...
⋮	⋮	⋮	⋮	⋮	⋮	⋮	⋮	⋮	...

Wallis clearly recognized Pascal's arithmetical triangle in his table. What he wanted was to be able to interpolate rows corresponding to $p = \frac{1}{2}, p = \frac{3}{2}, \ldots$ and columns corresponding to $n = \frac{1}{2}, n = \frac{3}{2}, \ldots$, from which he could find the desired value, which he wrote as $\square$, when both parameters equaled $\frac{1}{2}$. From his knowledge of Pascal's triangle, Wallis realized that in his table the relationship $a_{p,n} = [(p+n)/n]a_{p,n-1}$ holds, where

$$a_{p,n} = \binom{p+n}{n}$$

designates the entry in row p, column n. Using this same rule for the row $p = \frac{1}{2}$, he noted first that $a_{1/2,0} = 1$, because all other entries in column 0 were equal to 1. It followed that

$$a_{1/2,1} = \left(\frac{\frac{1}{2}+1}{1}\right) \cdot 1 = \frac{3}{2}, \qquad a_{1/2,2} = \left(\frac{\frac{1}{2}+2}{2}\right) \cdot \frac{3}{2} = \frac{5}{4} \cdot \frac{3}{2} = \frac{15}{8},$$

$$a_{1/2,3} = \frac{7}{6} \cdot \frac{5}{4} \cdot \frac{3}{2} = \frac{105}{48}, \ldots.$$

Similarly, since $a_{1/2,1/2} = \square$, he had

$$a_{1/2,3/2} = \left(\frac{\frac{1}{2}+\frac{3}{2}}{3/2}\right)\square = \frac{4}{3}\square, \qquad a_{1/2,5/2} = \frac{6}{5} \cdot \frac{4}{3}\square, \ldots$$

and the row $p = \frac{1}{2}$ was

$$1 \quad \square \quad \frac{3}{2} \quad \frac{4}{3}\square \quad \frac{15}{8} \quad \frac{8}{5}\square \quad \ldots$$

Wallis was able to fill in the remainder of the table analogously, but since he was interested in calculating $\Box$, he considered the ratios in this row. Because it was evident that the ratios of alternate terms continually decreased, that is, $a_{1/2,k+2} : a_{1/2,k} > a_{1/2,k+4} : a_{1/2,k+2}$ for all k, he made the assumption that this was true for ratios of adjoining terms as well. It followed that

$$\Box : 1 > \tfrac{3}{2} : \Box, \qquad \text{so} \qquad \Box > \sqrt{\tfrac{3}{2}};$$

that $$\tfrac{3}{2} : \Box > \tfrac{4}{3}\Box : \tfrac{3}{2}, \qquad \text{so} \qquad \Box < \tfrac{3}{2}\sqrt{\tfrac{3}{4}} = [(3 \times 3)/(2 \times 4)]\sqrt{\tfrac{4}{3}};$$

and that $$\Box > [(3 \times 3)/(2 \times 4)\sqrt{\tfrac{5}{4}}, \ \Box < (3 \times 3 \times 5 \times 5)/(2 \times 4 \times 4 \times 6)\sqrt{\tfrac{6}{5}}, \ldots.$$

Wallis was thus able to assert that $\Box$ (or $4/\pi$) could be calculated as an infinite product:

$$\Box = \frac{4}{\pi} = \frac{3 \times 3 \times 5 \times 5 \times 7 \times 7 \times \cdots}{2 \times 4 \times 4 \times 6 \times 6 \times 8 \times \cdots}.$$

11.2.5 The Area under the Sine Curve and the Rectangular Hyperbola

Although an infinite product was not perhaps the kind of area result Wallis had hoped for, other mathematicians of the period who were considering curves other than the power curves also had to be satisfied with answers not strictly arithmetical. For example, Pascal, around 1657, found the area under any portion of the sine curve and described the method in his short treatise entitled *Treatise on the Sines of a Quadrant of a Circle*. Consider the quadrant ABC of a circle, and let D be any point from which the sine DI is drawn to the radius AC (Fig. 11.9). Pascal then drew a "small" tangent EE' and perpendiculars ER, $E'R'$ to the radius. His claim was that "the sum of the sines of any arc of a quadrant is equal to the portion of the base between the extreme sines, multiplied by the radius." By the "sum of the sines," Pascal meant the sum of the infinitesimal rectangles formed by multiplying each sine by the infinitesimal arc represented by the tangent EE'. Hence, Pascal's theorem in modern terms is

$$\int_{\alpha}^{\beta} r \sin\theta \, d(r\theta) = r(r\cos\alpha - r\cos\beta).$$

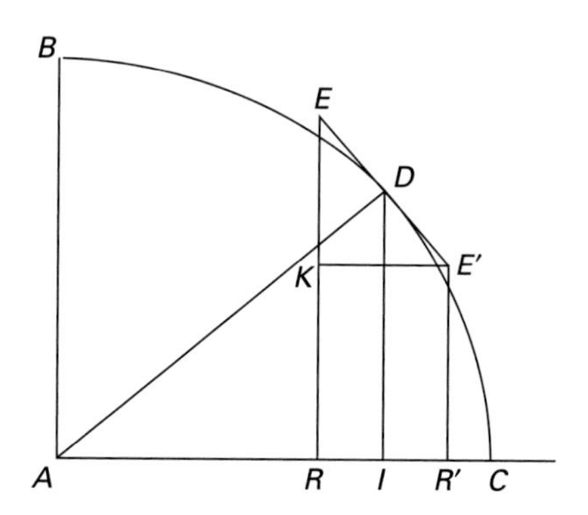

FIGURE 11.9 Pascal's area under the sine curve

For his proof, Pascal noted that triangles EKE' and DIA are similar, hence, $DI : DA = E'K : EE' = RR' : EE'$, and therefore, $DI \cdot EE' = DA \cdot RR'$. In other words, the rectangle formed from the sine and the infinitesimal arc (or tangent) is equal to the rectangle formed by the radius and the part of the axis between the ends of the arc, or $r \sin\theta d(r\theta) = r[r\cos(\theta + d\theta) - r\cos(\theta)] = r[d(r\cos\theta)]$. Adding these rectangles between the two given angles produced the result cited. Although this result proved important, and although Pascal generalized it immediately to give formulas for the integrals of powers of the sine, the most significant aspect of this proof was the appearance of the "differential triangle" EKE'. Leibniz's study of this particular work of Pascal was instrumental in his realization of the connection between the area problem and the tangent problem.

Our final example of mid–seventeenth-century work on the area problem starts with the 1647 *Opus geometricum* of the Belgian mathematician Gregory of St. Vincent (1584–1667). Gregory showed that if (x_i, y_i) for $i = 1, 2, 3, 4$ are four points on the rectangular hyperbola $xy = 1$ such that $x_2 : x_1 = x_4 : x_3$, then the area under the hyperbola over $[x_1, x_2]$ equals that over $[x_3, x_4]$ (Fig. 11.10). To prove this, divide the interval $[x_1, x_2]$ into subintervals at the points $a_i, i = 0, \ldots, n$. Because $x_2 : x_1 = x_4 : x_3$, it follows that $x_3 : x_1 = x_4 : x_2 = \nu$, or $x_3 = \nu x_1$, $x_4 = \nu x_2$. One can therefore conveniently subdivide the interval $[x_3, x_4]$ at the points $b_i = \nu a_i$, $i = 0, \ldots, n$. If rectangles are then inscribed in and circumscribed about the hyperbolic areas A_j over $[a_j, a_{j+1}]$ and B_j over $[b_j, b_{j+1}]$, it is straightforward to calculate the corresponding inequalities:

$$(a_{j+1} - a_j)\frac{1}{a_{j+1}} < A_j < (a_{j+1} - a_j)\frac{1}{a_j}$$

and

$$(b_{j+1} - b_j)\frac{1}{b_{j+1}} < B_j < (b_{j+1} - b_j)\frac{1}{b_j}.$$

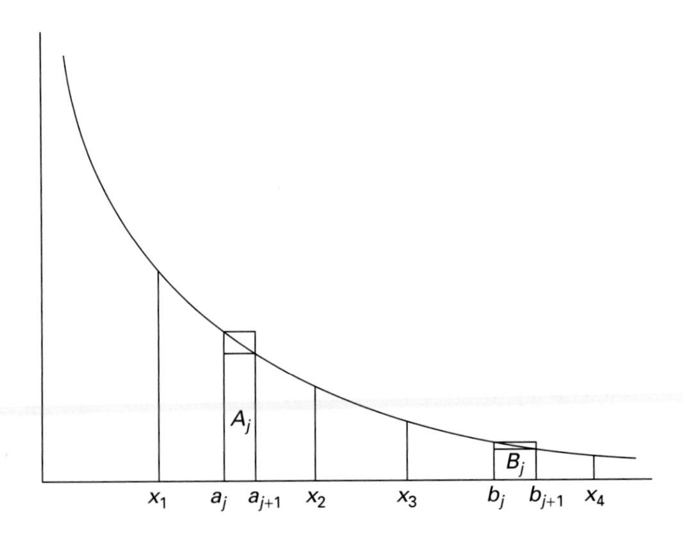

FIGURE 11.10 Gregory of St. Vincent's area under the hyperbola $xy = 1$

Substituting the values $b_j = va_j$ into the second set of inequalities gives

$$(a_{j+1} - a_j)\frac{1}{a_{j+1}} < B_j < (a_{j+1} - a_j)\frac{1}{a_j}.$$

Thus, both hyperbolic regions are squeezed between rectangles of the same areas. Because both intervals can be divided into subintervals as small as desired, it follows that the two hyperbolic areas are equal.

Gregory had not actually calculated the area under the hyperbola. However, when the Belgian Jesuit Alfonso Antonio de Sarasa (1618–1667) read Gregory's work in 1649, he noticed that this calculation implied that the area $A(x)$ under the hyperbola from 1 to x had the logarithmic property $A(\alpha\beta) = A(\alpha) + A(\beta)$. (Because the ratio $\beta : 1$ equals the ratio $\alpha\beta : \alpha$, the area from 1 to β equals the area from α to $\alpha\beta$. Because the area from 1 to $\alpha\beta$ is the sum of the areas from 1 to α and from α to $\alpha\beta$, the logarithmic property is immediate.) Thus, if one could calculate the areas under portions of the hyperbola $xy = 1$, one could calculate logarithms.

In 1668, Nicolaus Mercator (1620–1687) followed up on de Sarasa's hint. Having also learned the basic results for sums of powers, he decided to calculate $\log(1 + x)$ [the area A under the hyperbola $y = 1/(1 + x)$ from 0 to x]. To do this, he divided the interval $[0, x]$ into n subintervals of length x/n and approximated A by the sum

$$\frac{x}{n} + \frac{x}{n}\left(\frac{1}{1 + \frac{x}{n}}\right) + \frac{x}{n}\left(\frac{1}{1 + \frac{2x}{n}}\right) + \cdots + \frac{x}{n}\left(\frac{1}{1 + \frac{(n-1)x}{n}}\right).$$

Since each term $1/[1 + (kx/n)]$ is the sum of the geometric series $\sum_{j=0}^{\infty}(-1)^j(kx/n)^j$, it follows that

$$\begin{aligned} A &\approx \frac{x}{n} + \frac{x}{n}\sum_{j=0}^{\infty}(-1)^j\left(\frac{x}{n}\right)^j + \frac{x}{n}\sum_{j=0}^{\infty}(-1)^j\left(\frac{2x}{n}\right)^j + \cdots + \frac{x}{n}\sum_{j=0}^{\infty}(-1)^j\left(\frac{(n-1)x}{n}\right)^j \\ &= n\frac{x}{n} - \frac{x^2}{n^2}\sum_{i=1}^{n-1} i + \frac{x^3}{n^3}\sum_{i=1}^{n-1} i^2 + \cdots + (-1)^j\frac{x^{j+1}}{n^{j+1}}\sum_{i=1}^{n-1} i^j + \cdots \\ &= x - \frac{\sum_{i=1}^{n-1} i}{n \cdot n}x^2 + \frac{\sum_{i=1}^{n-1} i^2}{n \cdot n^2}x^3 + \cdots + (-1)^j\frac{\sum_{i=1}^{n-1} i^j}{n \cdot n^j}x^{j+1} + \cdots \end{aligned}$$

By the basic formulas for sums of integral powers derived by Fermat and Roberval, the coefficient of x^{k+1} in this expression is equal to $1/(k + 1)$ if n is infinite. Therefore,

$$\log(1 + x) = x - \frac{x^2}{2} + \frac{x^3}{3} - \frac{x^4}{4} + \cdots,$$

a power series in x, which enabled actual values of the logarithm to be calculated easily.

11.3 Rectification of Curves and the Fundamental Theorem

11.3.1 Van Heuraet and the Rectification of Curves

Another problem of interest in this time period was the rectification of curves—that is, the exact determination of their lengths. Although he was not the first to accomplish a rectification, Hendrick van Heuraet (1634–1660?) published the first general procedure for attacking the problem in his *On the Transformation of Curves into Straight Lines*, which appeared in van Schooten's 1659 Latin edition of Descartes's *Geometry*.

Van Heuraet began by showing that the problem of constructing a line segment equal in length to a given arc is equivalent to finding the area under a certain curve. Let P be an arbitrary point on the arc MN of the curve α (Fig. 11.11). The length PS of the normal line from P to the axis can be determined by Descartes's method. Taking an arbitrary line segment σ, van Heuraet defined a new curve α' by the ratio $P'R : \sigma = PS : PR$, where P' is the point on α' associated to P. (The σ is included so that both ratios are ratios of lines.) Drawing the differential triangle ACB with AC tangent to α at P, he noted that $PS : PR = AC : AB$. In modern notation, if $AC = ds$ and $AB = dx$, then van Heuraet's ratios yield $P'R : \sigma = ds : dx$ or $\sigma\, ds = P'R\, dx$. Because the sum of the infinitesimal tangents—or, equivalently, the infinitesimal pieces of the arc—over the curve MN gives the length of MN, van Heuraet concluded that $\sigma \cdot$ (length of MN) = area under the curve α' between M' and N'. Thus, if it is possible to derive the equation of α' from that of α and to calculate the area under it, the length of MN can also be calculated. Using modern notation, with $z = P'R = \sigma\, ds/dx = \sigma\sqrt{1 + (dy/dx)^2}$, van Heuraet's procedure can be written as

$$\sigma \cdot (\text{length of } MN) = \int_a^b z\,dx = \int_a^b \sigma\sqrt{1 + \left(\frac{dy}{dx}\right)^2}\,dx,$$

where a and b represent the x-coordinates of M and N. This is essentially the modern arc-length formula.

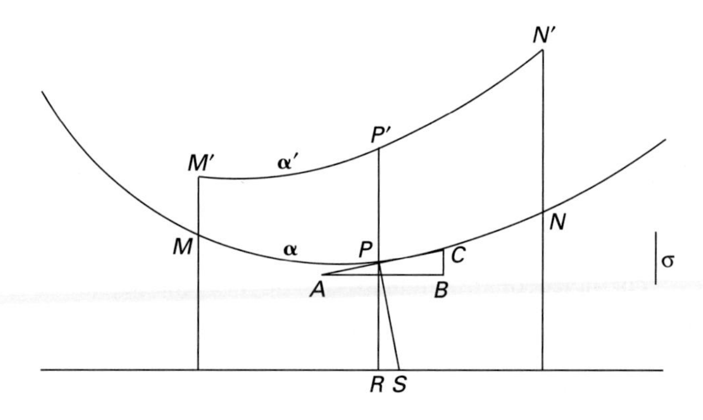

FIGURE 11.11 Van Heuraet's rectification of a curve

Van Heuraet illustrated his procedure with one of the few curves for which the area under the associated curve can actually be calculated, the semicubical parabola $y^2 = x^3$. Using Descartes's normal method, he calculated that the equation which must have a double root is $x^3 + x^2 - 2vx + v^2 - n^2 = 0$. Using Hudde's rule for finding the double root, he multiplied the terms of this equation by 3, 2, 1, 0 to get $3x^3 + 2x^2 - 2vx = 0$. Therefore, $v - x = 3x^2/2$, and $PS = \sqrt{\frac{9}{4}x^4 + x^3}$. Setting $\sigma = \frac{1}{3}$, van Heuraet defined the new curve α' by taking

$$z = P'R = \sigma \cdot \frac{PS}{PR} = \frac{1}{3}\frac{\sqrt{\frac{9}{4}x^4 + x^3}}{\sqrt{x^3}} = \sqrt{\tfrac{1}{4}x + \tfrac{1}{9}},$$

or, equivalently, $z^2 = \frac{1}{4}x + \frac{1}{9}$. Van Heuraet easily identified this curve as a parabola, the area under which he knew how to calculate. The length of the semicubical parabola from $x = 0$ to $x = b$ then equals this area divided by σ, that is,

$$\sqrt{\left(b + \tfrac{4}{9}\right)^3} - \tfrac{8}{27}.$$

After remarking that one can similarly explicitly determine the lengths of the curves $y^4 = x^5, y^6 = x^7, y^8 = x^9, \ldots$, van Heuraet concluded with the more difficult rectification of an arc of the parabola $y = x^2$, whose length depends on the determination of the area under the hyperbola $z = \sqrt{4x^2 + 1}$. That problem, in 1659, had not yet been satisfactorily solved. Nevertheless, van Heuraet's methods soon became widely known. In particular, the use of the differential triangle and the association of a new curve to a given curve helped to lead others to the ideas relating the tangent problem to the area problem.

11.3.2 Gregory and the Fundamental Theorem

Among the mathematicians who related the tangent problem to the area problem were Isaac Barrow (1630–1677) and James Gregory (1638–1675), both of whom decided to organize the material relating to tangents, areas, and rectification that they had gathered in their travels through France, Italy, and the Netherlands and to present it systematically. Not surprisingly, then, Barrow's *Geometrical Lectures* (1670) and Gregory's *Universal Part of Geometry* (1668) contained much of the same material presented in similar ways. In effect, both of these works were treatises on material today identified as calculus, but with presentations in the geometrical style each mathematician had learned in his university study. Neither was able to translate the material into a method of computation useful for solving problems.

As an example, consider how Gregory presented the fundamental theorem of calculus, the result linking the ideas of area and tangent. This result was the natural outcome of Gregory's study of the general problem of arc length, as discussed in the work of van Heuraet. Consider a monotonically increasing curve $y = y(x)$ with two other curves associated to it, the normal curve $n(x) = y\sqrt{1 + (dy/dx)^2}$ and $u(x) = cn/y = c\sqrt{1 + (dy/dx)^2}$, where c is a given constant. Now, constructing the differential triangle dx, dy, ds at a given point, Gregory argued from its similarity with the triangle formed by the ordinate y, the subnormal ν, and the normal n that $y : n = dx : ds = c : u$ and thus that both $u\,dx = c\,ds$

and $n\,dx = y\,ds$ (Fig. 11.12). Summing the first equation over the curve showed Gregory, as it had van Heuraet, that the arc length $\int ds$ can be expressed in terms of the area under the curve $(1/c)u(x)$. The sum of the second equation enabled Gregory to show that the area under $n = n(x)$ was equal, up to a constant multiple, to the area of the surface formed by rotating the original curve around the x-axis. Gregory proved both of these results by a careful Archimedean argument involving inscribed and circumscribed rectangles and a double *reductio ad absurdum*.

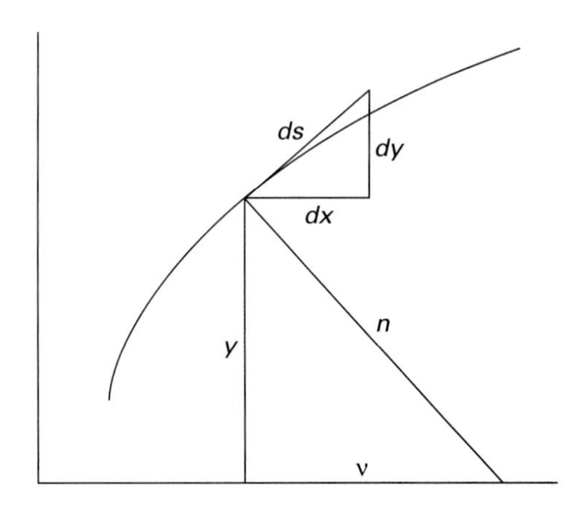

FIGURE 11.12 Gregory's differential triangle

Having shown that arc length can be found by an area, Gregory made a fundamental advance by asking the converse question: Can one find a curve $u(x)$ whose arc length s has a constant ratio to the area under a given curve $y(x)$? In modern notation, Gregory was asking whether it was possible to determine u such that

$$c\int_0^x \sqrt{1+\left(\frac{du}{dx}\right)^2}\,dx = \int_0^x y\,dx.$$

But this means that $c^2[1+(du/dx)^2] = y^2$ or that $du/dx = (1/c)\sqrt{y^2-c^2}$. In other words, Gregory had to determine a curve u, the slope of whose tangent was equal to a given function. Letting $z = \sqrt{y^2-c^2}$, Gregory simply defined $u(x)$ to be the area under the curve z/c from the origin to x. He then had to show that the slope of the tangent to this curve was given by z/c. What he in fact demonstrated, again by a *reductio* argument, was that the line connecting a point K on the u curve to the point on the axis at a distance cu/z from the x-coordinate of K is tangent to the curve at K.

Gregory's crucial advance, then, was the abstraction of the determination of area under a specific curve between two given x-values into the idea of area as a function of a variable. In other words, he constructed a new curve whose ordinate at any value x was equal to the area under the original curve from a fixed point up to x. Once this idea was conceived, it turned out that it was not difficult to construct the tangent to this new curve and to show that its slope at x was always equal to the original ordinate there.

11.3.3 Barrow and the Fundamental Theorem

Gregory had the idea of constructing a new curve for the particular purpose of finding arc length. Isaac Barrow, on the other hand, stated a more general version of part of the fundamental theorem as theorem 11 of lecture X of his *Geometrical Lectures*:

THEOREM. *Let ZGE be any curve of which the axis is AD, and let ordinates applied to this axis AZ, PG, DE, continually increase from the initial ordinate AZ. Also let AIF be a curve such that if any straight line EDF is drawn perpendicular to AD, cutting the curves in the points E, F and AD in D, the rectangle contained by DF and a given length R is equal to the intercepted space $ADEZ$. Also let $DE : DF = R : DT$ and join FT. Then TF will be tangent to AIF (Fig. 11.13).*

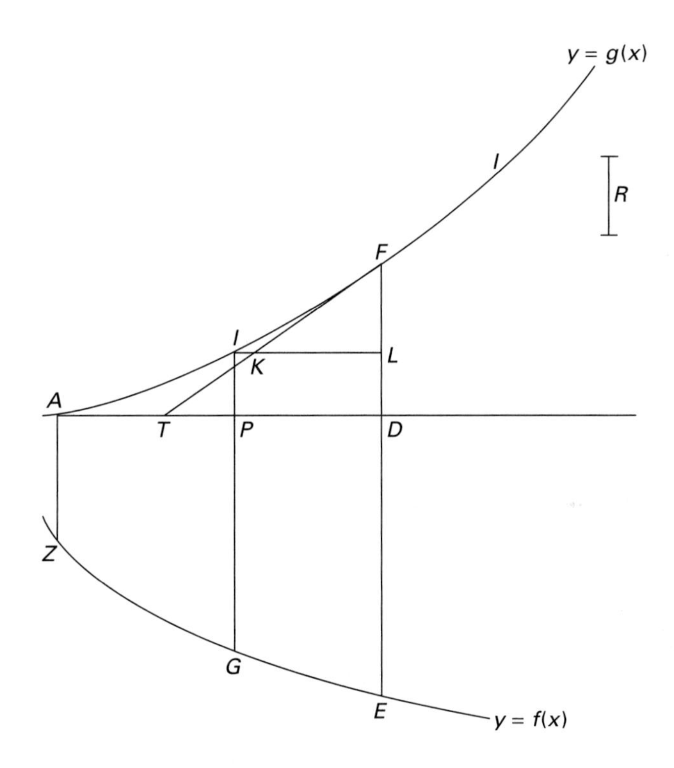

FIGURE 11.13 Barrow's version of the fundamental theorem

Like Gregory, Barrow began with a curve ZGE, written in modern notation as $y = f(x)$, and constructed a new curve $AIF = g(x)$ such that $Rg(x)$ is always equal to the area bounded by $f(x)$ between a fixed point and the variable point x. In modern notation, $Rg(x) = \int_a^x f(x)\,dx$. Barrow then proved that the length $t(x)$ of the subtangent to $g(x)$ is given by $Rg(x)/f(x)$, or that

$$g'(x) = \frac{g(x)}{t(x)} = \frac{f(x)}{R} \qquad \text{or} \qquad \frac{d}{dx}\int_a^x f(x)\,dx = f(x).$$

Barrow proved this result by showing that the line TF always lies outside the curve AIF. If I is any point on the curve $g(x)$ on the side of F toward A, and if IG is drawn parallel to AZ and KL parallel to AD, the nature of the curve shows that $LF : LK = DF : DT = DE : R$ or $R \cdot LF = LK \cdot DE$. Because $R \cdot IP$ equals the area of $APGZ$, it follows that $R \cdot LF$ equals the area of $PDEG$. Therefore, $LK \cdot DE =$ area $PDEG < PD \cdot DE$. Hence, $LK < PD$, or $LK < LI$, and the tangent line is below the curve at I. A similar argument applies for a point I on the side of F away from A.

In theorem 19 of lecture XI, Barrow proved the second part of the fundamental theorem, that

$$\int_a^b Rf'(x)\,dx = R(f(b) - f(a)),$$

by showing a correspondence between infinitesimal rectangles in the region under the curve $Rf'(x)$ and those in the (large) rectangle $R(f(b) - f(a))$.

Barrow had, therefore, discovered the inverse relationship of the tangent and the area problems. However, there is no indication in his *Geometrical Lectures* that he understood the fundamental nature of this relationship and the way it could be used to calculate areas. As Lucasian Professor of Mathematics at Cambridge University, however, Barrow did discuss some of his ideas with Isaac Newton. Newton, in fact, suggested a few improvements to Barrow's book—in particular, that it should include an algebraic method of calculating tangents based on the differential triangle. This method consists of drawing the differential triangle NMR at a point M on a given curve and calculating the ratio of $MP = y$ to $PT = t$ by using the corresponding ratio of $MR = a$ to $NR = e$ in the infinitesimal triangle (Fig. 11.14). Thus, if the curve is $y^2 = x^3$, Barrow replaced y by $y + a$, x by $x + e$ and found that $(y + a)^2 = (x + e)^3$ or $y^2 + 2ay + a^2 = x^3 + 3x^2e + 3xe^2 + e^3$. He then removed all terms containing a power of a or e or a product of the two, "for these terms have no value," and found that $y^2 + 2ay = x^3 + 3x^2e$, or $2ay = 3x^2e$. In the final step, he substituted y for a and t for e to get the ratio $y : t$. In this case, the result is $y : t = 3x^2 : 2y$. Barrow made no attempt to justify this method, a modification of Fermat's method, but only noted that he frequently used it in his own calculations.

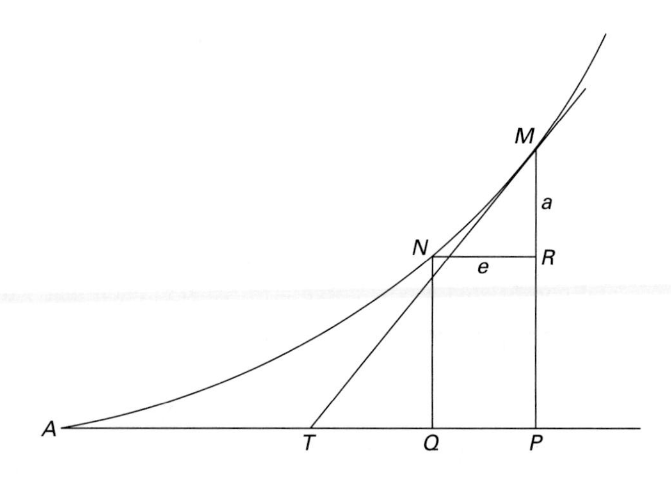

FIGURE 11.14 Barrow's differential triangle

The work of Barrow and Gregory can be thought of as a culmination of all of the seventeenth-century methods of calculating areas and tangents. But neither of these men developed these methods into a true computational and problem-solving tool. That process was accomplished first by Isaac Newton (1642–1727), during a few years in the mid–1660s, mostly in Cambridge, and by Gottfried Leibniz (1646–1716) between 1672 and 1676, in Paris.

11.4 Isaac Newton

Having mastered through self-study the entire achievement of seventeenth-century mathematics—in particular, the *Analytic Art* of Viète, the van Schooten edition of Descartes's *Geometry*, and Wallis's *Arithmetica infinitorum*—Isaac Newton spent the two years from late 1664 to late 1666 working out his basic ideas on calculus, partly in his rooms at Cambridge and partly back at his home in Woolsthorpe. More work followed in the next several years, and on at least three occasions, Newton wrote up his researches into a form suitable for publication. Unfortunately, for various reasons, Newton never published any of these three papers on calculus. Nevertheless, the so-called October, 1666 tract on fluxions, the *On Analysis by Equations with Infinitely Many Terms* of 1669, and the *Treatise on the Methods of Series and Fluxions* of 1671 all circulated to some extent in manuscript in the English mathematical community and demonstrated the great power of Newton's new methods.

11.4.1 Power Series

Newton clearly believed that he had expanded the "new analysis" that he had found in his readings. As he wrote at the beginning of the *Treatise on Methods*,

> Observing that the majority of geometers, with an almost complete neglect of the ancients' synthetical method, now for the most part apply themselves to the cultivation of analysis and with its aid have overcome so many formidable difficulties that they seem to have exhausted virtually everything apart from the squaring of curves, and certain topics of like nature not yet fully elucidated, I found not amiss, for the satisfaction of learners, to draw up the following short tract in which I might at once widen the boundaries of the field of analysis and advance the doctrine of series.

In particular, Newton intended to show that the doctrine of "series"—that is, what mathematicians today call **power series**—was central in "widening the boundaries" of analysis, thus enabling many curves to be "squared." Newton's discovery of power series came out of his reading of Wallis's work on determining the area of a circle. In considering areas, Wallis had always looked for a specific numerical value, or the ratio of two such values, because he wanted to determine the area under a curve between two fixed values, say, 0 and 1. Newton realized that further patterns could become evident if one calculated areas from 0 to an arbitrary value x—that is, if one considered area under a curve as a function of the varying endpoint of the interval. In modern notation, what Newton did was to look at the patterns in calculating

$$\int_0^x (1 - x^2)^n \, dx$$

for integral values of n and then to see what happened when n was a fraction, such as $\frac{1}{2}$.

Like Wallis, Newton noticed that Pascal's triangle appeared in his calculations. But Newton then applied Pascal's multiplicative formula for calculating the binomial coefficients,

$$\binom{n}{k} = \frac{n(n-1)(n-2)\cdots(n-k+1)}{k!},$$

to the case where n was not a positive integer. In fact, he decided that instead of limiting himself to expressions of the form $(1-x^2)^n$, he could achieve more generality by applying his results to $(a+bx)^n$. Recall that, for n a positive integer, the binomial theorem states that

$$(a+bx)^n = \binom{n}{0}a^n + \binom{n}{1}a^{n-1}bx + \binom{n}{2}a^{n-2}b^2x^2 + \cdots + \binom{n}{n}b^n x^n.$$

Newton in essence generalized this result to arbitrary rational values of n. For example, he calculated

$$\binom{\frac{1}{2}}{0} = 1,\ \binom{\frac{1}{2}}{1} = \frac{1}{2},\ \binom{\frac{1}{2}}{2} = \frac{\frac{1}{2}\left(\frac{1}{2}-1\right)}{2} = -\frac{1}{8},\ \binom{\frac{1}{2}}{3} = \frac{\frac{1}{2}\left(\frac{1}{2}-1\right)\left(\frac{1}{2}-2\right)}{6} = \frac{1}{16}, \ldots.$$

Realizing that for this case, unlike the case for positive integral n, there were infinitely many nonzero binomial coefficients, Newton was then able to express $(1+x)^{1/2}$ as an "infinite polynomial," or power series:

$$(1+x)^{1/2} = 1 + \tfrac{1}{2}x - \tfrac{1}{8}x^2 + \tfrac{1}{16}x^3 + \cdots.$$

But how did Newton know that his answer was correct? In this case, since he was claiming that the square root of $1+x$ was a power series, he simply squared the series. He found that the first two terms of the square were $1+x$, while the coefficients of all the other powers of x were 0, if he carried out the multiplication far enough. Similarly, he applied his new version of the binomial theorem to the case where $n = -1$. The result was

$$\begin{aligned}(1+x)^{-1} &= 1 + (-1)x + \frac{(-1)(-2)}{2!}x^2 + \frac{(-1)(-2)(-3)}{3!}x^3 + \cdots \\ &= 1 - x + x^2 - x^3 + \cdots.\end{aligned}$$

Since Newton also calculated the same power series by carrying out the long division of 1 by $1+x$, he again gained confidence that his result was correct.

Using his knowledge that the area under $y = x^n$ was $x^{n+1}/(n+1)$ and that the area under $y = 1/(1+x)$ was the logarithm of $1+x$, Newton found the power series for $\log(1+x)$ by integrating the above series term by term. It was, of course, the same series that Mercator had found:

$$\log(1+x) = x - \frac{x^2}{2} + \frac{x^3}{3} - \frac{x^4}{4} + \cdots.$$

Newton then proceeded to calculate the logarithms of 1 ± 0.1, 1 ± 0.2, 1 ± 0.01, and 1 ± 0.02 to over fifty decimal places. Using appropriate identities, such as $2 = (1.2 \times 1.2)/(0.8 \times 0.9)$ and $3 = (1.2 \times 2)/0.8$, as well as the basic properties of logarithms, Newton was able to calculate the logarithms of many small positive integers.

Knowledge of the binomial theorem let Newton deal with many other interesting series. For example, he worked out the series for $y = \arcsin x$ using a geometric argument: Suppose the circle AEC has radius 1 and $BE = x$ is the sine of the arc $y = AE$, or $y = \arcsin x$ (Fig. 11.15). The area of the circular sector APE is known to be $\frac{1}{2}y = \frac{1}{2}\arcsin x$. On the other hand, it is also equal to the area under $y = \sqrt{1-x^2}$ from 0 to x less $\frac{1}{2}x\sqrt{1-x^2}$. By his earlier calculation, Newton knew that

$$\sqrt{1-x^2} = 1 - \tfrac{1}{2}x^2 - \tfrac{1}{8}x^4 - \tfrac{1}{16}x^6 - \cdots.$$

It follows by integrating term by term and multiplying the above series by x that

$$y = \arcsin x = 2\int_0^x \sqrt{1-x^2}\,dx - x\sqrt{1-x^2} = x + \tfrac{1}{6}x^3 + \tfrac{3}{40}x^5 + \tfrac{5}{112}x^7 + \cdots.$$

Newton then applied a method he had developed of "inverting" series, that is, of solving an "equation" of this sort for x in terms of y. He found that

$$x = \sin y = y - \tfrac{1}{6}y^3 + \tfrac{1}{120}y^5 - \tfrac{1}{5040}y^7 + \cdots,$$

which is the same result the Indians had discovered several centuries earlier. Rather than calculating the cosine series analogously, however, Newton derived that series by calculating directly $\cos y = \sqrt{1-(\sin y)^2}$.

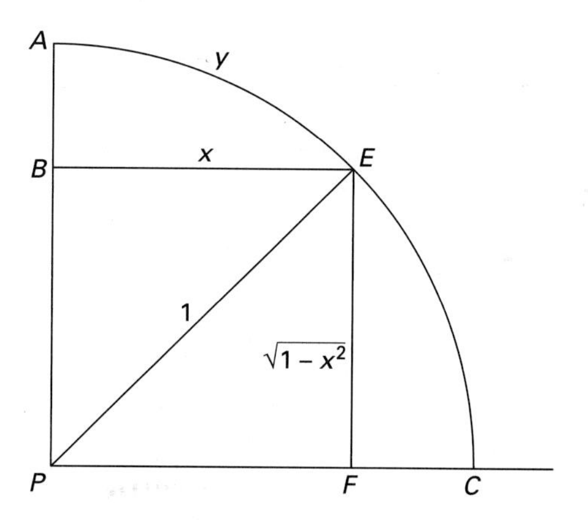

FIGURE 11.15 Newton's power series for $y = \arcsin x$

11.4.2 Algorithms for Calculating Fluxions and Fluents

Series were of fundamental importance to Newton's calculus. He used them in dealing with every algebraic or transcendental relation not expressible as a finite polynomial in one variable. Thus, given the basic rule for "integrating" a power of x, he was able to determine areas under all sorts of curves. In fact, he wrote out his three basic rules in *On Analysis*, and these can be summarized as follows:

1. If $y = ax^{m/n}$, then the area under y is $\frac{an}{m+n}x^{(m+n)/n}$.
2. If y is given by the sum of several such terms, even infinitely many, then the area under y is given by the sum of the areas of all the terms.
3. If the value of y is more complicated than in rule 2—that is, if the curve is expressed as $f(x, y) = 0$—then one must expand y as a (possibly infinite) sum of terms of the form $ax^{m/n}$ and then apply rules 1 and 2.

But Newton's calculus had far more to offer, beginning with the problems he only identified obliquely in the anagram he sent to Leibniz (presented at the opening of this chapter). More specifically, the two problems Newton considered as central to his new analysis, the solutions to which would resolve all the difficulties about curves faced by his predecessors, were

1. Given the length of the space continuously [that is, at every time], to find the speed of motion at any time proposed.
2. Given the speed of motion continuously, to find the length of the space described at any time proposed.

For Newton the basic ideas of calculus had to do with motion. Every variable in an equation was to be considered, at least implicitly, as a distance dependent on time. Of course, this idea was not new with Newton, but he did make the idea of motion fundamental: "I consider quantities as though they were generated by continuous increase in the manner of a space over which a moving object describes its course." The constant increase of time itself Newton considered virtually an axiom, for he gave no definition of time. What he did define was the concept of fluxion: The **fluxion** $\dot{x}$ of a quantity x dependent on time (called the **fluent**) was the speed with which x increased via its generating motion. In his early works, Newton did not attempt any further definition of speed. The concept of continuously varying motion was, Newton believed, completely intuitive.

Newton solved his first problem by a perfectly straightforward algorithm that determined the relationship of the fluxions $\dot{x}$ and $\dot{y}$ of two fluents x and y related by an equation of the form $f(x, y) = 0$: "Arrange the equation by which the given relation is expressed according to the dimensions of some fluent quantity, say x, and multiply its terms by any arithmetical progression and then by $\dot{x}/x$. Carry out this operation separately for each one of the fluent quantities and then put the sum of all the products equal to nothing, and you have the desired equation." As an example, Newton presented the equation $x^3 - ax^2 + axy - y^3 = 0$. First considering this as a polynomial of degree three in x, Newton multiplied using the progression 3, 2, 1, 0 to get $3x^2\dot{x} - 2ax\dot{x} + ay\dot{x}$. Next, considering the equation as a polynomial of degree three in y and using the same progression, he calculated $ax\dot{y} - 3y^2\dot{y}$. Putting the sum equal to nothing gave the desired relationship, $3x^2\dot{x} - 2ax\dot{x} + ay\dot{x} + ax\dot{y} - 3y^2\dot{y} = 0$. In terms of a ratio, this result is $\dot{x} : \dot{y} = (3y^2 - ax) : (3x^2 - 2ax + ay)$.

There are several important ideas to note in Newton's rule for calculating fluxions. First, Newton was not calculating derivatives, for he did not in general start with a function. What he did calculate is the differential equation satisfied by the curve determined by the given equation. In other words, given $f(x, y) = 0$ with x and y both functions of t, Newton's procedure produced what is today written as

$$\frac{\partial f}{\partial x}\frac{dx}{dt} + \frac{\partial f}{\partial y}\frac{dy}{dt} = 0.$$

Second, Newton used Hudde's notion of multiplying by an arbitrary arithmetic progression. In practice, however, Newton generally used the progression starting with the highest power of the fluent. Third, if x and y are considered as functions of t, the modern product rule for derivatives is built into Newton's algorithm. Any term containing both x and y is multiplied twice and the two terms are added.

Newton justified his rule, in effect, via infinitesimals. He first defined the **moment** of a fluent quantity to be the amount by which it increases in an "infinitely small" period of time. Thus, the increase of x in an infinitesimal time o is the product of the speed of x by o, or $\dot{x}o$. It follows that after this time interval, x will become $x + \dot{x}o$, and, similarly, y will become $y + \dot{y}o$. "Consequently, an equation which expresses a relationship of fluent quantities without variance at all times will express that relationship equally between $x + \dot{x}o$ and $y + \dot{y}o$ as between x and y; and so $x + \dot{x}o$ and $y + \dot{y}o$ may be substituted in place of the latter quantities, x and y, in the said equation."

Newton explained further through the example $x^3 - ax^2 + axy - y^3 = 0$, given earlier. Substituting $x + \dot{x}o$ for x and $y + \dot{y}o$ for y, gives the new equation

$$(x^3 + 3x^2\dot{x}o + 3x\dot{x}^2o^2 + \dot{x}^3o^3) - (ax^2 + 2ax\dot{x}o + a\dot{x}^2o^2)$$
$$+ (axy + ay\dot{x}o + ax\dot{y}o + a\dot{x}\dot{y}o^2) - (y^3 + 3y^2\dot{y}o + 3y\dot{y}^2o^2 + \dot{y}^3o^3) = 0.$$

"Now by hypothesis $x^3 - ax^2 + axy - y^3 = 0$, and when these terms are erased and the rest divided by o there will remain

$$3x^2\dot{x} + 3x\dot{x}^2o + \dot{x}^3o^2 - 2ax\dot{x} - a\dot{x}^2o$$
$$+ ay\dot{x} + ax\dot{y} + a\dot{x}\dot{y}o - 3y^2\dot{y} - 3y\dot{y}^2o - \dot{y}^3o^2 = 0.$$

But further, since o is supposed to be infinitely small so that it be able to express the moments of quantities, terms which have it as a factor will be equivalent to nothing in respect of the others. I therefore cast them out and there remains $3x^2\dot{x} - 2ax\dot{x} + ay\dot{x} + ax\dot{y} - 3y^2\dot{y} = 0$, as ... above."

Although this calculation is only an example and not a proof, Newton noted that it is immediately generalizable: "It is accordingly to be observed that terms not multiplied by o will always vanish, as also those multiplied by o of more than one dimension; and that the remaining terms after division by o will always take on the form they should have according to the rule." In other words, Newton assumed that the reader understood that the coefficient of $x^{n-1}\dot{x}o$ in the expansion of $(x + \dot{x}o)^n$ is n itself. But note also that Newton's only justification of his step of "casting out" any terms in which o appears was that they are "equivalent to nothing in respect of the others." There is no limit argument here. There is only the intuitive notion of the properties of these infinitesimal increments of time.

To solve his second problem of finding the relationship of fluents given the relationship of their fluxions, Newton simply reversed the above procedure: "Since this problem is the converse of the preceding, it ought to be resolved the contrary way: namely by arranging the terms multiplied by $\dot{x}$ according to the dimensions of x and dividing by $\dot{x}/x$ and then by the number of dimensions, ... by carrying out the same operation in the terms multiplied by ... $\dot{y}$, and, with redundant terms rejected, setting the total of the resulting terms equal to nothing." As his example, he took the same equation used before. Starting with $3x^2\dot{x} - 2ax\dot{x} + ay\dot{x} - 3y^2\dot{y} + ax\dot{y} = 0$, he divided the terms containing $\dot{x}$ by $\dot{x}/x$ (or, what amounts to the same thing, removed the $\dot{x}$ and raised the power of x by 1), then divided each term again by the new power of x to get $x^3 - ax^2 + axy$. Doing the analogous operation on the terms containing $\dot{y}$, he found $-y^3 + axy$. Noting that axy occurs twice, he removed one of these terms and produced the final equation, $x^3 - ax^2 + axy - y^3 = 0$.

Newton naturally realized that this procedure does not always work. He suggested, in fact, that one always check the result. But if the problem could not be solved by this simple "antiderivative" approach, Newton generally used the method of power series. Since the fluent equation determined by the fluxional equation $\dot{y} = x^n\dot{x}$, or $\dot{y}/\dot{x} = x^n$, is $y = x^{n+1}/(n+1)$, he suggested that when $\dot{y}/\dot{x}$ depends only on x, one should express the ratio by a power series and apply that rule to each term. For example, the equation $\dot{y}^2 = \dot{x}\dot{y} + x^2\dot{x}^2$ can be rewritten as $\dot{y}^2/\dot{x}^2 = \dot{y}/\dot{x} + x^2$. This quadratic equation in $\dot{y}/\dot{x}$ can be solved to give $\dot{y}/\dot{x} = \frac{1}{2} \pm \sqrt{\frac{1}{4} + x^2}$. By applying the binomial theorem, one gets the two series

$$\frac{\dot{y}}{\dot{x}} = 1 + x^2 - x^4 + 2x^6 - 5x^8 + \cdots \quad \text{and} \quad \frac{\dot{y}}{\dot{x}} = -x^2 + x^4 - 2x^6 + 5x^8 + \cdots.$$

The solutions to the original problem are then easily found to be

$$y = x + \tfrac{1}{3}x^3 - \tfrac{1}{5}x^5 + \tfrac{2}{7}x^7 + \cdots \quad \text{and} \quad y = -\tfrac{1}{3}x^3 + \tfrac{1}{5}x^5 - \tfrac{2}{7}x^7 + \cdots.$$

The solution method is more complicated if $\dot{y}/\dot{x}$ is given by an equation in both x and y, but even then Newton's basic idea was to express the given equation in terms of a power series.

Newton realized very early in his researches that the problem of finding the fluent, giving the fluxion, or of finding the distance, given the velocity, is equivalent to finding the area under a curve from its equation. As we have seen, he knew perfectly well how to do this in the case where the equation was a sum of powers of x. But Newton also discovered and used the fundamental theorem of calculus to solve area problems. For him, this theorem was virtually self-evident. Because he thought of the curve AFD as being generated by the motions of x and y, it followed that the area $AFDB$ was generated by the motion of the moving ordinate BD (Fig. 11.16). It was therefore obvious that the fluxion of the area was in fact the ordinate multiplied by the fluxion of BD. That is, if z represents the area under the curve, then $\dot{z} = y\dot{x}$, or $\dot{z}/\dot{x} = y$. This equation translates immediately into part of the modern fundamental theorem: If $A(x)$ represents the area under $y = f(x)$ from 0 to x, then $dA/dx = f(x)$. Newton noted that the area z can be found explicitly from the equation $\dot{z}/\dot{x} = y$ by using the techniques already discussed for finding fluents using series. But he also noted that "curves of this kind may sometimes be squared by means of finite equations also."

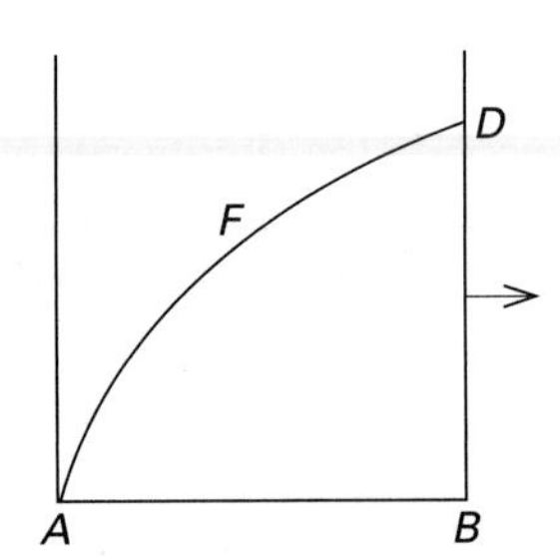

FIGURE 11.16 Newton and the fundamental theorem

To square curves by finite equations (that is, to find the area under them), one needs what is today referred to as a table of integrals. So Newton provided one. The first entry in the table he produced for the *Treatise on Methods* is a simple one—the area under $y = ax^{n-1}$ is $(a/n)x^n$—but the other entries are considerably more complex. In this brief excerpt from Newton's table, the function z on the right represents the area under the function y on the left:

$$y = \frac{ax^{n-1}}{(b+cx^n)^2} \qquad z = \frac{(a/nb)x^n}{b+cx^n}$$

$$y = ax^{n-1}\sqrt{b+cx^n} \qquad z = \frac{2a}{3nc}(b+cx^n)^{3/2}$$

$$y = ax^{2n-1}\sqrt{b+cx^n} \qquad z = \frac{2a}{nc}\left(-\frac{2}{15}\frac{b}{c}+\frac{1}{5}x^n\right)(b+cx^n)^{3/2}$$

$$y = \frac{ax^{2n-1}}{\sqrt{b+cx^n}} \qquad z = \frac{2a}{nc}\left(-\frac{2}{3}\frac{b}{c}+\frac{1}{3}x^n\right)\sqrt{b+cx^n}$$

In contrast to a modern table of integrals, Newton's table listed no transcendental functions, no sines or cosines, not even any logarithms. Although Newton knew the power series of these functions, he never treated them on an equal basis with algebraic functions. He did not operate with the sine, cosine, or logarithm algebraically, by combining them with polynomials and other algebraic expressions. Newton did, however, extend his table to functions whose integrals would be expressed today in terms of transcendental functions by expressing the integrals in terms of areas bounded by certain conic sections, areas that could be calculated using power series techniques. For example, given $y = x^{n-1}/(a+bx^n)$, he wrote that the area z can be expressed as $1/n$ times the area under the hyperbola $v = 1/(a+bu)$, where $u = x^n$.

Besides presenting the various algorithms, Newton showed, in the *Treatise on Methods*, how his solutions to the two basic problems can be applied to solve numerous other problems, such as determining maxima and minima, drawing tangent lines, finding the curvature of a curve, and determining arc length. He also discussed such methods as substitution in integrals and integration by parts. Thus, by the early 1670s, Newton had worked out virtually all of the important ideas found in the first several chapters of any modern calculus text. For reasons that we do not fully understand, however, Newton did not publish this material. Thus, it became known only to those who were fortunate enough to have access to one of the manuscripts. Newton, however, continued his researches in Cambridge and was soon to reveal publicly some of his discoveries.

11.4.3 The Synthetic Method of Fluxions and Newton's Physics

By the mid-1670s, Newton was somewhat unhappy with his use of "analysis" in developing the ideas of calculus. He had been studying the ancient Greek texts and believed that mathematical truth must be based on the tenets of proof that had been developed in Greece. Thus, although he had confidence in the efficacy of the algebraic methods of the "moderns," he began to reformulate his ideas more in keeping with the "geometry of the ancients." And therefore, when he began to compose his masterpiece, the *Philosophiae naturalis principia mathematica* (*Mathematical Principles of Natural Philosophy*) of 1687, the work in which he formulated his laws of motion and used them to derive the "system of the world," he decided to write his text in the form of a synthetic geometric treatise.

That he wrote the book at all is perhaps surprising, given Newton's aversion to publishing. But in the summer of 1684, the young English astronomer Edmond Halley (1656–1741) traveled to Cambridge to pose a critical question to Newton: "What would be the curve that would be described by the planets supposing the force of attraction towards the sun to be reciprocal to the square of their distance from it?" Newton's immediate answer was that he had already "calculated" the answer to this problem and that the curve would be an ellipse. When Halley pressed Newton for details, the Cambridge professor promised to send them along shortly. Several months went by, but in November of 1684 Halley received a ten-page treatise from Newton, not only purporting to answer the original question but also sketching a reformulation of astronomy in terms of forces. Halley was so impressed with this work, *De motu corporum in gyram* (*On the Motion of Bodies in an Orbit*), that he hurried back to Cambridge to attempt to persuade Newton to publish. Evidently, Halley did not have to work very hard at persuasion—Newton was already on the way to revising and expanding the treatise into the *Principia*.

The *Principia* begins with **Newton's laws of motion**:

1. Every body perseveres in its state of being at rest or of moving uniformly straight forward, except insofar as it is compelled to change its state by forces impressed.
2. A change in motion is proportional to the motive force impressed and takes place along the straight line in which that force is impressed.
3. To any action there is always an opposite and equal reaction.

In order to apply these laws, however, Newton needed to establish a mathematical framework for his physics. So he reformulated his ideas on fluxions from the analytic methods he used earlier into a more synthetic method that he called "the method of first and ultimate ratios." He used this method in proving eleven important lemmas in section 1 of the *Principia*. For example,

LEMMA 1. *Quantities, and also ratios of quantities, which in any finite time constantly tend to equality, and which before the end of that time approach so close to one another that their difference is less than any given quantity, become ultimately equal.*

The proof was obvious. If the quantities are ultimately unequal, they differ by a positive value D and therefore do not approach nearer to equality than D, a contradiction.

In lemma 2, Newton presented the situation where there is a set of inscribed rectangles in a curvilinear area and a corresponding set of circumscribed rectangles around that area. Newton claimed that the ultimate ratios that the inscribed figure, the circumscribed figure, and the curvilinear figure itself bear to one another as the width of the rectangles is diminished and their number increased indefinitely is the ratio of equality. Newton's proof is similar to one that would be given today. He showed that the difference between the areas of the circumscribed and inscribed figures is the area of a single rectangle, which, because its width "is diminished indefinitely, becomes less than any rectangle."

Newton demonstrated one of Galileo's important results in

LEMMA 10. *The spaces which a body describes when urged by any finite force, whether that force is determinate and immutable or is continually increased or continually decreased, are at the very beginning of the motion in the square ratio of the times.*

Newton's proof used his method of first and ultimate ratios on a geometric representation of distances as areas under velocity curves.

Newton believed that his method replaced the lengthy ancient proofs by *reductio ad absurdum*, but he realized that he had to convince his readers. So in the scholium to section 1, he wrote,

> I have preferred to make the proofs of what follows depend on the ultimate sums and ratios of vanishing quantities and the first sums and ratios of nascent quantities, that is, on the limits of such sums and ratios. . . . It may be objected that there is no such thing as an ultimate proportion of vanishing quantities, inasmuch as before vanishing the proportion is not ultimate, and after vanishing it does not exist at all. But by the same argument it could equally be contended that there is no ultimate velocity of a body reaching a certain place at which the motion ceases; for before the body arrives at this place, the velocity is not the ultimate velocity, and when it arrives there, there is no velocity at all. But the answer is easy; to understand the ultimate velocity as that with which a body is moving, neither before it arrives at its ultimate place and the motion ceases, nor after it has arrived there, but at the very instant when it arrives, that is, the very velocity with which the body arrives at its ultimate place and with which the motion ceases. And similarly the ultimate ratio of vanishing quantities is to be understood not as the ratio of quantities before they vanish or after they have vanished, but the ratio with which they vanish. . . . There exists a limit which their velocity can attain at the end of the motion, but cannot exceed. This is the ultimate velocity. . . . And since this limit is certain and definite, the determining of it is properly a geometrical problem. . . . Those ultimate ratios with which quantities vanish are not actually ratios of ultimate quantities [that is, there are no indivisibles], but limits which the ratios of quantities decreasing without limit are continually approaching, and which they can approach so closely that their difference is less than any given quantity, but which they can never exceed and can never reach before the quantities are decreased indefinitely.

Before discussing Newton's application of these ideas in the *Principia*, we will consider his use of them in defining and calculating fluxions. In his final tract on fluxions, the *De quadratura curvarum* (*On the Quadrature of Curves*) of 1691 (published in 1704), Newton wrote: "Fluxions are in the first ratio of the nascent augments or in the ultimate ratio of the evanescent part, but they may be expounded by any lines that are proportional to them." He then showed how to calculate the fluxion of x^n, where x flows uniformly:

> In the time that the quantity x comes in its flux to be $x + o$ [here o can be thought of as the "nascent augment"], the quantity x^n will come to be $(x + o)^n$, that is, by the method of infinite series,
>
> $$x^n + nox^{n-1} + \frac{n^2 - n}{2}o^2x^{n-2} + \cdots;$$
>
> and so the augments o and $nox^{n-1} + [(n^2 - n)/2]o^2x^{n-2} + \cdots$ are one to the other as 1 and $nx^{n-1} + [(n^2 - n)/2]ox^{n-2} + \cdots$. Now let those augments come to vanish [so now o is the "evanescent part"] and their ultimate ratio will be 1 to nx^{n-1}; consequently the fluxion of the quantity x is to the fluxion of the quantity x^n as 1 to nx^{n-1}.

This demonstration is not very different from the earlier calculation of fluxions, except that Newton does not write here of simply casting out terms that are "equivalent to nothing in respect of the others." In another manuscript, *Geometria curvilinea*, probably written a decade earlier but never published, however, Newton calculated the fluxions of the sine, tangent, and secant by the same method:

THEOREM. *In a given circle the fluxion of an arc is to the fluxion of its sine as the radius to its cosine; to the fluxion of its tangent as its cosine is to its secant; and to the fluxion of its secant as its cosine to its tangent.*

To demonstrate the result about the tangent, Newton considered a circle with center C and radius AC, where AB is the given arc and AT is a straight line tangent to the arc at A (Fig. 11.17). He then drew the secant CT, meeting the arc at B, and drew AS perpendicular to CT. Note that AS is the sine of AB, and CS is the cosine. Now let the arc and the tangent flow until they become Ab and At, respectively. Since the area of sector CBb is $\frac{1}{2}CA \times Bb$ and that of triangle CTt is $\frac{1}{2}CA \times Tt$, the ratio of arc Bb to segment Tt equals that of sector CBb to triangle CTt. To measure this ratio, Newton drew between the lines CT and Ct a new triangle Cpq, similar to triangle CTt but equal in area to sector CBb. The ratio of the two similar triangles is as the squares of their sides; thus, $Bb : Tt = Cp^2 : CT^2$. Now as the "augments" vanish, t and T will come together, as will p and q, and so Cp will become equal to CB. Thus, the ultimate ratio of Bb to Tt is CB^2 to CT^2. But $CT : CA = CA : CS$, and $CB = CA$, so $CB^2 = CT \cdot CS$. Thus, $CB^2 : CT^2 = (CT \cdot CS) : CT^2 = CS : CT$. In other words, the ratio of the fluxion of the arc to the fluxion of the tangent—that is, the ultimate ratio of Bb to Tt—is equal to the ratio of the cosine to the secant, as claimed. Newton proved the other two statements of the theorem by using this result and the basic trigonometric relationships.

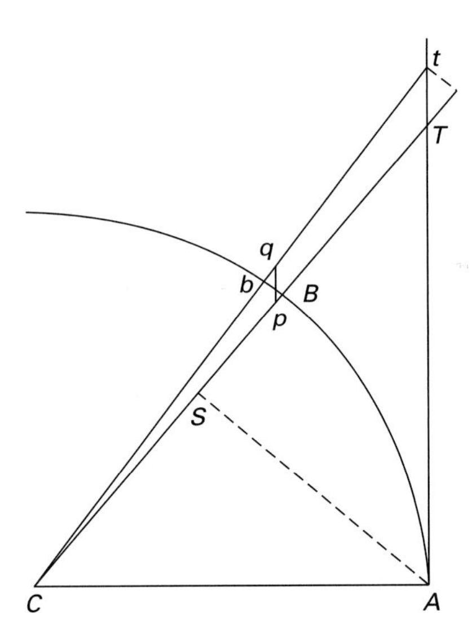

FIGURE 11.17 Newton's derivation of the fluxion of the tangent

We will now investigate how Newton used the ideas of limits in the *Principia*, in particular, how he used them in deriving versions of Kepler's laws of planetary motion. We begin with proposition 1 of section 2 of Book I:

PROPOSITION 1. *The areas which bodies made to move in orbits described by radii drawn to an unmoving center of forces . . . are proportional to the times.*

This result is, of course, Kepler's second law. Newton began his proof by dividing the time into equal finite parts. Suppose that in the first part, the body by its "inherent force" moves along the straight line segment AB (Fig. 11.18). If nothing were to impede the body, it would move in the next time interval along an equal segment Bc in the same direction, according to Newton's first law. Thus, if lines are drawn from A, B, and c to the center S, the triangles ASB and BSc will have equal areas. But since the body is being drawn to the center, Newton assumes that when it reaches B, the centripetal force acts and causes the body to change its path so that it moves in the direction BH. Now, a line is drawn from c parallel to BS, meeting BH at C. By the parallelogram law of combining forces, which Newton had worked out earlier, the body will be found at C at the end of the second time interval. If one now connects the center S to C and c, then triangle BSC is equal in area to triangle BSc and therefore to triangle ASB. Because this argument can be repeated for other equal time intervals, and because one can combine the equal triangles into larger regions, it follows that in this situation of force acting discretely, "any sums $SADS$ and $SAFS$ of the areas are to each other as the times of description." But Newton knew, of course, that the force acts continuously. Thus, he concluded his proof with the following: "Let the number of triangles be increased and their width decreased indefinitely, and their ultimate perimeter will be a curved line; and thus the centripetal force by which the body is continually drawn back from the tangent of this curve will act uninterruptedly, while any areas described, $SADS$ and $SAFS$, which are always proportional to the times of description, will be proportional to those times in this case."

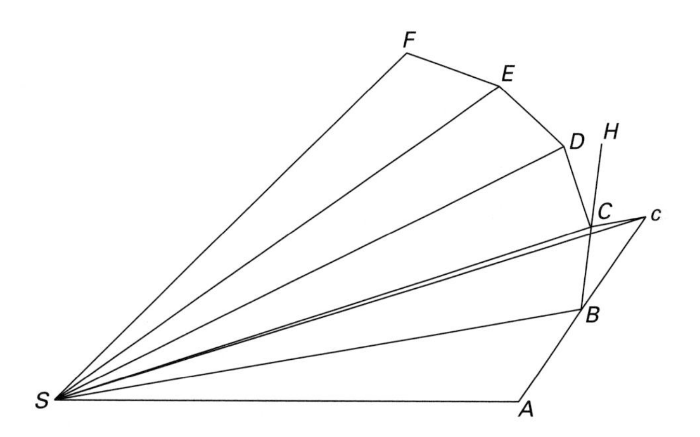

FIGURE 11.18 Newton's determination of the area law

To deal with the central forces by geometric methods, Newton needed a geometric representation of such a force, even when the force changes its magnitude and direction continuously. This he accomplished in proposition 6 and its corollaries, which describes a body orbiting about a center S in any curve (Fig. 11.19). If PX is tangent to the curve at P, if QT is perpendicular to PS at any other point Q on the orbit, and if QR is drawn to PX parallel to PS, then the centripetal force will be inversely as the solid $(SP^2 \times QT^2)/QR$, "provided that the magnitude of that solid is always taken as that which it has ultimately when the points P and Q come together." Recall that, by lemma 10, the distance a body

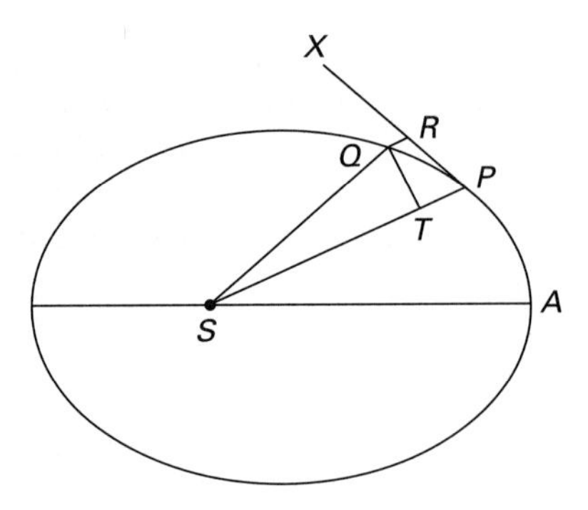

FIGURE 11.19 Determination of a geometrical method to measure centripetal force

travels under even a variable force is, at the very beginning of motion, as the square of the time. In this case, that distance is QR. But since QR also represents the change in motion of the body, it is proportional to the force (by Newton's second law). Therefore, QR is proportional to the force and the square of the time. But by proposition 1, the time is proportional to the area swept out, namely, that of the triangle SPQ, which is $\frac{1}{2}SP \cdot QT$. Thus, QR is proportional jointly to the centripetal force and to $(SP \cdot QT)^2$. Therefore, the force F is proportional to QR and inversely proportional to $(SP \cdot QT)^2$, as claimed. In modern terms, we can think of QR as a vector representing the acceleration caused by the initial force applied. The length of this vector is the magnitude of the acceleration, which, since $d = \frac{1}{2}at^2$, is itself proportional to the distance and inversely to the square of the time. Since force is proportional to acceleration and time to the area swept out, Newton's result follows.

With a geometric representation of the force at hand, Newton could calculate the force for various specific orbits. We will consider the most interesting case—Newton's proposition 11, in which the orbit is elliptical and the force is directed toward a focus S (Fig. 11.20). In

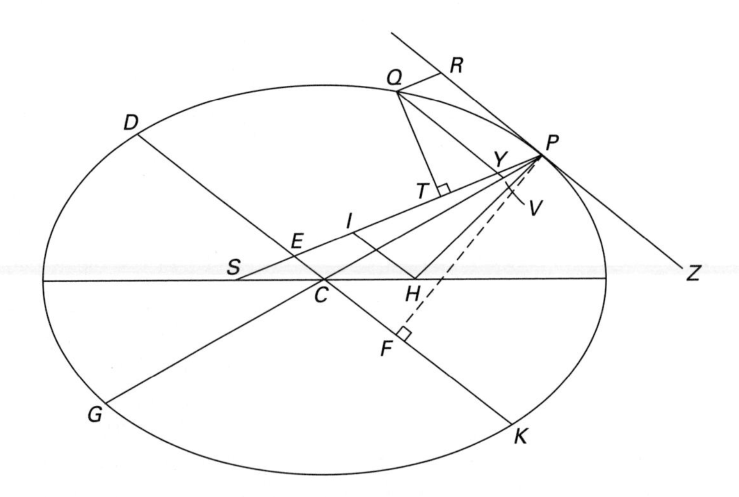

FIGURE 11.20 An elliptical orbit entails an inverse square force law

this case, using the same notation as in proposition 6, let DK, PG be conjugate diameters of the ellipse, where DK is parallel to PR, and let QV be drawn parallel to RP meeting PG at V. Furthermore, let SP cut DK at E and QV at Y, thus completing the parallelogram $QYPR$. As usual, let a and b represent the lengths of the semimajor and semiminor axes of the ellipse, respectively, and p represent the parameter. Newton first showed that $EP = a$. For if H is the second focus of the ellipse, and if HI is drawn parallel to EC, then $ES = EI$ and $\angle PIH = \angle YPR = \angle ZPH = \angle PHI$. Thus, $EP = (PS + PI)/2 = (PS + PH)/2 = 2a/2 = a$. He then set out five proportions:

$$p \times QR : p \times PV = QR : PV = PY : PV = PE : PC = a : PC \qquad \textbf{(11.1)}$$

$$p \times PV : GV \times PV = p : GV \qquad \textbf{(11.2)}$$

$$GV \times PV : QV^2 = PC^2 : CD^2 \qquad \textbf{(11.3)}$$

$$QV^2 : QY^2 = M : N \qquad \textbf{(11.4)}$$

$$QY^2 : QT^2 = EP^2 : PF^2 = a^2 : PF^2 = CD^2 : b^2 \qquad \textbf{(11.5)}$$

Proportion 11.1 follows from the similarity of triangles PVY and PCE and from the fact that $EP = a$, and proportion 11.2 is simply the cancellation law. Newton knew proportion 11.3 from his study of conic sections; it is Apollonius's proposition I-21 referred to a pair of conjugate axes. (See chapter 3, exercise 16d.) Proportion 11.4 is simply a definition of M and N, and proportion 5 depends on the similarity of triangles QTY and PFE and on Apollonius's VII-31 (that the rectangles constructed on any pair of conjugate diameters are equal). (See chapter 3, exercise 16e.) If one multiplies these proportions together, recalling that $b^2 = pa/2$, the result is $p \times QR : QT^2 = (2PC : GV) \times (M : N)$. But, as the points P and Q "come together," the two ratios on the right become ratios of equality. It follows that $p \times QR = QT^2$ and, on multiplying both sides by SP^2/QR, that

$$p \times SP^2 = \frac{SP^2 \times QT^2}{QR}.$$

Since, by proposition 6, the centripetal force is inversely proportional to the expression on the right, it is also inversely proportional to $p \times SP^2$. Since p is a constant, the force is inversely proportional to the square of the distance SP of a point from the focus.

After proving this result, Newton proved analogous results for hyperbolas (proposition 12) and parabolas (proposition 13) and then concluded with the following:

COROLLARY. *From the last three propositions it follows that if any body P departs from the place P along any straight line PR with any velocity whatever and is at the same time acted upon by a centripetal force that is inversely proportional to the square of the distance of places from the center, this body will move in some one of the conics having a focus in the center of forces; and conversely.*

This statement was criticized by many early readers of the *Principia*, since they wondered how Newton could conclude that an inverse square force law implied a conic section orbit from the converse of that result proved in propositions 11–13. So Newton added a brief argument to the corollary in the second and third editions (1713 and 1726) to the effect that a conic section can be constructed through a given point with a given focus, given tangent, and given curvature; that motion along this conic satisfies an inverse square force law; and that because the force and velocity together determine the curvature, this conic is the unique solution to the initial-value problem implied by the inverse square force law. Although sketchy, this argument is quite correct and can be expanded into a full formal proof of the corollary. In fact, Newton sketched other results in the *Principia* showing how to complete this proof, granting the solving of certain differential equations.

Proposition 11 and the corollary are closely related to Kepler's second law. Newton also proved a result closely related to Kepler's third law:

PROPOSITION 15. *The squares of the periodic times in ellipses are as the cubes of the major axes.*

Using the same notation and diagram as for proposition 11, by proposition 1, areas swept out are proportional to the time elapsed. Therefore, if Δt is the time taken in each ellipse to sweep out the infinitesimal area PSQ, the entire area of the ellipse is to the periodic time T as the area of triangle PSQ $(= \frac{1}{2}QT \cdot PS)$ is to Δt. Because the area of the ellipse is proportional to ab, thus ab is proportional to the product of T and $QT \cdot PS$. Also, for each of the elliptical orbits, the parameter p equals QT^2/QR. But QR, representing the force, is inversely proportional to SP^2. So p is proportional to $(QT \cdot PS)^2$, or $\sqrt{p}$ is proportional to $QT \cdot PS$. It follows that ab is proportional to $p^{1/2}T$. Since $b^2 = pa/2$, however, ab is also proportional to $a^{3/2}p^{1/2}$. It follows that T is proportional to $a^{3/2}$, or that T^2 is proportional to a^3, as claimed.

Kepler's laws and the inverse-square force law were to lead Newton to his law of universal gravitation, whose consequences were spelled out in Book III of the *Principia*. Newton's masterwork, read and commented on by many over the years, thus became the culminating document of the scientific revolution of the seventeenth century.

11.5 Gottfried Wilhelm Leibniz

The co-inventor of the calculus, Gottfried Leibniz, was brought to the frontiers of mathematical research by Huygens during his stay in Paris from 1672 to 1676. Like Newton, Leibniz read van Schooten's edition of Descartes's *Geometry*. He also studied the works of Pascal, which included the differential triangle. His subsequent investigations led to his invention of the differential and integral calculus by the end of his stay in Paris. It was only ten years later, however, that he began to publish his results in short notes in the *Acta eruditorum*, the German scientific journal that he helped to found.

11.5.1 Sums and Differences

Leibniz's idea, out of which his calculus grew, was the inverse relationship of sums and differences in the case of sequences of numbers. He noted that if A, B, C, D, E was an increasing sequence of numbers and L, M, N, P was the sequence of differences, then $E - A = L + M + N + P$; that is, "the sums of the differences between successive terms, no matter how great their number, will be equal to the difference between the terms at the beginning and the end of the series." It followed that difference sequences were easily summed.

Applying this result to geometry, Leibniz considered a curve defined over an interval divided into subintervals and erected ordinates y_i over each point x_i in the division. If one forms the sequence $\{\delta y_i\}$ of differences of these ordinates, its sum, $\sum_i \delta y_i$, is equal to the difference $y_n - y_0$ of the final and initial ordinates. Similarly, if one forms the sequence $\{\sum y_i\}$, where $\sum y_i = y_0 + y_1 + \cdots + y_i$, the difference sequence $\{\delta \sum y_i\}$ is equal to the original sequence of the ordinates. Leibniz extrapolated these two rules to handle the situation where there were infinitely many ordinates. He considered the curve as a polygon with infinitely many sides, at each intersection point of which an ordinate y is drawn to the axis. If the infinitesimal difference in ordinates is designated by dy, and if the sum of infinitely many ordinates is designated by $\int y$, the first rule translates into $\int dy = y$ and the second gives $d \int y = y$. Geometrically, the first means simply that the sum of the **differentials** (infinitesimal differences) in a segment equals the segment. (Leibniz assumed here that the initial ordinate equals 0.) The second rule does not have an obvious geometric interpretation, because the sum of infinitely many finite terms may well be infinite. So Leibniz replaced the finite ordinate y with an infinitesimal area $y\,dx$, where dx was the infinitesimal part of the x-axis determined by the intersection points of the sides of the infinite-sided polygon. Thus, $\int y\,dx$ could be interpreted as the area under the curve and the rule $d \int y\,dx = y\,dx$ simply meant that the differences between the terms of the sequence of areas $\int y\,dx$ are the terms $y\,dx$ themselves.

As part of his quest for the appropriate notation to represent ideas, Leibniz introduced the two notations d and $\int$ to represent his generalization of the idea of difference and sum. The latter is simply an elongated form of the letter S, the first letter of the Latin word *summa*, and the former is the first letter of the Latin word *differentia*. For Leibniz, both dy and $\int y$ were variables. In other words, d and $\int$ were operators which assigned an infinitely small variable and an infinitely large variable, respectively, to the finite variable y. But dy is always thought of as an actual difference, that between two neighboring values of the variable y, while $\int y$ is conceived of as a sum of all values of the variable y from a certain fixed value to the given one. Since dy is a variable, it too can be operated on by d to give a second-order differential, written as $d\,dy$, and even higher-order ones. It is perhaps somewhat difficult to conceive of these infinitesimal differences and infinite sums, but Leibniz and his followers became extremely adept at using these concepts in developing methods for solving many types of problems.

11.5.2 The Differential Triangle and the Transmutation Theorem

One of the earliest applications Leibniz made of the concept of a differential was to the idea of the differential triangle, a version of which he had seen in the work of Pascal and, perhaps, of Barrow. The differential triangle—the infinitesimal right triangle whose hypotenuse ds connects two neighboring vertices of the infinite-sided polygon representing a given curve—is similar to the triangle composed of the ordinate y, the tangent τ, and the subtangent t, so $ds : dy : dx = \tau : y : t$ (Fig. 11.21). Because ratios are involved in the idea of a tangent, Leibniz generally made one of these three differentials a constant. In other words, in choosing how to represent a curve as a polynomial with infinitely many sides, he made the polygon have equal sides (ds is constant, or $d\,ds = 0$), the projections of the sides on the x-axis be equal (dx is constant, or $d\,dx = 0$), or the projections of the sides on the y-axis be equal (dy is constant, or $d\,dy = 0$). In some sense, the variable chosen to have a constant differential can be thought of as the independent variable. In any case,

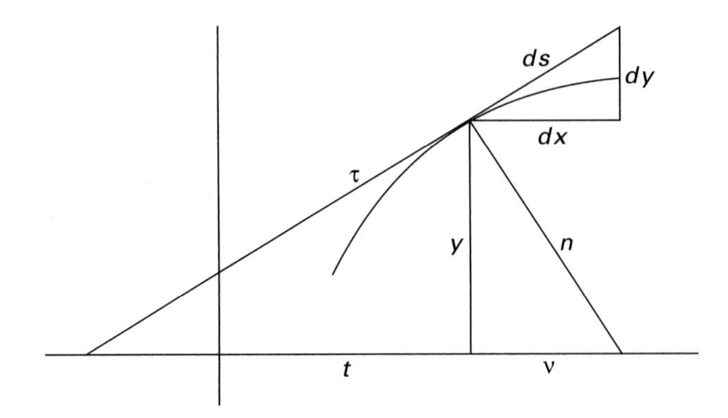

FIGURE 11.21 Leibniz's differential triangle

it was through manipulations of the differentials in the differential triangle, using his basic rules for manipulating with differentials, that Leibniz found the central techniques for his version of the calculus.

One of Leibniz's first discoveries through manipulation of differentials was his **transmutation theorem**, which led him to his arithmetical quadrature of the circle, a series expression for $\pi/4$. In the curve $OPQD$, where P and Q are infinitesimally close, he constructed the triangle OPQ (Fig. 11.22). Extending $PQ = ds$ into the tangent to the curve, drawing OW perpendicular to the tangent, and setting h and z as in the figure, he showed, using the similarity of triangle TWO to the differential triangle, that $dx : h = ds : z$, or $z\,dx = h\,ds$. The left side of the latter equation is the area under the rectangle $UVSR$, while the right side is twice the area of the triangle OPQ. It follows that the sum of all

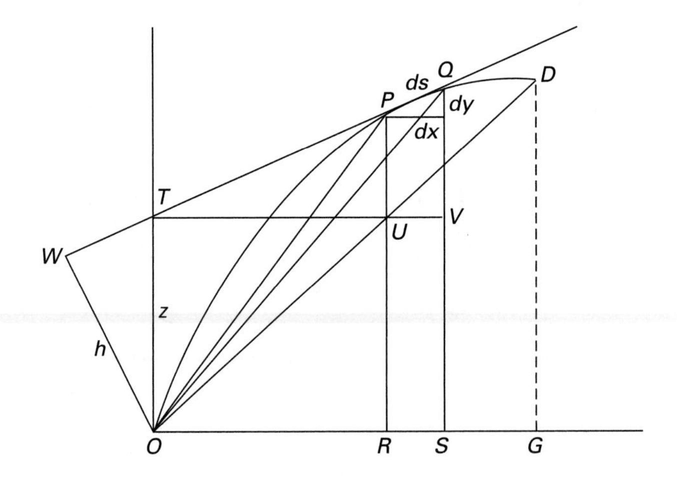

FIGURE 11.22 Leibniz's transmutation theorem

the triangles—namely, the area bounded by the curve $OPQD$ and the line OD—equals half the area under the curve whose ordinate is z, or $\int y\,dx - \frac{1}{2}(OG \cdot GD) = \frac{1}{2}\int z\,dx$. Denoting OG by x_0 and GD by y_0, Leibniz's transmutation theorem can be stated as

$$\int_0^{x_0} y\,dx = \frac{1}{2}\left(x_0 y_0 + \int_0^{x_0} z\,dx\right).$$

Because $z = y - PU = y - x(dy/dx)$, and because Leibniz could calculate tangents by using Hudde's rule, the transmutation theorem enabled him to find the area under the original curve, provided that $\int z\,dx$ was simpler to compute than $\int y\,dx$. For example, Leibniz applied this result to calculate the area of a quarter of the circle of radius 1 given by $y^2 = 2x - x^2$. In this case,

$$z = y - x\left(\frac{1-x}{y}\right) = \frac{x}{y} = \sqrt{\frac{x}{2-x}},$$

or

$$z^2 = \frac{x}{2-x} \qquad \text{or, finally,} \qquad x = \frac{2z^2}{1+z^2}.$$

By Leibniz's transmutation theorem, $\int y\,dx$ (or $\pi/4$) is equal to $\frac{1}{2}(1 + \int z\,dx)$. Since it is clear from Figure 11.23 that $\int z\,dx = 1 - \int x\,dz$, Leibniz concluded that

$$\int y\,dx = 1 - \int \frac{z^2}{1+z^2}\,dz.$$

By an argument analogous to Mercator's, he showed that

$$\frac{z^2}{1+z^2} = z^2(1 - z^2 + z^4 - z^6 + \cdots)$$

and hence that

$$\int y\,dx = 1 - \tfrac{1}{3}z^3 + \tfrac{1}{5}z^5 - \tfrac{1}{7}z^7 + \cdots.$$

Leibniz's formula for arithmetical quadrature, $\pi/4 = 1 - \frac{1}{3} + \frac{1}{5} - \frac{1}{7} + \cdots$, followed immediately.

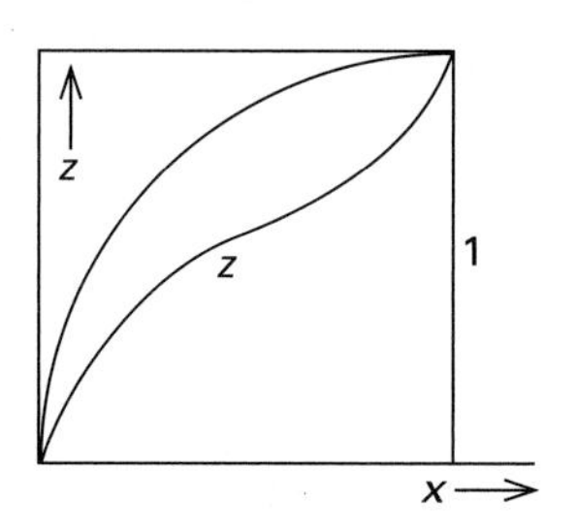

FIGURE 11.23 The transmutation function for the circle: $z^2 = x/(2-x)$ or $x = 2z^2/(1+z^2)$

11.5.3 The Calculus of Differentials

Leibniz discovered his transmutation theorem and the arithmetical quadrature of the circle in 1674. During the next two years, he discovered all of the basic ideas of his calculus of differentials. He first published some of these results in "A New Method for Maxima and Minima as well as Tangents, which is neither impeded by fractional nor irrational Quantities, and a remarkable Type of Calculus for them," a brief article that appeared in 1684 in the *Acta eruditorum*. In this paper, Leibniz was reluctant to define his differentials dx as infinitesimals because he believed there would be great criticism of these quantities, which had not been rigorously defined. Thus, he introduced dx as an arbitrary finite line segment. If y was the ordinate of a curve for which x was the abscissa, and if τ was the tangent to the curve at a point with t the subtangent, then dy was defined to be that line such that $dy : dx = y : t$. He then stated some basic rules of operation. If a is a constant, then $da = 0$; $d(v \pm y) = dv \pm dy$; $d(vw) = v\,dw + w\,dv$; and $d(v/y) = (\pm v\,dy \mp y\,dv)/y^2$. (The signs in the quotient rule depend, according to Leibniz, on whether the slope of the tangent line is positive or negative.)

Leibniz discovered the product and quotient rules in 1675. Here is his original proof of the first: "$d(xy)$ is the same thing as the difference between two successive xy's; let one of these be xy, and the other $(x + dx)(y + dy)$; then we have $d(xy) = (x + dx)(y + dy) - xy = x\,dy + y\,dx + dx\,dy$. The omission of the quantity $dx\,dy$, which is infinitely small in comparison with the rest, ... will leave $x\,dy + y\,dx$." He proved the quotient rule similarly.

In the 1684 paper, Leibniz continued by giving the power rule $d(x^n) = nx^{n-1}\,dx$ and the rule for roots $d\sqrt[b]{x^a} = (a/b)\sqrt[b]{x^{a-b}}\,dx$, noting that the first law includes the second if a root is written as a fractional power. The chain rule is almost obvious using Leibniz's notation. For example, to calculate the differential of $z = \sqrt{g^2 + y^2}$, where g is a constant, Leibniz set $r = g^2 + y^2$ and noted that $dr = 2y\,dy$ and $dz = d\sqrt{r} = dr/2\sqrt{r}$. Substituting the first equation into the second, he concluded that

$$dz = \frac{2y\,dy}{2z} = \frac{y\,dy}{z}.$$

To demonstrate the usefulness of his new calculus, Leibniz discussed how to determine maxima and minima. He noted that dv will be positive when v is increasing and negative when v is decreasing, since the ratio of dv to the always positive dx gives the slope of the tangent line. It follows that $dv = 0$ when v is neither increasing nor decreasing. At that place the ordinate will be a maximum (if the curve is concave down) or a minimum (if it is concave up). The tangent there will be horizontal. The question of concavity, Leibniz noted further, depends on the second differentials $d\,dv$: "When with increasing ordinates v its increments or differences dv also increase (that is, when dv is positive, $d\,dv$, the difference of the differences, is also positive, and when dv is negative, $d\,dv$ is also negative), then the curve is [concave up], in the other case [concave down]. Where the increment is maximum or minimum, or where the increments from decreasing turn into increasing, or the opposite, there is a *point of inflection*," that is, when $d\,dv = 0$.

At the end of his 1684 paper, Leibniz presented the problem of finding a curve whose subtangent is a given constant a. If y is the ordinate of the proposed curve, the differential equation of the curve is $y(dx/dy) = a$, or $a\,dy = y\,dx$. Leibniz set dx as constant, equivalent to having the abscissas form an arithmetical progression. The equation then can be written as $y = k\,dy$, where k is constant. It follows that the ordinates y are proportional to their increments dy, or that the y's form a geometric progression. Since the relationship of a geometric progression in y to an arithmetic progression in x is as numbers are to their logarithms, Leibniz concluded that the desired curve will be a "logarithmic" curve. (It is now called an *exponential curve*—but, after all, today's exponential and logarithmic curves are the same curves referred to different axes.) It follows from Leibniz's discussion that, since $x = \log\, y$, $d(\log\, y) = a(dy/y)$ where the constant a depends on the particular logarithm used.

Leibniz did not consider the logarithm further in the paper of 1684, but after discussion with Johann Bernoulli (1667–1748) some years later, he returned in 1695 to consideration of the differential of the logarithm as well as of the exponential function $z = y^x$. A direct calculation of the differential gives $dz = (y + dy)^{x+dx} - y^x$. Applying the binomial theorem and discarding powers of dy higher than the first as well as multiples $dx\,dy$ produces the equation $dz = y^{x+dx} + xy^{x+dx-1}\,dy - y^x$, a differential equation that is not homogeneous and that cannot apparently be simplified further, even in the special case where $y = b$ is constant and therefore $dy = 0$. To circumvent this difficulty, Leibniz attacked the problem differently by taking logarithms of both sides of the equation $z = y^x$ to get $\log\, z = x\, \log\, y$. The differential of this equation is then

$$a\,\frac{dz}{z} = xa\,\frac{dy}{y} + \log\, y\,dx.$$

It follows that

$$dz = \frac{xz}{y}\,dy + \frac{z\,\log y}{a}\,dx \qquad \text{or} \qquad d(y^x) = xy^{x-1}\,dy + \frac{y^x \log y}{a}\,dx.$$

If $x = r$ is constant, Leibniz noted, this rule reduces to the **power rule**: $d(y^r) = ry^{r-1}\,dy$.

11.5.4 The Fundamental Theorem and Differential Equations

Recall that Leibniz began his researches into what became his calculus with the idea that sums and differences are inverse operations. It followed that the fundamental theorem of calculus was completely obvious. In fact, "the general problem of quadratures can be reduced to the finding of a curve that has a given law of tangency." In other words, given the curve with ordinates y, if one can find a curve z such that $dz/dx = y$ (a curve with a given law of tangency), then $\int\, y\,dx = z$, or, in modern notation, assuming that $z(0) = 0$,

$$\int_0^b y\,dx = z(b).$$

But Leibniz was not as interested in finding areas as he was in solving differential equations, especially since it turned out that important physical problems could be expressed in terms of such equations. And Leibniz, like Newton, used power series methods to solve such equations. His technique, however, was different. For example, consider the equation expressing the relationship between the arc y and its sine x in a circle of radius 1, as discussed by Leibniz in 1693. The differential triangle with sides dy, dt, and dx is similar

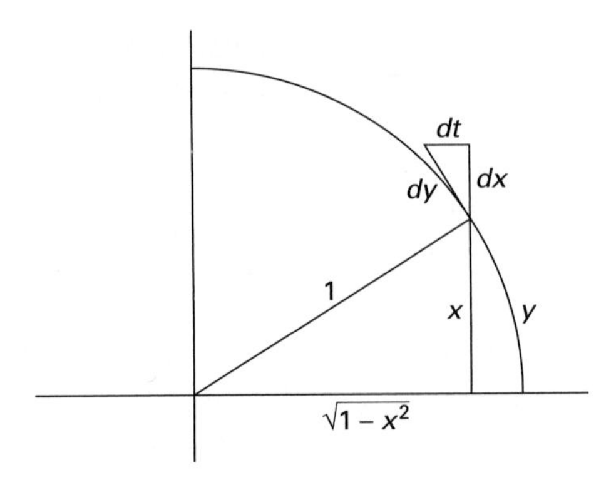

FIGURE 11.24 Leibniz's derivation of the differential equation for the sine

to the large triangle with corresponding sides 1, x, and $\sqrt{1-x^2}$ (Fig. 11.24), so

$$dt = \frac{x\,dx}{\sqrt{1-x^2}}.$$

By the Pythagorean theorem, $dx^2 + dt^2 = dy^2$. Substituting into this the value of dt and simplifying gave Leibniz the differential equation relating the arc and the sine: $dx^2 + x^2\,dy^2 = dy^2$. Considering dy as constant, he applied his operator d to this equation and concluded that $d(dx^2 + x^2\,dy^2) = 0$ or, using the product rule, that $2\,dx(d\,dx) + 2x\,dx\,dy^2 = 0$. Leibniz simplified this into the second-order differential equation

$$d^2x + x\,dy^2 = 0 \qquad \text{or} \qquad \frac{d^2x}{dy^2} = -x,$$

which is the familiar differential equation of the sine. (Note that Leibniz's method of manipulating with second-order differentials explains the seemingly strange placement of the 2's in the modern notation for the second derivative.)

Given the differential equation, Leibniz next assumed that x could be written as a power series in y: $x = by + cy^3 + ey^5 + fy^7 + gy^9 + \cdots$, with the coefficients to be determined. It was obvious to him that there could be no even-degree terms and that, since $\sin 0 = 0$, the constant term was also 0. Differentiating this series twice gives $d^2x/dy^2 = 2\cdot 3cy + 4\cdot 5ey^3 + 6\cdot 7fy^5 + 8\cdot 9gy^7 + \cdots$, a power series to be equated to the power series expressing $-x$. The identity of the coefficients then gives a series of simple equations:

$$2\cdot 3c = -b$$

$$4\cdot 5e = -c$$

$$6\cdot 7f = -e$$

$$8\cdot 9g = -f$$

$$\vdots$$

Setting $b = 1$ as the second initial condition, Leibniz solved these equations easily to get $c = -1/3!$, $e = 1/5!$, $f = -1/7!$, $g = 1/9!, \ldots$, and thus derived the sine series:

$$x = \sin y = y - \frac{1}{3!}y^3 + \frac{1}{5!}y^5 - \frac{1}{7!}y^7 + \frac{1}{9!}y^9 - \cdots.$$

Leibniz had by the early 1690s discovered most of the ideas present in current calculus texts but had never written out a complete, coherent treatment of the material. And, like Newton after the 1670s, he wanted to justify his work by appealing to Greek standards. He gave two separate justifications: First, he attempted to relate infinitesimals to Archimedean exhaustion: "For instead of the infinite or the infinitely small, one takes quantities as large, or as small, as necessary in order that the error be smaller than the given error, so that one differs from Archimedes' style only in the expression, which are more direct in our method and conform more to the art of invention." Leibniz's second justification made use of a law of continuity: "If any continuous transition is proposed terminating in a certain limit, then it is possible to form a general reasoning, which covers also the final limit." In other words, if one determined that a particular ratio is true in general, when, for example, the quantities dx, dy are finite, the same ratio will be true in the limiting case, when these quantities are themselves equal to 0. This justification is, in fact, very similar to Newton's own notion of limit. But justified or not, the technique of manipulating with these infinitesimal differentials became a very useful one, particularly for Leibniz's immediate followers, Johann Bernoulli and Jakob Bernoulli (1654–1705). In fact, they seemed to accept infinitesimals as actual mathematical entities and used them to achieve many important results both in calculus itself and in its applications to physical problems.

A few words about the priority controversy between Leibniz and Newton are in order here. It should be clear that although the two men discovered essentially the same rules and procedures which today are collectively called the calculus, their approaches to the subject were entirely different. Newton's approach was through the ideas of velocity and distance, whereas Leibniz's was through the ideas of differences and sums. Also, since Newton's work was well known in England well before it was published in the early eighteenth century, the successes of Leibniz and the Bernoulli brothers in applying their version caused certain English mathematicians to accuse Leibniz of plagiarism. Conversely, precisely because Newton had not published earlier, the Bernoullis accused him of plagiarizing from Leibniz. In 1711, the Royal Society, of which Newton was then the president, appointed a commission to look into the charges. Naturally, the commission found Leibniz guilty as charged. The unfortunate result of the controversy was that the interchange of ideas between English and continental mathematicians virtually ceased. As far as the calculus was concerned, the English all adopted Newton's methods and notation, while mathematicians on the continent used those of Leibniz. It turned out that Leibniz's notation and his calculus of differentials proved easier to work with. Thus, progress in analysis was faster on the continent. To its ultimate detriment, the English mathematical community deprived itself for nearly the entire eighteenth century of that great progress.

Exercises

1. Show that Fermat's two methods of determining a maximum or minimum of a polynomial $p(x)$ are both equivalent to solving $p'(x) = 0$.
2. Use one of Fermat's methods to find the maximum of $bx - x^3$. How would Fermat decide which of the two solutions to choose as his maximum?
3. Justify Fermat's first method of determining maxima and minima by showing that if M is a maximum of $p(x)$, then the polynomial $p(x) - M$ always has a factor $(x - a)^2$, where a is the value of x giving the maximum.
4. Use Fermat's tangent method to determine the relation between the abscissa x of a point B and the subtangent t that gives the tangent line to $y = x^3$.
5. Modify Fermat's tangent method to be able to apply it to curves given by equations of the form $f(x, y) = c$. Begin by noting that if $(x + e, \bar{y})$ is a point on the tangent line near to (x, y), then
$$\bar{y} = \frac{t+e}{t}y.$$
Then set
$$f(x, y) = f\left(x + e, \frac{t+e}{t}y\right).$$
Apply this method to determine the subtangent to the curve $x^3 + y^3 = pxy$.
6. Show that in modern notation, Fermat's method of finding the subtangent t to $y = f(x)$ determines t as $t = f(x)/f'(x)$. Show similarly that the modified method of exercise 5 is equivalent in modern terms to determining t as $t = -y(\partial f/\partial y)/(\partial f/\partial x)$.
7. Use Fermat's method to determine the subtangent to the ellipse $x^2/a^2 + y^2/b^2 = 1$.
8. Use Descartes's circle method to determine the subnormal to $y = x^{3/2}$.
9. Use Descartes's circle method to determine the slope of the tangent line to $y^2 = x$.
10. Use Hudde's rule applied to Descartes's method to show that the slope of the tangent line to $y = x^n$ at (x_0, x_0^n) is nx_0^{n-1}.
11. This problem is taken from Hudde's *De maximis et minimis*: Maximize $3ax^3 - bx^3 - (2b^2a/3c)x + a^2b$ using Hudde's rule.
12. Given that the volume of a cone is $\frac{1}{3}hA$, where h is the height and A the area of the base, use Kepler's method to divide up a sphere of radius r into infinitely many infinitesimal cones of height r, and then add up their volumes to get a formula for the volume of the sphere.
13. Use the result of exercise 12 of chapter 10 to derive formulas for the sums of integral squares and cubes from 1 to n. (It may be easier to rewrite that formula with $n + 1$ in place of n.) First, set $k = 3$ and expand the binomial coefficients on both sides. Rewrite the right side in terms of a sum of squares and a sum of integers. Replace the sum of integers by a formula for it, rearrange, and calculate the formula for the sum of squares. Next, set $k = 4$ and repeat. In this case, you will need to put in the formula for squares already calculated.
14. Repeat exercise 13 for $k = 5$ and discover the fifth-degree polynomial formula for the sum of integral fourth powers from 1 to n.
15. Fermat included the following result in a letter to Roberval of August 23, 1636: If the parabola with vertex A and axis AD is rotated around the line BD, the volume of this solid has the ratio 8 : 5 to the volume of the cone of the same base and vertex (Fig. 11.25). Prove that Fermat was correct, and show that this result is equivalent to the result on the volume of this same solid discovered by ibn al-Haytham and discussed in chapter 7.

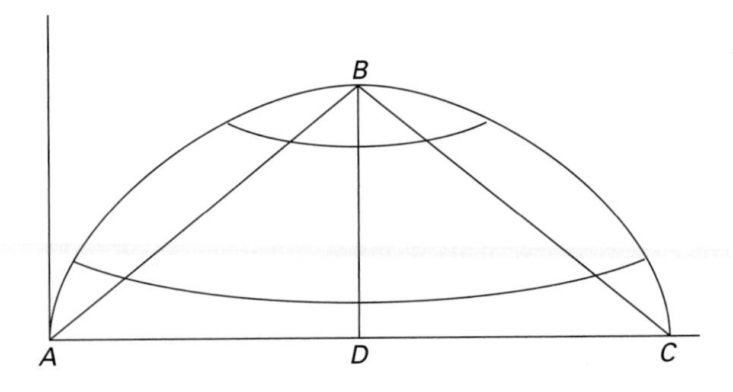

FIGURE 11.25 Fermat's problem of revolving a parabola around a line perpendicular to its axis

16. Determine the area under the curve $y = px^k$ from $x = 0$ to $x = x_0$ by dividing the interval $[0, x_0]$ into an infinite set of subintervals, beginning from the right with the points $a_0 = x_0$, $a_1 = (n/m)x_0$, $a_2 = (n/m)^2 x_0, \ldots$, where $n < m$, and proceeding as in Fermat's derivation of the area under the hyperbola.

17. Gregory derived various formulas for calculating the subtangents of curves composed of other curves by addition, subtraction, and the use of proportionals, and he used these in his calculations of power series. In particular, suppose that four functions are related by the proportion $u : v = w : z$. Show that the subtangent t_z is given by the formula

$$t_z = \frac{t_u t_v t_w}{t_u t_v + t_u t_w - t_v t_w}.$$

Derive the product and quotient rules for derivatives from this formula, given that if a function u is a constant, then its subtangent t_u is infinite.

18. Find the length of the arc of the curve $y^4 = x^5$ from $x = 0$ to $x = b$.

19. Show that to find the length of an arc of the parabola $y = x^2$, it is necessary to determine the area under the hyperbola $y^2 - 4x^2 = 1$.

20. Use Barrow's a, e method to determine the slope of the tangent line to the curve $x^3 + y^3 = c^3$.

21. Barrow was perhaps the first to calculate the slope of the tangent to the curve $y = \tan x$ using his a, e method. Suppose DEB is a quadrant of a circle of radius 1 and BX is the tangent line at B (Fig. 11.26). The tangent curve AMO is defined to be the curve such that if AP is equal to arc BE, then PM is equal to BG, the tangent to arc BE. Use the differential triangle to calculate the slope of the tangent to curve AMO as follows: Let $CK = f$ and $KE = g$. Since $CE : EK = \text{arc } EF : LK = PQ : LK$, it follows that $1 : g = e : LK$, or $LK = ge$ and $CL = f + ge$. Then $LF = \sqrt{1 - f^2 - 2fge} = \sqrt{g^2 - 2fge}$. Because $CL : LF = CB : BH$, one can transfer the ratio in the circle to that on the tangent curve. Demonstrate finally that

$$PT = t = \frac{BG \cdot CB^2}{CG^2} = \frac{BG \cdot CK^2}{CE^2},$$

and show that this result can be translated into the familiar formula $d(\tan x)/dx = \sec^2 x$. Given this result, can you say that Barrow differentiated a trigonometric function? Why or why not?

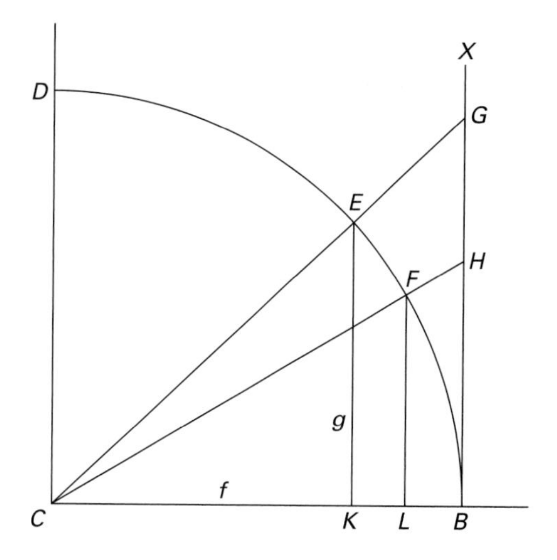

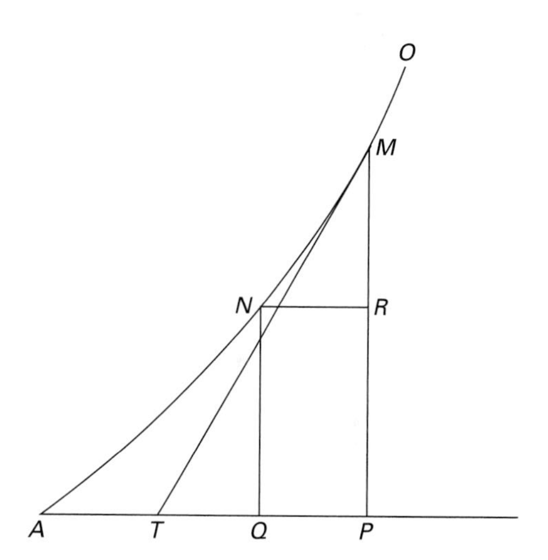

FIGURE 11.26 Barrow's calculation of the tangent line to the tangent function

22. Calculate a power series for $\sqrt{1 + x}$ by applying the square-root algorithm to $1 + x$.

23. Square Newton's power series for $(1 - x^2)^{1/2}$ and show that the resultant power series is equal to $1 - x^2$. (You need to convince yourself that every coefficient beyond that for x^2 is equal to 0.)

24. Calculate, using the power series for $\log(1 + x)$, the values of the logarithm of 1 ± 0.1, 1 ± 0.2, 1 ± 0.01,

1 ± 0.02 to eight decimal places. Using the identities presented in section 11.4.1 and others of your own devising, calculate a logarithm table of the integers from 1 to 10 accurate to eight decimal places.

25. Calculate the relationship of the fluxions of the equation $x^3 - ax^2 + axy - y^3 = 0$ using multiplication by the progression 4, 3, 2, 1. What do you notice? What would happen if you used a different progression?

26. Find the relationship of the fluxions using Newton's rules for the equation $y^2 - a^2 - x\sqrt{a^2 - x^2} = 0$. Put $z = x\sqrt{a^2 - x^2}$.

27. Solve the fluxional equation $\dot{y}/\dot{x} = 2/x + 3 - x^2$ by first replacing x by $x + 1$ and then using power series techniques.

28. Find the ratio of the fluxion of x to the fluxion of $1/x$ using Newton's "synthetic" method of fluxions.

29. Find the ratio of the fluxion of x to the fluxion of $1/x^n$ using Newton's "synthetic" method of fluxions.

30. Derive Newton's result that the fluxion of an arc to the fluxion of its secant is as its cosine to its secant. Use the result on the tangent demonstrated in section 11.4.3 and the fluxional relationship derived from $CT^2 = AT^2 + AC^2$ (Fig. 11.17), noting that AC, the radius of the circle, is fixed. Translate this result into a standard modern form of the derivative of the secant.

31. Derive Newton's result that the fluxion of an arc is to the fluxion of its sine as the radius of the arc to its cosine. Use the result on the tangent proved in section 11.4.3, the result of exercise 30, and fluxional relationships arising from the geometry of the situation.

32. Given the curve $y^q = x^p$ $(q > p > 0)$, use the transmutation theorem to show that

$$\int_0^{x_0} y\,dx = \frac{qx_0y_0}{p+q}.$$

Note that from $y^q = x^p$, it follows that $(q\,dy)/y = (p\,dx)/x$, and therefore that $z = y - x\,dy/dx = [(q-p)/q]y$.

33. Prove the quotient rule

$$d\left(\frac{x}{y}\right) = \frac{y\,dx - x\,dy}{y^2}$$

by an argument using differentials.

34. Derive the power series for the logarithm by beginning with the differential equation $dy = [1/(x+1)]\,dx$, assuming that y is a power series in x with undetermined coefficients, and solving simple equations to determine each coefficient in turn.

35. Derive the power series that determines the number $x + 1$, given its logarithm y, as Leibniz puts it—that is, the power series for the exponential function—by the method of undetermined coefficients. Begin with the differential equation $x + 1 = dx/dy$.

36. Compare and contrast the calculuses of Newton and Leibniz in terms of their notation, their ease of use, and their foundations.

37. Compare the efficacy of Fermat's tangent method and Descartes's circle method in determining the slope of the tangent line to the curve $y = x^n$. Note the kinds of calculations needed in each instance.

38. Outline a lesson introducing the concept of integration by applying the method of Fermat to curves whose equations are of the form $y = x^n$, for n a positive integer.

39. Outline a lesson introducing the determination of arc length using the method of van Heuraet. How does this differ from the method normally presented in calculus texts?

40. Compare Newton's analytic algorithm for fluxions with Barrow's procedure, evidently influenced by Newton. What are the similarities and the differences?

41. Is the notion of a differential as an infinitesimal a useful idea to present in a current calculus class, either a standard or a reform one? Would it make the derivation of the basic rules of calculus easier? Why or why not?

42. Why are Newton and Leibniz considered the inventors of the calculus rather than Fermat or Barrow, both of whom had methods for finding what are known today as derivatives and integrals?

References

Three works on the history of calculus together provide a fairly complete treatment of the material. The earliest is Carl Boyer, *The History of the Calculus and its Conceptual Development* (New York: Dover, 1959), which deals primarily with the central concepts underlying the calculus. Although this work generally considers ideas only as they prefigure modern ones, rather than as ideas influential in their own time, it is still an excellent treatment that covers most of the ideas discussed not only in this chapter but also (insofar as they are relevant to calculus) in several earlier and later chapters. Margaret E. Baron, *The Origins of the Infinitesimal Calculus* (Oxford: Pergamon Press, 1969) traces more of the methods actually used in the calculus up to the time of Newton and Leibniz. This book is very strong on the first part of the seventeenth century in particular and provides many examples that are useful in understanding how various mathematicians actually solved the problems they encountered. The third book, C. H. Edwards, *The Historical Development of the Calculus* (New York: Springer, 1979), also is devoted to showing exactly how mathematicians calculated, but unlike the previous work, it covers in detail the contributions of Newton and Leibniz as well as the work of their eighteenth- and nineteenth-century successors. A more general discussion of seventeenth-century mathematics is found in D. T. Whiteside, "Patterns of Mathematical Thought in the Later Seventeenth Century," *Archive for History of Exact Sciences* 1 (1960–62), 179–388.

Although few of the works cited in the first half of this chapter are available in their entirety in English, excerpts from many, including the works of Fermat, Torricelli, Pascal, and Wallis, are available in D. J. Struik, ed., *A Source Book in Mathematics, 1200–1800* (Cambridge: Harvard University Press, 1969). Michael Mahoney's biography of Fermat, cited in chapter 10, contains many details of Fermat's work on calculus. Descartes's work on normals is available in David Eugene Smith and Marcia L. Latham, trans., *The Geometry of René Descartes* (New York: Dover, 1954). Cavalieri's work is treated in Kirsti Andersen, "Cavalieri's Method of Indivisibles," *Archive for History of Exact Sciences* 31 (1985), 291–367. There is a detailed study of Wallis's *Arithmetica infinitorum*, with many excerpts, in J. F. Scott, *The Mathematical Work of John Wallis* (New York: Chelsea, 1981). The best treatment of van Heuraet is in J. A. van Maanen, "Hendrick van Heuraet (1634–1660?): His Life and Mathematical Work," *Centaurus* 27 (1984), 218–279. Gregory's work is analyzed in detail in H. W. Turnbull, ed., *James Gregory Tercentenary Memorial Volume* (London: G. Bell and Sons, 1939). Barrow's *Geometrical Lectures* are available in English in J. M. Child, *The Geometrical Lectures of Isaac Barrow* (Chicago: Open Court, 1916). Child, in his commentary, seems to credit Barrow with most of the invention of the calculus by translating his geometrical work into modern analytical terms, a step Barrow himself never took. Nevertheless, there is much of interest in Barrow's lectures. Isaac Newton's published treatises on fluxions, all published many years after they were written, are available in facsimile edition in Derek T. Whiteside, ed., *The Mathematical Works of Isaac Newton*, vol. 1 (New York: Johnson Reprint Corporation, 1964). Whiteside is also the editor of the excellent eight-volume set of all of Newton's surviving mathematical manuscripts (including the material that appears in the previous reference): *The Mathematical Papers of Isaac Newton* (Cambridge: Cambridge University Press, 1967–1981). Although the originals of many of the papers are in Latin, Whiteside has translated most of them into English. Newton's letters are in H. W. Turnbull, ed., *The Correspondence of Isaac Newton* (Cambridge: Cambridge University Press, 1960). This seven-volume set contains virtually all the extant letters to and from Newton, as well as other related material. A new translation of the *Principia* along with a guide to its study, appears in *Isaac Newton, The Principia: Mathematical Principles of Natural Philosophy*, trans. by I. Bernard Cohen and Anne Whitman (Berkeley: University of California Press, 1999). Excerpts from Leibniz's first published calculus treatises are in Struik's *A Source Book on Mathematics*, mentioned above. J. M. Child, *The Early Mathematical Manuscripts of Leibniz* (Chicago: Open Court, 1920) contains the edited translations of many of Leibniz's mathematical manuscripts up to about 1680. The commentaries must be read with care, because Child seems to be most interested in showing that much of Leibniz's work is derived from that of Barrow.

There are numerous works dealing with Newton. The best biography is Richard Westfall, *Never at Rest* (Cambridge: Cambridge University Press, 1980), which covers in stimulating detail not only Newton's mathematical achievements but also his work in various other areas of science. An excellent summary of Newton's mathematical achievements is V. Frederick Rickey, "Isaac Newton: Man, Myth, and Mathematics," *College Mathematics Journal* 18 (1987), 362–389. A summary of Newton's early work is found in two articles of Derek T. Whiteside: "Isaac Newton: Birth of a Mathematician," *Notes and Records of the Royal Society* 19 (1964), 53–62, and "Newton's

Marvelous Years: 1666 and All That," *Notes and Records of the Royal Society* 21 (1966), 32–41. For studies of the *Principia* and related works, see François De Gandt, *Force and Geometry in Newton's Principia* (Princeton: Princeton University Press, 1995); Dana Densmore, *Newton's Principia: The Central Argument; Translation, Notes, and Expanded Proofs* (Santa Fe: Green Lion Press, 1995); and Niccolò Guicciardini, *Reading the Principia: The Debate on Newton's Mathematical Methods for Natural Philosophy from 1687 to 1736* (Cambridge: Cambridge University Press, 1999). For a discussion of Newton's proof of the result that an inverse square force implies an elliptical orbit, see Bruce Pourciau, "Reading the Master: Newton and the Birth of Celestial Mechanics," *American Mathematical Monthly* 104 (1997), 1–19. The best works on Leibniz include Eric Aiton, *Leibniz, A Biography* (Bristol: Adam Hilger Ltd, 1985), a general work covering his entire scientific career, and Joseph E. Hofmann, *Leibniz in Paris, 1672–1676* (Cambridge: Cambridge University Press, 1974), covering in great detail the years in which Leibniz invented his version of the calculus. The controversy between Newton and Leibniz is discussed in A. R. Hall, *Philosophers at War: The Quarrel Between Newton and Leibniz* (Cambridge: Cambridge University Press, 1980).

CHAPTER TWELVE

Analysis in the Eighteenth Century

Jean Bernoulli, public professor of mathematics, pays his best respects to the most acute mathematicians of the entire world. Since it is known with certainty that there is scarcely anything which more greatly excites noble and ingenious spirits to labors which lead to the increase of knowledge than to propose difficult and at the same time useful problems through the solution of which . . . they may attain to fame and build for themselves eternal monuments among posterity, so I should expect to deserve the thanks of the mathematical world if . . . I should bring before the leading analysts of this age some problem upon which . . . they could test their methods, exert their powers, and, in case they brought anything to light, could communicate with us in order that everyone might publicly receive his deserved praise from us.

—Proclamation made public at Gröningen, the Netherlands, January, 1697

The driving force in the continued development of calculus in the eighteenth century was the desire to solve physical problems, the mathematical formulation of which was often in terms of equations among fluxions or among differentials. Although mathematicians in Britain as well as on the continent participated in this effort, the flexibility of Leibniz's notation seemed to give continental mathematicians an advantage, and the method of differential equations soon outstripped methods using fluxions. Thus, continental mathematicians thought it important to translate Newton's geometrical analysis of the *Principia* into the more algebraic analysis of differentials and thus derive many of Newton's results by their own methods. But mathematicians also posed and solved many new problems arising from applications of Newton's laws of motion. Gradually, the emphasis changed from the study of curves, which was central to both Newton's and Leibniz's mathematics, to the study of analytical expressions involving one or more variable quantities as well as certain constants, that is, functions of one or several variables. The relationship between the differentials of these variables and the variable dependent on them, determined by some physical situation, led to a differential equation whose solution explicitly determined the desired function. In fact, new classes of functions were discovered and analyzed through the differential equations that they satisfied.

The major figure in the development of analysis in the eighteenth century was the most prolific mathematician in history, Leonhard Euler. Much of this chapter will be devoted to his work, especially his work in formalizing mathematical analysis and in developing methods for solving ordinary and partial differential equations. The chapter will begin, however, with some of the problems set by the Bernoullis as challenges for the mathematicians of Europe, problems whose solutions helped to establish new ideas in mathematics that were later developed by Euler and others. It will conclude with a look at the foundations of calculus, beginning with George Berkeley's criticisms of both Newton's and Leibniz's justifications of their calculus and the responses to that criticism by Colin Maclaurin, Jean d'Alembert, and Joseph Louis Lagrange.

12.1 Differential Equations

It was the brothers Jakob and Johann Bernoulli (often known as Jacques and Jean, or James and John) who were among the first in Europe to understand the new techniques of Leibniz and to apply them to solve new problems. For example, in 1690, Jakob Bernoulli proved, using the calculus of differentials, that the **isochrone**—the curve along which an object descending under the influence of gravity would take the same amount of time to reach the bottom, from whichever point on the curve the descent began—was a **cycloid**, the curve traced by a point attached to the rim of a wheel rolling along a line.

Having succeeded in solving this problem, Jakob then proposed a new one: to determine the shape of the **catenary**, the curve assumed by a flexible but inelastic cord hanging freely between two fixed points. Although he was unable to solve the problem, his younger brother Johann published a solution in the *Acta eruditorum* of June, 1691. Johann's solution began with the differential equation $dy/dx = s/a$, derived from an analysis of the forces acting to keep the cord in position, where s represents arc length. Because $ds^2 = dx^2 + dy^2$, squaring the original equation gives

$$ds^2 = \frac{s^2\,dy^2 + a^2\,dy^2}{s^2}, \qquad \text{or} \qquad ds = \frac{\sqrt{s^2 + a^2}\,dy}{s},$$

or, finally,

$$dy = \frac{s\,ds}{\sqrt{s^2 + a^2}}.$$

An integration then shows that $y = \sqrt{s^2 + a^2}$, or $s = \sqrt{y^2 - a^2}$. Bernoulli concluded that

$$dx = \frac{a\,dy}{s} = \frac{a\,dy}{\sqrt{y^2 - a^2}}.$$

He was not able to express the integral of this equation in closed form, but he was able to construct the desired curve by making use of certain conic sections. In modern terminology, this equation can be solved in the form $x = a \ln\left(y + \sqrt{y^2 - a^2}\right)$ or in the form $y = a \cosh(x/a)$. In 1691, however, for Bernoulli, as for his contemporaries, an answer in terms of areas under, or lengths of, known curves was sufficient.

Over the next several years, both brothers posed other problems involving differential equations and, along with Leibniz, made much progress in developing methods of solution. In particular, in 1691, Leibniz found the technique of separating variables, that is, of rewriting a differential equation in the form $f(x)\,dx = g(y)\,dy$ and then integrating both sides to give the solution. He also developed the technique for solving the homogeneous equation $dy = f(y/x)\,dx$ by substituting $y = vx$ and then separating variables. By 1694, Leibniz had also solved the general first-order linear differential equation $m\,dx + ny\,dx + dy = 0$, where m and n are both functions of x. (In modern notation, this is the equation $dy/dx + ny = -m$.) He defined p by the equation $dp/p = n\,dx$ and substituted to get $pm\,dx + y\,dp + p\,dy = 0$. Because the last two terms on the left side are equal to $d(py)$, an integration gives $\int pm\,dx + py = 0$. This equation, giving the answer in terms of an area, provided Leibniz with the desired solution.

12.1.1 The Brachistochrone Problem

Another significant problem was proposed by Johann Bernoulli in 1696: "If two points A and B are given in a vertical plane, to assign to a mobile particle M the path AMB along which, descending under its own weight, it passes from the point A to the point B in the briefest time." Bernoulli noted that the required curve, the **brachistochrone**, or path of least time, was not a straight line, but a curve "well known to geometers." He first proposed the problem in an issue of *Acta eruditorum*, and later sent it, as noted in the chapter opening quotation, to mathematicians throughout Europe, including Isaac Newton—who, Bernoulli believed, had stolen Leibniz's methods and would not be able himself to solve this problem. Newton, however, solved it within twelve hours of his receipt of it and submitted his solution to the *Acta eruditorum*, where it was published in May, 1697 along with the solutions of Leibniz, Jakob Bernoulli, and Johann himself.

We will consider Johann Bernoulli's solution here. He began by noting that, according to Galileo, the velocity acquired by a falling body is proportional to the square root of the distance fallen. Second, he recalled Snell's law: When a light ray passes from a thinner to a denser medium, the ray is bent so that the sine of the angle of incidence is to the sine of the angle of refraction inversely as the densities of the media and therefore directly as the velocities in those media. This law had been derived by Fermat as an application of the principle that the path traversed by the light ray must take the least time. Bernoulli thus assumed that the vertical plane of the problem was composed of infinitesimally thick layers whose densities varied. The brachistochrone was therefore the curved path of a light ray whose direction changed continually as it passed from one layer to the next. At every point the sine of the angle between the tangent to the curve and the vertical axis was proportional to the velocity, and the velocity was in turn proportional to the square root of the distance fallen.

Now, denoting the desired brachistochrone curve by AMB and the curve representing the velocity at each point by AHE, let x and y be the vertical and horizontal coordinates, respectively, of the point M measured from the origin A and u the horizontal coordinate of the corresponding velocity at point H (Fig. 12.1). With m a point infinitesimally close to M, Bernoulli represented the infinitesimals Cc, Mm, and nm by dx, ds, and dy, respectively. From the fact that the sine of the angle of refraction nMm is $dy : ds$, which is in turn proportional to the velocity u, he derived the equation $dy : u = ds : a$, or $a\,dy = u\,ds$, or

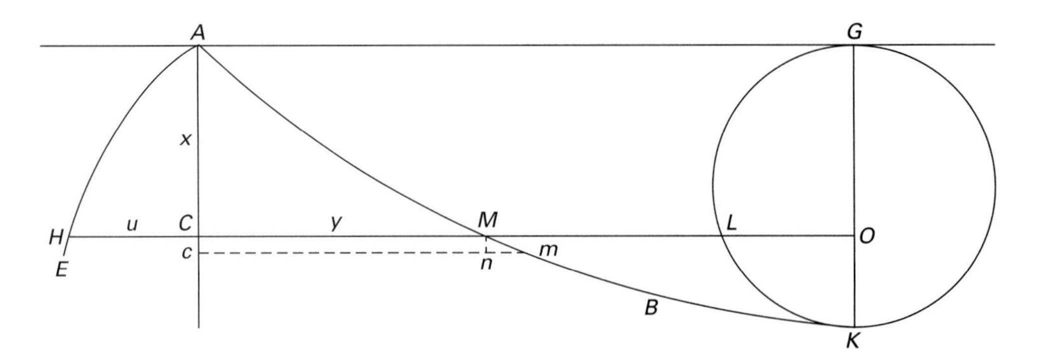

FIGURE 12.1 Johann Bernoulli's brachistochrone problem

$a^2\,dy^2 = u^2\,ds^2 = u^2\,dx^2 + u^2\,dy^2$, or, finally,

$$dy = \frac{u\,dx}{\sqrt{a^2 - u^2}}.$$

Because the curve AHE is a parabola with equation $u^2 = ax$, or $u = \sqrt{ax}$, substituting this value for u into the above equation produces the differential equation of curve AMB:

$$dy = dx\sqrt{\frac{x}{a-x}}.$$

Bernoulli immediately recognized that this equation defined a cycloid. To prove this analytically, he noted that

$$dx\sqrt{\frac{x}{a-x}} = \frac{a\,dx}{2\sqrt{ax - x^2}} - \frac{(a-2x)\,dx}{2\sqrt{ax - x^2}}.$$

Given that $y^2 = ax - x^2$ is the equation of a circle GLK, and that the first term on the right is the differential of arc length along this circle, an integration of the equation gives $CM = \text{arc}\,GL - LO$. Because $MO = CO - CM = CO - \text{arc}\,GL + LO$, and with the assumption that CO is equal to half the circumference of the circle, it follows that $MO = \text{arc}\,LK + LO$ or that $ML = \text{arc}\,LK$. It is then immediate that the curve AMK is a cycloid, as asserted.

12.1.2 Translating Newton's Synthetic Method of Fluxions into the Method of Differentials

Leibniz, the Bernoullis, and other continental mathematicians involved in the development of Leibniz's calculus of differentials read and analyzed Newton's *Principia*. They realized early on that Newton was using some of the same basic ideas as Leibniz, but from a different point of view. Therefore, to show that their methods were as good as, or better than, Newton's methods, mathematicians undertook a major effort to translate Newton's synthetic method of fluxions into the methods of the differential and integral calculus. (It should be noted that at the same time, British mathematicians, and even Newton himself, translated many of the propositions of the *Principia* into the analytical method of fluxions that Newton had developed in the 1660s.)

For example, we will consider the derivation of Kepler's area law by Jacob Hermann (1678–1733), a student of Jacob Bernoulli in Basel. Recall that Newton derived this law as proposition 1 of the *Principia*. Hermann proved it anew in his *Phoronomia* of 1716. He assumed that the trajectory of an orbiting body was a plane curve ANB, where the center of force is D, ds is the infinitesimal element of arc, and NC and nc are two tangents to the curve at N and n, respectively (Fig. 12.2). The line DC $(= p)$ is perpendicular to tangent NC $(= q)$. Lines ON and On are perpendicular to the tangents at the neighboring points N and n; thus, they meet at the center O of the **osculating circle**, the circle that best approximates the curve ANB near n. The radius $ON = \rho$ of that circle is what is known as the **radius of curvature** of curve ANB at N.

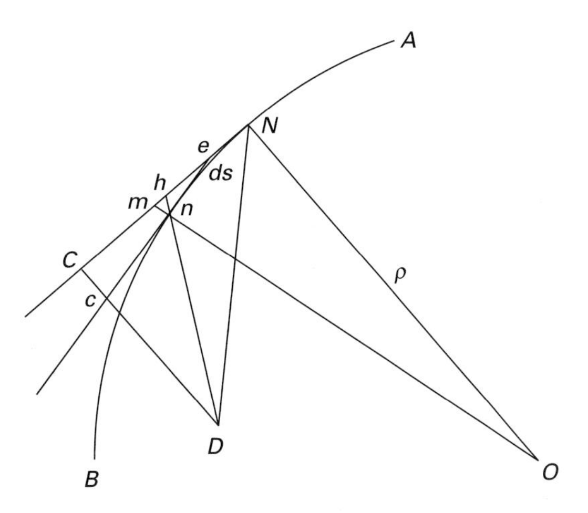

FIGURE 12.2 Hermann's proof of Newton's proposition 1

Hermann began with two basic principles for an arbitrary force G acting on a unit mass. First, since under those circumstances, force and acceleration are equal, he knew that (1) $Gt = v$, where v is the velocity and t is the time. Second, he took Galileo's law in the form (2) $2\ell/G = t^2$, where ℓ is the distance fallen from rest in time t. Now, let the central force F of the problem be split into two components: F_T, the force acting along the tangent, and F_N, the force acting along the normal. (Note that all forces may be assumed constant in an infinitesimal time interval.) By (1), $F_T\,dt = -dv$, where dt is the infinitesimal time it takes the body to go from N to n and $-dv$ is the corresponding change in the velocity. Therefore, $F_T v\,dt = -v\,dv$, and, since $v\,dt = ds$, we get (3) $F_T\,ds = -v\,dv$.

Next, by (2), we have $2\,d\alpha/F = dt^2 = ds^2/v^2$, where $d\alpha = hn$ is an infinitesimal Galilean fall from the tangent to the curve. Thus, $d\alpha = ds^2F/2v^2$. Since triangles Nmn and QnN are similar (Fig. 12.3), we get $nm : Nn = Nn : NQ = Nn : 2ON$, or $d\beta = ds^2/2\rho$, where $d\beta = nm$ is the infinitesimal change in position along the radius. (Note that here nm is perpendicular to NC rather than being the prolongation of On, while hn is parallel to DN. This ambiguity, common in the work of Hermann and others calculating with

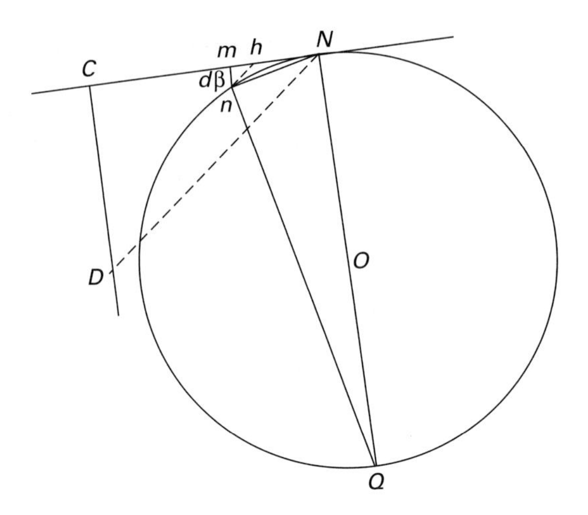

FIGURE 12.3 Hermann's proof of Newton's proposition 1, continued

infinitesimals, does not affect the final result, because the difference in the lengths of the infinitesimal line segments in question is a higher-order infinitesimal, which is neglected.) We now have

$$\frac{F}{F_N} = \frac{hn}{nm} = \frac{d\alpha}{d\beta} = \frac{ds^2 F}{2v^2} \cdot \frac{2\rho}{ds^2} = \frac{F\rho}{v^2},$$

so it follows that (4) $F_N \rho = v^2$, a standard result relating acceleration and velocity around a circle.

We know that $F_T : F_N = NC : DC = q : p$. Since triangles Cec and NOn are similar, again neglecting higher-order infinitesimals, we also have $Cc : Ce = Nn : NO$. But since Ce and CN differ by a higher-order infinitesimal, we can replace Ce by CN in this proportion. Since $Cc = dp$ and $Nn = ds$, we then have $dp/q = ds/\rho$. Dividing equation (3) by equation (4) and simplifying gives

$$\frac{F_T\, ds}{F_N \rho} = \frac{-v\, dv}{v^2}, \quad \text{or} \quad \frac{dv}{v} = -\frac{F_T\, ds}{F_N \rho} = -\frac{q}{p}\frac{dp}{q} = -\frac{dp}{p}.$$

It follows that $p\, dv + v\, dp = 0$, or $d(pv) = 0$, or, finally, $pv = 2k$, where k is a constant. But $p\, ds$ is twice the area of triangle DNn, or twice the infinitesimal area dA swept out by the line from the central force to the moving body. Therefore,

$$2k = pv = p\frac{ds}{dt} = \frac{p\, ds}{dt} = \frac{2\, dA}{dt},$$

and

$$\frac{dA}{dt} = k;$$

that is, the rate of change of area is constant. This is the content of Kepler's law of areas, so Hermann had now proved Newton's proposition 1.

Hermann also used differentials to prove the result that Newton stated in his corollary to propositions 11–13 of Book I of the *Principia*: that an inverse-square force law implies a conic section orbit. In Figure 12.4 (created by slightly modifying Fig. 11.19 by completing right triangles SPI and QRB and then drawing lines QH, QK, RG, and KP as shown), a body is moving along curve $APQL$ under the attraction of a central force F at S, which is inversely proportional to the square of the distance SP. We set S as the origin of the coordinate system. As in Figure 11.19, PQ is infinitesimal. We therefore set $PQ = ds$, $SI = x$, and $PI = y$. Then $SP = \sqrt{x^2 + y^2}$, $QH = PK = dx$, $PH = GB = dy$, and $KG = QB = -ddx = -d^2x$.

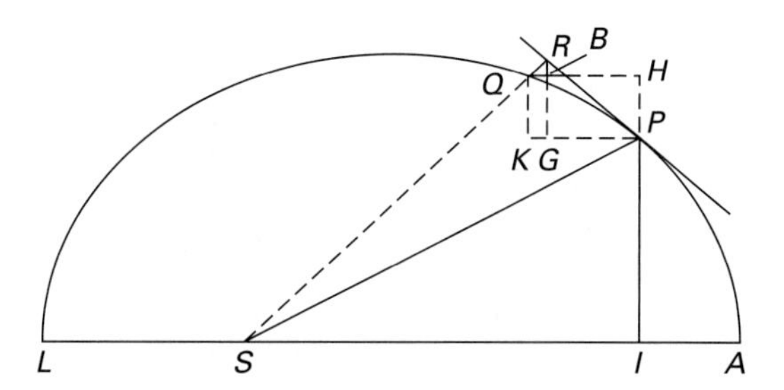

FIGURE 12.4 Hermann's proof of Kepler's first law

Because triangles QRB and SPI are similar, it follows that $QB : QR = SI : SP$, or $-d^2x : QR = x : \sqrt{x^2 + y^2}$. Therefore,

$$QR = \frac{-d^2x\sqrt{x^2 + y^2}}{x}.$$

Newton had shown that the force F is proportional to QR and inversely proportional to the square of the area of the infinitesimal triangle SQP, and Hermann now used this result. By standard techniques, that area is given by $\frac{1}{2}(y\,dx - x\,dy)$. Because F is also inversely proportional to $SP^2 = x^2 + y^2$, we get

$$\frac{-a\,d^2x\sqrt{x^2 + y^2}}{x(y\,dx - x\,dy)^2} = \frac{1}{x^2 + y^2}, \qquad \text{or} \qquad -a\,d^2x = \frac{x(y\,dx - x\,dy)^2}{(x^2 + y^2)^{3/2}},$$

as the second-order differential equation that Hermann needed to solve.

Now Hermann used the earlier area result, which implied that $y\,dx - x\,dy$ is constant. He therefore could perform two integrations. First, rewriting the differential equation in the form

$$-a\,d^2x = (y\,dx - x\,dy)\frac{xy\,dx - x^2\,dy}{(x^2 + y^2)^{3/2}},$$

he claimed that the integral was

$$-a\,dx = \frac{-y}{\sqrt{x^2 + y^2}}(y\,dx - x\,dy) = \frac{xy\,dy - y^2\,dx}{\sqrt{x^2 + y^2}}.$$

Although Hermann did not show how he found this integral, it is straightforward to show that the differential of this latter equation gives the original equation. Then, rewriting this result in the form

$$-\frac{ab\,dx}{x^2} = \frac{bxy\,dy - by^2\,dx}{x^2\sqrt{x^2+y^2}},$$

where b is a constant, Hermann integrated again to get

$$\frac{ab}{x} \pm c = \frac{b\sqrt{x^2+y^2}}{x}, \qquad \text{or} \qquad a \pm \frac{cx}{b} = \sqrt{x^2+y^2},$$

with c the arbitrary constant of integration. Hermann knew that this last equation was the equation of a conic section. In fact, it is a parabola if $b = c$, an ellipse if $b > c$, and a hyperbola if $b < c$. Hermann had therefore proved the result that Newton had only sketched in his corollary to propositions 11–13 of the *Principia*.

Interestingly enough, Johann Bernoulli criticized Hermann's result, because he thought it was not general enough in that Hermann did not introduce an arbitrary constant of integration in his first integral. But as others pointed out, such a constant would just have changed the axis of the curve, and Hermann simply made the assumption that that axis would be the x-axis. Bernoulli himself eventually gave another proof of the same result using differential calculus, but from a slightly different point of view.

12.1.3 Differential Equations and the Trigonometric Functions

In the 1730s, Leonhard Euler (1707–1783) developed some new techniques for solving linear differential equations, techniques that led to his invention of the modern notion of the sine and cosine functions. Recall that Newton had been able to calculate the fluxion of the sine and that Leibniz had derived the differential equation for the sine by a geometric argument and then had solved it to get its power series representation. But in the early years of the eighteenth century, physical problems that led to such equations were typically solved for time as a function of arc and thus did not involve the sine function directly. For example, the equation

$$dt = \frac{c\,ds}{\sqrt{c^2 - s^2}}$$

would be solved for t as $t = c \arcsin(s/c)$ rather than as $s = c \sin(t/c)$.

In contrast to the "missing" sine function, the exponential function was well known to Euler by 1730. In fact, he knew that the solution to $dy = ay\,dx$ was $y = e^{ax}$. Higher-order equations could also be solved by exponential functions, but it was not until 1739 that Euler realized that the sine function was also often necessary. On March 30 of that year, Euler presented a paper to the St. Petersburg Academy of Sciences in which he solved the differential equation of motion of a sinusoidally driven harmonic oscillator, that is, of an object acted on by two forces, one proportional to the distance and one varying sinusoidally with the time. The very statement of the problem is perhaps the earliest use of the sine as a function of time, and the resulting differential equation

$$2a\,d^2s + \frac{s\,dt^2}{b} + \frac{a\,dt^2}{g}\sin\frac{t}{a} = 0$$

(where s represents position and t time) is the earliest use of that function in such an equation. Two aspects of Euler's solution are of interest. First, as a special case, he deleted the sine term and solved the equation $2a\,d^2s + s\,dt^2/b = 0$, or, after multiplying through by $b\,ds$, the equation $2ab\,ds\,d^2s = -s\,ds\,dt^2$. An integration with respect to s then gave $2ab\,ds^2 = (C^2 - s^2)\,dt^2$, or

$$dt = \frac{\pm\sqrt{2ab}\,ds}{\sqrt{C^2 - s^2}},$$

the differential equation for the arcsine (with the positive sign) or the arccosine (with the negative sign). Euler, since he was interested in the motion rather than the time, solved the arccosine equation for s instead of t: $s = C\cos(t/\sqrt{2ab})$, the first such explicit analytic solution on record. Second, to solve the general case, Euler postulated a solution of the form $s = u\cos(t/\sqrt{2ab})$, where u is a new variable. He then substituted that solution into the equation and solved for u. This manipulation shows that Euler was already familiar with the basic differentiation rules for the sine and cosine.

There is more to the story of the sine and cosine. On May 5, 1739, Euler wrote to Johann Bernoulli, noting that he had solved in finite terms the third-order equation $a^3\,d^3y = y\,dx^3$. The solution was

$$y = be^{x/a} + ce^{-x/2a}\sin\frac{(f+x)\sqrt{3}}{2a}$$

where b, c, f are arbitrary constants arising from the three integrations. Euler did not reveal how he discovered this solution, but one reconstruction is based on the use of the known exponential solution $y = e^{x/a}$ to reduce the order of the equation. In this technique, which Euler had used earlier, one multiplies the original equation $a^3\,d^3y - y\,dx^3 = 0$ by $e^{-x/a}$ and assumes that this is the differential of $e^{-x/a}(A\,d^2y + B\,dy\,dx + Cy\,dx^2)$. It is then straightforward to show that a new solution of the original equation also satisfies the second-order equation $a^2\,d^2y + a\,dy\,dx + y\,dx^2 = 0$. To solve this latter equation requires a different Eulerian technique: guessing that a solution is of the form $y = ue^{\alpha x}$ and substituting this for y in the equation. Again, a bit of manipulation shows that the term $du\,dx$ can be eliminated by setting $\alpha = -\frac{1}{2}a$. The equation then reduces to $a^2\,d^2u + \frac{3}{4}u\,dx^2 = 0$, an equation of the same form as the one Euler had solved in March, 1739. In this case, the solution is $u = C\sin[(x+f)\sqrt{3}/2a]$, from which the general solution to the original third-order solution follows.

Since the sine and exponential functions had been used in the solution of the same differential equation, it was clear that Euler now considered the sine, and by extension, the other trigonometric functions, as functions in the same sense as the exponential function. But even more interesting is that it was the very introduction of these functions into calculus that led Euler to the solution method for the class of linear differential equations with constant coefficients, that is, equations of the form

$$y + a_1\frac{dy}{dx} + a_2\frac{d^2y}{dx^2} + a_3\frac{d^3y}{dx^3} + \cdots + a_n\frac{d^ny}{dx^n} = 0.$$

In a letter to Johann Bernoulli in September, 1739, Euler noted that his method was to replace the given differential equation by the algebraic equation

$$1 + a_1p + a_2p^2 + a_3p^3 + \cdots + a_np^n = 0$$

and then factor this "characteristic polynomial" into its irreducible real linear and quadratic factors. For each linear factor $1 - \alpha p$, one takes as the solution $y = Ae^{x/\alpha}$, and for each irreducible quadratic factor $1 + \alpha p + \beta p^2$, one takes as the solution

$$e^{-\alpha x/2\beta}\left(C \sin \frac{x\sqrt{4\beta - \alpha^2}}{2\beta} + D \cos \frac{x\sqrt{4\beta - \alpha^2}}{2\beta}\right).$$

The general solution is then a sum of the solutions corresponding to each factor. As an example, Euler solved the equation

$$y - k^4 \frac{d^4 y}{dx^4} = 0.$$

The corresponding algebraic equation $1 - k^4 p^4$ factors as $(1 - kp)(1 + kp)(1 + k^2 p^2)$. Thus, the solution is

$$y = Ae^{-x/k} + Be^{x/k} + C \sin \frac{x}{k} + D \cos \frac{x}{k}.$$

Euler did not say how he arrived at his algebraic solution method. But since he had discovered several months earlier that trigonometric functions were involved in the solution to the equation $y - a^3\, d^3y/dx^3 = 0$, one can surmise that he merely generalized that method. For, given one solution of the equation of the form $y = e^{x/a}$, the reduction procedure indicated in the earlier work provides in essence a factorization of the characteristic polynomial as $1 - a^3 p^3 = (1 - ap)(1 + ap + a^2 p^2)$. The general factorization method indicated in Euler's September letter would then have followed easily. In particular, it would have been clear that the sine and cosine terms come from the irreducible quadratic factors.

Johann Bernoulli was somewhat bothered by Euler's solution. He noted that the irreducible quadratic factors of the characteristic polynomial could be factored over the complex numbers, and thus that Euler's method led to the relation of complex roots of this polynomial to real solutions involving sines and cosines. Euler finally convinced Bernoulli that $2 \cos x$ and $e^{ix} + e^{-ix}$ were identical, because they satisfied the same differential equation, and therefore that using imaginary exponentials amounted to the same thing as using sines and cosines. It also followed that complex exponential functions were related to sines and cosines by the relationships

$$e^{ix} = \cos x + i \sin x \qquad \text{and} \qquad e^{-ix} = \cos x - i \sin x.$$

12.2 The Calculus of Several Variables

12.2.1 The Differential Calculus of Functions of Two Variables

We have already seen formal partial derivatives used in calculating differentials of equations, but the concept of a partial derivative was not yet developed. This concept first appeared explicitly, in fact, in terms of families $\{C_\alpha\}$ of curves, not, as one might expect from a modern perspective, in terms of surfaces defined by functions of two variables. Such families had been initially considered in the early 1690s by Leibniz. In the basic situation, there are two infinitesimally close curves from a given family, C_α and $C_{\alpha+d\alpha}$, intersected by a third curve D defined geometrically in terms of that family. For example, D may cut off equal arcs on all members of the family. In such a situation, to find the differential equation of D or construct its tangent, it was necessary to consider three different differentials of the ordinate y. Let P, P' be points on C_α and Q, Q' points on $C_{\alpha+d\alpha}$ (Fig. 12.5). One differential of y is that between two points on a single curve of the family, say $y(P') - y(P)$. This is the differential of y with x variable and α constant, designated by $d_x y$. A second differential is that between the y values of two corresponding points on the neighboring curves, say $y(Q) - y(P)$. This is the differential of y with α variable and x constant, designated by $d_\alpha y$. Differentiation using this differential was referred to as *differentiation from curve to curve*. Finally, there is a third differential, $y(Q') - y(P)$, denoted dy, which is the differential along the curve D.

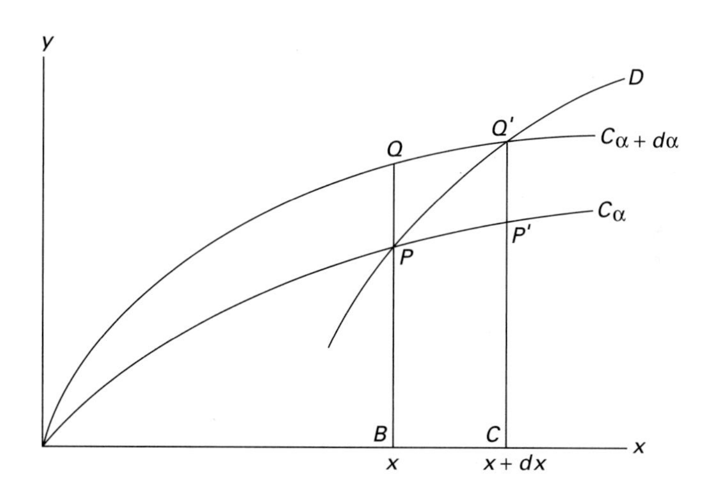

FIGURE 12.5 Partial differentiation in a family of curves

Leibniz had no difficulty in calculating either of the first two differentials, provided the curves were given algebraically. He could use his standard rules and treat either α or x as a constant, in effect taking the partial derivatives with respect to x or α, respectively. Unfortunately, many interesting curves $y(x, \alpha)$ were not given algebraically, but in terms of an integral. To solve this problem, Leibniz went back to his own basic ideas of the calculus, namely, that the differential is the extrapolation of a finite difference and the integral that of a finite sum. Since for two sets of finite quantities, the sum of the differences of the parts is equal to the difference of the sums of the parts, Leibniz discovered what is called the **interchangeability theorem** for differentiation and integration:

$$d_\alpha \int_b^x y(x, \alpha)\, dx = \int_b^x d_\alpha y(x, \alpha)\, dx.$$

Leibniz applied this result to determining the tangent to the curve D that cuts off equal arcs on the family of logarithmic curves $C_\alpha = \alpha \log x$. Because the arc length differential ds for the logarithmic curves is given by $ds = \sqrt{x^2 + \alpha^2}\, dx/x$, the desired curve D is determined by the condition

$$s(x, \alpha) = \int_1^x \frac{\sqrt{x^2 + \alpha^2}}{x}\, dx = K,$$

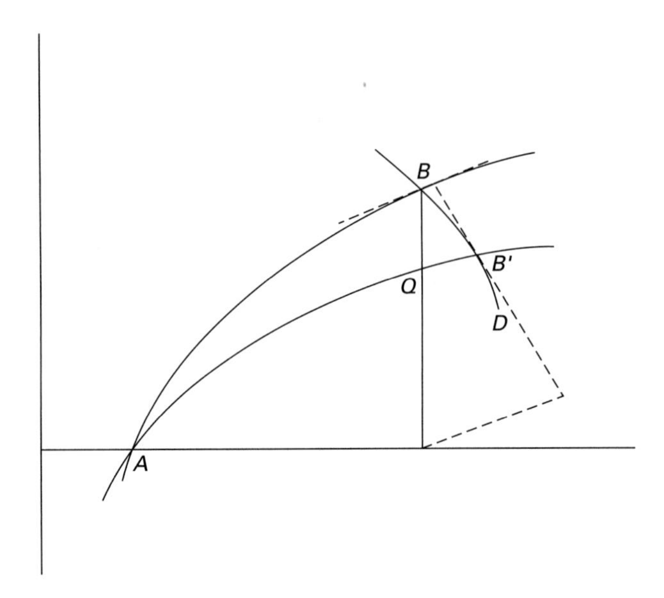

FIGURE 12.6 Finding the tangent to the curve D that cuts off equal arcs on the logarithmic curves $y = \alpha \log x$

where K is a constant. To determine the tangent, Leibniz needed to calculate the ratio of QB to QB' in the differential triangle of Figure 12.6. The first value is straightforward: $QB = d_\alpha \alpha \log x = \log x_0 \, d\alpha$, where x_0 is the x-coordinate of B. The arc QB' equals the difference $AB - AQ$, since $AB = AB'$. Therefore

$$QB' = d_\alpha s = d_\alpha \int_1^{x_0} \frac{\sqrt{x^2 + \alpha^2}}{x} \, dx.$$

By the interchangeability theorem, this integral is equal to

$$\int_1^{x_0} d_\alpha \frac{\sqrt{x^2 + \alpha^2}}{x} \, dx = \int_1^{x_0} \frac{\alpha}{x\sqrt{x^2 + \alpha^2}} \, d\alpha \, dx.$$

Because the limits of integration involve only x, however,

$$QB' = \alpha \, d\alpha \int_1^{x_0} \frac{dx}{x\sqrt{x^2 + \alpha^2}}.$$

Since QB' now is expressed by the area under a particular curve, it could be considered as known; hence, so was the desired ratio.

Two other important aspects of the calculus of functions of two variables, the notion of a total differential and the equality of the mixed second-order differentials, were discovered by Nicolaus Bernoulli (1687–1759), a nephew of both Johann and Jakob, who lived in the shadow of his more famous uncles and chose to publish very little in mathematics. In an article in the *Acta eruditorum* of June, 1719, Nicolaus asserted that the differential dy associated with a family of curves with parameter α is given by $dy = p \, dx + q \, d\alpha$, an expression he called the **complete differential equation** of the family, today known as the

total differential. In this equation, p and q are functions of x, y, and α. Although Nicolaus did not say how he derived this result, one possible method was via a geometric argument. The differential dy along the curve D crossing the given family can be expressed by

$$dy = y(Q') - y(P) = [y(Q') - y(Q)] + [y(Q) - y(P)]$$

$$= d_x(y + d_\alpha y) + d_\alpha y = d_x y + d_x\, d_\alpha y + d_\alpha y.$$

Since the middle term, a second-order differential, is infinitely small compared to the other terms, the result is $dy = d_x y + d_\alpha y$, or $dy = p\,dx + q\,d\alpha$ (see Fig. 12.5). It was Euler who eventually made the differential coefficients p and q themselves the fundamental concept of his calculus, replacing the differentials of Leibniz and the Bernoullis.

The equality of mixed second-order differentials follows from another geometric argument of Nicolaus Bernoulli, one in which the line segment $P'Q'$ (Fig. 12.5) is found in two different ways. On the one hand, $P'Q' = PQ + d_x PQ = d_\alpha y + d_x d_\alpha y$. On the other hand, $P'Q' = d_\alpha CP'$, which is in turn equal to $d_\alpha(BP + d_x BP) = d_\alpha BP + d_\alpha d_x BP = d_\alpha y + d_\alpha d_x y$. Comparing the two expressions yields the result $d_x d_\alpha y = d_\alpha d_x y$, or, in modern notation, $\partial^2 y/\partial x\, \partial\alpha = \partial^2 y/\partial\alpha\, \partial x$.

To solve the homogeneous linear differential equation $P\,dx + Q\,dy = 0$, where P and Q are functions of x and y, required an idea of Alexis Calude Clairaut (1713–1765), described in 1739. If $P\,dx + Q\,dy$ is the total differential of a function $f(x, y)$, then the equality of the mixed second-order differentials shows that $\partial P/\partial y = \partial Q/\partial x$. More importantly, however, Clairaut demonstrated that this condition was sufficient for $P\,dx + Q\,dy$ to be a total differential. He asserted, in fact, that under that condition, the function $f(x, y)$ was given by $\int P\,dx + r(y)$, where r was a function of y to be determined. The differential of $\int P\,dx + r(y)$ was, by Leibniz's result, equal to $P\,dx + dy \int(\partial P/\partial y)dx + dr$. But since

$$\frac{\partial P}{\partial y} = \frac{\partial Q}{\partial x} \quad \text{and} \quad \int \frac{\partial Q}{\partial x}\,dx = Q + s(y),$$

the differential can be rewritten as $P\,dx + Q\,dy + dr + s\,dy$. Therefore, if r is chosen so that $dr = -s\,dy$, the differential becomes $P\,dx + Q\,dy$, as desired. Clairaut had thus reduced the original two-variable problem to an ordinary differential equation in one variable, an equation he assumed to be solvable. He also easily extended this result to homogeneous linear equations in more than two variables.

12.2.2 Multiple Integration

The subject of multiple integration had its beginnings late in the seventeenth century, when Leibniz calculated areas of various regions on a hemisphere. To do this, he integrated expressions involving products of two differentials by integrating first with respect to one variable, holding the second constant, and then with respect to the second. Although the Bernoullis and others solved similar problems, it was not until 1731 that a systematic attempt to calculate volumes of certain regions as well as the areas of their bounding surfaces was published by Clairaut in his *Research on Curves of Double Curvature*. Clairaut demonstrated that surfaces could in general be represented by a single equation in three variables, but he most often considered cylindrical surfaces generated by a curve in one of the coordinate planes. Thus, to calculate the volume of a region between two cylinders given by $y = f(x)$, $z = g(y)$, he showed that the element of volume was given by $dx \int z\,dy$ and then used his equations to rewrite z and dy in terms of x so that he could integrate $z\,dy$. With the volume element now given entirely in terms of x, he was able to integrate again to calculate the desired volume.

Other mathematicians also calculated double integrals informally over the years, but it was only in 1769 that Euler gave the first detailed explanation of the concept of a **double integral**. For Euler, as for Leibniz and the Bernoullis, an "integral" was what we call an *antiderivative*. The use of these for finding areas was just an application. So Euler began by generalizing this notion of an integral to two variables. Thus, $\int \int Z\,dx\,dy$ was to mean a function of two variables, which, when twice differentiated, first with respect to x alone and second with respect to y alone, gave $Z\,dx\,dy$ as a differential. For example, Euler showed how to integrate

$$\int \int \frac{dx\,dy}{x^2 + y^2}.$$

The first integration can be performed with respect to either variable. Integrating each way, Euler found the values

$$\int \frac{dx}{x} \arctan \frac{y}{x} + X \qquad \text{and} \qquad \int \frac{dy}{y} \arctan \frac{x}{y} + Y,$$

where X is a function of x and Y is a function of y. Because the only way to perform the second integration, in either case, is by writing the integrand as a power series, Euler did so and showed that both integrations lead to the same final result

$$\int \int \frac{dx\,dy}{x^2 + y^2} = X + Y - \frac{y}{x} + \frac{y^3}{9x^3} - \frac{y^5}{25x^5} + \cdots.$$

Given then the idea of a double integral as double antiderivative, Euler generalized the concept of finding area using a single integration to that of finding volumes using this double integration. His basic idea, like that of Leibniz, was to integrate with respect to one variable first, keeping the other constant, and then deal with the second. For example, he found the volume under one octant of the sphere of radius a whose equation is $z = \sqrt{a^2 - x^2 - y^2}$. Taking an element of area $dx\,dy$ in the first quadrant of the circle in the xy-plane, Euler noted that the volume of the solid column above that infinitesimal rectangle is $dx\,dy\sqrt{a^2 - x^2 - y^2}$ (Fig. 12.7). To determine this volume, Euler first integrated with respect to y, holding x constant, to get

$$\left[\frac{1}{2}y\sqrt{a^2 - x^2 - y^2} + \frac{1}{2}(a^2 - x^2) \arcsin \frac{y}{\sqrt{a^2 - x^2}}\right] dx$$

as the volume under that piece of the sphere over the rectangle whose width is dx and whose length is y. Replacing y by $\sqrt{a^2 - x^2}$, Euler calculated the volume of the same piece up to that value of y to be $(\pi/4)(a^2 - x^2)\,dx$. Integrating with respect to x then gives $(\pi/4)(a^2x - \frac{1}{3}x^3)$ as the volume from the y-axis to x, and replacing x by a gives the total volume of the octant as $(\pi/6)a^3$ and that of the whole sphere as $(4\pi/3)a^3$.

After showing further how to calculate volumes of solid regions bounded above by the sphere and below by various areas of the plane, Euler noted that a double integral may also be used to calculate surface area. He gave, without much discussion, the element of surface

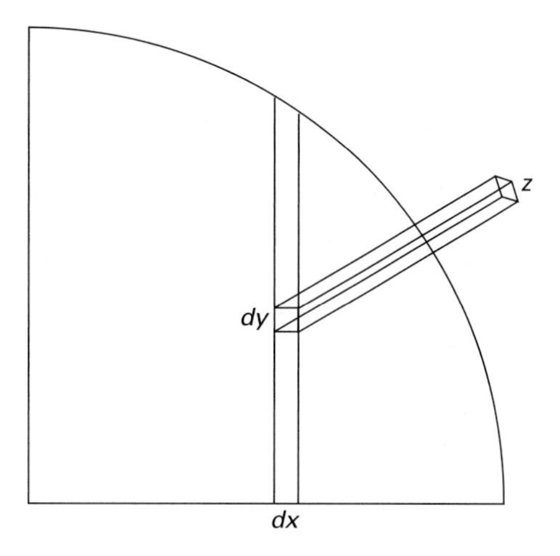

FIGURE 12.7 The volume of a sphere

area of the sphere as

$$\frac{a\,dx\,dy}{\sqrt{a^2 - x^2 - y^2}}.$$

He also noted that $\int\int dx\,dy$ over a region A was precisely the area of A.

12.2.3 Partial Differential Equations: The Wave Equation

Because many physical situations involved functions of more than one variable, it became necessary in the eighteenth century to deal with partial differential equations. Among the originators of this theory were Euler, Jean Le Rond d'Alembert (1717–1783), and Daniel Bernoulli (1700–1782). Here we will discuss only one particular type of partial differential equation—the wave equation—because the debate over the subject of vibrating strings, from which the equation was derived, led not only to certain methods of solution but also to a clarification of the idea of a function.

The discussion on the subject of vibrating strings began with a paper written by d'Alembert in 1747, in which he proposed a solution to the problem of the shape of a taut string undergoing vibration. Because the position of a point on the string varies both with its abscissa and with time, this shape is determined by a function $y = y(t, x)$ of two variables. D'Alembert considered the string to be composed of an infinite number of infinitesimal masses and then used Newton's laws to derive the partial differential equation for y, now called the **wave equation** and given in modern notation as

$$\frac{\partial^2 y}{\partial t^2} = c^2 \frac{\partial^2 y}{\partial x^2}.$$

D'Alembert then solved the equation in the form $y = \Psi(t + x) + \Gamma(t - x)$ for the special case $c^2 = 1$, where Ψ and Γ are arbitrary twice-differentiable functions. As d'Alembert pointed out, "this equation includes an infinity of curves." The elaboration of that statement led to much controversy.

D'Alembert himself first discussed the case where $y = 0$ when $t = 0$ for every x—that is, where the string is in the equilibrium position for $t = 0$. He further required that $y = 0$ for $x = 0$ and $x = l$ for all t—that is, that the string be held fixed at the two ends of the interval $[0, l]$. The first requirement shows that $\Psi(x) + \Gamma(-x) = 0$, while the second gives the results $\Psi(t) + \Gamma(t) = 0$ and $\Psi(t + l) + \Gamma(t - l) = 0$. It follows that $\Gamma(t - x) = -\Psi(t - x)$, or $y = \Psi(t + x) - \Psi(t - x)$; that $\Psi(-x) = -\Gamma(x) = \Psi(x)$, or Ψ is an even function; and that $\Psi(t + l) = \Psi(t - l)$, or Ψ is periodic of period $2l$. Furthermore, because the initial velocity is given by $\partial y/\partial t$ at $t = 0$, that is, by $v = \Psi'(x) - \Psi'(-x)$, and because the derivative of an even function is odd, d'Alembert concluded that the power series for the initial velocity must include only odd powers of x. In another paper, d'Alembert generalized the solution to the case where the initial position of the string was given by $y(0, x) = f(x)$ and the initial velocity by $v(0, x) = g(x)$. In this case, he obtained the result that the solution is only possible if $f(x)$ and $g(x)$ are odd functions of period $2l$ and (so that one could operate with these) if each function is given by a single analytic expression that is twice differentiable. For d'Alembert, then, a function was exactly that, an analytic expression, or what today is called a *formula*. Thus, even though $f(x)$ and $g(x)$ are given just on $[0, l]$, the function $y = \Psi(u)$ determining them must itself be given as a formula defined for all values of u. No other type of function could occur as a solution to a physical problem.

Two years after d'Alembert's initial paper, Euler published his own solution to the same problem, getting the same formal result as d'Alembert although with a somewhat different derivation. But Euler differed from d'Alembert in the kinds of initial position functions f he thought could be permitted. First, he announced that f could be any curve defined on the interval $[0, l]$, even one that was not determined by an analytic expression. It could be a curve drawn by hand. Thus, it was not necessary for the function to be differentiable at every point. Second, it was only the definition on the initial interval that was important. One could make the curve odd and periodic simply by defining it on $[-l, 0]$ by $f(-x) = -f(x)$ and then extending it to the entire real line by using $f(x \pm 2l) = f(x)$. After all, Euler reasoned, contrary to d'Alembert, as far as the physical situation was concerned, the initial shape of the string could be arbitrary. Even if there were isolated points where the function was not differentiable, one could still consider the curve to be a solution to the differential equation, because behavior at isolated points was not, Euler believed, relevant to the general behavior of a function over an interval. In fact, Euler thought that the consideration of such new kinds of functions would open "an entirely new field of analysis."

Both d'Alembert and Euler, although they carried on their analyses in terms of general "functions," always had in mind the examples of the sine and cosine. The former was odd and periodic, while the latter was even and periodic. In fact, in 1750, d'Alembert derived $y = (A \cos Nt)(B \sin Nx)$ as the solution to the wave equation by the technique of separation of variables, that is, by assuming $y = f(t)g(x)$ and then differentiating. Nevertheless, it was the third participant in the debate, Daniel Bernoulli, who explicitly referred to combinations of sines and cosines in his attempt to bring the debate back to the reality of physical strings.

Bernoulli, whose chief work at the University of Basel was in hydrodynamics and elasticity, began, in 1753, to explore the idea that a vibrating string potentially represents an infinity of tones, each superimposed on the others, and each being separately represented as a sine curve. It followed that the movement of a vibrating string can be represented by the function

$$y = \alpha \sin \frac{\pi x}{l} \cos \frac{\pi t}{l} + \beta \sin \frac{2\pi x}{l} \cos \frac{2\pi t}{l} + \gamma \sin \frac{3\pi x}{l} \cos \frac{3\pi t}{l} + \cdots,$$

where the sum is infinite. The initial position function, over which Euler and d'Alembert quarreled, is then represented by the infinite sum

$$y(0, x) = \alpha \sin \frac{\pi x}{l} + \beta \sin \frac{2\pi x}{l} + \gamma \sin \frac{3\pi x}{l} + \cdots.$$

Interestingly enough, Euler had written these series in 1750, probably intending only a finite sum, as an example of a possible solution to the equation. Bernoulli believed that the latter series could represent any arbitrary initial position function $f(x)$ with the appropriate choice of constants $\alpha, \beta, \gamma, \ldots$, but he could give no mathematical argument for the correctness of his view, only writing somewhat later that his representation provides an infinite number of constants, which can be used for adjusting the curve to pass through an infinite number of specified points. His view was challenged by Euler, who not only could not see any way of determining these coefficients but also realized that for a function to be represented by such a trigonometric series, it had to be periodic. By this argument, though, Euler showed himself to be caught between the older notion of function as formula and the newer view, which he was instrumental in helping evolve. Euler, after all, was willing to allow the arbitrary curve $f(x)$ defined over the interval $[0, l]$ to be extended by periodicity to the entire real line. But this was an example of what may be called *geometric periodicity*. It took no account of the algebraic expression by which f may be expressed. On the other hand, Euler's argument against Bernoulli was based on the algebraic periodicity of the trigonometric functions themselves on the whole real line. Euler had only an inkling of the modern notion of the domain of a function with the concomitant possibility that a function can be represented by different expressions on various parts of its domain.

The debate over the kinds of functions acceptable as solutions to the wave equation continued among these three mathematicians through the next decades, without any of them being convinced by any of the others. Although other mathematicians also entered the debate, it was not until the early years of the nineteenth century that a resolution of the problem was worked out through a complete analysis of the nature of trigonometric series.

12.3 The Textbook Organization of the Calculus

Many textbooks in calculus were written during the first century after Newton's and Leibniz's seminal work, both in Britain and on the continent. These books attempted to organize the topics just discussed, with the continental texts dealing with the calculus of differentials and the British texts with the calculus of fluxions. To some extent, the ideas were mutually translatable, but it gradually became clear that it was much easier both to calculate and to discover new ideas using the Leibnizian approach. We will look at the highlights of a few of these textbooks, concentrating on the ideas that were special in each one.

12.3.1 Textbooks in Fluxions

Among the earliest of the texts written in Britain was *A Treatise of Fluxions* by Charles Hayes (1678–1760), a book intended for teaching the new ideas to his own private students. Hayes gave a relatively clear treatment of Newton's analytic approach to fluxions but also included a study of the exponential (or logarithmic) curve $y = a^x$. Following Leibniz's procedure, he found the fluxion of y by taking the logarithm of both sides to get $\ell(y) = x\ell(a)$. Since he knew that the fluxion of the logarithm of any quantity is equal to the fluxion of the quantity divided by the quantity, he found $\dot{y}/y = \dot{x}\ell(a) + x\dot{\ell}(a)$. Because a is constant, the fluxion of its logarithm is 0, so $\dot{y} = y\dot{x}\ell(a) = a^x\ell(a)\dot{x}$. It follows that the subtangent $y(\dot{x}/\dot{y})$ of this curve is

$$y\frac{\dot{x}}{y\dot{x}\ell(a)} = \frac{1}{\ell(a)} = c,$$

a constant. Thus, to calculate the area under $y = a^x$, Hayes first noted that, in general, the fluxion of the area is $y\dot{x}$. Because in this case the subtangent is a constant c, it follows that $y\dot{x} = c\dot{y}$ and, therefore, that the fluxion of the area is $c\dot{y}$ and the area itself must be cy. Hayes's conclusion was that the area under the logarithmic curve between any two abscissas is proportional to the difference between the corresponding ordinates, a result not explicit in the work of either Leibniz or Bernoulli.

A more important British text, again written to supplement private instruction, was *A New Treatise of Fluxions*, published in 1737 by Thomas Simpson (1710–1761). Simpson's *Treatise* is basically Newtonian in approach, making much use of infinite series to solve, in particular, problems in integration. It is replete with problems, many of which have become familiar to today's students. Thus, in an early section on maxima and minima, Simpson showed how to find the greatest parallelogram inscribed in a triangle, the smallest isosceles triangle that circumscribes a given circle, and the cone of least surface area with a given volume. Also included in this section is perhaps the earliest solution to the problem of determining the maximum of a function of several variables, namely, $w = (b^3 - x^3)(x^2z - z^3)(xy - y^2)$. Although Simpson did not use the language of partial derivatives, he did calculate the relationships of the fluxions of w to that of each of x, y, and z separately, holding the other two variables constant, before setting $\dot{w}$ equal to 0 in each case and solving the resulting equations simultaneously.

Simpson is most famous today for the rule for numerical integration by parabolic approximation that bears his name. This rule appears not in his calculus text but in his *Mathematical Dissertations on a Variety of Physical and Analytical Subjects* of 1743. It is, however, not original to Simpson, having appeared in the works of other authors even in the seventeenth century.

The name of the Scottish mathematician Colin Maclaurin (1698–1746) is also known to today's students from a concept in the calculus text not original to him, the **Maclaurin series**. The series is found in Maclaurin's *A Treatise of Fluxions* (1742), where, in Book II, he demonstrated the rules of fluxions and their applications in an algebraic and algorithmic manner. Maclaurin provided details of the entire range of problems to which the calculus was being applied. He discussed maxima and minima and points of inflection; he found tangent lines and asymptotes; he determined curvature; and he gave a complete account of the brachistochrone problem. Maclaurin calculated areas under curves given by y in terms of x by showing that the fluxion of this area was $y\dot{x}$ and then using one of several methods to determine the fluent of this expression. Similarly, he calculated volumes and surface areas of solids of revolution by first determining their fluxions. He used an elementary form of multiple integration to study the gravitational attraction of ellipsoids. Finally, he dealt with the fluxions of exponential, trigonometric, and inverse trigonometric functions.

These latter functions appeared in the context of finding fluents of given fluxions. Thus, Maclaurin noted that the fluent of $\dot{y}/(a^2 + y^2)$ was the arc whose tangent was y in a circle of radius a, while the fluent of $a\dot{y}/\sqrt{a^2 - y^2}$ was the arc whose sine was y. More interestingly, however, Maclaurin also realized that a minor change in the function changed the fluent from a circular arc to a logarithm. For example, a change in the sign of y^2 in the first expression changes the fluent to $(1/2a)\log[(a + y)/(a - y)]$. It thus seemed that circular arcs could be represented by imaginary logarithms. Maclaurin was not, however, able to derive the consequences of this remark.

The series named for Maclaurin also occurs in Book II: Suppose that y is expressible as a series in z, say, $y = A + Bz + Cz^2 + Dz^3 + \cdots$. If $E, \dot{E}, \ddot{E}, \ldots$ are the values of y and its fluxions of various orders when z vanishes, then the series can be expressed in the form

$$y = E + \dot{E}z + \frac{\ddot{E}z^2}{1 \times 2} + \frac{\dddot{E}z^3}{1 \times 2 \times 3} + \cdots$$

(with the assumption that $\dot{z} = 1$). Maclaurin's proof was easy, given his assumption that y can be written in a power series. He first set $z = 0$ to get $A = E$. Next, he took the fluxion of the series and again set $z = 0$. It follows that $B = \dot{E}/\dot{z} = \dot{E}$. He continued to take fluxions and set $z = 0$ to complete the result. Maclaurin noted that this theorem had already been discovered and published by Brook Taylor (1685–1731).

Maclaurin worked out many examples of these series, including the series for the sine and the cosine in a circle of radius a. For example, if $y = \cos z$ (in a circle of radius a), then $\dot{y}/\dot{z} = \sqrt{a^2 - y^2}/a$. It follows that

$$\frac{\dot{y}^2}{\dot{z}^2} = \frac{a^2 - y^2}{a^2}$$

and that

$$\frac{2\dot{y}\ddot{y}}{\dot{z}^2} = -\frac{2y\dot{y}}{a^2}, \qquad \text{or} \qquad \frac{\ddot{y}}{\dot{z}^2} = -\frac{y}{a^2}.$$

Therefore, since $y = a$ when $z = 0$, we get $E = a$, $\dot{E} = 0$, and $\ddot{E} = -1/a$. The first three terms of the series for $y = \cos z$ are then $y = a + 0z - (1/2a)z^2$. More terms can be found easily.

Maclaurin also used his series for developing the standard derivative tests for determining maxima and minima: "When the first fluxion of the ordinate vanishes, if at the same time its second fluxion is positive, the ordinate is then a minimum, but is a maximum if its second fluxion is then negative." If the ordinate $AF = E$ and two values of the abscissa, one to the right of A (designated x) and one the same distance to the left (designated $-x$), are given (Fig. 12.8), the Maclaurin series shows that the corresponding ordinates are

$$PM = E + \dot{E}x + \frac{\ddot{E}x^2}{2} + \frac{\dddot{E}x^3}{6} + \cdots$$

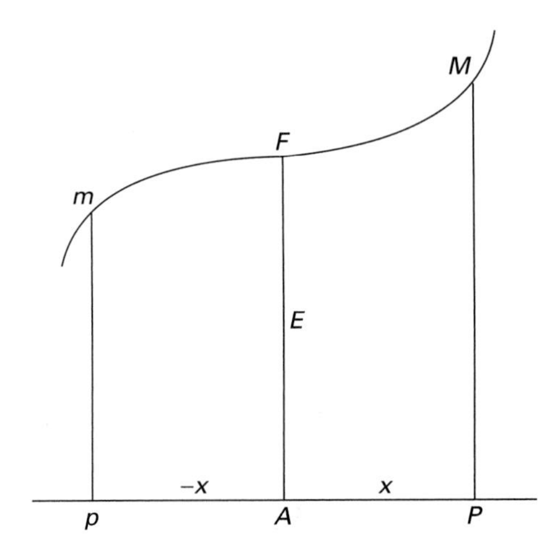

FIGURE 12.8 Maclaurin and the second derivative test

and

$$pm = E - \dot{E}x + \frac{\ddot{E}x^2}{2} - \frac{\dddot{E}x^3}{6} + \cdots.$$

Assuming that $\dot{E} = 0$ and that x is small enough, Maclaurin concluded that both of these ordinates will exceed the ordinate $AF = E$ when $\ddot{E}$ is positive (so that AF is a minimum) and both will be less than AF when $\ddot{E}$ is negative (so that AF is a maximum). Furthermore, Maclaurin concluded that if $\ddot{E}$ also vanishes and if $\dddot{E}$ does not, then either $PM > AF$ and $pm < AF$ or vice versa, so that AF is neither a maximum nor a minimum.

12.3.2 Textbooks in the Differential Calculus

The first textbook in the differential calculus was the *Analysis of Infinitely Small Quantities for the Understanding of Curves*, by Guillaume François l'Hospital (1661–1704), based on material provided to him by Johann Bernoulli. This very successful text provided a detailed treatment of the calculus of differentials and its application to numerous problems, but it is probably most famous as the source of **l'Hospital's rule** for calculating limits of quotients in the case where the limits of both numerator and denominator are zero:

PROPOSITION. *Let AMD be a curve ($AP = x$, $PM = y$, $AB = a$) such that the value of the ordinate y is expressed by a fraction, of which the numerator and denominator each become 0 when $x = a$, that is to say, when the point P corresponds to the given point B. It is required to find what will then be the value of the ordinate BD (Fig. 12.9).*

Supposing that $y = p/q$, l'Hospital simply noted that for an abscissa b infinitely close to B, the value of the ordinate y will be given by

$$y + dy = \frac{p + dp}{q + dq}.$$

But since this ordinate is infinitely close to y, and since both p and q are 0 at B, l'Hospital noted that $y = dp/dq$. L'Hospital did not offer trivial examples, his first being the function

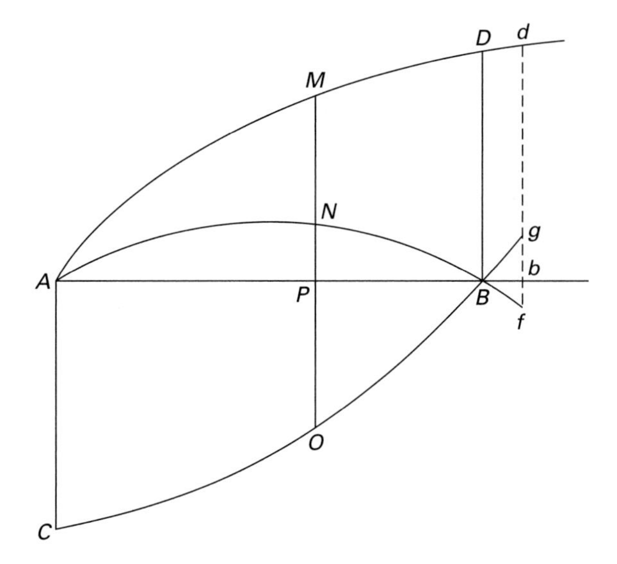

FIGURE 12.9 L'Hospital's diagram illustrating l'Hospital's rule. Notice that the function g is drawn below the x-axis, but the quotient function, represented by curve AMD, is above the x-axis. Think of all values of the functions involved as representing positive quantities.

$$y = \frac{\sqrt{2a^3x - x^4} - a\sqrt[3]{a^2x}}{a - \sqrt[4]{ax^3}},$$

where the value is to be found when $x = a$. A straightforward calculation of differentials gave him the answer $y = \frac{16}{9}a$.

The first important European successor to l'Hospital's text was the *Foundations of Analysis for the Use of Italian Youth*, by Maria Gaetana Agnesi (1718–1799). Agnesi's text, which appeared in 1748, was originally written to help her instruct her younger brothers in the subject. It thus explained concepts clearly and provided numerous examples. For example, in her section on maxima and minima, Agnesi presented such problems as that of cutting a line at a point so that the product of the length of one segment and the square of the other is maximal and that of finding the line segment of minimum length that passes through one vertex of a rectangle and intersects the extensions of both of the opposite sides.

Agnesi was especially thorough in her treatment of the logarithmic (exponential) curve. She noted that the ordinary rule for integration leads from $dx = ay^{-1}\,dy$ to $x = ay^{-1+1}/(-1+1)$, or $ay^0/0$, and that this "teaches us nothing." Thus, she dealt with this curve in other ways. She showed first that the curve whose ordinates increase geometrically while the abscissas increase arithmetically has the differential equation $dx = ay^{-1}\,dy$ and then that one can make computations by using appropriate infinite series. She also showed how to find the area under this curve, both over a finite interval and over an infinite one stretching to the left from a fixed abscissa x. She calculated this "improper integral" (today written as $\int_{-\infty}^{x} e^{t/a}\,dt$) to be ay, where y is the ordinate corresponding to x, and showed how to calculate the volume of the solid generated by revolving the curve around the x-axis.

Curiously, Agnesi's name is today attached to a small item in her book not even original to her. As an example in analytic geometry, she described geometrically a curve whose equation she determined to be $y = a\sqrt{a-x}/\sqrt{x}$, a curve that had earlier been named *la versiera*, derived from the Latin meaning "to turn." Unfortunately, the word *versiera* was also the abbreviation for the Italian word *avversiera*, meaning "wife of the devil." Because the English translator of Agnesi's work rendered this word as "witch," the curve has ever since been referred to as the "witch of Agnesi."

12.3.3 Euler's Textbooks

The most influential calculus textbooks of the eighteenth century were the three works of Euler, which appeared over a period of about twenty years: the *Introduction to Analysis of the Infinite* of 1748 (actually a precalculus text), the *Methods of the Differential Calculus* of 1755, and the *Methods of the Integral Calculus* of 1768–1770.

In the *Introduction*, Euler developed those topics "which are absolutely required for analysis" so that the reader "almost imperceptibly becomes acquainted with the idea of the infinite." In particular, since a critical tool of analysis was the notion of a power series, Euler developed these series for the most important transcendental functions, the trigonometric, exponential, and logarithmic functions. But he also introduced the notations and concepts that were to make obsolete all the discussions of such functions in earlier texts. All modern treatments of these functions are in some sense derived from those of Euler.

Euler defined exponential functions as powers in which exponents are variable and then—and this is a first—defined logarithms in terms of these. Namely, if $a^z = y$, Euler defined z to be the logarithm of y with base a. The basic properties of the logarithmic function are then derived from those of the exponential.

Euler then developed power series for the exponential and logarithmic functions to arbitrary base a by use of the binomial theorem. His technique made important use of both "infinitely small" and "infinitely large" numbers. Thus, he noted that since $a^0 = 1$, it followed that $a^\omega = 1 + \psi$, where both ω and ψ are infinitely small. Therefore, ψ must be some multiple of ω, depending on a, and

$$a^\omega = 1 + k\omega, \qquad \text{or} \qquad \omega = \log_a(1 + k\omega).$$

Euler noted next that for any j, $a^{j\omega} = (1 + k\omega)^j$ and, expanding the right side by the binomial theorem, that

$$a^{j\omega} = 1 + \frac{j}{1}k\omega + \frac{j(j-1)}{1\cdot 2}k^2\omega^2 + \frac{j(j-1)(j-2)}{1\cdot 2\cdot 3}k^3\omega^3 + \cdots.$$

If j is taken to equal z/ω, where z is finite, then j is infinitely large and $\omega = z/j$. The series now becomes

$$a^z = 1 + \frac{1}{1}kz + \frac{1(j-1)}{1\cdot 2j}k^2z^2 + \frac{1(j-1)(j-2)}{1\cdot 2j\cdot 3j}k^3z^3 + \cdots.$$

Because j is infinitely large, $(j-n)/j = 1$ for any positive integer n. The expansion then reduces to the series

$$a^z = 1 + \frac{kz}{1} + \frac{k^2z^2}{1\cdot 2} + \frac{k^3z^3}{1\cdot 2\cdot 3} + \cdots,$$

where k depends on the base a. Euler also noted that the equation $\omega = \log_a(1 + kw)$ implies that if $(1 + kw)^j = 1 + x$, then $\log_a(1 + x) = j\omega$. Since then $k\omega = (1 + x)^{1/j} - 1$, it follows that

$$\log_a(1+x) = \frac{j}{k}(1+x)^{1/j} - \frac{j}{k}.$$

Another clever use of the binomial theorem finally allowed him to derive the series

$$\log_a(1+x) = \frac{1}{k}\left(\frac{x}{1} - \frac{x^2}{2} + \frac{x^3}{3} + \cdots\right).$$

The choice of $k = 1$, or equivalently, $a = e$, gives the standard power series for e^z and $\ln z$.

Euler's treatment of "transcendental quantities which arise from the circle" is the first textbook discussion of the trigonometric functions that deals with these quantities as functions having numerical values, rather than as lines in a circle of a certain radius. Euler did not, in fact, give any new definition of the sine and cosine. He merely noted that he would always consider the sine and cosine of an arc z to be defined in terms of a circle of radius 1. All basic properties of the sine and cosine, including the addition and periodicity properties, were assumed known, although Euler did derive some relatively complicated identities. More importantly, he derived the power series for the sine and cosine through use of the binomial theorem and complex numbers.

From the identity $(\cos z \pm i \sin z)^n = \cos nz \pm i \sin nz$, Euler concluded that

$$\cos nz = \frac{(\cos z + i \sin z)^n + (\cos z - i \sin z)^n}{2}$$

and, by expanding the right side, that

$$\cos nz = (\cos z)^n - \frac{n(n-1)}{1 \cdot 2}(\cos z)^{n-2}(\sin z)^2 + \frac{n(n-1)(n-2)(n-3)}{1 \cdot 2 \cdot 3 \cdot 4}(\cos z)^{n-4}(\sin z)^4 + \cdots.$$

Again letting z be infinitely small, n infinitely large, and $nz = v$ finite, it follows from $\sin z = z$ and $\cos z = 1$ that

$$\cos v = 1 - \frac{v^2}{1 \cdot 2} + \frac{v^4}{1 \cdot 2 \cdot 3 \cdot 4} - \cdots.$$

There is much more about infinite processes in the *Introduction*, including infinite products as well as infinite series. For example, Euler considered the equality

$$\prod_p \frac{1}{1 - \dfrac{1}{p^n}} = \sum_m \frac{1}{m^n},$$

where the product is taken over all primes and the sum over all positive integers. This product and sum, both generalized to the case where n is any complex number s, are today called the *Riemann zeta function* of the variable s, the study of which has led to much new mathematics.

Euler began his *Differential Calculus* with a definition of the subject: "[It] is the method of determining the ratios of the evanescent increments which functions receive to those of the evanescent increments of the variable quantities, of which they are functions." Euler then gave a definition of the concept of function, influenced by his work on the wave equation: "When quantities depend on others in such a way that [the former] undergo changes themselves when [the latter] change, then [the former] are called functions of [the latter]; this is a very comprehensive idea which includes in itself all the ways in which one quantity can be determined by others." Thus, from now on, functions were to play the primary role in calculus, not curves, as in the time of Newton and Leibniz.

After developing rules for dealing with finite increments (or finite differences) for functions, Euler moved on to the "analysis of infinitesimals," where the differences or increments are now taken as "infinitely small." His rules for calculating with these infinitely small quantities, the differentials, produce the standard formulas of the differential calculus. For example, if $y = x^n$, then $y' = (x + dx)^n = x^n + nx^{n-1}\,dx + [n(n-1)/(1\cdot 2)]x^{n-2}\,dx^2 + \cdots$. Thus, $dy = y' - y = nx^{n-1}\,dx + [n(n-1)/(1\cdot 2)]x^{n-2}\,dx^2 + \cdots$. "But in this expression the second term with the rest of the sequence will vanish in comparison with the first." Thus, $d(x^n) = nx^{n-1}\,dx$. It should be noted here that Euler intended his argument to apply not just to positive integral powers of x, but to arbitrary powers. The binomial theorem, after all, applies to all powers. Thus, the expansion of $(x + dx)^n$ does not necessarily represent a finite sum; it may well represent an infinite series. Euler therefore noted immediately that $d(1/x^m) = -(m\,dx)/x^{m+1}$ and, more generally, that $d(x^{\mu/\nu}) = (\mu/\nu)x^{(\mu-\nu)/\nu}\,dx$.

Euler did not give an explicit statement of the modern chain rule, but dealt with special cases as the need arose. For example, if p is a function of x whose differential is dp, then $d(p^n) = np^{n-1}\,dp$. Euler's derivation of the product rule was virtually identical to that of Leibniz, but his derivation of the quotient rule was more original. He expanded $1/(q + dq)$ into the power series

$$\frac{1}{q+dq} = \frac{1}{q}\left(1 - \frac{dq}{q} + \frac{dq^2}{q^2} - \cdots\right),$$

neglected the higher-order terms, and then wrote

$$\frac{p+dp}{q+dq} = (p+dp)\left(\frac{1}{q} - \frac{dq}{q^2}\right) = \frac{p}{q} - \frac{p\,dq}{q^2} + \frac{dp}{q} - \frac{dp\,dq}{q^2}.$$

It follows, since the second-order differential $dp\,dq$ vanishes with respect to the first-order ones, that

$$d\left(\frac{p}{q}\right) = \frac{p+dp}{q+dq} - \frac{p}{q} = \frac{dp}{q} - \frac{p\,dq}{q^2} = \frac{q\,dp - p\,dq}{q^2}.$$

The differential of the logarithm required the power series developed in the *Introduction*. If $y = \ln x$, then

$$dy = \ln(x+dx) - \ln(x) = \ln\left(1 + \frac{dx}{x}\right) = \frac{dx}{x} - \frac{dx^2}{2x^2} + \frac{dx^3}{3x^3} - \cdots.$$

Dispensing with the higher-order differentials immediately gave Euler the formula $d(\ln x) = dx/x$.

Euler's approach to the arcsine function was through complex numbers. Substituting $y = \arcsin x$ into the formula $e^{iy} = \cos y + i\,\sin y$ gave $e^{iy} = \sqrt{1-x^2} + ix$. It followed that $y = (1/i)\ln(\sqrt{1-x^2} + ix)$ and therefore that

$$dy = d(\arcsin x) = \frac{1}{i}\frac{1}{\sqrt{1-x^2}+ix}\left(\frac{-x}{\sqrt{1-x^2}} + i\right)dx = \frac{dx}{\sqrt{1-x^2}}.$$

Finally, Euler's derivation of the differential of the sine function began with the calculation $d(\sin x) = \sin(x + dx) - \sin x = \sin x \cos dx + \cos x \sin dx - \sin x$. Euler then recalled his series expansions of the sine and the cosine and, again rejecting higher-order terms, noted that $\cos dx = 1$ and $\sin dx = dx$. It followed that $d(\sin x) = \cos x\, dx$, as desired.

Euler's chapter on the differentiation of functions of two or more variables does not record any of the struggles in the development of this idea that we discussed earlier. He merely noted that in dealing with such a function, the variables can change independently. The central concepts in this chapter, as in the case of functions of one variable, are that of the differential and the differential coefficient. In particular, Euler showed that if V is a function of the two variables x and y, then dV, the change in V resulting from the changes x to $x + dx$ and y to $y + dy$, is given by $dV = p\, dx + q\, dy$, where p, q are the differential coefficients resulting from leaving y and x constant, respectively.

Euler's *Differential Calculus* dealt with many other topics, including an introduction to differential equations (in which he showed how to generate these from a given equation in two variables), a discussion of the Taylor series, a chapter on various methods of converting functions to power series, an extensive discussion on finding the sums of various series (including those for the sums of the various powers of the integers), a variety of ways of finding the roots of equations numerically, and various methods for finding maxima and minima, in both the one-variable and the multi-variable cases.

The final part of Euler's trilogy in analysis, the *Integral Calculus*, begins with a definition of integral calculus as the method of finding, from a given relation of differentials of certain quantities, the quantities themselves. For Euler, as for Agnesi and Johann Bernoulli, integration was the inverse of differentiation rather than the determination of an area. Thus, this work deals with techniques for integrating (finding antiderivatives of) functions of various types as well as methods for solving differential equations. Euler began with such results as

$$\int ax^n\, dx = \frac{a}{n+1}x^{n+1} + C,$$

for $n \neq -1$, and

$$\int \frac{a\, dx}{x} = a \ln x + C = \ln cx^a,$$

and then gave a detailed discussion of partial fraction techniques, substitution, integration by parts, trigonometric integrals, and the use of infinite series. The bulk of the text dealt with differential equations. For example, Euler solved the general first-order linear equation $dy + Py\, dx = Q\, dx$ (or, in modern terms, $y' + Py = Q$) by separation of variables to get the general solution

$$y = e^{-\int P\, dx} \int e^{\int P\, dx} Q\, dx.$$

He showed how to integrate $P\, dx + Q\, dy$ in the "exact" case where $\partial P/\partial y = \partial Q/\partial x$ and how to find integrating factors in the case where $P\, dx + Q\, dy$ is not exact. He dealt with various cases of second-order and higher-order differential equations, including the linear case with constant coefficients. Finally, Euler concluded the book with a discussion of partial differential equations. But even though the original motivation for consideration of differential equations came largely from physical problems, Euler mentions nothing of these in the text.

The *Integral Calculus*, like the *Differential Calculus* and the *Introduction*, is a text in pure analysis, so much so that Euler did not even deal with applications to geometry. This fact perhaps explains the significant differences between Euler's works and a modern calculus text. In the *Differential Calculus* there are no tangent lines or normal lines, no tangent planes, no study of curvature—all topics with which Euler was fully conversant in 1740 but which appear only in some of his geometrical works. Even more surprisingly, there is no calculation of areas in the *Integral Calculus*, nor any material on lengths of curves, or volumes, or surface areas of solids. It follows that the fundamental theorem of calculus, central to modern textbooks, does not appear. There is not even a calculation of a definite integral. Euler was certainly familiar with using antiderivatives to calculate area and, in fact, used such ideas in various papers. On the other hand, since there does not appear in his work any clear notion of the area under a curve as a function, he did not consider its derivative.

Despite the gaps a modern reader might find in Euler's calculus texts, they proved influential to the end of the eighteenth century in presenting an organization and a clear explanation of the material that Euler and his predecessors had developed. It was, however, the new students entering the sciences after the upheavals of the French Revolution who inspired the writing of many new texts, which replaced those of Euler and were the direct ancestors of the texts of today.

12.4 The Foundations of the Calculus

As we have seen, the eighteenth century saw extensive development of the techniques of the calculus. But in the minds of some there was a nagging doubt as to the foundations of the subject. Most mathematicians regarded Euclid's *Elements* as the model of how mathematics should be done. Newton himself attempted to put calculus on a solid foundation, but it was hard for many mathematicians to understand his synthetic method of fluxions, while it was his analytic method that was spread by many of his followers in Britain. It was the foundations of that method, as well as the infinitesimals of Leibniz's followers, that were criticized by many.

12.4.1 George Berkeley's Criticisms and Maclaurin's Response

The most important criticism of both infinitesimals and fluxions was made by the Irish philosopher Bishop George Berkeley (1685–1753) in a 1734 tract entitled *The Analyst*. Berkeley agreed that the method of fluxions was helping mathematicians unlock the secrets of geometry and of nature. But the process of calculating fluxions appeared to him to be contradictory. As he wrote about the calculation of the fluxion of x^n:

> In the same time that x by flowing becomes $x + o$, the power x^n becomes $(x + o)^n$, i.e., by the method of infinite series
>
> $$x^n + nox^{n-1} + \frac{n^2 - n}{2}o^2x^{n-2} + \cdots,$$
>
> and the increments o and $nox^{n-1} + [(n^2 - n)/2]o^2x^{n-2} + \cdots$ are one to another as 1 to $nx^{n-1} + [(n^2 - n)/2]ox^{n-2} + \cdots$. Let now the increments vanish, and their last proportion will be 1 to

> nx^{n-1}. But it should seem that this reasoning is not fair or conclusive. For when it is said, let the increments vanish, i.e., let the increments be nothing, or let there be no increments, the former supposition that the increments were something, or that there were increments, is destroyed, and yet a consequence of that supposition, i.e., an expression got by virtue thereof, is retained.

Berkeley thus questioned how one can take a nonzero increment, do calculations with it, and then in the end set it equal to zero. He noted further that the methods of the continental mathematicians were no better. In particular, Berkeley claimed that he could not conceive of the infinitely small quantities they generally used. For example, "to conceive a part of such infinitely small quantity that shall be still infinitely less than it, and consequently though multiplied infinitely shall never equal the minutest finite quantity, is, I suspect, an infinite difficulty to any man whatsoever."

Berkeley's criticisms of the foundations of the calculus were valid. The question of when a value was zero and when it was not zero extended back even to the work of Fermat, and neither Newton nor Leibniz was ever quite able to resolve it. Nevertheless, several British mathematicians sprang to Newton's defense under Berkeley's attack. The most important response was that of Maclaurin in his *Treatise of Fluxions*, in which he attempted to found the theory of fluxions from a few basic axioms "after the manner of the ancients." But because the basic concept necessary to the theory of fluxions is instantaneous velocity, Maclaurin had to begin with a definition of this idea: "The velocity of a variable motion at any given term of time is not to be measured by the space that is actually described after that term in a given time, but by the space that would have been described if the motion had continued uniformly from that term." From a modern point of view, Maclaurin, by giving such a definition, missed the fundamental idea of instantaneous velocity as a limit of average velocities as the time interval approaches zero. Similarly, when he dealt with tangents to curves, he began with the ancient definition that a tangent to a curve is a straight line that touches the curve in such a way that no other straight line can be inserted between the curve and the line. Nevertheless, Maclaurin presented axioms for the use of these definitions and then proceeded to prove numerous theorems in the "manner of the ancients," using each time a double *reductio ad absurdum*.

Maclaurin was, however, well aware of Newton's use of the notion of "ultimate proportion of evanescent quantities," or "limits." Thus, he wrote about the ratio that is the limit of the various proportions that finite simultaneous increments of two variable quantities bear to one another as the two increments decrease until they vanish. He noted that to discover this limit, one must first determine the ratio of increments in general and then reduce to the simplest terms so that a part of the result might be independent of the value of the increments themselves. The desired limit then readily appears if one supposes the increments to "decrease until they vanish." For example, to find the ratio of the fluxion of x^2 to the fluxion of ax, Maclaurin calculated the ratio of the increments (as x increases to $x + o$) to be $2xo + o^2 : ao$, or $2x + o : a$. "This ratio of $2x + o$ to a continually decreases while o decreases and is always greater than the ratio of $2x$ to a while o is any real increment, but it is manifest that it continually approaches to the ratio of $2x$ to a as its limit."

Maclaurin thus vehemently denied Berkeley's contention that the method of first supposing a finite increment and then letting that increment vanish is contradictory. In fact, he noted, this method allows one to determine the ratio of the increments when the increments are finite and determine how the ratio varies with the increment. One can then easily determine what limit the ratio approaches as the increments are diminished. As a final response to Berkeley, Maclaurin even defined the tangent as a limit: "The tangent . . . is the

. . . line that limits the position of all the secants that can pass through the point of contact, though strictly speaking it be no secant, [just as] a ratio may limit the variable ratios of the increments, though it cannot be said to be the ratio of any real increments."

The problem with Maclaurin's treatment of the calculus, as noted by many of his contemporaries, was not that he failed rigorously to derive the rules, but that he did so in the "manner of the ancients." In particular, he used the method of exhaustion and its accompanying *reductio ad absurdum* argument. The use of such a method imposed a heavy toll on the reader. For example, the first 590 pages of this 754-page work do not contain any notation of fluxions. Every new idea is derived geometrically with great verbosity. And in the eighteenth century, few were willing to read through these detailed arguments. So, although Maclaurin's great efforts answered Berkeley's objections, they were not read or appreciated by most eighteenth-century mathematicians, people who saw themselves as breaking new ground rather than extending the methods of the ancients.

12.4.2 Euler and d'Alembert

On the continent, too, some justification of the procedures of the calculus was necessary. In his *Differential Calculus*, Euler developed the idea that the ratios involved in the calculation of derivatives were in fact simply versions of the ratio $0 : 0$. For Euler, infinitely small quantities were quantities actually equal to 0, because the latter is the only quantity smaller than any given quantity. But although two zeros are equal in such a way that their difference is always zero, Euler insisted that the ratio of two zeros, which depends on the origin of the quantities that are becoming zero, must be calculated in each specific case. As an example, he noted that $0 : 0 = 2 : 1$ is a correct statement because the first quantity on each side of the equal sign is double the second quantity. In fact, then, the ratio $0 : 0$ may be equal to any finite ratio at all. Therefore, "the calculus of the infinitely small is . . . nothing but the investigation of the geometric ratio of different infinitely small quantities."

Euler continued his discussion in this vein, noting further, for example, that "infinitely small quantities vanish in comparison with finite ones, and thus can be rejected insofar as those finite quantities are concerned." Similarly, the infinitely small quantity dx^2 will vanish with respect to dx and can thus be neglected because $(dx \pm dx^2)/dx = 1 \pm dx = 1$.

Interestingly enough, d'Alembert, in the article "Différentiel," which he wrote for the French compendium of knowledge, the *Encyclopédie*, in 1754, combined the ideas of Euler and Maclaurin. He agreed with Euler that there was no absurdity in considering the ratio $0 : 0$ because it may in fact be equal to any quantity at all. But the central idea of the differential calculus is that dy/dx is the limit of a certain ratio as the quantities involved approach 0. As an example of what he meant, d'Alembert calculated the slope of the line tangent to the parabola $y^2 = ax$ by first determining the slope of a secant through the two points (x, y) and $(x + u, y + z)$. This slope, the ratio $z : u$, is easily seen to be equal to $a : 2y + z$. "This ratio is always smaller than $a : 2y$; but the smaller z is, the greater the ratio will be and, since one may choose z as small as one pleases, the ratio $a : 2y + z$ can be brought as close to the ratio $a : 2y$ as we like. Consequently $a : 2y$ is the limit of the ratio $a : 2y + z$." It follows that $dy/dx = a/2y$. D'Alembert's wording is virtually identical to that of Maclaurin. He went somewhat further, however, by giving an explicit definition of the term *limit* in his *Encyclopédie* article on that notion: "One magnitude is said to be the **limit** of another magnitude when the second may approach the first within any given magnitude, however small, though the second magnitude may never exceed the magnitude it approaches." His idea, although apparently geometric rather than arithmetic, was not followed up by his eighteenth-century successors. Through the remainder of the century, most works on calculus attempted to explain the basis of the subject in terms of infinitesimals, fluxions, or the ratios of zeros.

12.4.3 Lagrange and Power Series

It was Joseph Louis Lagrange (1736–1813) who, near the end of the eighteenth century, attempted to give a precise definition of the derivative by eliminating all reference to infinitesimals, fluxions, zeros, and even limits, all of which he believed lacked proper definitions. He sketched his new ideas about derivatives in a paper of 1772 and then developed them in full in his text of 1797, the full title of which expressed what he intended to do: *The Theory of Analytic Functions, containing the principles of the differential calculus, released from every consideration of the infinitely small or the evanescent, of limits or of fluxions, and reduced to the algebraic analysis of finite quantities.* How could Lagrange accomplish the reduction of calculus purely to algebraic analysis? He did so by formalizing the idea used by most of his predecessors without question, the idea that any function can be represented as a power series. For Lagrange, if $y = f(x)$ is any function, then $f(x+i)$, where i is an indeterminate, could "by the theory of series" be expanded into a series in i:

$$f(x+i) = f(x) + pi + qi^2 + ri^3 + \cdots,$$

where $p, q, r \ldots$ are new functions of x independent of i. Lagrange then showed that the ratio dy/dx can be identified with the coefficient $p(x)$ of the first power of i in this expansion. He therefore had a new definition of this basic concept of the calculus. Since the function p is "derived" from the original function f, Lagrange named it a *fonction dérivée* (from which comes the English word *derivative*) and used the notation $f'(x)$. Similarly, the derivative of f' is written f'', that of f'' is written f''', and so on. Lagrange easily showed that $q = \frac{1}{2}f'', r = \frac{1}{6}f''', \ldots$.

Lagrange's argument for the expansion of a function f—generally thought of as a formula—began with the assertion that $f(x+i) = f(x) + iP$, where $P(x,i)$ is defined by

$$P(x,i) = \frac{f(x+i) - f(x)}{i}.$$

Lagrange assumed further that one can separate out from P that part p that does not vanish at $i = 0$. Thus, $p(x)$ is defined as $P(x,0)$ and then

$$Q(x,i) = \frac{P(x,i) - p(x)}{i},$$

or $P = p + iQ$. It follows that $f(x+i) = f(x) + ip + i^2Q$. Repeating the argument for Q, he wrote $Q = q + iR$ and substituted again. As an example of the procedure, Lagrange took $f(x)$ to be $1/x$. Since $f(x+i) = 1/(x+i)$, he calculated

$$P = \frac{1}{i}\left(\frac{1}{x+i} - \frac{1}{x}\right) = -\frac{1}{x(x+i)} \qquad p = -\frac{1}{x^2}$$

$$Q = \frac{1}{i}\left(-\frac{1}{x(x+i)} + \frac{1}{x^2}\right) = \frac{1}{x^2(x+i)} \qquad q = \frac{1}{x^3}$$

$$\vdots \qquad\qquad \vdots$$

Thus, the series becomes

$$\frac{1}{x+i} = \frac{1}{x} - \frac{i}{x^2} + \frac{i^2}{x^3} - \frac{i^3}{x^4} + \cdots.$$

At each stage of the expansion, the terms $iP, i^2Q, \ldots$ can be considered as the error terms resulting from representing $f(x+i)$ by terms up to that point. Furthermore, Lagrange claimed, the value of i can always be taken so small that any given term of this series is greater than the sum of the remaining terms—that is, the remainders are always sufficiently small that in fact the function is represented by the series. In fact, this result is what Lagrange used often later on. He also used a somewhat different form of his expansion result containing what is now called the *Lagrange form of the remainder* in the Taylor series. That is, he showed that for any given positive integer n, one can write

$$f(x+i) = f(x) + if'(x) + \frac{i^2 f''(x)}{2} + \cdots + \frac{i^n f^{(n)}(x)}{n!} + \frac{i^{n+1} f^{(n+1)}(x+j)}{(n+1)!},$$

for some value j between 0 and i. Although this new form is, perhaps, no more convincing to the modern reader than his earlier one, Lagrange himself was satisfied that his principle of a power series representation for every function was correct. After all, he claimed, it enabled him to derive anew all of the basic results of the calculus without any consideration of infinitesimals, fluxions, or limits.

One of these basic results is part of what is known today as the **fundamental theorem of calculus**, that if $F(x)$ represents the area under the curve $y = f(x)$ from a fixed ordinate, then $F'(x) = f(x)$. (It should be noted that Lagrange had no definition of area. He simply assumed that the area under a curve $y = f(x)$ was a well-determined quantity.) Lagrange began his proof by noting that $F(x+i) - F(x)$ represents that portion of the area between the abscissas x and $x+i$. Keeping to Euler's dictum that in a text on analysis one should not include diagrams, Lagrange nevertheless wrote that even without a figure one can easily convince oneself that if $f(x)$ is monotonically increasing, then

$$if(x) < F(x+i) - F(x) < if(x+i),$$

with the inequalities reversed if $f(x)$ is monotonically decreasing (Fig. 12.10). Expanding both $f(x+i)$ and $F(x+i)$, Lagrange determined that $f(x+i) = f(x) + if'(x+j)$ and $F(x+i) = F(x) + iF'(x) + (i^2/2)F''(x+j)$, where $0 < j < i$ (although the value of j may not be the same in both expansions). It follows that

$$if(x) < iF'(x) + (i^2/2)F''(x+j) < if(x) + i^2 f(x+j)$$

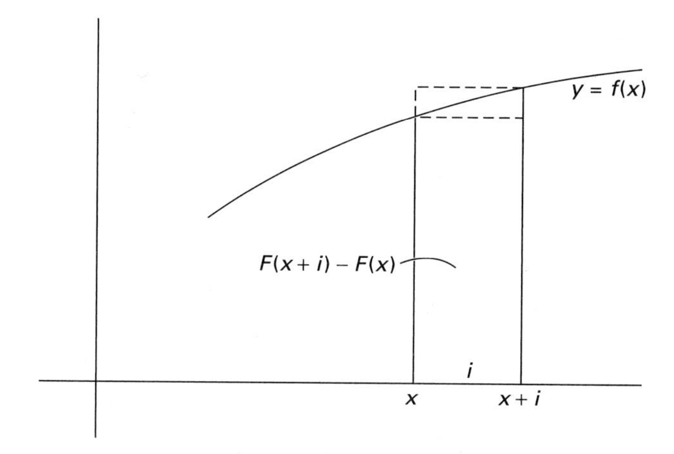

FIGURE 12.10 Lagrange and the fundamental theorem of calculus:
$if(x) < F(x+i) - F(x) < if(x+i)$

and therefore that

$$\left| i[F'(x) - f(x)] + \frac{i^2}{2}F''(x+j) \right| < i^2 f'(x+j),$$

where the absolute value sign is necessary to take care of both the increasing and the decreasing cases. Lagrange concluded that because the inequality holds no matter how small i is taken to be, it must be true that $F'(x) = f(x)$. He even calculated that if the conclusion were not true, the inequality would fail, for

$$i < \frac{F'(x) - f(x)}{f'(x+j) - \frac{1}{2}F''(x+j)}.$$

To finish his proof, Lagrange removed the condition that $f(x)$ must be monotonic on the original interval $[x, x+i]$. For if it is not, there is a maximum or a minimum of f on that interval and i can be chosen small enough so that the extreme value falls outside of the new interval $[x, x+i]$.

Most of the early objections to Lagrange's new foundation for the calculus were aimed at his new notations and the length of some of his calculations rather than at his assertion that any function can be expanded in a power series. Mathematicians in general continued to use the earlier differential methods, especially since Lagrange's book assured them that because there was a correct basis to the entire subject, any method that worked would be legitimate. Even Lagrange, in some of his other work, continued to employ the notation of differentials rather than that of derivatives. It was not until the second decade of the nineteenth century that various mathematicians pointed out that there exist differentiable functions that do not have a power series representation and thus that Lagrange's basic concept was not tenable. The story of the new attempts to supply a foundation to the ideas of the calculus will therefore be continued.

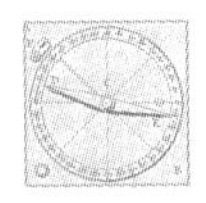

Exercises

1. Derive the equation of the catenary in closed form from Johann Bernoulli's differential equation

$$dx = \frac{a\,dy}{\sqrt{y^2 - a^2}}.$$

2. Derive Johann Bernoulli's original differential equation for the catenary as follows: Let the lowest point of the hanging cord be the origin of the coordinate system, and consider a piece of the cord of length s over the interval $[0, x]$. Let $T(x)$ be the (vector) tension of the cord at the point $P = (x, y)$. Let α be the angle that $T(x)$ makes with the horizontal and let ρ be the density of the cord. Show that the equilibrium of horizontal forces gives the equation $|T(0)| = |T(x)| \cos\alpha$, while that of the vertical forces gives $\rho s = |T(x)| \sin\alpha$. Since $dy/dx = \tan\alpha$, Bernoulli's equation can be derived by dividing the second equation by the first.

3. Translate Leibniz's solution of $m\,dx + ny\,dx + dy = 0$ into modern terms by noting that $dp/p = n\,dx$ is equivalent to $\ln p = \int n\,dx$ or to $p = e^{\int n\,dx}$. Solve $-3x\,dx + (1/x)y\,dx + dy = 0$ using Leibniz's procedure.

4. The modern way to derive Kepler's area law is to break the force into its radial and transverse components, rather than the tangential and normal components used by Hermann, and to use polar coordinates whose origin is the center of force. Assume then that the center of force is at the origin of a polar coordinate system. Using vector notation, set $\mathbf{u}_r = \mathbf{i}\cos\theta + \mathbf{j}\sin\theta$ and $\mathbf{u}_\theta = -\mathbf{i}\sin\theta + \mathbf{j}\cos\theta$. Show that $d\mathbf{u}_r/d\theta = \mathbf{u}_\theta$ and $d\mathbf{u}_\theta/d\theta = -\mathbf{u}_r$. Next, show that if $\mathbf{r} = r\mathbf{u}_r$, then the velocity $\mathbf{v}$ is given by $r(d\theta/dt)\mathbf{u}_\theta + (dr/dt)\mathbf{u}_r$. Show next that the radial component a_r and the transverse component a_θ are given respectively by

$$a_r = \frac{d^2r}{dt^2} - r\left(\frac{d\theta}{dt}\right)^2$$

and

$$a_\theta = r\frac{d^2\theta}{dt^2} + 2\frac{dr}{dt}\frac{d\theta}{dt}.$$

Since the force is central, $a_\theta = 0$. Multiply the differential equation expressing that fact by r and integrate to get $r^2\,d\theta/dt = k$, where k is a constant. Show finally that $r^2\,d\theta/dt = dA/dt$, where A is the area swept out by the radius vector. This proves Kepler's law of areas.

5. Show that the area of the infinitesimal triangle SQP of Figure 12.4 can be written as $\frac{1}{2}(y\,dx - x\,dy)$.

6. Convert Newton's geometrical description of the central force as proportional to QR and inversely proportional to the square of the area of triangle SQP (Fig. 12.4) into differentials, and compare the result with the component a_r of the force derived in exercise 4.

7. Show that the differential equation

$$\frac{d^2r}{dt^2} - r\left(\frac{d\theta}{dt}\right)^2 = -\frac{k}{r^2},$$

derived by assuming that the component a_r of the force from exercise 4 is inversely proportional to r^2, is equivalent to the differential equation Hermann derived using the inverse square property of the central force.

8. Show that the equation $a \pm cx/b = \sqrt{x^2 + y^2}$ is a parabola if $b = c$, an ellipse if $b > c$, and a hyperbola if $b < c$.

9. Show that the equation of any conic section may be written in the form $\sqrt{x^2 + y^2} = \alpha x + \beta$, where the origin is a focus of the conic. How do α and β determine the nature of the conic section?

10. Show that $y = e^{x/a}$ is a solution to the differential equation $a^3\,d^3y - y\,dx^3 = 0$. Next, assume that the product $e^{-(x/a)}(a^3\,d^3y - y\,dx^3)$ is the differential of $e^{-(x/a)}(A\,d^2y + B\,dy\,dx + Cy\,dx^2)$ and show that a new solution of the original equation must also satisfy $a^2\,d^2y + a\,dy\,dx + y\,dx^2 = 0$. (Hint: Calculate the differential and equate the two expressions. It may be easier if you rewrite the equations in modern notation using derivatives.)

11. Given that $y = e^x$ is a solution of $y''' - 6y'' + 11y' - 6y = 0$, show by a method analogous to that of exercise 10 that any other solution must satisfy $y'' - 5y' + 6y = 0$.

12. Solve the differential equation of exercise 11 by using Euler's procedure of factoring the characteristic polynomial.

13. Show that if $y = ue^{\alpha x}$ is assumed to be a solution of $a^2\,d^2y + a\,dy\,dx + y\,dx^2 = 0$, then if $\alpha = -\frac{1}{2}a$, one

can conclude that u is a solution to $a^2\, d^2u + \frac{3}{4}u\, dx^2 = 0$.

14. Solve $a^2\, d^2u + \frac{3}{4}u\, dx^2 = 0$. First multiply by du and integrate once to get $4a^2\, du^2 = (K^2 - 3u^2)\, dx^2$ or

$$dx = \frac{2a}{\sqrt{K^2 - 3u^2}}\, du.$$

Integrate a second time to get

$$x = \frac{2a}{\sqrt{3}} \arcsin \frac{\sqrt{3}u}{K} - f.$$

Rewrite this equation for u in terms of x as

$$u = C \sin\left(\frac{(x+f)\sqrt{3}}{2a}\right).$$

15. Solve the differential equation

$$(2xy^3 + 6x^2y^2 + 8x)dx \\ + (3x^2y^2 + 4x^3y + 3)\, dy = 0$$

using Clairaut's method.

16. Use Clairaut's technique of multiple integration to calculate the volume of the solid bounded by the cylinders $ax = y^2$, $by = z^2$ and the coordinate planes. First, determine the volume element $dx \int z\, dy$ by converting the integrand to a function of x and integrating. Then integrate the volume element with appropriate limits. Compare this method to the standard modern method.

17. Suppose that the solution to the wave equation $\partial^2 y/\partial t^2 = \partial^2 y/\partial x^2$ is given by $y = \Psi(t+x) - \Psi(t-x)$. Show that the initial conditions $y(0, x) = f(x)$, $y'(0, x) = g(x)$ and the condition $y(t, 0) = y(t, l) = 0$ for all t lead to the requirements that $f(x)$ and $g(x)$ are odd functions of period $2l$. (From d'Alembert)

18. Suppose that $y = F(t)G(x) = \Psi(t+x) - \Psi(t-x)$ is a solution to the wave equation $\partial^2 y/\partial t^2 = \partial^2 y/\partial x^2$. Show by differentiating twice that $F''/F = G''/G = C$, where C is some constant, and therefore that $F = ce^{t\sqrt{C}} + de^{-t\sqrt{C}}$ and $G = c'e^{x\sqrt{C}} + d'e^{-x\sqrt{C}}$. Apply the condition $y(t, 0) = y(t, l) = 0$ to show that C must be negative, and hence derive the solution $F(t) = A \cos Nt$, $G(x) = B \sin Nx$ for appropriate choice of A, B, and N. (d'Alembert)

19. Find the isosceles triangle of smallest area that circumscribes a circle of radius 1. (From Simpson)

20. Find the cone of least surface area with given volume V. (Simpson)

21. Calculate the first four nonzero terms of the power series for $y = \cos z$ using Maclaurin's technique, without explicitly using the derivatives of the cosine or the sine. Assume that the radius of the circle is 1.

22. Use l'Hospital's rule to show that

$$\lim_{x\to a} \frac{\sqrt{2a^3x - x^4} - a\sqrt[3]{a^2x}}{a - \sqrt[4]{ax^3}} = \frac{16}{9}a.$$

23. Show that $\int_{-\infty}^{x} e^{t/a}\, dt = ae^{x/a}$.

24. Given a rectangle, find the line of minimum length that passes through one vertex and through the extensions of the two opposite sides. (From Agnesi)

25. Sketch a particular example of the "witch of Agnesi," the curve given by $y^2 = 4(2-x)/x$. Show that it is symmetric about the x-axis and asymptotic to the y-axis. Show that the area between the curve and its asymptote is 4π.

26. Show that k in the series given in the text for a^z and $\log_a(1+x)$ is given by $k = \ln a$. (From Euler)

27. Derive the power series for $\ln(1+x)$ from the equation $\log(1+x) = j(1+x)^{1/j} - j$ by using the binomial theorem and assuming that j is infinitely large. (Euler)

28. If $y = \arctan x$, show that $\sin y = x/\sqrt{1+x^2}$ and $\cos y = 1/\sqrt{1+x^2}$. Then, if $p = x/\sqrt{1+x^2}$, show that $\sqrt{1-p^2} = 1/\sqrt{1+x^2}$. Since $y = \arcsin p$, it follows that $dy = dp/\sqrt{1-p^2}$ and that $dp = dx/(1+x^2)^{3/2}$. Conclude that

$$dy = \frac{dx}{1+x^2}. \qquad \text{(Euler)}$$

29. Calculate dy for $y = a^x$ by noting that $dy = a^{x+dx} - a^x = a^x(a^{dx} - 1)$ and then expanding $a^{dx} - 1$ into the power series $\ln a\, dx + [(\ln a)^2\, dx^2]/2 + \cdots$. (Euler)

30. Calculate dy for $y = \tan x$ by using the addition formula

$$\tan(x + dx) = \frac{\tan x + \tan dx}{1 - \tan x \tan dx}. \qquad \text{(Euler)}$$

31. Translate d'Alembert's definition of a limit into algebraic language, and compare your result with the modern definition of a limit.

32. Use Lagrange's technique to calculate the quantities p, q, r for the function $f(x) = \sqrt{x}$ and thus determine the first three terms of its power series representation.

33. Given that $f(x+i) = f(x) + pi + qi^2 + ri^3 + \cdots$, show that $p = f'(x)$, $q = f''(x)/2!$, $r = f'''(x)/3!, \ldots$.

34. Did eighteenth-century mathematicians use the fundamental theorem of calculus in the sense it is used today? What concepts must be defined before one can even consider this theorem? How were these concepts dealt with by eighteenth-century mathematicians? Did these mathematicians consider the fundamental theorem "fundamental"?
35. Compare Euler's trilogy of precalculus and calculus texts to a modern series of texts. What topics are common to Euler's texts and modern ones? What does Euler have that is missing in today's texts, and conversely? Could one use Euler's texts today?
36. Trace the development of the concept of the limit from Newton through Maclaurin to d'Alembert. How do their formulations agree? How do they compare with the modern formulation of this concept?
37. Develop several lessons to teach basic methods for solving various classes of differential equations using the formulations of Leibniz, Clairaut, and Euler.

References

Three books that cover aspects of the history of analysis in the eighteenth century are S. B. Engelsman, *Families of Curves and the Origins of Partial Differentiation* (Amsterdam: Elsevier, 1984); Umberto Bottazzini, *The Higher Calculus: A History of Real and Complex Analysis from Euler to Weierstrass* (New York: Springer-Verlag, 1986); and Ivor Grattan-Guinness, *The Development of the Foundations of Mathematical Analysis from Euler to Riemann* (Cambridge: MIT Press, 1970). Two good survey articles dealing with this material are H. J. M. Bos, "Calculus in the Eighteenth Century—The Role of Applications," *Bulletin of the Institute of Mathematics and Its Applications* 13 (1977), 221–227, and Craig Fraser, "The Calculus as Algebraic Analysis: Some Observations on Mathematical Analysis in the 18th Century," *Archive for History of Exact Sciences* 39 (1989), 317–336.

Johann Bernoulli's solution to the brachistochrone problem is available in David Eugene Smith, *A Source Book in Mathematics* (New York: Dover, 1959), vol. 2, 644–655. D. J. Struik, *A Source Book in Mathematics, 1200–1800* (Cambridge: Harvard University Press, 1969) provides translations of the important parts of the papers involved in the controversy over vibrating strings and also of l'Hospital's treatment of l'Hospital's rule. The book also includes selections from Berkeley's *The Analyst*, from Euler's discussion of the metaphysics of the calculus, and from Lagrange's *Theory of Analytic Functions*. Agnesi's *Foundations of Analysis* was translated into English as *Analytical Institutions*, John Colson, trans. (London: Taylor and Wilks, 1801). Euler's *Introduction* has been translated in two volumes as *Introduction to Analysis of the Infinite*, John D. Blanton, trans. (New York: Springer-Verlag, 1988, 1990). The first volume of Euler's *Differential Calculus* is also available in *Foundations of Differential Calculus*, John D. Blanton, trans. (New York: Springer-Verlag, 2000).

A new book detailing the work in Britain and on the continent, based on a study of Newton's *Principia*, is Niccolò Guicciardini, *Reading the Principia: The Debate on Newton's Mathematical Methods for Natural Philosophy from 1687 to 1736* (Cambridge: Cambridge University Press, 1999). Thomas Simpson's life and work is discussed by Frances Marguerite Clarke, *Thomas Simpson and His Times* (New York: Columbia University Press, 1929). For more details on Maclaurin's calculus and its influence, see Judith Grabiner, "Was Newton's Calculus a Dead End? The Continental Influence of Maclaurin's *Treatise of Fluxions*," *American Mathematical Monthly* 104 (1997), 393–410. There is no comprehensive mathematical biography of Leonhard Euler. But a good treatment of many details of his work is C. Truesdell, "The Rational Mechanics of Flexible or Elastic Bodies: 1638–1788," in Leonhard Euler, *Opera Omnia* (Leipzig, Berlin, and Zurich: Societas Scientarum Naturalium Helveticae, 1911–), (2) 11, part 2. In particular, this article is an introduction to a section of Euler's collected works, now totaling over 70 volumes, which have been appearing in four series for over 90 years and which are still not complete. A more detailed treatment of the invention of the calculus of the sine and cosine functions is found in Victor J. Katz, "The Calculus of the Trigonometric Functions," *Historia Mathematica* 14 (1987), 311–324. Various articles on Euler's work are found in a special journal issue on the bicentenary of his death: *Mathematics Magazine* 56 (5) (1983). Also, see E. A. Fellmann, ed., *Leonhard Euler 1707–1783: Beiträge zu Leben und Werk* (Boston: Birkhäuser, 1983). A discussion of the entire matter of vibrating strings and the wave equation can be found in Jerome R. Ravetz, "Vibrating Strings and Arbitrary Functions," a chapter in *The Logic of Personal Knowledge: Essays Presented to Michael Polanyi on his Seventieth Birthday, 11 March 1961* (London: Routledge and Paul, 1961), 71–88, as well as in chapter 1 of Bottazzini's *Higher Calculus* and in C. Truesdell's "Rational Mechanics."

CHAPTER THIRTEEN

Probability and Statistics in the Eighteenth Century

It seems that to make a correct conjecture about any event whatever, it is necessary only to calculate exactly the number of possible cases and then to determine how much more likely it is that one case will occur than another. But here at once our main difficulty arises, for this procedure is applicable to only a very few phenomena. . . . What mortal, I ask, could ascertain the number of diseases, counting all possible cases . . . and say how much more likely one disease is to be fatal than another . . . ? Or who could enumerate the countless changes that the atmosphere undergoes every day and from that predict today what the weather will be a month or even a year from now?

—Jakob Bernoulli, *Ars conjectandi*, 1713

In this chapter, we will consider eighteenth-century developments in probability and statistics. Jakob Bernoulli took up Huygens's work in probability and extended it, ultimately proving what is today known as the **Law of Large Numbers**. Abraham de Moivre carried this work even further by applying his knowledge of series to develop the curve of normal distribution and some of its properties. He was then able to apply his ideas to the pricing of annuities. Euler, Simpson, and others considered questions dealing with observational errors, questions whose eighteenth-century answers were superseded in the nineteenth century. Finally, Thomas Bayes and Pierre-Simon de Laplace attempted to answer questions of inverse probability, that is, how to determine probability from a consideration of certain empirical data. Their answers have proved somewhat controversial ever since.

13.1 Probability

The early work on probability discussed in Chapter 10 was chiefly concerned with the question of determining expectations and the associated probabilities in cases arising from various games or other gambling questions. But Pascal's idea of a "fair" distribution of the stakes in an interrupted game and Huygens's interest in "equitable" games show that probability in its beginnings was closely related to the notion of an **aleatory** contract, a contract providing for the exchange of a present certain value for a future uncertain one. Such contracts included annuities and maritime insurance policies, in which a certain sum of money is paid now in exchange for an unknown sum to be returned at a later date under certain conditions. For the contract to be "fair," the mathematicians argued, one needed to be able somehow to quantify the risk involved.

For certain types of games, early practitioners were able to work out efficient ways of counting successes and failures and thus to determine the expectation or probability *a priori*. In most realistic situations, however, it was much more difficult to quantify risk, that is, to determine the degree of belief that a "reasonable man" would have. How could one determine a "reasonable" price to pay for insurance? As the chapter opening quotation indicates, Jakob Bernoulli, in his study of the subject over some twenty years, wanted to be able to quantify risk in situations where it was impossible to enumerate all possibilities. To do this, he proposed to ascertain probabilities *a posteriori* by looking at the results observed in many similar instances, that is, by considering some statistics.

13.1.1 Jakob Bernoulli and the *Ars Conjectandi*

It seemed reasonably obvious to Bernoulli that the more observations one made of a given situation, the better one would be able to predict future occurrences. But he wanted to give a "scientific proof" of this principle, not only to show that increases in the number of observations enabled the actual probability of the event to be estimated to within any desired degree of accuracy, but also to show how to calculate exactly how many observations were necessary to ensure that the result was within a predetermined interval around the true answer. By the time of his death in 1705, Bernoulli had provided this scientific proof in his Law of Large Numbers. This was included in his important text on probability, the *Ars conjectandi* (*Art of Conjecturing*), a work not published until 1713.

The Law of Large Numbers appears in the fourth and last part of the *Ars conjectandi*. The first three parts are more in the spirit of earlier work on probability. In fact, part one is essentially a reprint of Huygens's 1657 work, with added commentary. Part two develops anew various laws of combinations, most of which were known in previous centuries, and part three applies these laws to solve more problems about games. Bernoulli did, however, generalize Pascal's ideas on the division of stakes in an interrupted game to the case where the two players' chances of winning a game are not equal, or, more generally, to the case of an experiment in which the chances of success or failure are not equal. Bernoulli showed that if the chance of success is a and the chance of failure is b (out of $a + b$ trials), then the probability of r successes in n trials is the ratio of $\binom{n}{n-r}a^r b^{n-r}$ to $(a + b)^n$. Similarly, the probability of at least r successes in n trials is the ratio of $\sum_{j=0}^{n-r} \binom{n}{j} a^{n-j} b^j$ to $(a + b)^n$.

Part four of the *Ars conjectandi* is entitled *On the Use and Applications of the Doctrine in Politics, Ethics, and Economics*. Although Bernoulli did not in fact present any practical applications, he did discuss various kinds of evidence seen in real life and how these pieces of evidence might be combined into a single probability statement. Realizing that in most real situations, absolute certainty (or probability equal to 1) is impossible to achieve, Bernoulli introduced the idea of **moral certainty**. He decided that for an outcome to be morally certain, it should have a probability no less than 0.999. Conversely, an outcome with probability no greater than 0.001 he considered to be morally impossible. It was to determine the moral certainty of the true probability of an event that Bernoulli formulated his theorem, the Law of Large Numbers.

To understand the discussion of the theorem, consider one of Bernoulli's examples. Suppose there is an urn containing 3000 white and 2000 black pebbles, although the observer does not know those numbers. The observer wants to determine the proportion of white to black by taking out, in turn, a certain number of pebbles and recording the outcome, at each step always replacing each pebble before taking out the next. Thus, in what follows, an observation is the removal of one pebble and a success is that the pebble is white. Assume then that N observations are made, that X of these are successes, and that $p = r/(r + s)$ is the (unknown) probability of a success. (Here r is the total of successful cases and s the total of unsuccessful ones. In the example, $p = 3/5$.) Bernoulli's theorem, in modern terminology, states that given any small fraction $\epsilon = 1/(r + s)$ and any large positive number c, a number $N = N(c)$ may be found so that the probability that X/N differs from p by no more than ϵ is greater than c times the probability that X/N differs from p by more than ϵ. In symbols, this result can be written as

$$P\left(\left|\frac{X}{N} - p\right| \leq \epsilon\right) > cP\left(\left|\frac{X}{N} - p\right| > \epsilon\right).$$

In other words, the probability that X/N is "close" to p is very much greater than the probability that it is not "close." Bernoulli's statement can easily be converted into the standard modern formulation: Given any $\epsilon > 0$ and any positive number c, there exists an N such that

$$P\left(\left|\frac{X}{N} - p\right| > \epsilon\right) < \frac{1}{c + 1}.$$

Because Bernoulli considered the basic statement of the theorem virtually intuitive, he felt his main contribution would be to determine the value $N(c)$ from which he could recover the true probability $p = r/(r + s)$ with "moral certainty," that is, with $c = 1000$. He in fact was able to accomplish this. He showed that if $t = r + s$, then $N(c)$ could be taken to be any integer greater than the larger of

$$mt + \frac{st(m - 1)}{r + 1} \qquad \text{and} \qquad nt + \frac{rt(n - 1)}{s + 1},$$

where m, n are integers such that

$$m \geq \frac{\log c(s - 1)}{\log(r + 1) - \log r} \qquad \text{and} \qquad n \geq \frac{\log c(r - 1)}{\log(s + 1) - \log s}.$$

In his example, Bernoulli calculated that for $r = 30$ and $s = 20$, the second expression was larger and therefore $N = 25{,}550$ for $c = 1000$. In other words, Bernoulli's result enabled him to know that 25,550 observations would be sufficient to achieve moral certainty that the relative frequency found would be within 1/50 of the true proportion 3/5.

Bernoulli's text ended with this calculation and similar ones for other values of c, perhaps because he was unhappy with this result. In the early eighteenth century, 25,550 was an enormous number, larger than the entire population of Basel, for example. What the result seemed to say was that nothing reliable could be learned in a reasonable number of observations. Bernoulli may have felt that he had failed in his quest to quantify the measure of uncertainty, especially since his intuition told him that 25,550 was much larger than necessary. He therefore did not include the promised applications of his method to politics and economics. Nevertheless, Bernoulli pointed the way toward a more successful attack on the problem by his slightly younger contemporary, Abraham De Moivre (1667–1754).

13.1.2 De Moivre and *The Doctrine of Chances*

De Moivre's major mathematical work was *The Doctrine of Chances*, first published in 1718, with new editions in 1738 and 1756. This probability text is much more detailed than the work of Huygens, partly because of the general advances in mathematics since 1657. For example, consider the dice problem of de Méré, solved as part of a more comprehensive problem:

PROBLEM III. *To find in how many trials an event will probably happen, or how many trials will be necessary to make it indifferent to lay on its happening or failing, supposing that a is the number of chances for its happening in any one trial and b the number of chances for its failing.*

De Moivre began his solution by noting that if there are x trials, then $b^x/(a+b)^x$ is the probability of the event failing x consecutive times. Since there are to be even odds as to whether the event happens at least once in x trials, this probability must equal $1/2$; that is, x must satisfy the equation

$$\frac{b^x}{(a+b)^x} = \frac{1}{2}, \qquad \text{or} \qquad (a+b)^x = 2b^x.$$

De Moivre easily solved this equation by taking logarithms:

$$x = \frac{\log 2}{\log(a+b) - \log b}.$$

Furthermore, he noted that if $a : b = 1 : q$, so that the odds against a success are q to 1, then the original equation can be rewritten in the form

$$\left(1 + \frac{1}{q}\right)^x = 2, \qquad \text{or} \qquad x \log\left(1 + \frac{1}{q}\right) = \log 2.$$

By expanding $\log[1 + (1/q)]$ in a power series, De Moivre concluded that if q is very large, then the first term $1/q$ of the series is sufficient and the solution can be written as $x = q \log 2$, or $x \approx 0.7q$. Thus, to solve de Méré's problem of finding how many throws of two dice are necessary to give even odds of throwing two 6's, De Moivre simply noted that $q = 35$, so $x = 24.5$. The required number of throws is therefore between 24 and 25, the same answer as Huygens found by a much lengthier calculation.

Like Bernoulli, De Moivre had as one of his important aims to determine the number of cases needed to estimate accurately the probability of an event. To accomplish this, he made detailed calculations of certain binomial coefficients. He initially restricted himself to equally likely occurrences and thus sought to find the probability of $n/2$ successes in n trials, that is, the ratio of the middle term $M = \binom{n}{n/2}$ of $(1+1)^n$ to the sum of all the terms, 2^n, for n large and even. He determined that this ratio $M : 2^n$ approached $2T(n-1)^n/(n^n\sqrt{n-1})$ as n increased without limit, where $T = e/\sqrt{2\pi}$. Since De Moivre knew that if n is large, then $(n-1)^n/n^n = [1-(1/n)]^n$ approximates e^{-1}, he concluded that the ratio $M : 2^n$ is equal to $2/\sqrt{2\pi n}$.

Next, by considering certain power series, DeMoivre found that if Q is a term of the binomial expansion $(1+1)^n$ at a distance t from the middle term M, then, for n large,

$$Q \approx Me^{-(2t^2/n)}.$$

In modern notation, we have

$$P\left(X = \frac{n}{2} + t\right) \approx P\left(X = \frac{n}{2}\right) e^{-(2t^2/n)} = \frac{2}{\sqrt{2\pi n}} e^{-(2t^2/n)}.$$

De Moivre thought of the various values of $Q = P(X = n/2 + t)$ as forming a curve, symmetric about $t = 0$ and with two inflection points. In fact, he calculated that the inflection points of this curve occurred at a distance $\frac{1}{2}\sqrt{n}$ from the maximum term. Thus, De Moivre had found what today is called the **normal curve**, here seen as an approximation to the binomial distribution.

Given his approximation to the individual terms of the binomial expansion and his representation of Q as a curve, De Moivre could approximate the sums of large numbers of such terms by integration. Thus, to find

$$\sum_{t=0}^{k} P\left(X = \frac{n}{2} + t\right),$$

he replaced this sum by

$$\frac{2}{\sqrt{2\pi n}} \int_0^k e^{-(2t^2/n)}\, dt$$

and evaluated the integral by writing the integrand as a power series and integrating term by term. For $k = \frac{1}{2}\sqrt{n}$, the series converged rapidly enough for him to conclude that the sum was equal to 0.341344. He therefore could conclude that, in modern terminology, for large n, the probability that a symmetric binomial experiment would fall within $\frac{1}{2}\sqrt{n}$ of the middle value $\frac{1}{2}n$ was 0.682688. De Moivre then calculated the corresponding values for various other multiples of $\sqrt{n}$. For De Moivre, $\sqrt{n}$ was the unit by which distances from the center were to be measured. He found, in fact, that the accuracy of a probability estimate increased as the square root of the number of experiments.

The discussion above applies only to cases where the chances of an event happening or failing are equal. But De Moivre did sketch a generalization of his method by showing how to approximate terms in $(a+b)^n$, where $a \neq b$. He concluded that if n is large and if M is the greatest term in the binomial expansion, then, first,

$$\frac{M}{(a+b)^n} \approx \frac{a+b}{\sqrt{2\pi abn}},$$

and, second, if Q is a term at distance t from M, then

$$Q \approx Me^{-[(a+b)^2/2abn]t^2}.$$

In modern notation, the first result means that

$$P(X = np) \approx \frac{1}{\sqrt{2\pi p(1-p)n}},$$

and the second means that

$$P(X = np + t) \approx P(X = np)e^{-[t^2/2np(1-p)]},$$

where X is a binomial distribution with n observations and probability of success $p = a/(a+b)$, with np assumed to be an integer.

De Moivre's ideas could be used to show that far fewer experiments are necessary to achieve the accuracy demanded in Bernoulli's example. For example, it can be shown that in the case where Bernoulli required 25,550 trials, De Moivre's method required just 6498. De Moivre himself, however, only gave examples in the equiprobable case. Thus, he showed, for example, that 3600 experiments will suffice to give the probability 0.682688 that an event will occur at least 1770 times and no more than 1830 times or the probability 0.99874 that an event will occur at least 1710 times and no more than 1890 times.

13.2 Applications of Probability to Statistics

13.2.1 Errors in Observations

An important question considered in the eighteenth century was how to deal with observational errors made in astronomy and other fields. It was certainly known that every observation was subject to error; thus, if one wanted to develop a theory, one had to understand the nature of the errors and how to compensate for them. For example, suppose one knows that a particular physical relationship is expressed by a linear function $y = a + bx$. One performs several observations of the phenomenon in question and finds the data points $(x_1, y_1), (x_2, y_2), \ldots, (x_k, y_k)$. Replacing x and y in the equation by each of these k pairs in turn gives k equations for the two unknown coefficients a, b. The system of k linear equations in two unknowns is thus overdetermined and, in general, has no exact solution. The idea, then, is somehow to determine the "best" approximation to a solution. In geometric terms, the problem is to find the straight line that is "closest," in some sense, to passing through the k observed points.

This problem of the combination of observations was discussed by various mathematicians in the eighteenth century, primarily in regard to astronomical observations. Among those who attempted solutions to the problem were Leonhard Euler in 1749, Tobias Mayer (1723–1762) in 1750, and Roger Boscovich (1711–1787) in 1760. Euler, in working on a problem involving the gravitational influence of Jupiter and Saturn on each other's orbits, ended up with a system of seventy-five equations in eight unknowns. He attempted to find the best solution by solving various small sets of his equations and combining the answers. Mayer, on the other hand, in looking at the detailed motion of the moon, had to solve a system of twenty-seven equations in three unknowns. He developed a systematic method of attack by dividing his equations into three groups of nine, adding the equations in each of the groups separately, and then solving the resulting system of three equations in three unknowns. What was not entirely clear was exactly what criteria should be used to divide up the equations. It was Boscovich, however, who made a significant advance in this problem as he dealt with a question involving the true shape of the earth. He stated the criteria that a method of determining the solutions to such systems of equations ought to satisfy, including the important one of minimizing the sum of the absolute values of the errors determined by substituting any particular set of values into the equations. A few years later, Laplace turned Boscovich's method into a detailed algebraic method in his own work on the same problem. Unfortunately, Laplace's method turned out to be difficult to work with and was replaced early in the nineteenth century by the method of least squares.

A related question also considered in the eighteenth century was to find a mathematical description of the error function itself. For example, Thomas Simpson in 1755 attempted to show that the error in observations would be diminished by taking the mean of several observations. He did this by assuming, for example, that the probability of errors of sizes $-5, -4, -3, -2, -1, 0, 1, 2, 3, 4, 5$ (in seconds) in a particular astronomical measurement was proportional to 1, 2, 3, 4, 5, 6, 5, 4, 3, 2, 1, respectively. Thus, the probability that a single error does not exceed 1 second is $16/36 = 0.444$ and that it does not exceed 2 seconds is $24/36 = 0.667$. On the other hand, Simpson calculated that the probability that the mean of six errors does not exceed 1 second is 0.725 and that it does not exceed 2 seconds is 0.967, thus showing the advantage of taking means.

Simpson attempted to generalize this result on taking means to more general error functions. But in the 1770s, Laplace made a more careful analysis by making explicit assumptions on the conditions an error function $\phi(x)$ should meet. These conditions were, first, that $\phi(x)$ should be symmetric about zero, assuming that it is equally probable that an observation is too big as that it is too small; second, that the curve must be asymptotic to the real axis in both directions, because the probability of an infinite error is 0; and third, that the total area under $\phi(x)$ should be 1, since the area under that curve between any two values represents the probability that the observation has error between those values. Unfortunately, there were many functions that satisfied Laplace's requirements. Through the use of various other arguments, Laplace settled on the function $\phi(x) = (m/2)e^{-m|x|}$, for some positive value m. Laplace soon found out, however, that calculations based on this error function led to great difficulties. A better answer was found in the nineteenth century by Gauss.

13.2.2 De Moivre and Annuities

To apply probability to the real world required knowledge of outcomes and their events. One particular application, which was of interest to De Moivre, was the application of probability to the pricing of annuities. Annuities had been sold for centuries, but were generally considered as a bet placed by the annuitant and a loan at interest extended by the seller. That is, the annuitant, who paid a fixed sum for a guarantee of regular payments until his death, was in effect betting that he would live long enough to collect all of his payment and more. The seller, on the other hand, considered the initial payment as a loan and the payoff to be interest, usually at a rate higher than the legal rate for lending money.

Before mathematicians thought about the subject, the pricing of annuities was set by either the experience of the parties involved or by a need of the seller for cash rather than by a consideration of statistics. For example, an English law of 1540 declared that a government annuity is worth seven years' purchase. In other words, the government would sell for A pounds an annuity that would guarantee the annuitant P pounds a year for life, where the P pounds per year was the amount necessary to pay back A pounds, including interest at a fixed rate, in seven years. Apparently, this contract was offered independently of the age or health of the buyer. The relationship between P and A was easy enough to determine, given an interest rate of r. Because $P/(1+r)$ pounds will give you P pounds in one year, $P/(1+r)^2$ will give you P pounds in two years, and so on, we find that

$$A = \frac{P}{1+r} + \frac{P}{(1+r)^2} + \cdots + \frac{P}{(1+r)^7} = \frac{P}{1+r}\sum_{i=0}^{6}\left(\frac{1}{1+r}\right)^i$$

$$= \frac{P}{1+r}\left(\frac{1-\left(\frac{1}{1+r}\right)^7}{1-\frac{1}{1+r}}\right) = \frac{P[(1+r)^7-1]}{r(1+r)^7}.$$

For example, if $P = 1$ and $r = .05$, we calculate that $A = 5.7864$. That is, one could buy a life annuity of 1 pound per year for 5.7864 pounds, assuming interest at 5%, regardless of one's age.

In modern terms, A is called the **present value of an annuity** of P pounds per year for seven years. Of course, it is easy enough to generalize this idea to an annuity for n years. That is, given an interest rate r, an annuity of P pounds per year for n years has a present value of

$$A = \frac{P[(1+r)^n - 1]}{r(1+r)^n}.$$

Again, if interest is at 5%, the present value of an annuity of 1 pound per year for 36 years is 16.5468.

What De Moivre wanted to do was to apply probabilities to this calculation to find a way of pricing annuities that was fairer to both buyer and seller, given that older annuitants were more likely to die sooner than younger ones. So the question for De Moivre was this: How much more likely were older annuitants to die sooner than younger ones? To answer this question, he needed mortality tables. Information for such tables was already being collected in the seventeenth century. For example, John Graunt (1620–1674) collected and analyzed information from the lists of deaths and their causes that began to be compiled in England in the sixteenth century, originally to keep track of the plague. Graunt published his material in *Natural and Political Observations on the Bills of Mortality* in 1662. Other studies of death rates had been done by Jan de Witt in the Netherlands in 1671 and Edmond Halley (using data from Breslau) in 1693. From looking at the data in these studies, De Moivre concluded that for purposes of pricing annuities he could assume, first, that the maximum age a person might live to was 86 years, and, second, that roughly the same number of people who were k years old now would die in each succeeding year. These assumptions can be expressed in terms of probability: If we let $n = 86 - k$, the number of possible years left, the probability of a k-year-old person dying in any given year of the n possible years left would be $1/n$. To put this another way, the probability of a k-year-old person living at least one year is $(n-1)/n$, at least two years is $(n-2)/n$, at least three years is $(n-3)/n$, and so on.

Given his assumptions, De Moivre claimed, in problem 1 of his *Treatise of Annuities on Lives* (1724, 1743, 1750), that the "fair" price of a life annuity of 1 pound per year for a person of age k, given an interest rate r, was

$$Q = \frac{1 - (sA/n)}{s - 1},$$

where $n = 86 - k$, $s = 1 + r$, and A is the present value of an annuity of 1 pound for n years. To prove this assertion, De Moivre began by noting that since the probability of having to pay back 1 pound at the end of the first year is $(n-1)/n$, we need only invest enough to pay back that amount, namely, $(n-1)/ns$. Similarly, the probability of having to pay back 1 pound after two years is $(n-2)/n$. The amount we need to invest today to do that is $(n-2)/ns^2$. Therefore, to be sure we have enough to pay back 1 pound per year for life, we need to invest the sum

$$\frac{n-1}{ns} + \frac{n-2}{ns^2} + \frac{n-3}{ns^3} + \cdots + \frac{n-(n-1)}{ns^{n-1}}.$$

This sum is then the present value Q of a life annuity of 1 pound per year for a person of age k years, where $n = 86 - k$.

To transform this formula into the closed form for Q, De Moivre took the present value formula for an annuity of 1 pound per year for n years:

$$A = \frac{1}{s} + \frac{1}{s^2} + \frac{1}{s^3} + \cdots + \frac{1}{s^n}.$$

Then he multiplied A by s/n and subtracted this value from 1:

$$1 - \frac{s}{n}A = \frac{n-1}{n} - \frac{1}{ns} - \frac{1}{ns^2} - \cdots - \frac{1}{ns^{n-1}}.$$

To get Q, De Moivre needed to multiply this formula by $1/(s-1)$, which he rewrote as

$$\frac{1}{s} + \frac{1}{s^2} + \frac{1}{s^3} + \cdots$$

He then multiplied in columns:

$$\begin{array}{cccccc}
\frac{n-1}{ns} & -\frac{1}{ns^2} & -\frac{1}{ns^3} & -\frac{1}{ns^4} & -\frac{1}{ns^5} & -\cdots \\
 & +\frac{n-1}{ns^2} & -\frac{1}{ns^3} & -\frac{1}{ns^4} & -\frac{1}{ns^5} & -\cdots \\
 & & +\frac{n-1}{ns^3} & -\frac{1}{ns^4} & -\frac{1}{ns^5} & -\cdots \\
 & & & +\frac{n-1}{ns^4} & -\frac{1}{ns^5} & -\cdots \\
 & & & & +\frac{n-1}{ns^5} & -\cdots
\end{array}$$

If the terms are added vertically, the result is

$$\frac{n-1}{ns} + \frac{n-2}{ns^2} + \frac{n-3}{ns^3} + \frac{n-4}{ns^4} + \frac{n-5}{ns^5} + \cdots,$$

thus proving De Moivre's result.

As an example, because we know that the present value of an annuity of 1 pound per year for 36 years at 5% interest is 16.5468, we can calculate the present value of a life annuity of 1 pound per year at 5% for a 50-year-old person. We set $A = 16.5468$ and $n = 86 - 50 = 36$. The present value Q is then

$$Q = \frac{1 - 1.05 \cdot 16.5468/36}{.05} = 10.3477.$$

Therefore, to get a life annuity of 1 pound per year at 5%, a 50-year-old person must pay 10.3477 pounds, based on de Moivre's mortality assumptions. De Moivre's treatise contained tables to calculate life annuities at ages from 1 to 86, with interest rates of 3%, $3\frac{1}{2}$%, 4%, 5%, and 6%.

13.2.3 Bayes and Statistical Inference

De Moivre's work on annuities was important, but his general work on probability was not applied to practical problems because it did not directly answer the question necessary for applications, the question of statistical inference: Given empirical evidence that a particular event happened a certain number of times in a given number of trials, what is the probability of this event happening in general? De Moivre (and Bernoulli) could only tell how likely it was that observed frequencies approximated a given probability. The first person to attempt a direct answer to the question of how to determine probability from observed frequencies was Thomas Bayes (1702–1761) in his *An Essay towards Solving a Problem in the Doctrine of Chances*, written toward the end of his life and not published until three years after his death.

Bayes began his essay with a statement of the basic problem: "Given the number of times in which an unknown event [i.e., an event of unknown probability] has happened and failed. Required the chance that the probability of its happening in a single trial lies somewhere between any two degrees of probability that can be named." In modern notation, if X represents the number of times the event has happened in n trials, x the probability of its happening in a single trial, and r and s the two given probabilities, Bayes's aim was to calculate $P(r < x < s|X)$, that is, the probability that x is between r and s, given X. Bayes proceeded to develop axiomatically, from a definition of probability, the two basic results about the probabilities of two events E and F that he would need. First, he showed that $P(E \cap F) = P(E)P(F|E)$, that is, the probability of both events happening is the product of the probability of E with the probability of F given E. Second, he proved what is today generally called **Bayes's theorem**, that $P(E|F) = P(E \cap F)/P(F)$. In words, the probability of E given that F has happened is the quotient of the probability of both happening divided by the probability of F alone. Bayes's basic problem, then, was the calculation of $P(E|F)$, where E is the event "$r < x < s$" and F is the event "X successes in n trials." To apply his theorem to this calculation, Bayes needed a way of determining the two probabilities $P(E \cap F)$ and $P(F)$.

Bayes naturally knew Bernoulli's result that if the probability of a success is a and that of a failure is b, then the probability of p successes and q failures in $n = p + q$ trials is $\binom{n}{q}a^p b^q$. To calculate sums of these terms, Bayes used De Moivre's approach of integration to attack the problem directly. Thus, he began by modeling the probabilities by a certain area:

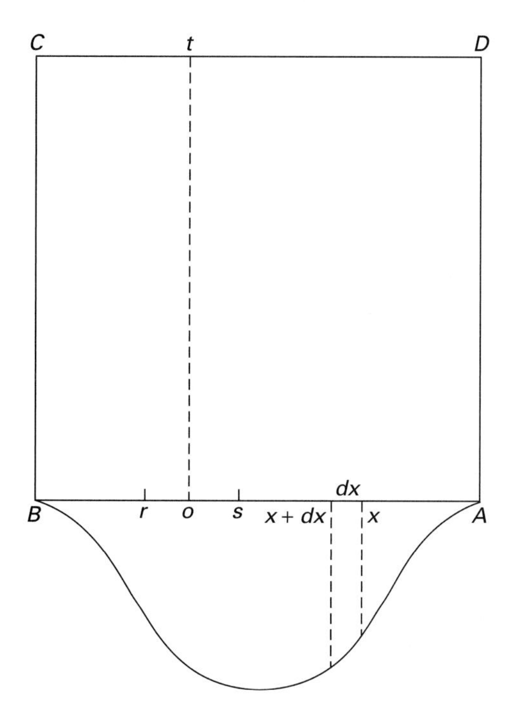

FIGURE 13.1 Bayes's theorem

> I suppose the square table ... $ABCD$ [of side 1] [Fig. 13.1] to be so made and leveled, that if either of the balls O or W be thrown upon it, there shall be the same probability that it rests upon any one equal part of the plane as another. ... I suppose that the ball W shall be first thrown, and through the point where it rests a line ot shall be drawn parallel to AD, and meeting CD and AB in t and o; and that afterwards the ball O shall be thrown $p+q$ or n times, and that its resting between AD and ot after a single throw be called the happening of the event M in a single trial.

In terms of the basic problem, the position of W determines the probability x. Bayes noted that the probability of the point o falling between any two points r and s is simply the length rs. Similarly, the probability of the event M given that W has been thrown is the length of Ao.

To calculate $P(E \cap F)$, Bayes first noted that any given probability range for the point o is represented by an interval on the axis AB, say, $[x, x+dx]$, measured from A. Because a particular x represents the probability of the ball landing to the right of ot, $1-x$ represents the probability of it landing to the left. The probability that the ball will land to the right p times in $p+q=n$ throws is thus given by $y = \binom{n}{q} x^p (1-x)^q = \binom{n}{p} x^p (1-x)^{n-p}$. Bayes drew the curve given by this function below the axis AB and used his first proposition to conclude that the probability of W lying above the coordinate interval $[x, x+dx]$ and the ball landing p times to the right of W is represented by the area under $[x, x+dx]$ and above the curve. It follows that $P(E \cap F) = P((r < x < s) \cap (X = p))$ is represented by the total area under the interval $[r, s]$ and above the curve, or, in modern notation, by

$$\int_r^s \binom{n}{p} x^p (1-x)^{n-p}\,dx.$$

Because $P(F) = P(X = p)$ can be thought of as $P((0 < x < 1) \cap (X = p))$, it follows from the above argument that $P(X = p)$ is represented by the entire area under the axis AB and above the curve, or by

$$\int_0^1 \binom{n}{p} x^p (1-x)^{n-p}\,dx.$$

Bayes's theorem then showed how to calculate $P(E|F)$:

$$P(E|F) = P((r < x < s)|(X = p)) = \frac{\displaystyle\int_r^s \binom{n}{p} x^p (1-x)^{n-p}\,dx}{\displaystyle\int_0^1 \binom{n}{p} x^p (1-x)^{n-p}\,dx}.$$

Although Bayes's problem was, in fact, formally solved, there were two obstacles to be overcome before one could consider the solution as a practical one. First, did Bayes's physical analogy of rolling balls on a table truly mirror the actual problems to which the theory would be applied? Could nature's choice of an unknown probability x really be the same as the rolling of a ball across a level table? Bayes answered this question by, in effect, restricting the application of the rule to just those circumstances in which, for any given number n of trials, all possible outcomes $X = 0$, $X = 1$, $X = 2, \ldots$ are equally likely. But is ignorance of the probabilities in a given situation equivalent to all possible outcomes being equally likely? This question has been debated extensively since Bayes's time.

Second, could the integrals in Bayes's formula actually be calculated? Bayes attempted to do so by expanding the integrands in power series. The integral in the denominator turned out to be $1/(n+1)$. The integral in the numerator, while not difficult to approximate when either p or $n - p$ is small, turned out to be very difficult otherwise. For example, if $p = n$, then the relevant quotient is

$$\frac{\displaystyle\int_r^s x^n\,dx}{\displaystyle\int_0^1 x^n\,dx} = s^{n+1} - r^{n+1}.$$

So, suppose nothing is known about an event M except that it has happened once. The chance that the unknown probability x of M is greater than $1/2$—that is, between $1/2$ and 1—is then $1^2 - (1/2)^2 = 3/4$. Similarly, if M has happened twice, the probability that x is greater than $1/2$ is $7/8$; in other words, the odds are 7 to 1 that there is more than an even chance of it happening. In this same situation, the odds are still better than even that the probability of x is greater than $2/3$.

13.2.4 The Calculations of Laplace

Bayes's formula did provide a start in answering the basic question of statistical inference. Further progress was made a few years later by Pierre-Simon de Laplace (1749–1827). In 1774, Laplace, using principles similar to those of Bayes, derived essentially the same result involving integrals for determining probability, given empirical evidence. Putting the question in terms of drawing tickets from an urn, he supposed that p white and q black tickets had been drawn from an urn containing an unknown proportion x of white tickets. Given any guessed value for x, Laplace showed how to calculate the probability that x differed from $p/(p+q)$ by as small a value ϵ as one wished. He was in fact able to demonstrate that

$$P\left(\left|x - \frac{p}{p+q}\right| \leq \epsilon | X = p\right) \cong \frac{2(p+q)^{3/2}}{\sqrt{2\pi}\sqrt{pq}} \int_0^{\epsilon} e^{-[(p+q)^3/2pq]z^2}\, dz$$

$$\cong \frac{2}{\sqrt{2\pi}} \int_0^{\epsilon/\sigma} e^{-(u^2/2)}\, du,$$

where $\sigma^2 = pq/(p+q)^3$. To show that this probability approached 1 as $p+q$ became large, whatever the value of ϵ, Laplace had to integrate $\int_0^\infty e^{-(u^2/2)}\, du$. Using a result of Euler's, he was in fact able to show that this integral equaled $\sqrt{\pi/2}$ and therefore established his result.

To go further in calculating, naturally, Laplace had to evaluate the integral $\int_0^T e^{-(u^2/2)}\, du$ for arbitrary T. This he did in 1785 by deriving two different series for this integral, one that converged rapidly for small T and one that did so for large T. He then applied his results to a genuine problem in statistical inference. During the 26-year period from 1745 to 1770, 251,527 boys and 241,945 girls had been born in Paris. Setting x as the probability of a male birth, Laplace made a straightforward calculation and demonstrated that the probability that $x \leq 1/2$ was 1.15×10^{-42}. He therefore concluded that it was "morally certain" that $x > 1/2$. He then extended his analysis using similar data from London to show that it was also "morally certain" that the probability of a male birth in London was greater than that in Paris.

Exercises

1. Suppose that a is the probability of success in an experiment and $b = 1 - a$ is the probability of failure. If the experiment is repeated three times, show that the probabilities of the number of successes S being 3, 2, 1, 0 are given, respectively, by $P(S = 3) = 1a^3$, $P(S = 2) = 3a^2b$, $P(S = 1) = 3ab^2$, and $P(S = 0) = 1b^3$.

2. Generalize exercise 1 to the case of n trials. Show that the probability of r successes is

$$P(S = r) = \binom{n}{n-r} a^r b^{n-r}.$$

3. Using the results of exercise 2, with $a = 1/3, b = 2/3$, and $n = 10$, calculate $P(4 \leq S \leq 6)$.

4. Complete Bernoulli's calculation of his example for the Law of Large Numbers by showing that if $r = 30$ and $s = 20$ (and therefore $t = 50$) and if $c = 1000$, then

$$nt + \frac{rt(n-1)}{s+1} > mt + \frac{st(m-1)}{r+1},$$

where m, n are integers such that

$$m \geq \frac{\log c(s-1)}{\log(r+1) - \log r}$$

and

$$n \geq \frac{\log c(r-1)}{\log(s+1) - \log s}.$$

Conclude that in this case the necessary number of trials is $N = 25{,}550$.

5. Use Bernoulli's formula to show that if one wants greater certainty in the problem in exercise 4, say, $c = 10{,}000$, then $N = 31{,}258$ trials are necessary.

6. Convert Bernoulli's statement of the Law of Large Numbers,
$$P\left(\left|\frac{X}{N} - p\right| \leq \epsilon\right) > cP\left(\left|\frac{X}{N} - p\right| > \epsilon\right),$$
into the modern statement,
$$P\left(\left|\frac{X}{N} - p\right| > \epsilon\right) < \frac{1}{c+1}.$$

7. Suppose that the probability of success in an experiment is 1/10. How many trials of the experiment are necessary to ensure even odds on it happening at least once? Calculate this by both De Moivre's exact method and his approximation.

8. In a lottery in which the ratio of the number of losing tickets to the number of winning tickets in 39 : 1, how many tickets should one buy to give oneself even odds of winning a prize?

9. Generalize De Moivre's procedure in problem III (of his text) to solve problem IV: To find how many trials are necessary to make it equally probable that an event will happen twice, supposing that a is the number of chances for its happening in any one trial and b the number of chances for its failing to happen. (Hint: Note that $b^x + xab^{x-1}$ is the number of chances in which the event may succeed no more than once, while $(a+b)^x$ is the total number of chances.) Approximate the solution for the case where $a : b = 1 : q$, with q large, and show that $x \approx 1.678q$.

10. De Moivre's result developing the normal curve implies that the probability P_ϵ of an observed result lying between $p - \epsilon$ and $p + \epsilon$ in n trials is given by
$$P_\epsilon = \frac{1}{\sqrt{2\pi np(1-p)}} \int_{-n\epsilon}^{n\epsilon} e^{-[t^2/2np(1-p)]}\, dt.$$
Change variables by setting $u = t/\sqrt{np(1-p)}$, and use symmetry to show that this integral may be rewritten as
$$P_\epsilon = \frac{2}{\sqrt{2\pi}} \int_0^{\sqrt{n}\epsilon/\sqrt{p(1-p)}} e^{-(1/2)u^2}\, du.$$
Calculate this integral for Bernoulli's example, using $p = .6$, $\epsilon = .02$, and $n = 6498$ and show that in this case $P_\epsilon = 0.999$, a value giving moral certainty. (Use a graphing utility.) Find a value for n that gives $P_\epsilon = 0.99$.

11. With interest at 4%, what is the present value of an annuity of 1 pound per year for 50 years?

12. With interest at 4%, what is the present value of a life annuity of 1 pound per year for someone aged 36?

13. Calculate $P(r < x < s | X = n-1)$ explicitly, using Bayes's theorem. In particular, suppose that you have drawn 10 white balls and 1 black ball from an urn containing an unknown proportion of white to black balls. If you now guess that this unknown proportion is greater than 7/10, what is the probability that your guess is correct?

14. Show that if an event of unknown probability happens n times in succession, the odds are $2^{n+1} - 1$ to 1 for more than an even chance of its happening again.

15. Imagine an urn with two balls, each of which may be either white or black. One of these balls is drawn and is put back before a new one is drawn. Suppose that in the first two draws, white balls have been drawn. What is the probability of drawing a white ball on the third draw?

16. The so-called St. Petersburg Paradox was a topic of debate among mathematicians studying probability theory in the eighteenth century. The paradox involves the following game between two players: Player A flips a coin until a tail appears. If it appears on his first flip, player B pays him 1 ruble. If it appears on the second flip, B pays 2 rubles, on the third flip, 4 rubles, ..., on the nth flip, 2^{n-1} rubles. What amount should A be willing to pay B for the privilege of playing? Show first that A's expectation, namely, the sum of the probabilities for each possible outcome of the game multiplied by the payoff for each outcome, is
$$\sum_{i=0}^{\infty} \frac{1}{2^i} 2^{i-1},$$
and then show that this sum is infinite. Next, play the game 10 times and calculate the average payoff. What would you be willing to pay to play? Why does the concept of expectation seem to break down in this instance?

17. Outline a lesson for a statistics course deriving Bayes's theorem and discussing its usefulness.

References

There are several good books on the early history of probability and statistics, including the works of Hacking and David cited in Chapter 10. A newer book that gives great detail on the mathematics of the first workers in probability, although often in modern language, is Anders Hald, *A History of Probability and Statistics and their Applications Before 1750* (New York: John Wiley, 1990). Another good treatment of the early history of statistics, which naturally also includes work on probability, is Stephen M. Stigler, *The History of Statistics* (Cambridge: Harvard University Press, 1986). A more philosophical treatment, concentrating on the ideas behind the notion of probability, is Lorraine Daston, *Classical Probability in the Enlightenment* (Princeton: Princeton University Press, 1988).

Brief selections from Jakob Bernoulli's *Ars conjectandi* are available in James Newman, ed., *The World of Mathematics* (New York: Simon and Schuster, 1956), vol. 3, 1452–1455, and D. J. Struik, *A Source Book in Mathematics, 1200–1800* (Cambridge: Harvard University Press, 1969), 316–320. A brief look at the entire *Ars conjectandi* is found in Ian Hacking, "Jacques Bernoulli's *Art of Conjecturing*," *British Journal for the History of Science* 22 (1971), 209–229. A reprint edition of Abraham De Moivre, *The Doctrine of Chances*, 3rd ed. (New York: Chelsea, 1967), is well worth perusing for its many problems and examples. Thomas Bayes, "An Essay towards Solving a Problem in the Doctrine of Chances," has been reprinted with a biographical note by G. A. Barnard in E. S. Pearson and M. G. Kendall, *Studies in the History of Statistics and Probability* (London: Griffin, 1970), 131–154. Two analyses of the question of the application of Bayes's results are presented in Stigler, *History of Statistics*, 122–131, and in Donald A. Gillies, "Was Bayes a Bayesian?" *Historia Mathematica* 14 (1987), 325–346. Laplace's earliest memoir on probability has been translated by Stephen Stigler as "Laplace's 1774 Memoir on Inverse Probability," *Statistical Science* 1 (1986), 359–378.

CHAPTER FOURTEEN

Algebra and Number Theory in the Eighteenth Century

All the pains that have been taken in order to resolve equations of the fifth degree, and those of higher dimensions, ... or, at least, to reduce them to inferior degrees, have been unsuccessful; so that we cannot give any general rules for finding the roots of equations which exceed the fourth degree.

—Leonhard Euler's *Algebra*, 1767

There were few major new developments in algebra in the eighteenth century, in contrast to other fields. The major effort, accomplished by mathematicians whose chief influence was felt in other areas, was a systematization of earlier material. For example, there were methods to solve systems of linear equations, as well as methods to solve algebraic equations of degree up to four. But these methods were *ad hoc*, so mathematicians sought more general procedures, for which some theoretical analysis was necessary. In this chapter, we will look at the general procedures that were developed to solve these two classes of equations. In addition, we will consider some of Euler's work in number theory, which was to have important consequences in subsequent years.

14.1 Systems of Linear Equations

The earliest systematic treatment of solutions of systems of linear equations is found in Maclaurin's *A Treatise of Algebra in Three Parts*, written in the 1730s but not published until 1748, after the author's death. Maclaurin began by showing that the solutions to two and three equations in the same number of unknowns can be found by solving for one unknown in terms of the others and substituting. He noted that if there are more unknowns than equations, there may be an infinite number of solutions, while in the opposite case, there may be no solutions at all; however, he did not give any examples of either situation.

Maclaurin did, however, present what he called a "general theorem" for eliminating unknowns in a system of equations, the method known today as **Cramer's rule**, after the Swiss mathematician Gabriel Cramer (1704–1752), who used it in his 1750 book *Introduction to the Analysis of Algebraic Curves*. If

$$\begin{aligned} ax + by &= c \\ dx + ey &= f \end{aligned},$$

then solving the first equation for x and substituting gives

$$y = \frac{af - dc}{ae - db},$$

and a similar answer for x. The system of three equations

$$\begin{aligned} ax + by + cz &= m \\ dx + ey + fz &= n \\ gx + hy + kz &= p \end{aligned}$$

is dealt with by first solving each equation for x, thus reducing the problem to a system in two unknowns, and then using the earlier rule to find

$$z = \frac{aep - ahn + dhm - dbp + gbn - gem}{aek - ahf + dhc - dbk + gbf - gec}.$$

In addition to giving the answer, Maclaurin noted the general rule that the numerator consists of the various products of the coefficients of x and y as well as the constant terms, each product consisting of one coefficient from each equation, while the denominator consists of products of the coefficients of all three unknowns. He also explained how to determine the sign of each term. Furthermore, he solved for y and x and showed that the general rule determining each of these values is analogous to the one for z. In particular, each of the three expressions has the same denominator. Maclaurin even extended the rule to systems of four equations in four unknowns, but did not discuss any further generalization.

The numerators and denominators in Maclaurin's solutions are, of course, what are known today as **determinants**. But the use of such combinations of coefficients as tools for solving systems of linear equations had appeared somewhat earlier. Leibniz had suggested a similar idea in a letter to l'Hospital in 1693 and had even devised a way of indexing the coefficients of the system by the use of numbers. In his book, Cramer generalized Maclaurin's work to systems of n linear equations in n unknowns, for any n, in connection with finding the coefficients of an algebraic curve passing through given points. In particular, he gave a clear description of how to construct the determinant of a given system.

Also in the mid–eighteenth century, Euler was exploring some more general ideas for solving systems of equations. In a paper in 1750, his concern was to resolve a paradox that Cramer had formulated, based on an earlier suggestion of Maclaurin. Cramer's paradox was based on two propositions that everyone in the early eighteenth century believed:

1. An algebraic curve of order n is uniquely determined by $n(n + 3)/2$ of its points.
2. Two algebraic curves of order n and m intersect in nm points.

The first result comes from elementary combinatorics. Basically, a curve of order n, one described by an nth-degree polynomial in two variables, has one coefficient of degree 0, two of degree 1, three of degree 2, four of degree 3, and so on. But since we can divide by any one coefficient, the total number of "independent" coefficients is

$$\sum_{i=1}^{n+1} i - 1 = \frac{(n+1)(n+2)}{2} - 1$$
$$= \frac{n(n+3)}{2},$$

and thus that many points are necessary to determine the curve. As for the second result, although it was known that the points of intersection of algebraic curves may be multiple or imaginary, examples were known where all mn points were real and distinct. The paradox then arose from consideration of the case $n \geq 3$, where it appears from the second proposition that there are n^2 points common to two algebraic curves of order n, while the first proposition implies that $n(n+3)/2$ (which is less than n^2) points should determine a unique curve.

Euler discussed the paradox and concluded that the first result, based on the fact that n linear equations in n unknowns should determine a unique n-fold solution, was not true without restriction. The conviction at the time that n equations determined n unknowns was so strong that no one earlier had taken the pains to discuss the cases where this did not happen. As noted, Maclaurin briefly discussed the situation where the number of equations was not equal to the number of unknowns, but he had not considered the possibilities in detail.

In his paper, Euler discussed various examples, without, however, being able to state a definite theorem. For example, he noted that $3x - 2y = 5, 4y = 6x - 10$ do not determine two unknowns, because if we solve for x and substitute, the equation for y becomes an identity, which does not allow us to determine a value. He also gave a system of four equations in four unknowns, which, after solving for two of the variables in terms of the others and substituting into the remaining two equations, again resulted in identities, so that the remaining two values were undetermined. Thus, the four equations do not determine four unknowns. So, he concluded, to the assertion that to determine n unknowns it is sufficient to have n equations, it is necessary to add the restriction that these equations must be so different that none of them is already "comprised" in the others. Although Euler did not explicitly define "comprise," it seems that, at least intuitively, he understood the concept of the rank of a system.

To resolve Cramer's paradox, Euler finally noted that "when two curves of fourth order meet in 16 points, as 14 points, when they lead to different equations, are sufficient to determine one curve of this order, these 16 points will always be such that three or more equations are already comprised in the others. In this way, these 16 points do not determine more than if they were 13 or 12 or even fewer points and in order to determine the curve entirely, one must add to these 16 points one or two others."

Although Euler had resolved the immediate paradox, it took over a century and a quarter for mathematicians to completely understand the ideas involved in undetermined or inconsistent systems.

14.2 Polynomial Equations

The second part of Maclaurin's algebra text is a treatise on solving polynomial equations, which presents all that had been discovered up to his time in well-organized form. Maclaurin included not only Cardano's rule for solving cubics and Ferrari's rule for quartics, but also Newton's method for approximating numerically the solution to an equation. He noted that the procedure by which equations are generated—multiplying together equations of the form $x - a = 0$ or other equations of degree smaller than the given one—showed that no equation can have more roots than the degree of the highest power. Furthermore, "roots become impossible [complex] in pairs" and therefore "an equation of an odd dimension has always one real root." He then discussed the general procedure for finding integral roots of monic polynomials: Check all divisors of the constant term as possible roots, and, if one such root α is found, divide the polynomial by $x - \alpha$ to reduce the degree.

Euler, in his *Complete Introduction to Algebra* of 1767, gave an even fuller treatment of the solution of such equations. In fact, he defined algebra itself as "the science which teaches how to determine unknown quantities by means of those that are known." Interestingly, he devoted considerable space in his text to explaining how to set up problems as equations. As he wrote, "in algebra, when we have a question to resolve, we represent the number sought by one of the last letters of the alphabet, and then consider in what manner the given conditions can form an equality between two quantities. This equality is represented by a kind of formula, called an *equation*, which enables us finally to determine the value of the number sought, and consequently to resolve the question." He then very systematically took his readers through the algebraic solution of equations of degrees 1, 2, 3, and 4, before concluding with the statement given in the opening of the chapter.

Lagrange, however, was not happy with leaving unresolved this question of how to solve higher-degree polynomial equations. Thus, in his *Reflections on the Algebraic Theory of Equations* of 1770, he began a new phase in the study of this subject by undertaking a detailed review of solution methods for cubics and quartics, to determine why they worked. Although he was not able to find analogous methods for higher-degree equations, he was able to sketch a new set of principles for dealing with these equations, which he hoped might ultimately succeed.

Lagrange began with a systematic study of the methods of solution of the cubic equation $x^3 + nx + p = 0$, starting essentially with Cardano's procedure. Setting $x = y - (n/3y)$ transforms this equation into the sixth-degree equation $y^6 + py^3 - (n^3/27) = 0$, which, with $r = y^3$, reduces in turn to the quadratic equation $r^2 + pr - (n^3/27) = 0$. This latter equation has two roots, r_1 and $r_2 = -(n/3)^2(1/r_1)$. But whereas Cardano took the sum of the real cube roots of r_1 and r_2 as his solution, Lagrange knew that each equation $y^3 = r_1$ and $y^3 = r_2$ had three roots. Thus, there were six possible values for y, namely, $\sqrt[3]{r_1}$, $\omega\sqrt[3]{r_1}$, $\omega^2\sqrt[3]{r_1}$, $\sqrt[3]{r_2}$, $\omega\sqrt[3]{r_2}$, and $\omega^2\sqrt[3]{r_1}$, where $\omega = \big(-1 + \sqrt{-3}\big)/2$ is a complex root of $x^3 - 1 = 0$ or of $x^2 + x + 1 = 0$. Lagrange could then show that the three distinct roots of the original equation were given by

$$x_1 = \sqrt[3]{r_1} + \sqrt[3]{r_2}$$

$$x_2 = \omega\sqrt[3]{r_1} + \omega^2\sqrt[3]{r_2}.$$

$$x_3 = \omega^2\sqrt[3]{r_1} + \omega\sqrt[3]{r_2}$$

Lagrange next noted that rather than consider x as a function of y, one could reverse the procedure, because the equation for y, which he called the *reduced equation*, was the one whose solutions enabled the original equation to be solved. The idea then was to express those solutions in terms of the original ones. Thus, Lagrange noted that any of the six values for y could be expressed in the form $y = \frac{1}{3}(x' + \omega x'' + \omega^2 x''')$, where (x', x'', x''') was some permutation of (x_1, x_2, x_3). It was this introduction of the permutations of the roots of an equation that provided the cornerstone not only for Lagrange's method but for the methods others were to use in the next century.

In the case of the cubic equation, there are several important ideas to note. First, the six permutations of the x_i lead to the six possible values for y and thus show that y satisfies an equation of degree 6. Second, the permutations of the expression for y can be divided into two sets, one consisting of the identity permutation and the two permutations that interchange all three of the x_i, and the second consisting of the three permutations that interchange just two of the x_i. (In modern terminology, the group of permutations of a set of three elements has been divided into two cosets.) For example, if $y_1 = \frac{1}{3}(x_1 + \omega x_2 + \omega^2 x_3)$, then the two nonidentity permutations in the first set change y_1 to $y_2 = \frac{1}{3}(x_2 + \omega x_3 + \omega^2 x_1)$ and $y_3 = \frac{1}{3}(x_3 + \omega x_1 + \omega^2 x_2)$, respectively. But then $\omega y_2 = \omega^2 y_3 = y_1$ and $y_1^3 = y_2^3 = y_3^3$. Similarly, if the results of the permutations of the second set are y_4, y_5, and y_6, it follows that $y_4^3 = y_5^3 = y_6^3$. Thus, because there are only two possible values for $y^3 = \frac{1}{27}(x' + \omega x'' + \omega^2 x''')^3$, the equation for y^3 is of degree 2. Finally, the sixth-degree equation satisfied by y has coefficients that are rational in the coefficients of the original equation. Lagrange considered several other methods of solution of the cubic equation but found in each case the same underlying idea. Each led to a rational expression in the three roots that took on only two values under the six possible permutations, thus showing that the expression satisfied a quadratic equation.

Lagrange next considered the solutions of the quartic equation. Ferrari's method of solving $x^4 + nx^2 + px + q = 0$ was to add $2yx^2 + y^2$ to each side, rearrange, and then determine a value for y such that the right side of the new equation,

$$x^4 + 2yx^2 + y^2 = (2y - n)x^2 - px + y^2 - q,$$

was a perfect square. After taking square roots of each side, he could then solve the resulting quadratic equations. The condition that the right side be a perfect square is that

$$(2y - n)(y^2 - q) = \left(\frac{p}{2}\right)^2, \qquad \text{or} \qquad y^3 - \frac{n}{2}y^2 - qy + \frac{4nq - p^2}{8} = 0.$$

The reduced equation is a cubic, which can, of course, be solved. Given the three solutions for y, Lagrange then showed, as in the previous case, that each is a permutation of a rational function of the four roots x_1, x_2, x_3, x_4 of the original equation. In fact, it turns out that $y_1 = \frac{1}{2}(x_1x_2 + x_3x_4)$ and that the twenty-four possible permutations of the x_i lead to only three different values for that expression: y_1, $y_2 = \frac{1}{2}(x_1x_3 + x_2x_4)$, and $y_3 = \frac{1}{2}(x_1x_4 + x_2x_3)$. The expression must therefore satisfy a third-degree equation, again one with coefficients rational in the coefficients of the original equation.

Having studied the methods for solving cubics and quartics, Lagrange was ready to generalize. First, as was clear from the discussion of cubic equations, the study of the roots of equations of the form $x^n - 1 = 0$ was important. For the case of odd n, Lagrange could show that all the roots could be expressed as powers of one of them. In particular, if n is prime and $\alpha \neq 1$ is one of the roots, then α^m for any $m < n$ can serve as a generator of all of the roots. Second, however, Lagrange realized that to attack the problem of equations of degree n, he needed a way of determining a reduced equation of degree $k < n$. Such an equation must be satisfied by certain functions of the roots of the original equation, functions that take on only k values when the roots are permuted by all $n!$ possible permutations. Because relatively simple functions of the roots did not work, Lagrange attempted to find some general rules for determining such functions and the degree of the equation they would satisfy.

Lagrange noted that if the values of the roots of the reduced equation are $f_1, f_2, \ldots, f_k$, where each f_i is a function of the n roots of the original equation, then the equation itself is given by $(t - f_1)(t - f_2)\cdots(t - f_k) = 0$. Although he could not prove that the degree of this equation in general is less than $n!$, he was able to show that its degree k, the number of different values taken by f under the permutations of the variables, always divided $n!$. One can read into this statement Lagrange's theorem to the effect that the order of any subgroup of a group divides the order of the group, but Lagrange never treated permutations as a "group" of operations. He did go on, however, to show how functions of the roots may be related. He proved that if all permutations of the roots that leave one such function u unchanged also leave another such function v unchanged, then v can be expressed as a rational function of u and the coefficients of the original equation. Furthermore, if u is unchanged by permutations that do change v, and if v takes on r different values for each value taken on by u, then v is the root of an equation of degree r whose coefficients are rational in u and the coefficients of the original equation. For example, in the cubic equation $x^3 + nx + p = 0$, the expression $v = \frac{1}{27}(x_1 + \omega x_2 + \omega^2 x_3)^3$ takes on two values under the six permutations of the roots, while $u = x_1 + x_2 + x_3$ is unchanged under those permutations. Then $v^2 + pv - (n^3/27) = 0$ is the equation satisfied by v. (Note here that $u = 0$.)

Lagrange presumably hoped to use this theorem to solve the general polynomial equation of degree n. He would start with a symmetric function of the roots, say, $u = x_1 + x_2 + \cdots + x_n$, which was unchanged under all $n!$ permutations, then find a function v that takes on r different values under these permutations. Thus, v would be a root of an equation of degree r with coefficients rational in the original coefficients (because the given symmetric function u was one of those coefficients). If that equation could be solved, then he could find a new function w that takes on, say, s values under the permutations that leave v unchanged. Thus, w would satisfy an equation of degree s. He would continue in this way until he reached the function x_1. Unfortunately, Lagrange was unable to find a general method of determining the intermediate functions such that they were of a form that could be solved by known methods. He was thus forced to abandon his quest. Nevertheless, his work did form the foundation on which all nineteenth-century work on the algebraic solution of equations was based.

14.3 Number Theory

The final part of Euler's *Algebra* is devoted to the solution of indeterminate equations. For example, Euler deals with techniques for finding solutions, in either rational numbers or integers, to equations of the form $p(x) = y^2$ where $p(x)$ is a polynomial of degree 2, 3, or 4. As a special case, he considered the solution in integers of the equation $Dx^2 + 1 = y^2$, which had been dealt with centuries earlier in India. Rather than presenting a general method of solution, however, Euler demonstrated a procedure to be applied in each case separately. He then concluded his discussion by presenting a table in which he listed solutions to the equation for values of D from 2 to 100. Although Euler did not prove that solutions exist for every D, such a proof was given by Lagrange in 1766 and included as an appendix in later editions of the *Algebra*. The text also included some other work on number theory, a subject Euler returned to frequently during his life.

14.3.1 Fermat's Last Theorem

As one of the final problems of his text, Euler presented his proof of the case $n = 3$ of **Fermat's Last Theorem**, a proof he had evidently found some fifteen years earlier:

THEOREM. *It is impossible to find any two cubes, whose sum, or difference, is a cube.*

Recall that Fermat, in virtually his only detailed proof of a number theory result, used the method of infinite descent to prove a theorem from which the case $n = 4$ of his theorem followed. Euler gave a detailed proof of this result and then applied the same technique to the case $n = 3$. That is, he began with the assumption that there were relatively prime integers x, y, z satisfying $x^3 + y^3 = z^3$ and showed that he could find smaller integers that satisfied the same equation. There is no loss of generality in assuming that both x and y are odd (and z is even), so Euler set $x + y = 2p$, $x - y = 2q$. Then $x = p + q$, $y = p - q$, and $x^3 + y^3 = 2p(p^2 + 3q^2)$. He then showed that if this latter expression were a cube, he could find a triple f, g, r less than x, y, z satisfying $f^3 + g^3 = r^3$.

If $2p(p^2 + 3q^2)$ were a cube, that cube would be even, so divisible by 8. Therefore, $\frac{1}{4}p(p^2 + 3q^2)$ is also an integral cube. Since the second factor in this expression is odd, so not divisible by 4, it follows that $\frac{1}{4}p$ is an integer. Euler then noted that the two factors must be relatively prime, except in the case where p is divisible by 3. So, assuming this is not the case, it follows that each factor must be a cube. Since $p^2 + 3q^2$ factors as $\left(p + q\sqrt{-3}\right)\left(p - q\sqrt{-3}\right)$, Euler asserted that each of these factors must itself be a cube. In other words, Euler evidently assumed without proof that the complex numbers of the form $a + b\sqrt{-3}$ form a unique factorization domain, a true statement, but one that is by no means obvious. Given this result, however, Euler wrote that $p \pm q\sqrt{-3} = \left(t \pm u\sqrt{-3}\right)^3$, so $p = t(t^2 - 9u^2)$ and $q = 3u(t^2 - u^2)$, with t odd and u even. In addition, since $\frac{1}{4}p$ is a cube, so also is $2p$. Thus, $2p = 2t(t + 3u)(t - 3u)$ is a cube and, since neither p nor t is divisible by 3, these three factors must be relatively prime. Thus, each of the factors is also a cube, say, $2t = r^3$, $t + 3u = f^3$, $t - 3u = g^3$. But then $f^3 + g^3 = r^3$, with each term less than the corresponding term in the original sum of cubes expression.

In the case where $p = 3r$, a similar argument shows that $\frac{2}{3}r = 2u(t + u)(t - u)$ is a cube that is the product of relatively prime factors. Thus, each factor is a cube, and we again get a new sum of cubes expression with smaller terms than the original one. Thus, the case $n = 3$ of Fermat's Last Theorem is proved by infinite descent. Euler believed, however, that his proof for $n = 3$ was sufficiently different from his proof for $n = 4$ that there was no hope of generalizing either into a proof of the entire theorem.

14.3.2 Residues

Probably around 1750, Euler began to write an elementary treatise on number theory, but he set it aside after completing sixteen chapters. The manuscript was discovered after his death and eventually published in 1849 under the title *Treatise on the Doctrine of Numbers*. The early chapters contain the calculation of such number theoretic functions as $\sigma(n)$, the number of divisors of an integer n, and $\phi(n)$, the number of integers prime to n and less than n. The most important part of the treatise, however, beginning in the fifth chapter, is Euler's treatment of the concept of congruence with respect to a given number d, now called the **modulus**. Euler defined the **residue of *a* with respect to *d*** as the remainder r on the division of a by d: $a = md + r$. He noted that there are d possible remainders, and that therefore all the integers are divided into d classes, each class consisting of those numbers having the given remainder. For example, division by 4 divides the integers into four classes, numbers of the form $4m$, $4m + 1$, $4m + 2$, and $4m + 3$. All numbers in a given class he regarded as "equivalent." Euler further showed that one can define operations on the classes such that if A and B are in the class of residues α and β, respectively, then $A + B$, $A - B$, nA, and AB are in the class of residues $\alpha + \beta$, $\alpha - \beta$, $n\alpha$, and $\alpha\beta$, respectively. Euler thus demonstrated, in modern terminology, that the function assigning an integer to its residue class is a ring homomorphism. In fact, it was out of such ideas that the theory of rings eventually developed.

Similarly, basic ideas of group theory are evident in Euler's discussion of residues of a series in arithmetic progression $0, b, 2b, \ldots$. Euler showed that if the modulus d and the number b are relatively prime, then this series contains elements from each of the d different residue classes. Therefore, b has an "inverse" with respect to d, a number p such that the residue of pb equals 1. On the other hand, if the greatest common divisor of d and b is g, then only d/g different residues appear and such an inverse does not exist. For example, the set of multiples of 2 contains elements from nine different residue classes with respect to modulus 9, and 5 is the inverse of 2, while the set of multiples of 3 contains elements from only three distinct residue classes with respect to modulus 9, and no inverse exists for 3.

Euler continued this line of investigation by considering the residues of a geometric series $1, b, b^2, b^3, \ldots$, where b is prime to d. The number n of distinct residues of this series can be no more than $\mu = \phi(d)$. Euler noted that this number n is the smallest number greater than 1 such that b^n has residue 1, because once this power is reached, all subsequent powers simply repeat the same remainders. To show that n is a factor of μ, he used an argument later to be standard in group theory by, in effect, considering the cosets of the subgroup of powers of b in the multiplicative group of residues of d relatively prime to d and showing that the order of the subgroup divides the order of the group. Euler first demonstrated that if r and s are residues, say, of b^ρ and b^σ, then rs is also a residue, of $b^{\rho+\sigma}$. Similarly, r/s is a residue. Thus, if r is a residue and $x < d$ is a nonresidue (a number prime to d that is not a residue of the series of powers), xr must also be a nonresidue. Therefore, if $1, \alpha, \beta, \ldots$ form the entire set of n residues, then $x, x\alpha, x\beta, \ldots$ form a set of n distinct nonresidues. Because any nonresidue not included in this latter list also leads to a set of n nonresidues, all distinct from the first list, Euler concluded that $\mu = mn$ for some integer

m. It follows that $b^\mu = b^{mn}$ has remainder 1 on division by d, or that $b^\mu - 1$ is divisible by d. A special case of this theorem, when d is a prime p, is **Fermat's Little Theorem**.

Euler spent much time calculating with residues between 1750 and his death in 1783. These calculations eventually led him to a statement of a theorem equivalent to the quadratic reciprocity theorem.

Euler called $p \neq 0$ a **quadratic residue** with respect to a prime q if there exist a and n such that $p = a^2 + nq$, that is, if $x^2 \equiv p \pmod{q}$ has a solution. Note that the condition of being a quadratic residue with respect to q depends only on the residue class of p with respect to q. For example, 1, 4, 9, $5 \equiv 4^2$ and $3 \equiv 5^2$ are quadratic residues with respect to 11, while 2, 6, 7, 8, and 10 are nonresidues. In a paper of 1783, Euler first proved that if $q = 2m + 1$ is an odd prime, then there are exactly m quadratic residues and therefore m nonresidues. Furthermore, he showed that the product and the quotient of two quadratic residues are again quadratic residues. He then determined that -1 is a residue with respect to q if q is of the form $4n + 1$, while it is a nonresidue if q is of the form $4n + 3$. At the end of the paper, however, after considering more examples, Euler presented four conjectures relating, for two different odd primes q and s, conditions under which each may or may not be a quadratic residue with respect to the other. These conditions may be written as follows:

1. If $q \equiv 1 \pmod 4$ and q is a quadratic residue with respect to s, then s and $-s$ are both quadratic residues with respect to q.
2. If $q \equiv 3 \pmod 4$ and $-q$ is a quadratic residue with respect to s, then s is a quadratic residue with respect to q and $-s$ is not.
3. If $q \equiv 1 \pmod 4$ and q is not a quadratic residue with respect to s, then s and $-s$ are both nonresidues with respect to q.
4. If $q \equiv 3 \pmod 4$ and $-q$ is not a quadratic residue with respect to s, then $-s$ is a quadratic residue and s is a nonresidue with respect to q.

Euler was not able to prove these results in 1783. They were restated in a somewhat different form by Adrien-Marie Legendre (1752–1833) in 1785 and 1798, but the first complete proof was given by Carl Friedrich Gauss in 1801 in his great work *Disquisitiones arithmeticae*.

Exercises

1. Derive Cramer's rule for three equations in three unknowns from the rule for two equations in two unknowns: Given the system

$$ax + by + cz = m$$

$$dx + ey + fz = n,$$

$$gx + hy + kz = p$$

solve each equation for x in terms of y and z, then form two equations in those variables and solve for z. Finally, determine y and x by substitution.

2. Find a cubic curve and a quadratic curve that intersect in six real points.

3. Consider the following system of linear equations given by Euler:

$$5x + 7y - 4z + 3v - 24 = 0$$

$$2x - 3y + 5z - 6v - 20 = 0$$

$$x + 13y - 14z + 15v + 16 = 0$$

$$3x + 10y - 9z + 9v - 4 = 0$$

Show that these four equations are "worth only two"—that they do not determine a unique 4-tuple as a solution.

4. Show that if n is prime, then the roots of $x^n - 1 = 0$ can all be expressed as powers of any such root $\alpha \neq 1$.

5. Let x_1, x_2 be the two roots of the quadratic equation $x^2 + bx + c = 0$. Since $t = x_1 + x_2$ is invariant under the two permutations of the two roots, while $v = x_1 - x_2$ takes on two distinct values, v must satisfy an equation of degree 2 in t. Find the equation. Similarly, x_1 is invariant under the same permutations as $x_1 - x_2$. Thus, x_1 can be expressed rationally in terms of $x_1 - x_2$. Find such a rational expression. Use the rational expression and equation you found to "solve" the original quadratic equation.

6. Determine the three roots x_1, x_2, x_3 of $x^3 - 6x - 9 = 0$. Use Lagrange's procedure to find the sixth-degree equation satisfied by y, where $x = y + 2/y$. Determine all six solutions of this equation and express each explicitly as $\frac{1}{3}(x' + \omega x'' + \omega^2 x''')$, where (x', x'', x''') is a permutation of (x_1, x_2, x_3) and ω is a complex root of $x^3 - 1 = 0$.

7. Show that the expression $x_1x_2 + x_3x_4$ takes on only three distinct values under the 24 permutations of four elements.

8. Let x_1, x_2, x_3, x_4 denote the four roots of the quartic equation $x^4 + ax^3 + bx^2 + cx + d = 0$. Set $\alpha = x_1x_2 + x_3x_4, \beta = x_1x_3 + x_2x_4, \gamma = x_1x_4 + x_2x_3$. Show that $\alpha + \beta + \gamma = b$, that $\alpha\beta + \alpha\gamma + \beta\gamma = ac - 4d$, and that $\alpha\beta\gamma = a^2d + c^2 - 4bd$. Show that these results imply that α, β, and γ are the roots of the cubic equation
$$y^3 - by^2 + (ac - 4d)y - (a^2d + c^2 - 4bd) = 0.$$

9. Use the results of exercise 8 to determine the reduced equation for the quartic equation $x^4 - 12x + 3 = 0$. Solve this reduced equation, a cubic. Use the values you obtain for α, β, and γ to solve the original quartic equation.

10. In Euler's proof of the case $n = 3$ of Fermat's Last Theorem, show that $\frac{1}{4}p$ and $p^2 + 3q^2$ are relatively prime if p is not divisible by 3. (Recall that $x = p + q$, $y = p - q$ are odd and relatively prime.)

11. In Euler's proof of the case $n = 3$ of Fermat's Last Theorem, consider the situation where $p = 3r$. In that situation, $\frac{3}{4}r(9r^2 + 3q^2) = \frac{9}{4}r(3r^2 + q^2)$ must be a cube. Show that the two factors in this expression are relatively prime. It follows that each must be a cube. In particular, $q^2 + 3r^2$ must be a cube. Factor this expression as in the text, using complex numbers of the form $a + b\sqrt{-3}$, and conclude that $q = t(t^2 - 9u^2)$, $r = 3u(t^2 - u^2)$, where t is odd and u is even. Also, since $\frac{9}{4}r$ is a cube, show that $\frac{2}{3}r = 2u(t + u)(t - u)$ is a cube where the factors are relatively prime. Conclude as in the case detailed in the text that one can now find three integers smaller than the original set satisfying the Fermat equation.

12. Calculate the distinct residues $1, \alpha, \beta, \ldots$ of $1, 5, 5^2, \ldots$ modulo 13. Then pick a nonresidue x of the sequence of powers and determine the coset $x, x\alpha, x\beta, \ldots$. Continue to pick nonresidues and determine the cosets until you have divided the group of all 12 nonzero residues modulo 13 into nonoverlapping subsets, the cosets of the group of powers of 5.

13. Determine the quadratic residues modulo 13.

14. Prove that -1 is a quadratic residue with respect to a prime q if and only if $q \equiv 1 \pmod 4$.

15. Outline a lesson for an algebra course that uses the technique of Maclaurin to teach the principles of Cramer's rule.

16. Prepare a report on the discovery of determinants by Seki Kowa, a Japanese mathematician of the seventeenth century.

References

Three books on the history of algebra have sections dealing with the eighteenth century: Luboš Nový, *Origins of Modern Algebra* (Prague: Academia Publishing House, 1973); Hans Wussing, *The Genesis of the Abstract Group Concept* (Cambridge: MIT Press, 1984); and B. L. van der Waerden, *A History of Algebra* (New York: Springer-Verlag, 1985). A good treatment of the early history of the theory of vector spaces is Jean-Luc Dorier, "A General Outline of the Genesis of Vector Space Theory," *Historia Mathematica* 22 (1995), 227–261. The early history of number theory is dealt with by André Weil in *Number Theory: An Approach through History from Hammurapi to Legendre* (Boston: Birkhäuser, 1984). A detailed discussion of the history of Fermat's Last Theorem up to the time of Dirichlet, including the work of Euler, is Harold Edwards, *Fermat's Last Theorem: A Genetic Introduction to Algebraic Number Theory* (New York: Springer, 1977).

The algebra treatise of Colin Maclaurin is *A Treatise of Algebra in Three Parts*, 2nd ed. (London: Millar and Nourse, 1756). Euler's original paper in which he discusses Cramer's paradox is "Sur une contradiction apparente dans la doctrine des lignes courbes," *Mémoires de l'Académie des Sciences de Berlin* 4 (1750), 219–223, in *Opera Omnia* (3), vol. 26, 33–45. Euler's algebra text, containing his ideas on the solution of polynomial equations and his proofs of the cases $n = 3$ and $n = 4$ of Fermat's Last Theorem, is *Elements of Algebra*, translated from the *Vollständige Anleitung zur Algebra* by John Hewlett (New York: Springer-Verlag, 1984). This Springer edition, a reprint of the original 1840 English translation, contains an introduction by C. Truesdell. Its many problems and ingenious methods of solution are worth a careful reading. Euler's incomplete manuscript on number theory, the *Tractatus de numerorum doctrina*, is in *Opera Omnia* (1), vol. 5, 182–283.

CHAPTER FIFTEEN

Geometry in the Eighteenth Century

I have principally sought such consequences of the [hypothesis of the acute angle] to see if it did not contradict itself. From them all I saw that this hypothesis would not destroy itself at all easily. I will therefore adduce some such consequences .... The most remarkable of such conclusions is that if the [hypothesis of the acute angle] holds, we would have an absolute measure of length for each line, for the content of each surface and each bodily space. Now this overturns a theorem that one can unhesitatingly count amongst the fundamentals of Geometry, and which up to now no one has doubted, namely that there is no such absolute measure.

—J. H. Lambert, *Theory of Parallel Lines*, 1766

Geometry in the eighteenth century was connected both to algebra, through the relationships codified under the term *analytic geometry*, and to calculus, through the application of infinitesimal techniques to the study of curves and surfaces. But there was also considerable interest in the continuing problem of Euclid's parallel postulate. In addition, Euler made several contributions to what would eventually become the subject of topology.

15.1 The Parallel Postulate

The eighteenth century saw renewed interest in the attempt to derive "rigorously" Euclid's parallel postulate from the other axioms and postulates and thus show that it was unnecessary for Euclid to have assumed his non-self-evident fifth postulate. Among those who wrote on this subject were Girolamo Saccheri and Johann Lambert.

15.1.1 Saccheri and the Parallel Postulate

Girolamo Saccheri (1667–1733) entered the Jesuit order in 1685 and subsequently taught philosophy in Genoa, Milan, Turin, and at the University in Pavia, near Milan, where he held the chair of mathematics until his death. In 1697, he published a work in logic containing a study of certain types of false reasoning, in which one begins with hypotheses that are incompatible with one another. Ultimately, he was led to the consideration of Euclid's postulates and the study of whether an alternative to Euclid's parallel postulate would be compatible or incompatible with the remaining axioms and postulates. It was this study that Saccheri finally published in 1733 in his *Euclid Freed of all Blemish*.

Saccheri's aim in the first part of his work, the only part to be considered here, was to free Euclid from the "blemish" of an unnecessary postulate by assuming that the parallel postulate is false and then deriving it as a logical consequence, thus finding a contradiction. Saccheri began with a consideration of the quadrilateral $ABCD$ with two equal sides CA and DB, both perpendicular to the base AB, the same quadrilateral considered some six hundred years earlier by al-Khayyāmī (Fig. 15.1). Using only Euclidean propositions not requiring the parallel postulate, Saccheri easily demonstrated that the angles at C and D are equal. There are then three possibilities for these angles: They are both right, both obtuse, or both acute. Saccheri called these possibilities the *hypothesis of the right angle*, the *hypothesis of the obtuse angle*, and the *hypothesis of the acute angle*, respectively. He then showed that these hypotheses are equivalent, respectively, to the line segment CD being equal to, less than, or greater than the line segment AB. It was "obvious" to Saccheri, as it was to all who considered the question in earlier times, that the only "true" possibility was the hypothesis of the right angle, since it is, in fact, implied by the parallel postulate. The other two hypotheses come from the assumption that the parallel postulate is false. Saccheri intended to derive the parallel postulate from each of these two "false" hypotheses, using only the "self-evident" axioms of Euclid, thus demonstrating that each possibility led to a contradiction.

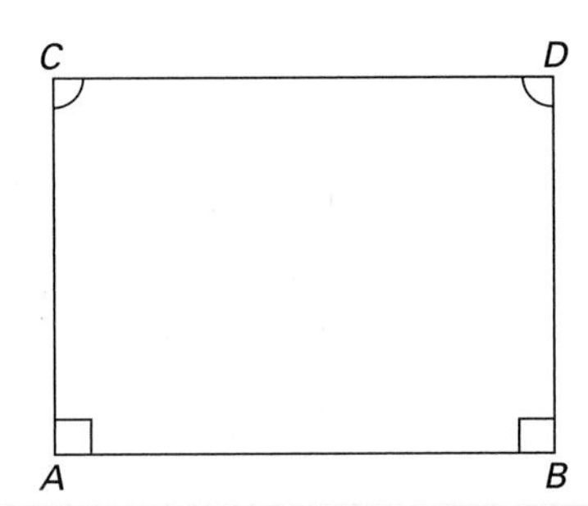

FIGURE 15.1 Saccheri's quadrilateral

Saccheri began by proving that if either of the hypotheses is true for one quadrilateral, then it is true for all. He then attacked the hypothesis of the obtuse angle. He was able to show that under that hypothesis, the two acute angles in any right triangle are together greater than a right angle. But he was also able to prove the parallel postulate, using

Elements I-17 to the effect that any two angles of a triangle are together less than two right angles. From the parallel postulate he could then prove, as did Euclid in *Elements* I, that the three angles of any triangle are together equal to two right angles. This result contradicted the first result, thus showing

PROPOSITION XIV. *The hypothesis of the obtuse angle is absolutely false, because it destroys itself.*

Saccheri next showed that the hypotheses of the right, obtuse, and acute angles are equivalent, respectively, to the results that the sum of the angles of any triangle is equal to, greater than, or less than two right angles and that the sum of the angles of a quadrilateral is equal to, greater than, or less than four right angles. He then proceeded to investigate in more detail the consequences of the hypothesis of the acute angle. Here, however, he was not able to derive the parallel postulate as a consequence. He did, however, derive other intriguing results. For example,

PROPOSITION XVII. *If the straight line AH is at right angles to any straight line AB, however small, I say that under the hypothesis of the acute angle it cannot be true that every straight line BD intersecting AB in an acute angle will ultimately meet AH produced (Fig. 15.2).*

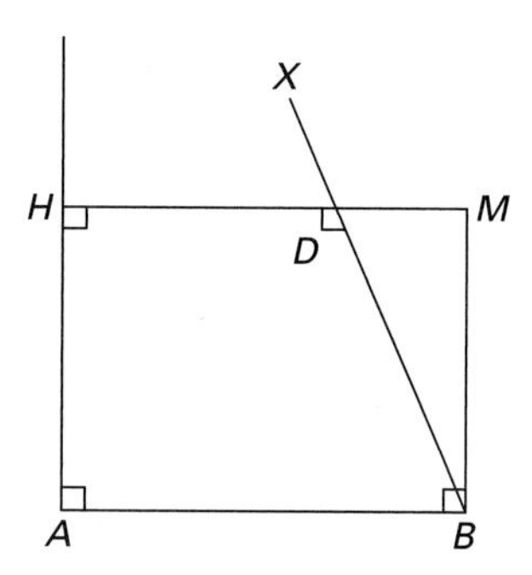

FIGURE 15.2 Saccheri's proposition XVII: BD and AH do not ultimately intersect

Suppose BM is also perpendicular to AB. Drop a perpendicular from M to AH intersecting that line at H. Because the sum of the angles of a quadrilateral is less than four right angles, it follows that angle BMH is acute. Similarly, if BX is drawn from B perpendicular to HM, intersecting that line at D, then angle XBA is also acute. But now BD extended cannot intersect AH extended, because, the angles at H and D both being right, this would contradict *Elements* I-17.

Because proposition XVII implies that there are two straight lines in the plane that do not meet, Saccheri could show in proposition XXIII that such lines either have a common perpendicular or else "they mutually approach ever more toward each other." Furthermore, in the latter case, the distance between the lines becomes smaller than any assigned length; that is, the lines are asymptotic. Saccheri was then able to prove in proposition XXXII that given a line BX perpendicular to a line segment AB, there is a certain acute angle

BAX such that the line AX "only at an infinite distance meets BX," while lines making smaller acute angles with BA intersect BX and those making larger ones all have a common perpendicular with BX (Fig. 15.3). Saccheri then concluded:

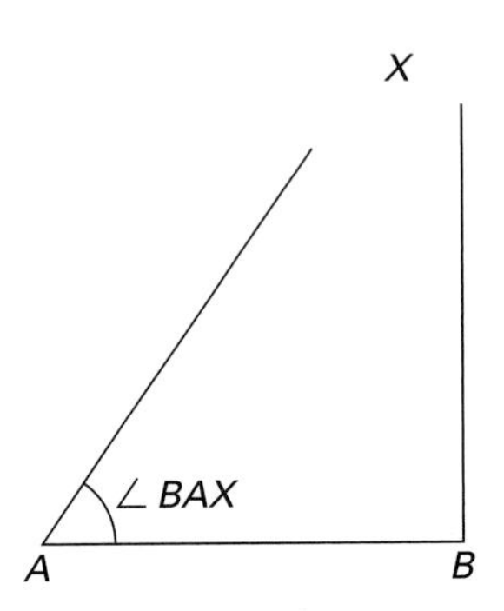

FIGURE 15.3 Saccheri's "angle of parallelism": AX and BX only meet at an "infinite" point X

PROPOSITION XXXIII. *The hypothesis of the acute angle is absolutely false, because it is repugnant to the nature of the straight line.*

Saccheri hardly gave a "proof" of this result. It appears, in fact, that he ended his quest with this proposition only because he had faith that the parallel postulate must be true. But then, apparently having second thoughts on the matter, he spent the next thirty pages attempting a further justification of his result.

15.1.2 Lambert and the Parallel Postulate

Johann Lambert (1728–1777), having studied at least a summary of Saccheri's work, attempted to improve on it. But his work on the parallel postulate, the *Theory of Parallel Lines*, finished by 1766, was never published, perhaps because Lambert was not completely happy with the conclusions. In the book, he considered a quadrilateral with three right angles and made three hypotheses as to the nature of the fourth angle, essentially the same three hypotheses Saccheri had proposed: It could be right, obtuse, or acute. He too was able to show that the first hypothesis implied the parallel postulate and that the second hypothesis gave a contradiction, but, as noted in the chapter opening, that the third hypothesis gave a long list of strange consequences.

The most surprising of Lambert's consequences was that in his fundamental quadrilateral the difference between 360° and the sum of the angles depended on the area of the quadrilateral—that is, the larger the quadrilateral, the smaller the angle sum. Consider the quadrilateral $ABCD$ with right angles at A, B, and C and an acute angle at D of measure β (Fig. 15.4). At point E between A and B, construct a perpendicular EF to AB. It follows that $\angle CFE$ is also acute. If its measure is α, then the measure of $\angle EFD$ is $180° - \alpha$. But the sum of the angles of quadrilateral $EBFD$ is less than 360°; thus, $90 + 90 + 180 - \alpha + \beta < 360$, or $\beta < \alpha$. Therefore, the angle sum of quadrilateral $ABCD$ is less than that of quadrilateral $AECF$, as stated.

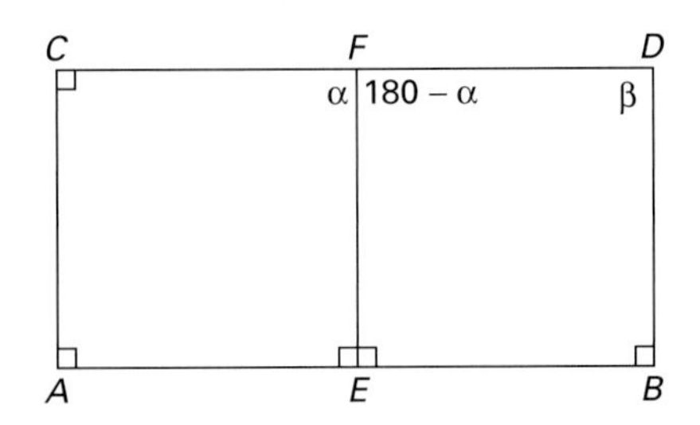

FIGURE 15.4 Lambert's proof that the angle of a quadrilateral decreases with the size of the quadrilateral

Lambert concluded from this result that the third consequence implied an absolute measure of length, area, and volume. For if we assume that quadrilateral $AEFC$ has $AE = AC$, then $\angle EFC$ is a determined acute angle, an angle that can fit no other such quadrilateral. Thus, the measure α of $\angle EFC$ may be taken as the absolute measure of the quadrilateral. Lambert was not able to deduce this absolute measure; that is, he could not determine what the angle would be if $AE = AC = 1$ foot, for example, but he did realize that this hypothesis destroyed entirely the notion of similar figures. It also implied that the difference between 180° and the sum of the angles of a triangle, the **defect** of the triangle, was proportional to the area of the triangle. Lambert realized that a similar result is true under his second hypothesis, with the defect being replaced by the excess of the angle sum over 180°. But he also knew that spherical triangles had this same property, that their angle sum was greater than 180° and that the excess was proportional to the area. He then argued by analogy: "I should almost therefore put forward the proposal that the third hypothesis holds on the surface of an imaginary sphere."

Lambert abandoned his study of Euclid's parallel postulate once he felt that he could not successfully refute the hypothesis of the acute angle, even though it seems he was convinced that Euclid's geometry was true of space. Nevertheless, he believed that because the geometry of the hypothesis of the obtuse angle was reflected in geometry on the sphere, the sphere of imaginary radius would perform the same function for the hypothesis of the acute angle. Although, by 1770, he had introduced the hyperbolic functions as complex analogs of the circular ones, in the sense that $\cosh ix = \cos x$ and $\sinh ix = i \sin x$, he was not able to apply these functions to develop a geometry on the imaginary sphere based on the hypothesis of the acute angle, nor could he give a construction in three-dimensional space of this imaginary sphere. It was only in the early nineteenth century, when analysis of this type could be brought to bear on the alternatives to the parallel postulate, that what is today called non-Euclidean geometry was developed.

15.2 Differential Geometry of Curves and Surfaces

The central thrust of eighteenth-century geometry was its connection to analysis. Thus, the second volume of Euler's *Introduction* provided a clear organization of the topic of plane curves. Euler began with a classification of quadratic and cubic curves and then gave a treatment of quartics that included 146 different forms. He next dealt with the various properties of curves in general, without using calculus, including asymptotes, curvature, and singular points. He concluded the volume with a discussion of quadric surfaces, showing that the general second-degree equation in three variables can be reduced by a change of coordinates to one of a limited number of canonical forms. He could then identify the type of surface: ellipsoid, elliptic or hyperbolic paraboloid, elliptic or hyperbolic hyperboloid (now called the *hyperboloids of one and two sheets*, respectively), cone, and parabolic cylinder.

15.2.1 Euler and Space Curves and Surfaces

It was not until 1775 that Euler applied calculus to the study of curves in space, expressing the curves parametrically via the arc length s. Thus, a curve was given by three equations $x = x(s)$, $y = y(s)$, $z = z(s)$. Taking differentials of each led Euler to the expressions $dx = p\,ds$, $dy = q\,ds$, $dz = r\,ds$, from which he derived the result $p^2 + q^2 + r^2 = 1$. The functions p, q, r, the derivatives of the coordinate functions with respect to arc length, are the components of the unit tangent vector to the curve. These components are also called the **direction cosines** of the tangent line (or of the curve itself) at the specified point.

Curvatures of plane curves at a point P had been defined by Newton, among others, as the reciprocal of the radius of the osculating circle meeting the curve at P. To define the curvature of a space curve, Euler used the unit sphere centered at a point $(x(s), y(s), z(s))$. If the "unit vectors" (p, q, r) at the neighboring parameter values s and $s + ds$, both considered to be emanating from that center, differ by an arc on the sphere equal to ds', then the **curvature** κ at that point was defined as $|ds'/ds|$, a value measuring how the curve at any point differs from a great circle on the sphere. Because the vector ds' is given by

$$\left(\frac{dx}{ds}(s+ds) - \frac{dx}{ds}(s),\ \frac{dy}{ds}(s+ds) - \frac{dy}{ds}(s),\ \frac{dz}{ds}(s+ds) - \frac{dz}{ds}(s)\right)$$
$$= \left(\frac{d^2x}{ds^2}\,ds,\ \frac{d^2y}{ds^2}\,ds,\ \frac{d^2z}{ds^2}\,ds\right),$$

it follows that

$$\kappa = \left|\frac{ds'}{ds}\right| = \sqrt{\left(\frac{d^2x}{ds^2}\right)^2 + \left(\frac{d^2y}{ds^2}\right)^2 + \left(\frac{d^2z}{ds^2}\right)^2}.$$

Euler next defined the **radius of curvature** ρ to be the reciprocal of the curvature:

$$\rho = \frac{ds^2}{\sqrt{(d^2x)^2 + (d^2y)^2 + (d^2z)^2}}.$$

It turned out, although it was not proved until the nineteenth century, that curvature is one of the two essential properties of a space curve. The other is the **torsion**, which measures the rate at which the curve deviates from being a plane curve. If the curvature and torsion are given as functions of arc length along a curve, then the curve is completely determined up to its actual position in space.

Fifteen years earlier, Euler had made a beginning in the differential geometry of surfaces, with a paper entitled *Research on the Curvature of Surfaces*. In that work, he noted that although the method of finding the curvature of a plane curve at a given point was well known, it was difficult even to define the curvature of a surface in space at a point.

Each section of a surface formed by a plane through the given point gives a different curve, and the curvatures of each of these sections may well be different, even if one restricts oneself to plane sections that are perpendicular to the surface. In the paper, Euler calculated these various curvatures and established some relationships among them. First, however, he needed to characterize planes perpendicular to the surface, that is, planes that pass through the normal line to the surface at the given point P. He showed that the plane with equation $z = \alpha y - \beta x + \gamma$ is perpendicular to the surface defined by $z = f(x, y)$ if $\beta(\partial z/\partial x) - \alpha\, \partial z/\partial y = 1$. Defining the principal plane to be the plane through P perpendicular both to the surface and to the xy-plane, Euler then demonstrated that if a given plane perpendicular to the surface makes an angle ϕ with the principal plane, the curvature of the section formed by that plane is given by $\kappa_\phi = L + M \cos 2\phi + N \sin 2\phi$, where L, M, N depend solely on the partial derivatives of z at P. Taking the derivative of this expression with respect to ϕ, Euler found that the maximum and minimum curvatures occur when $-2M \sin 2\phi + 2N \cos 2\phi = 0$ or when $\tan 2\phi = N/M$. But since $\tan(2\phi + 180°) = \tan 2\phi$, Euler concluded that if a maximum curvature occurs for a given value of ϕ, the minimum occurs at $\phi + 90°$. He was finally able to show that if κ_1 is the maximum curvature and κ_2 the minimum, and if the minimum curvature occurs at the principal plane, then the curvature of any section made by a plane at angle ϕ to the principal plane is given by $\kappa_\phi = \frac{1}{2}(\kappa_1 + \kappa_2) - \frac{1}{2}(\kappa_1 - \kappa_2) \cos 2\phi$.

15.2.2 The Work of Monge

Gaspard Monge (1746–1818) systematized the basic results of both analytic and differential geometry and added much new material in several papers, beginning in 1771, and in two textbooks written for his students at the École Polytechnique at the end of the century. For example, in a paper published in 1784, Monge presented for the first time the **point-slope form of the equation of a line**: "If one wishes to express the fact that this line [with slope-intercept equation $y = ax + b$] passes through the point M of which the coordinates are x' and y', which determines the quantity b, the equation becomes $y - y' = a(x - x')$, in which a is the tangent of the angle which the straight line makes with the line of x's." Monge's 1799 text *Descriptive Geometry*, on the other hand, did not deal with algebra at all, but relied on the basic ideas of pure geometry. Monge outlined many techniques for representing three-dimensional objects in two dimensions. He systematically used projections and other transformations in space to draw in two dimensions various aspects of space figures. He described in detail such concepts as the tangent plane to a surface, the intersection of two surfaces, the notion of a developable surface (a surface that can be flattened out to a plane without distortion), and the curvature of a surface.

In his second text, the *Application of Analysis to Geometry* of 1807, which grew out of lecture notes dating from 1795, Monge used algebra and calculus extensively. The first part of this work, which used only algebra, contained the earliest detailed presentation of the analytic geometry of lines in two- and three-dimensional space as well as planes in three-dimensional space. Thus, Monge indicated that points in space are to be determined by considering perpendiculars to each of three coordinate planes. A line in space is determined by its projection onto two of these three planes, the equations of the projections onto the xy-plane, for example, being given in the slope-intercept form or in the point-slope form. Monge showed how to find the intersection of two lines, as well as how to find a line parallel to a given line through a given point and a line through two given points. He also noted that the lines in the plane with equations $y = ax + \alpha$ and $y = a'x + \alpha'$ are perpendicular if $aa' = -1$.

Monge wrote the equation of a plane both in the form $z = ax + by + c$, where a and b are the slopes of the lines of intersection of this plane with the xz-plane and the yz-plane, respectively, and in the symmetric form $Ax + By + Cz + D = 0$, where the coefficients A, B, and C determine the direction cosines of the angles between the plane and the coordinate planes. He then proceeded to discuss all of the familiar problems dealing with points, lines, and planes, such as finding the normal line to a plane passing through a given point, finding the shortest distance between two lines, and finding the angle between two lines or between a line and a plane.

The second part of Monge's text was devoted to the study of surfaces. Here he used the entire machinery of calculus to develop analytically all of the topics he had considered in his *Descriptive Geometry*. He considered in detail how to determine from various types of descriptions the partial differential equation that represents a given surface as well as how, in certain cases, to integrate that equation. To develop the equations of the tangent plane and normal line to a surface, Monge began by noting that the differential equation that represents the surface $z = f(x, y)$ near a point (x', y', z') is

$$dz = \frac{\partial z}{\partial x}\,dx + \frac{\partial z}{\partial y}\,dy,$$

where the partial derivatives are evaluated at x' and y'. On the other hand, the equation of any plane through (x', y', z') can be written as $A(x - x') + B(y - y') + C(z - z') = 0$. For this plane to be a tangent plane, any point on it infinitely near the given point must also be on the surface, that is, must satisfy the differential equation of the surface. So, taking $x - x'$ as dx, $y - y'$ as dy, $z - z'$ as dz, Monge noted that the equation $A\,dx + B\,dy + C\,dz = 0$ must be identical to $dz = (\partial z/\partial x)dx + (\partial z/\partial y)dy$. It follows that $A/C = -\partial z/\partial x$, $B/C = -\partial z/\partial y$, and that the equation of the tangent plane is

$$z - z' = (x - x')\,\frac{\partial z}{\partial x} + (y - y')\,\frac{\partial z}{\partial y}.$$

The equations of the normal line to the surface, that is, the normal line to the tangent plane, are then calculated to be

$$x - x' + (z - z')\,\frac{\partial z}{\partial x} = 0 \qquad \text{and} \qquad y - y' + (z - z')\,\frac{\partial z}{\partial y} = 0.$$

15.3 Euler and the Beginnings of Topology

It was in the mid-1730s that Euler became aware of a little problem coming out of the town of Königsberg, in East Prussia (now in Russia). In the middle of the river Pregel, which ran through the town, there were two islands. The islands and the two banks of the river were connected by seven bridges. The question asked by the townspeople was whether it was possible to plan a stroll which passes over each bridge exactly once. Euler, as usual, instead of considering this problem in isolation, attacked and solved the general problem of the existence of such a path whatever the number of regions and bridges. In a paper published in 1736, he noted first that if one labeled the regions with letters $A, B, C, D, \ldots,$ one could then label a path using a series of letters representing the successive regions passed through (Fig. 15.5). Thus, $ABDA$ would represent a path leading from region A to region B and then to D and back to A, regardless of the particular bridges crossed. It followed immediately that a complete path satisfying the desired conditions must contain one more letter than the number of bridges. In the Königsberg case, that number must be 8.

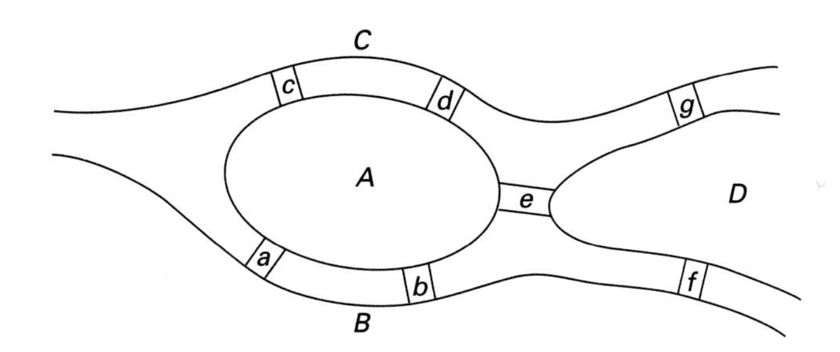

FIGURE 15.5 The seven bridges of Königsberg

Next, Euler realized that if the number k of bridges leading into a given region is odd, then the letter representing that region must occur $(k + 1)/2$ times. Thus, if there is one bridge leading to region A, then A will occur once; if there are three bridges, A will occur twice; and so on. It does not matter in this case whether the path starts in region A or in some other region. On the other hand, if k is even, then the letter representing that region will occur $k/2$ times if the path starts outside of the region and $k/2 + 1$ times if the path begins in the region. For example, if there are four bridges leading into region A, then A will occur twice if the route begins outside of A and three times if it begins in A. From the two different determinations of the number of letters that occurs in his representation of a particular path, Euler could determine whether a path passing over each bridge exactly once is possible:

> If the total of all the occurrences [calculated above for each region] is equal to the number of bridges plus one, the required journey will be possible, and will have to start from an area with an odd number of bridges leading to it. If, however, the total number of letters is one less than the number of bridges plus one, then the journey is possible starting from an area with an even number of bridges leading to it, since the number of letters will therefore be increased by one.

In the Königsberg case, an odd number of bridges, namely, 5, 3, 3, and 3, leads into each of the regions A, B, C, and D, respectively. The sum of the corresponding numbers, 3, 2, 2, and 2, is 9, which is more than "the number of bridges plus one." It follows that the desired path is impossible. In general, Euler noted that such a path will always be impossible if there are more than two regions approached by an odd number of bridges. If there are exactly two such regions, then the path is possible as long as it starts in one of those regions. Finally, if all the regions are approached by an even number of bridges, then a path crossing each bridge exactly once is always possible. Because once one knows whether a path is possible, the actual construction of it is straightforward, Euler had completely solved the problem he set.

Euler struggled a bit with another geometrical problem. In 1750, he wrote in a letter to Christian Goldbach (1690–1764), "I cannot yet give an entirely satisfactory proof of the following proposition: In every solid enclosed by plane faces the aggregate of the number of faces and the number of solid angles exceeds by two the number of edges." In other words, given any polyhedron with V vertices (solid angles), E edges, and F faces, then $V + F = E + 2$, or, more familiarly, $V - E + F = 2$. Eventually, Euler submitted a proof to the St. Petersburg Academy of Sciences in which he successively removed tetrahedron-shaped pieces from the given polyhedron in such a way that $V - E + F$ remained unchanged at each stage. Continuing in this process, Euler reached a single tetrahedron, for which $V = F = 4$ and $E = 6$, so that the desired relationship held. Unfortunately, it was not at all clear that Euler's dissection procedure could be carried out for an arbitrary polyhedron. A completely correct proof was given in 1794 by Adrien-Marie Legendre. On the other hand, Euler noted that the polyhedron problem as well as the Königsberg bridge problem were apparently part of a branch of geometry in which the relations depend on position alone and not at all on magnitudes. It was not until the late nineteenth and early twentieth centuries, however, that these facts and certain others were systematically studied and finally turned into the subject of topology.

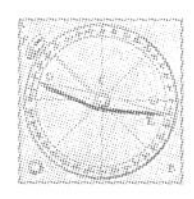

Exercises

1. Let $ABCD$ be a Saccheri quadrilateral, as in Fig. 15.1, with right angles at A and B. Show, using only Euclidean propositions not requiring the parallel postulate, that $\angle C = \angle D$.

2. Given the hypothesis of the acute angle, both Saccheri and Lambert showed that the sum of the angles of any triangle is less than two right angles. Let the difference between 180° and the angle sum of a triangle be the *defect* of the triangle. Suppose triangle ABC is split into two triangles by line BD (Fig. 15.6). Show that the defect of triangle ABC is equal to the sum of the defects of triangles ABD and BDC.

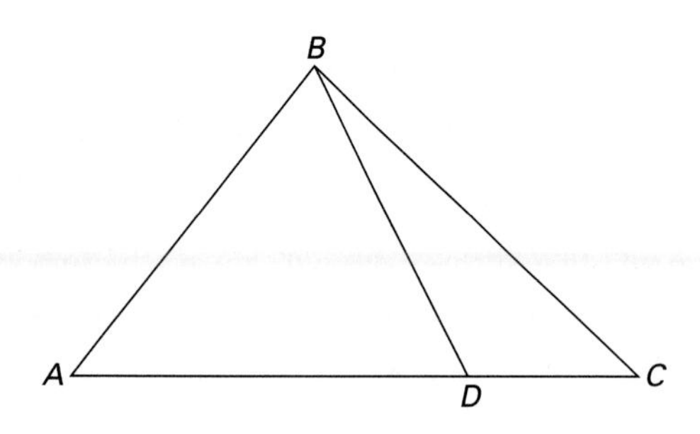

FIGURE 15.6 Calculation of the defect of triangle ABC

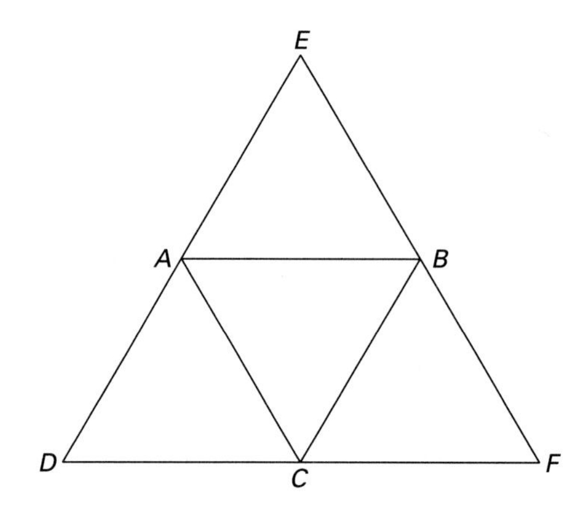

FIGURE 15.7 Calculation of the defect of triangle DEF

3. Assume that the defect of equilateral triangle DEF is $\beta > 0$. Bisect the sides at points A, B, C and form triangle ABC (Fig. 15.7). Show that triangles AEB, DAC, and CBF are all congruent isosceles triangles and that triangle ABC is equilateral. If the defect of triangle CBF is α and the defect of triangle ABC is γ, show that $\beta = 3\alpha + \gamma$. Also, show that, contrary to the situation in Euclidean geometry, triangle ABC

is not congruent to either of the triangles AEB, DAC, or CBF.

4. Given that the angle sum of a triangle made of great circle arcs on a sphere (a spherical triangle) is greater than two right angles, define the *excess* of a triangle as the difference between its angle sum and 180°. Show that if a spherical triangle ABC is split into two triangles by an arc BD from vertex B to the opposite side, then the excess of triangle ABC is equal to the sum of the excesses of triangles ABD and BDC.
5. Given the relationships $\cosh ix = \cos x$ and $\sinh ix = i \sin x$, determine $\cosh x$ and $\sinh x$ in terms of the cosine and sine functions, and show that

$$\cosh^2 x - \sinh^2 x = 1.$$

6. Show that Euler's result relating the curvature of any section of a surface made by a plane at angle ϕ to the principal plane is equivalent to the modern formulation $\kappa_\phi = \kappa_1 \sin^2 \phi + \kappa_2 \cos^2 \phi$.
7. Show that the plane $z = \alpha y - \beta x + \gamma$ is perpendicular to the surface $z = f(x, y)$ if

$$\beta\, \partial z/\partial x - \alpha(\partial z/\partial y) = 1.$$

(Show that the plane contains a normal line to the surface.)

8. Find the normal line to the plane

$$Ax + By + Cz + D = 0$$

that passes through the point (x_0, y_0, z_0).

9. Convert Monge's form of the equations of the normal line to the surface $z = f(x, y)$ into the modern vector equation of the line.
10. Show that an *Euler path* over a series of bridges connecting certain regions (a path that crosses each bridge exactly once) is always possible if there are either two or no regions approached by an odd number of bridges.
11. Construct Euler paths in the situations represented in Figure 15.8.

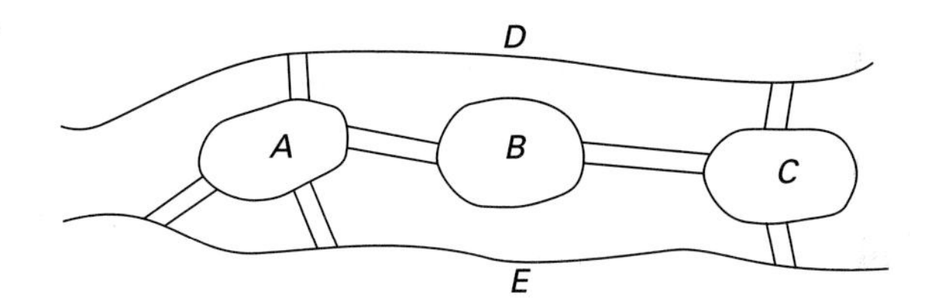

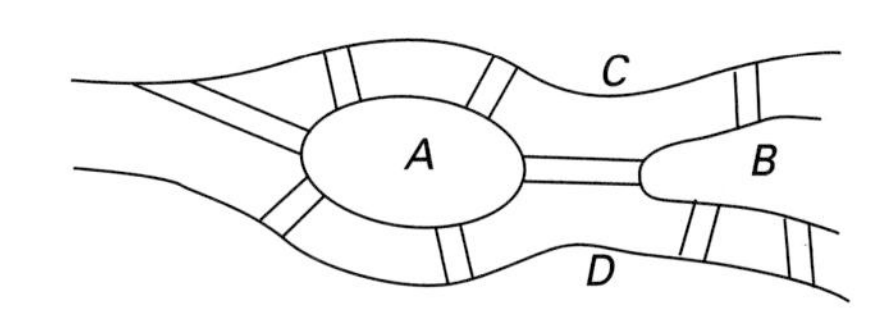

FIGURE 15.8 Bridge path problems

12. Develop a lesson in a course in three-dimensional analytic geometry that uses the work of Monge to derive the equation of the tangent plane to a surface.

References

The standard work on the history of non-Euclidean geometry is Roberto Bonola, *Non-Euclidean Geometry, A Critical and Historical Study of its Development*, H. S. Carslaw, trans. (New York: Dover, 1955). More recent works, both of which contain material on Saccheri and Lambert, are Jeremy Gray, *Ideas of Space: Euclidean, Non-Euclidean and Relativistic* (Oxford: Clarendon Press, 1989) and Boris A. Rosenfeld, *A History of Non-Euclidean Geometry: Evolution of the Concept of a Geometric Space*, Abe Shenitzer, trans. (New York: Springer-Verlag, 1988). Briefer surveys of the subject include Jeremy Gray, "Non-Euclidean Geometry—A Re-interpretation," *Historia Mathematica* 6 (1979), 236–258 and Torkil Heiede, "The History of Non-Euclidean Geometry," in Victor Katz, ed., *Using History to Teach Mathematics: An International Perspective* (Washington: MAA, 2000), 201–211. A geometry text containing much historical material on non-Euclidean geometry is Marvin Greenberg, *Euclidean and Non-Euclidean Geometries: Development and History*, 3rd ed. (New York: Freeman, 1993). A general survey of the history of differential geometry is found in Dirk J. Struik, "Outline of a History of Differential Geometry," *Isis* 19 (1933), 92–120, and the history of analytic geometry, including material on the eighteenth century, is detailed in Carl Boyer, *History of Analytic Geometry* (New York: Scripta Mathematica, 1956).

Girolamo Saccheri's, *Euclides Vindicatus* is available in its entirety in English: G. B. Halstead, trans. (New York: Chelsea, 1986). This Chelsea edition is a reprint of the first English translation of 1920 with added notes by Paul Stäckel and Friedrich Engel. Selections from Lambert's work, *The Theory of Parallel Lines*, are found in John Fauvel and Jeremy Gray, eds., *The History of Mathematics: A Reader* (London: Macmillan, 1987), 517–520. Euler's early work in analytic geometry is in volume 2 of *Introduction to Analysis of the Infinite*, John D. Blanton, trans. (New York: Springer, 1990). Euler's first work on differential geometry is "Recherches sur la courbure des surfaces," *Mem. de l'Academie des Sciences de Berlin* 16 (1760), 119–143, also found in *Opera Omnia* (1) 28, 1–22. Monge's work is thoroughly discussed in René Taton, *L'oeuvre scientifique de Monge* (Paris: Presses Universitaires de France, 1951). Finally, Norman L. Biggs, E. Keith Lloyd, and Robin J. Wilson, *Graph Theory: 1736–1936* (Oxford: Clarendon Press, 1986) contains a complete translation of Euler's article on the problem of the seven bridges of Königsberg.

CHAPTER SIXTEEN

Algebra and Number Theory in the Nineteenth Century

It is greatly to be lamented that this virtue of the real integers that they can be decomposed into prime factors which are always the same for a given integer does not belong to the complex integers [of arbitrary cyclotomic number fields], for were this the case, the entire theory, which is still laboring under many difficulties, could be easily resolved and brought to a conclusion. For this reason, the complex integers we are considering are seen to be imperfect, and there arises the question whether other types of complex numbers can be found . . . which would preserve the analogy with the real integers with respect to this fundamental property.

—Ernst Kummer, in "De numeris complexis, qui radicibus unitates et numeris integris realibus constant," 1847

lgebra in 1800 meant the solving of equations. By 1900, the term encompassed the study of various mathematical structures—that is, sets of elements with well-defined operations, satisfying certain specified axioms. It is this change in the notion of algebra that we will explore in this chapter.

The nineteenth century opened with the appearance of the *Disquisitiones arithmeticae* of Carl Friedrich Gauss, in which the "Prince of Mathematicians" discussed the basics of number theory, not only proving the law of quadratic reciprocity, but also introducing various new concepts that provided early examples of groups and matrices. Gauss's study of higher reciprocity laws soon led to his study of the so-called Gaussian integers, complex numbers of the form $a + bi$, where a and b are ordinary integers. Attempts to generalize the properties of these integers to integers in other number fields led Ernst Kummer to the realization that some of the most important of these properties, including unique factorization, fail to hold. To recover this property, along with a reasonable new meaning of the term *prime*, Kummer created (by 1846) what he called "ideal complex numbers," the study of which led Richard Dedekind in the 1870s to define "ideals" in rings of algebraic integers. These ideals had the property of unique factorization into prime ideals.

Gauss's study of the solutions of cyclotomic equations in the *Disquisitiones* as well as Augustin-Louis Cauchy's detailed study of permutations in 1815 helped with a new attack on the problem of solving algebraic equations of degree higher than 4. It was Niels Henrik Abel who finally proved (in 1827) the impossibility of solving a general equation of degree 5 or higher in terms of radicals. Soon thereafter, Evariste Galois sketched the relationship between algebraic equations and groups of permutations of the roots, a relationship whose complete development transformed the question of the solvability of an equation into one of the study of subgroups and factor groups of the group of the equation. Galois's work was not published until 1846 and was not completely understood until somewhat later. In 1854, Arthur Cayley gave the earliest definition of an abstract group, but his work, too, was somewhat in advance of its time and was not fully developed until 1882 by Walther von Dyck and Heinrich Weber. Meanwhile, the study of the "numbers" determined by the solutions of algebraic equations led to the definition, formulated by both Leopold Kronecker and Richard Dedekind, of a field of numbers and soon after to Weber's abstract definition of a field.

Another aspect of what is today called modern algebra, the theory of matrices, also developed in the mid–nineteenth century. Determinants had been used as early as the seventeenth century, but it was only in 1850 that James Joseph Sylvester coined the term **matrix** to refer to a rectangular array of numbers. Soon thereafter, Cayley developed the algebra of matrices. The study of eigenvalues was begun by Cauchy early in the century, in his work on quadratic forms, and then fully developed by Georg Frobenius, among others. In particular, Frobenius organized the theory of matrices into essentially the form it has today.

16.1 Number Theory

Carl Friedrich Gauss (1777–1855) burst on the mathematical scene with his *Disquisitiones arithmeticae* (*Investigations in Arithmetic*), published in 1801 but written before he had even studied the works of his contemporaries. Although some of what he thought he had discovered was already known to Euler or Lagrange or Legendre, Gauss's work contained many new discoveries in the theory of numbers.

16.1.1 Gauss and Congruences

Gauss began in chapter 1 by presenting the modern definition and notation for congruence: The integers b and c are **congruent** relative to the **modulus** a if a divides the difference of b and c. Gauss wrote this as $b \equiv c \pmod{a}$, noting that he adopted the symbol $\equiv$ "because of the analogy between equality and congruence," and called b and c each a **residue** of the other. He then discussed the elementary properties of congruence. For example, Gauss showed that the linear congruence $ax + b \equiv c \pmod{m}$ is always solvable by use of the Euclidean algorithm, if the greatest common divisor of a and m is 1. He also showed how to solve the Chinese remainder problem and calculate the Euler function $\phi(n)$, which gives the number of integers less than n that are relatively prime to n. In chapter 3, Gauss, like Euler before him, considered residues of powers. Noting that if p is prime and a is any number less than p, then the smallest exponent m such that $a^m \equiv 1 \pmod{p}$ is a divisor of $p - 1$, Gauss went on to show that in fact "there always exist numbers with the property that no power less than the $(p-1)$st is congruent to unity." A number a satisfying this property is called a **primitive root** modulo p. If a is a primitive root, then the powers a, $a^2, a^3, \ldots, a^{p-1}$ all have different residues modulo p and thus exhaust all of the numbers $1, 2, \ldots, p-1$. In particular, $a^{(p-1)/2} \equiv -1 \pmod{p}$.

The central topic of chapter 4 of the *Disquisitiones* is the law of quadratic reciprocity. Recall that Euler had stated but not proved this law, which describes the conditions under which two odd primes are quadratic residues of each other. Gauss gave numerous examples and special cases derived from calculations before presenting a proof of the theorem. He showed first, following Euler, that -1 is a quadratic residue of primes of the form $4n+1$ and a nonresidue of primes of the form $4n+3$. He next dealt with 2 and -2 and concluded that for primes of the form $8n+1$, both 2 and -2 are quadratic residues; for primes of the form $8n+3$, -2 is a quadratic residue and 2 is not; for primes of the form $8n+5$, both 2 and -2 are nonresidues; and for primes of the form $8n+7$, 2 is a quadratic residue and -2 is not. After characterizing the primes for which 3 and -3 are quadratic residues as well as those for 5 and -5 and 7 and -7, he was able to state the general result:

QUADRATIC RECIPROCITY THEOREM. *If p is a prime number of the form $4n+1$, $+p$ will be a [quadratic] residue or nonresidue of any prime number which taken positively is a residue or nonresidue of p. If p is of the form $4n+3$, $-p$ will have the same property.*

In terms of Legendre's suggestive notation for quadratic residues, in which the symbol $\left(\frac{p}{q}\right)$ is equal to 1 if p is a quadratic residue modulo q and to -1 if not, where q is an odd prime, the theorem can be stated in the following elegant form:

$$\left(\frac{p}{q}\right)\left(\frac{q}{p}\right) = (-1)^{\frac{p-1}{2}\frac{q-1}{2}}.$$

Similar formulas can be written expressing quadratic residue properties of -1 and ± 2 modulo a prime p. Given the product rule for residues, that

$$\left(\frac{a}{p}\right)\left(\frac{b}{p}\right) = \left(\frac{ab}{p}\right),$$

where a, b are prime to p, as well as rules for determining the residue situation when two numbers have common factors, it is possible to determine for any two positive numbers P and Q whether Q is a residue or a nonresidue of P.

Over the next several decades, Gauss attempted to establish a law of quartic reciprocity—that is, a law determining when numbers are congruent to fourth powers modulo other numbers. To deal with the latter case, he studied what are now called the **Gaussian integers**, the complex numbers $a+bi$, where a and b are ordinary integers, and established certain analogies between them and ordinary integers. After noting that there are four units (invertible elements) among the Gaussian integers, $1, -1, i$, and $-i$, he defined the **norm** of an integer $a+bi$ to be the product a^2+b^2 of the integer with its complex conjugate $a-bi$. He then called an integer **prime** if it cannot be expressed as the product of two other integers, neither of them units. Because an odd real prime p can be expressed as $p = a^2+b^2$ if and only if it is of the form $4n+1$, it follows that such primes, considered as Gaussian integers, are composite: $p = (a+bi)(a-bi)$. Conversely, primes of the form $4n+3$ are still prime as Gaussian integers. Because $2 = (1+i)(1-i)$, 2 is also composite. He could then conclude that 2 and primes of the form $4n + 1$ split as the product of two Gaussian primes in the domain of Gaussian integers, while primes of the form $4n + 3$ remain prime there.

With primes defined in the domain of Gaussian integers, Gauss proved the analog of the fundamental theorem of arithmetic, using theorems similar to those in Euclid's *Elements*, Book VII. First, he easily showed that any integer can be factored as the product of primes. To complete the analogy with ordinary integers, he then proved that this factorization is unique, at least up to unit factors, by first demonstrating, using his description of primes, that if any Gaussian prime p divides the product $qrs \cdots$ of Gaussian primes, then p must itself be equal to one of those primes, or one of them multiplied by a unit. Having established the unique factorization property of the Gaussian integers, Gauss considered congruence modulo Gaussian integers and then quartic reciprocity, stated in terms not of ordinary integers but of Gaussian integers.

16.1.2 Fermat's Last Theorem and Unique Factorization

The question of factorization in various domains turned out to be related not only to reciprocity but also to the continued attempts to prove Fermat's Last Theorem. The next mathematician after Euler to make any progress on the theorem was Sophie Germain (1776–1831). In a letter to Gauss written in 1819, she outlined her basic strategy. First, she proved the following

LEMMA. *If $x^p + y^p = z^p$ is a solution to the Fermat equation, where p is an odd prime, and if θ is a prime number with no nonzero consecutive pth-power residues modulo θ, then θ necessarily divides one of the numbers x, y, or z.*

She then noted that if, for a fixed p, one could find infinitely many primes θ satisfying the condition stated, each of these would have to divide one of x, y, or z, and therefore one of these three numbers would be divisible by infinitely many primes, which is absurd. Unfortunately, Germain was not able to find infinitely many primes satisfying this condition, even for a single exponent. But she did, for example, find enough such primes for exponent $p = 5$ to show that any solution would have to have at least 30 decimal digits. In addition, she was able to prove a related result for what has come to be called case 1 of Fermat's Last Theorem, the situation where none of the three numbers x, y, z is divisible by p:

GERMAIN'S THEOREM. *If p is an odd prime, and if there exists a prime θ such that (1) p is not a pth power modulo θ and (2) the equation $r' \equiv r + 1 \pmod{\theta}$ cannot be satisfied for any two pth-power residues, then case 1 of Fermat's Last Theorem is true for p.*

Using this result, Germain was in fact able to prove case 1 of Fermat's Last Theorem for every odd prime up to 97. Unfortunately, her technique did not help to determine the truth of Fermat's Last Theorem in the case where one of x, y, z is divisible by p. In 1825, however, Legendre succeeded in proving the complete result for $p = 5$. And in 1839, Gabriel Lamé (1795–1870) succeeded for the case $p = 7$. These latter proofs were all very long, involved difficult manipulations, and did not appear capable of generalization. If the theorem were to be proved, it seemed that an entirely new approach would be necessary.

Such a new approach to proving Fermat's Last Theorem was announced with great fanfare by Lamé at the Paris Academy meeting of March 1, 1847. Lamé claimed he had solved this longstanding problem and gave a brief sketch of the proof. The basic idea began with the factorization of the expression $x^p + y^p$ over the complex numbers as

$$x^p + y^p = (x + y)(x + \alpha y)(x + \alpha^2 y) \cdots (x + \alpha^{p-1} y),$$

where α is a primitive root of $x^p - 1 = 0$. (Recall that Euler had also used factorization over the complex numbers in his proof for $p = 3$.) Lamé next planned to show that if x and y are such that the factors in this expression are all relatively prime, and if also $x^p + y^p = z^p$, then each of the factors must itself be a pth power. He would then use Fermat's technique of infinite descent to find a solution in smaller numbers. On the other hand, if the factors were not relatively prime, he hoped to show that they had a common factor. Dividing by this factor would then reduce the problem to the first case. When Lamé finished his presentation, Joseph Liouville (1809–1882) took the floor and cast some serious doubt on Lamé's proposal. Basically, he noted that Lamé's idea to conclude that each factor was a pth power because the factors were relatively prime and their product was a pth power depended on the theorem that any integer can be uniquely factored into a product of primes. It was by no means obvious, he concluded, that such a result was true for complex integers of the form $x + \alpha^j y$.

Over the next several weeks, Lamé tried without success to overcome Liouville's objection. Then, on May 24, Liouville read into the proceedings a letter from Ernst Kummer (1810–1893), which effectively ended the discussion. Kummer noted not only that unique factorization fails in some of the domains in question, but that three years earlier he had published an article, admittedly in a somewhat obscure publication, in which he had demonstrated this failure in the case of $p = 23$.

Having discovered nonunique factorization, Kummer spent the next several years fashioning an answer to the question posed at the opening of this chapter. Kummer's basic problem is illustrated by the following example: Let D be the set of complex numbers of the form $m + n\sqrt{-5}$, with m and n ordinary integers, and consider $a = 2$, $b = 3$, $b_1 = -2 + \sqrt{-5}$, $b_2 = -2 - \sqrt{-5}$, $d_1 = 1 + \sqrt{-5}$, and $d_2 = 1 - \sqrt{-5}$, all in D. It is straightforward to show that none of these numbers factors in D, but since $ab = d_1 d_2$, $b^2 = b_1 b_2$, and $ab_1 = d_1^2$, it follows that unique factorization does not hold. If we now suppose that all the numbers involved are rational integers, then, assuming that a and b are relatively prime and that b_1 and b_2 are also, there would necessarily exist integers α, γ, and δ such that $a = \alpha^2$, $b = \gamma\delta$, $b_1 = \gamma^2$, $b_2 = \delta^2$, $d_1 = \alpha\gamma$, and $d_2 = \alpha\delta$. Then, for example, $ab = \alpha^2\gamma\delta = d_1 d_2$. These numbers, however, do not exist in D, because the original numbers do not factor. Kummer's goal was to create substitutes for these new integers and thereby restore unique factorization. In fact, he created a new type of complex number, "ideal complex numbers," which would uniquely factor into "ideal" prime factors. Using his ideal numbers, Kummer defined the notion of a "regular" prime number and then showed that Fermat's Last Theorem was true for such primes. He then demonstrated that all primes less than 100 with the exception of 37, 59, and 67 were regular, thus proving the theorem for those primes. Unfortunately, it turned out that there were infinitely many "irregular" primes, and to this day it is not known whether there are infinitely many regular ones.

Kummer's ideal numbers were, however, not defined set-theoretically, but only in terms of divisibility. Richard Dedekind (1831–1916), however, worked out a new concept to restore unique factorization, the concept of an ideal in a set of algebraic integers. An

algebraic integer is a complex number θ that satisfies an algebraic equation over the rational numbers—that is, an equation of the form

$$\theta^n + a_1\theta^{n-1} + a_2\theta^{n-2} + \cdots + a_{n-1}\theta + a_n = 0,$$

where the a_i are ordinary integers. For example, $\theta = \frac{1}{2} + \frac{1}{2}\sqrt{-3}$ is an algebraic integer because it satisfies the equation $\theta^2 - \theta + 1 = 0$, even though it is not of the "obvious" form $a + b\sqrt{-3}$ with a, b integers. Dedekind then showed that the sum, difference, and product of algebraic integers are also algebraic integers. He defined divisibility in the standard way: An algebraic integer α is **divisible** by an algebraic integer β if $\alpha = \beta\gamma$ for some algebraic integer γ. To develop general laws of divisibility, however, Dedekind needed to restrict himself to only a part of the domain of all algebraic integers. Thus, given any algebraic integer θ that satisfies an irreducible equation of degree n, he defined the system of algebraic integers Γ_θ corresponding to θ to be the set of those algebraic integers of the form $x_0 + x_1\theta + x_2\theta^2 + \cdots + x_{n-1}\theta^{n-1}$, where the x_i are rational numbers. In any such system Γ_θ, Dedekind could now define what it meant for an integer to be prime or irreducible. Modifying Gauss's definition, he called an algebraic integer in a particular system **irreducible** if it could not be factored into two other integers in that system, neither of which is invertible, but **prime** if, whenever it divides a product, it divides one of the factors.

Dedekind noted that the Gaussian integers Γ_i are a system of algebraic integers and that Gauss had proved that unique factorization into primes is true in this system. On the other hand, Dedekind could show, as illustrated above, that unique factorization did not hold in $\Gamma_{\sqrt{-5}}$, the domain determined by a root of $x^2 + 5 = 0$.

Dedekind defined an **ideal** as a subset I of Γ_θ that satisfied the conditions that if α and β were in I and ω in Γ_θ, then $\alpha + \beta$ and $\omega\alpha$ were in I. Furthermore, he defined a **principal ideal** (α) as the set of all multiples of a given integer α. If α and β are both elements of Γ_θ, then the ideal consisting of all integers of the form $r\alpha + s\beta$, with r, s in Γ_θ, is denoted by (α, β). Dedekind's next task was to define divisibility of ideals. He noted that if α is divisible by β, or $\alpha = \mu\beta$, the principal ideal generated by α is contained in that generated by β. Conversely, if every multiple of α is also a multiple of β, then α is divisible by β. Dedekind therefore extended this definition to arbitrary ideals: An ideal I is **divisible** by an ideal J if every number in I is contained in J. An ideal P, different from Γ_θ, is said to be **prime** if it has no divisor other than itself and Γ_θ—that is, if it is contained in no other ideal except Γ_θ itself. For example, the principal ideal (2) in $\Gamma_{\sqrt{-5}}$, although it is generated by an irreducible element, is not prime; it is contained in the prime ideal $(2, 1 + \sqrt{-5})$.

Dedekind noted further that there was a natural definition of a **product** of two ideals, namely, that IJ consists of all sums of products of the form $\alpha\beta$ where α is in I and β is in J. It is then obvious that IJ is divisible by both I and J. To complete the relationship between the two notions of product and divisibility, however, he had to prove two further theorems.

THEOREM 1. *If the ideal C is divisible by an ideal I, then there exists a unique ideal J such that the product IJ is identical with C.*

THEOREM 2. *Every ideal different from Γ_θ either is a prime ideal or may be represented uniquely in the form of a product of prime ideals.*

These two theorems gave Dedekind his new way of restoring unique factorization to any domain of algebraic integers. Thus, returning to our original example $\Gamma_{\sqrt{-5}}$, the (nonexistent) prime factors α, γ, δ were replaced by prime ideals $A = (2, 1 + \sqrt{-5})$, $G = (3, 1 + \sqrt{-5})$, $D = (3, 1 - \sqrt{-5})$ so that the principal ideals $(a) = (2)$, $(b) = (3)$, $(d_1) = (1 + \sqrt{-5})$, $(d_2) = (1 - \sqrt{-5})$ factored as $(a) = A^2$, $(b) = GD$, $(d_1) = AG$, and $(d_2) = AD$. The nonunique integer factorization $ab = d_1 d_2$ could then be replaced by the unique factorization into prime ideals: $(a)(b) = A^2 GD = (d_1)(d_2)$.

16.2 Solving Algebraic Equations

Because solving equations was the central concern of algebra before the nineteenth century, it is not surprising that major features of the new forms that algebra took in that century grew out of new approaches to this problem. In fact, some of the central ideas of group theory grew out of these new approaches.

16.2.1 Cyclotomic Equations

Lagrange had studied in detail the solvability of equations of degree less than 5 and had indicated possible means of attack for equations of higher degree. In the final chapter of his *Disquisitiones arithmeticae*, Gauss discussed the algebraic solution of **cyclotomic equations**, equations of the form $x^n - 1 = 0$, and the application of these solutions to the construction of regular polygons. Because the solution of the equation for composite integers follows immediately from that for primes, Gauss restricted his attention to the case where n is prime, and because $x^n - 1$ factors as $(x - 1)(x^{n-1} + x^{n-2} + \cdots + x + 1)$, it was the equation

$$x^{n-1} + x^{n-2} + \cdots + x + 1 = 0$$

that provided the focus for his work.

Gauss's plan for the solution of this $(n - 1)$st-degree equation was to solve a series of auxiliary equations, each of degree a prime factor of $n - 1$, with the coefficients of each in turn being determined by the roots of the previous equation. Gauss knew that the roots of $x^{n-1} + x^{n-2} + \cdots + x + 1$ could be expressed as powers r^i $(i = 1, 2, \ldots, n - 1)$ of any fixed root r. Furthermore, he realized that if g is any primitive root modulo n, the powers $1, g, g^2, \ldots, g^{n-2}$ include all the nonzero residues modulo n. It follows that the $n - 1$ roots of the equation can be expressed as $r, r^g, r^{g^2}, \ldots, r^{g^{n-2}}$ or even as $r^\lambda, r^{\lambda g}, r^{\lambda g^2}, \ldots, r^{\lambda g^{n-2}}$ for any λ less than n. His method of determining the auxiliary equations involved constructing **periods**, certain sums of the roots r^j, which in turn were the roots of the auxiliary equations.

In the case $n = 19$, where the factors of $n - 1$ are 3, 3, and 2, Gauss began by determining three periods of six terms each, each period to be the root of an equation of degree 3. The periods are found by choosing a primitive root modulo 19, here 2, setting $h = 2^3$, and computing

$$\alpha_i = \sum_{k=0}^{5} r^{ih^k} \qquad \text{for} \quad i = 1, 2, 4.$$

In modern terminology, the permutations of the 18 roots of $x^{18} + x^{17} + \cdots + x + 1 = 0$ form a cyclic group G determined by the mapping $r \to r^2$, where r is any fixed root. The periods here are the sums that are invariant under the subgroup H of G generated by the mapping $r \to r^h$. These sums contain six elements because $h^6 = 2^{18} \equiv 1 \pmod{19}$; that is, H is a subgroup of G of order 6. Furthermore, for $i = 1, 2, 4$, those mappings of the form $r \to r^{ih^k}$ for $k = 0, 1, \ldots, 5$ are precisely the three cosets of H in the group G. For example, because $H = \{r \to r, r^8, r^{64}, r^{512}, r^{4096}, r^{32768}\}$, it follows that

$$\alpha_1 = r + r^8 + r^7 + r^{18} + r^{11} + r^{12},$$

where the powers are reduced modulo 19. Similarly,

$$\alpha_2 = r^2 + r^{16} + r^{14} + r^{17} + r^3 + r^5 \quad \text{and} \quad \alpha_4 = r^4 + r^{13} + r^9 + r^{15} + r^6 + r^{10}.$$

Gauss then showed that $\alpha_1, \alpha_2, \alpha_4$ are roots of the cubic equation $x^3 + x^2 - 6x - 7 = 0$.

The next step is to divide each of the three periods into three further periods of two terms each, where again the new periods will satisfy an equation of degree 3. These periods,

$$\beta_i = \sum_{k=0}^{1} r^{im^k} \quad \text{for} \quad i = 1, 2, 4, 8, 16, 13, 7, 14, 9,$$

where $m = 2^9$, are invariant under the subgroup M generated by the mapping $r \to r^m$. Because $m^2 \equiv 1 \pmod{19}$, M has order 2 and has nine cosets corresponding to the given values of i. For example,

$$\alpha_1 = \beta_1 + \beta_8 + \beta_7 = (r + r^{18}) + (r^8 + r^{11}) + (r^7 + r^{12}).$$

Given these new periods of length 2, Gauss showed that β_1, β_8, and β_7 are all roots of the cubic equation $x^3 - \alpha_1 x^2 + (\alpha_2 + \alpha_4)x - 2 - \alpha_2 = 0$ and that each of the other β_i can be expressed as a polynomial in β_1. Finally, Gauss broke up each of the periods with two terms into the individual terms of which it is formed and showed that, for example, r and r^{18} are the two roots of $x^2 - \beta_1 x + 1 = 0$. The sixteen remaining roots of the original equation are then simply powers of r or can be found by solving eight other similar equations of degree 2.

Because the equations of degree 3 involved in the above example are solvable by the use of radicals, as is the equation of degree 2, Gauss had demonstrated that the roots of $x^{19} - 1 = 0$ are all expressible in terms of radicals. His more general result, applicable to any equation $x^n - 1 = 0$, showed first that one could discover a series of equations, each of prime degree less than $n - 1$, whose solutions would determine the solution of the original equation. He then proved that the auxiliary equations involved in his solution of $x^n - 1 = 0$ for n prime can always be reduced to equations of the form $x^m - A$, equations solvable by radicals. Thus, the original equation is solvable by radicals.

Naturally, if $n - 1$ is a power of 2, all of the auxiliary equations are quadratic and no special proof is necessary. In this case, however, Gauss noted further that the solutions can be constructed geometrically by Euclidean techniques. Because the roots of $x^n - 1 = 0$ can be considered as the vertices of a regular n-gon (in the complex plane), Gauss had proved that such a polygon can be constructed whenever $n - 1$ is a power of 2. The only such primes known to Gauss, and even to us today, are 3, 5, 17, 257, and 65,537. Gauss concluded his discussion by claiming, without proof, that no other regular n-gons (with n prime) can be constructed. This assertion was finally proved by Pierre Wantzel (1814–1848) in 1837.

16.2.2 The Theory of Permutations

Recall that Lagrange had attempted to find a solution to the quintic equation by considering permutations of the roots. To consider the question of higher-degree equations in detail, therefore, it was necessary to understand the theory of permutations. Substantial work on this concept was accomplished early in the nineteenth century by Augustin-Louis Cauchy (1789–1857).

Up to Cauchy's time, the term *permutation* generally referred to an arrangement of a certain number of objects, generally letters. It was Cauchy who first considered the importance of the action of changing from one arrangement to another. He used the word **substitution** to refer to such an action, what would today be called a permutation, that is, a one-to-one function from the given (finite) set of letters to itself. In a series of papers on the subject nearly thirty years after his initial efforts of 1815, Cauchy used the words *substitutions* and *permutations* interchangeably to refer to such functions. To avoid confusion, the latter word will generally be used, in its modern sense, in what follows.

Besides focusing on the functional aspect of a permutation, Cauchy used a single letter, say S, to denote a given permutation and defined the product of two such permutations S, T, written ST, as the permutation determined by first applying S to a given arrangement and then applying T to the resulting one. He named the permutation that leaves a given arrangement fixed the **identity permutation**, noted that the powers $S, S^2, S^3, \ldots$ of a given permutation must ultimately result in the identity, and then defined the **degree** of a permutation S to be the smallest power n such that S^n is equal to the identity. Cauchy even defined what he called a **circular (cyclic) permutation** on the letters $a_1, a_2, \ldots, a_n$ as one that takes a_1 to a_2, a_2 to $a_3, \ldots, a_{n-1}$ to a_n, and a_n to a_1. In 1844, Cauchy introduced the notation $(a_1a_2 \cdots a_n)$ for such a permutation. At that time, he also defined the inverse of a permutation S in the obvious way, using the notation S^{-1}, and introduced the notation 1 for the identity permutation. Further, given any set of substitutions on n letters, he defined what he called the **system of conjugate substitutions** determined by these—today called the **subgroup** generated by the given set—as the collection of all substitutions formed from the original ones by taking all possible products. Finally, he showed that the order of this system (the number of elements in the collection) always divides the order $n!$ of the complete system of substitutions on n letters.

16.2.3 The Unsolvability of the Quintic

Although he could solve cyclotomic equations with radicals, Gauss was convinced that the general equation of degree higher than 4 could not be solved by radicals. This result was finally proved in the mid-1820s by Niels Henrik Abel (1802–1829).

Abel's unsolvability proof involved applying results on permutations to the set of the roots of the equation. It is worth noting, however, that after proving his unsolvability result, Abel continued his research to attempt to solve two problems: (1) to find all equations of any given degree that are algebraically solvable; and (2) to decide whether a given equation is algebraically solvable or not. Although he was not able in what remained of his life to solve either of these questions in its entirety, he did make progress with a particular type of equation. In a paper published in Crelle's *Journal* in 1829, Abel generalized Gauss's solution method for cyclotomic equations. In those equations, the group of permutations of the roots is cyclic. Abel dealt with the case where the group of permutations is commutative and showed, in a solution analogous to Gauss's, that one could always reduce the solution to that of auxiliary equations of prime degree. It is because of this result that commutative groups today are often referred to as *Abelian groups*.

16.2.4 The Work of Galois

Although Abel could not complete his research program, this was largely accomplished by another genius who died young, Evariste Galois (1811–1832). Galois's thoughts on the subject of solvability of algebraic equations by radicals are outlined in the manuscript he submitted to the French Academy in 1831 and in a letter he wrote to his friend Auguste Chevalier just before the duel that ended his life. In that manuscript, he began by clarifying the idea of rationality. Since an equation has coefficients in a certain domain—for example, the set of ordinary rational numbers—to say that an equation is solvable by radicals means that one can express any root using the four basic arithmetic operations and the operation of root extraction, all applied to elements of this original domain. It turns out, however, that it is usually convenient to solve an equation in steps, as Gauss did in the cyclotomic case. Therefore, once one has solved $x^n = \alpha$, for instance, one has available as coefficients in the next step these solutions, expressible as $\sqrt[n]{\alpha}$, $r\sqrt[n]{\alpha}$, $r^2\sqrt[n]{\alpha}, \ldots$, where r is an nth root of unity. Galois noted that such quantities are adjoined to the original domain and that any quantity expressible by the four basic operations in terms of these new quantities and the original ones can then also be considered as **rational**. (In modern terminology, one begins with a particular field, then constructs an extension field by adjoining certain quantities not in the original field.) Galois also discussed in his introduction the notion of a permutation, using the same somewhat ambiguous language as Cauchy, and uses the word *group*, although not in a strictly technical sense, sometimes to refer to a set of permutations that is closed under composition and other times just to refer to a set of arrangements of letters determined by applying certain permutations.

Galois's main result is expressed as

PROPOSITION I. *Let an equation be given of which a, b, c, ... are the m roots.* [Galois tacitly assumes that this equation is irreducible and that all the roots are distinct.] *There will always be a group of permutations of the letters a, b, c, ... which has the following property: (1) that every function of the roots, invariant under the substitutions of the group, is rationally known; (2) conversely, that every function of the roots which is rationally known is invariant under the substitutions.*

Galois called this group of permutations the **group of the equation**. In modern usage, one normally considers the group of the equation as a group of automorphisms acting on the entire field created by adjoining the roots of the equation to the original field of coefficients. Galois's result then is that the group of the equation is that group of automorphisms of the extension field that leaves invariant precisely the elements of the original field (the elements "rationally known"). Besides giving a brief proof of his result, Galois presented Gauss's example of the cyclotomic polynomial $(x^n - 1)/(x - 1)$ for n prime. In that case, supposing that r is one root and g is a primitive root modulo n, the roots can be expressed in the form $a = r$, $b = r^g$, $c = r^{g^2}, \ldots$, and the group of the equation is the cyclic group of $n - 1$ permutations generated by the cycle $(abc \ldots k)$. On the other hand, the group of the general equation of degree n—that is, of the equation with literal coefficients—is the group of all $n!$ permutations of n letters.

Having stated the main theorem, Galois explored its application to the solvability question. His second proposition shows what happens when one adjoins to the original field one or all of the roots of some auxiliary equation (or of the original equation). Because any automorphism that leaves the new field invariant certainly leaves the original field invariant, the group H of the equation over the new field is a subgroup of the group G over the original field. In fact, G can be decomposed either as $G = H + HS + HS' + \cdots$ or as $G = H + TH + T'H + \cdots$, where S, S', T, T', ... are appropriately chosen permutations. Galois explained this entire procedure and noted that ordinarily these two decompositions do not coincide. When they do, however—and this always happens when all the roots of an auxiliary equation are adjoined—he called the decomposition **proper**. In modern terminology, a proper decomposition occurs when the subgroup H is **normal**—that is, when the right cosets $\{HS\}$ coincide with the left cosets $\{TH\}$. In these circumstances, the question of solvability reduces to the solvability of two equations each having groups of order less than the original one.

Gauss had already shown that the roots of the polynomial $x^p - 1$ with p prime can be expressed in terms of radicals. It follows that if the pth roots of unity are assumed to be in the original field, then the adjunction of one root of $x^p - \alpha$ amounts to the adjunction of all of the roots. If G is the group of an equation, this adjunction therefore leads to a normal subgroup H of the group G such that the index of H in G (the quotient of the order of G by that of H) is p. Galois also proved the converse: that if the group G of an equation has a normal subgroup of index p, then one can find an element α of the original field (assuming that the pth roots of unity are in that field) such that adjoining $\sqrt[p]{\alpha}$ reduces the group of the equation to H. Galois concluded that an equation is solvable by radicals as long as one can continue the process of finding normal subgroups until all of the resulting indices are prime. Galois gave the details of this procedure in the case of the general equation of degree 4, showing that the group of the equation of order 24 has a normal subgroup of order 12, which in turn contains one of order 4, which contains one of order 2, which contains the identity. It follows that the solution can be obtained by first adjoining a square root, then a cube root, and then two more square roots. Galois noted that the standard solution to the quartic equation uses precisely those steps. He also showed that an irreducible equation of prime degree is solvable by radicals if and only if each of the permutations transforms a root x_k into a root $x_{k'}$ with $k' \equiv ak + b \pmod{p}$. Since in the case $p = 5$, this group of permutations has 20 elements, while the group of the general fifth-degree polynomial has 120 elements, the general quintic is not solvable.

16.2.5 Jordan and the Theory of Groups of Substitutions

After Galois's death, his manuscripts lay unread until they were finally published in 1846 by Liouville in his *Journal des mathématiques*. Within the next few years, several mathematicians included Galois's material in university lectures or published commentaries on the work. In 1870, Camille Jordan (1838–1922) published his monumental *Treatise on Substitutions and Algebraic Equations*, which contained a somewhat revised version of Galois theory, among much else.

It is in Jordan's text, and in some of his papers of the preceding decade (which are essentially incorporated in the *Treatise*), that many modern notions of group theory first appear, although always in the context of groups of permutations (substitutions). Thus, Jordan defined a **group** as a system of permutations of a finite set with the condition that the product (composition) of any two such permutations belongs to the system. He could then show that every group contains a unit element 1 and, for every permutation a, another permutation a^{-1} such that $aa^{-1} = 1$. Jordan defined the **transform** of a permutation a by a permutation b to be the permutation $b^{-1}ab$ and the transform of the group $A = \{a_1, a_2, \ldots, a_n\}$ by b to be the group $B = \{b^{-1}a_1b, b^{-1}a_2b, \ldots, b^{-1}a_nb\}$ consisting of all the transforms. If B coincides with A, then A is said to be **permutable** with b. Although Jordan did not explicitly define a normal subgroup of a group, he did define a **simple group** as one that contains no subgroup (other than the identity) that is permutable with all elements of the group. For a nonsimple group G, there must then exist a **composition series**, a sequence of groups $G = H_0, H_1, H_2, \ldots, \{1\}$ such that each group is contained in the previous one and is permutable with all its elements (that is, is normal), and that no other such group can be interposed in this sequence. Jordan further proved that if the order of G is n and the orders of the subgroups are successively $n/\lambda, n/\lambda\mu, n/\lambda\mu\nu, \ldots$, then the integers $\lambda, \mu, \nu, \ldots$, are unique up to order; that is, any other such sequence has the same **composition factors**.

Jordan investigated in particular the set of groups that today are referred to as the **classical linear groups**. These are groups of what Jordan called linear substitutions and what are now written as $n \times n$ matrices operating on vectors in n-space. In general, the field of coefficients for these linear substitutions is the finite field of p elements, although Jordan did not refer to this set as a field. Among the groups he studied are the groups $GL(n, p)$ of all invertible linear transformations on n variables modulo p (the general linear group); $SL(n, p)$, the group of all such transformations with determinant 1 (the special linear group); and $PSL(n, p)$, the quotient group of $SL(n, p)$ by its subgroup of multiples of the identity matrix (the projective special linear group). For example, $SL(2, 5)$ consists of 2×2 matrices with each entry an integer between 0 and 4 and the determinant $ad - bc \equiv 1 \pmod 5$, and $PSL(2, 5)$ consists of equivalence classes of elements from $SL(2, 5)$ modulo the subgroup consisting of the identity matrix I and the matrix $4I$. The order of $SL(2, 5)$ is 120, and that of $PSL(2, 5)$ is 60. Jordan was able to show that $PSL(2, p)$ is a simple group for $p > 3$ and that $PSL(n, p)$ is simple for all $n \geq 3$. It was in fact the existence of these simple groups that led twentieth-century mathematicians to try to find all possible finite simple groups.

Note that if in the two-dimensional space of vectors with coefficients in the field of p elements, we put $(x_1, y_1) \equiv (x_2, y_2)$ if $x_1/y_1 = x_2/y_2$ (where $x_1/0$ is defined to be ∞), the set of equivalence classes is called $P_1(p)$, the **one-dimensional projective space** over the field with p elements. In this situation, we can consider the group $PSL(2, p)$ as the group of linear fractional transformations of the form

$$z' \equiv \frac{az + b}{cz + d} \pmod p,$$

with $ad - bc = 1$, where $z = x/y$; that is, $PSL\ (2, p)$ is a transformation group acting on the projective space $P_1(p)$. This group of linear fractional transformations is often called the **modular group**. Of course, one can generalize this entire construction and consider $PSL(n, p)$ as a group acting on projective space of dimension $n - 1$.

Jordan used some of the group-theoretic concepts he developed to restate some of Galois's results. He defined a **solvable group** as one that belongs to an equation solvable by radicals. Thus, a solvable group is one that contains a composition series with all composition factors prime. Because a commutative group always has prime composition factors, Jordan could show that an **Abelian equation**, one whose group contains only substitutions that are interchangeable among themselves, is always solvable by radicals. On the other hand, because the alternating group on n letters, which has order $n!/2$, is simple for $n > 4$, it follows immediately that the general equation of degree n is not solvable by radicals. With Jordan's work clarifying that of Galois, it became evident that the theory of permutation groups was intimately connected with the solvability of equations.

16.3 Groups and Fields—The Beginning of Structure

Certain concepts of group theory were implicit in the early nineteenth-century developments in number theory and the solvability of equations by radicals, in both of which areas Gauss played a significant role. Gauss's work in the theory of quadratic forms was also important in bringing to the fore ideas that were ultimately to be part of the abstract theory of groups. In addition, the understanding grew among mathematicians that certain sets of geometric transformations could also be considered as "groups."

16.3.1 Gauss and Quadratic Forms

Gauss discussed the theory of quadratic forms—that is, functions of two variables x, y of the form $ax^2 + 2bxy + cy^2$, with a, b, c integers—in chapter 5 of his *Disquisitiones*. Gauss's primary aim in his discussion of forms was to determine whether a given integer can be represented by a particular form. As a tool in the solution of this problem, he defined equivalence of two forms. A form $f = ax^2 + 2bxy + cy^2$ is **equivalent** to a form $f' = a'x'^2 + 2b'x'y' + cy'^2$ if there exists a linear substitution $x = \alpha x' + \beta y',\ y = \gamma x' + \delta y'$, with $\alpha\delta - \beta\gamma = 1$, that transforms f into f'. An easy calculation shows that any two equivalent forms have the same discriminant $b^2 - ac$. On the other hand, two forms with the same discriminant are not necessarily equivalent. Gauss was able to show that for any given value D of the discriminant there were finitely many classes of equivalent forms. In particular, there was a distinguished class, the principal class, consisting of those forms equivalent to the form $x^2 - Dy^2$.

To investigate these classes, Gauss presented a rule of composition for forms. In other words, given forms f, f', with the same discriminant, Gauss defined a new form F composed of f, f' (written $F = f + f'$), which had certain desirable properties. First, Gauss showed that if f and g are equivalent and if f' and g' are equivalent, then $f + f'$ is equivalent to $g + g'$. Therefore, the composition operation is an operation on classes. Gauss next showed that the operation of composition is both commutative and associative. Finally, Gauss showed that if any class K is composed with the principal class, the result will be the class K itself; that for any class K there is a class L (opposite to K) such that the composite of the two is the principal class; and that given any two classes K, L of the same discriminant, there is a class M with the same discriminant such that L is composed of M and K.

With the addition sign as the sign of operation, Gauss designated the composite of a class C with itself by $2C$, the composite of C with $2C$ as $3C$, and so on. Gauss then proved that for any class C, there is a smallest multiple mC that is equal to the principal class and that, if the total number of classes is n, then m is a factor of n. Naturally, this result reminded him of earlier material in the *Disquisitiones*. "The demonstration of the preceding theorem is quite analogous to the demonstrations [on powers of residue classes] and, in fact, the theory of the [composition] of classes has a great affinity in every way with the subject treated [earlier]." He could therefore assert, without proof, various other results that came from this analogy, in terms of what is now the theory of Abelian groups.

16.3.2 Kronecker and the Structure of Abelian Groups

Gauss, although he recognized the analogy between his two treatments, did not attempt to develop an abstract theory of groups. This development took many years and encompassed several other examples. It was Leopold Kronecker (1823–1891) who finally saw that an abstract theory could be developed out of these analogies. In a paper of 1870 in which he developed certain properties of the number of classes of Kummer's ideal complex numbers, Kronecker recalled Gauss's work on quadratic forms and noted that the principles he used occur in many situations, including the one with which Kronecker was dealing. Thus, he abstracted these: "Let $\theta', \theta'', \theta''', \ldots$ be finitely many elements such that to each pair there is associated a third by means of a definite procedure." Kronecker went on to require that this association, written as $\theta'\,\theta'' = \theta'''$, be commutative and associative and that, if $\theta'' \neq \theta'''$, then $\theta'\theta'' \neq \theta'\theta'''$. From the finiteness assumption, Kronecker then deduced the existence of a unit element 1 and, for any element θ, the existence of a smallest power n_θ such that $\theta^{n_\theta} = 1$.

Finally, Kronecker developed the fundamental theorem of Abelian groups, that there exists a finite set of elements $\theta_1, \theta_2, \ldots, \theta_m$ such that every element θ can be expressed uniquely as a product of the form $\theta_1^{h_1}\theta_2^{h_2}\cdots\theta_m^{h_m}$, where for each i, $0 \leq h_i < n_{\theta_i}$. Furthermore, the θ_i can be arranged so that each n_θ is divisible by its successor and the product of these numbers is precisely the number of elements in the system. Having proved the abstract theorem, Kronecker interpreted the elements in various ways, noting that analogous results in each case had been proved previously by others.

16.3.3 Groups of Transformations

Another source of the development of the group concept was the study of transformations of a geometric space. It was Felix Klein (1849–1925) who detailed this idea in his *Erlanger Programm* of 1872. This paper discussed the notion that the various geometrical studies of the nineteenth century could all be unified and classified by viewing geometry in general as the study of those properties of figures that remained invariant under the action of a particular group of transformations on the underlying space. (For Klein, a "group" of transformations was a set in which the composition of any two transformations was also in the set.)

Klein provided several examples of geometries and their associated groups. Ordinary Euclidean geometry in two dimensions corresponded to what Klein called the **principal group**, the group composed of all rigid motions of the plane along with similarity transformations and reflections. It is the invariants under these transformations that form the object of study in classical Euclidean geometry. Projective geometry consists of the study of those figures that are left unchanged by projections, transformations that take lines into lines. Since projections can also be thought of as transformations of two-dimensional projective space, these transformations can be expressed analytically as linear fractional transformations, namely, transformations of the form

$$x' = \frac{a_{11}x + a_{12}y + a_{13}}{a_{31}x + a_{32}y + a_{33}}, \qquad y' = \frac{a_{21}x + a_{22}y + a_{23}}{a_{31}x + a_{32}y + a_{33}},$$

where $\det(a_{ij}) \neq 0$. Because the principal group can be expressed analytically as the set of transformations of the form

$$x' = ax - by + c, \qquad y' = bex + aey + d,$$

with $a^2 + b^2 \neq 0$ and $e = \pm 1$, it is clear that it is a subgroup of the projective group. It then follows that there are fewer invariants of the latter than of the former, and so any theorem of projective geometry remains a theorem in Euclidean geometry, but not conversely.

16.3.4 Axiomatization of the Group Concept

The first mathematician to attempt an abstract definition of the group concept was Arthur Cayley (1821–1895), in his "On the Theory of Groups" of 1854. Cayley noted that the idea of a group of permutations was due to Galois and immediately proceeded to generalize it to any set of operations, or functions, on a set of quantities. He used the symbol 1 to represent the function that leaves all quantities unchanged and noted that for functions, there is a well-defined notion of composition that is associative, although not, in general, commutative. But then Cayley abstracted the basic ideas out of the concrete notion of operations and defined a **group** to be a "set of symbols, 1, α, β, ..., all of them different, and such that the product of any two of them (no matter in what order), or the product of any one of them into itself, belongs to the set ...." From a modern point of view, Cayley left out a significant portion of the definition, but evidently he was thinking of groups of permutations on finite sets and therefore assumed that the set of symbols was finite, that the product was associative, and that every symbol had an inverse, another symbol whose product with the first was 1.

To study different abstract groups, Cayley introduced the group table

	1	α	β	...
1	1	α	β	...
α	α	α^2	$\beta\alpha$	...
β	β	$\alpha\beta$	β^2	...
⋮	⋮	⋮	⋮	

in which each row and each column contains all the symbols of the group. He noted that if there are n elements in the group, then each element θ satisfies the symbolic equation $\theta^n = 1$. Cayley then showed by a familiar argument that if n is prime, the group is necessarily of the form $1, \alpha, \alpha^2, \ldots, \alpha^{n-1}$. If n is not prime, there are other possibilities. In particular, he displayed the group tables of the two possible groups of four elements and the two possible groups of six elements. In a paper of 1859, he described all five groups of order 8 by giving a list of their elements and defining relations as well as the smallest power (index) of each element that equals 1. For example, one of these groups contains the elements

$1, \alpha, \beta, \beta\alpha, \gamma, \gamma\alpha, \gamma\beta, \gamma\beta\alpha$ with the relations $\alpha^2 = 1$, $\beta^2 = 1$, $\gamma^2 = 1$, $\alpha\beta = \beta\alpha$, $\alpha\gamma = \gamma\alpha$, and $\beta\gamma = \gamma\beta$. Each element in this group, except the identity, has index 2.

Unfortunately, Cayley's work generated no interest either in England or on the continent for over two decades. But beginning in 1879, many mathematicians began to realize that it was worthwhile to combine Kronecker's and Cayley's definitions into a single abstract group concept. This process culminated in 1882 in work by both Walter Dyck (1856–1934) and Heinrich Weber (1842–1913). Dyck was familiar with the group notion as it had occurred in algebraic equations, number theory, and geometry and could identify the critical ideas of the theory. He knew that the group operations must obey the associative but not necessarily the commutative property and that each group element must have an inverse. But rather than give an axiomatic definition, he showed how to create what we call finitely generated groups by beginning with a finite number of elements $A_1, A_2, \ldots, A_m$ and considering all possible products of powers of these elements and their inverses. This group is today called the **free group on $\{A_i\}$**. Dyck then specialized to other groups by assuming various relations among the A_i.

But it was Weber, in a paper on quadratic forms, who gave a complete axiomatic description of a finite group without any reference to the nature of the elements composing it:

> A system G of h elements of any sort, $\theta_1, \theta_2, \ldots, \theta_h$, is called a group of order h if it satisfies the following conditions:
> I. Through some rule, which is called composition or multiplication, one derives from any two elements of the system a new element of the same system. In signs, $\theta_r\theta_s = \theta_t$.
> II. Always $(\theta_r\theta_s)\theta_t = \theta_r(\theta_s\theta_t) = \theta_r\theta_s\theta_t$.
> III. From $\theta\theta_r = \theta\theta_s$ and from $\theta_r\theta = \theta_s\theta$, there follows $\theta_r = \theta_s$.

From the given axioms and the finiteness of the group, Weber derived the existence of a unique unit element and, for each element, the existence of a unique inverse. He further defined a group to be an **Abelian group** if the multiplication is commutative and then proved the fundamental theorem of Abelian groups by essentially the same method used by Kronecker.

Although the use of the abstract group concept became more common over the next several years, it was not until 1893 that Weber published a definition that included infinite groups. He repeated his three conditions of 1882 and noted that if the group is finite, these conditions suffice to ensure that if any two of the three group elements A, B, C are known, there is a unique solution to the equation $AB = C$. On the other hand, this conclusion is no longer valid for infinite groups. In that case, one must assume the existence of unique solutions to $AB = C$ as a fourth axiom. This fourth axiom, even without finiteness, implies a unique identity and unique inverses for every element of the group.

Weber produced many examples of groups, including the additive group of vectors in the plane, the group of permutations of a finite set, the additive group of residue classes modulo m, the multiplicative group of residue classes modulo m relatively prime to m, and the group of classes of binary quadratic forms of a given discriminant under Gauss's law of composition. With the publication of this material, and its incorporation in Weber's 1895 *Textbook on Algebra*, the abstract concept of a group can be considered to have become part of the mathematical mainstream.

The notion of an abstract group having been defined, many mathematicians attempted to find general theorems about the structure of groups or to determine all finite groups of a given order. For example, Georg Frobenius (1849–1917) re-proved abstractly the theorem that if p^n is the largest power of a prime p that divides the order of a finite group, then the group has at least one subgroup of order p^n and, in fact, the number of such subgroups divides the order of the group and is congruent to 1 modulo p. This theorem had originally been proved by Ludvig Sylow (1832–1918) for permutation groups. Also, Otto Hölder (1859–1937), in his work on Galois theory in 1889, defined the notion of a **factor group** (or **quotient group**) and showed how these are Galois groups of the auxiliary equations that may come up in the process of solving a particular solvable equation. He further showed that in the composition series one gets in accomplishing the solution, the actual factor groups are unique, and not just their order. This result is now called the **Jordan-Hölder theorem**. Then in several papers in the early 1890s, Hölder studied finite groups of various orders and worked out some structure theorems. In particular, he determined the possible groups of orders p^2, p^3, p^4, pq, pq^2, and pqr, where p, q, r are distinct primes. The cases p^2 and pq had been proved earlier for permutation groups, but Hölder redid the proofs in a more abstract manner. For example, given a group of order pq with $p > q$, the abstract version of the Sylow theorem shows that there is one subgroup of order p and, if q does not divide $p - 1$, one subgroup of order q. In that case, the group is cyclic. On the other hand, if q does divide $p - 1$, then there is a second group that is generated by two elements S and T, with $S^q = 1$, $T^p = 1$, and $S^{-1}TS = T^r$, with $r \not\equiv 1 \pmod{p}$ and $r^q \equiv 1 \pmod{p}$.

16.3.5 The Concept of a Field

The story of field theory is much simpler to tell than that of group theory. The notion of field is certainly implicit in Galois's work around 1830. Recall that Galois discussed what it meant for quantities to be rational and how to adjoin a new element to a given set of rational quantities. For Galois, the notions of the rational number field Q and of an extension field $Q(\alpha)$ generated by either a transcendental quantity or a root of a given equation were intuitively obvious, and there was no need to name this concept. It was Kronecker, beginning in the 1850s, who tried to be more specific by actually constructing these fields.

Kronecker began with the integers and their quotients, the rational numbers. He was then able to construct the field we call $Q(\sqrt{2})$, for example, by considering the remainders of polynomials with rational coefficients on division by $x^2 - 2$. Because two polynomials with the same remainder are considered equal, it was straightforward to define the basic operations on this set of remainders and thereby construct a new set with the same basic properties as the rationals. Another way of looking at this construction is simply to consider the new set as being formed by adding a new element α to the rationals as well as all rational functions of α, with the condition that α^2 is always replaced by 2.

Dedekind, also beginning in the 1850s, was more concerned with the set of elements itself than with the process of adjunction. Recall that Dedekind was interested in the arithmetic of algebraic integers, complex numbers that could be expressed as roots of algebraic equations. Thus, Dedekind gave the following definition in his supplement to the 1871 edition of *Lectures on Number Theory* by Peter Lejeune-Dirichlet (1805–1859): "A system A of real or complex numbers α is called a *field* if the sum, difference, product and quotient of every pair of these numbers α belongs to the same system A." (He noted that zero cannot be a denominator in any such quotient and that a field must contain at least one number other than zero.) The smallest such system, of course, is the field of rational numbers, which is contained in every field, while the largest such system is the field of complex numbers, which contains every field. Thus, for Dedekind, unlike Kronecker, the adjunction of an algebraic element to a field always takes place in the field of complex numbers.

For both Dedekind and Kronecker, every field contained the field of rational numbers. Neither attempted to extend his definition to other types of fields, even though as far back as 1830 Galois had published a brief paper that in essence described finite fields. Galois's aim in that paper was to generalize Gauss's ideas in solving congruences of the form $x^2 \equiv a \pmod{p}$. Galois asked what would happen if, when a solution did not exist, one created a solution, exactly as one created the solution i to $x^2 + 1 = 0$. Thus, designating a solution to an arbitrary congruence $F(x) \equiv 0 \pmod{p}$ by the symbol i (where $F(x)$ is a polynomial of degree n and no residue modulo p is itself a solution), Galois considered the collection of p^n expressions $a_0 + a_1 i + a_2 i^2 + \cdots + a_{n-1} i^{n-1}$ with $0 \leq a_j < p$ and noted that these expressions can be added, subtracted, multiplied, and divided in the familiar manner. Galois next noted that if α is any of the nonzero elements of his set, some smallest power n of α must be equal to 1; then, by arguments analogous to those of Gauss in the case of residues modulo p, he showed that all such elements satisfy $\alpha^{p^n - 1} \equiv 1$ and that there is a primitive root β such that every nonzero element is a power of β. Galois concluded the paper by showing that for every prime power p^n, one can find an irreducible nth-degree congruence modulo p, a root of which generates what is today called the **Galois field of order p^n**. The simplest way to find such a polynomial, Galois remarked, is by trial and error. As an example, he showed that $x^3 - 2$ is irreducible modulo 7 and therefore that the set of elements $\{a_0 + a_1 i + a_2 i^2\}$, with i a zero of that polynomial and $0 \leq a_j < 7$ for $j = 0, 1, 2$, forms the field of order 7^3.

It was Heinrich Weber who combined the Dedekind-Kronecker version of a field with the finite systems of Galois into an abstract definition of a field in the same paper of 1893 in which he gave an abstract definition of a group. In fact, he used the notion of group in his definition: A **field** is a set with two forms of composition, addition and multiplication, under the first of which it is a commutative group and under the second of which the set of nonzero elements is a commutative group. Furthermore, the two forms of composition are related by the following rules: $a(-b) = -ab$; $a(b + c) = ab + ac$; $(-a)(-b) = ab$; and $a \cdot 0 = 0$. Weber further noted that in a field, a product can only be zero when one of the factors is zero. He then gave several examples of fields, including the rational numbers, the finite fields (of which he only cited the residue classes modulo a prime), and the fields of rational functions in one or more variables over a given field F. As in the case of groups, the notion of a field was incorporated into Weber's algebra text, thus enabling a new generation of students to understand this abstract concept.

16.4 Matrices and Systems of Linear Equations

The idea of a matrix has a long history, dating at least from its use by Chinese scholars of the Han period for solving systems of linear equations. In the eighteenth century, and even somewhat earlier, mathematicians calculated and used determinants of square arrays of numbers, often in the solution of systems of linear equations, even though the square arrays themselves were not singled out for attention. Other work in the nineteenth century led to more formal computations with such arrays and, by mid-century, to a definition of a matrix and the development of the algebra of matrices. But alongside this formal work, there was a more theoretical side to the development of the theory of matrices, namely, the work growing out of Gauss's study of quadratic forms, which ultimately led to the concepts of similarity, eigenvalues, and diagonalization.

16.4.1 Basic Ideas of Matrices

Recall that Gauss, in his theory of binary quadratic forms, considered the idea of a linear substitution that transforms one form into another; namely, if $F = ax^2 + 2bxy + cy^2$, then the substitution

$$x = \alpha x' + \beta y'$$
$$y = \gamma x' + \delta y'$$

converts F into a new form F' whose coefficients depend on the coefficients of F and those of the substitution. Gauss noted that if F' is transformed into F'' by a second linear substitution

$$x' = \epsilon x'' + \zeta y''$$
$$y' = \eta x'' + \theta y'',$$

then the composition of the two substitutions gives a new substitution transforming F into F'':

$$x = (\alpha\epsilon + \beta\eta)x'' + (\alpha\zeta + \beta\theta)y''$$
$$y = (\gamma\epsilon + \delta\eta)x'' + (\gamma\eta + \delta\theta)y''$$

The coefficient matrix of the new substitution is the product of the coefficient matrices of the two original substitutions. Gauss performed an analogous computation in his study of ternary quadratic forms $Ax^2 + 2Bxy + Cy^2 + 2Dxz + 2Eyz + Fz^2$, which in effect gave the rule for multiplying two 3×3 matrices. But although he wrote the coefficients of the substitution in a rectangular array and even used a single letter S to refer to a particular substitution, Gauss did not explicitly refer to this idea of composition as a multiplication.

In 1815, Cauchy published a fundamental memoir on the theory of determinants, in which he not only introduced the name **determinant** to replace several older terms, but also used the abbreviation $(a_{1,n})$ to stand for what he called the "symmetric system"

$$\begin{matrix} a_{1,1} & a_{2,2} & \cdots & a_{1,n} \\ a_{2,1} & a_{2,2} & \cdots & a_{2,n} \\ \vdots & \vdots & \ddots & \vdots \\ a_{n,1} & a_{n,2} & \cdots & a_{n,n} \end{matrix}$$

to which the determinant is associated. Although many of the basic results on calculating determinants had been known earlier, Cauchy's memoir gave the first complete treatment of these, including such ideas as the array of minors associated to a given array (the **adjoint**) and the procedure for calculating a determinant by expanding on any row or column. In addition, he followed Gauss in explicitly recognizing the idea of composing two systems

$(\alpha_{1,n})$ and $(a_{1,n})$ to get a new system $(m_{1,n})$, which is defined by the familiar law of multiplication

$$m_{i,j} = \sum_{k=1}^{n} \alpha_{i,k} a_{k,j}.$$

He then showed that the determinant of the new system was the product of those of the two original ones.

In a paper of 1844, Ferdinand Gotthold Eisenstein (1823–1852), a student of Gauss, introduced the explicit notation $S \times T$ to denote the substitution composed of S and T in his discussion of ternary quadratic forms, perhaps because of Cauchy's product theorem for determinants. Although Eisenstein realized that a calculation algorithm could be based on this notation, keeping in mind that "multiplication" was not commutative, he never developed fully his idea because of his untimely death at the age of 29. That development was carried out in England by Arthur Cayley and James Joseph Sylvester (1814–1897) in the 1850s.

In 1850, Sylvester coined the term **matrix** to denote "an oblong arrangement of terms consisting, suppose, of m lines and n columns" because out of that arrangement "we may form various systems of determinants." (The English word *matrix* meant "the place from which something else originates.") Sylvester himself made no use of the term at the time. It was his friend Cayley who put the terminology to use in papers of 1855 and 1858. In the earlier paper, Cayley noted that the use of matrices is very convenient for the theory of linear equations. Thus, he wrote

$$(\xi, \eta, \zeta, \ldots) = \begin{pmatrix} \alpha, & \beta, & \gamma, & \cdots \\ \alpha', & \beta', & \gamma', & \cdots \\ \alpha'', & \beta'', & \gamma'', & \cdots \\ \vdots & \vdots & \vdots & \vdots \end{pmatrix} (x, y, z, \ldots)$$

to represent the square system of equations

$$\begin{aligned} \xi &= \alpha x + \beta y + \gamma z + \cdots \\ \eta &= \alpha' x + \beta' y + \gamma' z + \cdots \\ \zeta &= \alpha'' x + \beta'' y + \gamma'' z + \cdots \\ \vdots & \quad \vdots \quad\quad \vdots \quad\quad \vdots \end{aligned}$$

He then determined the solution of this system using what he called the inverse of the matrix:

$$(x, y, z, \ldots) = \begin{pmatrix} \alpha, & \beta, & \gamma, & \cdots \\ \alpha', & \beta', & \gamma', & \cdots \\ \alpha'', & \beta'', & \gamma'', & \cdots \\ \vdots & \vdots & \vdots & \vdots \end{pmatrix}^{-1} (\xi, \eta, \zeta, \ldots).$$

This representation came from the basic analogy of the matrix equation to a simple linear equation in one variable. Cayley, however, knowing Cramer's rule, then described the entries of the inverse matrix in terms of fractions involving the appropriate determinants.

In 1858, Cayley introduced single-letter notation for matrices and showed not only how to multiply but also how to add and subtract them. He then exploited this idea, making constant use of the analogy between ordinary algebraic manipulations and those with matrices, but carefully noting where this analogy failed. Thus, using the formula for the inverse of a 3×3 matrix, he wrote that "the notion of the inverse ... matrix fails altogether when the determinant vanishes; the matrix is in this case said to be indeterminate .... It may be added that the matrix zero is indeterminate; and that the product of two matrices may be zero, without either of the factors being zero, if only the matrices are one or both of them indeterminate."

It was perhaps Cayley's use of the notational convention of single letters for matrices that suggested to him the result known as the **Cayley-Hamilton theorem**. For the case of a 2×2 matrix

$$M = \begin{pmatrix} a & b \\ c & d \end{pmatrix},$$

Cayley stated this result explicitly as

$$\det \begin{pmatrix} a - M & b \\ c & d - M \end{pmatrix} = 0.$$

Cayley first communicated this "very remarkable" theorem in a letter to Sylvester in November of 1857. In 1858, he proved it by simply showing that the determinant $M^2 - (a + d)M^1 + (ad - bc)M^0$ equaled zero (where M^0 is the identity matrix). Stating the general version in essentially the modern form that M satisfies the equation in λ, $\det(M - \lambda I) = 0$, the **characteristic equation**, Cayley noted that he had "verified" the theorem in the 3×3 case, but wrote further that "I have not thought it necessary to undertake the labour of a formal proof of the theorem in the general case of a matrix of any degree." It was Frobenius who took advantage of Cayley's notational innovation to give a complete proof some twenty years later.

16.4.2 Eigenvalues and Eigenvectors

Deeper results about matrices were based, not on matrix manipulation, but on spectral theory, the results surrounding the concept of an eigenvalue. In modern terminology, an **eigenvalue** of a matrix is a solution λ either of the matrix equation $AX = \lambda X$, where A is an $n \times n$ matrix and X is an $n \times 1$ matrix, or of $XA = \lambda X$, where A is $n \times n$ and X is $1 \times n$. An **eigenvector** corresponding to the eigenvalue λ is a vector X that satisfies the same equation. These concepts, in their origins and later development, were independent of matrix theory *per se*; they grew out of a study of various ideas that ultimately were included in that theory. Thus, the context within which the earliest eigenvalue problems arose during the eighteenth century was that of the solution of systems of linear differential equations with constant coefficients.

In works dating from 1743 to 1758, d'Alembert, motivated by the consideration of the motion of a string loaded with a finite number of masses (here restricted for simplicity to three), considered the system

$$\frac{d^2 y_i}{dt^2} + \sum_{k=1}^{3} a_{ik} y_k = 0 \qquad i = 1, 2, 3.$$

To solve this system, he multiplied the ith equation by a constant v_i for each i and added the equations together to obtain

$$\sum_{i=1}^{3} v_i \frac{d^2 y_i}{dt^2} + \sum_{i,k=1}^{3} v_i a_{ik} y_k = 0.$$

If the v_i are then chosen so that $\sum_{i=1}^{3} v_i a_{ik} + \lambda v_k = 0$ for $k = 1, 2, 3$, that is, if (v_1, v_2, v_3) is an eigenvector corresponding to the eigenvalue $-\lambda$ for the matrix $A = (a_{ik})$, the substitution $u = v_1 y_1 + v_2 y_2 + v_3 y_3$ reduces the original system to the single differential equation

$$\frac{d^2 u}{dt^2} + \lambda u = 0,$$

an equation which, after Euler's work on differential equations, could be easily solved and led to solutions for the three y_i. A study of the three equations in which it appears shows that λ is determined by a cubic equation with three roots. D'Alembert realized that for the solutions to make physical sense they had to be bounded as $t \to \infty$. This, in turn, would only be true provided that the three values of λ were distinct, real, and positive.

It was Cauchy who first solved the problem of determining in a special case the nature of the eigenvalues from the nature of the matrix (a_{ik}) itself. In all probability, he was influenced not by d'Alembert's work on differential equations, but by the study of quadric surfaces, a study necessary as part of the analytic geometry that Cauchy was teaching from 1815 at the École Polytechnique. A quadric surface (centered at the origin) is given by an equation $f(x, y, z) = K$, where f is a ternary quadratic form. To classify such surfaces, Cauchy wanted to find a transformation of coordinates under which f is converted to a sum or difference of squares. In geometric terms, this problem amounts to finding a new set of orthogonal axes in three-dimensional space by which to express the surface. But Cauchy then generalized the problem to quadratic forms in n variables, the coefficients of which can be written as a symmetric matrix. For example, the binary quadratic form $ax^2 + 2bxy + cy^2$ determines the symmetric 2×2 matrix

$$\begin{pmatrix} a & b \\ b & c \end{pmatrix}.$$

Cauchy's goal was to find a linear substitution on the variables such that the matrix resulting from this substitution was diagonal, a goal he achieved in a paper of 1829. Because the details in the general case are somewhat involved and because the essence of Cauchy's proof is apparent in the two-variable case, it is that case that we will consider here.

To find a linear substitution that converts the binary quadratic form $f(x, y) = ax^2 + 2bxy + cy^2$ into a sum of squares, it is necessary to find the maximum and minimum of $f(x, y)$ subject to the condition that $x^2 + y^2 = 1$. The point at which such an extreme value of f occurs is then a point on the unit circle that also lies on the end of one axis of one member of the family of ellipses (or hyperbolas) described by the equations $f(x, y) = k$. If one takes the line from the origin to that point as one of the axes and the perpendicular to that line as the other, the equation in relation to those axes will only contain the squares of the variables. By the principle of Lagrange multipliers, the extreme value occurs when the ratios $f_x/2x$ and $f_y/2y$ are equal. Setting each of these equal to λ gives the two equations

$$\frac{ax+by}{x} = \lambda \qquad \text{and} \qquad \frac{bx+cy}{y} = \lambda,$$

which can be rewritten as the system

$$(a-\lambda)x + by = 0$$
$$bx + (c-\lambda)y = 0.$$

Cauchy knew that this system has nontrivial solutions only if its determinant equals 0, that is, if $(a-\lambda)(c-\lambda) - b^2 = 0$. In matrix terminology, this equation is the characteristic equation $\det(A - \lambda I) = 0$, the equation that Cayley dealt with some thirty years later.

To see how the roots of the characteristic equation allow one to diagonalize the matrix, let λ_1 and λ_2 be those roots and (x_1, y_1), (x_2, y_2) be the corresponding solutions for x and y. Thus,

$$(a-\lambda_1)x_1 + by_1 = 0 \qquad \text{and} \qquad (a-\lambda_2)x_2 + by_2 = 0.$$

If one multiplies the first of these equations by x_2, the second by x_1, and subtracts the second product from the first, the result is the equation

$$(\lambda_2 - \lambda_1)x_1x_2 + b(y_1x_2 - x_1y_2) = 0.$$

Similarly, starting with the two equations involving $c - \lambda_i$, one arrives at the equation

$$b(y_2x_1 - y_1x_2) + (\lambda_2 - \lambda_1)y_1y_2 = 0.$$

Adding these two equations gives $(\lambda_2 - \lambda_1)(x_1x_2 + y_1y_2) = 0$. Therefore, if $\lambda_1 \neq \lambda_2$—and this is surely true in the case being considered, unless the original form is already diagonal—then $x_1x_2 + y_1y_2 = 0$. Because (x_1, y_1), (x_2, y_2) are only determined up to a constant multiple, one can arrange to have $x_1^2 + y_1^2 = 1$ and $x_2^2 + y_2^2 = 1$. In modern terminology, the linear substitution

$$x = x_1u + x_2v$$
$$y = y_1u + y_2v$$

is orthogonal. It is easy to compute that the new quadratic form arising from this substitution is $\lambda_1u^2 + \lambda_2v^2$, as desired. That λ_1 and λ_2 are real follows from assuming, on the contrary, that they are complex conjugates of one another. In that case, x_1 would be the conjugate of x_2 and y_1 that of y_2, and $x_1x_2 + y_1y_2$ could not be zero. Cauchy had therefore shown that all eigenvalues of a symmetric matrix are real and, at least in the case where they are all distinct, that the matrix can be diagonalized by use of an orthogonal substitution.

The basic arguments of Cauchy's paper provided the beginnings of an extensive theory dealing with the eigenvalues of various types of matrices. In general, however, throughout the middle of the nineteenth century, these results were all written in terms of forms, not in terms of matrices. For example, quadratic forms lead to symmetric matrices. The more general case of bilinear forms, functions of $2n$ variables of the form

$$\sum_{i,j=1}^{n} a_{ij} x_i y_j,$$

lead to general square matrices.

Although much of the theory of forms was worked out by Camille Jordan in his *Treatise on Substitutions*, it was Frobenius who in 1878 combined the ideas of his various predecessors into the first complete monograph on the theory of matrices. In particular, Frobenius dealt with various types of relations among matrices. For example, he defined two matrices A and B to be **similar** if there were an invertible matrix P such that $B = P^{-1}AP$ and **congruent** if a P existed with $B = P^t AP$, where P^t is the transpose of P. He showed that when two symmetric matrices were similar, the transforming matrix P could be taken to be **orthogonal**, that is, one whose inverse equaled its transpose. Frobenius then made a detailed study of orthogonal matrices and showed, among other things, that their eigenvalues were complex numbers of absolute value 1. Frobenius concluded his paper by showing the relationship between his symbolic matrix theory and the theory of quaternions. This theory, worked out by William Rowan Hamilton (1805–1865) in the 1840s, dealt with the four-dimensional algebra of quantities of the form $a + bi + cj + dk$, where a, b, c, d are real numbers and i, j, k satisfy the basic multiplication laws $i^2 = j^2 = k^2 = ijk = -1$ and the derived rules $ij = k$, $ji = -k$, $jk = i$, $kj = -i$, $ki = j$, and $ik = -j$. These rules can be extended to all the quaternions by use of the distributive law. The quaternions were an early example of a system with a noncommutative multiplication. Frobenius put these into the context of matrices by determining four 2×2 matrices whose algebra was precisely that of the quantities 1, i, j, and k of quaternion algebra.

16.4.3 Solutions of Systems of Equations

Frobenius was also responsible for clarifying the question of the nature of the set of solutions to a system of linear equations, a special case of which Euler had considered many years earlier. Recall that Euler had been bothered because a particular system did not determine a specific value for each of the unknowns. He did realize, of course, that it was the vanishing of the determinant of the matrix of coefficients of the system of n equations in n unknowns that prevented that system from having a unique solution. By the middle years of the nineteenth century, mathematicians were asking different questions; they wanted not only to determine when a system of m linear equations in n unknowns had solutions, but also the size of the set of solutions. From experience with determinants, they learned that if one extracted from such a system a subsystem of k equations with a nonvanishing $k \times k$ determinant in its matrix of coefficients, that subsystem could be solved. There were then other conditions on the determinants of the original system that would determine the nature of the solution set to that system, or indeed whether it could be solved at all. In the case where there were more equations than unknowns, it was understood that, in general, there would not be solutions; the system was overdetermined. So the major concern was with systems in which $n \geq m$, a system we will write in Cayley's notation in the form $AX = B$, with A an $m \times n$ matrix.

Although many mathematicians dealt with various aspects of this theory in the late nineteenth century, it was Charles L. Dodgson (1832–1898), more commonly known today as Lewis Carroll, who worked out the basic rules for such a system in his *An Elementary Treatise on Determinants* of 1867. There he discussed conditions on both the $m \times n$ matrix A of the linear system $AX = B$ and the $m \times (n+1)$ **augmented matrix** $(A|B)$ of the system that determined whether the system was consistent or inconsistent. Furthermore, he stated and proved a very general theorem that specified the nature of the set of solutions of an arbitrary system:

DODGSON'S THEOREM. *If there are m equations, containing n variables ($n \geq m$), and if there are among them r equations which have a nonvanishing order r determinant of their unaugmented matrix; and if when these r equations are taken along with each of the remaining equations successively, each set of $r+1$ equations has every order $r+1$ determinant of its augmented matrix equal to zero, then the equations are consistent. If any nonvanishing order r determinant of the system of r equations is selected, then the $n-r$ variables whose coefficients are not contained in it may have arbitrary values assigned to them. For each such set of arbitrary values, there is only one set of values for the other variables, and the remaining equations are dependent on these r equations.*

Dodgson's proof of this result was very constructive, and he proceeded to give several examples. Thus, consider the system of four equations in five unknowns:

$$\begin{aligned} u + v - 2x + y - z &= 6 \\ 2u + 2v - 4x - y + z &= 9 \\ u + v - 2x &= 5 \\ u - v + x + y - 2z &= 0 \end{aligned}$$

Dodgson noted that there is a nonvanishing order 2 determinant for the first two equations, that there is no nonvanishing order 3 determinant for the first three equations, but that there is a nonvanishing order 3 determinant for the system consisting of equations 1, 2, and 4. Thus, he concluded, those equations are consistent, and, since they contain five variables, there are $5 - 3 = 2$ variables to which one can assign arbitrary values. In addition, equation 3 is dependent on equations 1 and 2.

The notion of rank is implicit in Dodgson's theorem, but it was Frobenius who was able to abstract this concept from the work of his predecessors in 1879: "If in a determinant all minors of order $r+1$ vanish, but not all of those of order r are zero, then I call r the **rank** of the determinant." A few years earlier, he had also defined the notion of **linear independence**, both for equations and for n-tuples representing solutions to a system: In a homogeneous system, the solutions $(x_{11}, x_{12}, \ldots, x_{1n})$, $(x_{21}, x_{22}, \ldots, x_{2n})$, $\ldots$, $(x_{k1}, x_{k2}, \ldots, x_{kn})$ are independent when $c_1x_{1j} + c_2x_{2j} + \cdots + c_kx_{kj}$ cannot be zero for $j = 1, 2, \ldots, n$ without all of the c_i being zero. To define independence for equations, Frobenius set up a duality relationship. To a given system of linear homogeneous equations he associated a new system, the coefficients of the equations of which constituted a basis for the solutions of the original system. Thus, n-tuples and equations were similar objects seen from two different points of view. He then demonstrated that if the rank of a system of m equations in n variables was r, one could find a set of $n-r$ independent solutions. Reversing the roles of coefficients and coordinates of the solution set, he then found the associated system, which had rank $n-r$, and showed that this new system itself had an associated system with the same solutions as the original system.

For example, we have already seen that the homogeneous system

$$\begin{aligned} u + v - 2x + y - z &= 0 \\ 2u + 2v - 4x - y + z &= 0 \\ u + v - 2x &= 0 \\ u - v + x + y - 2z &= 0 \end{aligned}$$

has rank 3. Thus, it has a set of two independent solutions, and a basis for the set of solutions can be taken to be $(1, 3, 2, 0, 0)$ and $(1, -1, 0, 2, 2)$. The associated system is then

$$\begin{aligned} u + 3v + 2x &= 0 \\ u - v + 2y + 2z &= 0 \end{aligned}.$$

This system has rank 2, with a basis for the set of solutions given by $(1, 1, -2, 0, 0)$, $(3, -1, 0, -2, 0)$, $(3, -1, 0, 0, -2)$. It is then straightforward to show that the system associated to these solutions, namely,

$$\begin{aligned} u + v - 2x &= 0 \\ 3u - v - 2y &= 0, \\ 3u - v - 2z &= 0 \end{aligned}$$

has the same solutions as the original system.

16.4.4 Systems of Linear Inequalities

Although the theory of linear equations was worked out in detail in the nineteenth century, there was only limited progress in what today is considered a related subject—systems of linear inequalities. The first one to deal with these ideas in some detail was Joseph Fourier (1768–1830) in the 1820s. Fourier was interested in various types of problems in which inequalities appeared, including problems in mechanics, probability, elections, and the minimization of errors in a statistical context. In this context, Fourier worked out both algebraic and geometric methods of finding the region of solutions. For example, using a system of six inequalities in two variables, he found the convex polygon of feasible solutions (Fig. 16.1). In addition, Fourier gave indications of an interest in what is today called *linear programming*: Some of his problems require the finding of not only the "feasible region" but also an optimal point of some sort. In one example, in the case of three variables, where the linear inequalities defined half-planes in space, he first found the point of intersection furthest from the origin, then descended down from plane to plane, edge by edge, until he found the desired maximum or minimum value. Unfortunately, although Fourier's work was noted by others over the next few years, the technical difficulties involved in actually solving these linear programming problems at the time caused the theory to die out until the twentieth century.

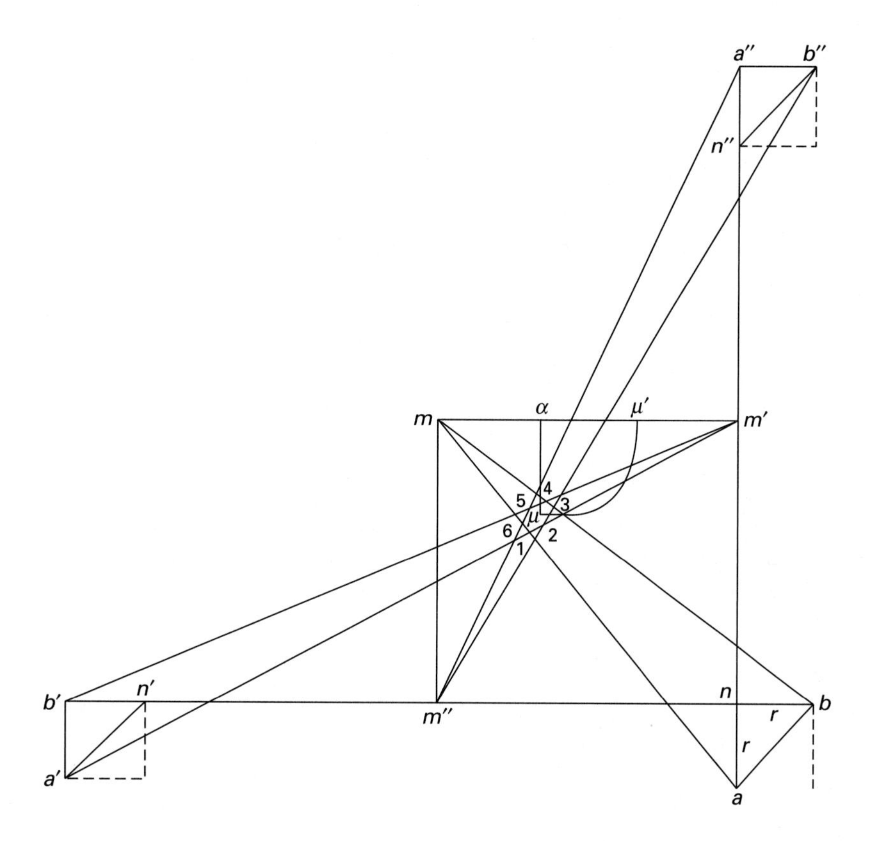

FIGURE 16.1 Polygon of feasible solutions to a system of six inequalities in two variables

Exercises

1. Prove that if p is prime and $0 < a < p$, then the smallest exponent m such that $a^m \equiv 1 \pmod{p}$ is a divisor of $p - 1$.
2. For the prime $p = 7$, calculate for each integer a with $1 < a < 7$ the smallest exponent m such that $a^m \equiv 1 \pmod{7}$. Show that the theorem in exercise 1 holds for all a.
3. Determine the primitive roots of $p = 13$; that is, determine numbers a such that $p - 1$ is the smallest exponent such that $a^{p-1} \equiv 1 \pmod{p}$.
4. Show that if any Gaussian prime p divides the product $abc \cdots$ of Gaussian primes, then p must equal one of those primes, or one of them multiplied by a unit. (Hint: Take norms of both sides.)
5. Factor $3 + 5i$ as a product of Gaussian primes.
6. Use Germain's theorem to show that there are no case 1 solutions to the Fermat equation for exponent 3 by showing, first, that 3 is not a cube modulo 7 and, second, that no two nonzero third-power residues differ by 1.
7. Show that a prime complex integer is irreducible.
8. Show that in the domain of complex integers of the form $m + n\sqrt{-5}$, the integers 2, 3, $-2 + \sqrt{-5}$, $-2 - \sqrt{-5}$, $1 + \sqrt{-5}$, $1 - \sqrt{-5}$ are all irreducible.
9. Show that in the domain of complex integers of the form $a + b\sqrt{-5}$, the principal ideal (2) is equal to A^2, where A is the ideal $(2, 1 + \sqrt{-5})$.

10. Use Gauss's method to solve the cyclotomic equation $x^6 + x^5 + x^4 + x^3 + x^2 + x + 1 = 0$.

11. In the example dealing with Gauss's solution to $x^{19} - 1 = 0$, show that α_1, α_2, and α_4 are roots of the cubic equation $x^3 + x^2 - 6x - 7 = 0$. (Hint: Use the relationship between the coefficients of this equation and the symmetric functions of the roots.)

12. In the example dealing with Gauss's solution to $x^{19} - 1 = 0$, show that β_1, β_8, and β_7 are roots of the cubic equation $x^3 - \alpha_1 x^2 + (\alpha_2 + \alpha_4)x - 2 - \alpha_2 = 0$, where the α's and β's are as in the text.

13. In the example dealing with Gauss's solution to $x^{19} - 1 = 0$, show that r and r^{18} are both roots of $x^2 - \beta_1 x + 1 = 0$, where r and β_1 are as in the text.

14. Wantzel showed that any construction problem that does not lead to an irreducible polynomial equation with degree a power of two and with constructible coefficients cannot be accomplished using straightedge and compass. Use this result to show that one cannot double a cube or trisect an arbitrary angle using straightedge and compass.

15. Calculate the Galois group G of the equation $x^3 + 6x = 20$ over the rational numbers. Show that this group has a normal subgroup H such that both H and the index of H in G are primes.

16. Write out the group tables for the two distinct groups of order 6. Show why there can be no other groups of that order.

17. Describe the five distinct groups of order 8.

18. Show that the order of the group $SL(2, p)$ is $p(p^2 - 1)$ and that of $PSL(2, p)$ is $\frac{1}{2}p(p^2 - 1)$.

19. Show that the group $PSL(2, p)$ can be considered as the group of linear fractional transformations $z' \equiv (az + b)/(cz + d) \pmod{p}$, with $ad - bc \equiv 1 \pmod{p}$, acting on the one-dimensional projective space $P_1(p)$.

20. Show that there are exactly two groups of order p^2, both of which are Abelian.

21. Given a group of order pq, $(p > q)$, with $S^q = 1$ and $T^p = 1$, show that if $S^{-1}TS = T^r$, then $r^q \equiv 1 \pmod{p}$.

22. Show that the Galois group of the equation $x^5 - 2$ over the rational numbers can be expressed as the group of substitutions of the form $x' \equiv ax + b \pmod{5}$ and that therefore it has 20 elements.

23. Create a field of order 5^3 by finding a third-degree irreducible congruence modulo 5.

24. Show that if the substitution $x = \alpha x' + \beta y'$, $y = \gamma x' + \delta y'$ with $\alpha\delta - \beta\gamma = 1$ transforms the quadratic form $F = ax^2 + 2bxy + cy^2$ into the form $F' = a'x'^2 + 2b'x'y' + c'y'^2$, then there is an "inverse" substitution of the same form that transforms F' into F.

25. Prove that if the product of two matrices is the zero matrix, then at least one of the factors has determinant 0.

26. Show explicitly the truth of the Cayley-Hamilton theorem that a matrix A satisfies its characteristic equation $\det(A - \lambda I) = 0$ in the case where A is a 2×2 matrix.

27. Using d'Alembert's method, determine explicitly the solution to the system of differential equations

$$\frac{d^2 y_i}{dt^2} + \sum_{k=1}^{3} a_{ik} y_k = 0 \qquad \text{for} \qquad i = 1, 2, 3.$$

Show why the three eigenvalues of the matrix (a_{ik}) must be distinct, real, and positive for the solution to make physical sense.

28. Use Cauchy's technique to find an orthogonal substitution that converts the quadratic form $2x^2 + 6xy + 5y^2$ into a sum or difference of squares.

29. Let $\alpha = 3 + 4i + 7j + k$ and $\beta = 2 - 3i + j - k$ be quaternions. Calculate $\alpha\beta$ and $1/\beta$.

30. Define the modulus $|\alpha|$ of a quaternion

$$a + bi + cj + dk$$

by $$|\alpha| = a^2 + b^2 + c^2 + d^2.$$

Show that $|\alpha\beta| = |\alpha||\beta|$.

31. Solve explicitly the system of linear equations

$$\begin{aligned} 2u + v + 2x + y + 3z &= 0 \\ 5u + 3v - 4x + 3y - 6z &= 0. \\ u + v - 8x + y - 12z &= 0 \end{aligned}$$

First determine the order of the maximal nonvanishing determinant in the matrix of coefficients.

32. Determine the rank of the matrix of coefficients of the system of equations in exercise 31. Find a basis for the set of solutions of this system.

33. Using the result of exercise 32, determine a system of linear equations that is associated to the system of exercise 31. Find a basis for the set of solutions for this system and then show that the system associated to the new system has the same solution set as the system of exercise 31.

34. Compare Weber's definition of a group with the standard modern definition. Show that they are equivalent.
35. Compare Weber's definition of a field with the standard modern definition. Can some of Weber's axioms be proved from other ones?
36. Design a lesson introducing the concept of a group through each of the following:
    (a) the notion of a permutation of a finite set;
    (b) the notion of composition of quadratic forms;
    (c) the notion of the residue classes modulo a prime p.
37. Compare the advantages and disadvantages of introducing a class to algebraic number fields using Kronecker's method of construction and Dedekind's method of considering subfields of the complex numbers.
38. Outline a lesson on manipulation of matrices following Cayley's 1858 treatment of the subject.

References

General references on the history of algebra include three works mentioned in the references to chapter 14: the books of Nový, Wussing, and van der Waerden. Other useful works include two books by Harold Edwards: *Fermat's Last Theorem. A Genetic Introduction to Number Theory* (New York: Springer-Verlag, 1977) and *Galois Theory* (New York: Springer-Verlag, 1984). An excellent survey of the history of group theory is Israel Kleiner, "The Evolution of Group Theory: A Brief Survey," *Mathematics Magazine* 59 (1986), 198–215. The history of the theory of matrices is presented in three articles by Thomas Hawkins: "Cauchy and the Spectral Theory of Matrices," *Historia Mathematica* 2 (1975), 1–29; "Another Look at Cayley and the Theory of Matrices," *Archives Internationales d'Histoire des Sciences* 26 (1977), 82–112; and "Weierstrass and the Theory of Matrices," *Archive for History of Exact Sciences* 17 (1977), 119–163. A reading of these articles gives a very detailed picture of the development of matrix theory and related areas of mathematics.

Gauss's *Disquisitiones* is available in English as Carl Friedrich Gauss, *Disquisitiones Arithmeticae*, Arthur A. Clarke, trans. (New York: Springer-Verlag, 1986). A perusal of the entire book is well worth the trouble. Some of Germain's work on Fermat's Last Theorem has been translated into English and appears in Reinhard Laubenbacher and David Pengelley, *Mathematical Expeditions: Chronicles by the Explorers* (New York: Springer, 1999), chapter 4. Dedekind's work on ideal theory is in Richard Dedekind, *Theory of Algebraic Integers*, John Stillwell, trans. (Cambridge: Cambridge University Press, 1996). Galois's "Memoir on the Conditions for Solvability of Equations by Radicals," translated by Harold Edwards, is found in Edwards, *Galois Theory*. Arthur Cayley's first paper on group theory is "On the Theory of Groups, as Depending on the Symbolic Equation $\theta^n = 1$," *Philosophical Magazine* (4) 7, 40–47. This paper is also found in Cayley, *The Collected Mathematical Papers* (Cambridge: Cambridge University Press, 1889–1897), vol. 2, 123–130. His major work on matrices is "A Memoir on the Theory of Matrices," *Philosophical Transactions of the Royal Society of London* 148 (1858), reprinted in Cayley, *Collected Mathematical Papers*, vol. 2, 475–496. Dodgson's work on linear equations is Charles L. Dodgson, *An Elementary Treatise on Determinants with their Application to Simultaneous Linear Equations and Algebraical Geometry* (London: Macmillan, 1867).

The story of the Paris Academy meeting of March 1, 1847 is told in more detail in Edwards, *Fermat's Last Theorem*, 76–80. For a detailed discussion of divisors and of Kummer's work on Fermat's Last Theorem, consult chapter 4 of Edwards, *Fermat's Last Theorem*. For more on ideal theory, see H. M. Edwards, "The Genesis of Ideal Theory," *Archive for History of Exact Sciences* 23 (1980), 321–378, and H. M. Edwards, "Dedekind's Invention of Ideals," in Esther Phillips, ed., *Studies in the History of Mathematics* (Washington: MAA, 1987), 8–20. A detailed discussion of Galois's propositions and their proofs is in Edwards, *Galois Theory*. For more information on the history of Galois theory, especially after the time of Galois, see B. Melvin Kiernan, "The Development of Galois Theory from Lagrange to Artin," *Archive for History of Exact Sciences* 8 (1971), 40–154. A discussion of Fourier's early work on the solution of inequalities is in Ivor Grattan-Guinness, "A New Type of Question: On the Prehistory of Linear and Non-Linear Programming, 1770–1940," in Eberhard Knobloch and David E. Rowe, eds., *The History of Modern Mathematics*, vol. 3 (Boston: Academic Press, 1993), 43–89.

CHAPTER SEVENTEEN

Analysis in the Nineteenth Century

In the above mentioned work of M. Cauchy [Cours d'analyse de l'École Polytechnique] ... one finds the following theorem: "If the different terms of the series $u_0 + u_1 + u_2 + u_3 + \cdots$ *are functions of one and the same variable quantity* x*, and indeed are continuous functions with respect to that variable in the neighborhood of a particular value for which the series converges, then the sum* s *of the series, in the neighborhood of this particular value, is also a continuous function of* x*." But it appears to me that this theorem admits exceptions. For example, the series* $\sin x - \frac{1}{2}\sin 2x + \frac{1}{3}\sin 3x - \cdots$ *is discontinuous for each value* $(2m+1)\pi$ *of* x*, where* m *is an integer. It is well known that there are many series with similar properties.*

—Niels Henrik Abel, in "Untersuchungen über die Reihe $1 + (m/1)x + (m/1)[(m-1)/2]x^2 + (m/1)[(m-1)/2][(m-2)/3]x^3 + \cdots$," 1826

Toward the end of the eighteenth century, with the growing necessity for mathematicians to teach rather than just do research, there was an increasing concern with how mathematical ideas should be presented to students and a concomitant increase in concern for rigor. Recall that Lagrange attempted to base all of calculus on the notion of a power series. And although texts were written using Lagrange's method, it was soon discovered that not all functions could be expressed by such series.

It was Augustin-Louis Cauchy, the most prolific mathematician of the nineteenth century, who first established the calculus on the basis of the limit concept so familiar today. Although the notion of limits had been discussed much earlier, even by Newton, Cauchy was the first to translate the somewhat vague notion of a function approaching a particular value into arithmetic terms by means of which one could actually prove the existence of limits. Cauchy used his notion of limit in defining continuity (in the modern sense) and convergence of sequences, both of numbers and of functions.

One of Cauchy's important results, that the sum of an infinite series of continuous functions is continuous, assuming this sum exists, turned out not to be true. Counterexamples were discovered as early as 1826 in connection with the series of sine and cosine functions now known as Fourier series. These series, although considered briefly by Daniel Bernoulli in the middle of the eighteenth century, were first studied in detail by Joseph Fourier in his work on heat conduction in the early nineteenth century. Fourier's work stimulated Peter Lejeune-Dirichlet to study in more detail the notion of a function and Bernhard Riemann to develop the concept known today as the Riemann integral.

Some unresolved questions in the work of Cauchy and Bernard Bolzano as well as the study of points of discontinuity growing out of Cauchy's wrong theorem led several mathematicians in the second half of the nineteenth century to consider the structure of the real number system. In particular, Richard Dedekind and Georg Cantor each developed methods of constructing the real numbers from the rational numbers and, in this connection, began the detailed study of infinite sets.

In his calculus texts, Cauchy defined the integral as a limit of a sum rather than as an antiderivative, as had been common in the eighteenth century. His extension of this notion of the integral to the domain of complex numbers led him by the 1840s to begin the development of complex analysis. Cauchy's ideas were further developed and extended by Riemann in the middle of the century. Because integration in the complex domain can be thought of as integration over a real two-dimensional plane, Cauchy was also able to state and Riemann to prove the theorem today known as Green's theorem, which relates integration around a closed curve to double integration over the region it bounds. Similar theorems relating integrals over a region to integrals over the boundary of the region were discovered by Mikhail Ostrogradsky and William Thomson. These theorems, today known as the divergence theorem and Stokes's theorem, turned out to have frequent applications in physics.

17.1 Rigor in Analysis

In 1797, Silvestre-Francois Lacroix (1765–1843) published the first of the three volumes of his *Treatise on Differential and Integral Calculus*. Lacroix intended this work to be a survey of the methods of calculus developed since the time of Newton and Leibniz. Thus, he not only presented Lagrange's view of the derivative of a function $f(x)$ as the coefficient of the first-order term in the Taylor series for f, but also dealt with the definition of dy/dx as a limit in the style of d'Alembert and as a ratio of infinitesimals in the manner of Euler. Lacroix was proud of the book's comprehensiveness and hoped that the true metaphysics of the subject would be found in what the various methods had in common.

For his teaching at the École Polytechnique in Paris, however, Lacroix wrote a shortened, one-volume version of his text entitled *Elementary Treatise on Differential and Integral Calculus*, whose continued popularity is attested by its appearance in nine editions between 1802 and 1881. In this work, Lacroix decided to base the differential calculus initially on the notion of a limit, defined in the process of his determination of the limit of a differential quotient. Thus, he showed that if $u = ax^2$ and $u_1 = a(x+h)^2$, then $2ax$ is "the **limit** of the ratio $(u_1 - u)/h$, or, is the value towards which this ratio tends in proportion as the quantity h diminishes, and to which it may approach as near as we choose to make it." After calculating several other limits of ratios, Lacroix explained that in fact "the differential calculus is the finding of the limit of the ratios of the simultaneous increments of a function and of the variables on which it depends."

Despite Lacroix's decision in the *Elementary Treatise* to begin differential calculus with limits, he rapidly moved to establish the Taylor series of a function. Like Lagrange, he believed that all functions could be expressed as series except perhaps at isolated points. He then proceeded to use the Taylor series representation to develop the differentiation formulas for various transcendental functions and the methods for determining maxima and minima.

In spite of the appeal of Lagrange's method, Cauchy found that it was lacking in "rigor." Cauchy, in fact, was not satisfied with what he believed were unfounded manipulations of algebraic expressions, especially infinitely long ones. Equations involving these expressions were only true for certain values, those values for which the infinite series was convergent. In particular, Cauchy discovered that the Taylor series for the function $f(x) = e^{-x^2} + e^{-(1/x^2)}$ does not converge to the function. Thus, from 1813 on, when he was teaching at the École Polytechnique, Cauchy began to rethink the basis of the calculus entirely. In 1821, at the urging of several of his colleagues, he published his *Cours d'analyse de l'École Royale Polytechnique*, in which he introduced new methods into the foundations of the calculus, and then its sequel of 1823, *Résumé des leçons donnees a l'École Royale Polytechnique sur le calcul infinitesimal*. We will study Cauchy's ideas on limits, continuity, convergence, derivatives, and integrals as they appeared in these texts, for they were to provide the model for calculus texts for the remainder of the century.

17.1.1 Limits

Cauchy's definition of limit occurs near the beginning of his *Cours d'analyse*: "If the successive values attributed to the same variable approach indefinitely a fixed value, such that finally they differ from it by as little as one wishes, this latter is called the **limit** of all the others." As an example, Cauchy noted that an irrational number is the limit of the various fractions that approach it. He also defined an **infinitely small quantity** as a variable whose limit is zero. Note that Cauchy was not defining the modern concept $\lim_{x\to a} f(x) = b$, for that concept involves two different variables. He seems to suppress the role of the independent variable entirely. Furthermore, it may appear that Cauchy's definition of a limit is little different from that of d'Alembert. To see, however, what Cauchy meant by his verbal definition and to discover the difference between it and the definitions of his predecessors, it is necessary to consider his use of the definition to prove certain specific results on limits. In fact, Cauchy not only dealt with both the dependent and independent variables, but also translated his statement arithmetically by use of the language of inequalities. As an example, consider Cauchy's proof of the following

THEOREM. *If, for increasing values of x, the difference $f(x+1) - f(x)$ converges to a certain limit k, the fraction $f(x)/x$ converges at the same time to the same limit.*

Cauchy began by translating the hypothesis of the theorem into an arithmetic statement: Given any value ϵ, as small as one wants, one can find a number h such that if $x \geq h$, then $k - \epsilon < f(x+1) - f(x) < k + \epsilon$. He proceeded to use this translation in his proof. Because each of the differences $f(h+i) - f(h+i-1)$ for $i = 1, 2, \ldots, n$ satisfies the inequality, so does their arithmetic mean,

$$\frac{f(h+n)-f(h)}{n}.$$

It follows that

$$\frac{f(h+n)-f(h)}{n}=k+\alpha,$$

where $-\epsilon<\alpha<\epsilon$, or, setting $x=h+n$, that

$$\frac{f(x)-f(h)}{x-h}=k+\alpha.$$

But then $f(x)=f(h)+(x-h)(k+\alpha)$, or

$$\frac{f(x)}{x}=\frac{f(h)}{x}+\left(1-\frac{h}{x}\right)(k+\alpha).$$

Because h is fixed, Cauchy concluded that as x gets large, $f(x)/x$ approaches $k+\alpha$, where $-\epsilon<\alpha<\epsilon$. Because ϵ is arbitrary, the conclusion of the theorem holds. Cauchy also proved the theorem for the cases $k=\pm\infty$ and then used it to conclude, for example, that as x gets large, $\log x/x$ converges to 0 and a^x/x $(a>1)$ has limit ∞.

17.1.2 Continuity

Given the definition of a limit, Cauchy could now define the crucial concept of continuity. The geometric notion of a continuous curve, one without any breaks, was generally understood, but Cauchy gave an analytic definition expressing this idea for functions: "The function $f(x)$ will be, between two assigned values of the variable x, a **continuous function** of this variable if for each value of x between these limits, the numerical [absolute] value of the difference $f(x+\alpha)-f(x)$ decreases indefinitely with α. In other words, the function $f(x)$ will remain continuous with respect to x between the given values if, between these values, an infinitely small increment of the variable always produces an infinitely small increment of the function itself." Note that Cauchy presented both an arithmetic definition and one using the more familiar language of infinitely small quantities. But because Cauchy had already defined such quantities in terms of limits, the two definitions meant the same thing. Cauchy demonstrated how to use his definition by showing, for example, that $\sin x$ is continuous (on any interval). For $\sin(x+\alpha)-\sin x=2\sin\frac{1}{2}\alpha\cos(x+\frac{1}{2}\alpha)$, and the right side clearly decreases indefinitely with α.

From a modern point of view, of course, Cauchy did not define continuity at a point, but over an interval. It would appear, however, that one can read the definition as defining continuity at each point in the interval. That is, given a particular value x and an $\epsilon>0$, one can find a $\delta>0$ such that $|f(x+\alpha)-f(x)|<\epsilon$ whenever $\alpha<\delta$. As we will see in the next section, though, Cauchy was not entirely clear on what quantities δ depended. This lack of clarity would lead him to an incorrect result.

17.1.3 Convergence

Cauchy's definition of convergence of series appears in chapter 6 of his *Cours d'analyse*: "Let $s_n = u_0 + u_1 + u_2 + \cdots + u_{n-1}$ be the sum of the first n terms [of a series], n designating an arbitrary integer. If, for increasing values of n, the sum s_n approaches indefinitely a certain limit s, the series will be called **convergent**, and the limit in question will be called the **sum** of the series. On the contrary, if, as n increases indefinitely, the sum s_n does not approach any fixed limit, the series will be **divergent** and will not have a sum." To clarify the definition, Cauchy stated what has become known as the **Cauchy criterion for convergence**. He realized that for a series to be convergent, it was necessary that the individual terms u_n must decrease to zero. This condition, however, was not sufficient. Convergence could only be assured if the various sums $u_n + u_{n+1}, u_n + u_{n+1} + u_{n+2}, u_n + u_{n+1} + u_{n+2} + u_{n+3}, \ldots$ "taken, from the first, in whatever number one wishes, finish by constantly having an absolute value less than any assignable limit." Cauchy offered no proof of this sufficiency condition, because without some arithmetical definition of the real numbers no such proof is possible. He did, however, offer examples. For example, he showed that the geometric series $1 + x + x^2 + x^3 + \ldots$, with $|x| < 1$, converges, because the sums $x^n + x^{n+1}, x^n + x^{n+1} + x^{n+2}, \ldots$, respectively equal to

$$x^n \frac{1-x^2}{1-x}, x^n \frac{1-x^3}{1-x}, \ldots,$$

are always between x^n and $x^n/(1-x)$ and the latter values both converge to zero with increasing n.

Interestingly, Cauchy had been preceded in his statement of the Cauchy criterion by Bernard Bolzano (1781–1848), a Czech mathematician. Bolzano's work, published in what was then a distant corner of Europe, had little contemporary influence. But he did show that the Cauchy criterion implied the **least upper bound principle**, eventually seen to be one of the defining properties of the real number system. Bolzano's convergence definition and his statement of the Cauchy criterion (applied to a series of functions rather than to constants) are contained in a

THEOREM. *If a series of quantities $F_1(x), F_2(x), F_3(x), \ldots, F_n(x), \ldots$ [where each $F_i(x)$ can be thought of as representing the sum of the first i terms of a series] has the property that the difference between its nth term $F_n(x)$ and every later term $F_{n+r}(x)$, however far from the former, remains smaller than any given quantity if n has been taken large enough, then there is always a certain constant quantity, and indeed only one, which the terms of this series approach, and to which they can come as close as desired if the series is continued far enough.*

Bolzano's proof of uniqueness of the limit is straightforward, but his proof of the existence for each x of a number $X(x)$ to which the series converges is faulty because Bolzano, like Cauchy, had no way of defining an arbitrary real number X. Nevertheless, he did show how to determine the X to within any degree of accuracy d. If n is taken sufficiently large so that $F_{n+r}(x)$ differs from $F_n(x)$ by less than d for every r, then $F_n(x)$ is the desired approximation to X.

Bolzano could now also prove the least upper bound property of the real numbers:

THEOREM. *If a property M does not belong to all values of a variable x, but does belong to all values which are less than a certain u, then there is always a quantity U which is the greatest of those of which it can be asserted that all smaller x have property M.*

Bolzano's proof of the existence of this least upper bound U of all numbers having the property M involves the creation of a series to which the convergence criterion can be applied. Because M is not valid for all x, there must exist a quantity $V = u + D$ such that it is false that M is valid for all x smaller than V. Now Bolzano considers the quantities $V_m = u + D/2^m$ for each positive integer m. If for all m, it is false that M is valid for all x less than V_m, then u itself must be the desired least upper bound. On the other hand, suppose that M is valid for all x less than $u + D/2^m$ but not for all x less than $u + D/2^{m-1}$. The difference between those two quantities is $D/2^m$, so Bolzano next applies this bisection technique to the interval $[u + D/2^{m-1}, u + D/2^m]$ and determines the smallest integer n such that M is valid for all x less than $u + D/2^m + D/2^{m+n}$ but not for all x less than $u + D/2^m + D/2^{m+n-1}$. Continuing this procedure, Bolzano constructs a sequence u, $u + D/2^m$, $u + D/2^m + D/2^{m+n}, \ldots$, which satisfies his Cauchy criterion and therefore must converge to a value U, which he can easily prove satisfies the conditions of the theorem.

The least upper bound principle easily implies the **intermediate value theorem**, that if $f(x)$ is continuous and if $f(\alpha)$, $f(\beta)$ have opposite signs, then there is a value of x between α and β for which $f(x) = 0$. Suppose $f(\alpha) < 0$ and $f(\beta) > 0$. Without loss of generality, we can also assume that $f(x) < 0$ for all $x < \alpha$. Then the property M that $f(x) < 0$ is not satisfied by all x but is satisfied by all x smaller than a certain $u = \alpha + \omega$, where $\omega < \beta - \alpha$ (because f is assumed continuous). It follows that there is a value U that is the largest such that $f(x) < 0$ for all $x < U$. It is straightforward to show that $f(U)$ can be neither positive nor negative. Thus, $f(U) = 0$, and the theorem is proved.

Cauchy's own proof of the same result did not make use of the Cauchy criterion. Instead, it relied implicitly on another axiom of the real number system, that any bounded monotone sequence has a limit, and explicitly on a result he had proved earlier, that if $f(x)$ is continuous and the sequence $\{a_i\}$ converges to a, then the sequence $\{f(a_i)\}$ converges to $f(a)$. Cauchy's procedure came from a standard approximation procedure used to approximate the solution to a polynomial equation $f(x) = 0$. Thus, if $f(\alpha) < 0$ and $f(\beta) > 0$, then setting $h = \beta - \alpha$ and m an arbitrary positive integer, Cauchy considered the signs of $f(\alpha + ih/m)$ for $i = 1, 2, \ldots, m$. Because there is some pair of consecutive values, say α_1 and β_1, for which $f(\alpha_1) < 0$ and $f(\beta_1) > 0$, with $\alpha < \alpha_1 < \beta_1 < \beta$, Cauchy next divided the interval $[\alpha_1, \beta_1]$ of length h/m into subintervals of length h/m^2 and repeated the argument. Continuing, he obtained an increasing sequence $\alpha, \alpha_1, \alpha_2, \ldots$ and a decreasing sequence $\beta, \beta_1, \beta_2, \ldots$, each of which must converge to the same limit a. It follows that both sequences, $f(\alpha), f(\alpha_1), f(\alpha_2), \ldots$ and $f(\beta), f(\beta_1), f(\beta_2), \ldots$, converge to the common limit $f(a)$. Because the values of the first sequence are all negative and those of the second positive, it must be that $f(a) = 0$, and the theorem is proved.

Given the Cauchy criterion for the convergence of a series, Cauchy developed in his text various tests by which one could demonstrate convergence in particular cases, beginning with tests for series of positive terms, say $u_0 + u_1 + u_2 + \cdots$. Cauchy used the **comparison test**, that if a given series is term-by-term bounded by a convergent series, then the given series is itself convergent, without any particular comment. His most common comparison was to a geometric series with ratio less than 1, a series whose convergence Cauchy proved initially by use of the Cauchy criterion. In fact, he used the comparison test to demonstrate the validity of many of his other tests. For example, Cauchy proved the **root test**: that if the limit of the values $\sqrt[n]{u_n}$ is a number k less than 1, then the series converges. Choosing a number U such that $k < U < 1$, Cauchy noted that for n sufficiently large, $\sqrt[n]{u_n} < U$ or $u_n < U^n$. It then follows by comparison with the convergent geometric series $1 + U + U^2 + U^3 + \cdots$ that the given series also converges. Similarly, if the limit of the roots is a number k greater than 1, then the series diverges.

Cauchy used a similar proof to demonstrate the **ratio test**: If, for increasing values of n, the ratio u_{n+1}/u_n converges to a fixed limit k, the series u_n will be convergent if $k < 1$ and divergent if $k > 1$.

For series involving positive and negative terms, Cauchy dealt with the idea of absolute convergence (although he did not use that terminology), adapted the root and ratio test to that case, proved the alternating series test, and showed how to calculate the sum and product of two convergent series. He also adapted his various tests to sequences of functions. In particular, he showed how to find the interval of convergence of a power series. Although some of the individual results on series had been known previously, Cauchy was the first to organize them into a coherent theory that allowed him and others to generalize to the case of series of complex numbers and functions.

There is one significant result in the *Cours d'analyse*, however, which, as Abel noted in 1826, is incorrect as stated:

THEOREM 6-1-1. *When the different terms of the series [$\sum_{n=0}^{\infty} u_n$] are functions of the same variable x, continuous with respect to that variable in the neighborhood of a particular value for which the series is convergent, the sum s of the series is also, in the neighborhood of this particular value, a continuous function of x.*

Cauchy's "proof" of this result is quite simple. We will present his argument in words, as he did, and then translate the words into modern symbols. Writing s_n as the sum of the first n terms of the series and s as the sum of the entire series, Cauchy denoted by r_n the remainder $s - s_n$. (Here s, s_n, and r_n are all functions of x.) To prove continuity of s, he needed to show that an infinitely small increment in x leads to an infinitely small increment in $s(x)$; that is, given $\epsilon > 0$,

$$\exists \delta \quad \text{such that} \quad \forall a,\ |a| < \delta \Rightarrow |s(x+a) - s(x)| < \epsilon. \tag{17.1}$$

Although, in certain earlier proofs, Cauchy actually calculated appropriate values for δ, in this case, he just attempted an argument using arbitrary infinitely small quantities. Thus, he wrote that an infinitely small increment α of x leads to an infinitely small increment of $s_n(x)$ because the latter is continuous for every n, or

$$\exists \delta \quad \text{such that} \quad \forall a,\ |a| < \delta \Rightarrow |s_n(x+a) - s_n(x)| < \epsilon. \tag{17.2}$$

Next, because the series converges for any x, r_n will itself be infinitely small for n large enough and will remain so for an infinitely small increment of x, or

$$\exists N \quad \text{such that} \quad \forall n,\ n > N \Rightarrow |r_n(x)| < \epsilon \quad \text{and} \quad |r_n(x+a)| < \epsilon. \tag{17.3}$$

Because the increment of s is the sum of those of s_n and r_n, Cauchy concluded that this increment is also infinitely small and therefore that s is itself continuous. We would write

$$
\begin{aligned}
|s(x+a) - s(x)| &= |s_n(x+a) + r_n(x+a) - s_n(x) - r_n(x)| \\
&\leq |s_n(x+a) - s_n(x)| + |r_n(x+a)| + |r_n(x)| \leq \epsilon + \epsilon + \epsilon = 3\epsilon.
\end{aligned}
\quad \textbf{(17.4)}
$$

What Cauchy failed to notice was that the δ in equation 17.2 depends on ϵ, x, and n, while the N in equation 17.3 depends on ϵ, x, and a. Unless we know that there is some value N that will work in equation 17.3 for all a with $|a| < \delta$, we cannot assert the truth of equation 17.4 (or equation 17.1). Cauchy's arguments with infinitely small increments obscured the needed relationships among the various quantities involved. For the proof to be valid, a notion of uniform convergence was necessary; that is, one needed the additional hypothesis that the number N can be chosen independently of x, at least in some fixed interval.

17.1.4 Derivatives

Cauchy's *Cours d'analyse* provided a treatment of the basic ideas of functions and series. In his 1823 text *Résumé des leçons donnees a L'École Royale Polytechnique sur le calcul infinitesimal*, Cauchy applied his new ideas on limits to the study of the derivative and the integral, the two basic concepts of the infinitesimal calculus.

After beginning this text with the same definition of continuity as in the earlier one, Cauchy proceeded in lesson 3 to define the derivative of a function as the limit of $[f(x+i) - f(x)]/i$ as i approaches the limit of 0, as long as this limit exists. Just as he did in his definition of continuity, Cauchy defined the concept of a derivative over an interval, in fact, an interval in which the function f is continuous. He noted that this limit will have a definite value for each value of x and, therefore, is a new function of that variable, a function for which he used Lagrange's notation $f'(x)$. The definition itself, although it can be thought of as expressing the quotient of infinitesimal differences as in Euler's work, is more directly taken from that section of Lagrange's *Analytic Functions* in which Lagrange, as part of his power series expansion of f, showed that $f(x+i) = f(x) + if'(x) + iV$, where V is some function that goes to zero with i. Cauchy was able to translate this theorem about derivatives into a definition of the derivative. He then calculated the derivatives of several elementary functions. For example, if $f(x) = \sin x$, then, using the identity $\sin(x+\alpha) - \sin x = 2\sin(\alpha/2)\cos(x+\alpha/2)$, he reduced the quotient of the definition to

$$\frac{\sin(i/2)}{i/2}\cos(x + i/2),$$

whose limit $f'(x)$ is seen to be $\cos x$.

There was, of course, nothing new about Cauchy's calculations of derivatives. Nor was there anything particularly new about the theorems Cauchy was able to prove about derivatives. Lagrange had derived the same results from his own definition of the derivative. But because Lagrange's definition of a derivative rested on the false assumption that any function could be expanded into a power series, the significance of Cauchy's works lies in his explicit use of the modern definition of a derivative, translated into the language of inequalities through his definition of limit, to prove theorems. The most important of these results, in terms of its later use, was in lesson 7:

THEOREM. *If the function $f(x)$ is continuous between the values $x = x_0$ and $x = X$, and if we let A be the smallest, B the largest value of the derivative $f'(x)$ in that interval, then* *the ratio of the finite differences*

$$\frac{f(X) - f(x_0)}{X - x_0}$$

must be between A and B.

Cauchy's proof of this theorem is the first to use the δ and ϵ so familiar to today's students. Cauchy began by choosing $\epsilon > 0$ and then choosing δ so that for all values of i with $|i| < \delta$ and for any value of x in the interval $[x_0, X]$, the inequality

$$f'(x) - \epsilon < \frac{f(x+i) - f(x)}{i} < f'(x) + \epsilon$$

holds. That such values exist follows from Cauchy's definition of the derivative as a limit. Note, however, that Cauchy used the fact, implicit in his definition of derivative on an interval rather than at a point, that, given ϵ, the same δ works for every x in the interval. In any case, Cauchy next interposed $n - 1$ new values $x_1 < x_2 < \cdots < x_{n-1}$ between x_0 and $x_n = X$, with the property that $x_i - x_{i-1} < \delta$ for each i, and applied the above inequality to the subintervals determined by each successive pair of values. It follows that for $i = 1, 2, \ldots, n$,

$$A - \epsilon < \frac{f(x_i) - f(x_{i-1})}{x_i - x_{i-1}} < B + \epsilon.$$

Cauchy then used an algebraic result to conclude that the sum of the numerators divided by the sum of the denominators also must satisfy the same inequality, that is,

$$A - \epsilon < \frac{f(X) - f(x_0)}{X - x_0} < B + \epsilon.$$

Because this result is true for every ϵ, the conclusion of the theorem follows.

As an immediate consequence of this theorem, Cauchy derived the **mean value theorem for derivatives**. Assuming that $f'(x)$ is continuous in the given interval (an assumption which of course justifies the assumption that it has a smallest value A and a largest value B), Cauchy used the intermediate value theorem to conclude that $f'(x)$ takes on every value between A and B, with x between x_0 and X. In particular, $f'(x)$ takes on the value in this theorem. Thus, there is a value θ between 0 and 1 such that

$$\frac{f(X) - f(x_0)}{X - x_0} = f'(x_0 + \theta(X - x_0)).$$

Using this mean value theorem, Cauchy then proved that a function with positive derivative on an interval is increasing there, one with negative derivative is decreasing, and one with zero derivative is constant.

17.1.5 Integrals

Although it used his new definition of limits, Cauchy's treatment of the derivative was closely related to the treatments in the works of Euler and Lagrange. Cauchy's treatment of the integral, on the other hand, broke entirely new ground. Recall that in the eighteenth century, integration was defined simply as the inverse of differentiation. Although Leibniz had developed his integral notation to remind us that the integral was an infinite sum of infinitesimal areas, the problems inherent in the use of infinities convinced eighteenth-century mathematicians to take the notion of the indefinite integral, or antiderivative, as their basic notion for the theory of integration. Mathematicians of course recognized that one could evaluate areas not only by use of antiderivatives but also by various approximation techniques. But it was Cauchy who first took these techniques as fundamental and proceeded to construct a theory of definite integrals upon them.

There are probably several reasons why Cauchy felt compelled to define the integral as the limit of a sum rather than in terms of antiderivatives. First, there were many situations where it was clear that an area under a curve made sense even though it could not be calculated by evaluating an antiderivative at the endpoints of an interval; such was the case in particular for certain piecewise continuous functions that showed up in Fourier's work on series of trigonometric functions. A second reason may well have arisen out of Cauchy's work in developing a theory of integrals of complex functions. Finally, Cauchy may have realized in the course of organizing his material for lectures at the École Polytechnique that there is no guarantee that an antiderivative exists for every function. Cauchy's own explanation of his reason for choosing to define an integral in terms of a sum, however, was that it works.

In the second part of his *Resumé*, Cauchy presented the details of a rigorous definition of the integral using sums. Cauchy probably took his definition from work on approximations of definite integrals by Euler and by Lacroix. But rather than consider their method a way of approximating an area (presumably understood intuitively to exist), Cauchy made their approximation into a definition. Thus, supposing that $f(x)$ is continuous on $[x_0, X]$, he took $n-1$ new intermediate values $x_1 < x_2 < \cdots < x_{n-1}$ between x_0 and $x_n = X$ and formed the sum

$$S = (x_1 - x_0)f(x_0) + (x_2 - x_1)f(x_1) + \cdots + (X - x_{n-1})f(x_{n-1}).$$

Cauchy noted that S depends both on n and on the particular values x_i selected. But, he wrote, "it is important to observe that if the numerical values of the elements $[x_{i+1} - x_i]$ become very small and the number n very large, the method of division will have only an insensible influence on the value of S."

To prove this result, Cauchy noted that if one chose a new subdivision of the interval by subdividing each of the original subintervals, the corresponding sum S' could be rewritten in the form

$$S' = (x_1 - x_0)f(x_0 + \theta_0(x_1 - x_0)) + (x_2 - x_1)f(x_1 + \theta_1(x_2 - x_1)) + \cdots$$
$$+ (X - x_{n-1})f(x_{n-1} + \theta_{n-1}(X - x_{n-1})),$$

where each θ_i is between 0 and 1. By the definition of continuity, this expression can be rewritten as

$$\begin{aligned} S' &= (x_1 - x_0)[f(x_0) + \epsilon_0] + (x_2 - x_1)[f(x_1) + \epsilon_1] + \cdots + (X - x_{n-1})[f(x_{n-1}) + \epsilon_{n-1}] \\ &= S + (x_1 - x_0)\epsilon_1 + (x_2 - x_1)\epsilon_2 + \cdots + (X - x_{n-1})\epsilon_{n-1} \\ &= S + (X - x_0)\epsilon' \end{aligned}$$

where ϵ' is a value between the smallest and largest of the ϵ_i. Cauchy then argued that if the subintervals are sufficiently small, the ϵ_i, and consequently ϵ', will be very close to zero, so that the taking of a subpartition does not change the value of the sum appreciably. Given any two sufficiently small subdivisions, one can take a third that subdivides each. The value of the sum for this third is then arbitrarily close to the values for each of the original two. It follows that "if we let the numerical values of [the lengths of the subdivisions] decrease indefinitely by increasing their number, the value of S ultimately becomes sensibly constant or, in other words, it will end by attaining a certain limit that will depend uniquely on the form of the function $f(x)$ and the extreme values x_0, X attributed to the variable x. This limit is what we call the definite integral [written $\int_{x_0}^{X} f(x)\,dx$]." This definition thus used a generalization of Cauchy's criterion for convergence to sequences not necessarily indexed by the natural numbers.

With the integral now defined in terms of a limit of sums, it was not difficult for Cauchy to prove the **mean value theorem for integrals**, that

$$\int_{x_0}^{X} f(x)\,dx = (X - x_0) f[x_0 + \theta(X - x_0)], \qquad \text{where } 0 \leq \theta \leq 1,$$

and also the additivity theorem for integrals over intervals. He then easily demonstrated the

FUNDAMENTAL THEOREM OF CALCULUS. *If $f(x)$ is continuous in $[x_0, X]$, if $x \in [x_0, X]$, and if*

$$F(x) = \int_{x_0}^{x} f(x)\,dx,$$

then $F'(x) = f(x)$.

To prove the theorem, Cauchy used the mean value theorem and additivity to get

$$F(x + \alpha) - F(x) = \int_{x}^{x+\alpha} f(x)\,dx = \alpha f(x + \theta\alpha).$$

If one divides both sides by α and passes to the limit, the conclusion follows from the continuity of $f(x)$. This version of the fundamental theorem can be considered the first one to meet modern standards of rigor, because it was the first in which $F(x)$ was clearly defined through an existence proof for the definite integral.

17.1.6 Fourier Series and the Notion of a Function

Abel's counterexample to Cauchy's false result on convergence was a **Fourier series**, a series of trigonometric functions of the type Euler and Daniel Bernoulli argued about in the middle of the eighteenth century. Fourier made a detailed study of these series in connection with his investigation of heat diffusion early in the nineteenth century. He first presented his work to the French Academy in 1807 and later reworked and expanded it into his *Analytic Theory of Heat* in 1822. Fourier began by considering the special case of the temperature distribution $v(t, x, y)$ at time t in a rectangular lamina infinite in the positive x direction, of width 2 in the y direction, with the edge $x = 0$ being maintained at a constant temperature 1 and the edges $y = \pm 1$ being kept at temperature 0. By making certain assumptions about the flow of heat, Fourier was able to show that v satisfies the partial differential equation

$$\frac{\partial v}{\partial t} = \frac{\partial^2 v}{\partial x^2} + \frac{\partial^2 v}{\partial y^2}.$$

Fourier then solved this equation under the condition that the temperature of the lamina had reached equilibrium, that is, that $\partial v/\partial t = 0$. Assuming that $v = \phi(x)\psi(y)$ (the method of separation of variables), he differentiated twice with respect to each variable to get $\phi''(x)\psi(y) + \phi(x)\psi''(y) = 0$, or

$$\frac{\phi(x)}{\phi''(x)} = -\frac{\psi(y)}{\psi''(y)} = A$$

for some constant A. The obvious solutions to these equations are $\phi(x) = \alpha e^{mx}$, $\psi(y) = \beta \cos ny$, where $m^2 = n^2 = 1/A$. Physical reasoning dictates that m be negative (for otherwise the temperature would tend toward infinity for large x), so Fourier concluded that the general solution of the original partial differential equation is a sum of functions of the types $v = ae^{-nx} \cos ny$. By using the boundary conditions $v = 0$ when $y = \pm 1$, Fourier showed that n must be an odd multiple of $\pi/2$ and therefore that the general solution is given by the infinite series

$$v = a_1 e^{-(\pi x/2)} \cos\left(\frac{\pi y}{2}\right) + a_2 e^{-(3\pi x/2)} \cos\left(\frac{3\pi y}{2}\right) + a_3 e^{-(5\pi x/2)} \cos\left(\frac{5\pi y}{2}\right) + \cdots$$

To determine the coefficients a_i, Fourier used the additional boundary condition that $v = 1$ when $x = 0$. Setting $u = \pi y/2$, that implied that the a_i satisfied the equation

$$1 = a_1 \cos u + a_2 \cos 3u + a_3 \cos 5u + \cdots,$$

a single equation for infinitely many unknowns, which Fourier turned into infinitely many equations by differentiating it infinitely often and each time setting $u = 0$. By noting the patterns determined by solving the first several of these equations, Fourier was able to determine that $a_1 = 4/\pi$, $a_2 = -4/3\pi$, $a_3 = 4/5\pi, \ldots$ and therefore to solve the partial differential equation. But in the usual spirit of mathematicians, once the original problem was solved, he began to consider the mathematical ramifications of his new type of solution. First, he noted that his values for the coefficients implied that

$$\cos u - \frac{1}{3}\cos 3u + \frac{1}{5}\cos 5u - \cdots = \frac{\pi}{4}$$

with $u \in (-\pi/2, \pi/2)$. But the same series clearly represents 0 for $u = \pi/2$ and represents $-\pi/4$ for $u \in (\pi/2, 3\pi/2)$. Fourier realized that this result would not be immediately believable to his readers, however, because it seemed to imply that the infinite cosine series represented the square wave of Figure 17.1. To a modern reader, it is not clear why

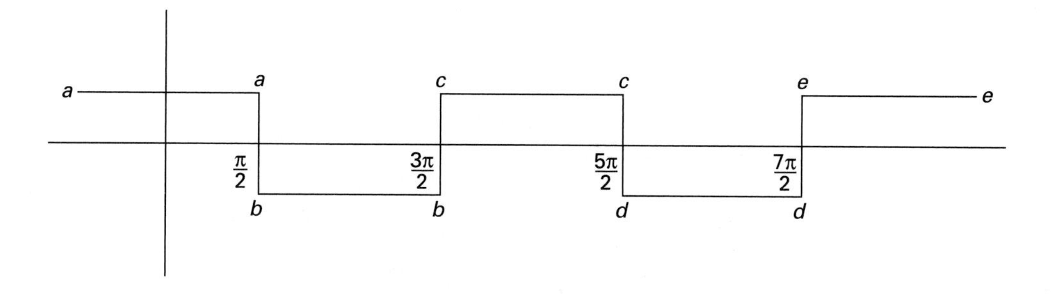

FIGURE 17.1 Fourier's square wave: $y = \cos u - \frac{1}{3}\cos 3u + \frac{1}{5}\cos 5u - \frac{1}{7}\cos 7u \cdots$

Fourier drew in the line segments perpendicular to the abscissa, for the value of the series at $u = k\pi/2$, for k odd, is always 0. Fourier, however, realizing that the partial sums of this series could always be represented by a curve without breaks, thought that the infinite sum too should be represented by such a curve. The questions of whether this curve represented a function in the modern sense or whether the series represented a continuous function in Cauchy's sense were not relevant to Fourier's work. He was interested in a physical problem and probably conceived of this solution in geometrical terms, which allowed him to draw a continuous curve without worrying about whether it represented a function.

Fourier next investigated the representation of various functions by series of trigonometric functions, the most general being the series of the form $c_0 + c_1 \cos x + c_2 \cos 2x + \cdots + d_1 \sin x + d_2 \sin 2x + \cdots$. Usually, however, Fourier limited himself to either a sine series or a cosine series. Furthermore, because he wanted to convince his readers of the validity of his methods, he determined the coefficients of his original series and others by new methods. For example, he showed that if one writes the function $\frac{1}{2}\pi f(x)$ as a sine series,

$$\tfrac{1}{2}\pi f(x) = a_1 \sin x + a_2 \sin 2x + a_3 \sin 3x + \cdots,$$

one can multiply both sides by $\sin nx\,dx$ for each integer n in turn and integrate over the interval $[0, \pi]$. Because

$$\int_0^\pi \sin mx \sin nx\,dx = \begin{cases} 0, & \text{if } m \neq n \\ \pi/2, & \text{if } m = n \end{cases},$$

it follows that

$$a_k = \int_0^\pi f(x) \sin kx\,dx,$$

as long as the integrals representing the coefficients make sense—that is, as long as the area under $f(x) \sin kx$ is well defined.

What kinds of functions, then, could be represented by these trigonometric series? To begin to answer this question, Fourier defined what he meant by the term **function**: "In general, the function $f(x)$ represents a succession of values or ordinates each of which is arbitrary. An infinity of values being given to the abscissa x, there is an equal number of ordinates $f(x)$. All have actual numerical values, either positive or negative or null. We do not suppose these ordinates to be subject to a common law; they succeed each other in any manner whatever, and each of them is given as if it were a single quantity." Despite this modern-sounding definition, Fourier never considered what today are called arbitrary functions. His examples show that he only intended to consider piecewise continuous functions. And, of course, Fourier only asserted that the series represented the given arbitrary function on the interior of a particular finite interval, such as $[0, \pi]$. The value of the series at the endpoints could easily be calculated separately, while the periodicity properties of the sine function enabled one to extend the original function geometrically to the entire real line.

Fourier attempted in certain cases to prove that his expansion actually represented the function by using trigonometric identities to rewrite the partial sum of the first n terms in closed form and then considering the limit as n increased. In general, however, he believed that his explicit calculation of the coefficients in his proposed expansion of an arbitrary function in terms of integrals that represented real areas was a convincing enough argument that the expansion was valid. Thus, for example, he calculated the expansion given later by Abel:

$$\tfrac{1}{2}x = \sin x - \tfrac{1}{2}\sin 2x + \tfrac{1}{3}\sin 3x - \tfrac{1}{4}\sin 4x + \cdots.$$

This series represented $\frac{1}{2}x$ on $[0, \pi/2)$ and the function of Figure 17.2 over the entire real line. Abel realized that not only did this function violate Cauchy's result on the sum of a series of continuous functions, but also that Fourier's attempts at a proof that the Fourier series converged to the original function were not sufficient.

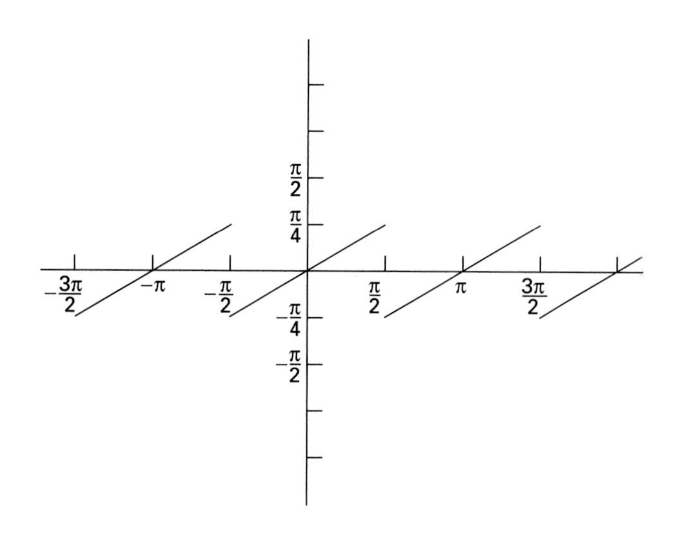

FIGURE 17.2 Fourier's graph of $y = \sin x - \frac{1}{2}\sin 2x + \frac{1}{3}\sin 3x - \frac{1}{4}\sin 4x \cdots$

Cauchy himself attempted a new proof of Fourier's assertion in 1826, but in that proof he assumed that if $(u_k - v_k) \to 0$ and $k \to \infty$, and if $\sum u_k$ is convergent, then the same is true of $\sum v_k$. Dirichlet noted that this assumption was erroneous in a paper of 1829, which contained the counterexample

$$u_k = \frac{(-1)^k}{\sqrt{k}} \qquad v_k = \frac{(-1)^k}{\sqrt{k}} + \frac{1}{k},$$

where it is obvious that the first series converges while the second diverges. Dirichlet then successfully attacked the problem of the convergence of Fourier series with a proof in the style of Cauchy's own analysis.

Rather than try to show that the Fourier series for an arbitrary function converged, Dirichlet lowered his sights drastically and found sufficient conditions on the function to assure this convergence. In particular, he showed that if a function $f(x)$ defined on $[-\pi, \pi]$ was continuous and bounded on that interval, except perhaps for a finite number of finite discontinuities, and in addition had only a finite number of turning values in that interval, then the Fourier series converged at each x in $(-\pi, \pi)$ to the value

$$\lim_{\epsilon \to 0} \tfrac{1}{2}[f(x - \epsilon) + f(x + \epsilon)].$$

[This value is equal to $f(x)$ if f is continuous at x.] The result held at the endpoints as well if, like Fourier, one interpreted $f(x)$ as being geometrically periodic outside the given interval. Dirichlet chose his conditions in order to be able to integrate products of the given function with trigonometric functions over certain intervals. He realized that although Cauchy's formulation of the definite integral enabled one to integrate functions with finitely many discontinuities, it would be quite difficult to extend this result to functions with infinitely many discontinuities in a given interval. In fact, he provided an example of a function that did not satisfy his original conditions, one that is continuous nowhere: the function that equaled a particular constant c at rational values of the variable x and a different constant d at irrational values of x. Although the function is defined for all x, one could not integrate the functions necessary to give the Fourier series.

17.1.7 The Riemann Integral

In 1853, Georg Bernhard Riemann (1826–1866) attempted to generalize Dirichlet's result by first determining precisely which functions were integrable according to Cauchy's definition of the integral, $\int_a^b f(x)\,dx$. He began, in fact, by changing the definition somewhat. Like Cauchy, he divided the interval $[a, b]$ into n subintervals $[x_{i-1}, x_i]$ for $i = 1, 2, \ldots, n$. Setting $\delta_i = x_i - x_{i-1}$, he now considered the sum

$$S = \sum_{i=1}^{n} \delta_i f(x_{i-1} + \epsilon_i \delta_i),$$

where each ϵ_i is between 0 and 1. This sum is more general than Cauchy's, because Riemann allowed the argument of the function f to take any value in the relevant subinterval. He then defined the integral to be the limit to which S tends, provided it exists, no matter how the δ_i and ϵ_i are taken. Riemann next asked a question that Cauchy had not: In what cases is a function integrable and in what cases not? Cauchy himself had only shown that a certain class of functions was integrable but had not tried to find all such functions. Riemann, on the other hand, formulated a necessary and sufficient condition for a finite function $f(x)$ to be integrable: "If, with the infinite decrease of all the quantities δ, the total size s of the intervals in which the variations of the function $f(x)$ are greater than a given quantity σ always becomes infinitely small in the end, then the sum S converges when all the δ become infinitely small," and conversely. (The **variation** of a function in an interval is the difference between the maximum and minimum values of the function on that interval.) As an example of a function defined on [0, 1] that does not satisfy Cauchy's criterion for integrability but is Riemann-integrable, Riemann gave

$$f(x) = \sum_{n=1}^{\infty} \frac{\phi(nx)}{n^2},$$

where $\phi(x)$ is defined to be x minus the nearest integer, or, if there are two equally close integers, to be equal to 0. It turns out that f is continuous everywhere except at the infinitely many points $x = p/2n$, with p and n relatively prime. But, because the variation of f near such a point is equal to $\pi^2/8n^2$, there are only finitely many points near which the variation is greater than any $\sigma > 0$, and the function satisfies Riemann's integrability criterion.

With a new class of functions now proved to be integrable, Riemann was able to extend Dirichlet's result on convergence of Fourier series. But rather than determine sufficient conditions on a function that ensure the convergence of its series, Riemann attacked the problem in reverse by beginning with a function that is representable by a trigonometric series and attempting to determine what consequences this representation has for the behavior of the function. Riemann was able in this way to find many types of functions that were representable by trigonometric series, but he never answered the entire question to his own satisfaction. Probably this was the reason that Riemann never published the manuscript in which this material appears.

17.1.8 Uniform Convergence

The work of Dirichlet and Riemann made it absolutely clear that a Fourier series can represent a discontinuous function and that therefore Cauchy's theorem on the sum of a series of continuous functions had to be modified. This was accomplished by several mathematicians, but it was Karl Weierstrass (1815–1897) who realized how to ensure that the sum function was continuous over an entire interval, as Cauchy had originally concluded. In the course of his lectures at the University of Berlin beginning in the 1850s, Weierstrass made a careful distinction between convergence of a series of numbers and that of a series of functions, a distinction that Cauchy had glossed over. He was then able to identify a crucial property of convergence of functions, that of uniform convergence over an interval: An infinite series $\sum u_n(x)$ is **uniformly convergent** in $[a, b]$ when, given any $\epsilon > 0$, there exists an N (dependent on ϵ) such that $|r_n(x)| < \epsilon$ for every $n > N$ and for every x in the interval $[a, b]$.

Given Weierstrass's definition, it was simple enough to correct Cauchy's proof for the case where the convergence of the series was uniform. But this definition also had a deeper influence. Not only did Weierstrass make absolutely clear how certain quantities in his definition depended on other quantities, he also completed the move away from the use of terms such as "infinitely small." Henceforth, all definitions involving such ideas were given completely arithmetically. For example, Eduard Heine (1821–1881), a professor at Halle who spent much time in Berlin discussing mathematics with Weierstrass, not only gave a definition of continuity at a point in a paper of 1872, but also reworked Cauchy's definition of continuity over an interval into the following: "A function $f(x)$ is called ... **uniformly continuous** from $x = a$ to $x = b$ if for any given quantity ϵ, however small, there exists a positive quantity η_0 such that for all positive values η which are less than η_0, $f(x \pm \eta) - f(x)$ remains less than ϵ. Whichever value one may give to x, assuming only that x and $x \pm \eta$ belong to the interval from a to b, the same η_0 must effect the required [property]." Heine then went on to prove that a function continuous over a closed interval is uniformly continuous in that interval, making implicit use of what is today called the **Heine-Borel theorem**. He also gave the earliest published proof of the theorem that a continuous function on a closed interval attains a maximum and a minimum.

Because Weierstrass himself did not publish many of his new ideas, it was through the efforts of his followers and his students that his concepts became the standard in mathematical analysis, a standard still in place today.

17.2 The Arithmetization of Analysis

Even with the new definitions of convergence by Bolzano and Cauchy, it was apparent by the middle of the nineteenth century that a crucial step was missing in their proofs of such results as the intermediate value theorem and the existence of a limit of a bounded increasing sequence. Although the new definitions enabled mathematicians to show that a certain sequence of numbers satisfied the Cauchy criterion, there was no way to assert the existence of a limit if one could not specify in advance what type of "number" this limit would be. Cauchy, among others, understood intuitively what the real numbers were. He had even asserted that an irrational number can be considered as a limit of a certain sequence of rational numbers. But he was thereby asserting the *a priori* existence of such a number, without any argument as to how that assertion could be justified.

17.2.1 Dedekind Cuts

By the middle of the century, several mathematicians were actively considering this matter of exactly what an irrational number is. They were no longer content to assume the existence of such objects, as their eighteenth-century predecessors had, particularly because they needed to explain some of the basic concepts in their courses. Among these mathematicians was Dedekind, who set forth a program of research that led to what is now called the *arithmetization of analysis*: the deduction of the theorems of analysis first from the basic postulates defining the integers and then from the principles of the theory of sets. Dedekind worked out the results of his investigations in 1858, but they were not published until 1872, in a brief work, *Continuity and Irrational Numbers*.

To find a "real definition of the essence of continuity" in a "purely arithmetic manner," Dedekind began by considering the order properties of the set R of rational numbers. The ones he considered most important were first, that if $a > b$ and $b > c$, then $a > c$; second, that if $a \neq c$, then there are infinitely many rational numbers lying between a and c; and third, that any rational number a divides the entire set R into two classes A_1 and A_2, with A_1 consisting of those numbers less than a and A_2 consisting of those numbers greater than a, the number a itself being assigned to either of the two classes. These two classes have the obvious property that every number in A_1 is less than every number in A_2. Although Dedekind had in mind the correspondence between rational numbers and points on the line, he realized that one could not use the geometrical line to define numbers arithmetically. Dedekind's aim, therefore, was "the creation of new numbers such that the domain of numbers shall gain the same completeness, or . . . the same continuity, as the straight line."

To create these new numbers, Dedekind decided to transfer to the domain of number the property he considered the essence of continuity of the straight line: that "if all points of the straight line fall into two classes such that every point of the first class lies to the left of every point of the second class, then there exists one and only one point which produces this division." Thus, Dedekind generalized his third property of the rationals into a new definition: "If any separation of the system R into two classes A_1, A_2 is given which possesses only this characteristic property that every number a_1 in A_1 is less than every number a_2 in A_2, then . . . we shall call such a separation a **cut** and designate it by (A_1, A_2)." Every rational number a determines a cut for which a is either the largest of the numbers in A_1 or the smallest of those in A_2, but there are certainly cuts not produced by rational numbers, for example, the cut determined by defining A_2 to be the set of all positive rational numbers whose square is greater than 2 and A_1 to be the set of all other rational numbers. As he wrote, "In this property that not all cuts are produced by rational numbers consists the incompleteness or discontinuity of the domain R of all rational numbers. Whenever, then, we have to do with a cut (A_1, A_2) produced by no rational number, we create a new, an *irrational* number α, which we regard as completely defined by this cut (A_1, A_2); we shall say that the number α corresponds to this cut, or that it produces this cut." Dedekind thus considered α to be a new creation of the mind which corresponded to the cut. Other mathematicians, however, believed that it was better to define the real number α to *be* the cut.

In any case, the collection of all such cuts determines the system $\mathcal{R}$ of real numbers. Dedekind was able to show that this system possessed a natural ordering $<$ that satisfies the same properties of order satisfied by the rational numbers. He then proved that the system $\mathcal{R}$ also possessed the attribute of continuity; that is, if it is broken up into two classes $\mathcal{A}_1$, $\mathcal{A}_2$ such that every number α_1 in $\mathcal{A}_1$ is less than every number α_2 in $\mathcal{A}_2$, then there exists one and only one real number α that is either the greatest number in $\mathcal{A}_1$ or the smallest number in $\mathcal{A}_2$. In fact, α is the real number corresponding to the cut (A_1, A_2) where A_1 consists of all rational numbers in $\mathcal{A}_1$ and A_2 consists of all rational numbers in $\mathcal{A}_2$.

Dedekind completed his essay by defining the standard arithmetic operations in his new system $\mathcal{R}$ and then proving some basic theorems. One of these theorems was that every bounded increasing sequence $\{\beta_i\}$ of real numbers has a limit. Letting $\mathcal{A}_2$ be the set of all numbers γ such that $\beta_i < \gamma$ for all i, and $\mathcal{A}_1$ all remaining numbers, Dedekind easily showed that the cut $(\mathcal{A}_1, \mathcal{A}_2)$ determines the number α that is the least number in $\mathcal{A}_2$ as well as the required limit.

17.2.2 Cantor and Fundamental Sequences

Dedekind's work on cuts appeared in print in 1872. In the same year, Georg Cantor (1845–1918) published an alternative arithmetic derivation of the real numbers. Cantor came at the problem of creating the real numbers from a point of view different from Dedekind's. He was interested in the old problem of the convergence of Fourier series and took up the question of whether a trigonometric series that represents a given function is necessarily unique. In 1870, he managed to prove uniqueness under the assumption that the trigonometric series converged for all values of x. But then he succeeded in weakening the conditions. First, in 1871, he showed that the theorem was still true if either the convergence or the representation failed to hold at a finite number of points in the given interval. Second, in the following year, he was able to prove uniqueness even if the number of these exceptional points was infinite, provided that the points were distributed in a specified way. To describe accurately this distribution of points, Cantor realized that he needed a new way of describing the real numbers.

Beginning, like Dedekind, with the set of rational numbers, Cantor introduced the notion of a **fundamental sequence**, a sequence $a_1, a_2, \ldots, a_n, \ldots$ with the property that "for any positive rational value ϵ there exists an integer n_1 such that $|a_{m+n} - a_n| < \epsilon$ for $n \geq n_1$, for any positive integer m." Such a sequence, now called a **Cauchy sequence**, satisfies the criterion Cauchy set out in 1821. For Cauchy, it was obvious that such a sequence converged to a real number b. Cantor, on the other hand, realized that to say this was to commit a logical error, for that statement presupposed the existence of such a real number. Therefore, Cantor used the fundamental sequence to *define* a real number b. In other words, Cantor associated a real number to every fundamental sequence of rational numbers. The rational number r was itself associated to a sequence, the sequence $r, r, \ldots, r, \ldots$, but there were also sequences that were not associated to rationals. For example, the sequence 1, 1.4, 1.41, 1.414, ..., generated by a familiar algorithm for calculating $\sqrt{2}$, was such a fundamental sequence.

Realizing that two fundamental sequences could well converge to the same real number, Cantor went on to define an equivalence relation on the set of such sequences. Thus, the number b associated to the sequence $\{a_i\}$ was said to be equal to the number b' associated to the sequence $\{a'_i\}$ if for any $\epsilon > 0$, there existed an n_1 such that $|a_n - a'_n| < \epsilon$ for $n > n_1$. The set B of real numbers was then the set of equivalence classes of fundamental sequences. It was not difficult to define an order relationship on these sequences as well as to establish the basic arithmetic operations. But Cantor wanted to show that the set he had defined was in some sense the same as the number line. It was clear to Cantor that every point on the line corresponded to a fundamental sequence, but he realized that the converse required an axiom: that to every real number (equivalence class of fundamental sequences) there corresponds a definite point on the line.

Having defined real numbers, Cantor returned to his original question in the theory of trigonometric series. By using his identification of the real numbers with the points on the line, he defined the **limit point** of a point set P to be "a point of the line so placed that in every neighborhood of it we can find infinitely many points of P.... By the neighborhood of a point should here be understood every interval which has the point in its interior. Thereafter it is easy to prove that a [bounded] point set consisting of an infinite number of points always has at least one limit point."

17.2.3 The Theory of Sets

In dealing with various sets of points, Cantor was led to numerous questions about sets. For example, he knew that the rational numbers were dense in the real numbers, in the sense that every interval contained infinitely many rationals, but they were not continuous. It would seem, therefore, that in some sense there should be more real numbers than rational numbers. In November, 1873, he posed this question in a letter to Dedekind: "Take the collection of all positive whole numbers n and denote it by (n); then think of the collection of all real numbers x and denote it by (x); the question is simply whether (n) and (x) may be corresponded so that each individual of one collection corresponds to one and only one of the other? ... As much as I am inclined to the opinion that (n) and (x) permit no such unique correspondence, I cannot find the reason."

Dedekind could not answer Cantor's question, but only a month later Cantor was able to show that such a correspondence was impossible. His original proof was somewhat complicated, so we will give his later "diagonal" proof, a proof that turned out to have many further uses. Suppose the real numbers of interval (0, 1) could be placed in one-to-one correspondence with the natural numbers. Then there is a listing $r_1, r_2, r_3, \ldots$ of the real numbers in the interval. Write each such number in its infinite decimal form:

$$r_1 = 0.a_{11}a_{12}a_{13}\ldots$$

$$r_2 = 0.a_{21}a_{22}a_{23}\ldots$$

$$r_3 = 0.a_{31}a_{32}a_{33}\ldots$$

Now define a number b by choosing $b = 0.b_1b_2b_3\ldots$, where $b_1 \neq a_{11}$, $b_2 \neq a_{22}$, $b_3 \neq a_{33}, \ldots$. Since b differs from r_1 in the first decimal place, from r_2 in the second, and so on, b cannot be anywhere in the list. Thus, a one-to-one correspondence between the real numbers in (0, 1) and the positive integers cannot exist.

Cantor was also able to prove in 1873 that the set of all algebraic numbers can be put into one-to-one correspondence with the set of natural numbers, where the **algebraic numbers** are those real numbers satisfying a polynomial equation with rational coefficients. It followed that there were an infinite number of **transcendental numbers**, numbers that were not algebraic. More importantly, however, Cantor had established a technique of counting infinite collections and had determined a clear difference in the size (or cardinality) of the continuum of real numbers on the one hand and the set of rational or algebraic numbers on the other.

Cantor soon realized that his concept of a one-to-one correspondence could be placed at the foundation of a new theory of sets. In 1879, he used the concept to begin the study of the cardinality of an infinite set. Two sets A and B were defined to be of the same **power** if there was a one-to-one correspondence between the elements of A and the elements of B. Cantor initially singled out the two special cases he had already considered: sets of the same power as the set N of natural numbers—these he called *denumerable sets*—and sets of the power of the set of real numbers. In his further attempt to understand the properties of the continuum, Cantor was led to establish over the next two decades a detailed theory of infinite sets.

Cantor's *Contributions to the Founding of Transfinite Set Theory* of 1895 began with a definition of a set as "any collection into a whole M of definite and separate objects m of our intuition or our thought." "Every set M has a definite 'power,' which we will also call its 'cardinal number,' " he continued. "We will call by the name 'power' or 'cardinal number' of M the general concept which, by means of our active faculty of thought, arises from the set M when we make abstraction of the nature of its various elements m and of the order in which they are given." By this "abstraction" Cantor meant that the cardinality of an infinite set was the generalization of the concept of "number of elements" for a finite set. Thus, the set of natural numbers and the set of real numbers have different cardinal numbers. The cardinality of the set of natural numbers, the set of smallest transfinite cardinality, Cantor called "aleph-zero," written $\aleph_0$; the cardinality of the real numbers he denoted by $\mathcal{C}$. Two sets are **equivalent**, or have the same cardinality, if there is a one-to-one correspondence between them. Cantor also defined the notion of $<$ for transfinite cardinals: The cardinality $\bar{\bar{M}}$ of a set M is less than that of a set N if there is no part of M that is equivalent to N but there is a part of N that is equivalent to M. It is then clear that for two sets M and N, no more than one of the relations $\bar{\bar{M}} = \bar{\bar{N}}$, $\bar{\bar{M}} < \bar{\bar{N}}$, or $\bar{\bar{N}} < \bar{\bar{M}}$ can occur. But Cantor could not show that at least one of these relations had to occur.

Because $\aleph_0 < \mathcal{C}$, Cantor posed the question whether any other cardinalities were possible for subsets of the real numbers. In 1878, he thought he had answered this question in the negative, but later realized that his proof was faulty. In fact, the conjecture that every subset of the real numbers has cardinality either $\aleph_0$ or $\mathcal{C}$, called the **Continuum hypothesis**, was eventually shown to be unprovable by use of any reasonable collection of axioms for the theory of sets.

17.2.4 Dedekind and Axioms for the Natural Numbers

Cantor had developed some of the more advanced ideas of set theory and had shown, along with Dedekind, how to construct the real numbers starting from the rational numbers. But it was Dedekind who completed the process of arithmetizing analysis by characterizing the natural numbers, and therefore the rational numbers, in terms of sets. In the work in which he accomplished this task—*What Are the (Natural) Numbers and What Do They Mean?*—developed over a fifteen-year period but only published in 1888, he also provided an introduction to the basic notions of set theory.

To characterize the natural numbers, Dedekind began with the notion that the natural numbers form a set of things, or "objects of our thought." Therefore, Dedekind defined the term *systeme*, here translated as *set*: "It very frequently happens that different things, $a, b, c, \ldots$ for some reason can be considered from a common point of view, can be associated in the mind, and we say that they form a set S.... Such a set S as an object of our thought is likewise a thing; it is completely determined when with respect to every thing it is determined whether it is an element of S or not." Given this necessarily somewhat vague definition, Dedekind proceeded to describe various simple relations involving sets. For example, a set A is a **part** of a set S when every element of A is also an element of S. Also, the set **compounded** out of any sets $A, B, C, \ldots$, denoted $\mathcal{M}(A, B, C, \ldots)$ consists of those elements that are in at least one of the sets $A, B, C, \ldots$, while the set of elements common to $A, B, C, \ldots$, is denoted $\mathcal{G}(A, B, C, \ldots)$. In modern terminology, Dedekind's "part" is a **subset**, $\mathcal{M}(A, B, C, \ldots)$ is the **union** of the sets $A, B, C, \ldots$, and $\mathcal{G}(A, B, C, \ldots)$ is their **intersection**.

A fundamental property of the natural numbers is that each number has a unique successor. In other words, there is a function ψ from the set N of natural numbers to itself given by $\psi(n) = n + 1$. Because different elements of N have different successors, Dedekind was led to the notion of a similar, or **injective**, transformation, one for which "to different elements a, b of the set S there always correspond different transforms $a' = \phi(a)$, $b' = \phi(b)$." In this case, there is an inverse transformation $\bar{\phi}$ of the system $S' = \phi(S)$, defined by assigning to every element s' of S' the unique element s that was transformed into it by ϕ. Two sets R and S are then said to be **similar** to one another if there exists an injective transformation ϕ defined on R such that $S = \phi(R)$.

The natural numbers also have the property that the image of N under the successor transformation is a proper subset of N itself, with the only element not belonging to that image being the element 1. In fact, Dedekind defined a set to be infinite precisely in the case where it is similar to a proper subset of itself. But do infinite sets exist at all? Dedekind was hesitant to prove results about such sets without an argument that they exist, so he gave one: "The totality S of all things which can be objects of my thought is infinite. For if s signifies an element of S, then is the thought s', that s can be object of my thought, itself an element of S." Given the transformation from S to itself defined by $s \rightarrow s'$, for which it is clear that the image is not all of S and that the transformation is injective, Dedekind concluded that the set S does satisfy the requirements of his definition.

Dedekind realized that the properties that N was a set possessing an injective successor function whose image was a proper subset of N did not characterize N uniquely. There may well be extraneous elements in any set S that satisfied these properties, elements that are not natural numbers. For example, the set of positive rational numbers satisfies all of the properties. So Dedekind added one more property, that an element belongs to N if and only if it is an element of every subset K of S having the property that 1 belongs to K and that the successor of every element of K is also in K. In other words, N is characterized by being the intersection of all sets satisfying the original properties. N thus contains a base element 1, the successor $\phi(1)$ of 1, the successor $\phi(\phi(1))$ of that element, and so forth, but no other elements.

From his characterization of the natural numbers, Dedekind was able to derive the principle of mathematical induction as well as to give a definition and derive the properties of the order relationship on N and the operations of addition and multiplication. It was this work, together with the work of Weierstrass and his school, that enabled the calculus to be placed on a firm foundation beginning with the basic notions of set theory. Thus, Dedekind, Weierstrass, and others showed that calculus has an existence independent of the physical world of motion and curves, the world used by Newton to create the subject in the first place.

17.3 Complex Analysis

Mathematicians had been using complex numbers since the sixteenth century, primarily in a formal sense, and did not generally conceive of them abstractly. It was the geometrical representation of these numbers, first published by the Norwegian surveyor Caspar Wessel (1745–1818) in an essay in 1797, that ultimately became the basis for a new way of thinking about these numbers, which soon convinced mathematicians that they could use them without undue worry.

17.3.1 Geometrical Representation of Complex Numbers

Wessel's aim in his *On the Analytical Representation of Direction* was not initially related to complex numbers as such. He felt that certain geometrical concepts could be more clearly understood if there was a way to represent both the length and direction of a line segment in the plane by a single algebraic expression. Wessel made clear that these expressions had to be capable of being manipulated algebraically. In particular, he wanted a way of expressing an arbitrary change of direction algebraically that was more general than the simple use of a negative sign to indicate the opposite direction.

Wessel began by dealing with addition: "Two straight lines are added if we unite them in such a way that the second line begins where the first one ends and then pass a straight line from the first to the last point of the united lines. This line is the sum of the united lines." Thus, whatever the algebraic expression of a line segment was to be, the addition of two had to satisfy this obvious property drawn from Wessel's conception of motion. In other words, he conceived of line segments as representing vectors. It was multiplication, however, that provided Wessel with the basic answer to his question of the representation of direction. To derive this multiplication, he established a number of properties that he felt were essential. First, the product of two lines in the plane had to remain in the plane. Second, the length of the product line had to be the product of the lengths of the two factor lines. Finally, if all directions were measured from the positive unit line, which he called 1, the angle of direction of the product was to be the sum of the angles of direction of the two factors. Designating by ϵ the line of unit length perpendicular to the line 1, he easily showed that his desired properties implied that $\epsilon^2 = (-\epsilon)^2 = -1$ or that $\epsilon = \sqrt{-1}$. A line of unit length making an angle θ with the positive unit line could now be designated by $\cos\theta + \epsilon \sin\theta$

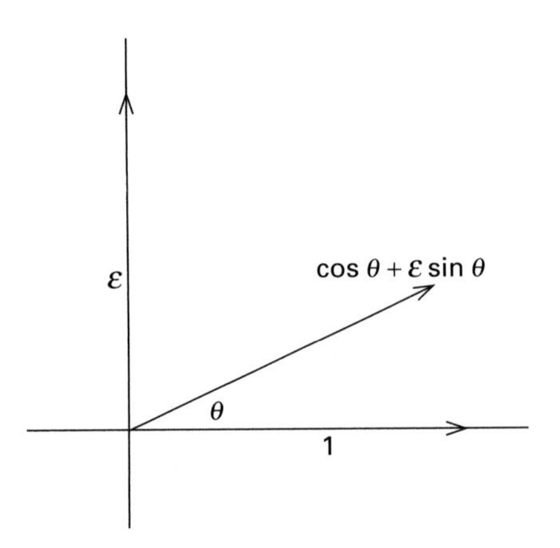

FIGURE 17.3 Wessel's geometric interpretation of complex numbers

and, in general, a line of length A and angle θ by $A(\cos\theta + \epsilon \sin\theta) = a + \epsilon b$, where a and b are chosen appropriately (Fig. 17.3). Thus, from Wessel's algebraic interpretation of a geometric line segment arose the geometric interpretation of the complex numbers. The obvious algebraic rule for addition satisfied Wessel's requirements for that operation, while the multiplication $(a + \epsilon b)(c + \epsilon d) = ac - bd + \epsilon(ad + bc)$ satisfied his axioms for multiplication. Wessel also easily derived from his definitions the standard rules for division and root extraction of complex numbers.

Unfortunately, Wessel's essay remained unread in most of Europe for many years after its publication. It was only because Gauss used the same geometric interpretation of the complex numbers in his proofs of the fundamental theorem of algebra (that every polynomial with real coefficients has a real or complex root) and in his study of quartic residues that this interpretation gained acceptance in the mathematical community.

17.3.2 Complex Functions

By the second decade of the nineteenth century, Gauss, with his geometric understanding of the meaning of complex numbers, began the development of the theory of complex functions. In particular, he defined the integral of a complex function $\phi(x)$ between two points in the complex plane as an infinite sum of the values $\phi(x)\,dx$, as x moves along a curve connecting the two points, where each dx is an infinitesimal complex increment of x. Gauss went on to assert the "very beautiful theorem" that as long as $\phi(x)$ is never infinite within the region enclosed by two different curves connecting the starting and ending points of this integral, then the value of the integral is the same along both curves. Although he did not express himself in those terms, Gauss was considering $\phi(x)$ as an analytic function. In any case, he never published a proof of this result, which is today known as **Cauchy's integral theorem**.

Although Cauchy discussed functions of a complex variable earlier, in terms of two functions of two real variables, it was not until 1825, with his new definition of a definite integral, that Cauchy was able to deal with complex functions in their own right. In his *Memoir on Definite Integrals Taken Between Imaginary Limits*, he explicitly defined the definite complex integral

$$\int_{a+ib}^{c+id} f(z)\,dz$$

to be the "limit or one of the limits to which the sum of products of the form $[(x_1 - a) + i(y_1 - b)]f(a + ib)$, $[(x_2 - x_1) + i(y_2 - y_1)]f(x_1 + iy_1), \ldots, [(c - x_{n-1}) + i(d - y_{n-1})]f(x_{n-1} + iy_{n-1})$ converges when each of the two sequences $a, x_1, x_2, \ldots, x_{n-1}$, c and $b, y_1, y_2, \ldots, y_{n-1}, d$ consist of terms that increase or decrease from the first to the last and approach one another indefinitely as their number increases without limit." In other words, Cauchy made Gauss's vague definition explicit by directly generalizing his own definition of a real definite integral and taking partitions of the two intervals $[a, b]$ and $[c, d]$. Cauchy realized, however, as had Gauss, that there were infinitely many different paths of integration beginning at $a + ib$ and ending at $c + id$. To show then that this definition made sense, Cauchy proved his integral theorem through an argument based on the calculation of variations. But since this argument required the existence and continuity of the derivative of f, and Cauchy had not explicitly defined what was meant by the derivative of a complex function, we will consider Riemann's definitions and his proof of the Cauchy integral theorem in his doctoral dissertation.

Riemann's dissertation of 1851, *Foundations for a General Theory of Functions of One Complex Variable*, began with a discussion of an important distinction between real and complex functions. Although the definition of function, "to every one of [the] values [of a variable quantity z] there corresponds a single value of the indeterminate quantity w," can be applied both to the real and to the complex case, Riemann realized that in the latter case, where $z = x + iy$ and $w = u + iv$, the limit of the ratio dw/dz defining the derivative could well depend on how dz approaches zero. Because for functions defined algebraically one could calculate the derivative formally and not have this problem, Riemann decided to make this existence of the derivative the basis for the concept of a complex function: "The complex variable w is called a function of another complex variable z when its variation is such that the value of the derivative dw/dz is independent of the value of dz."

As a first application of this definition, Riemann showed that such a complex function considered as a mapping from the z-plane to the w-plane preserves angles. For suppose p' and p'' are infinitely close to the origin P in the z-plane, with their images q', q'' infinitely close to the image Q of P. Writing the infinitesimal distance from p' to P both as $dx' + i\,dy'$ and as $\epsilon' e^{i\phi'}$, and the distance from q' to Q as both $du' + i\,dv'$ and $\eta' e^{i\psi'}$, with similar notations for the other infinitesimal distances, Riemann noted that his condition on the function implies that

$$\frac{du' + i\,dv'}{dx' + i\,dy'} = \frac{du'' + i\,dv''}{dx'' + i\,dy''}$$

or that

$$\frac{du' + i\,dv'}{du'' + i\,dv''} = \frac{\eta'}{\eta''} e^{i(\psi' - \psi'')} = \frac{dx' + i\,dy'}{dx'' + i\,dy''} = \frac{\epsilon'}{\epsilon''} e^{i(\phi' - \phi'')}.$$

It follows that $\eta'/\eta'' = \epsilon'/\epsilon''$ and that $\psi' - \psi'' = \phi' - \phi''$, or, in other words, that the infinitesimal triangles $p'Pp''$ and $q'Qq''$ are similar. Such an angle-preserving mapping is called a **conformal mapping**. Given this result, Riemann was able to demonstrate the

Riemann mapping theorem, that any two simply connected regions in the complex plane can be mapped conformally on each other by means of a suitably chosen complex function.

Riemann next determined what the existence of the derivative means in terms of the two functions u and v:

$$\frac{dw}{dz} = \frac{du + i\,dv}{dx + i\,dy} = \frac{\frac{\partial u}{\partial x}dx + \frac{\partial u}{\partial y}dy + i\left(\frac{\partial v}{\partial x}dx + \frac{\partial v}{\partial y}dy\right)}{dx + i\,dy}$$
$$= \frac{\left(\frac{\partial u}{\partial x} + i\frac{\partial v}{\partial x}\right)dx + \left(\frac{\partial v}{\partial y} - i\frac{\partial u}{\partial y}\right)i\,dy}{dx + i\,dy}.$$

If this value is independent of how dz approaches zero, then setting dx and dy in turn equal to zero and equating the real and imaginary parts of the two resulting expressions shows that

$$\frac{\partial u}{\partial x} = \frac{\partial v}{\partial y} \quad \text{and} \quad \frac{\partial v}{\partial x} = -\frac{\partial u}{\partial y}.$$

These equations are known as the **Cauchy-Riemann equations**, because Cauchy had derived them earlier, using an idea of Euler's. But Riemann went further than his predecessors, making them the center of his theory of complex functions. He noted, for example, that if these equations are satisfied, the derivative dw/dz is easily calculated to be $\partial u/\partial x + i\,\partial v/\partial x$, a value independent of dz.

Riemann then used the Cauchy-Riemann equations to give a detailed proof of the Cauchy integral theorem, based on the result today known as **Green's theorem** [named after George Green (1793-1841), although originally stated by Cauchy]. Riemann stated the theorem in the following form:

THEOREM. *Let X and Y be two functions of x and y continuous in a finite region T with infinitesimal area element designated by dT. Then*

$$\int_T \left(\frac{\partial X}{\partial x} + \frac{\partial Y}{\partial y}\right) dT = -\int_S (X\cos\xi + Y\cos\eta)\,ds,$$

where the latter integral is taken over the boundary curve S of T and ξ, η designate the angles the inward pointing normal line to the curve makes with the x-axis and y-axis, respectively.

Riemann proved this by using the fundamental theorem of calculus to integrate $\partial X/\partial x$ along lines parallel to the x-axis, getting values of X where the lines cross the boundary of the region. Because $dy = \cos\xi\,ds$ at each of those points, he could integrate with respect to y to get

$$\int\left[\int \frac{\partial X}{\partial x}dx\right]dy = -\int X\,dy = -\int X\cos\xi\,ds.$$

The other half of the theorem is proved similarly. Riemann then noted that

$$\frac{dx}{ds} = \pm\cos\eta \quad \text{and} \quad \frac{dy}{ds} = \mp\cos\xi,$$

where the sign depends on whether one gets from the tangent line to the inward normal line by traveling counterclockwise or clockwise. It follows that Green's theorem can be rewritten as

$$\int_T \left(\frac{\partial X}{\partial x} + \frac{\partial Y}{\partial y}\right) dT = \int_S \left(X\frac{dy}{ds} - Y\frac{dx}{ds}\right) ds.$$

To get the Cauchy integral theorem, we write $w(z) = u(z) + iv(z)$ and $dz = dx + i\,dy$. Instead of integrating $w(z)$ over two curves connecting the given points, we integrate it around a closed curve S, one half of which is the first curve and the other half of which is the second curve taken in the opposite direction. To show that the integrals over the two curves are equal is then the same as to show that the integral around the closed curve is zero. We first get

$$\int_S w(z)\,dz = \int_S (u + iv)(dx + i\,dy) = \int_S u\,dx - v\,dy + i\int_S v\,dx + u\,dy.$$

By parameterizing S as $x = x(s)$, $y = y(s)$, this integral becomes

$$\int \left(u\frac{dx}{ds} - v\frac{dy}{ds}\right) ds + i\int \left(v\frac{dx}{ds} + u\frac{dy}{ds}\right) ds$$

which, by Green's theorem, is equal to

$$-\int_T \left(\frac{\partial u}{\partial y} + \frac{\partial v}{\partial x}\right) dT + i\int_T \left(\frac{\partial u}{\partial x} - \frac{\partial v}{\partial y}\right) dT.$$

The Cauchy-Riemann equations then imply that these last two integrals are each zero, thus proving the Cauchy integral theorem.

17.3.3 The Riemann Zeta Function

One complex function that Riemann studied extensively has had major importance since his time; this function is known today as the **Riemann zeta function**. This function had its start in Euler's formula

$$\sum_{n=1}^{\infty} \frac{1}{n^s} = \prod_p \frac{1}{\left(1 - \frac{1}{p^s}\right)}, \qquad \textbf{(17.5)}$$

where the product on the right ranges over all prime numbers p. The formula results from expanding each factor on the right as

$$\frac{1}{\left(1 - \frac{1}{p^s}\right)} = 1 + \frac{1}{p^s} + \frac{1}{(p^2)^s} + \frac{1}{(p^3)^s} + \cdots$$

and then noting that their product is a sum of terms of the form

$$\frac{1}{\left(p_1^{n_1} p_2^{n_2} \cdots p_k^{n_k}\right)^s},$$

where the p_i are distinct primes and the n_i are positive integers. Because every positive integer can be expressed uniquely as a product of primes, the sum of all such expressions is exactly the left side of Euler's formula. Euler used this formula generally for integer values of s. In fact, letting $\zeta(s) = \sum_{n=1}^{\infty} 1/n^s$, he was able to show that $\zeta(2) = \pi^2/6$ and $\zeta(4) = \pi^4/90$ and also to provide a general method for calculating $\zeta(2n)$. Dirichlet, somewhat later, extended $\zeta(s)$ to real values $s > 1$ and was able to prove formula 17.5 rigorously for that case.

By rewriting the expression for $\zeta(s)$ in terms of integrals, Riemann was able in 1859 to extend its domain to the entire complex plane. He showed, in fact, that $\zeta(s)$ was finite for all values of s except $s = 1$ and also that $\zeta(-2n) = 0$ for every positive integer n. Riemann used $\zeta(s)$ in his attempts to find an analytic expression for $\pi(x)$, the number of primes less than x. Although he was not completely successful in this, he mentioned in passing that it was "very likely" that all the complex zeros of $\zeta(s)$ had their real part equal to $\frac{1}{2}$, but that he was unable to prove this result. This statement, that all the zeros have real part equal to $\frac{1}{2}$, has become known as the **Riemann hypothesis**. Although many mathematicians have attempted to prove it since Riemann's time, and recent computer calculations have shown that the 1,500,000,000 complex zeros closest to the real line all have their real part equal to $\frac{1}{2}$, no proof of the hypothesis nor, for that matter, any counterexample has yet been found. Its proof is still a major mathematical challenge, whose achievement would have far-reaching implications in number theory and other areas of mathematics.

17.4 Vector Analysis

Riemann stated Green's theorem in 1851 in terms of the equality of a double integral with an integral along a curve taken with respect to the curve element ds. It was the use in physics of integrals over curves to represent work done along the curves that seems to have inspired a change in notation that occurred in the 1850s, in which the curve integral was replaced by a line integral, an integral of the form $\int p\,dx + q\,dy$. Although this notation had been used in complex integration, physicists converted it into an expression involving vectors. Physicists were also interested in surface integrals, that is, integrals of functions and vector fields over two-dimensional regions. The notion of a surface integral was first introduced by Lagrange in 1811. He noted that if the tangent plane at dS makes an angle γ with the xy-plane, then simple trigonometry allows one to rewrite $dx\,dy$ as $\cos\gamma\,dS$. It followed that if A is a function of three variables, then $\int A\,dx\,dy = \int A\cos\gamma\,dS$, the second integral being taken over a region in the surface, the first over the projection of that region in the plane. Similarly, if β is the angle the tangent plane makes with the xz-plane and α the angle the tangent plane makes with the yz-plane, then $dx\,dz = \cos\beta\,dS$ and $dy\,dz = \cos\alpha\,dS$. Lagrange noted that α, β, and γ could also be considered as the angles that a normal to the surface element makes with the x, y, and z axes, respectively.

17.4.1 Surface Integrals and the Divergence Theorem

Lagrange used surface integrals in dealing with fluid dynamics. In 1813, Gauss used the same concept in considering the gravitational attraction of an elliptical spheroid. But Gauss went further than Lagrange in showing how to calculate an integral with respect to dS in the case where the surface S is given parametrically by three functions $x = x(p, q)$,

$y = y(p, q)$, $z = z(p, q)$. Using a geometrical argument, he demonstrated that

$$dS = \left[\left(\frac{\partial(y,z)}{\partial(p,q)}\right)^2 + \left(\frac{\partial(z,x)}{\partial(p,q)}\right)^2 + \left(\frac{\partial(x,y)}{\partial(p,q)}\right)^2\right]^{1/2} dp\,dq$$

and hence that any integral with respect to dS can be reduced to an integral of the form $\int f\,dp\,dq$, where f is either explicitly or implicitly a function of the two variables p, q.

Gauss used his study of integrals over surfaces to prove certain special cases of what is today known as the **divergence theorem**. The general case of this theorem was, however, first stated and proved in 1826 by Mikhail Ostrogradsky (1801–1861), a Russian mathematician who was studying in Paris in the 1820s. In his paper entitled "Proof of a Theorem in Integral Calculus," which came out of his study of the theory of heat, Ostrogradsky considered a surface with surface element ϵ bounding a solid region with volume element ω. With p, q, and r being three differentiable functions of x, y, z and with the angles α, β, and γ as defined above, Ostrogradsky stated the divergence theorem in the form

$$\int_V \left(\frac{\partial p}{\partial x} + \frac{\partial q}{\partial y} + \frac{\partial r}{\partial z}\right)\omega = \int_S (p\cos\alpha + q\cos\beta + r\cos\gamma)\epsilon,$$

where S is the boundary surface of the solid region V. Today, the theorem is generally written, by use of Lagrange's idea, in the form

$$\int\int\int_V \left(\frac{\partial p}{\partial x} + \frac{\partial q}{\partial y} + \frac{\partial r}{\partial z}\right) dx\,dy\,dz = \int\int_S p\,dy\,dz + q\,dz\,dx + r\,dx\,dy.$$

This result, like Green's theorem, is a generalization of the fundamental theorem of calculus, so Ostrogradsky's proof uses that theorem. To integrate $(\partial p/\partial x)\omega$ over a "narrow cylinder" with cross-sectional area $\bar{\omega}$ going through the solid in the x-direction, he used the fundamental theorem to express the integral as $\int (p_1 - p_0)\bar{\omega}$, where p_0 and p_1 are the values of p on the pieces of surface where the cylinder intersects the solid (Fig. 17.4). Because $\bar{\omega} = \epsilon_1 \cos\alpha_1$ on one section of the surface and $\bar{\omega} = -\epsilon_0\cos\alpha_0$ on the other, where α_1 and α_0 are the angles made by the normal at the surface elements ϵ_1, ϵ_0, respectively, Ostrogradsky had demonstrated that

$$\frac{\partial p}{\partial x}\omega = \int p_1\epsilon_1\cos\alpha_1 + \int p_0\epsilon_0\cos\alpha_0 = \int (p\cos\alpha)\epsilon,$$

where the left integral is over the cylinder and the right ones are over the two pieces of surface. Adding up the integrals over all such cylinders gives one-third of the desired result, the other two-thirds being done similarly.

17.4.2 Stokes's Theorem

The divergence theorem relates an integral over a solid to one over the bounding surface, while Green's theorem relates an integral over a region in the plane to one over the boundary curve. A similar result comparing an integral over a surface in three dimensions to one around the boundary curve, a result now known as **Stokes's theorem**, first appeared in print in 1854. George Stokes (1819–1903) had for several years been setting the Smith's

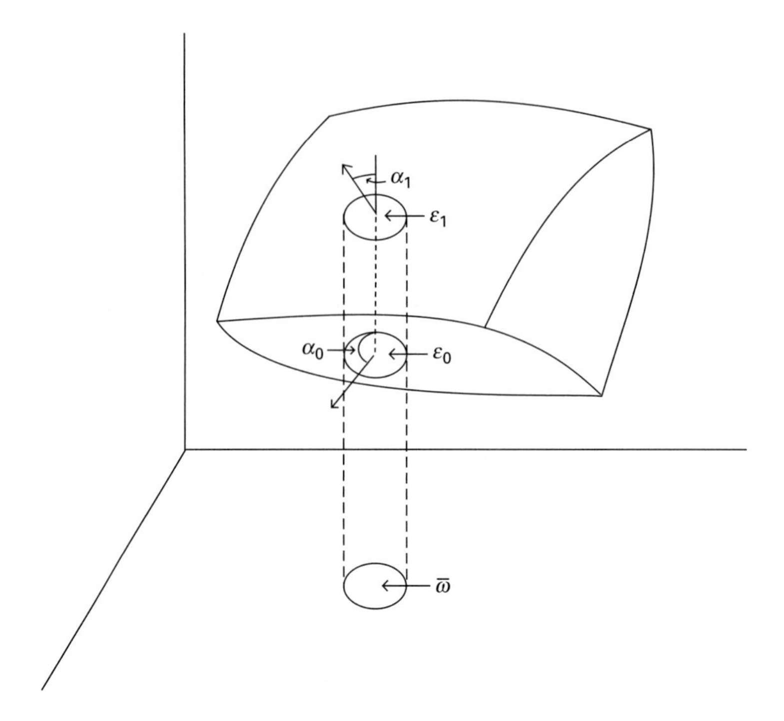

FIGURE 17.4 Ostrogradsky's proof of the divergence theorem

Prize Exam at Cambridge University and, in the February, 1854 examination, posed the following

PROBLEM 8. *If X, Y, Z be functions of the rectangular coordinates x, y, z; dS an element of any limited surface; ℓ, m, n the cosines of the inclinations of the normal at dS to the axes; ds an element of the boundary line, shew that*

$$\int\int\left[\ell\left(\frac{\partial Z}{\partial y}-\frac{\partial Y}{\partial z}\right)+m\left(\frac{\partial X}{\partial z}-\frac{\partial Z}{\partial x}\right)+n\left(\frac{\partial Y}{\partial x}-\frac{\partial X}{\partial y}\right)\right]dS$$
$$=\int\left(X\frac{dx}{ds}+Y\frac{dy}{ds}+Z\frac{dz}{ds}\right)ds$$

... the single integral being taken all around the perimeter of the surface.

It is not known whether any of the students proved the theorem, although the physicist Clerk Maxwell (1831–1879) did sit for the exam. The first published proof of the result, at least for the case where the surface is given explicitly as a function $z = z(x, y)$, appeared in 1861 in a monograph by Hermann Hankel (1839–1873). Hankel substituted the value of z and $dz = (\partial z/\partial x)\,dx + (\partial z/\partial y)\,dy$ into the right-hand integral, thus reducing it to an integral in two variables, then used Green's theorem to convert it to a double integral easily seen to be equal to the surface integral on the left.

Both Stokes's theorem and the divergence theorem appear in the opening chapter of Maxwell's *Treatise on Electricity and Magnetism* and are used often in the remainder of the work. Because Maxwell was an advocate of quaternion notation in physics, he wrote out these theorems in quaternion form, using the fact that if the vector operator $\nabla = (\partial/\partial x)i + (\partial/\partial y)j + (\partial/\partial z)k$ is applied to the vector $\sigma = Xi + Yj + Zk$, the resulting quaternion can be written

$$\nabla\sigma = -\left(\frac{\partial X}{\partial x} + \frac{\partial Y}{\partial y} + \frac{\partial Z}{\partial z}\right) + \left(\frac{\partial Z}{\partial y} - \frac{\partial Y}{\partial z}\right)i + \left(\frac{\partial X}{\partial z} - \frac{\partial Z}{\partial x}\right)j + \left(\frac{\partial Y}{\partial x} - \frac{\partial X}{\partial y}\right)k.$$

Maxwell named the scalar part of $\nabla\sigma$ (the part not including i, j, k) the **convergence** of σ and the vector part (the part including those terms) the **curl** of σ, because of his interpretation of their physical meaning. Maxwell's convergence is the negative of what is today called the **divergence** of σ.

The pure vector form of these theorems finally appeared in the work of Josiah Willard Gibbs (1839–1903) near the end of the century. Gibbs had realized, after reading the work of Maxwell, that it was basically just the two types of products of vectors that Maxwell needed, namely, the dot product and the cross product, rather than the full machinery of quaternions. Thus, in lectures at Yale during the 1880s, Gibbs introduced the modern concepts and notations for $A \cdot B$ and $A \times B$, where A and B are vectors in 3-space. In the context of the divergence theorem and Stokes's theorem, Gibbs set $dV = dx\,dy\,dz$ to be the element of volume, $d\mathbf{a} = dy\,dz\,i + dz\,dx\,j + dx\,dy\,k$ the element of surface area, and $d\mathbf{r} = i\,dx + j\,dy + k\,dz$. He could then write the divergence theorem in the form

$$\int\!\!\int\!\!\int \nabla \cdot \sigma\, dV = \int\!\!\int \sigma \cdot d\mathbf{a}$$

and Stokes's theorem as

$$\int\!\!\int (\nabla \times \sigma) \cdot d\mathbf{a} = \int \sigma \cdot d\mathbf{r},$$

or, alternatively as

$$\int\!\!\int\!\!\int \mathbf{div}\,\sigma\, dV = \int\!\!\int \sigma \cdot d\mathbf{a} \quad \text{and} \quad \int\!\!\int \mathbf{curl}\,\sigma \cdot d\mathbf{a} = \int \sigma \cdot d\mathbf{r}.$$

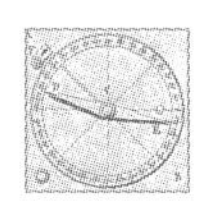

Exercises

1. Prove the theorem of Cauchy:

$$\text{If } \lim_{x\to\infty} f(x+1) - f(x) = \infty,$$
$$\text{then } \lim_{x\to\infty} \frac{f(x)}{x} = \infty.$$

2. Use the theorem of exercise 1 and the analogous theorem in the text to show that

$$\lim_{x\to\infty} \frac{a^x}{x} = \infty \quad \text{and} \quad \lim_{x\to\infty} \frac{\log x}{x} = 0.$$

3. Use the modern definition of continuity and Cauchy's trigonometric identity $\sin(x + \alpha) - \sin x = 2\sin\frac{1}{2}\alpha\cos(x + \frac{1}{2}\alpha)$ to show that $\sin x$ is continuous at any value of x.

4. Prove the following theorem of Cauchy: If $f(x)$ is positive for sufficiently large values of x and if the ratio $f(x+1)/f(x)$ converges to k as x increases indefinitely, then $[f(x)]^{1/x}$ also converges to k as x increases indefinitely.

5. Use the theorem of exercise 4 to show that

$$\lim_{x\to\infty} x^{1/x} = 1.$$

6. Use the Cauchy criterion to show that the series
$$1 + \frac{1}{1!} + \frac{1}{2!} + \frac{1}{3!} + \cdots$$
converges.

7. Show that the series $\{u_k(x)\}$, where $u_1(x) = x$ and $u_k(x) = x^k - x^{k-1}$ for $k > 1$, satisfies the hypotheses of Cauchy's theorem 6-1-1 in a neighborhood of $x = 1$ but not the conclusion. Analyze Cauchy's proof for this case to see where it fails.

8. By putting $a^i = 1 + \beta$, use Cauchy's definition of the derivative to show that the derivative of $y = a^x$ is $y' = a^x / \log_a(e)$.

9. Show that if $f(x)$ is continuous on $[a, b]$ and if $a = x_0 < x_1 < \cdots < x_n = b$ is a partition of $[a, b]$ into subintervals, then the sum
$$f(x_0)(x_1 - x_0) + f(x_1)(x_2 - x_1) + \cdots + f(x_{n-1})(x_n - x_{n-1})$$
is equal to $(b - a)f(x_0 + \theta(b - a))$ for some θ between 0 and 1.

10. Complete Bolzano's proof of the least upper bound criterion by showing that the value U to which the constructed sequence converges is the least upper bound of all numbers having the property M.

11. Let A be the set of numbers in $(\frac{3}{5}, \frac{2}{3})$ that have decimal expansions containing only finitely many zeros and sixes after the decimal point and no other integer. Find the least upper bound of A.

12. Suppose M is the property that $x < 0$ or that $x^3 < 3$. Since this property does not belong to all x but does belong to all x less than 1, it satisfies the conditions of Bolzano's least upper bound theorem. Beginning with the quantity $V = 1 + 1$, for which M is not valid for all x smaller than it, use Bolzano's proof method to construct an approximation to $\sqrt[3]{3}$ accurate to three decimal places.

13. Show that $\phi(x) = \alpha e^{mx}$, $\psi(y) = \beta \cos ny$, with $m^2 = n^2 = 1/A$, are solutions to
$$\frac{\phi(x)}{\phi''(x)} = -\frac{\psi(y)}{\psi''(y)} = A.$$
Conclude that $v = ae^{-nx} \cos ny$ is a solution to
$$\frac{\partial^2 v}{\partial x^2} + \frac{\partial^2 v}{\partial y^2} = 0.$$

14. Use Fourier's method of integration to calculate the Fourier series for $\phi(x) = \frac{1}{2}x$ referred to in the text. Check the correctness of this result for $x = \pi/2$ by using other known series sums.

15. Prove Cauchy's theorem on the continuity of the sum of a series of continuous functions under the additional assumption that the series converges uniformly.

16. Let
$$u_k = \frac{1}{k(k+1)}$$
and
$$v_k(h) = u_k + \frac{2h}{((k-1)h + 1)(kh + 1)},$$
where h is a positive real variable. Show that $\lim_{h\to 0} v_k(h) = u_k$, that $\sum u_k$ converges, and that $\sum v_k(h)$ converges for sufficiently small h. Show also that
$$\lim_{h\to 0} \sum v_k(h) \neq \sum u_k.$$

17. Define a natural ordering $<$ on Dedekind's set of real numbers $\mathcal{R}$ defined by the notion of cuts. That is, given two cuts $\alpha = (\mathcal{A}_1, \mathcal{A}_2)$, and $\beta = (\mathcal{B}_1, \mathcal{B}_2)$, define $\alpha < \beta$. Show that this ordering $<$ satisfies the same basic properties on $\mathcal{R}$ as on the set of rational numbers.

18. Define an addition on Dedekind's cuts. Show that $\alpha + \beta = \beta + \alpha$ for any two cuts α and β.

19. Prove the theorem that every bounded increasing sequence of real numbers has a limit number, using Dedekind's cuts and also using Cantor's fundamental sequences. Which proof is easier?

20. Show that if $\{a_i\}$ and $\{b_i\}$ are fundamental sequences with $\{b_i\}$ not defining the limit 0, then $\{a_i/b_i\}$ is also a fundamental sequence.

21. Define the product AB of two fundamental sequences $A = \{a_i\}$ and $B = \{b_i\}$ as the sequence consisting of the products $\{a_i b_i\}$. Show that this definition makes sense and that if $AB = C$, then $B = C/A$, where division is defined as in exercise 20.

22. Let the complex function $w(z)$ be given as the sum $u(x, y) + iv(x, y)$. Suppose that the Cauchy-Riemann equations are satisfied, that is, that $\partial u/\partial x = \partial v/\partial y$ and $\partial v/\partial x = -\partial u/\partial y$. Show that the derivative dw/dz is equal to $\partial u/\partial x + i\, \partial v/\partial x$.

23. Use Stokes's theorem to show that if $\operatorname{curl} \sigma = 0$, then $\int_C \sigma \cdot d\mathbf{r}$ is independent of the curve C but depends only on its endpoints. Similarly, show that if

div $\sigma = 0$, then $\int_S \sigma \cdot d\mathbf{a}$ is independent of the surface C but depends only on its boundary curve.

24. What did Cauchy mean by his statement that an irrational number is the limit of the various fractions that approach it? What did Cauchy understand by the term *irrational number* (or even by the term *number*)?
25. Explain the differences between Cauchy's definition of continuity on an interval and the modern definition of continuity at a point. Does a function that satisfies Cauchy's definition satisfy the modern one for every point in the interval? Does a function that satisfies the modern definition for every point in an interval satisfy Cauchy's definition?
26. Develop a lesson plan for teaching the concept of uniform convergence by beginning with Cauchy's incorrect theorem and proof.

References

Among the best general works on the history of analysis in the nineteenth century are Ivor Grattan-Guinness, ed., *From the Calculus to Set Theory* (London: Duckworth, 1980), a collection of essays by experts on the various topics covered; Umberto Bottazzini, *The Higher Calculus: A History of Real and Complex Analysis from Euler to Weierstrass* (New York: Springer, 1986); and Ivor Grattan-Guinness, *The Development of the Foundations of Mathematical Analysis from Euler to Riemann* (Cambridge: MIT Press, 1970). The best treatment of Cauchy's work on calculus is Judith V. Grabiner, *The Origins of Cauchy's Rigorous Calculus* (Cambridge: MIT Press, 1981).

Cauchy's *Cours d'analyse de l'École Royale Polytechnique* is reprinted in Cauchy, *Oeuvres complète d'Augustin Cauchy* (Paris: Gauthier-Villars, 1882–), series 2, vol. 3; the *Résumé des leçons donnees a l'École Royale Polytechnique sur le calcul infinitesimal* is in series 2, vol. 4. Bolzano's paper on the intermediate value theorem is available in S. B. Russ, "A Translation of Bolzano's Paper on the Intermediate Value Theorem," *Historia Mathematica* 7 (1980), 156–185. A part of Abel's paper in which he gave a counterexample to Cauchy's theorem on convergence is translated in G. Birkhoff, *A Source Book in Classical Analysis* (Cambridge: Harvard University Press, 1973), 68–70. Fourier's *Analytical Theory of Heat* is available in an English translation (New York: Dover, 1955). Dedekind's work on Dedekind cuts is in *Continuity and Irrational Numbers*, which appears in *Essays on the Theory of Numbers*, Wooster Beman, trans. (La Salle, Ill.: Open Court, 1948). The same volume also contains a translation of *What Are the (Natural) Numbers and What Do They Mean?* Cantor's work on set theory is available as *Contributions to the Founding of the Theory of Transfinite Numbers*, P. E. B. Jourdain, trans. (Chicago: Open Court, 1915). Wessel's "On the Analytical Representation of Direction," is translated in David Smith, *A Source Book in Mathematics* (New York: Dover, 1959), 55–66. Part of Riemann's *Foundations for a General Theory of Functions of One Complex Variable* is translated in Garrett Birkhoff, *A Source Book in Classical Analysis* (Cambridge: Harvard University Press, 1973), 48–50.

For a brief overview of the work of Cauchy and others on the notion of continuity, see Judith V. Grabiner, "Who Gave You the Epsilon? Cauchy and the Origins of Rigorous Calculus," *American Mathematical Monthly* 90 (1983), 185–194. For more details on the development of the concept of the derivative, see Judith V. Grabiner, "The Changing Concept of Change: The Derivative from Fermat to Weierstrass," *Mathematics Magazine* 56 (1983), 195–203. For more discussion of the work of Bolzano, see I. Grattan-Guinness, "Bolzano, Cauchy and the 'New Analysis' of the Early Nineteenth Century," *Archive for History of Exact Sciences* 6 (1970), 372–400. Grattan-Guinness claims that Cauchy took the central ideas of his definitions of continuity and convergence from Bolzano. But see also H. Freudenthal, "Did Cauchy Plagiarize Bolzano?" *Archive for History of Exact Sciences* 7 (1971), 375–392 for a contrary view. Joseph Dauben's biography of Cantor, *Georg Cantor: His Mathematics and Philosophy of the Infinite* (Princeton: Princeton University Press, 1979), provides a detailed study of Cantor's work and how it relates to the mathematics and philosophy of his day. More information on the divergence theorem, as well as on Green's theorem and Stokes's theorem, can be found in Victor J. Katz, "The History of Stokes' Theorem," *Mathematics Magazine* 52 (1979), 146–156.

CHAPTER EIGHTEEN

Statistics in the Nineteenth Century

Observations and statistics agree in being quantities grouped about a Mean; they differ, in that the Mean of observations is real, of statistics is fictitious. The mean of observations is a cause, as it were the source from which diverging errors emanate. The mean of statistics is a description, a representative quantity put for a whole group, the best representative of the group, that quantity which . . . minimizes the error unavoidably attending such practice.

—Francis Edgeworth in "Observations and Statistics: An Essay on the Theory of Errors of Observation and the First Principles of Statistics," 1885

The nineteenth century saw the beginning of the application of statistical methods in various fields, particularly agriculture and the social sciences. It was, in fact, these applications that led to the development of various standard statistical techniques in the nineteenth and twentieth centuries. In this chapter, we will begin with one of the earliest statistical methods, that of least squares; then, after a brief look at Laplace's survey of the entire field of probability and statistics, we will consider the new interpretation of the normal curve in the middle of the nineteenth century, look at some of the developments in statistical procedures in the final decades of that century, and conclude with a brief glance at some types of statistical graphs.

18.1 The Method of Least Squares

Perhaps the most important statistical development of the nineteenth century was the **method of least squares**. Developed to give a procedure for combining observations, this method proved more effective than the eighteenth-century methods discussed earlier. Legendre was the first to publish this method, in 1805, but Gauss gave a much better justification of it a few years later.

18.1.1 The Work of Legendre

Although it is not known what influenced Legendre to develop the method of least squares, he discussed it in 1805 in an appendix to a work on the determination of cometary orbits. He began his discussion by outlining the problem, presented here in modern notation: There is a system of m equations $V_j(\{x_i\}) = a_{j0} + a_{j1}x_1 + a_{j2}x_2 + \cdots + a_{jn}x_n = 0$ $(j = 1, 2, \ldots, m)$ in n unknowns $(m > n)$ for which one wants to find the "best" approximate solutions $\bar{x}_1, \bar{x}_2, \ldots, \bar{x}_n$. For each equation, the value $V_j(\{\bar{x}_i\}) = E_j$ is the error associated with that solution. Legendre's aim, like those of his predecessors, was to make all the E_i small: "Of all the principles which can be proposed for that purpose, I think there is none more general, more exact, and more easy of application, than that of which we have made use in the preceding researches, and which consists of rendering the sum of the squares of the errors a minimum. By this means there is established among the errors a sort of equilibrium which, preventing the extremes from exerting an undue influence, is very well fitted to reveal that state of the system which most nearly approaches the truth."

To determine the minimum of the squares of the errors, Legendre applied the tools of calculus. For the sum of the squares $E_1^2 + E_2^2 + \cdots + E_m^2$ to have a minimum when x_1 varies, its partial derivative with respect to x_1 must be zero:

$$\sum_{j=1}^{m} a_{j1}a_{j0} + x_1 \sum_{j=1}^{m} a_{j1}^2 + x_2 \sum_{j=1}^{m} a_{j1}a_{j2} + \cdots + x_n \sum_{j=1}^{m} a_{j1}a_{jn} = 0.$$

Because there are analogous equations for $i = 2, 3, \ldots, n$, Legendre noted that he now had n equations in the n unknowns x_i and therefore that the system could be solved by "established methods." Although he offered no derivation of the method from first principles, he did observe that his method was a generalization of the method of finding the ordinary mean of a set of observations of a single quantity. For in that case (the special case where $n = 1$ and $a_{j1} = -1$ for each j), if we set $b_j = a_{j0}$, then the sum of the squares of the errors is $(b_1 - x)^2 + (b_2 - x)^2 + \cdots + (b_m - x)^2$. The equation for making that sum a minimum is $(b_1 - x) + (b_2 - x) + \cdots + (b_m - x) = 0$, so that the solution

$$x = \frac{b_1 + b_2 + \cdots + b_m}{m}$$

is just the ordinary mean of the m observations.

18.1.2 Gauss and the Derivation of the Method of Least Squares

Within ten years of Legendre's publication, the method of least squares was a standard method in solving astronomical and geodetical problems throughout the continent. In particular, it appeared in 1809 in Gauss's *Theory of Motion of the Heavenly Bodies*. Gauss, however, did not quote Legendre. In fact, Gauss claimed that he had been using the principle himself since 1795. Gauss's statement led to a pained reaction from Legendre, who noted that priority in scientific discoveries could only be established by publication. And, in fact, whether or not Gauss used the method privately, there is no evidence that he discussed it with anyone else before Legendre's own publication.

The priority dispute notwithstanding, Gauss went further with the method than Legendre had. First, he realized that it was not enough to say that one can use "established methods" to solve the system of n equations in n unknowns produced by the method of least squares. In real applications, there are often many equations and the coefficients are not integers but real numbers calculated to several decimal places. Cramer's rule in these cases would require enormous amounts of calculation. Gauss therefore devised a systematic method of elimination for systems of equations, a method of multiplying the equations by appropriately chosen values and then adding these new equations together. The procedure, now known as the **method of Gaussian elimination** and virtually identical with the method used by the Han Chinese 1800 years earlier, results in a triangular system of equations, that is, a system in which the first equation involves but one unknown, the second two, and so on. Thus, the first equation can be easily solved for its only unknown, the solution substituted in the second to get the value for the second unknown, and so on until the system is completely solved. Gauss's procedure was improved somewhat later in the century by the German geodesist Wilhelm Jordan (1842–1899), who used the method of least squares to deal with surveying problems. Jordan devised a method of substitution, once the triangular system had been found, to further reduce the system to a diagonal one in which each equation only involved one unknown. This **Gauss-Jordan method** is the one typically taught in modern linear algebra courses as the standard method for solving systems of linear equations.

Second, Gauss developed a much better justification for the method of least squares than the somewhat vague "general principle" of Legendre. He derived the method from his prior discovery of a suitable function $\phi(x)$ describing the probability of an error of magnitude x in the determination of an observable quantity, a function different from the ones worked out in the previous century. Gauss's criteria were the same as Laplace's earlier: that $\phi(x)$ should be symmetric about zero, that the curve must be asymptotic to the real axis in both directions, and that the total area under $\phi(x)$ should be 1. Gauss joined these criteria to the original problem of Legendre of determining the values of m linear functions $V_1, V_2, \ldots, V_m$ of n unknowns $x_1, x_2, \ldots, x_n$. Supposing that the observed values of these were $M_1, M_2, \ldots, M_m$, with corresponding errors $\Delta_1, \Delta_2, \ldots, \Delta_m$, Gauss noted that because the various observations were all independent, the probability of all these errors occurring was $\Omega = \phi(\Delta_1)\phi(\Delta_2)\cdots\phi(\Delta_m)$. To find the most probable set of values meant maximizing Ω, but to do this required a better knowledge of ϕ. Thus, Gauss made the further assumption that "if any quantity has been determined by several direct observations, made under the same circumstances and with equal care, the arithmetical mean of the observed values affords the most probable value." Taking each V_i as the simplest linear function of one variable, namely, $V_i = x_1$, he determined ϕ by supposing that $x_1 = (1/m)(M_1 + M_2 + \cdots + M_m)$, the mean of the observations, gives the maximum value of Ω. That maximum occurs when $\partial\Omega/\partial x_i = 0$ for all i. After some calculation, he finally concluded that

$$\phi(\Delta) = \frac{h}{\sqrt{\pi}} e^{-h^2\Delta^2}.$$

That this was the "correct" error function followed for Gauss because he was able easily to derive from it the method of least squares. After all, given this function ϕ, the product Ω in the general case is given by

$$\Omega = h^m \pi^{-(1/2)m} e^{-h^2(\Delta_1^2+\Delta_2^2+\cdots+\Delta_m^2)}.$$

To maximize Ω, therefore, it is necessary to minimize the $\sum \Delta_i^2$, that is, to minimize the sum of the squares of the errors, the very procedure that Legendre had developed.

That the distribution of errors is "normal"—that is, is determined by Gauss's function—gained even more credence because it was soon supported by much empirical evidence. In particular, Friedrich Bessel (1784–1846) made three sets of measurements of star positions for several hundred stars and compared the theoretical number of errors between given limits, according to the normal law, with the actual values. The comparison showed very close agreement. Meanwhile, Laplace gave a new theoretical derivation of the normal law in a paper of 1810. His result was based on what today is called the **central limit theorem**, to the effect that any mean, not just the total number of successes in m trials, will, if the number of terms becomes large, be approximately normally distributed. This was a generalization of De Moivre's calculations of the previous century involving the terms of the binomial theorem. The work of Laplace soon established the function $y = Ae^{-kx^2}$ as that representing error distributions and, in general, probability distributions, in a wide variety of situations.

In his book *Analytic Theory of Probability*, published in 1812, Laplace collected all the material so far developed in probability theory, beginning with the definition of probability of an event as "the ratio of the number of cases favorable to it, to the number of possible cases, when there is nothing to make us believe that one case should occur rather than any other." He included the statement and proof of the central limit theorem and its application to the question of the inclinations of the orbits of comets, a problem he had considered for many years. Furthermore, he dealt with the applications of the theory of probability to such topics as insurance, demographics, decision theory, and the credibility of witnesses. In fact, it was Laplace's view that through the theory of probability, mathematics could be brought to bear on the social sciences, just as calculus was the major tool in mathematizing the physical sciences. Laplace's prediction began to come true well before the end of the century.

18.2 Statistics and the Social Sciences

Recall that the normal curve $y = Ae^{-kx^2}$ was first developed by De Moivre from computations of probabilities in a binomial experiment and that it turned out, in its formulation by Gauss and then Laplace, to represent a distribution of errors. But by the middle of the nineteenth century, this curve had been seized on by Adolphe Quetelet (1796–1874), a Belgian mathematician, astronomer, meteorologist, sociologist, and statistician, as the key to developing his concept of the "average man." By compiling vast numbers of statistics covering not only physical characteristics such as height and weight, but also "moral" characteristics such as the propensities for individuals to commit crimes or to become drunk, he proposed to be able to develop the idea of the representative individual in a given society at a given time. The concept of the average man—and there were of course different average men (and, perhaps, women) for various ages and classes in each country—was to be a device for smoothing the manifold variations among people and somehow revealing the regular laws of society, a "social physics."

Quetelet noticed that many of the characteristics he gathered could be plotted in terms of a normal curve. That is, there was a mean value and "errors" from the mean, which were distributed in the same way as errors of measurement. In 1846, he wrote a letter to the Grand Duke of Saxe Coburg expressing his belief in the use of the normal curve for social analysis. Suppose, he said, that one wanted to make a thousand copies of a particular statue. These copies would naturally be subject to a wide variety of errors, but, in fact, the errors would combine in a very simple fashion. In fact, he wrote, "the experiment has been already made. Yes, surely, more than a thousand copies have been measured of a statue . . . which in all cases differs little from it. These copies were even living ones." Quetelet's "statues" were Scottish soldiers; he had compiled measures of the chest circumferences of 5732 of them and noted that the measurements were distributed normally around a mean of about 40 inches. His conclusion was that because the measurements were distributed as they would be if nature were aiming for an ideal type, this must be the case. Thus, his distribution showed that there was an "average" Scottish soldier, and the deviations from the average were simply due to a combination of accidental causes.

In his use of the normal curve in this situation and others, Quetelet's unit of deviation from the mean was the probable error, rather than the now common standard deviation. A data point is one **probable error** from the mean if it has a percentile rank of 25 or 75. That is, in a normal distribution, a particular value is as likely to be within one probable error of the mean as not. This measure of deviation had been introduced early in the century in connection with the theory of errors.

Certainly many disagreed with Quetelet's program of looking for normal distributions everywhere; nevertheless, the idea of a normal distribution became central in many arguments involving statistics. Francis Galton (1822–1911), an English statistician, used Quetelet's ideas in his own work in biology in trying to mathematize Charles Darwin's theory of evolution by looking at the inheritance of variation. Galton was curious as to why the same normal curve persisted generation after generation, and why the variability did not increase over the years. One experiment that he conducted in 1875 was on the size of the seeds in a particular type of sweet pea. He took an equal number of seeds of each of seven sizes and studied their offspring. It turned out that the sizes of seeds in each set of progeny were normally distributed, but that the variability in each group—that is, the spread of the data—was essentially the same.

But what Galton also found is that the size of all the seeds in the generation of offspring, looked at in its totality, were normally distributed. To explain why all the small normal distributions added together gave a "large" one, Galton created what he called the "quincunx." The quincunx was a device with a glass face and a funnel at the top. Very small steel balls were poured through the funnel and moved through an array of pins, each ball striking one pin at each level and, in principle, falling to the left or right with equal probability. The balls were then collected in compartments at the bottom. The resulting distribution was binomial, so it resembled a normal curve. But then Galton modified the device by intercepting the balls at an intermediate level (Fig. 18.1). The result at that level was again approximately normal. If the balls were released from any one compartment at the intermediate level, the distribution of balls at the bottom was approximately normal, with mean directly under the compartment that was opened up. Of course, releasing a compartment nearer the center produced a higher curve. Releasing all the compartments at the intermediate level produced a mixture of these curves of varying sizes, but since this gave the same result as when the balls were not intercepted, the resulting curve was again normal. Thus, this quincunx enabled Galton to show that a normal mixture of normal distributions was itself normal.

In his experiments with peas, Galton also observed that although the distribution of the seed sizes of the offspring from each parent was normal, the mean of each group was not the same as that of the parental seed sizes. In fact, the means were linearly related to those of the

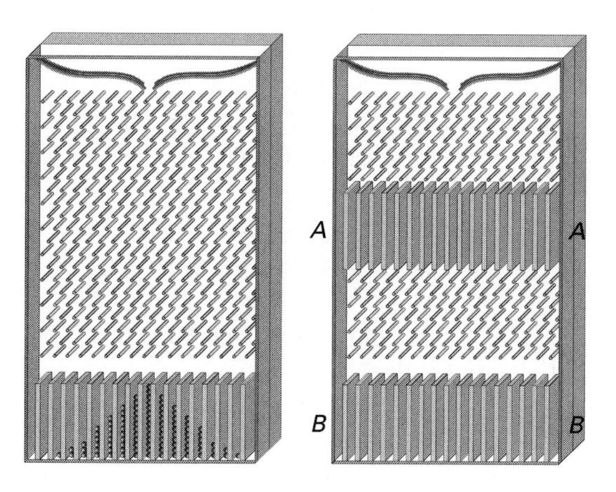

FIGURE 18.1 Galton's quincunx

parent seeds, but with the slope of this line being $\frac{1}{3}$. In other words, the second generation "regressed" to the overall mean. It was this regression combined with the overall normality that preserved the variability in the entire population over the generations. Thus, Galton originated the statistical study of **regression** (or *reversion*, as he first called it). Interestingly, to calculate the slope of this regression line, Galton simply plotted the points and estimated the slope of that straight line that "best" connected these points.

Galton carried out several other studies involving inheritance, including an extensive one on the heights of children as related to the heights of their parents (Table 18.1). In this case, he noted that both the heights of the children of parents of a given height and the heights of the parents of children of a given height were normally distributed. He then found that if one scaled the two sets of data by units of probable error, the slopes of the regression lines were the same. This slope became the coefficient of "corelation" of the two variables, later to become the **correlation coefficient**. This value, which Galton could generally only roughly approximate, could be used to measure the strength of the relationship of the two variables. Galton realized, of course, that a strong correlation does not necessarily imply a causal relationship.

Other English statisticians tried to clarify and extend the work of Galton and come to grips with the basic philosophy behind looking at normal curves representing various types of distributions. Francis Edgeworth (1845–1926), quoted at the beginning of the chapter, commented on the difference between *observations*, as used in astronomy, and *statistics*, as collected in the social sciences. Thus, Edgeworth realized that the use of statistics in the social sciences would not have the "objective" character that the use of the theory of errors had in measurements in astronomy. Nevertheless, he worked to develop this new tool as best he could.

In particular, in his work of 1885, he developed a basic significance test. Given two "means," he first estimated what he called their "fluctuations," or, in modern terms, twice the variance. Thus, if $\bar{x}$ was a sample mean, then the fluctuation was $2\sum(x_i - \bar{x})^2/n^2$. If c_1^2 and c_2^2 were estimates of the fluctuations of the two means, then $\sqrt{c_1^2 + c_2^2}$ estimated the modulus of the difference of the means. (In modern terms, this modulus is $\sqrt{2}$ times

TABLE 18.1

Galton's Data on Heights

Note that "height of mid-parent" means the average height of the parents, with mother's height scaled up by 1.08.

Height of the mid-parent (in inches)	Height of the Adult Child														Total no. of Adult Children	Total no. of mid-parents	Medians
	< 61.7	62.2	63.2	64.2	65.2	66.2	67.2	68.2	69.2	70.2	71.2	72.2	73.2	> 73.7			
> 73.0	—	—	—	—	—	—	—	—	—	—	—	1	3	—	4	5	—
72.5	—	—	—	—	—	—	—	1	2	1	2	7	2	4	19	6	72.2
71.5	—	—	—	—	1	3	4	3	5	10	4	9	2	2	43	11	69.9
70.5	1	—	1	—	1	1	3	12	18	14	7	4	3	3	68	22	69.5
69.5	—	—	1	16	4	17	27	20	33	25	20	11	4	5	183	41	68.9
68.5	1	—	7	11	16	25	31	34	48	21	18	4	3	—	219	49	68.2
67.5	—	3	5	14	15	36	38	28	38	19	11	4	—	—	211	33	67.6
66.5	—	3	3	5	2	17	17	14	13	4	—	—	—	—	78	20	67.2
65.5	1	—	9	5	7	11	11	7	7	5	2	1	—	—	66	12	66.7
64.5	1	1	4	4	1	5	5	—	2	—	—	—	—	—	23	5	65.8
< 64.0	1	—	2	4	1	2	2	1	1	—	—	—	—	—	14	1	—
Totals	5	7	32	59	48	117	138	120	167	99	64	41	17	14	928	205	—
Medians	—	—	66.3	67.8	67.9	67.7	67.9	68.3	68.5	69.0	69.0	70.0	—	—	—	—	—

the standard deviation.) Edgeworth was extremely conservative, however, in coming to conclusions. He would only call the difference between two means significant if it exceeded two moduli. This is equivalent to using a two-sided test of significance at a level of 0.005, a much stronger requirement than the levels of 0.05 or 0.01 often used today.

In the last decade of the nineteenth century, two other English statisticians, Karl Pearson (1857–1936) and his student George Udny Yule (1871–1951), did further work in showing how to use statistics to come to definite conclusions about the relationships between several quantities. Pearson not only introduced the standard deviation in 1893 but also developed the chi-square statistic as one way of measuring the relationship between two quantities. And Yule showed how to calculate the regression equation by using, in essence, Gauss's least squares technique to find the line of best fit. But all of the procedures of these statisticians were only designed to show relationships in quantities already tabulated. A significant use of statistics today is in the design and analysis of experimental procedures that will enable, for example, farmers to determine the effectiveness of different types of fertilizer on the yield of crops and doctors to determine the efficacy of differing treatments for a particular disease. Furthering the study of statistics so that it could be applied to such problems was the work of the twentieth-century.

18.3 Statistical Graphs

One major innovation of the nineteenth century was the use of graphs to represent data. Many types of graphs were developed, some by several different people. We consider here some examples of the more important ones.

It was William Playfair (1759–1823) who developed a number of graphical designs so that he could replace tables of numbers with a visual representation. As he wrote, "a man who has carefully investigated a printed table, finds, when done, that he has only a very faint and partial idea of what he has read; and that like a figure imprinted on sand, is soon totally erased and defaced." Thus, Playfair drew charts of all sorts. One example, which was both a line graph and a bar chart, shows the relationship between wages and the price of wheat over a 250-year period (Fig. 18.2). Another example, which uses areas of circles to represent quantity (and incorporates a pie chart), shows the relationship between population and revenue of many of the countries of Europe (Fig. 18.3). The area of each circle represents the area of the country. The vertical line at the left of each circle represents population; the one on the right represents revenue. The slope of the lines connecting these shows whether, in proportion to its population, a country is burdened with heavy taxes.

Florence Nightingale (1820–1910), who was a nurse during the Crimean war, drew several very effective pie charts giving the monthly death rates during the war. In Figure 18.4 (page 472), each wedge is divided into three sections representing different causes of death. The innermost section represents death due to wounds, the middle section represents death due to "other causes," and the large outer section shows death due to preventable disease. Her graphs, when published in England, led to a great outcry and caused the War Department to improve the sanitary conditions at field hospitals.

Histograms, bar charts with the horizontal axis denoting a continuous variable, were first used by A. M. Guerry in France in 1833. But they were named by Karl Pearson, who used them in his *The Mathematical Theory of Evolution* of 1895. Figure 18.5 (page 473) is an example from that book, from which it is easy to see that there are fewer examples of flowers with a greater number of petals.

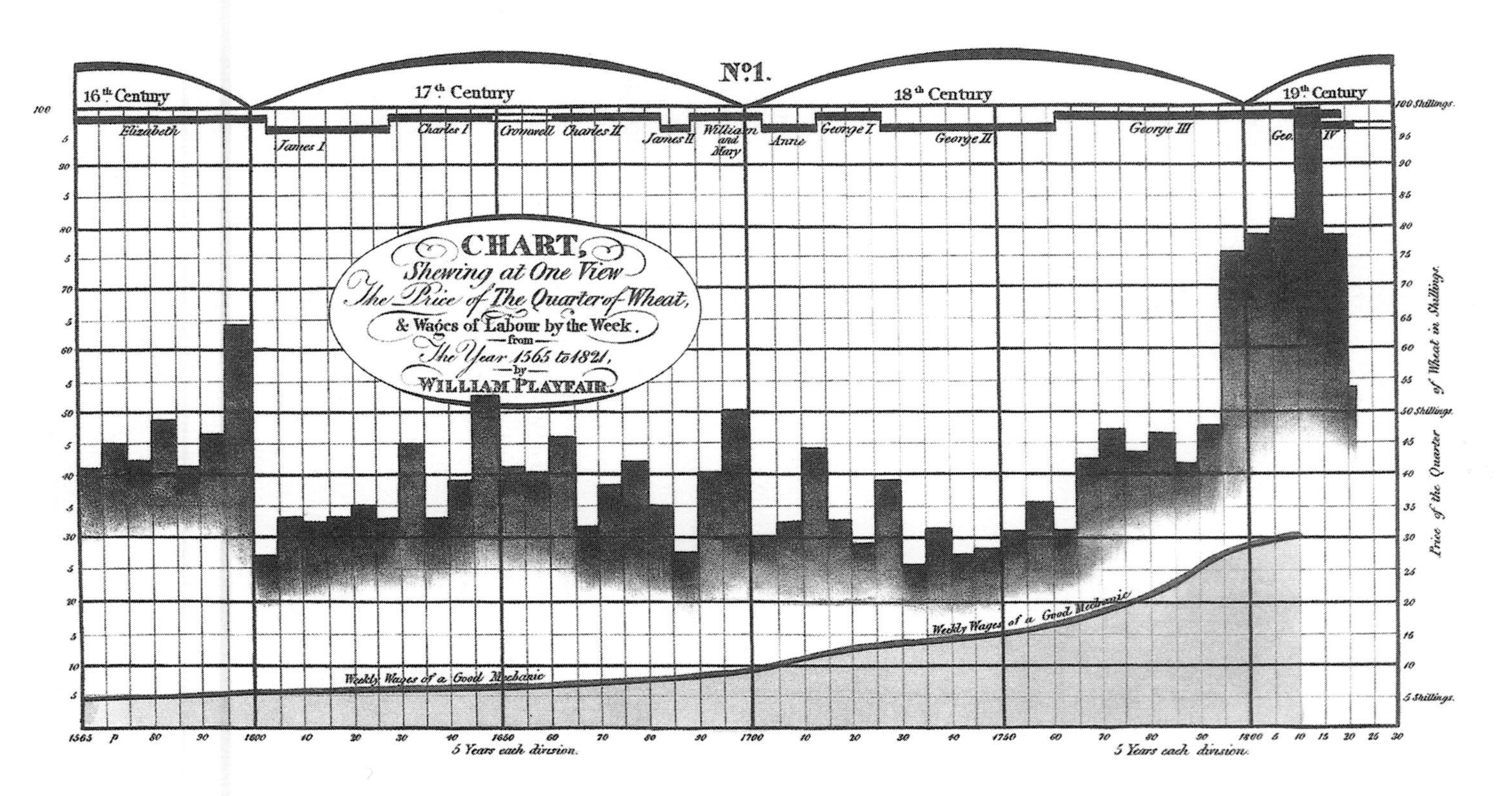

FIGURE 18.2 Playfair's graph of wages and the price of wheat

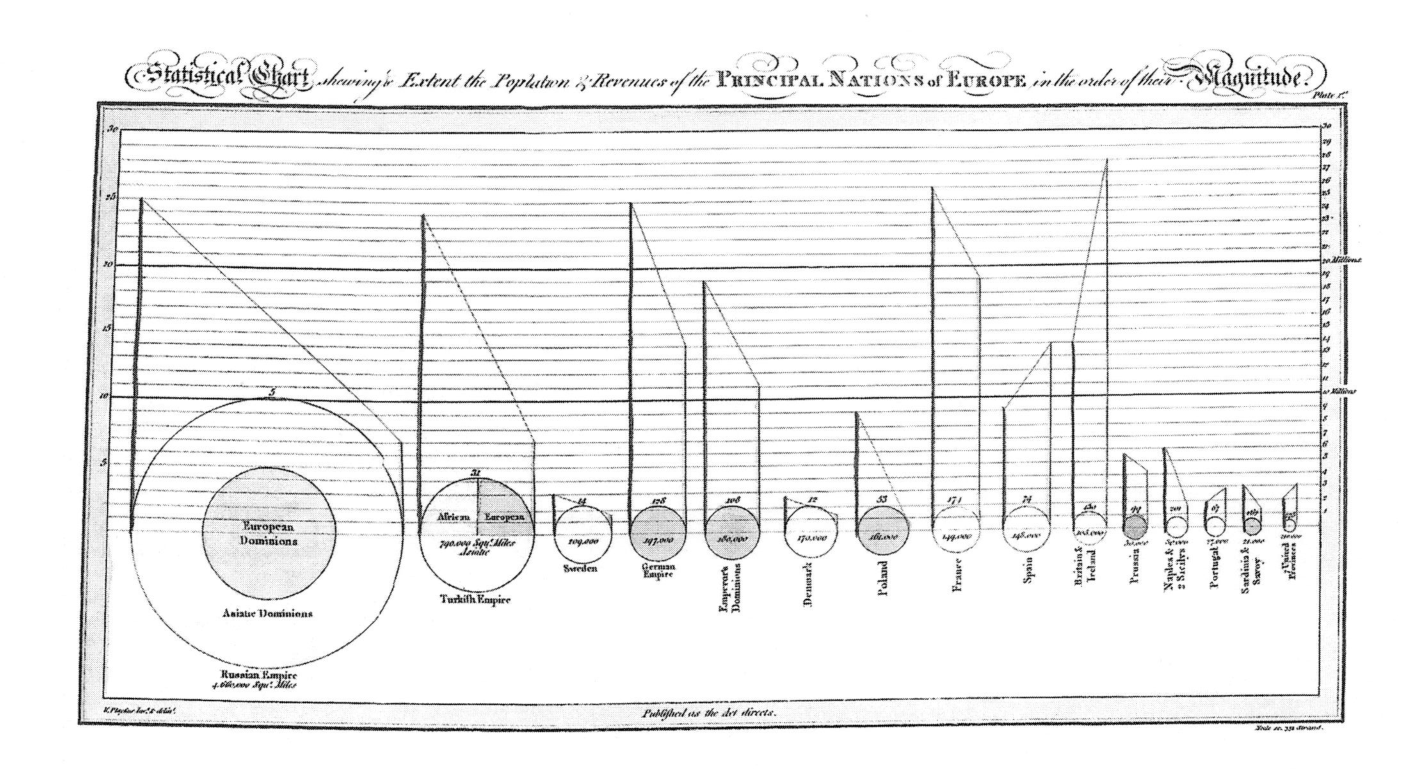

FIGURE 18.3 Playfair's graph of population and revenue, in which areas of circles represent the areas of the countries

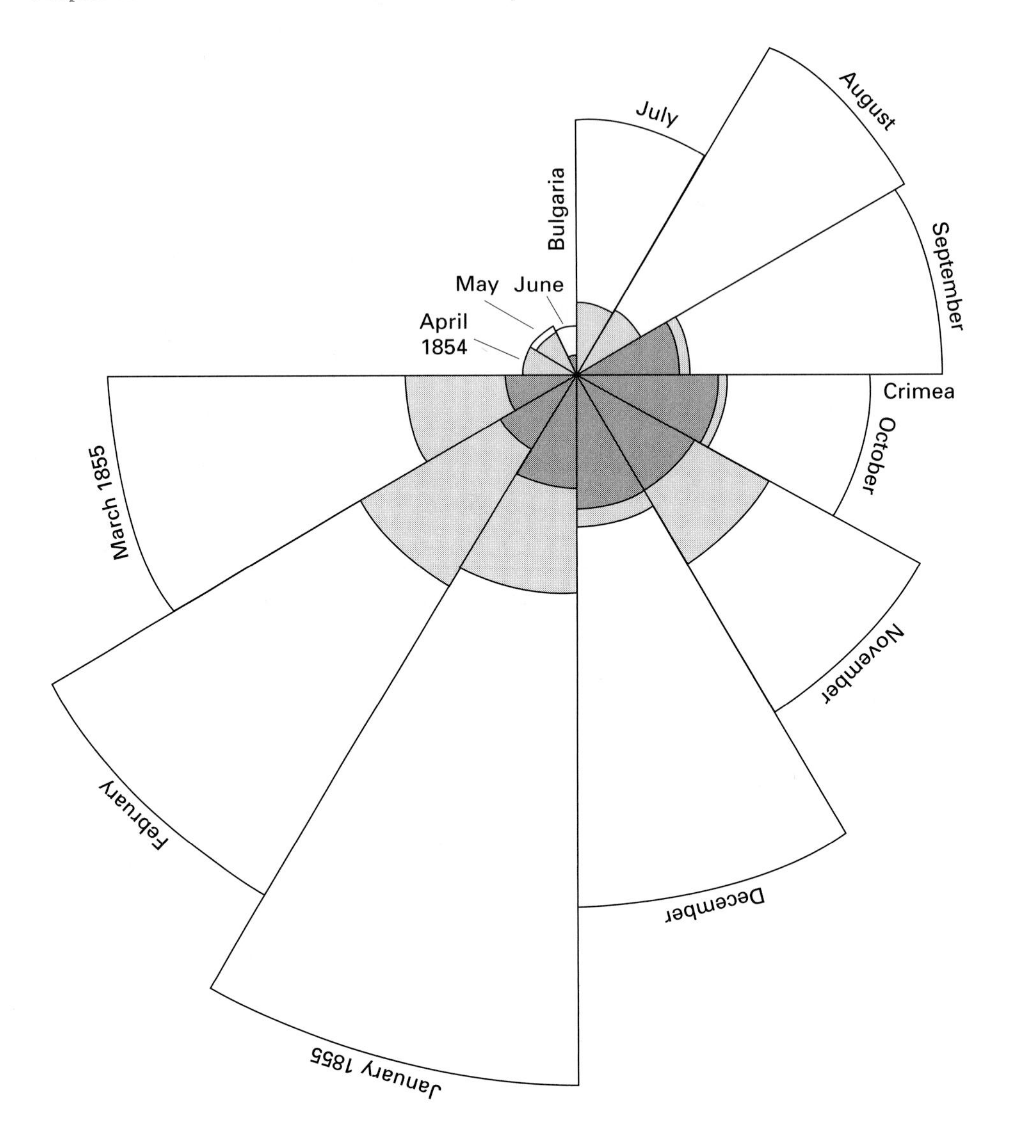

FIGURE 18.4 Nightingale's pie chart of death rates during the Crimean war

One final graph is the **ogive**, a graph of a cumulative frequency distribution, invented by Galton around 1875. In the example shown in Figure 18.6, Galton was graphing the strength of pull in pounds of 519 males ages 23 to 26. He plotted the strength along the vertical axis and the percentile rank along the horizontal and drew the curve representing strength as a function of percentile rank. For example, those at the 70th percentile can pull a weight of about 80 pounds. Since the population is normally distributed in this situation, Galton's function is now called the inverse normal cumulative distribution. Today, however, on such a graph, the horizontal and vertical axes are generally interchanged to determine, for a given weight, the percentage of the population that can pull no more than that.

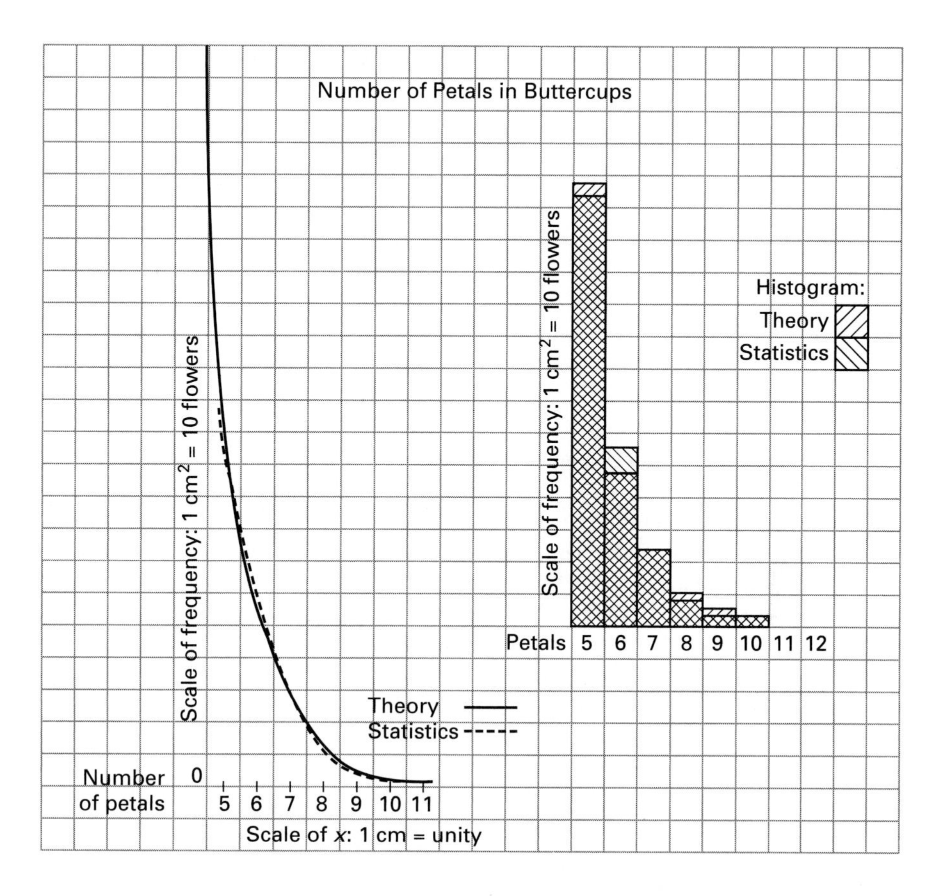

FIGURE 18.5 A histogram from Pearson's *The Mathematical Theory of Evolution*

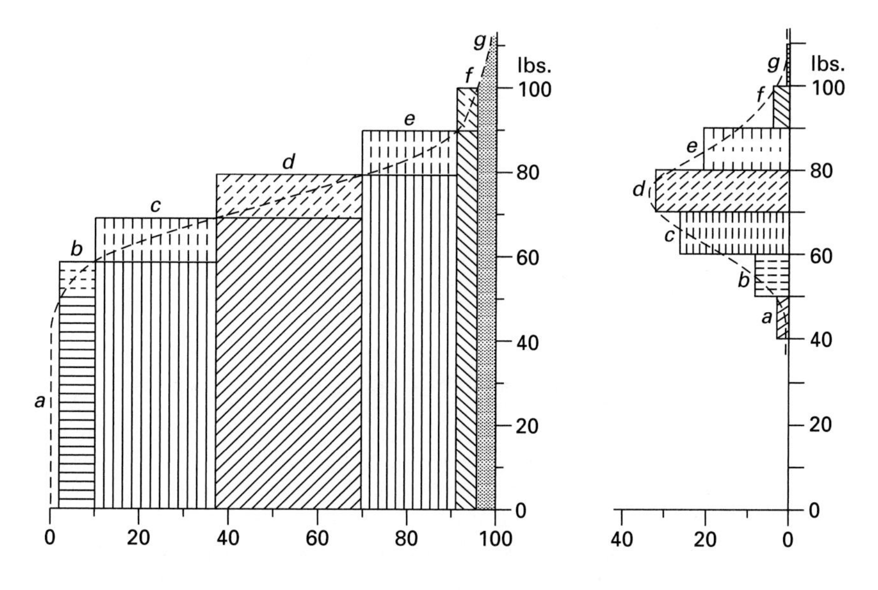

FIGURE 18.6 Galton's ogive of male strength of pull

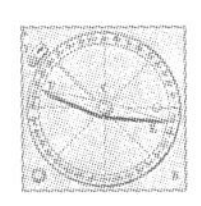

Exercises

1. Given four measured values for the independent variable x as $x_1 = 2.0$, $x_2 = 4.0$, $x_3 = 5.0$, $x_4 = 6.0$ and four corresponding measured values for the dependent variable y as $y_1 = 2.5$, $y_2 = 4.5$, $y_3 = 7.0$, $y_4 = 8.5$, use the method of least squares to determine the constants a and b that give the best linear function $y = ax + b$ that represents this measured relationship.
2. Given that the x value determining one standard deviation in the normal curve $y = (1/c\sqrt{\pi})e^{-(x^2/c^2)}$ occurs at an inflection point of the curve, show that this value is given by $x = c/\sqrt{2}$.
3. Show that in a normal distribution with standard deviation σ, a measurement is one probable error from the mean if it is approximately 0.675σ from the mean and is one modulus from the mean (the value c from exercise 2) if it is $\sqrt{2}\sigma$ from the mean. (Use a table of the normal curve to solve this problem.)
4. For this exercise and the next one, use Galton's table (Table 18.1). Use a histogram to graph the heights of adult children of parents of height 68.5 inches. Then graph the heights of parents of children of height 69.2 inches. Show that these graphs are graphs of an approximately normal distribution. Find the medians in each case.
5. Use separate histograms to graph the heights of all adult children (use the totals column) and the heights of all parents. Show that these distributions are approximately normal. Find the medians in each case and calculate the standard deviations. (Use statistical software if necessary.)
6. Using Figure 18.2, show how Playfair concluded that "never at any former period [i.e. before 1820] was wheat so cheap, in proportion to mechanical labour, as it is at the present time [1821]."
7. Using Figure 18.3, show how Playfair concluded that the people of Great Britain (the sixth circle from the right) were excessively taxed in relation to the people of other European countries. (Note that the scales of population on the left and revenue on the right are not particularly related to one another.)
8. In Figure 18.3, the areas of the circles are proportional to the areas of the countries. The second circle from the left represents the Turkish empire. Note that it is divided into three sections representing the African, European, and Asian domains of that empire. Thus, that circle is in the form of a pie chart. Estimate the relative sizes of those three sections of the Turkish empire from the pie chart.
9. In Pearson's histogram (Fig. 18.5), ten flowers are represented by one unit in the vertical direction. The histogram represents the number of buttercups found with a given number of petals. How many buttercups had seven petals? How many had five petals?
10. Redraw Galton's ogive (Fig. 18.6) to represent the distribution of males who are able to pull a given number of pounds. Show that this distribution is approximately normal. Then redraw the ogive as a cumulative frequency distribution, with strength of pull on the horizontal axis and the numbers on the vertical axis.
11. Describe why the publication of Nightingale's pie charts would have led to an outcry in Britain. Find out what happened in the hospitals after these were published and write a brief report.

References

Among the best general works on the history of statistics is Stephen M. Stigler, *The History of Statistics* (Cambridge: Harvard University Press, 1986). This work provides a detailed study of the development of the field of statistics from its beginnings through 1900. Another work with more mathematical details is Anders Hald, *A History of Mathematical Statistics from 1750 to 1930* (New York: Wiley, 1998). Graphs of various sorts, including some of the early ones illustrated in the text, are discussed in Edward R. Tufte, *The Visual Display of Quantitative Information*, 2nd ed. (Cheshire, Conn.: Graphics Press, 2001).

Legendre's early work on the method of least squares is available in David Smith, *A Source Book in Mathematics* (New York: Dover, 1959), 576–579. Gauss's initial discussion of the method of least squares is translated by C. H. Davis in *Theory of Motion of the Heavenly Bodies Moving about the Sun in Conic Sections* (Boston: Little, Brown, 1857). Adolphe Quetelet's *Letters Addressed to H.R.H. the Grand Duke of Saxe Coburg and Gotha, on the Theory of Probabilities, as Applied to the Moral and Political Sciences* was translated by O. G. Downes and published in London by Charles and Edwin Layton in 1849. Galton's initial work on regression is in "Regression Towards Mediocrity in Hereditary Stature," *Journal of the Anthropological Institute* 15 (1886), 246–263. Many of his conclusions about statistical methods are included in *Natural Inheritance* (London: Macmillan, 1889). Edgeworth's essay, in which occurs the quotation at the opening of this chapter, is "Observations and Statistics: An Essay on the Theory of Errors of Observation and the First Principles of Statistics," *Transactions of the Cambridge Philosophical Society* 14 (1885), 13–169.

For more details on the method of least squares, see R. L. Plackett, "The Discovery of the Method of Least Squares," *Biometrika* 59 (1972), 239–251. This paper is reprinted in M. G. Kendall and R. L. Plackett, eds., *Studies in the History of Statistics and Probability*, vol. 2 (New York: Macmillan, 1977), 279–291. This volume, and its companion, E. S. Pearson and M. G. Kendall, eds., *Studies in the History of Statistics and Probability*, vol. 1 (Darien, Conn.: Hafner, 1970), provide a valuable collection of essays on the history of probability and statistics, most of which originally appeared in *Biometrika*. See also William C. Waterhouse, "Gauss's First Argument for Least Squares," *Archive for History of Exact Sciences* 41 (1991), 41–52. For more details on Gauss-Jordan reduction, see Steven C. Althoen and Renate McLaughlin, "Gauss-Jordan Reduction: A Brief History," *The American Mathematical Monthly* 94 (1987), 130–142 and Victor J. Katz, "Who Is the Jordan of Gauss-Jordan?" *Mathematics Magazine* 61 (1988), 99–100.

CHAPTER NINETEEN

Geometry in the Nineteenth Century

I am ever more convinced that the necessity of our geometry cannot be proved—at least not by human reason for human reason. It is possible that in another lifetime we will arrive at other conclusions on the nature of space that we now have no access to. In the meantime we must not put geometry on a par with arithmetic that exists purely a priori but rather with mechanics.

—Carl Friedrich Gauss, in an 1817 letter to Heinrich Olbers (1758–1840)

Several important developments occurred in geometry during the nineteenth century. One of them arose out of the continued debate over the parallel postulate, whose truth implied that the sum of the angles of a plane triangle was equal to two right angles. Toward the end of his life, Gauss noted that he had long been convinced that the parallel postulate could not be proved and that the acceptance of alternatives could well lead to new and interesting geometries, the "truth" of which for the physical world could only be established by experiment. Nevertheless, Gauss never published any of his ideas on the subject. It was Nikolai Lobachevsky and János Bolyai, who in the 1820s published, independently of each other, the first full treatments of a non-Euclidean geometry.

It took nearly forty years, however, for the ideas of non-Euclidean geometry to make an impression in the mathematical community. It was only with the work of Hermann von Helmholtz and others in the 1860s on the general notion of a geometrical manifold of arbitrary dimension that the meaning of these new ideas for the study of geometry took hold. Shortly thereafter, various models of non-Euclidean geometries in Euclidean space were introduced, thus convincing the mathematical community that the non-Euclidean geometries were as valid as the Euclidean one from a logical standpoint and that the question of the "truth" of Euclidean geometry for the real world no longer had an obvious answer.

Although there were other significant developments in geometry during the century, we will limit ourselves in this chapter to two. First, we will consider the expansion of geometry to spaces of dimension greater than 3. Although several mathematicians experimented with this idea by simply generalizing their formulas and theorems, it was Hermann Grassmann who, beginning in 1844, first attempted a detailed study of n-dimensional vector spaces from a geometric point of view. Unfortunately, Grassmann's work, like that of Bolyai and Lobachevsky, was not appreciated until the end of the century, when Giuseppe Peano gave a set of axioms for a finite-dimensional vector space to provide a basis for the study of higher-dimensional geometry.

A second area of geometry we will deal with is the theory of graphs, a theory whose growth was influenced by the four-color problem. It turned out that this theory had other important applications as well.

19.1 Non-Euclidean Geometry

Recall that in the eighteenth century both Saccheri and Lambert attempted to prove Euclid's parallel postulate by assuming that it was false and trying to derive a contradiction. Saccheri believed that he had succeeded in this endeavor, but Lambert realized that his attempt was a failure. Both attacked the problem through synthetic means, trying to use the methodology of Euclid to show that he had assumed an unnecessary postulate. The nineteenth century, however, with its increasing use of analysis to solve all sorts of problems, provided a new approach to this one as well. And interestingly enough, it was the hyperbolic functions of Lambert that were called into service to make the connection between analysis and a new geometry, a connection that Lambert himself had missed.

19.1.1 Taurinus and Log-Spherical Geometry

Lambert had noted that the hypothesis of the acute angle would seem to hold on the surface of a sphere of imaginary radius, but it was Franz Taurinus (1794–1874), a man of independent means who pursued mathematics as a hobby, who made this connection explicit in a work of 1826. Taurinus began with a formula of spherical trigonometry connecting the sides and an angle of an arbitrary spherical triangle on a sphere of radius K, a formula we have already seen in the work of Islamic mathematicians:

$$\cos\frac{a}{K} = \cos\frac{b}{K}\cos\frac{c}{K} + \sin\frac{b}{K}\sin\frac{c}{K}\cos A,$$

where the triangle has sides a, b, c and opposite angles A, B, C (Fig. 19.1). Replacing K by iK—that is, making the radius of the sphere imaginary (whatever that means)—and

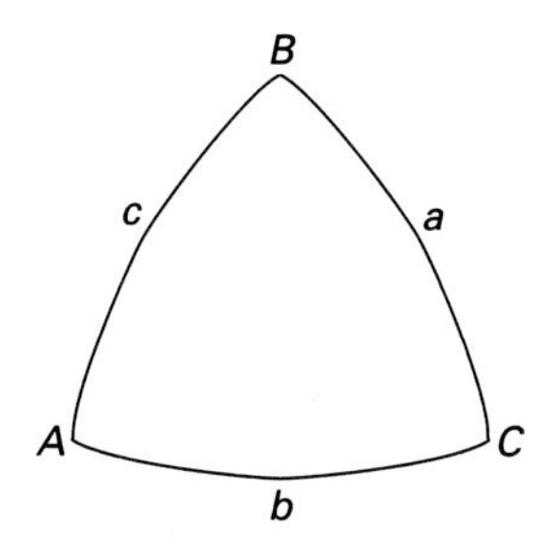

FIGURE 19.1 Spherical triangle on sphere of radius K

recalling that $\cos ix = \cosh x$ and $\sin ix = i \sinh x$, Taurinus derived the new formula

$$\cosh \frac{a}{K} = \cosh \frac{b}{K} \cosh \frac{c}{K} - \sinh \frac{b}{K} \sinh \frac{c}{K} \cos A. \qquad \textbf{(19.1)}$$

Taurinus called the geometry defined by this formula "log-spherical geometry," but he realized that this geometry was not possible in the plane. Exploring the consequences of the formula gives some idea of the properties of this geometry, however. For example, if the triangle is equilateral ($a = b = c$), the formula becomes

$$\cosh \frac{a}{K} = \cosh^2 \frac{a}{K} - \sinh^2 \frac{a}{K} \cos A$$

or

$$\cos A = \frac{\cosh^2 \frac{a}{K} - \cosh \frac{a}{K}}{\sinh^2 \frac{a}{K}} = \frac{\left(\cosh \frac{a}{K}\right)\left(\cosh \frac{a}{K} - 1\right)}{\cosh^2 \frac{a}{K} - 1} = \frac{\cosh \frac{a}{K}}{\cosh \frac{a}{K} + 1}.$$

Because $\cosh(a/K) > 1$, it follows that $\cos A > \frac{1}{2}$ and therefore that $A < 60°$. In other words, the sum of the angles of an equilateral triangle in this geometry is less than 180°. On the other hand, it is easy to see that as either the sides get smaller or the radius K gets larger, the angle A approaches 60° and the geometry approaches Euclid's geometry. In fact, one can also show (by using appropriate power series expansions) that in the limit as K approaches infinity, Taurinus's formula 19.1 reduces to the Euclidean law of cosines: $a^2 = b^2 + c^2 - 2bc \cos A$.

A second important formula of spherical trigonometry, which connects the angles and a side of a spherical triangle, is

$$\cos A = -\cos B \cos C + \sin B \sin C \cos \frac{a}{K}.$$

On replacing K by iK, this formula becomes a formula of log-spherical geometry:

$$\cos A = -\cos B \cos C + \sin B \sin C \cosh \frac{a}{K}. \qquad \textbf{(19.2)}$$

For the special case where $A = 0°$ and $C = 90°$, formula 19.2 reduces to

$$\cosh \frac{a}{K} = \frac{1}{\sin B}.$$

Naturally, a triangle with right angle at C and an angle of zero degrees at A does not exist in Euclid's geometry. Recall, however, that Saccheri had realized that the hypothesis of the acute angle led to the concept of asymptotic straight lines. Thus, Taurinus's triangle must be thought of as one in which two sides are asymptotic (Fig. 19.2). The angle B and the length a of the third side are then related through the formula $\sin B = \text{sech}(a/K)$. This formula can be rewritten in the form

$$\tan \frac{B}{2} = e^{-a/K},$$

a formula that was to become fundamental in the work of Lobachevsky.

Formula 19.2 also shows that if one constructs an isosceles right triangle and splits it into two right triangles by drawing the altitude a, then the relationship between a and the base angle A of the original triangle is given by $\cosh(a/K) = \sqrt{2} \cos A$ (Fig. 19.3). It follows

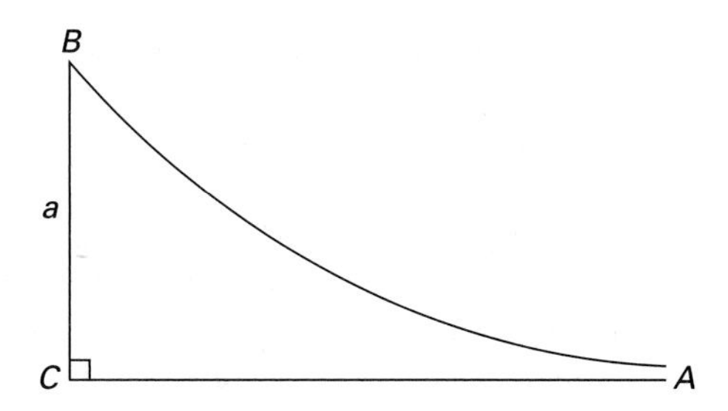

FIGURE 19.2 Triangle in which $\angle C = 90^\circ$ and $\angle A = 0^\circ$

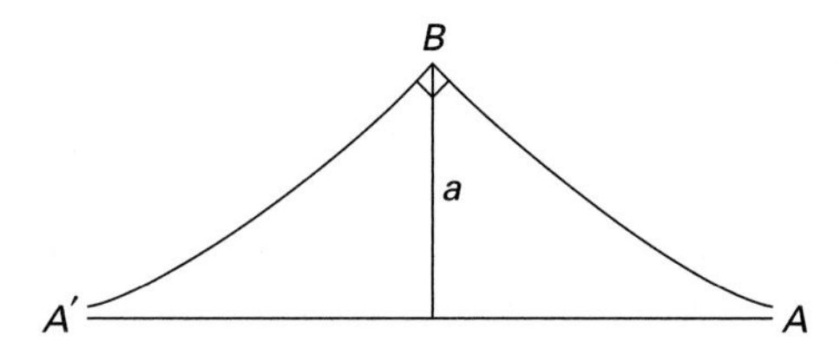

FIGURE 19.3 Isosceles right triangle with both base angles equal to 0°

that the maximum possible altitude h of an isosceles right triangle occurs when $A = 0^\circ$, that is, when the two legs of the triangle are both asymptotic to the hypotenuse. In that case, $\cosh(h/K) = \sqrt{2}$, or

$$K = \frac{h}{\ln(1 + \sqrt{2})}.$$

Taurinus noted further that the area of a triangle is proportional to its defect (as Lambert had already discovered), the length of the circumference of a circle of radius r is $2\pi K \sinh(r/K)$, and the area of a circle of radius r is $2\pi K^2[\cosh(r/K) - 1]$.

19.1.2 The Non-Euclidean Geometry of Lobachevsky and Bolyai

Despite his analytic results, Taurinus was not convinced that his geometry on an imaginary sphere was applicable to any "real" situation. The formulas were simply a collection of pretty results with no real content. But because neither Saccheri nor Lambert had succeeded in their attempts to refute the hypothesis of the acute angle, other mathematicians began to believe that a plane geometry in which that hypothesis was valid could exist. And that geometry would have as its analytic basis the formulas of Taurinus. The two mathematicians who first had confidence enough in their new ideas to publish them were the Russian Nikolai Ivanovich Lobachevsky (1792–1856) and the Hungarian János Bolyai (1802–1860). Both began work on the problem of parallels determined to find a correct refutation of the hypothesis of the acute angle. And both gradually changed their minds.

In 1826, Lobachevsky gave a lecture at Kazan University in which he outlined a geometry that allowed more than one parallel to a given line through a given point; he later published, in both Russian and German, versions of his geometric researches, the most important being his 1840 survey in German entitled *Geometrical Investigations on the Theory of Parallel Lines*. Bolyai, who also did most of his creative work in the 1820s, published his material (in Latin) in 1831 as an appendix to a geometric work of his father Farkas Bolyai, called *Appendix exhibiting the absolutely true science of space, independent of the truth or falsity of Axiom XI [the Parallel Postulate] of Euclid, that can never be decided a priori.* Because the ideas of Lobachevsky and Bolyai turned out to be remarkably similar, we will concentrate on the work of the former.

Lobachevsky began with a summary of certain geometric results that were true independent of the parallel postulate. He then stated clearly his new definition of parallels: "All straight lines which in a plane go out from a point can, with reference to a given straight line in the same plane, be divided into two classes—*cutting* and *not-cutting*. The *boundary lines* of the one and other class of those lines will be called **parallel** to the given line." Thus, if BC is a line, A a point not on the line, and AD the perpendicular from A to BC, one can first draw a line AE perpendicular to AD (Fig. 19.4). The line AE does not meet BC. Lobachevsky then assumed that there may be other lines through A, such as AG, that also do not meet BC, however far they are prolonged. In passing from a cutting line, such as AF, to a not-cutting line, such as AG, there must be a line AH that is the boundary between these two sets. It is AH that is the parallel to BC. The angle HAD between AH and the perpendicular AD, an angle dependent on the length p of AB, is what Lobachevsky called the **angle of parallelism**, written $\Pi(p)$. If $\Pi(p) = 90°$, then there is only one line through A parallel to BC and the situation is the Euclidean one. If, however, $\Pi(p) < 90°$, then there will be a corresponding line AK on the other side of AD from AH that also makes the same angle $\Pi(p)$ with AD. It is thus always necessary in this non-Euclidean situation to distinguish two different sides in parallelism. In any case, on each side of AD, under the non-Euclidean assumption, there are infinitely many lines through A that do not meet BC.

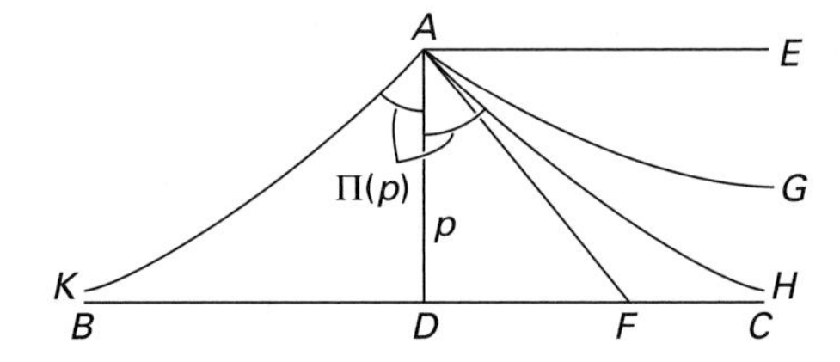

FIGURE 19.4 Lobachevsky's angle of parallelism

From the non-Euclidean assumption, Lobachevsky derived many results, some of which were in essence known to Saccheri and/or Lambert. For example, he showed that the property that $\Pi(p) < 90°$ for any p is equivalent to the property that the angle sum of every triangle is less than 180°. In that case, not only is the equation $\Pi(p) = \alpha$ solvable for any α less than a right angle, but also parallel lines are asymptotic to one another.

Lobachevsky's most interesting results, which were unknown to Saccheri and Lambert, involved the trigonometry of the non-Euclidean plane. Through a complex argument involving spherical triangles and triangles in the non-Euclidean plane, he was able to evaluate explicitly the function $\Pi(x)$ in the form

$$\tan \tfrac{1}{2}\Pi(x) = e^{-x},$$

essentially the same result obtained by Taurinus. It followed that $\Pi(0) = \pi/2$ (or that for small values of x, the geometry is close to the Euclidean one) and that $\lim_{x\to\infty} \Pi(x) = 0$. Lobachevsky could then derive new relationships among the sides a, b, c and the opposite angles A, B, C of an arbitrary non-Euclidean triangle:

$$\sin A \cot \Pi(b) = \sin B \cot \Pi(a), \tag{19.3}$$

$$\cos A \cos \Pi(b) \cos \Pi(c) + \frac{\sin \Pi(b) \sin \Pi(c)}{\sin \Pi(a)} = 1, \tag{19.4}$$

$$\cot A \sin C \sin \Pi(b) + \cos C = \frac{\cos \Pi(b)}{\cos \Pi(a)}, \tag{19.5}$$

$$\cos A + \cos B \cos C = \frac{\sin B \sin C}{\sin \Pi(a)}. \tag{19.6}$$

Lobachevsky's formulas imply standard Euclidean formulas when the sides of the triangle are small. The explicit formula for $\Pi(x)$ implies that

$$\cot \Pi(x) = \sinh x \qquad \cos \Pi(x) = \tanh x \qquad \sin \Pi(x) = \frac{1}{\cosh x}.$$

One then approximates the values of the hyperbolic functions by the terms of their power series up to degree 2 to get $\cot \Pi(x) = x$, $\cos \Pi(x) = x$, and $\sin \Pi(x) = 1 - \frac{1}{2}x^2$. Substituting these approximations into the four formulas and neglecting terms of degree higher than 2 gives the results

$$b \sin A = a \sin B,$$

$$a^2 = b^2 + c^2 - 2bc \cos A,$$

$$a \sin(A + C) = b \sin A,$$

$$\cos A + \cos(B + C) = 0.$$

The first two results are the familiar laws of sines and cosines, respectively, while the last two, when combined with the first two, are equivalent to the result that $A + B + C = \pi$. It also follows that if one replaces the sides a, b, c of the triangle by ia, ib, ic, respectively, Lobachevsky's results transform into standard results of spherical trigonometry. Thus, Lobachevsky's geometry is essentially the same as Taurinus's log-spherical geometry on the sphere of imaginary radius.

Lobachevsky saw in his trigonometric formulas "a sufficient foundation for considering the assumption of [non-Euclidean] geometry as possible. Hence," he concluded, "there is no means, other than astronomical observations, to use for judging of the exactitude which pertains to the calculations of the ordinary geometry." As noted in the chapter opening, Gauss, too, realized that the creation of a new and apparently valid geometry in which Euclid's parallel postulate did not hold showed that there was no "necessity" to Euclid's geometry and that one could not automatically conclude that Euclid's geometry held in the world in which we live. It was necessary to experiment to decide whether the geometry of the physical universe was Euclidean or not. Lobachevsky actually attempted such an experiment, using data on star positions, but the results were inconclusive.

The work of Bolyai and Lobachevsky, although responding to an age-old question about the parallel postulate, drew very little response from the mathematical community before the 1860s. There are several reasons for this, including the fact that some (though not all) of their articles were published in somewhat obscure sources and not in the major languages of the day, as well as the general difficulty of getting acceptance for an entirely new idea in mathematics. But it would appear that the most important reason that the discoveries of the Hungarian and the Russian did not immediately become part of the mainstream of mathematics was that few could really understand what a non-Euclidean plane was. Although the arguments of the founders were correct and logically coherent, and although they displayed what appeared to be reasonable mathematical formulas involving known functions, the "reality" of this new geometry simply was not accepted. Until non-Euclidean geometry could be seen as part of a more general system of geometry and be connected via this system to Euclidean geometry, it was not to be anything more than a curious sidelight.

19.1.3 Models of Non-Euclidean Geometry

Among those who attempted to develop a more general system of geometry in which to imbed Lobachevsky's geometry was Hermann von Helmholtz (1821–1894). In a paper of 1869 entitled "On the Facts Which Lie at the Foundations of Geometry," he attempted to list a set of hypotheses that would provide the basis for any reasonable study of geometry. First, he assumed that a space of n dimensions is a **manifold**. In other words, he assumed the existence of n independent coordinates near every point, at least one of which varies continuously as the point moves. It seems clear from his examples that he did not require that the same set of coordinates always applied in the entire manifold. Helmholtz's second axiom was that rigid bodies exist. This assumption permits one to equate two different spatial objects by superposition. Third, Helmholtz asserted that rigid bodies can move freely. This implied that the curvature of the manifold was constant. These hypotheses led Helmholtz to his own concept of the physical space in which we live—namely, as a three-dimensional manifold of constant curvature. It follows that there are three possibilities for the curvature of physical space: positive, negative, or zero. The third option leads to Euclidean geometry. Contrariwise, if the curvature is positive, we have **spherical space**, in which straight lines return upon themselves and there are no parallels. Such a space, it turns out, has no boundaries, but is finite in extent, like the surface of a sphere. If the curvature is negative, we get what Helmholtz termed a **pseudospherical space**, in which straight lines run out to infinity, and a pencil of straight lines may be drawn through a point on a flat surface in the space which does not intersect another given straight line in that surface. This latter geometry is basically Lobachevsky's non-Euclidean geometry. Furthermore, spherical geometry also turned out to be a non-Euclidean geometry, one in which no parallel lines exist. Thus, the two possible negations of Euclid's parallel postulate could both lead to possible geometries of our physical space.

Because Lobachevsky's geometry appeared to be valid on a pseudosphere, a surface of constant negative curvature, Eugenio Beltrami (1835–1900), an Italian mathematician who held chairs in mathematics in Bologna, Pisa, Pavia, and finally Rome, attempted to construct this surface. It turned out that he could construct only a portion of the surface in

Euclidean three-dimensional space. Nevertheless, Beltrami succeeded in determining the appropriate metric on this surface, that is, an expression for the length ds of an infinitesimal element of a curve, which is the square root of a positive definite quadratic form in the infinitesimal coordinate elements. He was then able to show the connection between this metric and Lobachevsky's trigonometric laws for non-Euclidean space. In an 1868 article, he began by parametrizing the sphere of radius k (and curvature $1/k^2$) situated in Euclidean three-dimensional space by

$$x = \frac{uk}{\sqrt{a^2+u^2+v^2}}, \qquad y = \frac{vk}{\sqrt{a^2+u^2+v^2}}, \qquad z = \frac{ak}{\sqrt{a^2+u^2+v^2}},$$

for some value a. It is then straightforward to calculate the metric form ds^2 on the sphere by substitution into the Euclidean form $ds^2 = dx^2 + dy^2 + dz^2$:

$$ds^2 = k^2 \frac{(a^2+v^2)\,du^2 - 2uv\,du\,dv + (a^2+u^2)\,dv^2}{(a^2+u^2+v^2)^2}.$$

To transform this result into one on a pseudosphere of curvature $-1/k^2$, Beltrami simply replaced u by iu and v by iv. The resulting metric,

$$ds^2 = k^2 \frac{(a^2-v^2)\,du^2 + 2uv\,du\,dv + (a^2-u^2)\,dv^2}{(a^2-u^2-v^2)^2},$$

turned out to have the required properties.

On the pseudosphere, the curves $u = c$ and $v = c$ are geodesics orthogonal to $v = 0$ and $u = 0$, respectively, for any constant $c < a$. Thus, Beltrami could consider a right triangle with one vertex at the origin, one leg along the curve $v = 0$, one leg along a curve $u = c$, and the hypotenuse along a geodesic through the origin that makes an angle θ with $v = 0$ (Fig. 19.5). He calculated the lengths of these three sides by integration of the appropriate

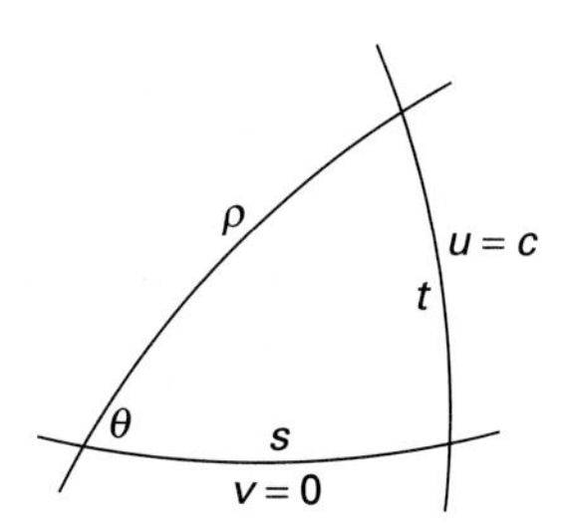

FIGURE 19.5 Beltrami's calculations on the pseudosphere:

$$\rho = \frac{1}{2}k \ln \frac{a+r}{a-r}$$

$$s = \frac{1}{2}k \ln \frac{a + r\cos\theta}{a - r\cos\theta}$$

$$t = \frac{1}{2}k \ln \frac{\sqrt{a^2-u^2}+v}{\sqrt{a^2-u^2}-v}$$

metric forms. For the hypotenuse, set $u = r\cos\theta$ and $v = r\sin\theta$. It follows that

$$ds = \frac{ka\,dr}{a^2 - r^2}.$$

This element of arc is easily integrated from 0 to r to get the length ρ of the hypotenuse:

$$\rho = \frac{1}{2}k\ln\frac{a+r}{a-r}.$$

Similarly, along the curve $v = 0$,

$$ds = \frac{a\,du}{a^2 - u^2},$$

and the length s of this leg of the triangle up to a given u is

$$s = \frac{1}{2}k\ln\frac{a+u}{a-u} = \frac{1}{2}k\ln\frac{a + r\cos\theta}{a - r\cos\theta}.$$

Finally, the metric along $u = c$ is given by

$$ds = \frac{k\sqrt{a^2-u^2}}{a^2-u^2-v^2}\,dv,$$

a differential whose integral up to a particular v is

$$t = \frac{1}{2}k\ln\frac{\sqrt{a^2-u^2}+v}{\sqrt{a^2-u^2}-v}.$$

With a bit of algebraic manipulation on the values for ρ, s, and t, Beltrami showed that

$$\frac{r}{a} = \tanh\frac{\rho}{k}, \qquad \frac{r}{a}\cos\theta = \tanh\frac{s}{k}, \qquad \frac{v}{\sqrt{a^2-u^2}} = \tanh\frac{t}{k}.$$

It then follows that

$$\cosh\frac{s}{k}\cosh\frac{t}{k} = \cosh\frac{\rho}{k}.$$

This result, identical with Taurinus's equation (19.1) and Lobachevsky's equation (19.4) for the case of a right triangle, shows that Beltrami's surface with its associated metric gives the same geometry as Lobachevsky's non-Euclidean plane. In other words, Beltrami's calculations showed that Taurinus's apparently mysterious use of a sphere of imaginary radius was equivalent to the introduction of a new metric on an appropriate two-dimensional manifold.

A different way of looking at Lobachevsky's geometry is simply to consider the imaginary sphere to be projected onto the interior of the circle $u^2 + v^2 = a^2$, where u and v are the parameters given above. It then turns out that straight lines in the Lobachevskian plane are represented by chords in the circle (Fig. 19.6a). Parallel straight lines are those whose intersection is at the circumference of the circle, with the circumference itself representing points at "infinity." Chords that do not intersect inside the circle represent lines that do not intersect at all in the Lobachevskian plane. A similar model of Lobachevskian geometry in the interior of a circle was developed in 1882 by Henri Poincaré (1854–1912). In this model, straight lines are represented by arcs of circles that are orthogonal to the boundary circle. Parallel lines are then represented by circular arcs that intersect at the boundary.

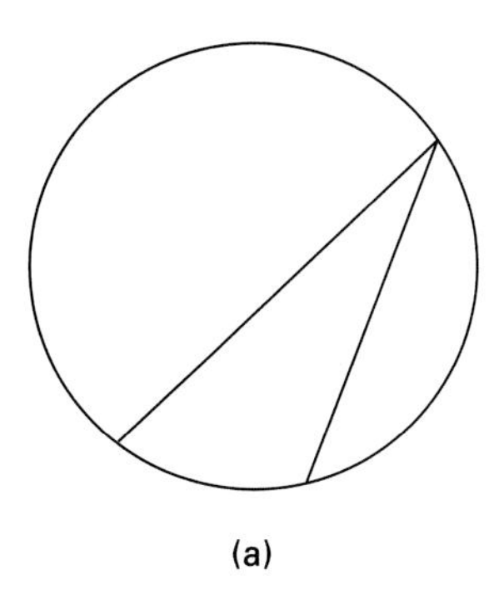

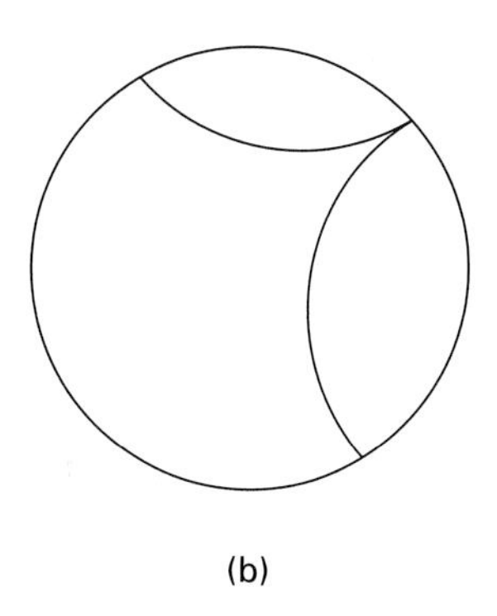

FIGURE 19.6 Two models of the Lobachevskian plane

This model has the advantage that angles between circles are measured in the Euclidean way. Figure 19.6b then shows why the angle sum of a triangle is less than π.

It was the use of models of Lobachevskian geometry as subsets of the ordinary Euclidean plane that helped to convince mathematicians by the end of the century that non-Euclidean geometry was as valid as Euclid's. Any contradiction in the former geometry would, by translation to the model, lead to a contradiction in the latter. Saccheri's attempts to "vindicate" Euclid had failed. With the work of Lobachevsky, Bolyai, Beltrami, and Poincaré, among others, it was now clear that Euclid was truly vindicated. He had been completely correct in his decision 2200 years earlier to take the parallel postulate as a postulate. Because the Lobachevskian alternative to Euclid's parallel postulate led to a geometry as valid as Euclid's, it was impossible to prove that postulate as a theorem.

19.2 Geometry in n Dimensions

The Greek limitation of geometry to three dimensions had been overcome by various mathematicians in the early nineteenth century. For example, Ostrogradsky in the 1830s had generalized his divergence theorem to n dimensions almost casually (simply by adding three dots to the end of various formulas), while Cauchy even earlier had dealt with geometric objects of arbitrary dimensions in explaining his version of the diagonalization of a symmetric matrix. The actual phrase "geometry of n dimensions" seems to have first appeared in 1843 in the title of a paper by Cayley. The article itself, however, was purely algebraic and only touched on geometry in passing.

19.2.1 Grassmann and the *Ausdehnungslehre*

The first mathematician to present a detailed theory of spaces of dimension greater than three was Hermann Grassmann (1809–1877), a German mathematician and philologist whose brilliant work was unfortunately not recognized during his lifetime. Grassmann's aim in his *The Science of Linear Extension* of 1844 and in its reworking in 1862 was to develop a systematic method of expressing geometric ideas symbolically, beginning with the notion of geometrical multiplication.

Four years before he wrote his first detailed work discussing n-dimensional spaces, Grassmann was already able to deal with the multiplication of vectors in two- and three-dimensional spaces in a paper devoted to a new explanation of the theory of the tides. He defined the **geometrical product of two vectors** to be "the surface content of the parallelogram determined by these vectors" and the **geometrical product of three vectors** to be the "solid (a parallelepiped) formed from them." Defining in an appropriate way the sign of such products, he was able to show that the geometrical product of two vectors is distributive and anticommutative and that the geometrical product of three vectors all lying in the same plane is zero. Because the area of the parallelogram that is the geometrical product of two vectors is equal to the product of the lengths of the two vectors and the sine of the angle between them, this product is identical in numerical value to the length of the modern cross product. The difference, of course, is in the geometrical nature of the object produced by multiplying. Rather than the product being a new vector, it is a two-dimensional object. But because there is a one-to-one correspondence between Grassmann's parallelograms (two of which are considered equal if they have the same area and lie in the same or parallel planes) and modern vectors in three-dimensional space, determined by associating to each parallelogram the normal vector whose length equals the parallelogram's area, the two multiplications are essentially identical. The advantage of Grassmann's product, however, is that it, unlike the cross product, is generalizable to higher dimensions. It is that generalization that is the basis of Grassmann's major works of 1844 and 1862.

Grassmann began the discussion in his text, particularly in the clearer 1862 version, with the notion of a vector as a straight line segment with fixed length and direction. Two vectors are to be added in the standard way by joining the beginning point of the second vector to the end point of the first. Subtraction is simply the addition of the negative, that is, the vector of the same length and opposite direction. Vectors are the simplest examples of what Grassmann called an **extensive quantity**. In general, such a quantity is a linear combination of a set of n independent "units," $\epsilon_1, \epsilon_2, \ldots, \epsilon_n$. Addition of extensive quantities is by the obvious method: $\sum \alpha_i \epsilon_i + \sum \beta_i \epsilon_i = \sum (\alpha_i + \beta_i)\epsilon_i$. Similarly, one can multiply an extensive quantity by a scalar. Grassmann noted that the basic laws of algebra hold for his extensive quantities, and he defined the **space** of the quantities $\{\epsilon_i\}$ to be the set of all linear combinations of them.

Grassmann next defined multiplication of extensive quantities by use of the distributive law: $(\sum \alpha_i \epsilon_i)(\sum \beta_j \epsilon_j) = \sum \alpha_i \beta_j [\epsilon_i \epsilon_j]$, where each quantity $[\epsilon_i \epsilon_j]$ is called a **quantity of the second order**. Because this new sum must be an extensive quantity, it too must be expressible as a linear combination of units. Thus, Grassmann needed to define second-order units. Assuming that the multiplication rules defined on units extend to the same rules on any extensive quantities, he demonstrated that there are only four basic possibilities for defining second-order units: First, all of the quantities $[\epsilon_i \epsilon_j]$ could be independent. Second, one could have $[\epsilon_i \epsilon_j] = [\epsilon_j \epsilon_i]$. (This multiplication satisfies all of the ordinary algebraic multiplication rules.) Third, one could have $[\epsilon_i \epsilon_j] = -[\epsilon_j \epsilon_i]$. (This implies that $[\epsilon_i \epsilon_i] = 0$ for all i.) Finally, one could have all products $[\epsilon_i \epsilon_j] = 0$. It is the third form of multiplication, called **combinatory multiplication**, that Grassmann considered in detail in the remainder of his work. According to his condition, then, for any first-order extensive quantities A and B, the multiplication rule $AB = -BA$ holds.

With the combinatory product of two first-order units defined, it was straightforward for Grassmann to define products of three or more first-order units using the same basic rules.

For example, if there are three first-order units $\epsilon_1, \epsilon_2, \epsilon_3$, then there are three second-order units $[\epsilon_1\epsilon_2], [\epsilon_2\epsilon_3], [\epsilon_3\epsilon_1]$ and one third-order unit $[\epsilon_1\epsilon_2\epsilon_3]$. (Any other product of three first-order units would have two factors in common and would therefore be equal to 0.) If there are four first-order units, then there are six second-order units, four third-order units, and one fourth-order unit. Grassmann noted further that the product of n linear combinations of n first-order units, $(\sum \alpha_{1i}\epsilon_i)(\sum \alpha_{2i}\epsilon_i)\cdots(\sum \alpha_{ni}\epsilon_i)$, is equal to $\det(\alpha_{ij})[\epsilon_1\epsilon_2\cdots\epsilon_n]$, where the bracketed expression is the single unit of nth order. Grassmann's combinatory product determines, in modern terminology, the exterior algebra of a vector space.

The ideas in his *Science of Linear Extension* came from Grassmann's desire to express various geometric concepts symbolically. So, in particular, he thought of his second-order quantities as parallelograms and his third-order quantities as parallelepipeds. But even though there was no specifically geometric interpretation of higher-order quantities, Grassmann saw that the symbolic manipulations did not require the limitation to any particular number of dimensions. Not only did he construct the exterior algebra, but he also developed many of the important ideas relating to vector spaces, that is, to the space of all linear combinations of n units. As early as 1840, he had developed the notion of the inner product of two vectors as the algebraic product of one vector multiplied by the perpendicular projection of the second onto it and showed that in coordinate form, the inner product was given by $(\sum \alpha_i\epsilon_i)(\sum \beta_j\epsilon_j) = \sum \alpha_i\beta_i$. And in his text, he developed the notions of linear independence and basis, showed that any vector can be uniquely expressed as a linear combination of the elements in a basis, and proved that in an n-dimensional space, any vector in a basis can be replaced by another vector independent of the remaining $n-1$ vectors. He demonstrated that an orthogonal system of quantities is linearly independent (where two vectors are orthogonal if their inner product is zero) and proved the well-known result that for two subspaces U, W of a space V,

$$\dim(U + W) = \dim U + \dim W - \dim(U \cap W).$$

19.2.2 Vector Spaces

The basic notions of linear algebra, including those of linear independence and linear combinations, were used in many areas of mathematics during the nineteenth century, generally without reference to Grassmann's work, but it was not until the end of the century that an abstract definition of a vector space was formulated. The first mathematician to give such a definition was Giuseppe Peano (1858–1932) in his *Geometric Calculus* of 1888. Peano's aim in the book, as the title indicates, was the same as Grassmann's, namely, to develop a calculus of geometric objects. Thus, much of the book consists of various calculations dealing with points, lines, planes, and solid figures. But in chapter IX, Peano gave a definition of what he called a **linear system**. Such a system consists of quantities provided with operations of addition and scalar multiplication. The addition must satisfy the commutative and associative laws (although these laws were not cited as such by Peano), while the scalar multiplication satisfies two distributive laws, an associative law, and the law that $1v = v$ for every quantity v. In addition, Peano included as part of his axiom system the existence of a zero quantity satisfying $v + 0 = v$ for any v as well as $v + (-1)v = 0$. Peano also defined the **dimension** of a linear system as the maximum number of linearly independent quantities in the system. In connection with this idea, Peano noted that the set of polynomial functions in one variable forms a linear system, but that there is no such maximum number of linearly independent quantities and therefore the dimension of this system must be infinite.

Peano's work, like that of Grassmann, had no immediate effect on the mathematical world. His definition was forgotten, although mathematicians continued to use the basic concepts involved. For example, in 1893, as part of his work on algebraic number fields, Dedekind defined a space Ω as the set of all linear combinations of an independent set of n algebraic numbers with coefficients in a field. He noted that the numbers of this space satisfy the basic properties now attributed to a vector space, without referring to any such definition elsewhere. And he proved, using induction, the important result that any $n + 1$ numbers in Ω are dependent. Although he did not state explicitly that no smaller set of generators would determine the space, his definition essentially assured this, and thus he had shown that the dimension of a (finite-dimensional) vector space is well-defined.

Aspects of vector space theory continued to appear in the mathematical literature, but it was not until the twentieth century that a fully axiomatic treatment of the subject entered the mathematical mainstream.

19.3 Graph Theory and the Four-Color Problem

A **graph** in modern terminology consists of a non-empty set V, whose elements are called **vertices**, and a set E whose elements are **edges**, where each edge consists of a pair of vertices. In geometric terminology, the edges of the graph are arcs that join pairs of vertices. Graph theory in the West has its origins in Euler's 1736 solution of the problem of the seven bridges of Königsberg, discussed earlier. Although Euler himself solved the problem algebraically, it is easy enough to draw a graph with the vertices representing the regions of the city and the edges representing the bridges. Another interesting problem that later became part of graph theory was found by William Rowan Hamilton in 1856. In fact, Hamilton turned the problem into a game, which was marketed in 1859. This *Icosian* game consisted of a graph with twenty vertices on which pieces were to be placed in accordance with various conditions, the overriding consideration being that a piece was always placed at the second vertex of an edge on which the previous piece had been placed (Fig. 19.7). The first set of extra conditions Hamilton proposed was, given pieces placed on five initial points, to cover the board with the remaining pieces in succession such that the last piece placed is adjacent to the first. In more modern terminology, Hamilton's problem was to discover a cyclic path that passed through each vertex exactly once, in contrast to Euler's problem of discovering a path traversing each edge exactly once. Hamilton gave several examples of ways in which this could be accomplished but gave no general method for determining in cases other than his special graph whether or not such a path could be constructed.

The earliest purely mathematical consideration of a special class of graphs appeared in an article by Arthur Cayley in 1857. Cayley, inspired by a consideration of possible combinations of differential operators, defined and analyzed the general notion of a **tree**, a connected graph that has no cyclic paths and therefore whose number of edges is one fewer than the number of vertices. In particular, Cayley dealt with the notion of a **rooted tree**, a tree in which one particular vertex is designated as the root. He exhibited the possible rooted trees with two, three, and four vertices (which Cayley called **knots**) or, equivalently, with one, two, or three edges (**branches**, to continue the botanical analogy). By a clever combinatorial argument, Cayley then developed a recursive formula for determining the

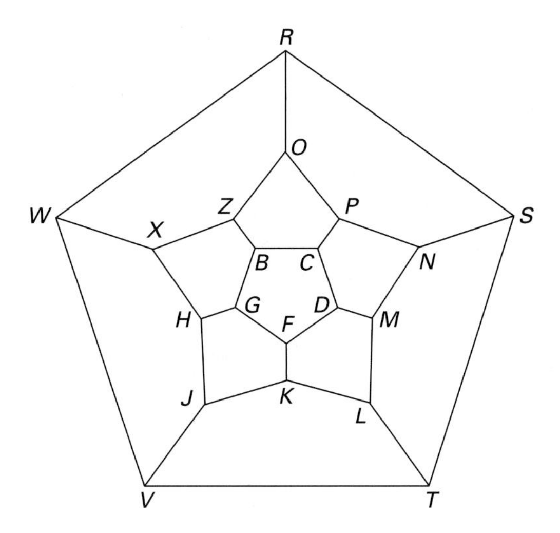

FIGURE 19.7 The Icosian game of William Rowan Hamilton

number A_r of different trees with r branches (where "different" is appropriately defined) (Fig. 19.8). He showed, for example, that $A_1 = 1$, $A_2 = 2$, $A_3 = 4$, $A_4 = 9$, and $A_5 = 20$. In 1874, Cayley applied his results to the study of chemical isomers and a few years later succeeded in developing a formula for counting the number of unrooted trees with a given number of vertices.

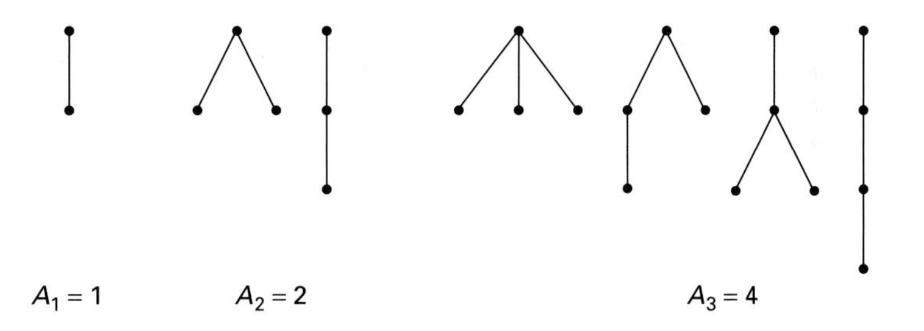

FIGURE 19.8 Different trees with r branches, for $r = 1, 2, 3$

The problem that has played the greatest role in the development of modern graph theory, a problem that was addressed by many mathematicians from its first formulation in 1852, is the **four-color problem**. The problem was described in a letter from Augustus De Morgan (1806–1871) to Hamilton on October 23 of that year: "A student of mine [Frederick Guthrie] asked me today to give him a reason for a fact which I did not know was a fact—and do not yet. He says that if a figure be anyhow divided and the compartments differently coloured so that figures with any portion of common boundary line are differently coloured—four colours may be wanted, but not more. . . . My pupil says he guessed it in colouring a map of England. The more I think of it, the more evident it seems." De Morgan was not able to think of a case of a map where five colors were required, and although he thought the sufficiency of four colors was "evident," he could give no proof of that either. Hamilton was not interested in the problem, but in the following two decades Cayley and others spent much time in a futile search for a proof.

A few basic results were worked out in this period, however, that would be essential to any proof of the **four-color theorem**, the result that Guthrie had asserted. First, we recall Euler's formula relating the number of vertices, edges, and faces on a convex polyhedron: $F - E + V = 2$. If we imbed the polyhedron in a sphere, then project it onto a plane by lines through the north pole, we get a map in the plane (Fig. 19.9). In fact, Cauchy was able to prove Euler's formula by reducing it to a consideration of these plane maps. In any case, the resulting plane map satisfies the relation $F - E + V = 1$, because an exterior region (or face) has been "lost" in the projection.

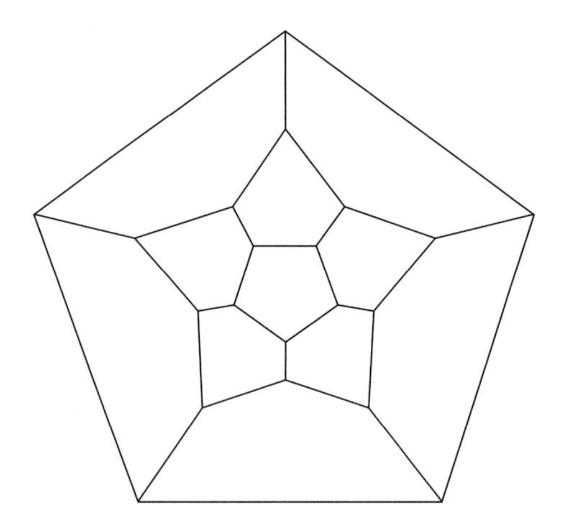

FIGURE 19.9 Map of a convex polyhedron projected onto a plane

We can now show, using Euler's formula, that every map has at least one country with five or fewer neighbors. Let us assume the map has F countries, E boundary lines, and V vertices (or meeting points). We can assume that at least three boundary lines meet at each vertex. Thus, it appears that there are at least $3V$ boundary lines in total. However, since each boundary line has two ends, we can conclude that there are at least $\frac{3}{2}V$ boundary lines. That is, $E \geq \frac{3}{2}V$, or $V \leq \frac{2}{3}E$. Now, let us assume that there is no country with five or fewer neighbors—that is, that every country has at least six neighbors. It follows that there are at least $6F$ boundary lines, except that, again, each boundary line is counted twice, because there is a country on each side. So we get $E \geq 3F$, or $F \leq \frac{1}{3}E$. But then $F - E + V \leq \frac{1}{3}E - E + \frac{2}{3}E = 0$, contradicting Euler's formula. Hence, the result follows.

Now suppose that there are maps that require at least five colors. Pick such a map with the minimal number of countries. We note that such a map cannot contain a two-sided country (a digon) or a three-sided country (a triangle). For example, if the map contained a triangle, we could shrink the triangle to a point, thus removing one country, color the new map with four colors (which implies that around that point there are only three colors), then reinstate the triangle and use the fourth color to color it (Fig. 19.10). The proof in the digon case is similar. Unfortunately, this argument does not extend to minimal maps that contain a square or a pentagon. So a different argument would be needed there to complete the proof of the four-color theorem.

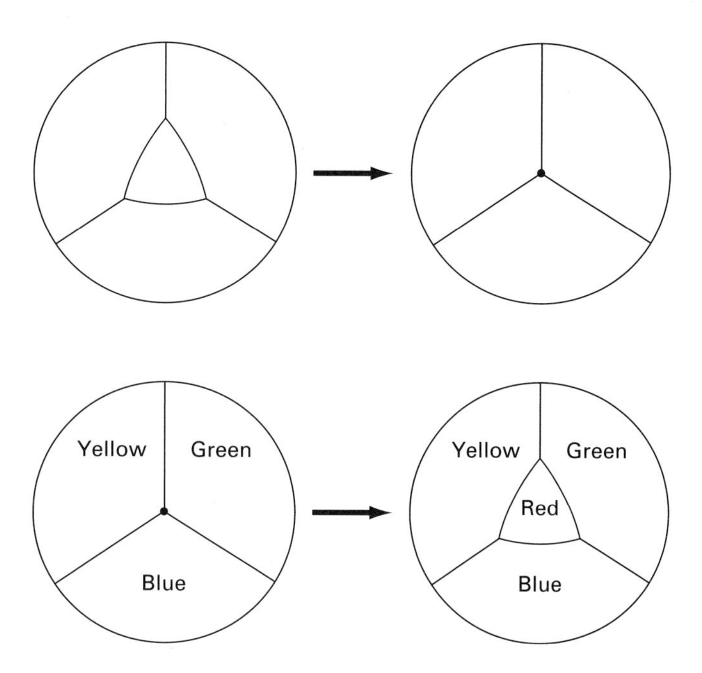

FIGURE 19.10 Coloring a minimal map containing a triangle

In 1879, Alfred Kempe (1849–1922) published a proof of the theorem that seemed to solve the problem for the square and the pentagon. His procedure for coloring any map is the following: First, find a country with five or fewer neighbors. Cover this country with a blank piece of paper (a patch) of the same shape but slightly larger, and extend all the boundaries that meet this patch into the center. This amounts to shrinking the country to a point, thus reducing the number of countries by one. Repeat this procedure until there is just one country left. This country can then be colored with one of the four colors. Then reverse the process, stripping off each patch in turn. At each stage, color the remaining country with any available color until the entire map is colored. The tricky part of Kempe's argument is, of course, the last, in which he claims one of the four colors will always be available when a country is restored. If the restored country is a digon or triangle, there is no problem. For the cases where it is a square or a pentagon, Kempe gave a complicated argument involving chains of countries colored with two colors, beginning with two that border the restored country, an argument that was accepted as correct when Kempe's paper was published. However, ten years later, Percy Heawood (1861–1955) discovered a flaw in Kempe's argument for the pentagon, which could not easily be remedied. Still, enough of the argument was true that Heawood could prove that five colors are always sufficient for coloring a map. Nevertheless, it was necessary for some new ideas to emerge before the four-color problem could eventually be solved.

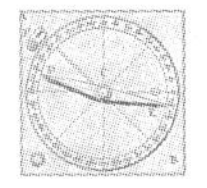

Exercises

1. Derive the formula

$$\cos a = \cos b \cos c + \sin b \sin c \cos A$$

for an arbitrary spherical triangle with sides a, b, c and opposite angles A, B, C on a sphere of radius 1 by dividing the triangle into two right triangles and applying the formulas of chapter 3.

2. Show that the formula in exercise 1 changes to

$$\cos \frac{a}{K} = \cos \frac{b}{K} \cos \frac{c}{K} + \sin \frac{b}{K} \sin \frac{c}{K} \cos A$$

if the sphere has radius K, where a, b, c are expressed in a linear measure.

3. By using power series, show that Taurinus's log-spherical formula,

$$\cosh \frac{a}{K} = \cosh \frac{b}{K} \cosh \frac{c}{K} - \sinh \frac{b}{K} \sinh \frac{c}{K} \cos A,$$

reduces to the law of cosines as $K \to \infty$.

4. Show that Taurinus's formula for an asymptotic right triangle on a sphere of imaginary radius i, namely, $\sin B = 1/\cosh x$, is equivalent to Lobachevsky's formula for the angle of parallelism, $\tan(B/2) = e^{-x}$.

5. Given that $\tan \frac{1}{2}\Pi(x) = e^{-x}$, where $\Pi(x)$ is Lobachevsky's angle of parallelism, derive the formulas

$$\sin \Pi(x) = \frac{1}{\cosh x} \quad \text{and} \quad \cos \Pi(x) = \tanh x,$$

and show that their power series expansions up to degree 2 are $\sin \Pi(x) = 1 - \frac{1}{2}x^2$ and $\cos \Pi(x) = x$, respectively.

6. Substitute the results of exercise 5 into Lobachevsky's formulas,

$$\sin A \tan \Pi(a) = \sin B \tan \Pi(b)$$

and

$$\cos A \cos \Pi(b) \cos \Pi(c) + \frac{\sin \Pi(b) \sin \Pi(c)}{\sin \Pi(a)} = 1,$$

to derive the laws of sines and cosines when the sides a, b, c of the non-Euclidean triangle are "small."

7. Show that if ABC is an arbitrary triangle with sides a, b, c, then the formulas $a \sin(A + C) = b \sin A$ and $\cos A + \cos(B + C) = 0$, along with the law of sines, imply that $A + B + C = \pi$.

8. Describe geometrically Beltrami's parametrization of the sphere of radius k given by

$$x = \frac{uk}{\sqrt{a^2 + u^2 + v^2}}, \qquad y = \frac{vk}{\sqrt{a^2 + u^2 + v^2}},$$

$$z = \frac{ak}{\sqrt{a^2 + u^2 + v^2}}.$$

9. Show that replacing u, v by iu, iv, respectively, transforms the sphere of exercise 8 with curvature $1/k^2$ to a pseudosphere with curvature $-1/k^2$.

10. Show that Beltrami's formulas for the lengths ρ, s, t of the sides of a right triangle on his pseudosphere transform into

$$\frac{r}{a} = \tanh \frac{\rho}{k}, \qquad \frac{r}{a} \cos \theta = \tanh \frac{s}{k},$$

$$\frac{v}{\sqrt{a^2 - u^2}} = \tanh \frac{t}{k},$$

and then show that

$$\cosh \frac{s}{k} \cosh \frac{t}{k} = \cosh \frac{\rho}{k}.$$

11. Letting i, j, k be first-order units in three-dimensional space, determine the combinatory product of $2i + 3j - 4k$, $3i - j + k$, and $i + 2j - k$.

12. Show that in Grassmann's combinatory multiplication,

$$\left(\sum \alpha_{1i}\epsilon_i\right)\left(\sum \alpha_{2i}\epsilon_i\right)\cdots\left(\sum \alpha_{ni}\epsilon_i\right) = \det(\alpha_{ij})[\epsilon_1\epsilon_2\cdots\epsilon_n],$$

where each linear combination is of a given set of n first-order units and where $[\epsilon_1\epsilon_2\cdots\epsilon_n]$ is the single unit of nth order.

13. Find a path in Hamilton's Icosian game (Fig. 19.7) that passes through each vertex exactly once and returns to the starting point.

14. Get an outline map of the continental United States and color it with four colors.

15. Show that a map with the minimal number of countries that requires at least five colors cannot contain a digon (a country with only two boundary edges).

16. Is the analytic form of non-Euclidean geometry as presented by Taurinus, Lobachevsky, and Beltrami a bet-

ter way of presenting the subject than the synthetic form? How can one make sense of a sphere of imaginary radius?

17. Is it simply by chance that both of the mathematicians who first published accounts of non-Euclidean geometry were from countries not in the mainstream of nineteenth-century mathematics, or might there be substantive reasons?

18. Outline a lesson using trigonometry to show the relationships between spherical geometry and Lobachevsky's non-Euclidean geometry.

References

Some of the important works on various aspects of the history of geometry in the nineteenth century include Jeremy Gray, *Ideas of Space: Euclidean, Non-Euclidean and Relativistic* (Oxford: Clarendon Press, 1989); B. A. Rosenfeld, *A History of Non-Euclidean Geometry* (New York: Springer-Verlag, 1988); Roberto Bonola, *Non-Euclidean Geometry. A Critical and Historical Study of Its Development* (New York: Dover, 1955); Julian Lowell Coolidge, *A History of Geometrical Methods* (New York: Dover, 1963); and Michael J. Crowe, *A History of Vector Analysis* (New York: Dover, 1985). A history of graph theory through annotated reprints of original works on the subject is Norman Biggs, E. Keith Lloyd, and Robin J. Wilson, *Graph Theory, 1736–1936* (Oxford: Clarendon, 1976). The history of the four-color problem itself is well covered in Robin Wilson, *Four Colors Suffice* (Princeton: Princeton University Press, 2002).

Bonola's book contains translations of Lobachevsky's *Geometrical Investigations on the Theory of Parallel Lines* and Bolyai's *Appendix exhibiting the absolutely true science of space*. Helmholtz's "On the Origin and Significance of Geometrical Axioms" is in James Newman, ed., *The World of Mathematics* (New York: Simon and Schuster, 1956), vol. 1, 646–668. Beltrami's papers in which he deals with the metric for the pseudosphere and its relation to non-Euclidean geometry have been translated into English and appear in John Stillwell, *Sources of Hyperbolic Geometry* (Providence: American Mathematical Society, 1996). The 1844 version of the *Ausdehnungslehre* has been translated into English and appears, with other works of Grassmann, in *A New Branch of Mathematics*, Lloyd C. Kannenberg, trans. (Chicago: Open Court, 1995). The 1862 version has also been translated by Kannenberg and is available as *Extension Theory* (Providence: American Mathematical Society, 2000). Kempe's "proof" of the four-color theorem appeared as "On the Geographical Problem of the Four Colours," *American Journal of Mathematics* 2 (part 3) (1879), 193–200. It is reprinted in Biggs, Lloyd, and Wilson, *Graph Theory*, 96–102.

Grassmann's *Ausdehnungslehre* is discussed in detail in J. V. Collins, "An Elementary Exposition of Grassmann's *Ausdehnungslehre*, or Theory of Extension," *American Mathematical Monthly* 6 (1899), 193–198, 261–266, 297–301, and 7 (1900), 31–35, 163–166, 181–187, 207–214, 253–258. An article on Grassmann's creation of linear algebra is Desmond Fearnley-Sander, "Hermann Grassmann and the Creation of Linear Algebra," *American Mathematical Monthly* 86 (1979), 809–817. For more details on the origins of the theory of vector spaces and abstract linear algebra, see Jean-Luc Dorier, "A General Outline of the Genesis of Vector Space Theory," *Historia Mathematica* 22 (1995), 227–261 and Gregory H. Moore, "The Axiomatization of Linear Algebra: 1875–1940," *Historia Mathematica* 22 (1995), 262–303. Studies of the four-color problem include Oystein Ore, *The Four-Color Problem* (New York: Academic Press, 1967) and Thomas Saaty and Paul Kainen, *The Four-Color Problem: Assaults and Conquest* (New York: Dover, 1986).

CHAPTER TWENTY

Aspects of the Twentieth Century

[Emmy Noether] had great stimulating power, and many of her suggestions took final shape only in the works of her pupils or co-workers .... She could just utter a far-seeing remark like this, "Norm rest symbol is nothing else than cyclic algebra," in her prophetic lapidary manner, out of her mighty imagination that hit the mark most of the time and gained in strength in the course of years; and such a remark could then become a signpost to point the way for difficult future work.... She originated above all a new and epoch-making style of thinking in algebra.

—From an address in memory of Emmy Noether by Hermann Weyl, 1935

In the winter of 1899-1900, David Hilbert (1862–1943) was invited to make one of the major addresses at the second International Congress of Mathematicians, to be held in Paris in August, 1900. Hilbert took many months to decide on the topic of his address, finally deciding by July that, because a new century was just beginning, he would discuss some central open problems of mathematics, problems he believed would set the tone for mathematics in the twentieth century.

Hilbert began his talk with a discussion of the criteria for a significant problem: Its statement must be clear and easily comprehended; it should be difficult but not completely inaccessible; and its solution should have significant consequences. The twenty-three problems Hilbert presented to his audience encompassed virtually all of the branches of mathematics. For example, from the foundations of mathematics, Hilbert called for a proof of the continuum hypothesis as well as an investigation of the consistency of the axioms of arithmetic. From number theory came the questions as to whether α^β, where α is algebraic and β is irrational, always represents a transcendental, or even just an irrational, number, and whether it can always be determined if a Diophantine equation is solvable. From analysis came the Riemann hypothesis, whether all complex zeros of the Riemann zeta function have real part $\frac{1}{2}$, and whether one can always solve boundary-value problems in the theory of partial differential equations.

Hilbert's problems did in fact prove to be central in twentieth-century mathematics. Many were solved, and significant progress was achieved on the remainder, although the Riemann hypothesis is still unproven. (In fact, a prize of $1 million has been offered for its proof.) In this chapter, we will touch on only two of Hilbert's problems, but we will discuss other highlights of the twentieth century, looking in particular at some of the nineteenth-century ideas that were more fully developed as well as at certain old questions that were answered. For those who want to explore more extensively the mathematics of the twentieth century, the library shelves are open, and there are many resources available as a guide.

20.1 The Growth of Abstraction

By the end of the nineteenth century, mathematicians began to understand that the axioms that Euclid had given for geometry were not entirely adequate. In fact, it was clear that Euclid had made various assumptions in his proofs that were not explicitly mentioned in his list of axioms. Thus, Hilbert developed a new set of axioms for geometry and showed that they were consistent—that is, that no contradictions could be deduced from them, at least on the assumption that arithmetic was consistent. As the twentieth century began, the axiomatic approach gained favor. Thus, once a particular construct appeared to be important, mathematicians attempted to discover sets of axioms for the system and to derive as much as possible from the axioms without any appeal to a particular concrete manifestation of them.

20.1.1 The Axiomatization of Vector Spaces

Although Peano had given a set of axioms for vector spaces in 1888, this had been largely ignored. In 1918, Hermann Weyl (1885–1955) made a new attempt to give an axiomatic treatment of the subject in his *Space-Time-Matter*, as a basis for his development of the theory of relativity from basic principles. Although there is no indication that Weyl was familiar with the work of Peano, his axiom system was virtually the same. The only difference was that, unlike his Italian predecessor, Weyl insisted that vector spaces be finite dimensional. Thus, his final axiom stated that "there are n linearly independent vectors, but every $n + 1$ vectors are linearly dependent." Unfortunately, Weyl's work was even less influential than that of Peano. The notion of a vector space needed a third discovery, this time in the context of analysis.

Several mathematicians in the 1920s were studying the metric properties of spaces of many dimensions—that is, the properties that depend on a distance function in the space. In particular, Stefan Banach (1892–1945), in his dissertation of 1920, introduced the notion of what is now called a **Banach space**, a vector space possessing a distance function under which all Cauchy sequences converged. As Banach wrote in the published version of his dissertation in 1922, "the aim of the present work is to establish certain theorems valid in different functional domains. . . . Nevertheless, in order not to have to prove them for each particular domain, which would be painful, I have chosen to take a different route; that is, I will consider in a general sense the sets of elements of which I will postulate certain properties. I will deduce from them certain theorems and then I will prove for each specific functional domain that the chosen postulates are true." Thus, Banach began with a set of thirteen axioms characterizing the notion of a vector space over the real numbers. And since Banach was interested in spaces of functions, his vector spaces were not limited to a finite dimension. Although Banach's axiom system contained more axioms than necessary, his paper had great influence. And by the time of the publication ten years later of his *Theory of Linear Operators*, in which the axioms were repeated, the abstract notion of a vector space had become part of the mathematical vocabulary.

20.1.2 The Theory of Rings

The vector spaces that Dedekind studied in the late nineteenth century were also fields and thus possessed a reasonable multiplication operation in addition to the additive one. A mathematical structure R with two operations, addition and multiplication, is today called a **ring** if R is a commutative group with respect to addition, if multiplication is associative, and if the distributive laws hold. The first detailed study of such objects, under the name "linear associative algebras," was carried out by the American mathematician Benjamin Peirce (1809–1880) around 1870. By a **linear associative algebra** (today called simply an **algebra**), Peirce meant a ring that is at the same time a finite dimensional vector space over a field F. (Peirce limited the field of coefficients to the field of real numbers.) Peirce's chief aim in his work was to describe all possible algebras of dimensions 1 through 5 and some of dimension 6, by considering the possible multiplication tables for the basis elements. In the course of this work, however, he introduced two important definitions: A nonzero element a of a ring is **nilpotent** if some power a^n is zero, whereas it is **idempotent** if $a^2 = a$. Peirce was then able to demonstrate the

THEOREM. *In every algebra there is at least one idempotent or at least one nilpotent element.*

Because an algebra is finite dimensional, any nonzero element A of the algebra must satisfy an equation of the form

$$\sum_{i=1}^{n} a_i A^i = 0$$

for some n. This equation can be rewritten in the form $BA + a_1 A = 0$, or $(B + a_1)A = 0$, where B is a linear combination of powers of A. It follows that $(B + a_1)A^k = 0$ for every $k > 0$ and therefore that $(B + a_1)B = 0$, or $B^2 + a_1 B = 0$. It is immediate from the last equation that if $a_1 \neq 0$, then

$$\left(-\frac{B}{a_1}\right)^2 = -\frac{B}{a_1}$$

and $-B/a_1$ is an idempotent element. If $a_1 = 0$, then $B^2 = 0$ and B is a nilpotent element.

Several other mathematicians in the last quarter of the nineteenth century studied special algebras, in particular **simple algebras**, those having no nontrivial two-sided ideals. (A **two-sided ideal** in an algebra, or in any ring R, is a subset I such that if α and β belong to I so does $\alpha + \beta$, $r\alpha$, and αr, for any r in R. This notion generalizes the definition of Dedekind for ideals in a ring of algebraic integers.) In particular, in a paper of 1907, Joseph Henry Maclagan Wedderburn (1882–1948) proved that any simple algebra is a matrix algebra, not necessarily over a field, but over a division algebra. (A **division algebra** is an algebra with multiplicative identity such that every nonzero element has a multiplicative inverse.) Frobenius had already proved that over the field of real numbers there were only three division algebras: the real numbers, the complex numbers, and the quaternions. Wedderburn himself proved in 1909 that the only finite division algebras were the finite fields themselves, and these were well known. Division algebras over certain other fields were classified by Emmy Noether (1882–1935) and others in 1932.

Noether accomplished much else dealing with the theory of rings in the 1920s. In particular, she was able to develop a decomposition theory for ideals analogous to Dedekind's prime factorization but applicable to rings more general than the rings of integers in algebraic number fields. These more general rings are now called **Noetherian rings**, commutative rings with identity that satisfy the ascending chain condition, that every chain $I_1, I_2, \ldots, I_k, \ldots$ of ideals in the ring such that $I_k \subset I_{k+1}$ breaks off after a finite number of terms. For these rings, Noether was able to derive decomposition results that were somewhat weaker than unique prime factorization of ideals. Noether was also able to characterize those rings R for which the entire Dedekind theory of prime factorization of ideals holds. In particular, she showed not only that all domains of integers in algebraic number fields possess unique factorization of ideals into primes, but also that the "integral elements" in fields of algebraic functions in one variable have this property.

As indicated at the opening of this chapter, Noether had great influence on her coworkers, especially in emphasizing the structural rather than the computational aspects of algebra. In fact, the second volume of probably the most important algebra text of the first half of the twentieth century, *Modern Algebra*, by B. L. van der Waerden (1903–1996), is based largely on her ideas. A comparison of this text to texts only a few years earlier shows the great changes Noether initiated.

20.1.3 The Axiomatization of Set Theory

Georg Cantor raised many questions about the theory of infinite sets in his work of the late nineteenth century, some of which he could not answer. Other mathematicians also attacked these questions around the turn of the twentieth century.

Recall that Cantor had shown that for two sets M and N, with cardinality $\bar{\bar{M}}$, $\bar{\bar{N}}$, respectively, no more than one of the relations $\bar{\bar{M}} = \bar{\bar{N}}$, $\bar{\bar{M}} < \bar{\bar{N}}$, or $\bar{\bar{N}} < \bar{\bar{M}}$ can occur. It seemed obvious to him as far back as 1878, furthermore, that exactly one of these relations should hold; that is, if the two sets did not have the same cardinality, then one set must have the same cardinality as a subset of the other. It was only later that Cantor realized that it was not a trivial matter to deny the existence of two sets, neither one of which was equivalent to a subset of the other. In fact, in his *Contributions* of 1895, he made explicit mention of this **trichotomy principle** and stressed not only that he did not have a proof of it, but that its proof must surely be difficult. He therefore carefully avoided using this principle in other proofs in his theory.

Cantor also realized that this question of trichotomy was closely related to another principle, that every set can be **well-ordered**. This principle states that for any set A, there exists an order-relation $<$ such that each nonempty subset B of A contains a least element, an element c such that $c < b$ for every other b in B. The natural numbers are well-ordered under their natural ordering. The real numbers, on the other hand, are not well-ordered under their natural ordering, but Cantor in 1883 thought it nearly self-evident that a well-ordering existed. By the mid-1890s, however, he began to realize that this result too needed a proof, but again he knew that such a proof would be difficult.

Another troubling aspect of set theory at the beginning of the twentieth century was the appearance of a number of seeming paradoxes. One of the earliest is today called **Russell's paradox**, because it was published by Bertrand Russell (1872–1970) in 1903. Recall that Dedekind and Cantor believed that virtually any description of "objects of our thought" would define a set. However, Russell, and Ernst Zermelo (1871–1953) two years earlier, determined that defining sets that contain themselves as elements would lead to contradictions. For suppose such a set M containing each of its subsets $m, m', \ldots$ as elements exists. Then consider those subsets m that do not contain themselves as elements. These constitute a set M_0. We can prove of M_0 (1) that it does not contain itself as an element, (2) that it contains itself as an element. First, M_0, being a subset of M, is itself an element of M, but not an element of M_0. For otherwise M_0 would contain as an element a subset of M (namely, M_0 itself) that contains itself as an element, and that would contradict the notion of M_0. Second, it follows that M_0 itself is a subset of M that does not contain itself as an element. Thus, it must be an element of M_0. Russell himself published several other versions of this paradox, the simplest being the barber paradox: A barber in a certain town has stated that he will cut the hair of all those persons and only those persons in the town who do not cut their own hair. Does the barber cut his own hair?

By early in the twentieth century, it was therefore clear that a new approach to set theory was necessary, one that narrowed Dedekind's vague definition of a set and that also axiomatized the subject. This new approach was taken by Zermelo when he published a set of axioms for set theory in 1908. Zermelo's method of axiomatization began with a collection of unspecified objects and a relation among them that was defined by the axioms. In other words, Zermelo started with a domain $\mathcal{B}$ of objects and a relation $\in$ of membership between some pairs of these objects. An object is called a **set** if it contains another object (except as specified by axiom 2). To say that $A \subseteq B$ meant that if $a \in A$, then also $a \in B$. Zermelo's seven axioms were as follows (with the names he gave them):

1. (Axiom of extensionality) If, for the sets S and T, $S \subseteq T$ and $T \subseteq S$, then $S = T$.
2. (Axiom of elementary sets) There is a set with no elements, called the empty set, and for any objects a and b in $\mathcal{B}$, there exist sets $\{a\}$ and $\{a, b\}$.
3. (Axiom of separation) If a propositional function $P(x)$ is definite (see below) for a set S, then there is a set T containing precisely those elements x of S for which $P(x)$ is true.
4. (Power set axiom) If S is a set, then the power set $\mathcal{P}(S)$ of S is a set. (The power set of S is the set of all subsets of S.)
5. (Axiom of union) If S is a set, then the union of S is a set. (The union of S is the set of all elements of the elements of S.)
6. (Axiom of choice) If S is a disjoint set of nonempty sets, then there is a subset T of the union of S that has exactly one element in common with each member of S.
7. (Axiom of infinity) There is a set Z containing the empty set such that for any object a, if $a \in Z$, then $\{a\} \in Z$.

Zermelo never discussed exactly why he chose the particular axioms he did. But one can surmise the reasons for most of them. The first axiom merely asserts that a set is determined by its members, and the second axiom was probably motivated by Zermelo's desire to have the empty set as a legitimate set and also to distinguish between an element and the set consisting solely of that element. Similarly, the power set axiom and the axiom of union were designed to make clear the existence of certain types of sets constructed from other types that were used in many arguments. The axiom of separation is Zermelo's method of correcting Cantor's definition of a set as defined by any property, thereby eliminating Russell's paradox. By this axiom, there must be, first, a given set S to which the function describing the property applies and, second, a definite propositional function, a function defined in such a way that the membership relation on $\mathcal{B}$ and the laws of logic always determine whether $P(x)$ holds for any particular x in S. The axiom of infinity was designed by Zermelo to clarify Dedekind's argument as to the existence of infinite sets. That argument had been met with disapproval by many mathematicians, partly because it seemed to be a psychological rather than a mathematical argument. Zermelo thus proposed his own axiom, which asserts that an infinite set can be constructed.

Finally, the axiom of choice, which allows one somehow to "choose" an element from each subset of a given set, was introduced to enable Zermelo to give a proof of both the well-ordering theorem and the trichotomy principle. In addition, Zermelo noted that this axiom had already been applied frequently, without explicit mention, in numerous mathematical arguments. Notwithstanding the fact that the axiom had been used for over thirty years, its publication by Zermelo soon raised a storm of controversy. The essence of the controversy was whether the use of infinitely many arbitrary choices was a legitimate procedure in mathematics. This question soon became part of the broader question as to what methods were permissible in mathematics at all, and if all methods must be constructive. And then of course arose the questions of what constituted a construction and what it meant to say that a mathematical object existed. Mathematicians had rarely debated such points before, but the use of the seemingly innocuous principle of making choices now led to the proof of a result, the well-ordering theorem, of which many mathematicians were skeptical. There was also a wide diversity of opinion in the mathematical community about Zermelo's choice of axioms and therefore about the validity of Zermelo's results. In particular, Zermelo was criticized for not proving his axioms to be consistent. Zermelo admitted that he could not prove consistency, but felt that it could be done eventually. He was convinced, however, that his system was complete in the sense that from it all of Cantorian set theory could be derived.

For any consensus to be possible, two changes had to take place in Zermelo's system. First, the axioms themselves needed to be somewhat modified. On the suggestion of several mathematicians, Zermelo himself in 1930 introduced a new system, now called **Zermelo-Fraenkel set theory** [after Abraham Fraenkel (1891–1965)]. The major change from Zermelo's original system was the introduction of a new axiom, the axiom of replacement, intended to insure that the set $\{\mathbf{N}, \mathcal{P}(\mathbf{N}), \mathcal{P}(\mathcal{P}(\mathbf{N})), \ldots\}$ exists in the Zermelo theory, where $\mathbf{N}$ is the set of natural numbers. This axiom states "If M is a set and if M' is obtained by replacing each member of M with some object of the domain [$\mathcal{B}$], then M' is also a set." As a second change, the nature of a "definite" propositional function had to be clarified, because this was essential to the axiom of separation. It turned out that this clarification had more to do with logic than with set theory, and ultimately it became the accepted view that axiomatic set theory needed to be embedded in the field of logic. For various reasons, there are even today certain schools of mathematicians who do not accept one or more of

Zermelo's axioms. But it is fair to say that the successes achieved within mathematics on the basis of these axioms have convinced the great majority of working mathematicians that the axioms form a workable basis for the theory of sets.

The axiom of choice itself, although probably the most controversial of Zermelo's axioms, turned out to have numerous applications throughout mathematics. For example, it was used to prove that every vector space has a basis and that in a commutative ring every proper ideal can be extended to a maximal ideal. Also, Max Zorn (1906–1993) derived from the axiom of choice a maximal principle, now known as **Zorn's lemma**: If $\mathcal{A}$ is a family of sets that contains the union of every chain $\mathcal{B}$ contained in it, then there is a set A^* in $\mathcal{A}$ that is not a proper subset of any other $A \in \mathcal{A}$. (A **chain** $\mathcal{B}$ means a set of sets such that for every two sets B_1, B_2 in $\mathcal{B}$, either $B_1 \subseteq B_2$ or $B_2 \subseteq B_1$.) Zorn's aim in stating this axiom was, in fact, to replace the well-ordering theorem in various proofs in algebra. He claimed that the latter, although equivalent to his own axiom, did not belong in algebraic proofs because it was somehow a transcendental principle. In any case, Zorn's lemma soon became an essential part of the mathematician's toolbox.

Even though the axiom of choice proved useful, some of its consequences were unsettling and totally unexpected. Among the most surprising of these results was the **Banach-Tarski paradox** first noted in 1924 by Stefan Banach and Alfred Tarski (1901–1983). They proved, using the axiom, that any two spheres of different radii are equivalent under finite decomposition. One can take a sphere A of radius 1 inch and a sphere B the size of the earth and partition each into the same number of pieces, $A_1, A_2, \ldots A_m$ and $B_1, B_2, \ldots B_m$, respectively, such that A_i is congruent to B_i for each i. With results such as this provable using the axiom of choice, there was great interest in clarifying its exact status with regard to the other axioms of set theory. It was certainly not clear that the axiom could not lead to a contradiction.

Zermelo realized that a proof of the consistency of his axioms would be extremely difficult. Although he and others worked on this problem through the 1920s, it was not until 1931 that Kurt Gödel (1906–1978), an Austrian mathematician who spent most of his life at the Institute for Advanced Study in Princeton, showed in essence that there could be no such proof. In fact, he showed that in any system that contains the axioms for the natural numbers—Dedekind's axioms, for example, could be proved in Zermelo-Fraenkel set theory—it was impossible to prove the consistency of the axioms within that system. Gödel also showed that such a system was incomplete, that is, that there are propositions expressible in the system such that neither they nor their negations are provable. Given these results, the only hope for dealing with the axiom of choice was to prove that it was relatively consistent, that is, that its addition to the set of axioms did not lead to any contradictions that would not already have been implied without it. Gödel was able to give such a proof by the fall of 1935. Within the next three years, he also succeeded in showing that the continuum hypothesis was relatively consistent within Zermelo-Fraenkel set theory.

A final result in the determination of the relationship of the axiom of choice to Zermelo-Fraenkel set theory was completed by Paul Cohen (b. 1934) in 1963. Cohen, using entirely new methods, was able to show that both the axiom of choice and the continuum hypothesis are independent of Zermelo-Fraenkel set theory (without the axiom of choice). In other words, it is not possible to prove or to disprove either of those axioms within set theory, and, furthermore, one is free to assume the negation of either one without fear of introducing any new contradictions to the theory. With these and other more recent results, it seems to be the case in set theory, as has already been shown in geometry, that there is not one version but many different possible versions, depending on one's choice of axioms. Whether this will be good or bad for the progress of mathematics is a matter for history to decide.

20.2 Major Questions Answered

The twentieth century saw the answers to several major questions first raised much earlier. In particular, mathematicians were able to prove Fermat's Last Theorem, classify completely all simple groups, and, with the help of computers, prove the four-color theorem. The stories of each of these results will be sketched here.

20.2.1 The Proof of Fermat's Last Theorem

As we have seen, Kummer's idea enabled Fermat's Last Theorem to be proved for many prime exponents p. And during the twentieth century, various other techniques were used to show that the theorem was true for all primes less than 125,000 and that case 1 of the theorem was true for all primes less than 3,000,000,000. But there is a big difference between proving the theorem for finitely many primes and proving it for *all* primes. It was becoming clear that an entirely new approach was necessary.

The new approach involved the idea of an **elliptic curve**. This is a curve of the form $y^2 = ax^3 + bx^2 + cx + d$, where a, b, c, and d are rational and where the cubic polynomial in x has distinct roots. Recall that Diophantus had determined rational solutions to at least one equation of this form, and Euler had studied the situation more generally. By late in the nineteenth century, it was known that there was a definition of "addition" on the rational points on an elliptic curve (including the "point at infinity"), that is, on the set of pairs of rational numbers that solved the equation, which turned this set into an Abelian group $E(\mathbf{Q})$. In the 1920s, Louis Mordell (1888–1972) proved that this group was a finitely generated Abelian group (that is, that all its elements could be written as sums of multiples of finitely many elements); later, Carl Siegel (1896–1981) proved that the set of integral points on such a curve was finite.

Earlier, we considered the so-called modular group of linear fractional transformations $f(z) = (az + b)/(cz + d)$, where $ad - bc = 1$, as transformations on projective space over a finite field. But more generally, the name **modular group** is given to the group of these transformations with integral coefficients acting on the upper half of the complex plane $\{z = x + iy | y > 0\}$. [It is not difficult to check that in this situation $f(z)$ is also in the upper half of the complex plane.] These modular groups and their subgroups had been studied extensively beginning in the late nineteenth century, but it was the Japanese mathematicians Goro Shimura (b. 1928) and Yutaka Taniyama (1927–1958) who first saw a connection between the modular group and elliptic curves. Although a discussion of the exact connection is beyond the scope of this book, Shimura and Taniyama conjectured that every elliptic curve comes in a very definite manner from a modular form, a function from the upper half-plane to the complex numbers that is as invariant as possible under certain subgroups of the modular group. The **Taniyama-Shimura conjecture**, which was made somewhat more precise by André Weil (1906–1998) in the 1960s, is frequently stated simply as "every elliptic curve is modular."

During the late 1960s, Robert Langlands (b. 1936), at the Institute for Advanced Study in Princeton, began to believe that the unification of modular forms and elliptic curves implied by the Taniyama-Shimura conjecture was only one part of a much greater scheme of unification of aspects of number theory and analysis, many other elements of which he began to conjecture. In fact, Langlands proposed what is now known as the **Langlands program**, a concerted effort to prove these unifying conjectures one by one, leading ultimately to a great unification of mathematics. Although the Langlands program is still in its beginning stages, its first great triumph would be the proof of the Taniyama-Shimura conjecture.

In the early 1980s, Gerhard Frey (b. 1944) noted the close relationship between the Taniyama-Shimura conjecture and Fermat's Last Theorem. Suppose a solution $a^p + b^p = c^p$ of the Fermat equation existed for $p > 3$, where we may as well assume that b is even and that $a \equiv -1 \pmod 4$. Frey then considered the elliptic curve $y^2 = x(x - a^p)(x + b^p)$ and, by studying various functions defined on the curve, came to believe that it was impossible for it to exist. In particular, it seemed to him that the existence of the curve would contradict the Taniyama-Shimura conjecture. The exact links between this curve and the conjecture were clarified by 1986 by Jean-Pierre Serre (b. 1926) and Kenneth Ribet. Basically, the three mathematicians established that the Frey elliptic curve was not modular. Thus, the truth of the Taniyama-Shimura conjecture would establish that the Frey curve could not exist, or that Fermat's Last Theorem was true.

Over the next seven years, Princeton mathematician Andrew Wiles (b. 1953), who had been fascinated by Fermat's Last Theorem since reading about it while growing up in England, worked in secret in his attic study to try to establish the Taniyama-Shimura conjecture. Finally, by May of 1993, having developed numerous new techniques in number theory, Wiles believed he had proved the conjecture, at least for a certain class of elliptic curves to which the Frey curve belonged. He therefore arranged to give a series of three lectures at a number theory conference in Cambridge, England, in June of that year. Although he did not state the goal of his lectures at the beginning, the mathematicians in attendance soon understood that a major result would be announced by the end. Thus, when Wiles concluded the third lecture on June 23 by writing the statement of Fermat's Last Theorem on the board and saying "I think I'll stop here," the audience burst into sustained applause.

The dramatic nature of Wiles's announcement notwithstanding, it turned out that there was a flaw in the proof, which was only discovered during the review process of his manuscript. Wiles labored mightily over the next year to correct the flaw, finally enlisting the help of Richard Taylor, one of his former graduate students. And then on September 19, 1994, Wiles had a brilliant new insight and all the pieces of the puzzle came together. Two new manuscripts were soon prepared, one co-authored with Taylor; the normal review process took place, and the May, 1995 issue of *Annals of Mathematics* contained the complete proof of Fermat's Last Theorem.

Wiles's work, besides leading to a proof of a very old conjecture, opened new doors to many topics in number theory, and since 1995, other mathematicians have used his ideas to push ahead. In fact, in 1999, Taylor and others published a proof of the complete Taniyama-Shimura conjecture. Although it is still a mystery as to what proof of the theorem Fermat had in mind, Wiles's twentieth-century proof is certainly one whose ideas will have ramifications into the twenty-first century and beyond.

20.2.2 The Classification of the Finite Simple Groups

Ever since Camille Jordan showed that the alternating group A_n for $n \geq 5$ and certain groups of matrices with coefficients in the field with p elements were simple groups, mathematicians attempted to find other families of such groups. By the end of the nineteenth century, Leonard Dickson (1874–1954) and others had generalized some of Jordan's results. For example, Dickson showed that the projective special linear group $PSL(n, p^k)$ over the field with p^k elements was simple for $n > 1$ except in a few trivial cases. Similarly, Jordan had studied other subgroups of the general linear group over a field of p elements, subgroups defined by their leaving invariant certain bilinear forms. He was able to show that quotients of these groups by their centers, the subgroups that commuted with every element of the group, were also simple. Dickson generalized these results to analogous groups defined over finite fields of order p^k.

The simple groups that Dickson studied, now called the **projective symplectic, orthogonal,** and **unitary groups**, had analogs, the so-called classical groups, when the coefficients of the matrices were allowed to be complex numbers. But there were other families of classical groups of matrices over the complex numbers as well. Dickson was able to show that one of these families of exceptional groups had a finite analog, each of whose members was simple. It turned out that the others too had finite analogs, although these were only discovered by Claude Chevalley (1909–1984) in the 1950s. Meanwhile, near the turn of the twentieth century, Frank N. Cole (1861–1927) and George A. Miller (1863–1951) showed that five groups first discovered by Emile Mathieu (1835–1890) around 1860 were simple groups that were not part of any of the known families. (Such groups are now called **sporadic groups**.) Mathieu, naturally, defined these as permutation groups. For example, one of these groups, now referred to as M_{12}, is the group generated by the following three permutations on a set of 12 elements: $A = (1, 2, 3, 4, 5, 6, 7, 8, 9, 10, 11)$, $B = (5, 6, 4, 10)(11, 8, 3, 7)$, and $C = (1, 12)(2, 11)(3, 6)(4, 8)(5, 9)(7, 10)$. It turns out that this group has order 95,040.

In 1963, Walter Feit (b. 1930) and John Thompson (b. 1932) made a major advance in group theory by showing, in a massive paper in the *Pacific Journal of Mathematics*, that every group of odd order was solvable, and thus that there were no simple groups of odd order besides the cyclic groups of prime order. Using some of the new techniques developed by Feit and Thompson in their paper, group theorists began a major attack on the problem of simple groups. Between 1965 and 1974, twenty-one new so-called **sporadic simple groups** were discovered, the largest being of order approximately 10^{54}. Mathematicians began to wonder, in fact, whether more would continue to be discovered. By 1972, however, Daniel Gorenstein (1923–1992) began to believe the opposite: that there were only finitely many sporadic groups and that a complete classification of the finite simple groups was possible. In fact, he laid out a program for accomplishing this goal. And it was only nine years later, in February, 1981, that he was able to announce that the classification of the finite simple groups was complete. This result, which involved the combined efforts of several hundred mathematicians and whose proof probably covered some 10,000 pages, stated that there were four basic classes of these groups: the cyclic groups of order p, the alternating groups on $n \geq 5$ letters, 16 infinite families of classical groups, and 26 sporadic groups.

The theorem asserting the classification of the finite simple groups is different from a normal mathematical theorem, in that its proof is impossible for any one person to absorb and verify. Nevertheless, since the individual pieces have been checked, and since the arguments made in the various papers that comprise the proof often overlap, most mathematicians now believe in the truth of the result. Still, such a proof stretches our notion of what a proof should be. Further stretching is necessary for the proof of the four-color theorem.

20.2.3 The Proof of the Four-Color Theorem

Once the initial disappointment over the failure of Kempe's attempted proof of the four-color theorem had worn off, mathematicians attacked the problem with increasing vigor in the twentieth century. In fact, it is said that nearly every mathematician in the first half of the century tried to solve this problem. Among the many ideas that soon emerged as crucial to the solution were those of an unavoidable set of regions and a reducible configuration. An **unavoidable set** is a set of regions, at least one of which must always appear in any map. For example, we have seen that the set consisting of a digon, a triangle, a square, and a pentagon is unavoidable. Every map must contain one of these. Another unavoidable set consists of a digon, a triangle, a square, two adjacent pentagons, or a pentagon adjacent to a hexagon. A **reducible configuration** is any arrangement of regions that cannot occur in a minimal map requiring at least five colors. For if a map contains such a configuration, then any coloring of the remainder of the map with four colors can be extended to a coloring of the entire map. We have seen that a digon, a triangle, and a square are all reducible configurations and that the failure of Kempe's argument lay in his inability to prove that a pentagon was reducible.

One of the many mathematicians who sought reducible configurations was George David Birkhoff (1884–1944), who spent most of his career at Harvard. Among many other results on the problem, he showed that the configuration of Figure 20.1(a) was reducible. Actually, by his time, mathematicians had generally converted the map-coloring problem to a problem in graph theory. In other words, a map was converted to a graph whose vertices were a set in one-to-one correspondence with the regions of the map, such that two vertices were joined by an edge if and only if the two corresponding regions had a common boundary arc. (Compare parts a and b of Fig. 20.1.) If we define a coloring of a graph to be an assignment of colors to the vertices so that no two vertices lying on a common edge have the same

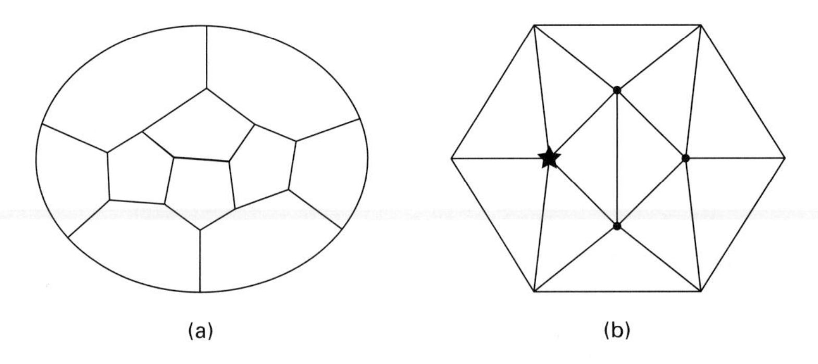

FIGURE 20.1 (a) Reducible configuration of the four-color theorem. (b) Graph derived from the map in part a

color, then the four-color theorem for maps is equivalent to a four-color theorem for graphs. It was thus possible to apply many of the results of the emerging study of graphs to this old problem. In fact, Birkhoff actually stated his reducibility result as "A five-vertex with three consecutive five-neighbors is reducible." [The five-vertex is the vertex in Fig. 20.1(b) marked by a star; the three five-neighbors are marked by dots.]

Given the two notions of unavoidable sets and reducible configurations, the goal toward which Birkhoff and others were working was to find an unavoidable set of reducible configurations, for such a set would provide a proof of the theorem. Every map would have one member of the set, and each of these members would be reducible and therefore could not exist in a minimal map requiring five or more colors. During the first half of the twentieth century, much of the work toward this goal was piecemeal, as attempts to find unavoidable sets and reducible configurations were mostly independent. The first person who advocated a systematic search for an unavoidable set of reducible configurations was Heinrich Heesch (1906–1995). Heesch invented a method for showing that a particular configuration was unavoidable, but initially he feared that an unavoidable set of reducible configurations would contain about ten thousand configurations, each of which would have to be checked. Nevertheless, he began work on the problem and, by 1948, when Wolfgang Haken (b. 1928) heard him lecture on the subject, had checked five hundred.

Nearly twenty years later, Heesch was still working on the problem, having checked thousands more configurations. Haken then suggested that computers might be useful in working through all the detailed calculations necessary. Unfortunately, Haken was not a programming expert, and it soon was clear that without efficient programs, computers in the late 1960s were not fast enough to make substantial progress on the problem. But in 1972, after Haken gave a lecture in which he stated he would quit working on the problem, Kenneth Appel (b. 1932), who was a programming expert, suggested that a computer attack on the problem was in fact feasible at the time. Over the next four years, the pair worked through the problem, first proving that there was an unavoidable set of reducible configurations, each of which would be amenable to a computer check, and then actually finding and checking the set. It helped that early in 1976 Haken's university, the University of Illinois, acquired a faster computer and allowed the pair all the computer time they needed. When Appel and Haken announced in July, 1976 that they had completed the proof, their unavoidable set contained 1936 configurations, but by the time they published the following year, they had succeeded in reducing the number to 1482.

A proof by an essential use of a computer was an entirely new phenomenon in mathematics. Computers, since their introduction, have been used to help mathematicians make conjectures, but the proof of the four-color theorem was the first in which a computer was actually used for the detailed work of constructing a formal proof. As was to be expected, this proof has generated much controversy since its appearance. Many mathematicians still do not accept the proof as valid, because the general standard of acceptance of a proof has always been that it can be checked by many members of the mathematical community. And although a computer program itself can be checked, there is no way for mathematicians to check the various details of the work actually performed by the computer. How this debate over the use of the computer will be resolved is not at all clear. It is possible (though it seems unlikely in the foreseeable future) that the four-color theorem will one day be proved in the traditional way, but in any case, Appel and Haken's proof initiated a new debate over what constitutes a mathematical proof.

20.3 Growth of New Fields of Mathematics

20.3.1 The Statistical Revolution

In general, all of the statistical procedures that mathematicians worked on in the nineteenth century were designed to show relationships in quantities already collected and tabulated. Today, of course, statisticians design experiments to test certain hypotheses. The methods for doing this were developed in the early twentieth century. For example, how do we prove or disprove the claim of the lady tasting tea who says she can taste the difference between tea made by first putting in the milk and tea made by adding the milk afterwards? Or, to take a more serious example, how do we determine whether a certain type of fertilizer will produce bigger crop yields? After all, we know from experience that crop yields vary normally as a result of circumstances that we are often not entirely aware of. So what kinds of experiments can we devise? This question was considered in detail by Ronald Fisher (1890–1962) in his position as chief statistician at a British agricultural experiment station.

As an example of what Fisher did, we will consider his methodology in answering the question about fertilizer. First, we need to design a reasonable experiment. So we divide a field into two strips and each strip into, say, ten blocks. That is, we have ten pairs of blocks, each at essentially the same place in the field, in order to try to remove as many other variables from the problem as possible. Then we treat one block in each pair with fertilizer and leave the adjacent block untreated. After the growing season, we measure the yields. We thus have ten pairs of numbers. Let us subtract the yield without fertilizer from the yield with fertilizer (so we probably get positive numbers). We then have a sample of ten numbers. The question is whether this set is significantly different from a set we would get by taking a random sample from a set of numbers normally distributed with mean zero. The analysis then proceeds by calculating the so-called **Student's *t*-statistic**. This is so named because it was essentially developed by William Gosset (1876–1937) of the Guinness Brewery in Dublin, who used the pseudonym "Student" in his publications on the analysis of experimental data. This statistic enabled reasonable results to be estimated with the use of only a small sample, the types of samples Gosset had to deal with at the brewery.

Having calculated the t-statistic for the fertilizer case, we consult a table calculated from the curves of the t-statistic, which are nearly (but not quite) normal curves. (The larger the sample size, the more closely the t-curves approach the normal curve.) We want to know if our value is sufficiently improbable under the assumption that it comes from a normally distributed set with mean zero. The definition of "sufficiently improbable" is something we have to decide on in advance, but today it usually means with probability less than 0.05 or 0.01. If our calculated value meets this standard, we say that we reject the null hypothesis that there is no significant difference in the yields of the two types of plots. In this case, we conclude that the fertilizer is effective.

The case of the lady tasting tea is a bit simpler. Fisher proposed the following test: The lady will be given eight cups of tea in a random order, four of which are made in one way and four in the other. The lady's task is to divide the eight cups into the correct two sets of four. Fisher noted that there are

$$\binom{8}{4} = 70$$

ways of selecting a set of four out of a set of eight. A person who could not discriminate between the two processes of making tea would have only a 1 in 70 (approximately 0.014) chance of picking the correct set. Thus, if the null hypothesis is that the lady cannot discriminate among cups of tea, then if she does pick the correct set, it is sufficiently improbable (probability less than 0.05) that she has done this by accident that we reject the null hypothesis. Note that if the lady picks three cups correctly, the chances of this happening without her having the claimed ability to discriminate are 16 out of 70, or approximately 0.23. That number is sufficiently large that we would not reject the null hypothesis in that case. Fisher noted that there are other ways of conducting this test, particularly if the lady claims not that she can always distinguish, but that she can do it more often than not. For example, if there are twelve cups, then the probability is again less than 0.05 that, without any discriminating ability, the lady could pick either five or six cups correctly.

In his two major books, *Statistical Methods for Research Workers* (1925) and *The Design of Experiments* (1935), Fisher laid out careful instructions on how to deal with all the major parts of the design and analysis of a statistical experiment. As we noted in the two examples, there are three major parts of this analysis. First, we need a null hypothesis, basically a statement that there is no difference between two situations. This hypothesis must allow us to specify a unique distribution function for the test statistic. Then we have to work out the observations and order them somehow to show their relative deviation from the null hypothesis. How we do this must be determined by experience. In fact, the entire process of formulating a null hypothesis and making observations to determine whether to accept it is an art and cannot be reduced to a mechanical process. As Fisher wrote, "It is, I believe, nothing but an illusion to think that this process can ever be reduced to a self-contained mathematical theory of tests of significance. Constructive imagination, together with much knowledge based on experience of data of the same kind, must be exercised before deciding on what hypotheses are worth testing, and in what respects. Only when this fundamental thinking has been accomplished can the problem be given a mathematical form." Finally, we need a measure of how far these observations differ from the null hypothesis. This measure is usually expressed as a probability that this particular observation could occur, given the null hypothesis. In general, we reject the null hypothesis when this probability is less than a certain predetermined value.

Interestingly, Fisher himself realized that rejecting the null hypothesis does not necessarily prove the efficacy of the particular cause in question. One generally cannot do that on the basis of one experiment, no matter how well designed it may be. But the result of one experiment does give indications that can be confirmed by repeated experiments. As Fisher wrote, "No isolated experiment, however significant in itself, can suffice for the experimental demonstration of any natural phenomenon. In relation to the test of significance, we may say that a phenomenon is experimentally demonstrable when we know how to conduct an experiment which will rarely fail to give us a statistically significant result."

An alternative to Fisher's methods was worked out by Egon Pearson (1895–1980) and Jerzy Neyman (1894–1981) during the late 1920s and the early 1930s. Egon Pearson was the son of Karl Pearson and worked at his father's laboratory at University College, London. Neyman originally worked at the University of Warsaw, before coming to London and later moving to the United States. What they decided was that looking at a single null hypothesis could not give a significant answer. They wanted a statistical test to provide a choice between alternatives. And to give a reasonable answer, they wanted a test that rarely led to error.

As Pearson and Neyman understood Fisher, there was only one kind of statistical error—rejecting the null hypothesis when it is in fact true. This they called an error of the first kind. But they also wanted to consider an error of the second kind: accepting a hypothesis that is false. In their basic use of their method, they considered two hypotheses, say, H_1 and H_2. The assumption is that one of these hypotheses is true. Then one decides on a so-called critical region R and conducts observations to determine whether they fall in R or not. That is, if one makes an observation in R, one rejects H_1 and accepts H_2. If one makes an observation outside R, one accepts H_1 and rejects H_2. Then the probability $P(R|H_1)$ is the probability that one will reject H_1 when it is true. That is the probability of an error of the first kind, and the idea is to make this value small, say, less than 0.05 or 0.01. The probability $1 - P(R|H_2)$ is then the probability of rejecting H_2—and therefore accepting H_1—when H_1 is false. This is an error of the second kind. The value $P(R|H_2)$ is called the **power** of the test, and we want that value as close to 1 as possible. The question is always how to choose the alternative hypothesis H_2 and a critical region. These often involve the choice of a sample size.

For example, in the case of the lady tasting tea, Fisher's null hypothesis, which we will call H_1, is that the probability p of the lady's being able to tell the type of a given cup of tea is equal to $\frac{1}{2}$. This is, of course, equivalent to the statement that the lady cannot discriminate. Neyman and Pearson would then assert an alternative hypothesis H_2, that the lady can discriminate, or that $p > \frac{1}{2}$. In fact, they would also conduct the test differently. In their analysis of the situation, they would give the lady n pairs of cups of tea, with one cup in each pair prepared with tea first and the other with milk first. We will then agree with the lady's claim if the number of pairs correctly identified is at least as great as a given value t specified in advance. Suppose we say that we will agree with the lady's claim if she makes no more than two errors in identifying ten pairs of cups of tea. That is, an observation will be in the critical region R if the number X of correct pairs is equal to 8, 9, or 10. If our observation is in R, then we reject H_1 and accept H_2. We first calculate the probability of an error of the first kind, that is, of rejecting H_1 when it is true. This value is calculated to be

$$P\left(X = 8 \,\middle|\, p = \frac{1}{2}\right) + P\left(X = 9 \,\middle|\, p = \frac{1}{2}\right) + P\left(X = 10 \,\middle|\, p = \frac{1}{2}\right)$$
$$= \binom{10}{8}\left(\frac{1}{2}\right)^{10} + \binom{10}{9}\left(\frac{1}{2}\right)^{10} + \binom{10}{10}\left(\frac{1}{2}\right)^{10}$$
$$= (45 + 10 + 1)\frac{1}{2^{10}} = 0.054688.$$

This probability is not less than 0.05, but it is close. In this case, we may want to accept the result anyway, because of our calculation of an error of the second kind.

To calculate the error of the second kind, we calculate the power $P(R|H_2)$ as a function of the actual probability p of the lady's correctly identifying the tea preparation. In the case of the given critical region, this function is

$$\binom{10}{8}p^8(1-p)^2 + \binom{10}{9}p^9(1-p) + p^{10}.$$

To then determine the power for a given p, it is best to graph this function (Fig. 20.2). We see, for example, that if $p = 0.9$, the power is approximately 0.96. In other words, the probability of rejecting H_2 if in fact H_1 is false is about 0.04. That is, if the lady does

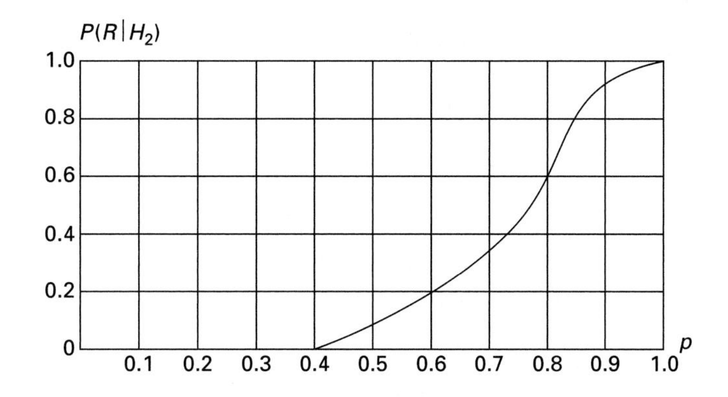

FIGURE 20.2 Graph of $\binom{10}{8}p^8(1-p)^2 + \binom{10}{9}p^9(1-p) + p^{10}$

not identify at least eight pairs correctly, and we therefore accept H_1, the probability of the lady's actually having the claimed discriminatory ability is only 0.04. Since, in this situation, the probabilities of both errors are relatively small, Neyman and Pearson would accept the validity of this hypothesis test. Of course, it is worth noting that the calculation of the power depends on p, and thus we must be very clear about exactly what we are testing. There is rarely a straightforward way to design an experiment with small errors of both kinds.

What happened in the 1930s was that neither Fisher nor Neyman really understood the other. Each published polemics in which he accused the other of reaching false conclusions with his methods. Part of the problem seems to have been that the men's different approaches were applicable to different types of statistical questions. So sometimes one method appeared to work better than the other, but the two sides were reluctant to recognize that. What happened later is that the textbooks over the past fifty years or so merged the two approaches and did not even try to discuss the differences. In fact, most current textbooks state that a researcher must specify the level of significance before conducting the experiment, which was what Neyman and Pearson required; must not draw conclusions from a nonsignificant result, which follows Fisher; and must include the errors of the first and second kinds. But an experimenter who does meet the specified significance level will make a definite conclusion. And Fisher himself warned that one experiment does not allow one to make a conclusion. He even suggested that nonsignificant results should be published as well as significant ones, so that the literature will fairly reflect what has been happening. But this has turned out to be impossible, in particular because of the decisions of journal editors.

A major impetus for the development of better statistical procedures was the requirements of the Second World War. Statisticians were brought in to help make predictions, thus providing a basis for action, in such areas as quality control, personnel selection, gunnery and bombing, and weather forecasting. Among the new techniques developed was the idea of sequential analysis. Instead of deciding between two alternatives, as Neyman and Pearson suggested, the statistician can test sequentially by deciding at various times to make more observations. Thus, sample size is not fixed in advance. The major developer of this theory, couched in terms of statistical decision functions, was Abraham Wald

(1902–1950), who had escaped from Austria in 1938. Among others who worked in this area was David Blackwell (b. 1919), the first African American elected to the National Academy of Sciences. In fact, he systematized the entire field of decision theory in his 1954 textbook *Theory of Games and Statistical Decisions*.

By the 1950s, it was becoming clear that statistical computations were increasingly to be performed on computers. And once this process was well underway, it became possible to analyze larger and larger data sets. Fisher had worked on agricultural problems and developed his methods based on the relatively small data sets that could be dealt with by hand. But with computerization, statisticians had an opportunity to apply statistical methods to the huge data sets being generated in biological and biomedical research. One example of such a data set was in the genome project of the late 1990s, which succeeded in mapping the human genome. In the future, it may well be possible to use the results of this analysis to develop personalized cures for various diseases in individuals.

But even though the use of computers in statistics requires the use of algorithms to perform the analyses, it is still true that, as both Fisher and Neyman suggested, there is always a need for judgment. One cannot write down mechanical rules and then come to definite conclusions by following them. To infer the truth of a statement via statistics is something that can only be done through a combination of experiments, repeated as accurately as possible over a long period of time. Nevertheless, the theoretical advances made by the developers of the statistical methods have allowed mathematicians, using their best judgment, to come to conclusions they have every reason to believe are correct.

20.3.2 Linear Programming

Linear programming deals with the problem of maximizing or minimizing a linear function $a_1x_1 + a_2x_2 + \cdots + a_nx_n$, subject to constraints that are linear inequalities in the variables x_i. Strangely enough, although methods for solving systems of linear equations have been studied for over 2000 years, prior to the Second World War very little attention had been paid to the study of systems of linear inequalities and even less to the study of solutions that maximized a linear function, except for some initial steps by Fourier and a few others in the nineteenth century.

Modern work on linear programming stems from two major sources, military and economic. Among the mathematicians who dealt with economic questions was the Russian Leonid V. Kantorovich (1912–1986), who in 1939 wrote a book entitled *Mathematical Methods in the Organization and Planning of Production*. Kantorovich believed that one way to increase productivity in a factory or an entire industrial organization was to improve the distribution of the work among individual machines, the orders to various suppliers, the different kinds of raw materials, the different types of fuels, and so on. Kantorovich was the first to recognize that these problems could all be put into the same mathematical language and that the resulting mathematical problems could be solved numerically. For various reasons, however, Kantorovich's work was not followed up by Soviet economists or mathematicians, so that the general mathematical solutions to problems such as the ones he explored were first discovered in the United States.

It was the requirements of the Air Force during the war that led to the consideration of linear programming problems in the United States. The specific problems to be dealt with concerned such matters as deployment of particular units to particular theaters of war, scheduling of training for technical personnel, and supply and maintenance of equipment. It soon became clear that the efficient coordination of the various aspects of these problems required new mathematical techniques, which were only developed around 1947. In that year, the Air Force set up a working group called Project SCOOP (Scientific Computation of Optimum Programs), among the principal members of which was George Dantzig (b. 1914). It was he who worked out the basic ideas of the **simplex method** for solving linear programming problems.

The first step is to determine the feasible set of solutions, or the convex polyhedron in the appropriate dimensional space that contains all solutions to the set of linear inequalities. The next step, to move along the edges of this polyhedron from one vertex to another to maximize the linear function, was first rejected by Dantzig as inefficient, but was ultimately seen in fact to provide the most efficient way of determining the desired solution, which is always achieved at one of the vertices. However, in most problems of interest, there were a large number of variables and equations; thus, Dantzig's process required some sort of machine computation. The first test of the simplex method on a major problem was accomplished in the fall of 1947 on an early computer. Over the next several years, various computational techniques were worked out to enable the newly developed computers to be employed in solving linear programming problems with several hundred variables and equations. In fact, the applications of linear programming have grown rapidly over the past decades in parallel with the increasing speed and computational power of modern computers.

20.4 Computers and Mathematics

When a non–technically educated person thinks of mathematics today, the most obvious aspect of the subject that comes to mind is the use of computers. Mathematicians themselves are only gradually accepting the entrance of machine computation into their field. For many mathematicians, pencil and paper are still the most important tools. Yet the rapid advances in computing power since the 1950s have brought the computer into the mainstream of mathematics, and a growing number of mathematicians now make use of it not only in generating examples but also in constructing proofs. In fact, as we have seen, the proof of the four-color theorem, as well as the growth in statistics and linear programming, are heavily dependent on the computer. Thus, we will conclude this chapter with a sketch of the most important aspects of the history of the computer.

20.4.1 The Prehistory of Computers

The dream of mechanical calculation must have occurred as far back as Greek times when Ptolemy was probably forced to use a large number of human "computers" to generate the various tables that appear in his *Almagest*. Some Islamic scientists in the Middle Ages in fact did use certain instruments to help in their own calculations, particularly those related to astronomy. The calculation of astronomical tables was important in Europe as well by the early seventeenth century, and logarithms were invented in part to help in this regard. In short order, two Englishmen, Richard Delamain (first half of the seventeenth century) and William Oughtred, independently created a physical version of a logarithm table in the form of a slide rule, a circular (later rectilinear) arrangement of movable numerical scales, which enabled multiplications and divisions as well as computations involving trigonometric functions to be performed easily. And somewhat later, both Pascal and Leibniz constructed machines to aid in computation. As Leibniz wrote, "it is unworthy of excellent men to lose hours like slaves in the labor of calculation, which could be safely relegated to anyone else if the machine were used."

Unfortunately, neither Leibniz's machine nor the various improved models built by others during the following century and a half were actually used to any extent in the way Leibniz envisaged. The mathematical practitioners themselves continued to do calculations by hand, probably because the machines, operated manually, provided little advantage in speed. For complicated calculations, naturally, tables were used, particularly tables of logarithms and trigonometric functions, even though these tables, originally calculated by hand, frequently contained errors. It was not until around 1821, when the Industrial Revolution was in full swing in England and the steam engine had been invented, that the brilliant mind of Charles Babbage (1792–1871) conceived the idea of using this new technology to drive a machine that would increase the speed as well as the accuracy of numerical computation.

Babbage realized that the calculation of the values of a polynomial function of degree n could be effected by using the fact that the nth-order differences were always constant. To take a simple example, consider Table 20.1 for the function $f(x) = x^2$. Note that in this short table the second differences—that is, the differences of the first differences of the values—are all 2. Thus, to calculate the values of $f(x)$, it is only necessary to perform additions, working backwards from the second difference column to the first difference column to the desired tabular values. (Naturally, one must begin with certain given values, say, $2^2 = 4$ and the first first difference 3.) This idea was the principle behind Babbage's original machine, his Difference Engine (Fig. 20.3). The plans for the machine called for seven axes, representing the tabular values and the first six differences, each axis containing wheels that could be set to represent numbers of up to twenty decimal digits. The axes would be interconnected so that the constant set up in one of the difference axes would add to the number set up in the next lower difference axis, and so on, until the tabular axis was reached. By repeating the process continually, the desired tabular values for polynomial functions of degree up to 6 could be calculated for as many values of the variable as desired.

TABLE 20.1

1st and 2nd Order Differences of a Quadratic Function

x	$f(x)$	First Difference	Second Difference
1	1		
2	4	3	
3	9	5	2
4	16	7	2
5	25	9	2
6	36	11	2

FIGURE 20.3 A modern model of Babbage's Difference Engine. (Source: Neuhart Donges Neuhart Designers, Inc.)

Babbage realized too that any continuous function could be approximated in an appropriate interval by a polynomial and therefore that the machine could be used to calculate tables for virtually any function of interest to scientists of the day. His aim, in fact, was to attach the machine to a device for making printing plates so that the tables could be printed without any new source of error being introduced. Unfortunately, although Babbage succeeded in convincing the British government to provide him with a grant to help in the building of the Difference Engine, a complete model was never constructed, not only because of various difficulties in developing machine parts of sufficient accuracy, but also because ultimately the government lost interest in the project and Babbage himself became interested in a new project, the development of a general-purpose calculating machine, his Analytical Engine.

Babbage began his new project in 1833 and had elaborated the basic design by 1838. His new machine contained many of the features of today's computers. Constructed again of numerous toothed wheels on axes, it was to consist of two basic parts, the **store** and the **mill**. The store was the section in which numerical variables were kept until they were to be processed and where the results of the operations were held; the mill was the section in which the various operations were performed. To control the operations, Babbage devised a system of punched cards, which were to contain both the numerical values and the instructions for the machine. Although Babbage never wrote out a complete description of his Analytical Engine and, in fact, never had the financial resources actually to construct it, he did leave for posterity some 300 sheets of engineering drawings, each about 2 by 3 feet, and many thousands of pages of detailed notes on his ideas. Modern scholars have concluded by examining these papers that the technology of the time was probably sufficient to construct the engine, but because there was insufficient interest on the part of the British government to finance such a massive project, the engine remained only a theoretical construct.

In 1840, Babbage gave a series of seminars on the workings of the Analytical Engine to a group of Italian scientists assembled in Turin, one of whom summarized the seminars in a published article. The seventeen-page article was translated into English in 1843 and supplemented by an additional forty pages of notes by Ada Byron King, Countess of Lovelace (1815–1852). In her notes, Lovelace not only expanded on various parts of the article about the detailed functioning of the engine, but also gave explicit descriptions of how it would solve specific problems. Thus, she described, for the first time in print, what would today be called a computer program, in her case a program for computing the Bernoulli numbers, the numbers that appear as the coefficients B_i in the expansion

$$\frac{x}{e^x - 1} = 1 - \frac{x}{2} + B_2\frac{x^2}{2} + B_4\frac{x^4}{4!} + B_6\frac{x^6}{6!} + \cdots.$$

By a bit of algebraic manipulation using the power series expansion for e^x, Lovelace rewrote this equation in the form

$$0 = -\frac{1}{2}\frac{2n-1}{2n+1} + B_2\left(\frac{2n}{2!}\right) + B_4\left(\frac{2n(2n-1)(2n-2)}{4!}\right) + B_6\left(\frac{2n(2n-1)\cdots(2n-4)}{6!}\right) + \cdots + B_{2n},$$

a form from which the various B_i can be calculated recursively. Thus, to calculate B_{2n}, one needs three numerical values 1, 2, n, as well as the values B_i for $i < 2n$, which presumably have already been calculated. Instruction cards are then needed to get the machine to multiply n by 2, subtract 1 from that result, add 1 to that result, divide the two last results, multiply the result by $-\frac{1}{2}$, divide $2n$ by 2, multiply the result by B_2, and so on. The results of certain of these calculations, such as $2n - 1$, are used several times during the calculations and therefore need to be moved to various registers where the calculations will take place. At certain stages in the calculation, the machine is instructed to subtract an integer from $2n$ and then decide on the next step, depending on whether the result is positive or 0. If it is 0, the equation for B_{2n} is complete and the machine easily solves it; if the result is positive, the machine repeats many of the preceding steps. It is not difficult to see that some of the basic concepts of modern-day programming, including loops and decision steps, are included in Lovelace's description. And to clarify this description, Lovelace wrote out a flow chart giving a diagram of the above program.

Besides discussing the basic functioning of the Analytical Engine, Lovelace described what kinds of jobs it could do and noted explicitly that it could perform symbolic algebraic operations as well as arithmetic ones. But, she noted, "the Analytical Engine has no pretensions whatever to *originate* anything. It can do whatever we *know how to order it* to perform. It can *follow* analysis; but it has no power of *anticipating* any analytical relations or truths.... It is however pretty evident, on general principles, that in devising for mathematical truths a new form in which to record and throw themselves out for actual use, views are likely to be induced, which should again react on the more theoretical phase of the subject." A better description of the computer's limitations and its implications for the development of mathematics could hardly be written today.

20.4.2 Turing and Computability

One of the reasons that Babbage's ideas were not brought to fruition with an actual Analytical Engine was that even in mid–nineteenth-century England, there was no perceived societal need that made it worth the enormous resources that would have been necessary for its construction. And although various computational devices and analog computers were devised in the century after Babbage's design work, devices generally adapted to solving specific mathematical problems that otherwise would require enormous amounts of manual computation, it was military necessities during the two world wars, especially the second, that were to lead to the actual construction of the first electronic computers, in many essentials based on Babbage's ideas. Still, there were other theoretical ideas worked out in the years immediately before the Second World War that were to be fundamental in the development of these computers. One of these was the idea of computability in the work of Alan Turing (1912–1954).

Computability, for Turing, was the problem of determining a reasonable but precise answer to the questions of what a computation is and whether a given computation can in fact be carried out. To answer these questions, Turing extracted from the ordinary process of computation the essential parts and formulated these in terms of a theoretical machine, now known as a **Turing machine**. Furthermore, he showed that there is a "universal" Turing machine, a machine that can calculate any number or function that can be calculated by any special machine, provided it is given the appropriate instructions.

Turing's machine, presented in a major paper of 1936, was formed from three basic concepts: a finite set of states, or configurations, $\{q_1, q_2, \ldots, q_k\}$; a finite set of symbols $\{a_0, a_1, a_2, \ldots, a_n\}$, which are to be read and/or written by the machine (where a_0 is taken as a blank symbol); and a process of changing both the states and the symbols to be read. To accomplish its job, the machine is supplied with instructions by means of an (infinite) tape running through it; the tape is divided into squares, with a finite number of these squares bearing a nonblank symbol. At any given time, there is just one square, say the rth, in the machine, bearing a symbol S_r. To simplify matters further, the possible instructions given by a symbol are limited to replacing the symbol on the square by a new one, moving the tape one square to either the right or the left, and changing the state of the machine. Thus, at any given moment, the pair (q_i, S_r) will determine the behavior of the machine according to the particular functional relationship that defines the behavior. If the function is not defined on a particular pair (q_i, S_r), the machine simply halts. It is the symbols printed by the machine, or at least a determined subset of them, which represent the number to be computed. Turing's contention, backed up by many arguments in his paper, is that the operations above are all those necessary actually to compute a number.

As an example, Turing constructed a machine to compute the sequence 010101 . . . . This machine has four states, q_1, q_2, q_3, q_4, and is capable of printing two symbols, 0 and 1. The tape for this machine is entirely blank initially, and the machine begins in state q_1. The instructions that the machine uses are as follows:

1. If it is in state q_1 and reads a blank square, it prints 0, moves one square to the right, and changes to state q_2.
2. If it is in state q_2 and reads a blank square, it moves one square to the right and changes to state q_3.
3. If it is in state q_3 and reads a blank square, it prints a 1, moves one square to the right, and changes to state q_4.

4. If it is in state q_4 and reads a blank square, it moves one square to the right and changes to state q_1.

It is easy enough to see that this machine does accomplish what is desired, although Turing for technical reasons arranged the printing so that there are figures only on alternate squares. And although this example does not demonstrate it, the reason for the motion of the tape in either direction is to give the machine a memory. Thus, the machine can reread a particular square and act in different ways depending on its particular state at the time. In this way the machine can "remember" numbers written earlier and use them in subsequent computations.

It is the possibility of memory that leads to perhaps the most surprising part of Turing's paper, his proof of the existence of a single machine that can compute any computable number. Turing's idea for this was to take the set of instructions for any given machine, like those written above, and turn that set systematically into a series of symbols, called the **standard description** of the machine. The universal machine is then supplied with a tape containing this standard description followed by the symbols on the input originally supplied to that machine. Turing was in fact able to give a rather explicit description of the behavior of this universal machine in terms of a functional relationship, as described earlier. The main idea is that the machine acts in cycles, each cycle representing, first, a look at the standard description of the particular machine, second, a look at one square of that machine's input, and, finally, a corresponding action. Although Turing did not at the time attempt the physical construction of a machine with the capabilities he proved could exist, it was his idea that led directly to the concept of an all-purpose computer that could be programmed to do any desired computation. Naturally, there are physical limits to the size of a machine and the length of a program—limits that did not exist in Turing's theoretical model with its infinite tape. Modern technology, however, seems to extend these physical limits so often that today's computers are better and better approximations to Turing's universal machine with each passing year.

20.4.3 Von Neumann's Computer

The work of Turing was only one facet of the many theoretical and applied problems that had to be solved before the modern computer could be constructed, and there were numerous people who worked on these problems, particularly during the 1930s and 1940s. One was Claude Shannon (1916–2001), who developed the basic algebra of switching circuits to enable a machine actually to perform additions and multiplications. But the man most responsible for the shape of the ultimate result was probably John von Neumann (1903–1957), who immediately after the Second World War gathered a brilliant group of scientists and engineers at the Institute for Advanced Study in Princeton. Their task was to take the experience developed during the war years in the development of two early computers, the ENIAC and the EDVAC, and combine it with recently developed theoretical knowledge to develop what one of the project's backers called "the most complex research instrument now in existence. . . . Scholars have already expressed great interest in the possibilities of such an instrument and its construction would make possible solutions of which man at the present time can only dream."

The group under von Neumann decided to organize the computer into four main sections: an arithmetic unit, a memory, a control, and an input-output device, the first two being quite analogous to Babbage's mill and store, respectively. The arithmetic unit, now generally called the **central processing unit**, is the place where the machine performs the elementary operations, those operations the designers have decided should not be reduced any further. These elementary operations are essentially wired into the machine, while any other operation is built out of the elementary ones by a set of instructions. Recall that the number system of Babbage's analytical engine was decimal. But with the advent of electronic, rather than mechanical, devices for representing numbers, it turned out that it was simpler to represent numbers in binary form, so that any particular device holding a digit would only need to have two states, on and off, to represent the two possibilities of 1 and 0. Von Neumann was in fact instrumental in designing efficient sets of decimal-binary and binary-decimal conversion instructions so that an operator could enter numbers in the normal decimal mode and receive answers in that mode as well, without compromising the speed and ease of construction of the machine.

The memory unit of the machine needed to be able to take care of two different tasks: storing the numbers that were to be used in the calculations and storing the instructions by which the calculations were to be made. But because instructions themselves could be stored in appropriate numerical code, the machine only needed to be able to distinguish between the actual numbers and coded instructions. Moreover, in order to compromise between the "infinite" memory desired by the user of the machine and the finite memory constructible by the engineer, it was decided to organize the memory in hierarchies, such that some limited amount of memory was immediately accessible while a much larger amount could be accessed at a somewhat slower rate. It was also decided that in order to achieve a sufficiently large memory in a reasonable physical space, the units that stored an individual digit needed to be microscopic parts of some large piece.

The control unit was the section where the instructions to the machine resided, the orders that the machine actually obeyed. Again, compromises had to be worked out between the desire for simplicity of the equipment and the desirability, in terms of speed, of a large number of different types of orders. In any case, one of the more important aspects of the control procedure, an aspect of which even Lady Lovelace was aware, was the ability of the machine to use a given sequence of instructions repeatedly. But because the machine must be made aware of when the repetition should end, it was also necessary to design a type of order that lets the machine decide when a particular iteration is complete. Furthermore, the control unit needed to have a set of instructions that integrated the input and output devices into the machine. Von Neumann was particularly interested, in fact, in assuring that the latter devices would allow for both printed and graphical outputs, because he realized that some of the more important results of a particular computation may best be explored graphically.

The computer eventually constructed at the Institute for Advanced Study, based on von Neumann's design and finished in 1951, proved to be the model for the more advanced computers built in succeeding years. Technological achievements in regard to computers over the remainder of the twentieth century both increased their capacity and decreased their size by factors probably undreamed of by members of the working group of the late 1940s. Computers have now become so much a part of everyday life that we can scarcely imagine how we would accomplish many common tasks without them. And even mathematicians have grown accustomed to using computers to generate new examples and to help verify conjectures. It is clear that in the twenty-first century, computers will play an essential role in the continued development of mathematical ideas.

Exercises

1. Show that each of the following multiplication tables of two basis elements i, j determines an associative algebra of degree 2 over the real numbers. Are there any other tables that do so?

	i	j
i	i	j
j	j	0

	i	j
i	i	j
j	0	0

	i	j
i	j	0
j	0	0

2. Find a nilpotent element in the algebra of 2×2 matrices over the rational numbers.

3. Find an idempotent element (that is not a diagonal matrix) in the algebra of 2×2 matrices over the rational numbers.

4. Find the multiplicative inverse of the element $1 + 2i - 3j + 4k$ in the division ring of the quaternions.

5. The following is the **Richard paradox**, named after its originator, Jules Richard (1862–1956): Arrange all two-letter combinations in alphabetic order, then all three-letter combinations and so on, and then eliminate all combinations that do not define a real number. (For example, "six" defines a real number, but "sx" does not.) Then the set of real numbers definable in a finite number of letters forms a denumerable, well-ordered set $E = \{p_1, p_2, \ldots\}$. Now define the real number $s = .a_1a_2\ldots$ between 0 and 1 by requiring a_n to be one more than the nth decimal of p_n if this decimal is not 8 or 9 and equal to 1 otherwise. Although s is defined by a finite number of letters, it is not in E, a contradiction. How can one resolve this paradox? How is this paradox related to the barber paradox or to Zermelo's original paradox?

6. Show that the trichotomy law follows from Zermelo's well-ordering theorem.

7. Show that Zermelo's axiom of separation resolves Russell's barber paradox as well as the Richard paradox (exercise 5) in the sense that it excludes certain "sets" from discussion.

8. Let $f(z) = (az + b)/(cz + d)$, where a, b, c, d are integers and $ad - bc = 1$. Show that if $z = x + iy$ is in the upper half of the complex plane, then so is $f(z)$.

9. Consider the elliptic curve given by the equation $y^2 = x^3 + 17$. Turn the set of rational points into an Abelian group as follows: If P_1 and P_2 are rational points, first construct the line connecting them. Next, determine the point P_3' where the line intersects the curve again. Finally, let the sum $P_1 + P_2$ be the point P_3, which is the reflection of P_3' in the x-axis. (If $P_1 = P_2$, then take the tangent line at that point to begin the process.) The additive identity for this group will be the point P_0 at infinity. Using this addition, show that the sum of $P_1 = (2, 5)$ and $P_2 = (4, 9)$ is $(-2, 3)$. Find the point that is double $(-2, 3)$.

10. Consider the elliptic curve given by the equation $y^2 = x^3 - 43x + 166$. Using the description of addition on elliptic curves given in exercise 9, calculate all multiples of the rational point $(3, 8)$. Determine the order of this point.

11. Determine the order of the group $PSL(2, 8)$. [Recall that $PSL(n, p^k)$ is the quotient group of $SL(n, p^k)$ by its subgroup consisting of multiples mI of the identity matrix, where $m^n \equiv 1 \pmod{p^k}$.]

12. Determine the order of the group $PSL(3, 4)$. Show that this group has the same order as A_8, the alternating group on eight letters. (It turns out that the two groups are not isomorphic.)

13. In the Mathieu group M_{12} described in the text, calculate AB, BA, AC, and CA.

14. Convert the map of Figure 19.9 into a graph whose vertices are in one-to-one correspondence with the regions of the map and in which two vertices are joined by an edge whenever the corresponding regions have a common boundary.

15. Suppose we modify Fisher's method for verifying the claim of the lady tasting tea so that she has to taste twelve cups and pick out the set of six of one type. Show that the probability is less than 0.05 that she could pick five or six cups correctly if she was totally without any discriminating ability.

16. In the Neyman-Pearson version of the test for the lady tasting tea, show that

$$P\left(X = 8 \,\middle|\, p = \frac{1}{2}\right) = \binom{10}{8}\left(\frac{1}{2}\right)^{10}.$$

17. In the Neyman-Pearson version of the test for the lady tasting tea, verify that the power $P(R|H_2)$ is given as a function of the probability p as

$$\binom{10}{8}p^8(1-p)^2 + \binom{10}{9}p^9(1-p) + p^{10}.$$

18. Show that for a polynomial of degree n, the nth-order differences are always constant.
19. Construct a difference table that would enable Babbage's Difference Engine to calculate the pyramidal numbers, the numbers that are the sums of the triangular numbers and can be considered as representing, say, the number of cannon balls contained in triangular pyramids of given height.
20. Use Lovelace's equation to calculate B_2, B_4, and B_6.
21. Consider a Turing machine with two states q_1, q_2 capable of printing two symbols 0 and 1. Suppose it is defined by the following instructions:
    (a) If the machine is in state q_1 and reads a 1, it prints 1, moves one square to the right and remains in state q_1.
    (b) If the machine is in state q_1 and reads a blank, it prints 1, moves one square to the right and changes to state q_2.

    (Note that there is no instruction for the machine when it is in state q_2.) Suppose the machine begins in state q_1 with a tape whose first square is blank, whose next two squares to the right have 1's, and all of the rest of whose squares to the right are blank, and suppose further that the leftmost 1 is the initial square to be read. Show that the final configuration of the tape will be the same as the initial one except that it will have three 1's instead of two. Interpreting a tape with n 1's as representing the number $n + 1$, show that this Turing machine will calculate the function $f(n) = n + 1$ for n any non-negative integer.
22. Determine a Turing machine that computes the function $f(n) = 2n$.
23. Look up the Banach-Tarski paradox in a text on set theory and discuss its meaning and implications. Do you believe the result? How does this relate to your belief in the truth of the axiom of choice?
24. Find a copy of an early edition of Van der Waerden's *Modern Algebra* text. Compare it with your current abstract algebra text.
25. Discuss the implications of a computer proof of the four-color theorem. Do you believe that a proof using computers is rigorous?

References

Although there is no easily accessible general history of twentieth-century mathematics, each of the topics covered in this chapter is dealt with by one or more excellent works. The history of the theory of vector spaces can be found in Gregory H. Moore, "The Axiomatization of Linear Algebra: 1875–1940," *Historia Mathematica* 22 (1995), 262–303. A good part of twentieth-century algebra, including particularly the theory of algebras, is covered in part three of B. L. Van der Waerden, *A History of Algebra from al-Khwarizmi to Emmy Noether* (New York: Springer, 1985). Gregory H. Moore, *Zermelo's Axiom of Choice: Its Origins, Development, and Influence* (New York: Springer, 1982) deals with the development of set theory early in the century. Two histories of Fermat's Last Theorem are Simon Singh, *Fermat's Enigma: The Epic Quest to Solve the World's Greatest Mathematical Problem* (New York: Walker and Company, 1997) and Amir Aczel, *Fermat's Last Theorem: Unlocking the Secret of an Ancient Mathematical Problem* (New York: Four Walls Eight Windows, 1996). Since neither of these books goes into many technical details, a reader interested in those should consult David Cox, "Introduction to Fermat's Last Theorem," *American Mathematical Monthly* 101 (1994), 3–14, and Fernando Gouvêa, "A Marvelous Proof," *American Mathematical Monthly* 101 (1994), 203–222, as well as the references in both of these articles. There are numerous surveys of the work leading up to the theorem on the classification of finite groups. Among these are Joseph Gallian, "The Search for Finite Simple Groups," *Mathematics Magazine* 49 (1976), 163–179, and Ron Solomon, "On Finite Simple Groups and Their Classification," *Notices of the American Mathematical Society* 42 (1995), 231–239. The history of the four-color theorem can be found in Robin Wilson, *Four Colors Suffice: How the Map Problem Was Solved* (Princeton: Princeton University Press, 2002) and, with more mathematical details, in Thomas L. Saaty and Paul C. Kainen, *The Four-Color Problem: Assaults and Conquest* (New York: Dover, 1986). Two of the best studies of twentieth-century statistics are Gerd Gigerenzer et al., *The Empire of Chance: How Probability Changed Science and Everyday Life* (Cambridge: Cambridge University Press, 1989) (which also contains some work on earlier centuries) and David Salsburg, *The Lady Tasting Tea* (New York: Freeman, 2001). See George B. Dantzig, *Linear Programming and Extensions* (Princeton: Princeton University Press, 1963) for more information on linear programming and its historical development. That work also contains numerous references to original papers in the field. Among the books that provide histories of various facets of computing are B. V. Bowden, ed., *Faster Than Thought* (London: Pitman, 1953), which includes a copy of Lady Lovelace's translation and notes as well as material on British computers of the immediate postwar period; Herman Goldstine, *The Computer from Pascal to von Neumann* (Princeton: Princeton University Press, 1972), which deals primarily with work in which the author was involved in the 1940s and 1950s; and N. Metropolis et al., eds., *A History of Computing in the Twentieth Century* (New York: Academic Press, 1980), which contains the proceedings of a 1976 conference at which many of the pioneers of computing gave presentations.

Hermann Weyl's text, which contains his definition of a vector space, is *Space–Time–Matter*, Henry Brose, trans. (New York: Dover Publications, 1952). The original texts of many of the articles on set theory discussed in this chapter can be found in Jean van Heijenoort, ed., *From Frege to Gödel: A Source Book in Mathematical Logic, 1879–1931* (Cambridge: Harvard University Press, 1967). Many of the original papers on aspects of graph theory, including those important in the four-color problem, are available in Norman Biggs, E. Keith Lloyd, and Robin J. Wilson, *Graph Theory, 1736–1936* (Oxford: Clarendon, 1976). The two major works of Ronald Fisher mentioned in this chapter are *Statistical Methods for Research Workers* (Edinburgh: Oliver and Boyd, 1925) and *The Design of Experiments* (Edinburgh: Oliver and Boyd, 1935). Both books went through numerous editions. Jerzy Neyman's approach to statistics is set forth in his own text, *Probability and Statistics* (New York: Holt, 1950). Ada Lovelace's "Note G" to her translation of L. F. Menabrea, "Sketch of the Analytical Engine Invented by Charles Babbage," is in Philip and Emily Morrison, eds., *Charles Babbage: On the Principles and Development of the Calculator and Other Seminal Writings by Charles Babbage and Others* (New York: Dover, 1961), 225–297.

More information on Noether's mathematics and the developments it inspired is found in James W. Brewer and Martha K. Smith, eds., *Emmy Noether: A Tribute to Her Life and Work* (New York: Marcel Dekker, 1981) and in Auguste Dick, *Emmy Noether, 1882–1935* (Boston: Birkhäuser, 1981). For a discussion of the philosophical implications of the computer-assisted proof of the four-color theorem, see two articles by Thomas Tymoczko: "Computers, Proofs and Mathematicians: A Philosophical Investigation of the Four-Color Proof," *Mathematics Magazine* 53 (1980), 131–138, and "The Four-Color Problem and its Philosophical Significance," *Journal of Philosophy* 76 (1979), 57–83.

APPENDIX

Using This Textbook in Teaching Mathematics

One of the primary goals of learning the history of mathematics is to make use of it in teaching the subject, whether at the primary, secondary, or college level. This textbook includes the history of most of the mathematical topics taught at those levels, so this appendix will present various tools for making use of that history in teaching. First, for each of the standard mathematics courses taught in secondary school or in the undergraduate years, there is a list of the topics generally covered and the section of this text that covers the history of that topic. Of course, in any given school situation, the topics may be somewhat different, or topics listed in one course may be taught in a different one. Nevertheless, it should be simple to match your courses with what is listed here. This list will also be useful in designing a history of mathematics course by theme. Next, various ideas on how to use the history of mathematics in teaching mathematics are presented. Finally, there is a time line of the history of mathematics along with major world events, to help teachers and their students relate the history of mathematics to world history in general.

Courses and Topics

Pre-Algebra

Number systems, place value, decimals	1.1.2, 1.2.2, 6.1, 7.1, 9.3.1
Arithmetic algorithms	1.1.2, 1.2.2, 1.2.4, 4.1, 5.1, 5.3.1, 6.1, 9.1.1
Lengths, areas, and volumes	1.1.4, 1.2.3, 3.1.1, 8.1.1
Pythagorean theorem	1.2.4, 5.2.1, 6.2

Elementary Algebra

Algebraic symbolism	4.1, 9.1.1, 9.1.2, 10.1.1
Linear equations, proportions	1.1.3, 1.2.5
Systems of linear equations	1.2.5, 5.3.1

Algebraic manipulations, laws of exponents	2.2.2, 4.1.2, 7.2.3, 9.1.1, 9.1.2
Quadratic equations	1.2.5, 2.2.2, 6.3, 7.2.1, 7.2.2

Geometry

Figurate numbers, Pythagorean triples	2.1.1
Logical arguments and proof	2.1.2, 3.1.4, 4.2
Axioms for geometry	2.2.1, 7.4.1
Constructions	2.2.1, 2.2.2
Theorems on triangles and parallelograms	2.2.1
Theorems on circles	2.2.3
Ratio and proportion, similarity	2.2.4
Areas and volumes	5.2.2, 6.2
Regular solids	2.2.6

Intermediate Algebra

Algebraic symbolism	9.1.5
Quadratic equations and systems	1.2.5, 8.3.1, 8.3.2, 9.1.5
Polynomial algebra	7.2.3
Solving polynomial equations	5.3.2, 7.2.5, 9.1.1, 9.1.3, 9.1.5, 10.1.2, 10.2.2
Systems of linear equations	14.1
Irrational numbers	2.1.1, 2.2.4, 2.2.6
Elementary number theory	2.2.5
Complex numbers	9.1.4, 17.3.1
Elementary combinatorics	6.4, 7.3.1, 7.3.2, 8.2.1, 8.2.2

Trigonometry

Trigonometric functions	3.3.2, 3.3.3, 6.5, 7.5.1, 7.5.3, 8.1.2
Solving plane triangles	3.3.4, 9.2.3
Solving spherical triangles	3.3.5, 7.5.2, 9.2.3
Problem solving using trigonometry	3.3.1, 3.3.5, 8.1.2, 9.2.3

Elementary Probability and Statistics

Basic laws of probability	10.3.1, 13.1.1
Probability calculations	10.3.2, 13.1.2, 13.2.2
Binomial distribution and normal curve	13.1.2, 13.2.1, 13.2.4, 18.1.2, 18.2
Mean and standard deviation	18.2

Precalculus

Calculus

Linear Algebra

Differential Equations

Solutions of ordinary differential equations	11.5.4, 12.1.1, 12.1.3, 12.2.1
Applications to physics	12.1.2
Partial differential equations	12.2.3, 17.1.6
Fourier series	12.2.3, 17.1.6

Abstract Algebra

Congruences	5.4, 6.3, 14.3.2, 16.1.1
Basic number theory	10.4
Groups	16.2.2, 16.2.5, 16.3.1, 16.3.2, 16.3.3, 16.3.4, 20.2.2
Fermat's Last Theorem	10.4, 14.3.1, 16.1.2, 20.2.1
Galois theory	14.2, 16.2.1, 16.2.3, 16.2.4
Algebraic numbers and fields	16.1.1, 16.1.2, 16.3.5
Rings and ideals	16.1.2, 20.1.2

Modern Geometry

Attempted proofs of the parallel postulate	7.4.1, 15.1.1, 15.1.2
Non-Euclidean geometry	19.1.1, 19.1.2, 19.1.3
Graph theory	15.3, 19.3, 20.2.3

Advanced Calculus

Set theory	17.2.3, 20.1.3
Development of integers and real numbers	17.2.1, 17.2.2, 17.2.4
Functions	17.1.6
Limits and continuity	17.1.1, 17.1.2
Convergence	17.1.3, 17.1.8
Riemann integral	17.1.7

Complex Analysis

Complex functions	17.3.2
Cauchy-Riemann equation	17.3.2
Cauchy integral theorem	17.3.2
Riemann zeta function	17.3.3

Sample Lesson Ideas for Incorporating History

There are numerous ways to incorporate history into the teaching of mathematics. Four general ways have been described by Man-Keung Siu in an article entitled "The ABCD of Using History of Mathematics in the (Undergraduate) Classroom" in Victor Katz, ed., *Using History to Teach Mathematics: An International Perspective* (Washington: MAA, 2000), 3–9. And although Siu writes about the "undergraduate" classroom, his ideas are certainly worthwhile for the secondary classroom as well. His four ways are A (Anecdotes), B (Broad Outline), C (Content), and D (Development of Mathematical Ideas). To illustrate what he means, here are some examples.

Anecdotes

There are numerous anecdotes about mathematicians that add spice to a class and a little entertainment. They may introduce a human element or forge some links with cultural history or underline some particular concept. Some of these anecdotes are described in this text, but many others are available in online sources or in the *Mathematical Circles* books of Howard Eves. For example, there is the story of Galois's duel and how he wrote down a lot of his theory in a letter to a friend the night before he was killed. There is the story of the mathematical contests in fifteenth century Italy during the period of discovery of the solution method for cubic equations. There is the story of Archimedes jumping out of his bath and running through the streets of Syracuse shouting "Eureka" upon his discovery of the laws of hydrostatics. And there is the story of Wiles and his seven-year-long quest in his attic to find the proof of Fermat's Last Theorem.

Broad Outline

History can be used to sketch a broad outline of a subject, before you plunge into the teaching of it. For example, before teaching trigonometry, you might sketch its origins in Greece in the search for ways to solve spherical triangles in order to predict heavenly phenomena. Then you can look at how trigonometry traveled to India, where new trigonometric functions were developed, then to the Islamic world, where all the six functions were tabulated and many better methods were devised to solve plane and spherical triangles. Finally, you can note that trigonometry was reintroduced to Europe in the work of several mathematicians, all of whom learned significant ideas from their Islamic predecessors.

As another example, before teaching quadratic equations, you can discuss the geometric algebra of the Babylonians and the Greeks as they learned to solve problems that we consider "quadratic." You can then move on to Islam and al-Khwarizmi's first text in the subject. You could also discuss how later Islamic writers used al-Khwarizmi's algorithms applied to new kinds of numbers. Finally, you can look at Descartes's treatment of the geometric meaning of the solution of quadratic equations as well as at Galileo's use of quadratic methods in his study of the paths of projectiles.

Or, before beginning the study of calculus, you might describe some of the methods developed in Greece, India, and Islamic countries to solve problems that were later clarified and extended by Europeans in the seventeenth century. Students should become aware of the numerous problems that were solved earlier and know why the algorithms of calculus made the frequently ad hoc methods of the past obsolete.

Content

The history of a particular mathematical topic can help students understand some subtle mathematical ideas. For example, Augustin-Louis Cauchy's incorrect proof of the theorem that the limit of a sequence of continuous functions is continuous provides an ideal opportunity for students in advanced calculus to learn that even great mathematicians could be wrong. If you ask the students to read that proof, and then work through an example, such as a Fourier series, showing that the result cannot be true in general, they will gain a greater appreciation of the necessity for more careful definitions. They may even figure out on their own a way to repair Cauchy's argument.

Euler's discussion of Cramer's paradox provides an ideal entry into the notion of the "rank" of a system of equations. Students can try to figure out how the paradox can be explained and what this means for the solution of certain systems. They can then work through the theorem of Dodgson and some relevant examples to try to figure out how to tell the nature of the solution set of a particular system. With enough experience in solving systems, students will find the notion of rank and the notions of linear dependence and independence much more natural.

On a more elementary level, the development of the quadratic formula is an ideal place to use history. Many students have more success understanding the formula once they realize that a quadratic equation is a statement about real squares and rectangles. Thus, if they work their way in detail through a Babylonian argument (or an Islamic argument) for the solution of a type of quadratic equation by manipulating geometric objects, under your guidance, they may be able to construct at least a version of the quadratic formula on their own.

As a final example, the introduction of complex numbers is often difficult for students, because in one year they are told that negative numbers do not have square roots and in the next year they are told that they do. It may seem to them that mathematical rules are arbitrary. But if you develop the cubic formula and show students how complex numbers appear naturally in the solutions of certain equations with real coefficients and real solutions, they may begin to believe that complex numbers are necessary. Then if you further develop the geometric notion of a complex number, as done originally by Wessel, students may, like mathematicians of the nineteenth century, start to believe that complex numbers are perfectly "real" objects with their own geometric rules of manipulation, rules that can be turned into arithmetic rules as well.

Development of Mathematical Ideas

For some subjects, it can prove worthwhile to organize a course or a large chunk of a course by bearing in mind the historical evolution of the subject (assuming that we can find out—at least in outline—what that is). To do this, we need to look at the history of the subject, noting the high points and the developments that seemed necessarily to precede other developments; to keep in mind what concepts have historically been especially difficult to understand; and then to try to order the topics of the course so that they agree as closely as possible with this history.

As one example, consider trigonometry. In a standard trigonometry course, one begins by defining the sine and cosine, then calculates these values for 30°, 45°, and 60° via some elementary geometry, and then tells the students to use their calculators if they want to calculate the sine of 37°. Thus, when the sum and difference formulas or the half-angle formula are discussed later, they have no meaning for the students. It would seem more reasonable to try to develop the subject based on its development by Ptolemy 1850 years ago.

To do this, we begin with the basic trigonometry definitions (noting that here we do differ from Ptolemy, who used chords rather than sines, cosines, and tangents). But then we calculate the sine values by use of geometry. We do not just calculate values for 30, 45, and 60, but develop them for 72, 18, 36, and 54, by use of similarity principles and the quadratic formula. We next develop, again using geometry, the half-angle formula and the sum and difference formulas and use these to work out trigonometry tables. Now, given that students have calculators, we can have them use those, but just to do arithmetic, including square roots. And it is useful to ask them to carry out all the calculations to eight or ten decimal places, since that is what their calculators do anyway. It becomes clear to the students that, first, the basic formulas are developed to help them calculate trigonometric values, and, second, that they cannot find the sine of 1 degree via this method. Since it would be nice to be able to find such a value, we use approximation, beginning with the observation that the sine is nearly linear for small values. Students can then easily get a six-place approximation for the sine of 1 degree and then, in theory, can complete a table of sines.

The next stage of trigonometry involves solving plane triangles. This is done rather as in any trigonometry text, by first dealing with right triangles and then developing the laws of sines and cosines to do other triangles. Various examples of practical reasons for solving triangles can be given. But in keeping with the idea of historical development, one then includes a section on spherical trigonometry, where, depending on the class, one can either derive the basic formulas for right triangles or just state them. These can then be applied to two types of problems: the astronomical problems of determining the time of sunrise and sunset and the geographical problems of determining distances and angles on the surface of the earth. For this latter problem, it is necessary to be able to solve non-right spherical triangles. So students can break up the triangles into right triangles and use the earlier formulas or develop new formulas. In particular, students should understand that in the spherical case, knowing the three angles determines the triangle. That is, there is no concept of "similarity" for spherical triangles. With this development, students have a firm idea of the ideas of trigonometry and their importance.

If one wants to go further, one can also develop some analytic trigonometry, as it was done in the seventeenth and eighteenth centuries. It is not particularly difficult even to develop the power series for the sine and the cosine, assuming a knowledge of the binomial theorem and a willingness to use infinitely large and infinitely small numbers.

To take one other example, consider the case of abstract algebra on the undergraduate level. Most such courses start with the definition of a group, even though, historically, this notion was only developed after mathematicians had spent many decades studying examples. Thus, a course that focuses on the historical evolution of the subject should begin with examples. One place to start is with the idea of the complex numbers as developed through the cubic formula of Cardano. As a first example of a group, one can have students consider the groups of complex roots of unity. Many of the basic theorems on Abelian groups can be proved first using those particular groups. One can next move on to modular arithmetic, as developed by Gauss and others. Students can study the basic properties of the fields with a prime number of elements. Then, as Galois did, they can look at polynomials over such fields. The "imaginary" roots of these polynomials enable students to construct the finite fields of prime power order. Again, these provide substantial examples of various concepts of group theory.

Since the examples so far are all commutative, it is necessary to study other examples. So we can work our way through Lagrange's analysis of the solution methods for cubic and quartic polynomial equations, and develop the notion of a permutation of the roots of an equation. We can then have students study permutations in general, and again prove important results, many of which will be essentially the same as results proved in the commutative examples studied earlier. One other useful object to study at this point is the set of transformations of a geometric object, such as a square, a tetrahedron, or even a dodecahedron. The entire set of such transformations can be thought of as a "sub-object" of the set of permutations of the vertices. But these are nice concrete geometric examples that students can hold in their hands.

It is only after a thorough study of the various examples that one should introduce the abstract notion of a (finite) group. By this point, the students will be familiar enough with the basic theorems in several cases that they will understand how these results can be applied to the general case of an abstract finite group. At this point, one can proceed as in standard abstract algebra courses by considering, for example, the Sylow theorems and seeing how these determine the possible groups of various orders. Although what has been outlined here is only, perhaps, a one-semester course, there are many ways to follow the historical development of other abstract algebra concepts in a second semester. [Note that Saul Stahl's text *Introductory Modern Algebra: A Historical Approach* (New York: Wiley, 1996) embodies this approach.]

I hope that the examples given will enable teachers and prospective teachers to work out many other specific ways of using the history of mathematics in teaching mathematics. There are many other suggestions to think about in the discussion questions in the text, and there will be further examples available on the web site for the text, www.aw.com/katz.

Time Line

The time line on pages 529–535 lists significant world political and cultural events that occurred in the centuries from 3000 BCE to the twentieth century, as well as the mathematical ideas that developed as those events were taking place.

Date	Political Events			Cultural Events	Mathematical Ideas	
	Asia	Africa	Europe		Africa/Asia	Europe
3000 BCE	Sumerian kingdoms emerge in what is now Iraq; Harappan civilization exists in India.	Upper and Lower Egypt united (c. 3100 BCE).		Pyramids built at Giza; writing develops in Mesopotamia and Egypt.	Base-60 place value system develops in Mesopotamia; Egyptians develop base-10 grouping system for numbers.	
2000	Aryan tribes move into India from the northwest; Shang dynasty established in China.	Egyptian empire expands into western Asia.	Major civilization develops on Crete, later destroyed by earthquakes and invasions; Trojan war fought c. 1200 BCE.	Hammurapi establishes law code in Babylon (c. 1790 BCE).	Egypt: Rhind and Moscow papyri; unit-fraction calculations; areas and volumes. Babylonia: Pythagorean triples; quadratic equations.	
1000	King David establishes capital in Jerusalem (c. 1000 BCE); Persians conquer and rule Southwest Asia and Egypt; Zhou dynasty in China.		Greece settled by invaders from the north; Roman republic founded.	Confucius flourishes c. 600 BCE.	India: Pythagorean theorem; circle measurements. China: Pythagorean theorem.	Thales, first to write about mathematical proof; Pythagoras and "all is number"
500			Greek city-states flourish.	Vedas codified in India; classical age of Greece.		Hippocrates and quadrature of lunes, duplication of cube; discovery of incommensurability

Date	Political Events			Cultural Events	Mathematical Ideas	
	Asia	Africa	Europe		Africa/Asia	Europe
400		Carthage is a major power in North Africa.	Alexander the Great (356–323 BCE) conquers Egypt and Southwest Asia.	Foundation of Roman legal system		Plato, Theaetetus, Eudoxus, Aristotle
300	Seleucids rule Babylonia; Qin unifies China; Ashoka popularizes Buddhism throughout India.	Ptolemies rule in Egypt.	Punic wars begin.	Library and Museum founded in Alexandria.	Euclid; Apollonius	Archimedes
200	Han dynasty begins in China.		Romans extend rule throughout the Mediterranean.		Hipparchus and beginning of trigonometry	
100		Cleopatra is defeated (30 BCE); Egypt becomes Roman province.	Roman empire established.	Birth of Jesus; paper made in China.	*Nine Chapters* compiled in China.	
0	Fall of second temple in Jerusalem (70).		Roman empire reaches its greatest extent.	Paul's missions transform Christianity into a widespread religion.	Heron and applied mathematics	
100 CE				New Testament canonized.	Ptolemy and the *Almagest*	

200			Barbarian invasions begin.		Diophantus and the *Arithmetica*; Liu Hui and surveying
300	Gupta dynasty in India		Christianity becomes state religion in Roman empire; division of Roman empire into West and East.	Council of Nicaea (325)	Pappus
400			Fall of Rome		Death of Hypatia; Aryabhata and sine tables
500				Birth of Muhammad	Volume of sphere in China
600	Muslims conquer Middle East.	Muslims conquer North Africa.			Brahmagupta and indeterminate equations
700	Baghdad founded as capital of Islamic empire (762).		Muslims conquer Spain; Battle of Tours (732)		Development of decimal place value system in India
800			Charlemagne crowned Holy Roman emperor (800).	House of Wisdom established in Baghdad.	Al-Khwarizmi and algebra

Date	Political Events			Cultural Events	Mathematical Ideas	
	Asia	Africa	Europe		Africa/Asia	Europe
900			Beginnings of English and French kingdoms	Russians converted to Christianity.	Abu Kamil and algebra; Abu'l-Wafa and spherical trigonometry	
1000	First crusade establishes Christian rule in Jerusalem (1095–1099).		William the Conqueror conquers England.		Egypt: Ibn al-Haytham, sums of powers, and volumes of paraboloids. Persia: Omar Khayyam and solution of cubics. China: Pascal triangle developed.	
1100	Northern India conquered by Muslims; Christian kingdoms established in Middle East.	Islam expands into sub-Saharan Africa.		Gothic art and architecture flourish.	Bhaskara and Pell equation; al-Samaw'al and development of decimals and polynomial algebra; Sharaf al-Din al-Tusi and cubic equations	Translations of mathematical works from Arabic into Latin; Abraham ibn Ezra and combinatorics

1200	Constantinople sacked by Crusaders (1202–1204); Genghis Khan conquers much of Asia; Marco Polo reaches China; Muslim sultanate founded in India; Muslims reestablish rule in Middle East.		Christian Kings defeat Muslims in Spain.	Magna Carta (1215) establishes foundation of English constitutional rights; paper produced in Italy; Universities founded in Paris, Cambridge, and Oxford.	Persia: Nasir al-Din al-Tusi and trigonometry. China: Chinese remainder theorem; solution of polynomial equations. Morocco: development of combinatorics.	Leonardo of Pisa
1300	Ming dynasty founded in China.	Timbuktu flourishes as intellectual center in Mali.	Hundred Years' War between England and France	Black Death ravages Europe.		Levi ben Gerson and combinatorics; Mertonian school and kinematics
1400	Ottomans overthrow Byzantine empire; voyages of Chinese to India and Africa	Bartholomew Dias rounds Cape of Good Hope.	Beginning of Renaissance in Italy; voyages of Columbus	Leonardo da Vinci; Gutenberg prints Bible (1456).	Power series for sine and cosine in India; al-Kashi and decimal calculations	Luca Pacioli and the *Summa*
1500	Portuguese establish trading posts in India.		Spanish colonize the Americas; defeat of Spanish Armada	Luther and the Protestant Reformation; Portuguese explorers reach China; William Shakespeare		Solution of cubic equations; Copernicus and heliocentrism; numerous algebra texts written; Viète and algebraic symbolism

Date	Political Events			Cultural Events	Mathematical Ideas	
	Asia	Africa	Europe		Africa/Asia	Europe
1600			English colonies founded in North America; English Civil War and Glorious Revolution; Royal Society and Académie des Sciences founded in London and Paris.			Galileo and physics; Kepler and his laws; Newton and the *Principia*; Napier and logarithms; development of algebraic symbolism; analytic geometry and calculus; early work in probability
1700			American and French Revolutions	Founding of École Polytechnique; invention of steam engine (1769)		Euler; ordinary and partial differential equations; calculus of several variables; Bayes, Laplace, and statistical inference
1800	British rule India.	Africa carved up into European colonies.	Napoleon conquers Europe, then is defeated at Waterloo (1815); Congress of Vienna; revolutions of 1848; founding of German empire	Railroads established; steamships sail the oceans; telephone and telegraph invented.		Rigor in analysis; groups and fields; Galois theory; non-Euclidean geometry; growth of statistics

1900	Chinese overthrow emperor; Ottoman empire collapses; Europeans carve up Middle East; India gains independence; Communist rule in China begins; Korean and Vietnamese wars; Israeli-Arab conflict begins and continues.	Colonies gain independence.	First World War; German, Austrian, Russian empires collapse; Bolshevik revolution in Russia; Nazi period in Germany; Second World War; collapse of Soviet Union	Invention of airplane; artificial satellites; travel to moon		Abstraction in algebra; statistical methodology; computer revolution

Answers to Selected Problems

CHAPTER ONE

1. $\begin{smallmatrix}|||\\||\end{smallmatrix}$ ∩∩𓍢; YY $\begin{smallmatrix}YYY\\YY\end{smallmatrix}$

3. $99\,\overline{2}\,\overline{4}$

5. $5 \div 13 = \overline{4}\,\overline{13}\,\overline{26}\,\overline{52}$; $6 \div 13 = \overline{4}\,\overline{26}\,\overline{52} = \overline{4}\,\overline{8}\,\overline{13}\,\overline{104}$; $8 \div 13 = 2(4 \div 13) = \overline{2}\,\overline{13}\,\overline{26}$

7. 9

13. $18 \leftrightarrow 3{,}20$; $32 \leftrightarrow 1{,}52{,}30$; $54 \leftrightarrow 1{,}6{,}40$; $1{,}04 \leftrightarrow 56{,}15$. If the only prime divisors of n are 2, 3, and 5, then n is a regular sexagesimal.

17. The correct formula in the first case gives $V = 56$, while the Babylonian version gives $V = 60$, for an error of 7%. In the second case, the correct formula gives $V = \frac{488}{3} = 162\,\frac{2}{3}$, while the Babylonian formula gives $V = 164$, for an error of 0.8%.

21. $12\,\overline{\overline{3}}\,\overline{15}\,\overline{24}\,\overline{32} = 12\,\frac{129}{160}$; $\left(12\,\frac{129}{160}\right)^2 = 164.0000391$

23. $(67319, 72000, 98569)$

25. $\ell = 32$; $w = 24$

27. 1 *mina*, $15\frac{5}{6}$ *gin*

31. $x = 3\frac{1}{2}$, $y = 2\frac{1}{3}$

CHAPTER TWO

5. $8 \cdot \dfrac{n(n+1)}{2} + 1 = 4n^2 + 4n + 1 = (2n+1)^2$

19. 9

21. For $46 : 6$, the calculation is $46 = 7 \cdot 6 + 4$; $6 = 1 \cdot 4 + 2$; $4 = 2 \cdot 2$. For $23 : 3$, the calculation is $23 = 7 \cdot 3 + 2$; $3 = 1 \cdot 2 + 1$; $2 = 2 \cdot 1$.

27. Since BC is the side of a decagon, triangle EBC is a 36-72-72 triangle. Thus, $\angle ECD = 108°$. Since CD, the side of a hexagon, is equal to the radius CE, it follows that triangle ECD is an isosceles triangle with base angles equal to 36°. Thus, triangle EBD is a 36-72-72 triangle and is similar to triangle EBC. Therefore, $BD : EC = EC : BC$, or $BD : CD = CD : BC$, and the point C divides the line segment BD in extreme and mean ratio.

29. Circumference = 250,000 *stades* = 129,175,000 ft = 24,465 miles; diameter = 79,577.5 *stades* = 41,117,680 ft = 7787 miles

CHAPTER THREE

3. Let d be the diameter of the circle, t_i the length of one side of the regular inscribed polygon of $3 \cdot 2^i$ sides, and u_i the length of the other leg of the right triangle formed from the diameter and the side of the polygon. Then

$$\frac{{t_{i+1}}^2}{d^2} = \frac{t_i^2}{t_i^2 + (d+u_i)^2}$$

or

$$t_{i+1} = \frac{dt_i}{\sqrt{t_i^2 + (d+u_i)^2}} \qquad u_{i+1} = \sqrt{d^2 - {t_{i+1}}^2}.$$

If P_i is the perimeter of the ith inscribed polygon, then $\dfrac{P_i}{d} = \dfrac{3 \cdot 2^i t_i}{d}$. So let $d = 1$. Then $t_1 = \dfrac{d}{2} = 0.5$ and $u_1 = \dfrac{\sqrt{3}d}{2} = 0.8660254$. Repeated use of the algorithm gives $P_9 = 3.141590016$.

5. Let the equation of the parabola be $y = -x^2 + 1$. Then the tangent line at $C = (1, 0)$ has the equation $y = -2x + 2$. Let the point O have coordinates $(-a, 0)$. Then $MO = 2a + 2$, $OP = 1 - a^2$, $CA = 2$, $AO = 1 - a$. So $MO : OP = (2a+2) : (1 - a^2) = 2 : (1 - a) = CA : AO$.

7. Since $BOAPC$ is a parabola, we have $DA : AS = BD^2 : OS^2$, or $HA : AS = MS^2 : OS^2$. Thus, $HA : AS =$ (circle in cylinder) : (circle in paraboloid). So the circle in the cylinder, placed where it is, balances the circle in the paraboloid placed with its center of gravity at H. Since the same is true whatever cross section line MN is taken, Archimedes can conclude that the cylinder, placed where it is, balances the paraboloid, placed with its center of gravity at H. If we let K be the midpoint of AD, then K is the center of gravity of the cylinder. Thus, $HA : AK =$ cylinder : paraboloid. But $HA = 2AK$. So the cylinder is double the paraboloid. But the cylinder is also triple the volume of the cone ABC. Therefore, the volume of the paraboloid is $\frac{3}{2}$ the volume of the cone ABC, which has the same base and same height.

11. Suppose the cylinder P has diameter d and height h, and suppose the cylinder Q is constructed with the same volume but with its height and diameter both equal to f. It follows that $d^2 : f^2 = f : h$, or that

$f^3 = d^2h$. It further follows that one must construct the cube root of the quantity d^2h, and this can be done by finding two mean proportionals between 1 and d^2h, or, alternatively, two mean proportionals between d and h (the first of which will be the desired diameter f).

13. Focus of $y^2 = px$ is at $\left(\frac{p}{4}, 0\right)$. Length of latus rectum is therefore $2\sqrt{p \cdot \frac{p}{4}} = 2 \cdot \frac{p}{2} = p$.
19. If the two parallel lines are $x = 0$ and $x = k$ and the perpendicular line is the x-axis, then the equation of the curve satisfying the problem is $y^2 = px(k - x)$ or $y^2 = kpx - px^2$.
21. crd $30° = 31; 03, 30$; crd $15° = 15; 39, 47$; crd $7\frac{1}{2}° = 7; 50, 54$
23. crd $12° = 12; 32, 36$; crd $6° = 6; 16, 49$;
crd $3° = 3; 08, 29$; crd $1\frac{1}{2}° = 1; 34, 15$;
crd $\frac{3}{4}° = 0; 47, 07$
25. $\lambda = 60°$, $\rho = 35°47'$; $\lambda = 90°$, $\rho = 63°46'$
29. $40°53'$
31. $\lambda = 45°$: sun is $28°23'$ from zenith; $\lambda = 90°$: sun is $21°9'$ from zenith
33. $29°59'$ north of east

CHAPTER FOUR

1. 84
3. $72\frac{1}{4}$, $132\frac{1}{4}$
5. 13, 3
11. Assume that the theorem is true, and that line AB is cut at C. Then $AB^2 + BC^2 = 3AC^2$. But since $AB = AC + BC$, we have $(AC + BC)^2 + BC^2 = 3AC^2$. This reduces to $AC^2 + 2AC \cdot BC + 2BC^2 = 3AC^2$ or $AC \cdot BC + BC^2 = AC^2$. This in turn implies that $BC(AC + BC) = AC^2$ or that $AB \cdot BC = AC^2$. But this is precisely the statement that AB is cut in extreme and mean ratio at C.
13. 336
15. $A: 15\frac{5}{7}$, $B: 18\frac{4}{7}$

CHAPTER FIVE

1. [illegible]
3. $51\frac{41}{109}$, $32\frac{12}{109}$, $16\frac{56}{109}$
5. $10\frac{15}{16}$ pounds
7. The side of the square is 250 *pu*.
13. $\frac{9}{25}$, $\frac{7}{25}$, $\frac{4}{25}$
17. 23
19. 9

CHAPTER SIX

7. $x = 2$, $y = 1000$, $N = 3000$
9. $x = 2$, $y = 731$; $x = 20$, $y = 7310$
11. 59
13. $m = 12$, $n = 53$
15. $x = 9$, $y = 82$
17. $x = 180$, $y = 649$
21. $\frac{1}{14}$ of a day; $\frac{2}{14}$, $\frac{3}{14}$, $\frac{4}{14}$, $\frac{5}{14}$
23. One solution is that the first traveler had 11 coins, the second 13 coins, and that the purse had 30 coins.
25. The interpolation scheme gives 948, while the algebraic formula gives 953.

CHAPTER SEVEN

1. Al-Khwārizmī's rule for solving $bx + c = x^2$ translates to the formula

$$x = \sqrt{\left(\frac{b}{2}\right)^2 + c} + \frac{b}{2}.$$

The main point of the geometric proof is that rectangle $RBMN$ is equal to rectangle $NKTL$. Then, because $\left(\frac{b}{2}\right)^2$ is equal to rectangle $KHGT$, $\left(\frac{b}{2}\right)^2 + c$ is represented by rectangle $MAGL$, so that the square root in the formula is equal to the side of that square, namely, GA.

3. (a) 4; (b) 3
5. (a) $x = \sqrt{2\frac{1}{2} + \sqrt{1000}} - \sqrt{2\frac{1}{2}}$, $y = 10 + \sqrt{2\frac{1}{2}} - \sqrt{2\frac{1}{2} + \sqrt{1000}}$

(b) $x = 10 + \sqrt{50} - \sqrt{50 + \sqrt{20,000} - \sqrt{5000}}$,
$y = \sqrt{50 + \sqrt{20,000} - \sqrt{5000}} - \sqrt{50}$

15. If $y = bx^2 - x^3$, then $y' = 2bx - 3x^2$ and $y' = 0$ when $x = 0$ or when $x = \frac{2b}{3}$. The second deriva-

tive test shows that $x_0 = \frac{2b}{3}$ makes y maximal. If we now consider the graph of $f(x) = x^3 - bx^2 + d$, we note that it has a maximum at 0 (and $f(0) = d$) and a minimum at $\frac{2b}{3}$ $\left(\text{and } f\left(\frac{2b}{3}\right) = -\frac{4b^2}{27} + d\right)$. For this graph to cross the x-axis twice for x positive, this minimum value must be negative. Thus, there are two positive solutions to the cubic if $-\frac{4b^3}{27} + d < 0$ or if $4b^3 > 27d$; there is one positive solution if $-\frac{4b^3}{27} + d = 0$ or if $4b^3 = 27d$; and there are no positive solutions if $-\frac{4b^3}{27} + d > 0$ or if $4b^3 < 27d$.

19. 3460 miles
23. The angle opposite the 60° side is 56°; that opposite the 75° side is 112°23′; and that opposite the 31° side is 29°32′.

CHAPTER EIGHT

1. 3600, 2400, 1200
3. 6;23,21 = 6.389
5. 5.47
11. $\frac{15}{56}$ of a day, or approximately 6 hr 26 min
15. 1572 days
17. $x = 5,\ y = 4$ or $x = 6,\ y = 3$
19. $x = 3,\ y = 6$

CHAPTER NINE

1. They will meet in $3\frac{15}{16}$ days; the courier from Rome will have traveled $140\frac{5}{8}$ miles and the one from Venice will have traveled $109\frac{3}{8}$ miles.
3. First: 10 days; second: 24 days; third: $17\frac{1}{7}$ days
5. Approximately 2.5 *denarii* per *lira* per month
9. $5 - 2\sqrt{3} - 2\sqrt{2},\ 5 + 2\sqrt{3} - 2\sqrt{2}$
11. 57
13. 15 dukes, 450 earls, 27,000 soldiers
15. $\frac{3 \pm \sqrt{5}}{2}, -3$
17. $x = \sqrt[3]{\sqrt{26} + 5} - \sqrt[3]{\sqrt{26} - 5}$
23. 8, 2
25. Assuming, as in Figure 9.1, that $d = 3.5b$, the pole heights are $p, \frac{7}{9}p, \frac{7}{11}p, \frac{7}{13}p, \ldots$
27. $AB = 8.46$; $AG = 14.10$
29. 25°40′ and 14°20′
35. 1.88 years = 687 days
39. Suppose the initial velocity of the projectile is v_0. The horizontal distance $x(t)$ is given by $x = (v_0 \cos\alpha)t$, while the vertical distance $y(t)$ is given by $y = -at^2 + (v_0 \sin\alpha)t$. If we solve the first equation for the parameter t and substitute in the second, we get $y = -a\left(\frac{x}{v_0 \cos\alpha}\right)^2 + (\tan\alpha)x$. Since y is a quadratic function of x, the graph of the function is a parabola, as claimed.
41. From exercise 39, we have the equation expressing y as a function of x. This quadratic function achieves its maximum when $x = \frac{-\tan\alpha}{-2a/(v_0^2\cos^2\alpha)} = \frac{1}{2a}v_0^2 \tan\alpha \cos^2\alpha$. The y value at this point gives the maximum value, and that is $y = \frac{1}{4a}v_0^2 \tan^2\alpha \cos^2\alpha = \frac{1}{4a}v_0^2 \sin^2\alpha$. When $\alpha = 45°$, this expression is equal to 5000. So $\frac{1}{4a} \cdot \frac{1}{2}v_0^2 = 5000$ and $\frac{1}{4a}v_0^2 = 10,000$. To find the maximum height when $\alpha = 30°$, we calculate $y = 10,000 \sin^2(30°) = 10,000 \cdot \frac{1}{4} = 2500$. To find the maximum height when $\alpha = 60°$, we calculate $y = 10,000 \sin^2(60°) = 10,000 \cdot \frac{3}{4} = 7500$. We note that Galileo's results are slightly in error.

CHAPTER TEN

1. $x = -9 \pm \sqrt{57}$
3. The asymptotes are $x = s$ and $y = r$.
7. $(e - cg)y^2 + (de + fgc - bcg)xy + bcfgx^2 + (dek - fg\ell c)y - bcfg\ell x = 0$
15. The probability of no six in four throws is $\left(\frac{5}{6}\right)^4 = \frac{625}{1296}$. Thus, the odds in favor are $(1296 - 625) : 625 = 671 : 625$.
17. 42 : 22
19. First: $\frac{37}{72}$; second: $\frac{35}{72}$
21. 9 : 6 : 4

CHAPTER ELEVEN

5. $t = \dfrac{pxy - 3y^3}{3x^2 - py}$

7. $t = -\dfrac{a^2y^2}{b^2x} = -\dfrac{a^2 - x^2}{x}$

11. The maximum occurs when $x = -\sqrt{\dfrac{2b^2a/3c}{9a - 3b}}$ and the minimum occurs when x is the positive square root of that expression.

19. If $y = x^2$, then $y' = 2x$. The arc length formula gives
$$L = \int_a^b \sqrt{1 + (2x)^2}\,dx = \int_a^b \sqrt{1 + 4x^2}\,dx.$$
Thus, to calculate this integral, it is necessary to find the area under the curve $y = \sqrt{1 + 4x^2}$, or $y^2 - 4x^2 = 1$, a hyperbola.

27. $y = 4x - 2x^2 + \dfrac{x^3}{3} - \dfrac{x^4}{2} - \dfrac{2x^5}{5} - \cdots$

33. We have
$$d\left(\frac{x}{y}\right) = \frac{x + dx}{y + dy} - \frac{x}{y} = \frac{xy + y\,dx - xy - x\,dy}{y(y + dy)}$$
$$= \frac{y\,dx - x\,dy}{y^2}.$$
We can ignore the $y\,dy$ in the denominator because it is infinitesimally small with respect to the y^2 term.

CHAPTER TWELVE

3. $y = x^2 + \dfrac{k}{x}$

13. Suppose $y = ue^{\alpha x}$ satisfies the differential equation $a^2\,d^2y + a\,dy\,dx + y\,dx^2 = 0$. We calculate that $dy = e^{\alpha x}\,du + \alpha u e^{\alpha x}\,dx$ and that $d^2y = e^{\alpha x}\,d^2u + \alpha e^{\alpha x}\,dx\,du + \alpha e^{\alpha x}\,du\,dx + \alpha^2 u e^{\alpha x}\,dx^2$. Therefore, $a^2\,d^2y + a\,dy\,dx + y\,dx^2 = a^2e^{\alpha x}\,d^2u + (2\alpha a^2 e^{\alpha x} + ae^{\alpha x})\,du\,dx + (a^2\alpha^2 u e^{\alpha x} + a\alpha u e^{\alpha x} + ue^{\alpha x})dx^2$. To eliminate the term in $du\,dx$, we must have $2\alpha a^2 + a = 0$, or $\alpha = -\dfrac{1}{2a}$. Then the coefficient of the dx^2 term is $e^{\alpha x}\left(a^2\dfrac{1}{4a^2} - a\dfrac{1}{2a} + 1\right)u$, or $\frac{3}{4}ue^{\alpha x}$. Dividing through by $e^{\alpha x}$, we find that u must be a solution to $a^2\,d^2u + \frac{3}{4}u\,dx^2 = 0$.

15. $x^2y^3 + 2x^3y^2 + 4x^2 + 3y = k$

19. Base $= 2\sqrt{3}$, side $= 2\sqrt{3}$

25. To find the area bounded by the curve and the y-axis, it is easiest first to solve for x and then integrate. Thus, we rewrite the equation of the curve in the form $xy^2 = 8 - 4x$ and solve this as $x = \dfrac{8}{y^2 + 4}$. Then the area is given by
$$A = \int_{-\infty}^{\infty} \frac{8}{y^2 + 4}\,dy = 8\left[\frac{1}{2}\arctan\frac{y}{2}\right]_{-\infty}^{\infty}$$
$$= 4\frac{\pi}{2} - 4\frac{-\pi}{2} = 4\pi.$$

CHAPTER THIRTEEN

3. $\frac{24{,}864}{59{,}049} = 0.42$

7. $x = 6.6$, while the approximation is 6.3. With 7 trials, the odds are better than even, while with 6 they are less than even.

11. 21.4822

13. 0.91

15. 9/10

CHAPTER FOURTEEN

5. If $v = x_1 - x_2$, then $v^2 = (x_1 + x_2)^2 - 4x_1x_2 = t^2 - 4c$. Thus, v satisfies a quadratic equation in t. It is simpler to rewrite this equation in the form $v^2 = b^2 - 4c$, so $v = \pm\sqrt{b^2 - 4c}$. We also have $x_1 = \dfrac{(x_1 - x_2) - b}{2} = \dfrac{v - b}{2}$. It follows that $x_1 = \dfrac{\sqrt{b^2 - 4c} - b}{2}$ and the second solution x_2 is given by $\dfrac{-\sqrt{b^2 - 4c} - b}{2}$.

7. The three distinct values are $x_1x_2 + x_3x_4$, $x_1x_3 + x_2x_4$, and $x_1x_4 + x_2x_3$.

9. The reduced equation is $y^3 - 12y - 144 = 0$. Its solutions are $\alpha = 6$, $\beta = -3 + \sqrt{-15}$, $\gamma = -3 - \sqrt{-15}$. The solutions to the original equation are then $\frac{1}{2}\sqrt{6} \pm \sqrt{\sqrt{6} - 1\frac{1}{2}}$ and $\frac{1}{2}\sqrt{6} \pm \sqrt{-\sqrt{6} - 1\frac{1}{2}}$.

13. Quadratic residues modulo 13 are 1, 3, 4, 9, 10, 12.

CHAPTER FIFTEEN

3. Since $< ABC = 60 - \gamma/3$ and $< F = 60 - \beta/3$, it follows that $< FBC = < FCB = 60 + \gamma/6$. The sum of the angles of triangle CBF is then $180 + \gamma/3 - \beta/3$ and also $180 - \alpha$. Therefore, $\alpha = \beta/3 - \gamma/3$, or $\beta = 3\alpha + \gamma$.

5. $\cosh x = \cos ix$, $\sinh x = -i \sin ix$
7. The normal line to the surface $z = f(x, y)$ is a line in the direction of the gradient, namely, $\left(\frac{\partial z}{\partial x}, \frac{\partial z}{\partial y}, -1\right)$. The normal vector to the plane $z = \alpha y - \beta x + \gamma$ is $(\beta, -\alpha, 1)$. Thus, the plane will contain the normal line if these two vectors are perpendicular, that is, if their dot product is zero. This amounts to the condition $\beta\frac{\partial z}{\partial x} - \alpha\frac{\partial z}{\partial x} - 1 = 0$, as stated.
11. (a) $EADCBAEC$; (b) $CBDBADACAC$

CHAPTER SIXTEEN

3. 2, 6, 7, 11
5. $3 + 5i = (1 - 4i)(-1 + i)$
9. If $A = (2, 1 + \sqrt{-5})$, then A^2 consists of sums of multiples of $2 \times 2 = 4$, $2(1 + \sqrt{-5}) = 2 + 2\sqrt{-5}$, and $(1 + \sqrt{-5})(1 + \sqrt{-5}) = -4 + 2\sqrt{-5}$. Each of these numbers is a multiple of 2 (by some integer in the domain), so $A^2 \subset (2)$. But also $2 = (-1)[4 + (-4 + 2\sqrt{-5}) + (-1)(2 + 2\sqrt{-5})]$, so $2 \in A^2$, and therefore $(2) \subset A^2$. Therefore $(2) = A^2$.
13. From the equation $x^2 - \beta_1 x + 1 = 0$, where $\beta_1 = r + r^{18}$, we have $x = \frac{1}{2}\left(\beta_1 \pm \sqrt{\beta_1^2 - 4}\right) = \frac{1}{2}\left(r + r^{18} \pm \sqrt{r^2 - 2 + r^{17}}\right) = \frac{1}{2}\left[r + r^{18} \pm \sqrt{(r - r^{18})^2}\right] = \frac{1}{2}\left[(r + r^{18}) \pm (r - r^{18})\right]$. Thus, the two roots of this equation are r and r^{18}.
15. S_3
17. There are three Abelian groups of order 8. The first is the cyclic group of order 8. The second has two generators, α and β, with $\alpha^4 = 1 = \beta^2$ and $\alpha\beta = \beta\alpha$. The third has three generators α, β, and γ, with $\alpha^2 = \beta^2 = \gamma^2 = 1$, and with all the generators commuting with each other. There are also two non-Abelian groups of order 8. The first has two generators, α and β, with $\alpha^4 = 1 = \beta^2$ and with $\alpha\beta = \beta\alpha^3$. The second one has two generators, α and β, with $\alpha^4 = 1$, $\beta^2 = \alpha^2$, and $\alpha\beta = \beta\alpha^3$.
23. $x^2 + x + 1$ is irreducible modulo 5. Therefore, $\{a_0 + a_1\alpha + a_2\alpha^2\}$ is a field of order 5^3, where α satisfies $\alpha^3 = -\alpha - 1$ and $0 \le a_j < 5$, for $j = 0, 1, 2$.
29. $\alpha\beta = 12 - 9i + 18j + 24k$, $\frac{1}{\beta} = \frac{2}{15} + \frac{1}{5}i - \frac{1}{15}j + \frac{1}{15}k$
31. Order of maximal nonvanishing determinant is 2. One way of expressing the solution is $u = -10x - 15z$, $v = 18x - y + 27z$, with x, y, z arbitrary.
33. Associated system is $-10u + 18v + x = 0$, $-v + y = 0$, $-15u + 27v + z = 0$; a basis for the set of solutions is $(1, 0, 10, 0, 15)$, $(0, 1, -18, 1, 27)$.

CHAPTER SEVENTEEN

3. Let $\epsilon > 0$. We need to find $\delta > 0$ such that $|\sin(x + \alpha) - \sin x| < \epsilon$, whenever $|\alpha| < \delta$. We choose $\delta = \epsilon$. Then for $|\alpha| < \delta$, we have

$$\begin{aligned} |\sin(x + \alpha) - \sin x| &= \left|2 \sin \tfrac{1}{2}\alpha \cos\left(x + \tfrac{1}{2}\alpha\right)\right| \\ &\le 2 \cdot \tfrac{1}{2}|\alpha| \cdot 1 = |\alpha| < \delta = \epsilon. \end{aligned}$$

Therefore, $\sin x$ is continuous at x.
11. $\frac{2}{3}$
23. First part: Let C_1, C_2 be two paths connecting the points p_1, p_2 in 3-space. Let C be the closed path from p_1 to p_2, which goes along C_1 from p_1 to p_2 and then goes back to p_1 along C_2. Then, $\int_{C_1} \sigma \cdot dr - \int_{C_2} \sigma \cdot dr = \int_C \sigma \cdot dr = \int\int_A (\nabla \times \sigma) \cdot da = 0$. It follows that $\int_{C_1} \sigma \cdot dr = \int_{C_2} \sigma \cdot dr$.

CHAPTER EIGHTEEN

1. $a = 1.53$, $b = -0.87$
3. Using a table of the normal curve, we find that a percentile rank of 75 corresponds to a z-score of 0.675. Therefore, one probable error from the mean, which corresponds to a percentile rank of 75, is at distance approximately 0.675σ from the mean, where σ is the standard deviation. By the result of exercise 2, a distance c of one modulus from the mean corresponds to $\sqrt{2}\sigma$.
9. Approximately 24 flowers had seven petals and 133 had five petals

CHAPTER NINETEEN

1. Assuming that angle B is acute, we drop a perpendicular from angle B to the opposite side b, intersecting that side at point D. Designate BD by h, AD by r, and DC by $b - r$. In right triangle BDC, we have $\cos a = \cos h \cos(b - r)$. In right triangle BDA, we have $\cos A = \dfrac{\tan r}{\tan c}$ and $\cos c = \cos h \cos r$. Therefore, $\cos h = \dfrac{\cos c}{\cos r} = \dfrac{\cos a}{\cos(b - r)}$. So

$$\begin{aligned}\cos a &= \frac{\cos c \cos(b - r)}{\cos r}\\ &= \frac{\cos c(\cos b \cos r + \sin b \sin r)}{\cos r}\\ &= \cos c \cos b + \cos c \sin b \tan r\\ &= \cos c \cos b + \cos c \sin b \tan c \cos A\\ &= \cos b \cos c + \sin b \sin c \cos A.\end{aligned}$$

7. From $a \sin(A + C) = b \sin A$ and the law of sines, we have $\dfrac{\sin B}{b} = \dfrac{\sin A}{a} = \dfrac{\sin(A + C)}{b}$. Therefore, $\sin(A + C) = \sin B$, or, interchanging A and B, $\sin(B + C) = \sin A$. From $\cos A + \cos(B + C) = 0$, we have $\cos(B + C) = -\cos A$. The sine result implies that either $A = B + C$ or $A = \pi - (B + C)$. The cosine result shows that the second equation is the correct one. Thus, $A + B + C = \pi$.

11. $-18[ijk]$

13. $WRSTVJHGBCDFKLMNPOZXW$

CHAPTER TWENTY

1. One other example is the complex numbers: If the basis is taken to be i, j, then the four multiplications are $i \cdot i = i, i \cdot j = j \cdot i = j$, and $j \cdot j = -i$.

3. One example is $\begin{pmatrix} 1 & 1 \\ 0 & 0 \end{pmatrix}$.

9. $2 \times (-2, 3) = (8, -23)$

11. 504

13. $AB = (1, 2, 3, 8, 4, 11, 9, 10, 6, 5, 7)$,
$BA = (1, 2, 7, 3, 10, 8, 9, 5, 4, 6, 11)$,
$AC = (1, 12, 2)(3, 7, 11)(4, 9, 6)(5, 10, 8)$,
$CA = (1, 11, 12)(2, 6, 10)(3, 8, 5)(4, 9, 7)$

15. Since $\dbinom{12}{6} = 924$, if the woman has no discriminating ability, the probability of picking six cups correctly is $1/924$ and the probability of picking five cups correctly is 36/924. The total probability then is $37/924 = 0.04$.

19. The pyramidal numbers are numbers of the form $\dbinom{n}{3}$, $(n \geq 3)$, which can therefore be expressed in the form of a cubic polynomial: $\frac{1}{6}n^3 - \frac{1}{2}n^2 + \frac{1}{3}n$. These numbers are the numbers 1, 4, 10, 20, 35, ... Their first differences are the triangular numbers 3, 6, 10, 15, ... Their second differences are the integers 3, 4, 5, ..., and their third differences are all constantly 1. Thus, to calculate the pyramidal numbers, we start with the third differences and note that the first second difference is 3, the first first difference is 3, and the first pyramidal number is 1. We can then use the Difference Engine to calculate by finding in turn the integers, the triangular numbers, and the pyramidal numbers by repeated addition.

General References in the History of Mathematics

Each chapter of this text includes references to works that can provide further information on the material of that chapter. In general, however, to learn the history of a specific topic in mathematics, it pays to begin one's search in one of the following works:

- Ivor Grattan-Guinness, ed., *Companion Encyclopedia of the History and Philosophy of the Mathematical Sciences* (London: Routledge, 1994). This two-volume encyclopedia includes brief (sometimes too brief) articles on some 180 topics in the history and philosophy of mathematics, each written by an expert in the field. The encyclopedia is particularly strong on what are generally thought of as applied mathematical topics, such as mechanics, physics, engineering, and the social sciences, while it is relatively weak on some of the more standard topics in the history of mathematics. Nevertheless, it is an excellent source in which to begin research on a topic in the history of mathematics.
- Joseph W. Dauben, ed., *The History of Mathematics from Antiquity to the Present: A Selective Annotated Bibliography* (Providence: American Mathematical Society, 2000). This CD-ROM contains approximately 4800 entries by 38 experts, who have attempted to pick the "best" works in their respective fields. Although the CD format is not as easy to use as a printed book, this is probably the best place to start on a quest for historical articles on a particular topic.
- Morris Kline, *Mathematical Thought from Ancient to Modern Times* (New York: Oxford University Press, 1972). This book is the most comprehensive of the works in the history of mathematics and pays particular attention to the nineteenth and twentieth centuries. It provides chapter bibliographies for further help. However, it completely lacks any information about Chinese mathematics and provides only sketchy information about the mathematics of India and the Islamic world.
- Charles C. Gillispie, ed., *Dictionary of Scientific Biography* (New York: Scribners, 1970–1990). This 18-volume encyclopedia (including two recent supplementary volumes) is in essence a comprehensive history of science organized biographically. There are articles about virtually every mathematician mentioned in this text and, naturally, articles about many who are not mentioned. There are also special essays on topics in Egyptian, Babylonian, Indian, Japanese, and Mayan mathematics and astronomy. The encyclopedia includes an extensive index, which allows the reader to look up a mathematical topic and find references to all the mathematicians who considered it.

For readers interested in original sources, there are several collections of selections from important mathematical works, all translated into English. These include Henrietta O. Midonick, ed., *The Treasury of Mathematics* (New York: Philosophical Library, 1965); Ronald Calinger, ed., *Classics of Mathematics* (Englewood Cliffs, N.J.: Prentice Hall, 1995); D. J. Struik, ed., *A Source Book in Mathematics, 1200–1800* (Cambridge: Harvard University Press, 1969); Garrett Birkhoff, ed., *A Source Book in Classical Analysis* (Cambridge: Harvard University Press, 1973); David Eugene Smith, ed., *A Source Book in Mathematics* (New York: Dover, 1959); and John Fauvel and Jeremy Gray, eds., *The History of Mathematics: A Reader* (London: Macmillan, 1987).

Naturally, more research continues to be done in the history of mathematics, and there are many journals that publish articles on the subject. The most important of such journals, which can be found in most major university libraries, are *Historia Mathematica* and *Archive for History of Exact Sciences*. The former publishes in each issue a list of abstracts of recent articles in the history of mathematics. To keep up with current literature, however, it is best to consult *Mathematical Reviews*, published monthly by the American Mathematical Society, or the *Isis Current Bibliography of the History of Science and its Cultural Influences*, published every year as the fifth issue of *Isis*, the journal of the History of Science Society. This latter publication contains an extensive listing, by subject, of articles published during the previous twelve months on the history of science, including, of course, the history of mathematics. All of these sources are available online at many research libraries.

Finally, the Internet contains numerous sources on the history of mathematics, some of which are excellent and others that are best ignored. Among the best sites are the following:

- David Joyce's History of Mathematics Home Page (http://aleph0.clarku.edu/~djoyce/mathhist/)
  This starting point to a wealth of resources is provided by David Joyce of Clark University in Worcester, Massachusetts. There are pages on regional mathematics, subjects, books, journals, bibliography, history of mathematics texts, and the like, as well as an excellent list of web resources clearly categorized, an extensive chronology, and time lines.
- The Math Forum Internet Mathematics Library (http://mathforum.org/library/topics/history/)
  This site provides an extensive list of annotated links to other sites. The sites are ordered alphabetically and the collection can be viewed in outline or annotated form. A well-designed search engine allows for a variety of searches.
- St. Andrews MacTutor History of Mathematics Archive (http://www-history.mcs.st-and.ac.uk/history/)
  This site offers a collection of biographies of mathematicians and a variety of resources on the development of various branches of mathematics. This is an extremely rich and extensive site with some excellent pages, although the quality is not always consistent. In particular, the biographies should be read with a critical eye.
- Trinity College, Dublin, History of Mathematics Archive
  (http://www.maths.tcd.ie/pub/HistMath/HistMath.html)
  This site, created and maintained by David Wilkins, includes biographies of some seventeenth- and eighteenth-century mathematicians, material on Berkeley, Newton, Hamilton, Boole, Riemann, and Cantor, and an extensive directory of history of mathematics web sites.
- The British Society for the History of Mathematics (http://www.dcs.warwick.ac.uk/bshm/resources.html)
  This site, maintained by June Barrow-Green for the Society, contains an extensive list of history sites, carefully annotated and categorized. Anyone seriously interested in the history of mathematics should consider joining the BSHM; the Society has a wonderful newsletter and conducts numerous meetings on various topics in the history of mathematics.

Index and Pronunciation Guide*

PRONUNCIATION KEY

a	act, bat	j	just, fudge	œ	as in German schön or in French feu
ā	cape, way	k	keep, token	R	rolled r as in French rouge or in German rot
â	dare, Mary	KH	as in Scottish loch or in German ich	sh	shoe, fish
ä	alms, calm	N	as in French bon or un	th	thin, path
ch	child, beach	o	ox, wasp	u	up, love
e	set, merry	ō	over, no	û	urge, burn
ē	equal, bee	o͝o	book, poor	y	yes, onion
ə	like a in alone or e in system	o͞o	ooze, fool	z	zeal, lazy
g	give, beg	ô	ought, raw	zh	treasure, mirage
i	if, big	oi	oil, joy		
ī	ice, bite	ou	out, cow		

*As an aid to pronouncing the names of the various mathematicians discussed in the book, the phonetic pronunciation of many of them is included in parentheses after the name. Naturally, since many foreign languages have sounds that are not found in English, this pronunciation guide is only approximate. To get the exact pronunciations, it is best to consult a native speaker of the appropriate language.

St. Petersburg
Moscow
Königsberg
Cambridge
Berlin
Gröningen
Oxford
Leiden
Halle
London
Göttingen
Leipzig
Paris
Königsberg
Vienna
Basel
Tours
Brescia
Lyon
Milan
Orange
Bologna
Toulouse
Pisa
Florence
Rome
Barcelona
Istanbul
Toledo
Bithynia
Crotona
Chios
Seville
Athens
Samos
Miletus
Syracuse
Perge
Damascus
Baghdad
Behistun
Marrakech
Babylon
Alexandria
Uruk
Sus
Larsa
Cairo
Luxor
Mecca